J Enoksen

A good friend

Best Regards

Brian Rothschild

Oct 2009

The Future of Fisheries Science in North America

FISH & FISHERIES SERIES

VOLUME 31

**Series Editor: David L.G. Noakes, Fisheries & Wildlife Department,
Oregon State University, Corvallis, USA**

For other titles published in this series, go to
www.springer.com/series/5973

Richard J. Beamish • Brian J. Rothschild
Editors

The Future of Fisheries Science in North America

 Springer

Editors

Richard J. Beamish
Pacific Biological Station
Fisheries & Oceans Canada
3190 Hammond Bay Road
Nanaimo B.C. V9T 6N7 Canada
Richard.Beamish@dfo-mpo.gc.ca

Brian J. Rothschild
School for Marine Science and Technology
University of Massachusetts, Dartmouth
706 South Rodney French Boulevard
New Bedford, MA 02744-1221 USA
brothschild@umassd.edu

ISBN 978-1-4020-9209-1   e-ISBN 978-1-4020-9210-7

Library of Congress Control Number: 2008940866

Printed on acid-free paper

springer.com

Foreword

This volume and the preceding symposium, The Future of Fishery Science in North America, held during February 13–15, 2007, are part of the commemoration of the 50th Anniversary of the American Institute of Fishery Research Biologists (AIFRB). The symposium was a milestone in some ways. It was the first symposium of this kind co-sponsored by the National Marine Fisheries Service (NMFS), the Department of Fisheries and Oceans Canada (DFO), and AIFRB. It marked what we hope is the beginning of an ongoing series of symposia on topics of broad interest in fisheries science that the AIFRB will organize and co-sponsor.

The Symposium on the Future of Fishery Science in North America covered a wide range of topics and raised many questions that will need to be studied in the future. The contributors in this book took a stimulating look at what is to come in fishery science. The symposium also provided an opportunity to bring people with new perspectives into contact, sowing the groundwork for new perspectives. Each chapter is the author's vision of the future in a particular area of fisheries research. All authors had their own method of presenting their vision, but most looked back into the history and then projected into the future. There is some overlap in the broad spectrum of activities, but it is insightful how individuals saw differences in some common topics such as ecosystem-based management. Authors vary in experience from an undergraduate to scientists who have seen just about everything there is to see but have learned to expect the unexpected. Authors were invited to speculate more than is normally allowed in peer-reviewed publications, as it will be informative to future generations of researchers to read what they think will happen and needs to happen in fisheries science in North America. We hope the chapters in this volume pique the interest of readers and accelerate new ideas for the future of fishery science and conservation.

The Symposium was co-sponsored by the National Oceanic and Atmospheric Administration (NOAA) Fisheries and Sea Grant, Fisheries and Oceans Canada, and the North Pacific Research Board. In addition, there were a number of generous donors: School of Aquatic and Fisheries Sciences University of Washington, ECorp Consulting, Inc., Simrad Fisheries, International Pacific Halibut Commission, Census for Marine Life, West Palm Beach Fishing Club, Biosonics, Hydroacoustic Technology, Inc., Shakespeare Fishing Tackle, and Cabela Outfitters. I would like to thank Dr. Wendy Watson-Wright, Assistant Deputy Minister, Fisheries and

Oceans Canada, and Dr. Bill Hogarth, Assistant Administrator, NOAA Fisheries Service for attending and giving keynote addresses to open the Symposium.

Many AIFRB members worked very hard to make the Symposium a success. Although a simple thank you is inadequate, I do want to acknowledge all of these dedicated people: Brian Rothschild and President-elect Dick Beamish, the original instigators of the 50th Symposium and the creative forces for all that followed; our hard-working Treasurer, Allen Shimada; the Local Organizing Committee comprised of Nancy Davis, Bill Aron, Kristan Blackhart, Vince Gallucci, Rick Methot, Kate Myers, David Somerton, and Fred Utter; Ray Wilson for organizing the Poster Session; and the Fund Raising Committee comprised of Vidar Wespestad, chair, Bern Megrey, Membership Chair Tom Keegan, and John Jolley, Dick Beamish, and Brian Rothschild.

There was a tremendous amount of help from a host of people before, during, and after the symposium. The National Steering Committee developed the program concept: Dick Beamish, John Boreman, Bill Fox, Rick Methot, Joanne Morgan, Steve Murawski, Victor Restrepo, Andy Rosenberg, and Brian Rothschild. The Session Chairs recruited excellent speakers for the symposium. The Chairs were Jim Balsiger, Bill Fox, Mike Fogarty, Jake Rice, Ken Drinkwater, Anne Hollowed, Rick Methot, Bob Mohn, Van Holliday, and Ken Foote.

David Somerton and Vickie Lingwood, Alaska Fisheries Science Center (AFSC), National Oceanic and Atmospheric Administration Fisheries (NOAA), were responsible for the web pages for registration and online abstract submission with the generous support of Russ Nelson (AFSC). NOAA Fisheries staff was also critical contributors: Steve Murawski, John Boreman, Bill Zahner, Marty Golden, Christopher Moore (HQ), and Wendy Carlson (AFSC Graphics). Nancy Davis organized and recruited the many volunteers needed to assist during the symposium: Andrea Belgrano, Kristan Blackhart, John Brandon, Troy Buckley, Lisa Crosson, Keith Denton, Katy Doctor, Martin Dorn, Heather Gibbs, Marty Golden, Pete Haaker, Jim Hastie, Katrina Hoffman, Mary Hunsiker, Bob Lauth, Alan Lin, Jodie Little, Stacey Miller, Vija Pelekis, Mary Ramirez, Cheryl Ryder, Nancy Somerton, Ian Stewart, Nancy Utter, Mark Wilkins, John Williams, Matt Wilson, and Stephani Zador.

It takes tremendous effort to organize a successful symposium. On behalf of AIFRB, we thank all of these people and organizations for their time, money, and effort.

The AIFRB is a professional organization established in 1956 to promote the application of fishery science in the conservation and proper use of fishery resources. Our goal is to advance the theory, practice, and application of fishery biology and related sciences, promote high professional standards in fishery science, and to recognize excellence and outstanding achievements of scientists in fishery conservation and science. Members are individuals with formal training and demonstrate competence and achievement in fields of fishery science and research. As a member demonstrates higher levels of leadership, excellence, and achievement in the field through publications, professional responsibility, teaching, and editorial excellence, the member is advanced from Associate Member to Member to Fellow. Advancement in membership can also be based on an individual's role in promoting key fishery-related legislation or exceptional program planning and management in fishery organizations, or involvement in activities related to aquatic resource conservation. As a result, AIFRB has a strong cadre of leaders in various fishery and conservation research fields that can provide expertise on a wide range of topics. A new program for the Institute is using this expertise to mentor early career members by providing reviews of research manuscripts at an early stage in their preparation.

The Institute also recognizes excellence and leadership in fishery science through its awards. An Outstanding Achievement Award is given annually to an individual who has made significant contributions to the advancement of fishery science and conservation. The individual has demonstrated excellence in science, mentoring of students and colleagues, publications and leadership in fishery, and conservation science. One of the symposium organizers, Brian Rothschild, was selected for this award in 2004. An Outstanding Achievement Award is also given to groups of fishery scientists who as an organizational unit have demonstrated these same levels of achievement and excellence. Another AIFRB award, which was first announced during the symposium, is the Kasahara Early Career Award. This new award recognizes scientists early in their careers who have already demonstrated excellence and outstanding science and leadership. The W.F. Thompson Award is presented annually to the best student paper published that year. It recognizes excellence in research on fish or some aspects of aquatic science and conservation by a student.

The AIFRB also supports the training and education of the next generation of scientists in fields of fishery science and conservation through the Research Assistance Awards. Four awards are given each year to support student research or presentation of their research results at major symposia and meetings.

Linda Jones
President
American Institute of Fishery Research Biologists

Volume Foreword

Bill Ricker's advice to fishery scientists was to expect the unexpected. That advice remains true, and this volume goes to considerable lengths to anticipate what might be unexpected. A search of any book of quotations will produce numerous references to the future. Most of those quotations warn of the perils of making predictions about the future. I once read a book about the incredible predictions of the future made during the 1930s, most of which did not come true (Onosko 1979). That is the obvious danger of predictions, but an unlikely one for this volume. For this volume the most useful statement is likely the one attributed to Jason Kauffman, "The best way to predict the future is to create it."

There are few fishery scientists other than Brian Rothschild and Dick Beamish, who could, or would, take on the task of convening the symposium and editing the resulting volume to predict the future of fisheries science in North America. They are remarkably qualified individuals, with unmatched backgrounds in the field, and they have brought together contributions from individuals who are most likely to create that future. Andy Dixon, a person with an uncommon amount of common sense, was my teacher in Agricultural Science classes when I started high school. He once told us that you could understand everything of human behavior when you realize that we are motivated by the desire to be able to say, "I told you so." Perhaps. However, it is clear that the contributors to this volume are quite honestly trying to forecast the future for the sake of fishery science, not just to satisfy any self-serving needs. Most of the contributors have more than enough personal experience to realize the importance of their science.

The volume is directed to fishery science in North America, and so it will set the standard for others to follow. Of course there are specific details and peculiarities to North America because of simple facts of geography, history, and politics. The sweep of the coverage demonstrates that such a volume could hardly consider anything more than this geographic area. Were it to attempt to cover the rest of the world, the fabric would be stretched so thin as to be little more than vague generalities or simple platitudes. In contrast, this volume combines both breadth and depth, from molecular genetics to remote sensing techniques, from climate change to socioeconomics.

It is always useful to ask what elephants are in the room, and this volume does that in several ways. We are all familiar with the story of the blind men, each of whom attempts to describe an elephant, based on his own limited individual experience of one small part of the creature. Here we have a number of informed experts in the field, each commenting on individual perspectives, and from that combination we can assemble an image of the biggest elephant – fishery science. But there are also a number of smaller elephants already with us. Aquaculture and fish farming will not only remain a regional concern but will also increase in importance with proposals for marine aquaculture and the development of broader international marketing. Biodiversity seems to be an omnipresent consideration in North America, even though there is as yet limited agreement on how to apply or to even define the concept. Ecosystem management is on every agenda; however it might be applied. Climate change is probably better established in fishery science than many other areas through the contributions of Dick Beamish and his colleagues, but the implications for fishery science remain to be determined.

This volume should be read as a detailed advice column. It contains advice to young scientists in the form of recommendations to develop their expertise in risk assessment, spatially explicit models, statistics, interdisciplinary collaboration, and bioeconomic models of sustainability in addition to their obvious background in fishes and fish biology. The volume also has advice for politicians, senior scientists, resource managers and, most importantly, the general public. In some cases this advice takes the form of necessary "recipes," as for example detailed updates on techniques ranging from molecular genetics to remote sensing and quantitative ecological modeling. In other cases the advice is informed opinions from those who have reviewed their areas of expertise and developed clear predictions for the future. Coverage includes both marine and freshwater systems and a range of species from high profile examples to those yet to gain broad public attention

Several noteworthy themes emerge. It is clear that climate, biodiversity, remote sensing, and monitoring will be major focal points for fishery science. Ecosystem-based management is a given for almost everyone, and various contributors offer specific advice on how best to adjust to that paradigm. Fishery science will undoubtedly continue to grow and develop and to incorporate new technologies, methodologies, and applications. There are very clear messages for the growing importance of broader collaborations and interdisciplinary approaches for fishery scientists. Economics and conservation are two of the major areas that must be incorporated in the future.

If we start with advice from Bill Ricker perhaps we should consider his example. He certainly had an ecosystem view, and was an accomplished naturalist with an incredible diversity of expertise in botany, entomology, languages, literature, and astronomy, as well as a strongly quantitative approach (Ricker 2005). We need to realize that this volume is not a roadmap; it is a GPS unit.

Editor, Springer Fish and Fisheries Series   Dr. David L. G. Noakes
Professor of Fisheries and Wildlife
Senior Scientist, Oregon Hatchery Research Center
Oregon State University
Corvallis, Oregon, USA

References

Onosko T (1979) Wasn't the future wonderful? A view of trends and technology from the 1930's. E P Dutton, New York

Ricker WE (2006) One man's journey through the years when ecology came of age. Environ Biol Fish 75: 7–37

Contents

Chapter 1
On the Future of Fisheries Science

Brian J. Rothschild and Richard J. Beamish

Abstract The invited papers in this book provide a range of opinions about the future of fisheries science in North America. The ideas of each author are carefully thought out speculations of what will change in their field and how the changes may be used to improve the stewardship of fisheries. The collection of thoughts does not cover all areas of fisheries science, but there is sufficient diversity to stimulate readers to contemplate what changes they anticipate. This introductory chapter is our perspective on the contents of the book and on the future of our science. We hope that this chapter and the chapters of our colleagues signal the urgent need for change and for strong leadership.

Keywords Fisheries Management · North America · ecosystem approach

1.1 Introduction

The future can evolve in many directions. Our theme is that while fisheries science is generally thought of in the context of an applied science, the quality of this applied science is limited by fundamental knowledge. In other words, significant investments need to be made in fundamental research in order to improve the quality of information available to improve the quality of fishery management. Our visions of the future are remarkably similar considering it is seen through different issues in different oceans. The fundamental understanding of the biology of key species must improve if we are to forecast the impacts of changing physical and biological environments on

B.J. Rothschild
School for Marine Science and Technology, University of Massachusetts Dartmouth,
706 South Rodney French Boulevard, New Bedford, MA 02744-1221, USA
e-mail: brothschild@umassd.edu

R.J. Beamish
Pacific Biological Station, Fisheries & Oceans Canada, 3190 Hammond Bay Road,
Nanaimo B.C. V9T 6N7 Canada
e-mail: Richard.Beamish@dfo-mpo.gc.ca

R.J. Beamish and B.J. Rothschild (eds.), *The Future of Fisheries Science in North America*,   1
Fish & Fisheries Series, © Springer Science + Business Media B.V. 2009

recruitment. For too long fisheries science has been used to manage fisheries without the necessary understanding of the processes that regulate the size of year classes.

1.2   Motivation

Fisheries science has focused on relatively simple goals. These goals are changing rapidly. A number of chapters set the stage for these changes that essentially specify or provide underpinning for the new requirements of fishery management. For example, Rice exposes us to the convention on biological diversity and a "new" requirement to include biodiversity in management. Benson discusses the conservation of biodiversity and its application to fishery management. O'Boyle describes the paradigm shift driven by the ecosystem approach to management, which has been anointed by the acronym EAM. Kaplan and Levin point out the challenges generated by EAM. Stringer et al. provide a perspective on the interaction between changing demands on the scientific community and the practice of scientific investigations in fisheries. Stokesbury et al. suggest application of the naturalist's approach, while Schnute and Richards recommend taking a perspective of high dimensionality that is thoughtful about the use of models. Timely and reliable assessments of multiple stocks require efficiencies beginning with the collection of data through to the presentation of results. Methot explores how this can be done by the development and acceptance of broadly applicable assessment methods. Not only will new concepts be required but, also, students will need to be trained to be responsive to the shifting focus. Peterman makes it abundantly clear that probabilistic approaches to fishery science will need shifts in training direction and also a capability to communicate and interact with wider audiences.

1.3   The Scope of Research

Ecosystem Approaches – Various workers are already considering responses to the new requirements. Some involve the application and evolution of existing approaches such as trophodynamics. Koen-Alonso considers how trophodynamic models can be applied to fisheries management. New approaches are also being developed. For example, Rochet and Trenkel give critical examination to the indicator concept. Norton et al. consider the cyclical interaction of California current ecosystem and the economics of the fishery. Saila reviews several ecosystem models and demonstrates how artificial life models can contribute to a fuller understanding of relevant ecosystem processes. Pepin makes the case that at the end of the day much will need to be done to develop a resolution of the recruitment problem. Watson-Wright identifies the commitment that is needed to take us back to an ecosystem approach to management.

Technology – Technology is our window to observing the ocean environment. It is our way of picturing the four-dimensional world that lies beneath the surface of the sea. There are at least two parts to technology. The first part is the sensors. The second part is the platforms that carry or form a chassis for the sensors. Holliday surveys the field and makes the critical observation that sensor technology may be the weakest link in developing a materially improved approach to ecosystem management. Traditionally one of the most used technologies in fisheries research is acoustics. The state of the art in acoustics is described by Foote. There is increasing interest in optical approaches. An optical approach that has interesting specific uses is light detection and ranging technology (LIDAR). LIDAR is a focus of the chapter by Churnside et al. Clarke et al. describe one of the more innovative ideas in platforms, the autonomous underwater vehicle (AUV). Godø gives a perspective on the new technologies that will be required to undertake observations required by the ecosystem approach.

Stock Structure and Genetics – One of the challenging problems in fisheries involves the fact that the same individual species may consist of groups of fish that do not mix freely. Sometimes these fish can be thought of as stocks, but sometimes they are genetically differentiated. How these stocks or genetically independent groups are differentiated and how the information on their structure applies to management is a continuing problem. To this end, Cadrin and Secor review the history and future of the incorporation of information on genetically or otherwise discrete stock of fish. The study of genetics is a rapidly growing field. Waples and Naish review the context of the genetic and evolutionary considerations in fishery management. Kochzius discusses recent developments in DNA-based species and stock identification.

Stock Assessment – The evolution of stock assessment concepts has been relatively slow. However, important new ideas are emerging. For example, the basic approach in stock assessment is to provide managers with an optimal magnitude of size- or age-specific fishing mortality. This seemingly simple problem becomes complex because several models can be used to estimate optimal mortality with different results. Which is best? Mohn considers the intercomparison of models – an issue that will be increasingly important as models begin to include assessment of several species simultaneously. A specific approach to interacting models is provided by Jiao, who considers Bayesian averaging. Observers of the process are framing new requirements for stock assessment. Methot develops the concept of operational stock assessment as an analogue to operational oceanography.

Application of Other Fields Such as Engineering – In a way, the science of fishery management relates almost to engineering and operations research techniques for managing inventories. Holt and de la Mare show how the theory of engineering control systems can be applied to the management of sockeye salmon. Recognizing that fisheries management is essentially a decision-making process, Goodman describes how applying decision theory, an important branch of applied mathematics, modifies the world view of traditional quantitative approaches.

Environment and Climate – The understanding of the relation of the environment to the variability in fish stocks is one of the biggest gaps. One practical example of the importance of the ocean environment relates to the idea that a depressed stock can generally be *rebuilt* evidently not recognizing that one interpretation of a decline in stock magnitude relates to its returning to normal abundance after a series of large year classes. Imagine taking an ecosystem approach to management without a better understanding of the ocean environment. The understanding of the environment is made complex in that the ocean environment is embedded in a hierarchy of space and time scales. A critical research issue is the way that the smaller local time and space scales are embedded in the broader and longer basin and climate scales. But understanding and adapting to new knowledge on the ocean environment – the response to surprises – will provide new insights to add understanding as described by Beamish and Riddell. A clear result of learning from surprises is the transition from hindcasting to forecasting and the dynamical shifts in approach that are necessary to make this transition as discussed by Hollowed and Bailey. Schwing et al. discuss the steps that need to be undertaken to enhance this understanding of climate variability. Fisheries science is no longer just about fish. A focus on ecosystems and the environment requires an improved communication with the public. Squires shows how social science research in general and economics in particular can improve fisheries management. The focus of economic research will shift from overfishing to addressing the sustainability of ecosystems, the loss of biodiversity and the changes in the ocean.

1.4   Our Perspective

It is important to focus on what is meant by fisheries science so that the investments can be focused. The process of observing and reporting nature were the roots of fisheries science. Early fisheries science started out trying to understand the population ecology of fish, which included their associated species and their environment. It was the push to go fishing after the Second World War that changed the emphasis to assessing how many fish could be harvested. Fisheries management and the science that supported management became oriented toward providing managers with a magnitude of size-specific fishing mortalities for each stock. The basic idea was to use yield per recruit theory, production theory, or stock and recruitment theory to determine optimal levels of fishing mortality. Then, once these were determined, estimates of the age- or size-specific abundance of a stock could be employed to estimate fishing mortality. If fishing mortality was greater than the optimum, it was reduced, and vice versa.

In recent years new objectives for fishery management are beginning to emerge. Unfortunately, these make a problem that is already costly and difficult even more costly and difficult. Instead of "simply" determining optimum yield, estimating actual fishing mortality, and determining whether the actual mortality is greater or less than optimal, the "new requirements" of fishery management additionally include the mantras: (1) *managing ecosystems*, (2) *managing habitat*, (3) *ending overfishing*, (4) *using the precautionary approach*, and (5) *rebuilding stocks*.

So now the nature of the investments can be clarified. But before this can be accomplished, it is necessary to further shape the requirements of future fisheries management by considering what we have learned and failed to learn over the past 10 decades during which fisheries management has been practiced.

Leading questions have been: What causes large- and small-year classes? What is overfishing? What is the effect of the ocean environment on the abundance of particular fish stocks? What is the effect of fishing intensity on recruitment, biomass, and production?

Additional questions may be needed for some species such as Pacific salmon (*Oncorhynchus* spp.). Hatchery production of Pacific salmon may now account for about one half of all Pacific salmon catches, which are currently at historic high levels. Current catches of pink (*O. gorbuscha*) and chum (*O. keta*) salmon represent about 88% of the total catch, and large numbers of these two species are produced in hatcheries. What are the impacts of these hatchery salmon on wild salmon? Because salmon are anadromous, what are the impacts of Pacific salmon produced in one country on the production from another country?

Over the past century parts of these questions have been answered. Turn-of-the-century curiosity on the causes of large- and small-year classes of cod and herring in Norway has still not been resolved. There is no unique definition of overfishing. The effect of the ocean environment on fish stocks is generally not understood. Interactions among species are not understood. While some advances have been made with regard to the effects of fishing intensity, we have learned to address the problem only partially.

These difficulties have arisen from a variety of circumstance that have related to the difficulty of the problem and the evolution of the body of knowledge. The problem of large- and small-year classes has evolved into the problem of understanding the intertwined influence of fishing intensity and the ocean environment on recruitment. This is a keystone problem in the sense that all of the problems previously cited would be solved if this problem was solved. To be specific, the term overfishing has been erroneously used to explain *any* decline in a fishery stock. Of course, stocks decline (or increase) because of a favorable or unfavorable ocean environment, or perhaps because of large increases in hatchery production. In fact, an unusual decline in a stock can be determined only after the fact, in the sense that our body of knowledge is not sufficient to determine whether the observation of a depressed year class is a chance event or whether it heralds a genuine long-term depression in the abundance of the stock.

So in shaping the problem we are quick to realize that these new objectives are in a sense a recasting of the old relatively simple approach into generally more complex, more difficult problems. This means that since in many instances the store of fundamental knowledge is exhausted, we need to think of new creative and innovative ways to shore up the fundamental knowledge base to support the applied research required by the new objectives.

1.5   Managing Ecosystems

The present state of the field is that each stock is managed independently of every other stock. While single species management is roundly criticized, techniques are generally not available to manage two interacting species, let alone an entire ecosystem. The difficulty lies at two levels. In the two-species case, there are the interactions between juveniles and adults of a stock of interest with the juveniles and adults of another stock of interest. One form of interaction relates to the trophodynamic interrelationship between the two stocks. Stomach analyses are used to study this interaction, but the problem is exceedingly difficult because of the changing size relationships between the species, and the nonlinear aspects of prey–predator interaction, and the fact that the interactions between the two species of interest exist in a setting that generally involves other stocks of fish and invertebrates. The second level relates to the survival of the eggs and larvae of the two species of interest. On one hand, the survival of eggs and larvae may be independent of explicit interactions at the egg and larval stage. On the other hand, the dynamics of egg and larval survival to some extent depend on density dependent relationships at the egg and larval stage, and these in turn depend to some extent on the production of eggs and the quality of the eggs of the two species of interest. In the case of anadromous fish such as Pacific salmon and river herring (for example), there is the added complexity of an anadromous life-history strategy that involves feeding in a vast ocean and returning to spawn sometimes in the exact area in freshwater that the fish was hatched. There are even further complications given a sometimes complex freshwater life history.

Understanding how marine ecosystems support fish and associated species is not easy. It is even more difficult now because we are changing the climate at a rate that is much greater than the rates of change that created the life-history strategies of the organisms in these ecosystems. It is not too much of an exaggeration to see the future as a crisis within fisheries science. In an oversimplified way, fisheries science finds itself trying to understand how the factors that created a particular life-history strategy will change the strategy as the factors change. Adding to this complexity, in many cases, are the changes in biology that may have occurred from fishing. Thus the complexity of the tasks of ecosystem management should be apparent to just about anyone who has thought about the situation. If progress is to be made, it is clear that a much improved understanding of these fundamental population dynamic issues needs to be acquired not only in the context of population dynamics per se, but also for population dynamics as forced by the ever-changing, multiscale physical environment.

1.6 Managing Habitat

It is generally recognized that habitat, or more precisely the quality of habitat, constrains the abundance of the harvested stocks. The habitat problem has many facets. The importance of freshwater habitat for Pacific and Atlantic salmon is well known. The impact of dams, logging, and the general destruction of spawning and rearing habitats has been extensively studied. Less well known are the impacts of changing flow and temperature patterns that are forecasted to occur in the future (see Chapter by Beechie et al.). For marine habitats, the estuarine/riverine aspect is perhaps the easiest to define. The degradation of estuaries is well-known. These degradations range from loss of productive areas, pollution, toxics, etc. These are well known, and it is up to the political system to resolve the issues, some of which might include the recognition of introduction of classes of substances not previously recognized, such as hormone inhibitors.

Moving to the open ocean, the habitat problem becomes more complex. There are two classes of problems: the benthic problem and the water-column problem.

As far as the benthic problem is concerned, most discussion relates to the effect of bottom-tending gear. There are views that bottom-tending gear should not be allowed to destroy structure. Beyond this belief is an important scientific issue, which is, how does bottom-tending gear modify the biological productivity of the bottom boundary layer? In some settings it is conceivable that bottom-tending gear decreases productivity; in other circumstances it is conceivable that it increases productivity. In other settings it is easy to imagine how continual application of bottom-tending gear could change the equilibrium species composition. The problem rapidly becomes complex because bottom-tending gear not only changes the productivity of the benthos per se through physical interaction with the bottom but also by changing the equilibrium abundance of the species of fish that feed on the benthos.

There is not much that can be said about water-column habitat except that it is obviously critical to the majority of species of concern. The scientific problems basically relate to the ecosystem, which is covered under ecosystem management. The two serious scientific issues relate to the greenhouse-gas-induced warming and acidification of open ocean habitats. The realization that the deep ocean is warming, as well as the surface waters, is a reminder that the biology of species known to be important forage for commercially important species is so poorly known that it may not be possible to forecast the impacts of a changing climate on these forage species. In the North Pacific the intensity of the winter Aleutian Low strongly affects the recruitment of many species. It is still not known if the long-term impact of greenhouse gas increases will weaken or strengthen the Aleutian Low. Natural cycles are important, as the California sardine (*Sardinops sagax*) literature shows. Thus, sorting out the impacts of natural climate trends and impacts resulting from greenhouse-gas-induced climate change is a major challenge.

1.7 Ending Overfishing

Ending overfishing is an admirable goal. However, the concept of overfishing has drifted in and out of a scientific definition. The term first arose in England in the eighteenth century. The concept was that increased fishing was correlated with the decline of the stock. As fishery science advanced, it became difficult to develop an operational definition of overfishing. As a consequence, in the early 1900s ICES (International Council for the Exploration of the Sea) formed an overfishing committee – Committee B – to study overfishing. Resulting reports did not yield a clear definition.

The subject was more or less fallow until the post-World War II era when Beverton and Holt, Ricker, and Schaefer developed various theories that linked stock size or fishing mortality and indices of stock production. These theories yield various optima or asymptotic relationships that gave the maximum production for some level of fishing mortality. Any magnitude of fishing mortality greater than the optimum could be described as overfishing. Unfortunately, there were three types of problems with these definitions. First, optima were not generally obtainable, data did not fit the theory and finally data were often not available for the most sensitive parts of the curves. Furthermore, the term stock overfishing was utilized with yield per recruit theory, and yield per recruit theory has no intrinsic optimality consequences. So it all boils down to understanding the productivity of the stock.

1.8   Precautionary Approach

A precautionary approach sounds like a good idea. In a decision-making setting, such as fishery management, a precautionary approach would suggest that we err on the side of caution. For example, if a normative fishing mortality is estimated to be $F = 0.5$, let's be cautious and set fishing mortality at $F = 0.4$.

The problem is that while it is easy to estimate $F = 0.5$, it is not easy to develop rules that set the degree of caution. For example, person A might be more cautious then person B, and so person A would set $F = 0.2$, and so on. In other words, different individuals have different degrees of risk proneness or risk aversion. An illustrative parable relates to three starving individuals waiting to cross an avenue streaming with high velocity traffic to reach a restaurant on the other side. The risk-prone individual doesn't look either way and crosses the street. He is hit by a car and killed. The risk neutral person looks both ways, carefully assesses the flow of traffic and navigates safely to the restaurant. The risk-adverse person is afraid to cross the street and as a consequence starves to death.

The subject of risk has been studied in detail in disciplines aligned with decision theory, utility theory, and risk analysis. An examination of these fields will reveal that they require higher quality and more informative data and understanding than presently available. Some of the ingredients require a definition of "risk." Risk of what? Risk of a stock becoming extinct? Recruitment failure for 1, 2, 3,… N years into the future? The stock falling below a particular level? In addition to defining the terms of risk, these techniques require probabilistic calculations of the effect of any management action or of the environment on the stock becoming extinct, etc. And finally there needs to be an assessment of the utility associated with the outcomes. As specified above, this problem becomes complex in risk-averse or risk-prone settings. Basically the state of knowledge is not well-tuned to dealing with these problems in a probabilistic setting.

1.9   Rebuilding Stocks

A common assumption in contemporary fisheries management is that all stock declines are caused by fishing and that if fishing mortality is reduced, then stock abundance would increase. There are many examples that show this assumption is not necessarily true. So how do we know when fishing mortality is coupled or uncoupled with changes in stock size?

It is important to recognize that there are major world fisheries that are healthy and at high levels of abundance. As previously mentioned, Pacific salmon catches are at historic high levels and have been at these levels since the mid-1990s, even though some stocks have declined substantially. Pacific halibut (*Hippoglossus stenolepis*) is a major fishery off the Pacific coast of North America, and by all accounts, it is in good shape. The United States fishery for walleye pollock (*Theragra chalcogramma*) in the Bering Sea remains as one of the largest fisheries in the world. A focus on rebuilding stocks is a major task for fisheries science, but it is also important that fisheries science recognizes and reports the reasons that some major North American fisheries are in very good shape.

Aquaculture is outside of the scope of this book, but aquaculture is an important component of efforts to rebuild and sustain stocks. Many people in North America have eaten *Tilapia*, farmed catfish, farmed salmon, farmed shrimp, or farmed shellfish. It is apparent that an increasing amount of "seafood" is no longer "wild" and that some aspects of fishing are turning into farming. The impact of the developing fish farming industry on the traditional fisheries remains to be determined, but there is an impact.

1.10 A Common Theme

In a way the new-found goals or mantras of fishery management exemplified by these goals have common threads. The common thread that relates to all of the problems is a basic understanding of single-species population dynamics in a multiple species setting involving physical forcing. The main issue in dealing with this problem is to define it in a manageable way, and to somehow simplify its very high dimensionality and multiscale nature. This is a major innovation that certainly includes the 5-decade-old theories of yield per recruit, production, and stock and recruitment, but at the same time recognizes that they explain only a small part of the variability in the data. Advance estimation techniques, including Bayesian analysis, refine inference. However, these advanced techniques, however useful, are still constrained by a lack of data and a 5-decade-old view of the problem (basically the effect of stock size or fishing mortality on the productivity of the stock independent of causal connections with other populations and forcing by the physical environment). With these constraints it is not surprising that existing techniques, while becoming more and more refined, do not yield basic insights into the new requirements of fishery management.

What is needed is a new theory and new observations to support the theory. It is not exactly a new idea to enhance the existing theories with environmental data. To be successful though, the enhancement needs to be more than correlational. It can readily be seen that the acquisition of new or enhanced biological theories and new data will be a complex task that cannot be accomplished with less than a critical mass of resources or within the confines of applied research. In other words, it is time to resort to the traditional scientific approach of developing new and enhanced theories, develop data sets that enable testing and amplifying the theories and then recreating this cycle.

Reliable scientific information comes slowly and is costly. We think that wise managers at all levels in our political systems now know this. Fisheries science will eventually produce reliable stock assessments that will clearly identify risk, but it is not clear when this will happen. The development of farmed and certifiably safe-cultured seafood may ease the pressure on fisheries managers and fish populations, but it is also not clear when we will get to this stage in fisheries management or how the nutritional requirements of aquacultured fish are coupled with wild fisheries. The immediate problem is that there will be dramatic changes among species in regional ecosystems around North America as climate and ecosystems change. Fisheries scientists will need to focus on understanding the dynamics of regional ecosystems. Fisheries managers will remain as a major user of science, but it will be the general public that needs to become more aware of the importance of understanding their own impacts on marine ecosystems. Thus, ecosystem management needs to include ecosystem understanding by the public because marine ecosystem health will become an index of human impacts on the planet. Our way of doing fisheries science needs to change, and change will only happen with strong teams and effective leadership. We recognize strong teams and leaders in competitive sports, but the team approach has not been popular in fisheries science. We have experimented with team approaches in the past such as GLOBEC, FOCI, and CalCOFI, and we need to use this experience to rethink how we carry out fisheries science.

We see the future of fisheries science in North America as centred on regional ecosystem-based stewardship. The best advice will come from the best teams that have the best leaders. There will be a need for more field observations and more and cheaper monitoring. A team approach effectively solves the age-old problem of data ownership. All of the skills needed to know what to measure, to make accurate measurements, to analyze the data correctly and to interpret the results would be by a team. Teams need to use the new electronic reporting technologies to provide information quickly to colleagues around the world. Thus fisheries science around North America needs to move away from the individual investigator approach and become an "ecosystem" of interdisciplinary marine stewards. Fisheries science entered a new regime when it became obvious that the changes in our climate were having dramatic impacts on the population ecology of the species in our fisheries. Fisheries science needs to move in a new direction, and the direction is beyond alarming the public. The tough task is advising what to do now. If we do not maintain fisheries, there is not much use in paying for a large number of fisheries scientists beyond the need for educators and the curious. What does a manager do tomorrow? Do banks continue to lend money to fishermen? What will happen to coastal communities that depend on fishing? It is important to fund a project that will tell us that the food base is changing, but it is equally important to recognize that the future of our fisheries may be in unchartered waters, and our clients cannot wait for years while we charter their future.

It is important that organizations such as the American Institute of Fishery Research Biologists continue to bring thoughtful people together to talk about changes, as Jones describes in her paper in this book. However, there is no organization in North America that currently represents our vision of fisheries science. There are a number of societies and organizations that offer annual opportunities for scientific exchange, but there is no focus for the regional marine fisheries science issues around North America. We believe that it is time to think about the equivalent of a North American ICES. We think that fisheries science needs a clearly thought out strategy that articulates large scales and regional priorities. The old saying that if you do not know where you are going, any road will get you there, probably applies to our current way of doing fisheries science. We need to excite fisheries science with a new spirit. For example, in the Pacific and the Atlantic we could have an "International Year of the Salmon." Fisheries scientists are drawn to the profession because of a passion for discovery and the satisfaction of working with other living things. The potential members of the potential scientific teams still have this passion. We now need the individuals to step up and lead the teams.

Chapter 2
Biodiversity, Spatial Management, and the Ecosystem Approach

Jake C. Rice

Abstract For most of its history, fisheries science has focused on the dynamics of single populations. Ecosystem considerations have gained prominence in recent decades, with attention given to predation and prey shortages as sources of mortality and environmental features as drivers of variation in recruitment, growth, and maturity. However, these are still population-based considerations, just linking the population dynamics to components of the ecosystem in which the species lives.

Keywords ecosystem approach · fisheries management · spatial management · biodiversity · economic incentives · ecosystem modelling

Endorsement of the Convention on Biological Diversity has given governments new responsibilities for conservation of biodiversity as well as the target species of fisheries. Adoption of an Ecosystem Approach in many marine policy sectors, including fisheries, has also broadened the responsibilities of our agencies. These changes have some important consequences for fisheries research in the future, two of which will be discussed in this talk.

The first consequence is that biodiversity conservation requires frameworks for assessment and accountability as rigorous and unambiguous as has been created for fisheries management. However, we lack the assessment and management decision tools for biodiversity that can function like SSB and F do for managing individual populations. The paper will also outline how the fisheries science community will need to approach the conceptual vagaries and/or quantitative intractability that characterises much of the current literature on conservation of biodiversity, if biodiversity is to be genuinely integrated into fisheries science and management.

The second consequence is that management, and the science which supports it, is going to have to shift focus from being fundamentally population based to being fundamentally spatially based. This entails much more than just looking at Marine Protected

J.C. Rice
Canadian Science Advisory Secretariat, Department of Fisheries and Oceans,
Ottawa, Canada
e-mail: Jake.Rice@dfo-mpo.gc.ca

R.J. Beamish and B.J. Rothschild (eds.), *The Future of Fisheries Science in North America*,   13
Fish & Fisheries Series, © Springer Science + Business Media B.V. 2009

Areas as the new magic bullet of fisheries management. The paper will outline these other considerations and what they imply for fisheries science in the future.

2.1   The Evolving Fisheries Agenda

Historically the business of fisheries science and management was to find ways to achieve sustainable use of the target species of the fishery. The notion of what constituted "sustainability" was vague and ill-defined for much of this period, but generally focused on keeping yields high and stable, keeping spawning biomasses large enough to avoid reduced recruitment, and finding new fishing opportunities (Smith 1994; Anderson 1998). The clients of fisheries science were fisheries management agencies, who needed advice on where to set output (and occasionally input) controls in the directed fisheries. The science support for these management tools consisted of evaluating the status and trends of the exploited populations, and of the factors that determined the productivity of those populations (Beverton and Holt 1957; Ricker 1975). These assessments were in turn supported by studies of the biology of those species, and development of increasingly sophisticated population modeling methods (Smith 1994; Rozwadowski 2002; Anderson 2000).

The concept of sustainability became more complex, but not necessarily clearer, as international conservation agreements added concepts of intergenerational options and precaution (United Nations 1992, 2002). Sustainability remained the central pillar of successful fisheries management, but with the additional recognition that sustainability requires success on at least three dimensions – ecological, economic, and social (Charles 2001; FAO 2002). Through the last quarter of the twentieth century the sustainability agenda was still largely restricted to choices among options measured by the status of and benefits from the target species of the fishery, but protection of species considered to be at risk of extinction became a more prominent consideration (FAO 1999, 2000). This expanded interpretation of sustainability led fisheries science to expand the basis for its support to management, to include multispecies assessment models (Pope et al. 1991), coupled bio-economic models (Clark 1990; Hannesson 1993), and more recently coupled biophysical models (Brander and Mohn 2004; Tjelmeland and Lindstrom 2005; King 2005), and formal management strategy evaluations (Butterworth and Punt 2003; Kell et al. 2005).

Even in this relatively narrow framework, success in conventional fisheries management was elusive. For stocks assessed by ICES and DFO, fewer than half of them are within precautionary conservation reference points for spawning biomass or fishing mortality, where such reference points have been defined (Rice in press). Even the US reports more than 30% of stock "overfished" or "experiencing over-fishing" despite legislation requiring action to remedy overfishing when it occurs (NOAA 2006a). In some extreme cases, such as the exploitation of sharks, largely as retained bycatch in mixed-species fisheries, FAO considered these fisheries to be sufficiently unsustainable that it has implemented an international plan of action for conservation and management of sharks (FAO 2002).

While fisheries science struggles to develop, test, and apply the more inclusive biophysical and bio-economic models, in the more demanding frameworks of management strategy evaluations, fisheries management has had to take on additional responsibilities as well. Accountability for performance on the social and economic dimensions of sustainability has escalated, with a rapidly expanding role for social sciences in fisheries management (Hanna et al. 1996; Durrenburger and King 2000; Neis and Felt 2001; NOAA 2003). Economic instruments are being used to augment input and output controls on fisheries (Grafton et al. 2006), revealing in the process intrinsic stresses between economic efficiencies and social dependencies of fisheries which are hard for management to resolve simultaneously (Shotten 2000; Copes and Charles 2004) Likewise accountability for the footprint of the fishery on the greater ecosystem has also escalated (Hall 1999; Barnes and Thomas 2005; Costanza 2006). Despite nearly 2 decades of focused research on the ecosystem effects of fishing (Rice 2005a; ICES 2006a), advances in management the ecosystem effects of fishing, are only slowly working into management. Spatial management measures are being implemented to address some types of impacts (Kruse et al. 2001; Wilen and Botsford 2005), and some forage species have had the needs of dependent predators considered explicitly in their exploitation strategies (Anonymous 1997; Carscadden et al. 2002; Cury et al. 2003). However, only in frameworks as complete and demanding as eco-certification is the ecosystem context of managing fisheries fully realized (FAO 2001; Peterman 2002; Marine Stewardship Council 2004).

2.2 The Global Biodiversity Agenda

While fisheries science and management has been struggling to address unsustainability of directed fisheries, accommodate environmental variability and change, achieve economic efficiency and social justice, and reduce its ecosystem impacts, the global agenda for conservation and sustainable use of the world's resources has moved on. The Rio Declaration included much more than reference to the precautionary approach, which was the product that fisheries science and management focused on as most relevant to their business. It laid the foundations for a global biodiversity policy agenda that was followed rapidly by the Convention on Biological Diversity (http://www. biodiv.org/convention/convention.shtml) and many more policy commitments on the conservation of biodiversity in WSSD (http://www.un.org/events/wssd/). The commitments made in these two settings alone constitute a framework for the use of biological resources that is far more comprehensive than how fisheries science and management has been interpreting "ecosystem approach" and the "precautionary approach." The explicit commitments in the Convention on Biological Diversity are summarized in Table 2.1. These stakes were raised by the explicit goals and commitments adopted with deadlines for achievement, some as soon as 2010, most by 2015, by WSSD, meetings of the Conference of Parties to the CBD, and supported by enabling provisions in other agreements pose major new challenges to fisheries (Table 2.2).

These commitments go beyond high-level policy statements. In fora such as the UN meeting on Biodiversity Beyond National Jurisdictions, which set the call for periodic Global Marine Assessments, and the Conference of Parties of the CBD, which adopted recommendations for developing the scientific basis for high seas marine protected areas, the agreed texts lay out specific workplans for marine biodiversity science (Table 2.3). The elements of these workplans look familiar to any fisheries scientist. They include:

Table 2.1 Selected provisions for the Convention on Biological Diversity of particular relevance to marine and freshwater fisheries science and management

Article	Provision of Convention paraphrased in fisheries context
6a	Develop national strategies ... for the conservation and sustainable use of biological diversity
6b	Integrate conservation and sustainable use of biological diversity into relevant sectoral policies
7a	Identify ecosystem components important for conservation and sustainable use of biological diversity
7b	Monitor the components of biological diversity
7c	Identify processes and categories of activities which (may) have significant adverse impacts on conservation and sustainable use of biological diversity
7d	Maintain and organize . . . data . . . from identification and monitoring activities
8a	Establish a system of protected areas or areas where special measures need to be taken to conserve biological diversity
8b	Develop guidelines for the selection, establishment and management of protected areas . . . to conserve biological diversity
8c	Promote the protection of ecosystems, natural habitats and the maintenance of viable populations of species in natural surroundings
8d	Rehabilitate and restore degraded ecosystems and promote the recovery of threatened species
8h	Prevent the introduction of, control or eradicate those alien species which threaten ecosystems, habitats or species
8i	Endeavor to provide the conditions needed for compatibility between present uses and the conservation of biological diversity and the sustainable use of its components
8j	Respect, preserve and maintain knowledge, innovations and practices of indigenous and local communities embodying traditional lifestyles relevant for the conservation and sustainable use of biological diversity
8l	Where a significant adverse effect on biological diversity has been determined pursuant to Article 7, regulate or manage the relevant processes and categories of activities and
10a	Integrate consideration of the conservation and sustainable use of biological resources into national decision-making
10b	Adopt measures relating to the use of biological resources to avoid or minimize adverse impacts on biological diversity
10c	Protect and encourage customary use of biological resources in accordance with traditional cultural practices . . . compatible with conservation or sustainable use
12a	Establish and maintain programmes for . . . education and training in measures for the identification, conservation and sustainable use of biological diversity
12b	Promote and encourage research which contributes to the conservation and sustainable use of biological diversity
12c	Promote and cooperate in the use of scientific advances in biological diversity research in developing methods for conservation and sustainable use of biological resources

Table 2.2 Explicit goals and objectives under the three pillars of the Convention for Biological Diversity: conservation of biological diversity; the sustainable use of its components; and, the fair and equitable sharing of benefits arising from the use of genetic resources, and related provisions of other agreements

Goal 1: Promote the conservation of the biological diversity of ecosystems, habitats and biomass

Target 1.1: At least 10% of each of the world's ecological regions effectively conserved

Convention on Biological Diversity (CBD) Decision VII, paragraph 21

World Parks Congress in recommendation 5.23

Target 1.2: Areas of particular importance to biodiversity such as tropical and cold water coral reefs, seamounts, hydrothermal vents, mangroves, seagrasses, spawning, aggregations and other vulnerable areas in marine habitats effectively protected

CBD Decision VII/5, paragraph 61

Fifth meeting of the Informal Consultative Process on Oceans and the Law of the Sea (ICP) Recommendation 6a

Third informal consultation on the United Nations Fish Stocks Agreement

WSSD paragraph 32c

Goal 2: Promote the conservation of species diversity

Target 2.1: Reduce the decline of, restore, or maintain, populations of species of selected taxonomic groups

WSSD paragraph 31a (at least B_{msy})

WSSD paragraph 32c

IUCN Red List – framework for comprehensive assessment of marine species

FAO Code of Conduct

FAO International Plans of Action (IPOA) for the Conservation and Management of Sharks

FAO IPOA for Reducing Incidental Catch of Seabirds in Longline Fisheries

FAO IPOA for the Management of Fishing Capacity

Target 2.2: Status of threatened species improved

World Parks Congress Recommendation 5.04

IUCN Red List for Marine Species

Convention on International Trade in Endangered Species (CITES)

Also targets in Regional Seas Conventions and Programmes

Goal 3: Promote the conservation of genetic diversity

Target 3.1: Genetic diversity of … harvested species of trees, fish and wildlife and other valuable species conserved, and associated indigenous and local knowledge maintained

CBD states that activities to reach this target should be implemented together with those associated with goals 1, 2, 4, 5, 7, and 8)

Goal 4: Promote sustainable use and consumption

Target 4.1: Biodiversity-based products derived from sources that are sustainably managed, and production areas managed consistent with the conservation of biodiversity

Target 4.1.1: All exploited fisheries products derived from sources that are sustainably managed, and unsustainable uses of other marine and coastal species minimized

WSSD paragraph 31a

WSSD paragraphs 31b–f and 32c

Article 7 of the FAO Code of Conduct for responsible fisheries

(continued)

Table 2.2 (continued)

Target 4.1.2: All mariculture facilities operated consistent with the conservation of biodiversity and social equity
Article 9 of the FAO Code of Conduct for responsible fisheries
WSSD paragraph 31h

Target 4.2: Unsustainable consumption of biological resources, or that impacts upon biodiversity reduced
Same paragraphs as subtargets 4.1.1 and 4.1.2

Target 4.3: No species of wild flora or fauna endangered by international trade
Marine Aquarium Council Certification Schemes (for ornamentals)
Convention on International Trade in Endangered Species (CITES)

Goal 5: Pressure from habitat loss, land use change and degradation, and unsustainable water use, reduced
Target 5.1: Rate of loss and degradation of natural habitats decreased
WSSD paragraph 32c

Goal 6: Control threats from invasive alien species
Target 6.1: Pathways for major potential alien invasive species controlled
International Convention on the Control and Management of Ship's Ballast Water and Sediments by the International Maritimes Organization

Target 6.2: Management plan in place for major alien species that threaten ecosystems, habitats or species
International Convention on the Control and Management of Ship's Ballast Water and Sediments by the International Maritimes Organization

Goal 7: Address challenges to biodiversity from climate change and pollution
Target 7.1: Maintain and enhance resilience of the components of biodiversity to adapt to climate change
Coral Bleaching Workplan
CBD Decision VII/5, appendix 1
CBD activities under goals 1–6 and 8

Target 7.2: Reduce pollution and its impacts on biodiversity
UNEP GEO Yearbook 2003
WSSD paragraphs 33 and 34
Global Programme of Action for the protection of the marine environment from land-based activities
International Maritime Organization (particularly sensitive sea areas)
Relevant components of the United Nations Convention on the Law of the Sea (UNCLOS), London and PoPs Conventions
CBD Operational Objective 1.2 in Decision VII/5 annex 1

Goal 8: Maintain capacity of ecosystems to deliver goods and services and support livelihoods
Target 8.1: Capacity of ecosystems to deliver goods and services maintained
WSSD para 29d

Target 8.2: Biological resources that support sustainable livelihoods, local food security and health care, especially of poor people, maintained
CBD Decision VII/5 – annex 1
Millennium Development Goals (MDG), activities under targets 1, 2, 4, 5 and 7

(continued)

Table 2.2 (continued)

Goal 9: Maintain socio-cultural diversity of indigenous and local communities

    Target 9.1: Protect traditional knowledge, innovations and practices

        Millennium Development Goal 9

        Rio Agenda 21

        Article 8j of the CBD

    Target 9.2: Protect the rights of indigenous and local communities over their traditional knowledge, innovations and practices, including their rights to benefits sharing

        Same as above 9.1

Goal 10: Ensure the fair and equitable sharing of benefits arising out of the use of genetic resources

    Target 10.1: All transfers of genetic resources are in line with the Convention on Biological Diversity, the International Treaty on Plant Genetic Resources for Food and Agriculture and other applicable agreements

        CBD Bonn Guidelines

        CBD Decision VII/9

        Relevant provisions of the United Nations Convention on the Law of the Sea (UNCLOS)

    Target 10.2: Benefit arising from the commercial and other utilization of genetic resources shared with the countries providing such resources

        Same as 10.1

Goal 11: Parties have improved financial, human, scientific, technical and technological capacity to implement the Convention

    Target 11. 1: New and additional financial resources are transferred to developing countries parties, to allow for the effective implementation of their commitments under the Convention, in accordance with Article 20

        Article 20 of the convention on biological diversity

    Target 11.2: Technology is transferred to developing country parties, to allow for the effective implementation of their commitments under the Convention, in accordance with Article 20, para 4

        Same as 11.1

Table 2.3a Charge in the UN General Assembly call for periodic Global Marine Assessments (http://www.unep.ch/regionalseas/partners/p_gma.htm)

"The marine assessment process should address all dimensions of marine ecosystems, including the physical and chemical environment, biota, and socioeconomic aspects

The assessments would address the state of marine ecosystems, causes of change, benefits derived from marine ecosystems, and threats and risks

The geographic scope of the assessments should span coastal and estuarine waters through ocean basins, taking account of terrestrial and atmospheric influences

As to the topics to be addressed in the assessment process, the Group of Experts suggested that a list of issues or activities could include:

> ➤ Intentional large-scale perturbations of the open ocean, such as deliberate fertilization and carbon sequestration;
> ➤ Effects of habitat degradation in the marine environment on fisheries
> ➤ Assessment of deep-sea and open-ocean conditions (e.g., biodiversity, productivity) integrated across all oceans;
> ➤ Increased atmospheric input of nitrogen to the oligotrophic open ocean;
> ➤ Review of methodologies for the economic valuation of marine ecosystem services;
> ➤ Implications of coastal degradation for human health and safety; and
> ➤ Best practices for particular emerging uses of the ocean."

…"The Group also recommended the establishment of a Global Scientific Assessment Panel."

(continued)

Table 2.3b Workplan adopted at the 2006 Conference of Parties to the Convention on Biological Diversity, with regard to providing a scientific foundation for establishing and managing Marine Protected Areas on the High Seas (http://www.biodiv.org/doc/meeting.aspx?mtg = COP-08)

Paragraph COPVIII/21/7

Requests the Executive Secretary, in collaboration with the United Nations Division
for Ocean Affairs and the Law of the Sea, and other relevant international organizations,
to further analyze and explore options for preventing and mitigating the impacts of some
activities to selected seabed habitats and report the findings to future meetings
of the subsidiary body on scientific, technical and technological advice

COPVIII/24/44

Requests the Executive Secretary to work actively with, and to take into account
scientific information available from, the range of relevant expertise available in governmental,
intergovernmental, non-governmental, regional and scientific institutions, expert scientific
processes and workshops, and, indigenous and local communities, where appropriate, to: . . .

(a) Synthesize, with peer review, the best available scientific studies on priority areas
for biodiversity conservation in marine areas beyond national jurisdiction, including
information on status, trends and threats to biodiversity of these areas as well as distribution
of seamounts, cold water coral reefs and other ecosystems, their functioning and the ecology
of associated species, and to disseminate this through the clearing-house mechanism

(b) Refine, consolidate and, where necessary, develop further scientific and ecological criteria
for the identification of marine areas in need of protection, and biogeographical and other
ecological classification systems, drawing on expertise and experience at the national and
regional scale

(c) Collaborate in the further development of spatial databases containing information
on marine areas beyond the limits of national jurisdiction, including the distribution
of habitats and species, in particular rare or fragile ecosystems, as well as the habitats
of depleted, threatened or endangered species, and data on national and regional marine
protected areas and networks

(d) As appropriate, facilitate work relating to scientific issues, including those raised in annex II
of the report of the Ad Hoc Informal Open-ended Working Group to study issues relating to
the conservation and sustainable use of marine biological diversity in areas beyond national
jurisdiction

(e) Collate information concerning customary use of biological resources in accordance with
traditional cultural practices that are compatible with the conservation and sustainable use
of biological diversity in marine areas beyond the limits of national jurisdiction

COP VIII/24/45. Urges Parties and other Governments to undertake and actively promote
scientific research and information exchange, and to cooperate with the Executive Secretary
on the activities proposed in paragraph 44 above

- Assessing status and trends of components of marine ecosystems
- Assessing the effects of environmental forcers on the dynamics of components of marine ecosystems
- Assessing the impacts of specific human activities on marine ecosystems and their components
- Assessing rates, forms, and consequences of habitat degradation
- Partitioning causation of observed trends among multiple potential causal factors

This type of workplan is exactly what fisheries scientists have done – or tried to do – for decades with exploited resources. Management, likewise, has tried to ensure sustainable uses of the exploited stocks based on the science as it comes available. As noted above, the path has not been smooth, easy, or fast to travel. Now those two communities are challenged to cover the same ground for the full spectrum of marine biodiversity – and do so in the next few years. For the rest of this paper I will consider each of the major cutting-edge tools we have been using recently in "target-resource-oriented management" (TROM – as per FAO 2003), and the science supporting TROM, with regard to the potential the current favored tools have for addressing the ambitious and imminent commitments and workplan posed by the global marine biodiversity agenda.

2.3 Evaluation of Fisheries Science and Management Tools for Addressing Biodiversity

2.3.1 The New Generation of Ecosystem Models

Multispecies assessment models have contributed to improved assessment advice, through including predator–prey interactions (including sometimes cannibalism) in the estimates of natural mortality of assessed species (Hollowed et al. 2000; Steffansson 2003). However, to perform reliably they are highly demanding of data on species' diets across a range of predator and prey sizes, seasons, and years. Like typical assessment models most multispecies assessment models also require catch at age data on the species being assessed together. These two data requirements greatly constrain the potential contribution that these models can make to assessing status and trends of the broad range of marine biodiversity.

More general trophodynamic models can to some extent circumvent the data demands of multispecies assessment models through additional assumptions about mass-balance or other system-level constraints (Kooijman 2000; Shannon et al. 2003). The substitution of system-wide constraints as structural assumptions in the models for actual data on who eats whom and how abundant the various species are during the parameterization period, allow these generalized trophodynamic models to investigate many scenarios with implications for the conservation of biodiversity. However, by working with relatively few firm data constraints these "ecosystem" models must either work at very aggregated scales of biological diversity, or if disaggregated into many species or even life-history stages of species, they necessarily provide substantial scope for self-validating judgment calls by the users. When the former strategy is followed these models can provide limited insight into all but the most coarse-scale questions about the conservation and sustainable use of biodiversity. When the latter strategy is followed, the highly structured but weakly constrained mass-balance models often perform substantially at variance with other models of the same systems (Plaganyi and Butterworth 2004; Koen-Alonzo and Yodzis 2005), suggesting that their quantitative results should be viewed with great caution.

The third class of new models from fisheries management are the coupled biophysical models. As noted earlier there are models which do well at capturing the effects of environmental forcing on stock dynamics (Brander and Mohn 2004; PICES 2005; Schrum et al. 2006). However, in general the successes are models linking ocean physics to the dynamics of single (or rarely two or three tightly interacting) species that have been subjected to intensive directed study (ICES 2007). These are significant advances, but neither scientific resources nor time is adequate to make this approach the standard for addressing the global biodiversity conservation issues. Answers are needed much sooner, and for biodiversity generally, not just for the few most intensively studied species.

If the fisheries ecosystem modeling community is going to become a major contributor to the new demands for conservation of biodiversity, they have to begin to investigate new questions. The research on what features of a species' productivity make it resilient to exploitation have to generalize into what properties of an aquatic ecosystem give it its integrity, resilience, and general ecosystem "health" (Rice 2003; Hewitt et al. 2005; Hughes et al. 2005)? This is a question that has been explored in depth for terrestrial ecosystems (Gundersen and Pritchert 2002; Elmqvist et al. 2003), but for reasons discussed later in the paper, the answers cannot all be assumed to generalize from terrestrial to marine systems without some adaptations of unknown but possibly large magnitude.

Also, just as ecosystem modelers working on traditional fisheries depend on field data, ecosystem modeling in a biodiversity context expands the information needed from monitoring aquatic systems. It will be necessary to quantify what are natural patterns of interannual variation in biodiversity components and properties of ecosystem. This has been challenging in single-species assessment and management, and is likely to be at least as difficult in the broader biodiversity context. It is also necessary to get a better understanding of how environmental forcing affects whole interacting systems. There is growing acceptance of the concept of ecosystem "regime shifts," but there is a very weak grasp of even what ought to be predictable and what is not predictable across regimes, although such questions are central to conservation and sustainable use of biodiversity (Beamish et al. 2004). Finally, a "fishing out" phase has long been an accepted practice in single species fisheries. Biodiversity conservation requires reconsidering how such "managed" reductions affect the ecosystem processes maintaining biodiversity. All these are questions which the fisheries ecosystem modeling community must address on a priority basis, if progress is to be made on the many commitments to conservation of biodiversity.

2.3.2 Management Strategy Evaluations (MSE) and Biodiversity

Over the past decade the modeling framework supporting fisheries has expanded to full management strategy evaluations. This framework has the capacity to include biodiversity considerations in all four core components – the objectives to be met by the strategies, the harvest control rules tested within the framework, the operating model being used in the simulation, and the types of robustness tests that are applied within the framework. The ways that biodiversity considerations might be incorporated in each component are discussed in ICES (2005, 2006a, b). However, it is also important to note that although the potential to add biodiversity concerns to the MSE framework are present, a review of 16 case histories being examined in ICES (2006b) found one indirect biodiversity-related objective, and no biodiversity considerations in any harvest control rules or operating models. In only one case was there even an environmental forcer in a harvest control rule, and only four cases where an environmental forcer was included in the operating model. None of the case histories tested the robustness of any control rules to any consideration other than the state of the target species and features of the directed fisheries.

As with ecosystem modeling, the MSE framework is capable of being used to address issues of conservation and sustainable use of biodiversity, but the potential is not being explored. Again, experts working in this framework have to begin to ask new sets of questions. One key is to be able to address seriously what biodiversity objectives should be made operational and part of management systems to begin with? It is easy to make long lists of ecosystem objectives, but making them operational within management systems is much more challenging (Sainsbury and Sumaila 2003; Rice and Rochet 2005; CSAS 2007). Making conceptual biodiversity objectives operational requires determining indicators that are not just scientifically interesting, but functional in management in ways that SSB and F can be used to directly guide management advice (ICES 2002; Nicholson and Jennings 2004; Greenstreet and Rogers 2006). That, in turn requires determining how Target and Limit Reference Points should be positioned on the practical biodiversity indicators. Finally, addressing biodiversity considerations in a MSE framework will require determining what control rules have been documented and predictable consequences for status on indicators relative to reference points, without their utility being compromised by "implementation uncertainty" (Richards and Rice 1996; Scott 1998). The scientific community is only at the preliminary stages of exploring these important questions.

2.3.3   Use of Economic Instruments to Address Biodiversity Considerations

Economic incentives have received so much recent attention as an essential part of sustainable fisheries because they have the advantage of bringing goals of harvesters and managers into accord. They make it possible for the individuals who pay the short-term price for conservation actions to also reap the longer-term benefits of their short-term sacrifices (Bromley 1991; Devlin and Grafton 1998). This is possible within a TROM framework because the currency used in accounting is consistent throughout; the value of catch foregone now is offset by increased value of greater catches in future. Unfortunately this consistency of accounting will not transfer directly to biodiversity considerations. Many of the benefits most desired by those advocating conservation of biodiversity are aspects of protecting ecosystem structure and function (Pikitch et al. 2004) and ecosystem services essential to healthy coastal ecosystem and coastal residents (Costanza 2006). Their diffuse nature means that they cannot be uniquely harvested or the benefits otherwise monopolized by those making the short-term (or not so short term) sacrifices. There may be ways to quantify the value of ecosystem benefits that would accrue from better protection of biodiversity. However, it is far from straightforward to see how these would be converted into currencies that would form economic incentives for fish harvesters to increase the selectivity of their gears, decreased effort, and overall footprint of their fisheries, and fund better surveillance and compliance – all documented consequences of at least some cases when economic instruments have been used in fisheries (Rice 2008).

Again, we find ourselves posing a new class of questions for the research community specializing in the use of economic instruments fisheries. First, how far is it possible to go with valuation of biodiversity goods and services? Proponents are enthusiastic about the concept (Jorgensen et al. 2005) but it needs a hard look in the context of marine biodiversity. More importantly, how can valued biodiversity goods and services be converted into direct economic incentives for fishers to restrain activity? This is much more demanding than just documenting that under some circumstances MPAs increase catches. The framework of economic incentives may indeed be appropriate for biodiversity conservation as well as for TROM, but it looks like it will require developing a system of accounting and financial transfers for which I know of no terrestrial counterparts, despite the much longer history of using economic instruments in the conservation of terrestrial biodiversity. Eco-certification may provide that accounting system, but it places all the costs on those in the fishery and requires only those choosing to purchase their products to provide all the financial incentives. It is society in general that benefits from – and increasingly is demanding – the conservation of biodiversity, not just those choosing to pay a market premium for certified fish products.

2.3.4   Stewardship, Co-management and Biodiversity

Co-management and shared stewardship are major pillars in the contemporary approach to fisheries management (Berkes et al. 2000; Olsson et al. 2004), reflecting that governments have finally learned that command-and-control management from the top rarely achieves the desired outcomes. For co-management to success it is essential that all parties agree on what is being managed or "stewarded," and that all the parties accept the legitimacy of the other players at the co-management table. In TROM, the former has usually been straightforward although not always easy (Grey 2005), whereas in fisheries with many sectors (many gears or commercial and recreational) the latter has been more challenging (cf. Browman and Stergiou 2006). Successes in co-management approaches to complex fisheries are possible, as the Puget Sound/Gulf of Georgia salmon initiative documents (NOAA 2005), but they require years to develop and major commitments of institutional resources and time from all parties (e.g., Harwood and Stokes 2003).

There is no question that the lessons learned about the need for participatory governance in fisheries will have to be transferred to the governance structures for conservation and sustainable use of biodiversity. However, from my personal experience in the Canadian Large Ocean Management Area (LOMA) initiatives (DFO 2006), the US external Ecosystem Task Team (NOAA 2006b), and the ICES Expert Group developing the guidance document for the EU Marine Strategy (European Community 2005; Rice et al. 2005), the entire fisheries governance community has been one of the most reluctant participants in the larger integrated management initiatives inherent in these moves to greater ecosystem focus in management. Meaningful progress on conservation and sustainable use of biodiversity requires much broader definitions of the problems that management has to solve. This reluctance to make decision-making in fisheries simply a part of a larger, more integrated management system is not an encouraging signal that progress will be swift in making biodiversity considerations central to fisheries governance.

To improve this situation, those working on cooperative governance of fisheries have to address several issues. First, there need to be clear guidelines for legitimacy of a very broad range of stakeholder in biodiversity, without the industry sector and fisheries managers necessarily having priority over other biodiversity interests. Because the current fisheries governance systems are unlikely to become simply one player in wider integrated management in the near future, the existing fisheries governance systems need to broaden their activity base and accountability quickly. The US regional fisheries management councils must start to give biodiversity goals priority over harvest goals, not just discuss them as an extra agenda item. In Canada the fishing industry and Fisheries Management need to become fully committed to the integrated management initiatives driven by the Oceans Act and the LOMAs. In Europe the gap must be bridged between implementation of the Species and Habitats Directives via DG Environment and OSPAR (European Community 1992) and the Common Fisheries Policy through the new DG Fisheries and the new Regional Advisory Councils (European Community 2007).

2.3.5   Spatial Management and Biodiversity

Marine Protected Areas (MPAs) are emerging as a preferred tool for addressing marine biodiversity considerations in management. Policy commitments are in place internationally (Aqorau 2003, Table 2.2), as well at national and regional scales. The policy commitments and their operational science support are intended to ensure both features of special ecological significance and representative slices of all major biogeographic marine areas are given protection (CBD 2006). Evidence is accumulating that in many cases MPAs can provide conservation benefits to species, habitats, and communities, and at least sometimes to fisheries on the margins of the MPAs (Roberts et al. 2002; Murawski et al. 2000; Wilen and Botsford 2005). Although the studies showing these benefits often fail to partition the effects of the spatial aspect of MPAs from their effect of reducing overall catch or effort (Pastoors et al. 2000; Hilborn et al. 2004), and often do not account for the ecological footprint of increased fishing effort in new areas where more effort may be needed to take the same catch in areas of lower fish density (Jennings et al. 2005), Marine protected areas at least partially address 6 of the 11 biodiversity targets in Table 2.2.

There is no question that MPAs will be an important part of management measures implemented for conservation and sustainable use of biodiversity. However, it is important to note that they are better suited to be used as a tool in some contexts than others. The enthusiasm for protected areas as a tool for conservation of biodiversity was imported from terrestrial ecology and management, where their effectiveness was founded on some features intrinsic to terrestrial ecosystems. As developed in detail in Rice (2005b) the key habitat features of terrestrial ecosystems are persistent and predictable (at least probabilistically) on local spatial scales (often down to meters or centimeters) and temporally out to scales of seasons or even longer. These features of persistence are the basis for the three-dimensional complexity and patchiness of terrestrial systems, as well to regulating the carrying capacity of terrestrial ecosystems for species or communities.

In marine ecosystems, similar scales of persistence and predictability of habitat features is found in benthic and estuarine habitats but not pelagic ones. There the key habitat processes of productivity, retention, and dispersal (Bakun 1996) generally are predictable only on spatial scales of tens of kilometers upward and on temporal scales of hours to a few weeks (Rice 2005b). These differences in scales greatly affect how competition and carrying capacity work in pelagic systems, and those differences require new thinking about if, where, and how to implement place-based management. It is particularly important to consider how to protect biodiversity through representative MPAs when habitat features are not stably associated with places.

The easy answer might be to just make the protected places bigger, so that integrated over the probabilistic distribution of the habitat features there is still a high likelihood of protecting functionally "representative" systems. Such proposals have been made, such as the one by Greenpeace in spring 2006 to place slightly more than 40% of the world's oceans in no-take MPAs (Greenpeace 2006), but global and national governance systems have not considered that scale of closures of the sea to even be a constructive starting point for dialogue (United Nations 2006). Yet another priority for fisheries science and management is to confront the special challenges that the fundamental dynamic processes of pelagic, and to some extent demersal, marine ecosystems pose for spatial management approaches. Then we will know if and how the concepts of MPAs can be adapted to those systems, and build on the lessons being learned about how to use MPAs to address biodiversity concerns in benthic and near-coastal systems.

2.4 Concluding Comments

The global biodiversity policy agenda is much more comprehensive than the global fisheries policy agenda, and it is gaining momentum and constituents at national and international scales. The agenda has been largely ignored by the core fisheries science and management community, who have focused their innovative thinking largely on placing fisheries on a broader basis both ecologically and in terms of fisheries governance. In this area there have been successes, many reviewed in other sections of this book. However, scientific progress has taken significant time and resources, and movement of scientific progress into management has been slow.

At the same time, the global biodiversity community has identified ambitious targets and timetables, that if achieved will overtake the pace of change in fisheries, and render much of fisheries science and management largely inconsequential in the big picture. This raises the spectre that policies and decisions which have much more scope and impact than fisheries policies will be made with little engagement by the fisheries science and management communities. Such changes are not certain to occur, but there are many indications that the global trends are in that direction. The implications and risks are too profound to ignore.

When the typical workplans of biodiversity initiatives are considered, the work truly is the business of fisheries research, just done on much broader seascapes. Fisheries science needs to move swiftly into that business, if we are to continue to have a leadership role in supporting marine policy and management. Most of the challenges we would have to take on are not different in kind from the current challenges we address, only different in scope. Ecosystem modelers need to ask new questions that will be more complex and data hungry. Management Strategy Evaluations need to start taking the environment and ecosystem seriously in objectives, operating models, decision rules, and robustness testing. Economic instruments need to address the concepts of rights and entitlements relative to biodiversity, as well as equity of distribution of costs and benefits on the various space and time scales. Co-management will be very complex as biodiversity empowers lots of new players, and the reluctance of fisheries to be part of integrated management governance systems needs to be confronted. The notion of habitat in pelagic and demersal ecosystems needs more serious thinking, so spatial management tools can be adapted for effective application there. Together this agenda promises the fisheries science and management community a future as exciting and interesting as the past 50 years. They just need to achieve it all in the next decade or less.

Acknowledgments I would like to thank the International Policy team at DFO, particularly L. Ridgeway, R. Sauve, and J. Mooney for bringing me into the international biodiversity policy deliberations as a science advisor, allowing me to broaden my perspective on what our jobs in fisheries science and management really are.

References

Anderson, E.A. (1998) The history of fisheries management and scientific advice: the ICNAF/ NAFO history from the end of World War II to the present. Journal of Northwest Atlantic Fisheries Science 23: 75–94.

Anderson, E.A. (ed) (2000) 100 Years of Science Under ICES: A symposium held in Helsinki 1–4 August 2000. ICES Marine Science Symposia # 215. ICES, Copenhagen.

Anonymous (1997) Forage Fishes in Marine Ecosystems. Lowell Wakefield Fisheries Symposium Series 14, Alaska Sea Grant AK-SG-97–01. Fairbanks, AK. 828 pp.

Aqorau, T. (2003) Obligations to protect marine ecosystems under international Conventions and other legal instruments. In Sinclair, M., Valdimarsson G. (eds) Responsible Fisheries in the Marine Ecosystem. FAO, Cambridge, pp 25–40.

Bakun, A. (1996) Patterns in the Ocean: Ocean Processes and Marine Population Dynamics. California Sea Grant, San Diego, CA, 323 pp.

Barnes, P.W., Thomas, J.P. (eds) (2005) Benthic Habitats and the Effects of Fishing. American Fisheries Society Symposium 41. AFS, Washington, DC, 890 pp.

Beamish, R.J., McFarlane, G.A., King, J.A. (2004) Regimes and the history of the major fisheries off Canada's west coast. Progress in Oceanography 60: 355–385.

Berkes, F., Colding, J., Folke, C. (2000) Rediscovery of traditional ecological knowledge as adaptive management. Ecological Applications 10: 1251–1262.

Beverton, R.J.H., Holt, S.J. (1957) On the Dynamics of Exploited Fish Populations. Fisheries Investigations Series II:19, UK Ministry of Fisheires, Lowestoft, UK.

Brander, K., Mohn, R. (2004) Effect of the North Atlantic oscillation on recruitment of Atlantic cod (Gadus morhua). Canadian Journal of Fisheries and Aquatic Sciences 61: 1558–1564.

Bromley, D.W. (1991) Environment and Economy: Property Rights and Public Policy. Blackwell, Cambridge.

Browman, H.L., Stergiou, K.I. (eds) (2006) Politics and socio-economics of ecosystem-based management of marine resources. Marine Ecology Progress Series 2006: 1–73.

Butterworth, D.S., Punt, A.E. (2003) The role of harvest control laws, risk and uncertainty and the precautionary approach in ecosystem-based management. In Sinclair, M., Valdimarsson G. (eds) Responsible Fisheries in the Marine Ecosystem. FAO, Rome, pp 311–319.

Carscadden, J.E., Montevecchi, W.A., Davoren, G.K., Nakashima, B.S. (2002) Trophic relationships among capelin (Mallotus villosus) and seabirds in a changing ecosystem. ICES Journal of Marine Science 59: 1027–1033.

CBD (2006) Convention on Biological Diversity Programs and Issues: Protected Areas. http:// www.biodiv.org/programmes/cross-cutting/protected/default.asp

Charles, A.T. (2001) Sustainable Fishery Systems. Fish and Aquatic Resources Series No. 5. Blackwell Science, Oxford, 370 pp.

Clark, C.W. (1990) Mathematical Bioeconomics – The Optimal Management of Renewable Resources. Wiley, New York.

Copes, P., Charles, A.T. (2004) Socioeconomics of individual transferable quotas and community-based fishery management. Agricultural and Resource Economics Review 17: 171–181.

Costanza, R. (2006) Thinking broadly about costs and benefits in ecological management. Integrated Environmental Assessment and Management 2: 166–173.

CSAS (2007) Guidance document on identifying conservation priorities and phrasing conservation objectives for large ocean management areas. CSAS Science Advisory Report 2007: 02.

Cury, P., Shannon, L., Shin, Y.-J. (2003) The functioning of marine ecosystems: a fisheries perspective. In Sinclair, M., Valdimarsson G. (eds) Responsible Fisheries in the Marine Ecosystem. FAO, Rome, pp 103–123.

Devlin, R.A., Grafton, R.Q. (1998) Economic Rights and Environmental Wrongs: Property Rights for the Common Good. Edward Elgar Press, Cheltenham.

DFO (2006) Ocean Action Plan: Integrated management. www.dfo-mpo.gc.ca/oceans/ management-gestion/integratedmanagement-gestionintegree.

Durrenburger, E.P., King, T.D. (eds) (2000) State and Community in Fisheries Management: Power, Policy, and Practice. Bergen and Garvey Press, Westport, CT.

Elmqvist, T., Folke, C., Nystrom, M., Peterson, G., Bengtsson J., Walker, B., Norberg, J. (2003) Response diversity, ecosystem change, and resilience. Frontiers in Ecology and Environment 1: 488–494.

European Community (1992) Council Directive 92/43 on the Conservation of Natural Habitats and of Wild Fauna and Flora. http://www.jncc.gov.uk/page = 1374

European Community (2005) Directive of the European Parliament and of the Council establishing a Framework for Community Action in the field of Marine Environmental Policy (Marine Strategy Directive) [SEC (2005) 1290], In COM(2005) 505 Final, pp. 31, Brussels.

European Community (2007) European Fisheries: General Information. http://ec.europa.eu/ fisheries/faq/general_en.htm

FAO (1999) Report of the Meeting of the FAO *AD HOC* Expert Group on Listing Criteria for Marine Species under CITES, Cape Town, South Africa, 20 November 1998. COFI/99/Inf.16 Part II. http://www.fao.org/docrep/meeting/x0702e.htm

FAO (2000) An appraisal of the suitability of the CITES criteria for listing commercially exploited aquatic species. FAO Fisheries Circular No. 954. 66 pp.

FAO (2001) Product certification and ecolabelling for sustainability. FAO Fisheries Technical Paper 442: 83.

FAO (2002) International Plan of Action for the Conservation and Management of Sharks. http:// www.fao.org/fi/website/FIRetrieveAction.do?dom = org&xml = ipoa_sharks.xml

FAO (2003) The Ecosystem Approach to Fisheries. FAO Technical Guidelines for Responsible Fisheries. No. 4, Suppl. 2, 112 pp. FAO, Rome. ftp://ftp.fao.org/ docrep/fao/005/y4470e/ y4470e00.pdf

Grafton, R.Q. et al. (2006). Incentive-based approaches to sustainable fisheries. Canadian Journal of Fisheries and Aquatic Science 63: 699–710.

Greenpeace (2006) Defending Our Oceans: A Map of Proposed High Seas Marine Protected Areas. http://oceans.greenpeace.org/en/documents-reports/roadmap

Greenstreet, S.P.R., Rogers, S.I. (2006) Indicators of the health of the North Sea fish community: identifying reference levels for an ecosystem approach to management. ICES Journal of Marine Science 63: 573–593.

Grey, T.S. (ed) (2005) Participation in Fisheries Governance. Kluwer/Springer, The Netherlands.

Gundersen, L.H., Pritchard, L. (eds) (2002) Resilience and the Behaviour of Large-Scale Systems. Island Press, Washington, DC.

Hall, S.J. (1999) The Effects of Fishing on Marine Ecosystems and Communities. Blackwell, Oxford, 396 pp.

Hanna, S., Folke, C., Maler, K.-G. (eds) (1996) Rights to Nature: Ecological, Economics, Cultural and Political Principles of Institutions for the Environment. Island Press, Washington, DC.

Hannesson, R. (1993) Bioeconomic Analysis of Fisheries an FAO Fishing Manual. Blackwell, Oxford, 144 pp.

Harwood, J., Stokes, K. (2003) Coping with uncertainty in ecological advice: lessons from fisheries. Trends in Ecology and Evolution 18: 617–622.

Hewitt, J.E., Anderson, M.J., Thrush, S.F. (2005) Assessing and *monitoring* ecological *community* health in marine systems. Ecological Applications 15: 942–953.

Hilborn, R.L., Garcia, S., et al. (2004) When can marine protected areas improve fisheries management? Ocean and Coastal Management 47: 197–205.

Hollowed, A.B., Bax, N., Beamish, R.J., Collie, J., Fogarty, M., Livingston, P., Pope, J.G., Rice, J.C. (2000) Are multispecies models an improvement on single species models for measuring fishing impacts on marine ecosystems?. ICES Journal of Marine Science 57: 707–719.

Hughes, T.P., Bellwood, D.R., Folke, C., Steneck, R.S., Wilson, J. (2005) New paradigms for supporting resilience of marine ecosystems. Trends in Ecology and Evolution 20: 380–386.

ICES (2002) Report of the Advisory Committee on Fisheries Management 2001. ICES Cooperative Research Report 255 (3 vols).

ICES (2005) Report of the Study Group on Management Strategies, ICES CM/2005/ ACFM:09. 72 pp. http//:www.ices.dk/reports/ACFM/2005/SGMAS/SGMAS05.pdf

ICES (2006a) Report of the Working Group on Ecosystem Effects of Fishing ICES CM 2006 Ace:05. http://www.ices.dk/reports/ACE/2006/WGECO06.pdf

ICES (2006b) Report of the Study Group on Management Strategies. ICES CM 2006/ ACFM:15, 165 pp.

ICES (2007) Report of the Working Group on Regional Ecosystem Descriptions ICES CM 2007:ACE/01, 160 pp.

Jennings, S., Freeman, S., Parker, R., Duplisea, D.E., Dinmore, T.A. (2005) Ecosystem consequences of bottom fishing disturbance. American Fisheries Society Symposium 41: 73–90.

Jorgensen, S.E., Costanza, R., Xu, F.-L. (eds) (2005) Ecological Indicators for Assessment of Ecosystem Health. CRC Press, Boca Raton, FL, 439 pp.

Kell, L.T., Pastoors, M.A., Scott, R.D., Smith, M.T., Van Beek, F.A., O'Brien, C.M., Pilling, G.M. (2005) Evaluation of multiple management objectives for Northeast Atlantic flatfish stocks: sustainability vs. stability of yield. ICES Journal of Marine Science 62: 1104–1117.

King, J.R. (ed) (2005) Report of the Study Group on the Fisheries and Ecosystem Responses to Recent Regime Shifts. PICES Science Report No. 28, 162 pp.

Koen-Alonso, M., Yodzis, P. (2005) Multispecies modelling of some components of the marine community of northern and central Patagonia, Argentina. Canadian Journal of Fisheries and Aquatic Sciences 62: 1490–1512.

Kooijman, S.A.L M. (2000) Dynamic energy and mass budgets in biological systems. Cambridge University Press, New York.

Kruse, G.H. et al. (2001) Spatial processes and management of marine populations. University of Alaska Sea Grant AK-SG-01–02. Fairbanks, AK, 720 pp.

Marine Stewardship Council (2004) MSC Principles and Criteria for Sustainable Fisheries. http:// www.msc.org/html/

Murawski, S.A., Brown, R., Lai, H.L., Rago, P.J., Hendrickson, L. (2000) Large-scale closed areas as a fishery management tool in temperate marine systems: the Georges Bank experience. Bulletin of Marine Science 66: 755–798.

Neis, B., Felt, L. (eds) (2001) Finding our sea legs: Linking fishery people and their knowledge with science and management. ISER Books, St. John's.

Nicholson, M.D., Jennings, S.J. (2004) Testing candidate indicators to support ecosystem-based management: the power of monitoring surveys to detect temporal trends in fish community metrics. ICES Journal of Marine Science 61: 35–42.

NOAA (2003) Social Science Research Within NOAA: Review and Recommendations. Final Report to the NOAA Science Advisory Board by the Social Science Review Panel, Washington, DC.

NOAA (2005) Puget Sound Salmon Recovery Plan. http://www.noaa.gov/salmon_recovery_planning/ RecoveryDomains/PugetSoundPlan/ExecSumm.pdf

NOAA (2006a) The Status of US Fisheries – 2006. http://www.nmfs.noaa.gov/sfa/statusoffisheries/ SOSmain.htm

NOAA (2006b) NOAA External Ecosystem Task Team Report. http://ecosystems.noaa.gov/docs/ EETt.pdf

Olsson, T., Folke, C., Berkes, F. (2004) Adaptive co-management for building resilience in social-ecological systems. Environmental Management 34: 75–90.

Pastoors, M.A., Rijnsdorp, A.D., Van beek, F.A. (2000) Effects of a partially closed area in the North Sea ("plaice box") on stock development of plaice. ICES Journal of Marine Science 57: 1014–1022.

Peterman, R.M. (2002) Eco-certification: an incentive for dealing effectively with uncertainty, risk, and burden of proof in fisheries. Bulletin of Marine Science 70: 669–681.

PICES (2005) Marine Ecosystems of the North Pacific (Bering Sea Chapter). PICES Special Publication 1. http://www.pices.int/publications/special_publications/NPESR/2005/ File_9_pp_153_176.pdf

Pikitch, E.K., Santora, C., Babcock, E.A., Bakun, A., Bonfil, R., Conover, D.O., Dayton, P., Doukakis, P., Fluharty, D., Heneman, B., Houde, E.D., Link, J., Livingston, P.A., Mangel, M., McAllister, M.K., Pope, J., Sainsbury, K.J. (2004) Ecosystem-based fishery management. Science 305: 346–347.

Plaganyi, E.E., Butterworth, D.S. (2004) A critical look at the potential of ECOPATH with ECOSIM to assist in practical fisheries management. South African Journal of Marine Science 26: 261–287.

Pope, J.G. (1991) The ICES Multispecies Assessment working Group: evolution, insights, and future problems. pp 22–33 in M. Sissenwine and N. Daan (eds) Multispecies Models relevant to Management of Living Marine Resources. ICES Marine Science Symposium 193.

Rice, J.C. (2003) Environmental health indicators. Ocean and Coastal Management 46: 235–259.

Rice, J.C. (ed) (2005a) Ecosystem effects of fishing: impacts, metrics, and management strategies. ICES Cooperative Research Report 272: 177.

Rice, J.C. (2005b) Understanding fish habitat ecology to achieve conservation. Journal of Fish Biology 67(Suppl B): 1–22.

Rice, J.C. (2008) An ecologist's view of economic instruments and incentives. International Journal of Global Environmental Issues 7: 191–204.

Rice, J.C., Rochet, M.-J. (2005) A framework for selecting a suite of indicators for fisheries management. ICES Journal of Marine Science 62: 516–527.

Rice, J.C., Trujillo, V., Jennings, S., Hylland, K., Hagstrom, O., Astudillo, A., Jensen, J.N. (2005) Guidance on the application of of the ecosystem approach to management of human activities in the European marine environment. ICES Cooperative Research Report 273: 22.

Richards, L.J., Rice J.C. (1996) A framework for reducing implementation uncertainty in fisheries management. North American Journal for Fisheries Management 16: 488–494.

Ricker, W.E. (1975) Computatition and Interpretation of Biological Statistics of Fish Populations. Bulletin of the Fisheries Research Board of Canada 191. 382 pp.

Roberts, C.M., Bohnsack, J.H., Gell, F., Hawkins, J.P., Goodridge, R. (2002) Effects of marine reserves on adjacent fisheries. Science 294: 1921–1923.

Rozwadowski, H.M. (2002) The Sea Knows No Boundaries: A Century of Marine Science Under ICES. University of Washington Press, Seattle, WA, 410 pp.

Sainsbury, K., Sumaila, U.R. (2003) Incorporating ecosystem objectives into management of sustainable marine fisheries, including 'best practice' reference points and use of marine protected areas. In Sinclair, M., Valdimarsson, G. (eds) Responsible Fisheries in the Marine Ecosystem. FAO, Cambridge, pp 343–362.

Schrum, C., Alekseeva, I., St. John, M. (2006) Development of a coupled physical-biological ecosystem model ECOSMO Part I: model description and validation for the North Sea. Journal of Marine Systems 61: 79–99.

Scott, A. (1998) Cooperation and quotas. In Pitcher, T.J. et al. (eds) Reinventing Fisheries Management. Fish and Fisheries Series # 23. Kluwer, Dordrecht, pp 201–214.

Shannon, L.J., Moloney, C.L., Jarre, A., Field, J.G. (2003) Trophic flows in the southern Benguela during the 1980s and 1990s. Journal of Marine Systems 39: 83–116.

Shotten, R. (ed) (2000) Current property rights systems in fisheries management; Proceedings of a Conference, Freemantle W.A.. FAO Fisheries Technical Paper 404: 684.

Smith, T.D. (1994) Scaling Fisheries: The Science of Measuring the Effects of Fishing 1855–1955. Cambridge Scientific Press, Cambridge, 392 pp.

Steffansson, G. (2003) Multi-species and ecosystem models in a management context. In Sinclair, M., Valdimarsson, G. (eds) Responsible Fisheries in the Marine Ecosystem. FAO, Cambridge, pp 171–188.

Tjelmeland, S., Lindstrom, U. (2005). An ecosystem element added to the assessment of Norwegian spring-spawning herring: implementing predation by minke whales. ICES Journal of Marine Science 62: 285–294.

United Nations (1992) United Nations Convention on the Environment and Development. http://www.unep.org/unep/partners/un/unced/home.htm

United Nations (2002) Plan of Implementation – World Summit on Sustainable Development. New York, 44 pp.

United Nations (2006) Report of the United Nations Open-Ended Informal Consultative Process on Oceans and the Law of the Sea. http://www.un.org/Depts/los/consultative_process/consultative_process_info.htm

Wilen J.E., Botsford, L.W. (2005) Spatial Management of Fisheries. California Sea Grant College Program. Fisheries Paper 05_01. http://repositories.cdlib.org/csgc/rcr/Fisheries05_01

Chapter 3
Biodiversity and the Future
of Fisheries Science

Ashleen J. Benson

*My duty is to protect biological diversity… This principle
is now clearly spelled out as a pillar of proper fisheries
management.*

Hon. Loyola Hearn, Minister of Fisheries and Oceans Canada
December 19, 2006

Abstract The political and economic influence of environmental nongovernmental agencies has major implications for fisheries science in the future. For example, the conservation of biodiversity has been adopted as a global benchmark for successful fisheries management by the Marine Stewardship Council, among others. While agreeable in principle, demonstrating progress toward biodiversity conservation is a significant challenge facing young scientists in the near future. Here I present a personal view on the future of biodiversity in fisheries management that is informed by an extensive review of the ecological and fisheries literature. I show that there is no agreement on the definitions and measures that form the basis for conservation initiatives. However, in contrast to the broader ecological literature, fisheries science has a sophisticated theory of resource exploitation that can be extended to include biodiversity-type objectives in order to achieve successful management.

Keywords biodiversity, fisheries management, management objectives, stock structure

3.1 Introduction

This paper is a personal perspective on the future of fisheries science. I am currently a Ph.D. student in fisheries science and management, and as such, my opinion does not benefit from years of experience. However, my formal education has

A.J. Benson
School of Resource and Environmental Management, Simon Fraser University,
Burnaby, British Columbia,V5A 1S6, Canada
e-mail: ajbenson@sfu.ca

R.J. Beamish and B.J. Rothschild (eds.), *The Future of Fisheries Science in North America*,   33
Fish & Fisheries Series, © Springer Science + Business Media B.V. 2009

been strongly interdisciplinary, encompassing the economic, sociopolitical, and ecological aspects of fisheries management. Therefore, while I do not have the advantage of hindsight, I have a breadth of understanding of the fisheries management problem that permits me to comment to a degree on the future as I see it.

An immediate challenge facing fisheries scientists is to expand the scope of our research beyond single species in order to achieve a holistic, ecosystem-based approach to management. Central to ecosystem-based management is the notion of biodiversity conservation, which is the primary concern of groups such as the Marine Stewardship Council and other environmental nongovernmental organizations. The influence of such groups can be expected to intensify as they gain political and economic power – ultimately their influence may set global-scale objectives and standards for fisheries science. In these early years of shifting political context, I believe that fisheries scientists must think critically about what they are being asked to do, and identify inconsistencies between what is politically desirable and what is scientifically feasible. In my opinion, the management of marine biodiversity is a very good example of one such inconsistency.

In this paper I attempt to define biodiversity based on a directed review of the ecological and fisheries literature, in order to clarify the ecosystem-level objectives of fisheries management. However, it appears that the ecological theory of biodiversity conservation is not, and may never be, well developed. Like many graduate students I initially approached ecosystem-based management with enthusiasm, but found I fell short in its application – how does one reconcile single-species stock assessment, harvest management, and "biodiversity" conservation? Based on the introductory quote, it appears that others do not share my confusion. Nonetheless, I believe that the willingness of fisheries scientists to set aside a sophisticated (albeit, single species) theory of resource management in favor of a grander (and more complicated) approach may require further consideration. After all, as discussed later in this paper, the theoretical basis of fishery science recognizes diversity *within* fish populations across space (e.g., Ricker 1973), and as such, single-species theory may be directly relevant to implementing biodiversity policy.

I expect that fisheries scientists in the future will revisit and adapt single-species theory to account for the fact that many fish populations are naturally spatially discontinuous and heterogeneous. This natural state is advantageous because it spreads exposure to external perturbations across space and time, enabling the population to maintain production in spite of changing environmental conditions (Hilborn et al. 2003). When viewed in this context, it is evident that important aspects of fish populations are not captured by existing, abundance-based assessment approaches, and some measure of spatial diversity within populations is required. This is the definition of biodiversity that requires the attention of fisheries scientists, and which will emerge as our understanding of marine populations evolves. By shifting emphasis to systems-level characteristics of fish populations such as spatially complex dynamics, interactions, and distribution, we may see marked improvements in fisheries management.

3.2   What Is Biodiversity?

In spite of prolific use of the term, there is no consensus on the definition or the value of biodiversity. The term "biodiversity" has been used to describe various aspects of biological populations, including the number of species present in a community, genetic variability, and diversity among ecological systems (Harper and Hawksworth 1995). The lack of consistency exists for two reasons: first, biodiversity is a pseudocognate term in that most people assume that the definition is intuitive and that others automatically share their understanding; and secondly, several perspectives on the meaning of biodiversity have developed over time (Gaston 1996). These perspectives can be classified as those in which biodiversity is approached as a concept, those that consider biodiversity to be an entity that can be measured, and those that consider it to be a sociopolitical construct (Gaston 1996).

The competing definitions of biodiversity all emphasize its multidimensional nature, and build on a basic theme that equates biodiversity with the "variety of life" (e.g., McNeely et al. 1990; Wilson 1992). A commonly cited definition is that of the Convention on Biological Diversity (UNEP 1992):

> ...the variability among living organisms from all sources, including *inter alia*, terrestrial, marine and other aquatic systems and the ecological complexes of which they are part; this includes diversity within species, between species and of ecosystems...

Although this characterization has certain intuitive appeal, it does not advance a precise understanding of the concept. This language unfortunately reflects the majority of biological studies on the topic. The "variety of life" is frequently deconstructed into smaller categories of genetic, species, or taxonomic diversity, and ecosystem diversity, in an effort to facilitate its study (Harper and Hawksworth 1995; McAllister 1991). Some authors divide the categories further into genes, populations, species, assemblages, and whole systems (Soule 1991). Regardless of the scheme, all classifications emphasize hierarchies both within and between levels in the system (e.g., Noss 1990). Biodiversity is commonly defined in terms of these hierarchical entities rather than the processes underlying the observed patterns (Smith et al. 1993). Viewed from this perspective, biodiversity is seen to be the singular end of the evolutionary process, rather than a by-product of evolution (Zeide 1997). In focusing exclusively on units such as species and genes, the basic concept of biodiversity is insufficient because it ignores the fact that biodiversity is the biological response to a variable environment (Hengeveld 1996).

Most biologists tend to avoid the problem of defining biodiversity and instead approach it as something that is generally understood to be both real and measurable. There are volumes written on the loss of biodiversity and how to rebuild and conserve what remains. Statements expressing concern over what biologists refer to as the sixth mass extinction on our planet are common (Perrings et al. 1995), as are biodiversity evaluations that are presented in terms of some numeric value (e.g., Wilson 1988).

Biodiversity is firmly tied to the notion of preserving the natural environment, and there is a growing public understanding that biodiversity is "good" and should therefore be maintained. In this sense, biodiversity is not a neutral scientific concept, rather, it is either imparted with a value or it is perceived as representing the value of nature (Gaston 1996). This is an important distinction because at its most basic level, the problem of nature conservation is fundamentally linked to human population growth and resource consumption. Thus, the conservation of nature is more of a political problem than an ecological one because it requires value judgements to be made regarding what biodiversity is good for and how best to allocate scarce conservation resources (Vane-Wright 1996).

3.3   The Ecological Justification for Conserving Biodiversity

The relationship between biodiversity and characteristics of ecosystems such as productivity, stability, resilience, and function, has been a key area of focus in ecological research for at least 4 decades, but relatively few unifying principles for management have emerged (Hughes and Petchey 2001). This may be because biodiversity is most often treated as an academic area of study instead of a broadly applied one. In addition, there is little research directed at predicting how human activities are likely to influence biodiversity, or at the importance of such an effect relative to other human impacts on ecosystems (Srivastava 2002). Much of the experimental work that has been conducted focuses on biodiversity in terrestrial plant communities, and there is substantial uncertainty as to how these results might be generalized for other species and ecosystems (Loreau et al. 2001). In addition, it is difficult to reconcile both the approach and the results of small-scale and large-scale observational experiments. However, taken in aggregate, the empirical evidence suggests that biodiversity is an important predictor of ecological proc- esses at small scales, and can be expected to decrease in importance at regional and ecosystem-level scales (Loreau et al. 2001).

3.4   Biodiversity and Stability

Our current understanding of the relationships between biodiversity and ecosys- tems emerged from two branches of ecology: community ecology, which focuses on the interrelationships between species (competition, predation, parasitism, etc.), and ecosystem ecology which focuses on energy flows and nutrient cycling (Holling et al. 1995). The majority of historical work on the relationship between biodiversity and ecosystems is firmly cast in the community ecology approach. Prior to the 1970s ecologists believed that stability increased with the number of species in an ecosystem (species richness) (Elton 1958). This conclusion was supported by repeated observations that the population densities within simplified terrestrial communities were more variable than those within complex, diverse communities (McCann 2000). However, subsequent work showed that increased diversity tends to either destabilize community dynamics, or is of little importance to stability (May 1973). These and other disparate results have come to be known as the "diversity-stability debate" and have created uncertainty within the scientific community, because natural systems are observed to be both highly unpredictable and diverse. The field of ecology is currently replete with studies searching for diversity–stability relationships, and the results appear to depend greatly on the species and scale of examination (Hughes and Petchey 2001; McCann 2000). Although ecologists have yet to identify a universal diversity–stability relationship that can be applied across scale, species, and ecosystem, it appears that high diversity on average yields greater ecosystem stability (Loreau et al. 2001; McCann 2000). However, the persistence of an ecosystem results not from its biodiversity per se, but rather from the fact that it is comprised of species that can respond differently to change. This idea is referred to as the "insurance hypothesis" – biodiversity provides a buffer against environmental variations because the mixed responses of different species mean that the aggregate characteristics of ecosystems are stabilized (Loreau et al. 2001).

Confusion also arises from the fact that the definition of stability used in many analyses focuses on systems behavior and the notion of achieving constancy, a tradition borrowed from classical physics and engineering (Holling 1973). In keeping with the classical approach, both the dynamics of populations and the physical environment are defined with respect to an equilibrium level, and the environment is seen to affect organisms but is not affected by them. In reality, this definition of stability is insufficient for marine systems which have been shown to shift rapidly between alternate states (e.g., Benson and Trites 2002).

3.5 Biodiversity and Resilience

The relationship between resilience and biodiversity can be represented in terms of stability landscapes in which species-rich environments have landscapes with deep pits implying high local stability whereas species-poor environments have landscapes with shallow pits and low local stability (Peterson et al. 1998). However, ecosystem resilience is not a simple function of the number of species present at a given time (Elmqvist et al. 2003); there are competing models of how increases in species richness relate to increases in local stability of ecosystems. The first model of how ecosystems function and organize around species is the "redundant species hypothesis," which proposes that there is a minimum level of biodiversity that is required to maintain ecosystem processes, and species additions or losses above this level have minor impacts (Walker 1992). Under this perspective a limited number of keystone organisms drive the critical ecosystem processes (drivers) and other species exist in the niches created by the keystone groups (passengers) (Holling et al. 1995). Most ecological function resides in the keystone species, whose presence or absence determines the state of an ecosystem (Walker 1995).

Empirical evidence highlights the importance of keystone species in both natural and managed systems – they maintain critical processes in ecosystems under stress, even as the species composition changes (Folke et al. 1996). However, identification of keystone species is problematic, because depending on the state of the ecosystem, species can be either a driver or a passenger (Lawton 1994; Lister 1998). The second model is the "rivet hypothesis" which suggests that all species contribute evenly to ecosystem function. Similar to rivets on an airplane wing, the effects of losing a small number of species (rivets) are buffered by the overlapping functions of other species, but continuous loss ultimately leads to collapse of the system (the airplane wing falls apart) (Ehrlich and Ehrlich 1981). In this model, an ecological function is not lost until all species performing that function are removed from the system (Peterson et al. 1998). As such, compensation by the remaining species masks degradation of the ecosystem and a loss of adaptive capacity. The third model is the "idiosyncratic response hypothesis," which proposes that it is not the number of species that is important, but the particulars of the species and the ecological history of a region that is important for ecosystem function (Lawton 1994; Kunin and Lawton 1996). However, ecosystem structure and function can be sustained at the regional scale, independent of the mix and relative abundance of species that are present (Schindler 1990). This suggests that groups of organisms are more critical than single species to the maintenance of ecosystem functions.

Ecosystems generally possess considerable functional redundancy (i.e., the number of alternative species that can provide a particular function following a disturbance), which acts to stabilize ecosystem processes. However the loss of a functional group (a decrease in functional diversity) can drastically alter ecosystem functioning (Folke et al. 2004; Peterson et al. 1998). In addition to diversity in the function of groups of species, the adaptability of the groups that are present (i.e., the diversity of responses to environmental change that exists within a functional group) plays a key role in determining the resilience of ecosystems (Elmqvist et al. 2003). This "response diversity" is governed by species and population diversity. For example, sockeye salmon in Bristol Bay, Alaska exhibit a broad range of population distributions and life history strategies whose productivity varies differently in relation to climate, such that in spite of high variability in relative abundance, the populations in aggregate have sustained high productivity over time (Folke et al. 2004). In short, the stock aggregate has maintained its resilience.

3.6   The Economic Justification for Conserving Biodiversity

Ecologists and economists both recognize that human beings cannot live in isolation – we are critically dependant on other organisms for our survival. In terms of value, plants and animals are important not only because they house the genetic library of the planet, but also because they provide food, timber, industrial resources, and medicines; in addition, they sustain the flow of ecological services (e.g., waste assimilation and filtering of pollutants) that are used by humans (Ehrlich and Ehrlich 1992; Kunin and Lawton 1996). In this sense, biodiversity serves both the direct and indirect needs of society, and ecosystems are fundamental "factors of production" that are being threatened by human activities (Barbier et al. 1994; Folke et al. 1996). The value of biodiversity from an economic perspective is therefore intimately linked with its functional role in maintaining ecosystem processes.

Economists view the preservation of biodiversity as a form of insurance against lost utility (in terms of use and services) of organisms at some future date (Williams and Humphries 1996). This notion is captured in the "option value" of species (Faith 1995). Option value corresponds not just to the unknown future values of known species, but also to the unknown values of unknown species. A general view of the option value of biodiversity is that it maximizes the capacity of humans to adapt to ecological change (Reid 1994). As such, the basis for biodiversity conservation from an economic perspective is to maximize species' abilities to adjust to a changing environment (Williams and Humphries 1996). The value of biodiversity therefore lies in the capacity of organisms to adapt through natural selection thereby maintaining options for future generations of humans. Because selection acts at the level of character (feature), the focus turns from attempts to value *species* to attempts to value the *features of species* (Williams and Humphries 1996). Fundamental to the idea of an option value is a high degree of uncertainty about both the environmental conditions and the realized value of species under those conditions (Faith 1995). At any given time, an individual species that contributes more novel features to a protected subset is of greater value than another that contributes fewer features (Faith 1995). However, uncertainty with respect to the future value of organisms greatly complicates the rationale for differentially ranking them for conservation. This point has sparked a substantial, and somewhat circular, debate on the point of weighting and non-weighting of both the features and the species that own them. Most valuation systems are inherently subjective (Reid 1994), and in order to avoid any subjectivity it is necessary to consider all species as equal (Wilson 1992). However, if the features of species at lower levels (e.g., genetic and phenotypic characteristics) are weighted equally, it is inevitable that species at higher levels (e.g., species assemblages) will be weighted differentially because certain species will contribute relatively more to the preserved set of features than others (Faith 1995; Williams and Humphries 1996).

3.7   "Biodiversity" as a Management Objective

The biodiversity objectives for ecosystem management are somewhat ambiguous. According to some authors the objective is to protect the species and ecosystems most at risk of extinction (Hansen et al. 1999), others promote maximizing species richness within a particular geographic region (Smith et al. 1993), while others focus on maintaining a range of ecosystem services (Folke et al. 1996). In addition to a lack of clear objectives, the preservation of biodiversity is thus far hampered by the absence of a well-defined unit of analysis that might be used to formulate an objective function for biodiversity management (Weitzman 1995).

In principle, biodiversity can be measured at a variety of hierarchical levels, from molecular and genetic (alpha diversity), to the ecosystem level (beta diversity) (Harper and Hawksworth 1995). However, there is no consensus on which is most important for assessment and conservation because a loss of biodiversity at any level may represent a loss of future opportunities (i.e., option value) (Agardy 2000). Conceptual and methodological problems exist even at the species level, where ecologists debate whether species counts (i.e., species richness) are sufficient, or whether each species should be weighted by its relative abundance in order to reveal the relative effects of rare and common species, as in Simpson's index and the Shannon-Weiner measure of "effective number" (Baumgartner 2004). These abundance-based indices do not explicitly account for the uniqueness of species. In contrast, indices that have been developed by economists such as Weitzman (1992, 1995), and Solow et al. (1993), stem from the notion of product diversity and focus on measuring the differences between species using metrics that incorporate species richness as well as the features of species (Baumgartner 2004; Faith 1995). It is important to note that measures of species richness are nonstationary because of the taxonomic inflation phenomenon wherein scientific progress drives the elevation of lower taxa to the rank of species (Knapp et al. 2005).

No approach to measuring biodiversity fully captures all of the characteristics of species within an area. As such, the choice of index requires a prior judgement on the purpose of biodiversity and the value of its components (Baumgartner 2004). Further problems with quantifying species diversity in a given area include the biasing effect of sampling effort on biodiversity measures (Gotelli and Colwell 2001), the effects of species movements through invasion, migration, and changes in distribution (Hawksworth 1995), and questions pertaining to the boundaries of the system of interest (Gaston and Williams 1993). The measurement problem is particularly pronounced in the marine environment, where organisms have wider and more variable geographic ranges than on land (May 1995), and where significantly less effort has been directed at generating species inventories, particularly for deep-sea environments (The Royal Society 2003). The enormity of the problem has led many researchers to propose proxy indicators for biodiversity. The use of indicators is based on the idea that small subsets of species, habitats, or ecosystems can be used to expedite biodiversity assessments, and may remove the necessity for a full census of global diversity (Pearson 1995). However, the search for appropriate indicators is problematic (e.g., Gaston and Williams 1993; Pearson 1995) and in some cases debate on this point threatens to eclipse the underlying issue. For example, the criteria for selection of indicator species are usually based more on legal and political imperatives rather than on scientific assessments (Pearson 1995). This turns the focus to the species itself rather than what it is supposed to be indicating. In addition to these fundamental limitations of the analytical approach, many researchers conclude that there is no single spatial or temporal scale at which to describe the natural system because the scale of interest depends greatly on the chosen species (Bunnell and Huggard 1999; Levin 1992).

It is important to recognize that most conservation and recovery efforts to date focus on those species that have already been identified as being under threat of extinction. This raises fundamental questions related to prioritizing conservation efforts, and the choice and identification of critical species for conservation (e.g., keystone species versus charismatic species) (Walker 1995). Given these constraints, it is reasonable to question the practicality of focusing on species when tackling the greater problem of biodiversity conservation. The species-based endeavor is not without merit; however, if one accepts that the "biodiversity crisis" extends beyond the massive extinction of species to include all ecologically and socially undesirable changes in the composition and functioning of ecosystems, then the scope of the traditional species-based approach is not sufficiently general to account for all objectives. In addition, the number and variety of definitions and measures of biodiversity that form the basis for conservation initiatives indicate that biodiversity remains a poorly conceived and immature concept (Hengeveld 1996). Furthermore, the level of biodiversity is constantly changing because it is bound to the state of the ecosystem. Therefore, even if there was an objective measure of biodiversity, fixing target levels for it would largely be an academic exercise (Kamppinen and Walls 1999).

3.8   A Fisheries Science Perspective

Objectives for biodiversity conservation are increasingly identified in terms of population viability and probabilities of extinction for closed, identifiable populations of rare or endangered species (e.g., Nicholson and Possingham 2006). Application of this approach is hampered by insufficient understanding of the process of extinction of marine populations, by a lack of data on population abundance over time, and by the fact that fish populations are seldom closed to migration (Wainwright and Waples 1998). This extinction-centric approach to management is somewhat at odds with traditional fisheries science, which is concerned with preserving stock productivity and the maintenance of future fishing opportunities for relatively abundant fish species. Adopting the extinction-centric approach to conserving biodiversity requires a concomitant shift in focus to the recovery of rare, depleted fish species and may not be necessary or appropriate for populations that continue to sustain fisheries. The challenge that remains for fisheries science is to develop a tractable interpretation of biodiversity that is relevant for the management of commercially exploited fish species. The scale of biodiversity management will be constrained in the near term by the limited application of genetic techniques to fish populations – the preservation of genetic or even phenotypic diversity is as yet beyond the scope of fisheries management. Similarly, both the theory and application of ecosystem-based management are under development. Therefore, in an operational sense, the unit of interest for fisheries management remains the population.

Defining a population for fisheries management is a difficult and research-intensive aspect of fisheries science. The "stock concept" has been debated for over 100 years, and a variety of definitions exist in the literature (Begg et al. 1999).

In general, the recent definitions admit that a stock is largely a construct of management, and does not represent a single, homogenous group of fish (Begg et al. 1999). The key uncertainty in applying the stock concept is the ecological implication of incorrectly treating a diverse group of fish as a homogenous unit for assessment and management. For example, the biological limits within which fisheries can operate are determined by the production function, or stock–recruitment relationship, of the underlying "stock" (Mace 2001). Stock assessments typically proceed on the assumption that this relationship is stationary across both space and time. However, for spatially structured populations, changes in productivity of the sub-stocks that result from harvesting or natural disturbance can lead to changes in the aggregate recruitment relationship that can never be fully understood or anticipated (Walters 1987). The assessment and management issues stemming from the stock aggregate situation parallel those of mixed-stock fisheries, where subpopulations that contribute to the total recruitment have a range of productivities. When treated as a unit stock, the lower productivity populations will invariably be overfished and decline in abundance, resulting in a lower weighting of the unproductive populations in the aggregate recruitment function. This leads to a pathology in which the recruitment function actually appears to show increasing productivity as the stock declines, leading to recommendations of increased exploitation rates (Ricker 1973).

Substructure within fish stocks is represented by either spatially discrete units or different life history types within a population. Species that demonstrate within-population structure include sockeye salmon (*Oncorhynchus nerka*) (Hilborn et al. 2003), Pacific herring (*Clupea pallasi*) (Hay and McCarter 1997), Arctic char (*Salvelinus alpinus*) (Secor 1999), Atlantic herring (*Clupea harengus*), and cod (*Gadus morhua*) (Smedbol and Stephenson 2001). This aspect of fish populations has been noted since the early nineteenth century, but the high cost of collecting data on the compositional complexity of a population tends to preclude explicit consideration of spatial and subpopulation-level impacts (Walters 1987). This omission presents a problem for fisheries, because management that does not account for population structure within and among stock complexes can lead to overexploitation of the stock components, an erosion of within-species diversity, and ultimate depletion of the productive potential of the stock aggregate (de la Mare 1996; Hilborn et al. 2003; Ruzzante et al. 2000; Stephenson 1999). Based on these arguments, preserving within-stock diversity (response diversity) is a biodiversity-type objective that can be operationalized, as demonstrated to a degree in certain fisheries (Hilborn et al. 2003), and is directly consistent with both the theory of fisheries science, and the current call to maintain the resilience of populations and ecosystems (Resilience Alliance and Santa Fe Institute 2004).

Recognition of the importance of stock identification is manifest in widespread interest in tagging and genetic research programs as well as an increasing focus on spatially explicit population models. However, in many cases it is unclear how this information can or should be incorporated into management. For example, does detailed information on stock structure easily translate to stock-specific management? To what extent does the management approach depend on the stock structure? What is the appropriate spatial scale of management? How much within-stock diversity is "enough"? The answers to these questions are likely to be case specific, and conditioned on the local goals of fisheries management.

Fisheries management involves making decisions that balance resource conservation and exploitation, given imperfect information about the resource, the environment, and the resource users. In recognition of the highly uncertain nature of fisheries, levels of "risk" and "precaution" are now frequently provided in stock assessments. However, the management implications of such measures tend to be highly uncertain themselves. The evaluation of management procedures addresses this problem by formally testing the implications of uncertainties on the quality of management decisions (de la Mare 1998). Fundamental to this approach is the specification of a minimally realistic model of the exploited population that incorporates uncertainties in the key biological processes. However, spatial and multispecies interactions are rarely explicitly incorporated into such models (Punt 2006). Stock structure can be added to the list of important characteristics that are seldom addressed in these evaluations. Given that many fisheries are now moving toward formalized management procedures, fisheries scientists have a strategic role to play in identifying the important aspects of exploited populations that govern the productive capacity of the stock, and in developing methods for including these features in models of "true" stock dynamics. The key population processes will likely include more than age-structure and apparent spawner–recruit relationships, particularly for population-"rich" species such as salmon, herring, rockfish, and cod, which exhibit numerous spawning populations within a single management unit (Stephenson 1999).

3.9   Science Requirements for Managing Within-Stock Diversity

Several challenges must be addressed before biodiversity conservation can be incorporated into the management of commercially exploited fish populations. Foremost is the identification of a metric that reflects relevant changes in the population. This requires knowledge of the aspects of population dynamics that support the ability of the subpopulations to respond differentially to disturbance (i.e., the maintenance of stock resilience). Fisheries ecologists emphasize the importance of the spatial distribution of the spawning stock, as it reflects evolutionary adaptations to the marine environment that maximize the survival of progeny (Sinclair 1988). In fact, recruitment success may be determined as much by spawner abundance as by the spawning distribution (DeYoung and Rose 1993). The spatial distribution of the spawning stock may therefore be a reasonable proxy for stock diversity, provided that a broad spawning distribution reflects the varied responses of the subpopulations to disturbance. Taken together with information on abundance and age-structure, this information can augment the stock assessment by adding a dimension of interpretation that does not exist under standard, nonspatial approaches. It is important to note that the metric should reflect within-stock diversity, but may not necessarily directly measure the mechanism that maintains diversity. For example, the population structure of many marine fish species is determined by dispersal and homing dynamics (McQuinn 1997). From a management perspective, measuring the details of dispersal (i.e., the age-specific rate of straying) may be less important than obtaining an accurate measure of the spatial complexity of the stock, in part because the relationship between straying rate and stock resilience may be less direct than the relationship between spawning distribution and stock diversity.

Methods for incorporating stock diversity into harvest decision rules present another challenge for managing the diversity of commercially important fish species. A fundamental question related to harvest decision planning is the optimal scale of stock assessment. As discussed earlier, there is a level of heterogeneity in fish stocks that is assumed not to exist for stock assessment and management, in large part because the subpopulation boundaries have not been reliably determined. Additional, practical considerations for conducting aggregate stock assessments include a reduction in data quality that would go along with disaggregating the data, and increased annual survey costs arising from the intensified sampling that would be required for fine-scale stock assessment. It is reasonable to assume that resources for moving in this direction will be scarce, particularly given the additional fisheries management expenditures that are required for implementing finer-scale assessment and harvest management (e.g., enforcement costs).

Given the high probability that subpopulations will continue to be assessed in aggregate, is there a way to include the measure of biodiversity in a harvest control rule? Adding a dimension of constraint for setting allowable catch is a minor problem when compared to the issue of setting the appropriate threshold for biodiversity. This is the point at which fisheries scientists must carefully consider how to apply the language of biodiversity conservation. For example, if the overarching objective is to maintain biodiversity, does this translate to a minimum number of spawning sites? How is this number related to the aggregate stock production? Many current fishery harvest control rules are based on relative measures of depletion – can a similar method be applied to stock diversity? Must the threshold be absolute? The difficulty in identifying a threshold level of diversity is matched by the problem of implementing harvest policies for spatially structured stocks. The management questions arising from this approach pertain to the appropriate allocation of fishing effort among the various subpopulations and the tractability of a spatial approach to management, which will be determined by the type of fishery (i.e., a terminal fishery versus a fishery that operates on migratory aggregations), the fishery dynamics, and ultimately, the increased costs of management.

3.10 Conclusion

A direct review of the literature indicates that the justifications for conserving biodiversity that have featured prominently in the development of international conventions and policies are characterized by wide-ranging hypotheses and disparate results. Nonetheless, these lines of argument have yielded both political and global economic imperatives that now set the agenda for fisheries management. Fisheries science has yet to feature prominently in the development of biodiversity policies and objectives, but given its unique position at the interface between ecological theory and application, I believe that fisheries science has much to offer in this regard. It has been observed that environmental problems arise from the negative net impact of many small decisions, including the focus on single-species management (Odum 1982). In my opinion, the apparent incompatibility of traditional fisheries science and biodiversity conservation does not arise from a myopic focus on single species per se; rather, problems arise from a preoccupation with population abundance, and insufficient consideration of spatial and subpopulation-level impacts of management policies. The imperative to conserve biodiversity is no longer up for debate: there are now political, ecological, and economic justifications to do so. The challenge that remains for fisheries science in the future is to develop a tractable interpretation of biodiversity that can be rendered operational in the management of commercial fisheries.

The preservation of within-stock diversity (response diversity) is a biodiversity-type objective that is immediately relevant for fisheries management. This aspect of fish stocks is seldom explicitly included in stock assessments and evaluations of management strategies, but it may determine the ability of the stock aggregate to maintain its productive capacity under exploitation. Managing for within-stock diversity requires a measure of diversity, a method for incorporating the measure into harvest control rules, and a spatial approach to management that accounts for differences in productivity among population components. Fisheries science has developed a sophisticated theory intended to address these types of questions as they relate to fish stock production. The theory should be extended to include within-stock diversity. The implications of and necessity for moving in this direction can be tested using existing methodology such as management procedure evaluations, which can be used to prioritize information requirements based on the key uncertainties in the fishery system. Additionally, by explicitly considering the feasibility of various spatial management regimes in light of the objectives of those interested in the fishery, this approach may provide a strategy for addressing the potential institutional mismatch that may exist between current management arrangements and those required to conserve population diversity.

Acknowledgments This work was funded through grants awarded to A.J. Benson by the Natural Sciences and Engineering Research Council of Canada and the Herring Research and Conservation Society of British Columbia. I am grateful for constructive input from Chris Wood, Sean Cox, Jaclyn Cleary, Jake Schweigert, Randall Peterman, Mariano Koen-Alonso, an anonymous reviewer, Richard Beamish, and Brian Rothschild.

References

Agardy T (2000) Effects of fisheries on marine ecosystems: A conservationist's perspective. ICES J Mar Sci 57: 761–765

Barbier EB, Burgess JC, Folke C (1994) Paradise lost? The ecological economics of biodiversity. Earthscan, London

Baumgartner S (2004) Measuring the diversity of what? And for what purpose? A conceptual comparison of ecological and economic measures of biodiversity. Paper presented at the XIII EAERE Annual Conference Budapest, June 2004

Beechie, T., Pess, G., Roni, P. and Giannico, G. (2008) Setting river restoration priorities: a review of approaches and a general protocol for identifying and prioritizing actions. *North American Journal of FIsheries Management* **28**, 891–905.

Begg GA, Friedland KD, Pearce JB (1999) Stock identification and its role in stock assessment and fisheries management: An overview. Fish Res 43: 1–8

Benson AJ, Trites AW (2002) Ecological effects of regime shifts in the Bering Sea and eastern North Pacific Ocean. Fish Fish 3: 95–113

Bunnell FL, Huggard DJ (1999) Biodiversity across spatial and temporal scales: Problems and opportunities. Forest Ecol Manage 115: 113–126

de la Mare WK (1996) Some recent developments in the management of marine living resources. Frontiers of Population Ecology. CSIRO, Melbourne, pp. 599–616

de la Mare WK (1998) Tidier fisheries management requires a new MOP (management-oriented paradigm). Rev Fish Biol Fish 8: 349–356

DeYoung B, Rose GA (1993) On recruitment and distribution of Atlantic cod (*Gadus morhua*) off Newfoundland. Can J Fish Aquat Sci 50: 2729–2741

Ehrlich PR, Ehrlich AH (1981) Extinction: The causes and consequences of the disappearance of species. Random House, New York

Ehrlich PR, Ehrlich AH (1992) The value of biodiversity. Ambio 21: 219–226

Elmqvist T, Folke C, Nystrom M, Peterson G, Bengtsson J, Walker B, Norberg J (2003) Response diversity, ecosystem change, and resilience. Front Ecol Environ 1: 488–494

Elton CS (1958) Ecology of invasions by animals and plants. Chapman & Hall, London

Faith DP (1995) Phylogenetic pattern and the quantification of organismal biodiversity. *In* Biodiversity measurement and estimation. *Edited by* DL Hawksworth. Chapman & Hall, London, pp. 45–58

Folke C, Holling CS, Perrings C (1996) Biological diversity, ecosystems, and the human scale. Ecol Appl 6: 1018–1024

Folke C, Carpenter S, Walker B, Scheffer M, Elmqvist T, Gunderson L, Holling CS (2004) Regime shifts, resilience, and biodiversity in ecosystem management. Annu Rev Ecol Syst 35: 557–581

Gaston KJ (1996) What is biodiversity?. *In* Biodiversity: A biology of numbers and difference. Chapter 1. *Edited by* KJ Gaston. Blackwell Science, Oxford

Gaston KJ, Williams PH (1993) Mapping the world's species – The higher taxon approach. Biodivers Lett 1: 2–8

Gotelli NJ, Colwell RK (2001) Quantifying biodiversity: Procedure and pitfalls in the measurement and comparison of species richness. Ecol Lett 4: 379–391

Hansen AJ, Rotella JJ, Kraska MPV, Brown D (1999) Dynamic habitat and population analysis: An approach to resolve the biodiversity manager's dilemma. Ecol Appl 9: 1459–1476

Harper JL, Hawksworth DL (1995) Preface. *In* Biodiversity measurement and estimation. *Edited by* DL Hawksworth. Chapman & Hall, London, pp. 5–12

Hawksworth DL (1995) Biodiversity measurement and estimation. Chapman & Hall, London

Hay DE, McCarter PB (1997) Larval distribution, abundance, and stock structure of british columbia herring. J Fish Biol 51: 155–175

Hengeveld R (1996) Measuring ecological biodiversity. Biodivers Lett 3: 58–65

Hilborn R, Quinn TP, Schindler DE, Rogers DE (2003) Biocomplexity and fisheries sustainability. PNAS 100: 6564–6568

Holling CS (1973) Resilience and stability of ecological systems. Annu Rev Ecol Syst 4: 1–23

Holling CS, Schindler DW, Walker B, Roughgarden J (1995) Biodiversity in the functioning of ecosystems: An ecological synthesis. *In* Biodiversity loss: Economic and ecological issues. *Edited by* CA Perrings, K Maler, C Folke, CS Holling, B-O Jansson. Cambridge University Press, New York, pp. 44–83

Hughes JB, Petchey OL (2001) Merging perspectives on biodiversity and ecosystem functioning. TREE 16: 222–223

Kamppinen M, Walls M (1999) Integrating biodiversity into decision making. Biodivers Conserv 8: 7–16

Knapp S, Lughadha EN, Paton A (2005) Taxonomic inflation, species concepts, and global species lists. TREE 20: 7–8

Kunin WE, Lawton JH (1996) Does biodiversity matter? Evaluating the case for conserving species. In Biodiversity: A biology of numbers and difference. Chapter 11. Edited by KJ Gaston. Blackwell Science, Oxford

Lawton JH (1994) What do species do in ecosystems? Oikos 71: 367–374

Levin SA (1992) The problem of pattern and scale in ecology. Ecology 73: 1943–1967

Lister NE (1998) A systems approach to biodiversity conservation. Environ Monit Assess 49: 123–155

Loreau M, Naeem S, Inchausti P, Bengtsson J, Grime JP, Hector A, Hooper DU, Huston MA, Raffaelli D, Schmid B, Tilman D, Wardle DA (2001) Biodiversity and ecosystem functioning: Current knowledge and future challenges. Science 294: 804–808

Mace PM (2001) A new role for MSY in single-species and ecosystem approaches to fisheries stock assessment and management. Fish Fish 2: 2–32

May RM (1973) Stability and complexity in model ecosystems. Princeton University Press, Princeton, NJ

May RM (1995) Conceptual aspects of the quantification of the extent of biological diversity. In Biodiversity measurement and estimation. Edited by DL Hawksworth. Chapman & Hall, London, pp. 13–20

McAllister DE (1991) What is biodiversity? Can Biodivers 1: 4–6

McCann KS (2000) The diversity-stability debate. Nature 405: 228–233

McNeely JA, Miller KR, Reid WV, Mittermeier RA, Werner TB (1990) Conserving the world's biodiversity. IUCN/WRI/CI/WWF/World Bank, Washington, DC.

McQuinn IH (1997) Metapopulations and the Atlantic herring. Rev Fish Biol Fish 7: 297–329

Nicholson E, Possingham HP (2006) Objectives for multiple-species conservation planning. Conserv Biol 20: 871–881

Noss RF (1990) Indicators for monitoring biodiversity: A hierarchical approach. Conserv Biol 4(4): 355–364

Odum WE (1982) Environmental degradation and the tyranny of small decisions. BioScience 32(9): 728–729

Pearson DL (1995) Selecting indicator taxa for the quantitative assessment of biodiversity. In Biodiversity measurement and estimation. Edited by DL Hawksworth. Chapman & Hall, London, pp. 75–80

Perrings CA, Maler K, Folke C, Holling CS, Jansson B-O (1995) Biodiversity loss: Economic and ecological issues. Cambridge University Press, New York

Peterson G, Allen Craig R, Holling CS (1998) Ecological resilience, biodiversity, and scale. Ecosystems 1: 6–18

Punt AE (2006) The FAO precautionary approach after almost 10 years: Have we progressed towards implementing simulation-tested feedback-control management systems for fisheries management? Nat Res Model 19: 441–464

Reid WV (1994) Setting objectives for conservation evaluation. In Systematics and conservation evaluation. Systematics Association Special Volume 50. Edited by PL Forey, CJ Humphries, RI Vane-Wright. Clarenden Press, Oxford, pp. 1–13

Resilience Alliance and Santa Fe Institute (2004) Thresholds and alternate states in ecological and social-ecological systems. Resilience Alliance, Online posting:http://resalliance.org/ev_en.php?ID=2497_201&ID2=DO_TOPIC.

Ricker WE (1973) Two mechanisms that make it impossible to maintain peak period yields from stocks of pacific salmon and other fishes. J Fish Res Board Can 30: 1275–1286

Ruzzante DE, Taggart CT, Lang S, Cook D (2000) Mixed-stock analysis of Atlantic cod near the gulf of St. Lawrence based on microsatellite DNA. Ecol Appl 10: 1090–1109

Schindler DW (1990) Experimental perturbations of whole lakes as tests of hypotheses concerning ecosystem structure and function. Oikos 57: 25–41

Secor DH (1999) Specifying divergent migrations in the concept of stock: The contingent hypothesis. Fish Res 43: 13–34

Sinclair M (1988) Marine populations: An essay on population regulation and speciation. University of Washington Press, Seattle, WA

Smedbol RK, Stephenson R (2001) The importance of managing within-species diversity in cod and herring fisheries of the north-western Atlantic. J Fish Biol 59(Suppl A): 109–128

Smith TB, Bruford MW, Wayne RK (1993) The preservation of process: The missing element of conservation programs. Biodivers Lett 1: 164–167

Solow A, Polasky S, Broadus J (1993) On the measurement of biological diversity. J Environ Econ Manage 34: 60–68

Soule ME (1991) Conservation tactics for a constant crisis. Science 253: 744–749

Srivastava DS (2002) The role of conservation in expanding biodiversity research. Oikos 98(2): 351–360

Stephenson RL (1999) Stock complexity in fisheries management: A perspective of emerging issues related to population sub-units. Fish Res 43: 247–249

The Royal Society (2003) Measuring biodiversity for conservation. Policy Document 11/03. Online posting: www.royalsoc.ac.uk

UNEP (1992) Convention on biological diversity, June 1992. United Nations Environment Programme, Environmental Law and Institutions Programme Activity Centre, Nairobi

Vane-Wright RI (1996) Identifying priorities for the conservation of biodiversity: Systematic biological criteria within a socio-political framework. *In* Biodiversity: A biology of numbers and difference. Chapter 12. *Edited by* KJ Gaston. Blackwell Science, Oxford

Wainwright TC, Waples RS (1998) Prioritizing pacific salmon stocks for conservation: Response to Allendorf et al. Conserv Biol 12: 1144–1147

Walker B (1992) Biological diversity and ecological redundancy. Conserv Biol 6: 18–23

Walker B (1995) Conserving biological diversity through ecosystem resilience. Conserv Biol 9(4): 747–752

Walters CJ (1987) Nonstationarity of production relationships in exploited populations. Can J Fish Aquat Sci 44: 156–165

Weitzman M (1992) On diversity. Quart J Econ 107: 363–405

Weitzman M (1995) Diversity functions. *In* Biodiversity loss: Economic and ecological issues. *Edited by* CA Perrings, K Maler, C Folke, CS Holling, B-O Jansson. Cambridge University Press, New York, pp. 21–43

Williams PH, Humphries CJ (1996) Comparing character diversity among biotas. *In* Biodiversity: A biology of numbers and difference. Chapter 3. *Edited by* KJ Gaston. Blackwell Science, Oxford

Wilson EO (1988) The current state of biological diversity. *In* BioDiversity. *Edited by* EO Wilson. National Academy Press, Washington, DC, pp. 3–18

Wilson EO (1992) The diversity of life. Allen Lane/Penguin Press, London.

Zeide B (1997) Assessing biodiversity. Environ Monit Assess 48: 249–260

Chapter 4
The Implications of a Paradigm Shift in Ocean Resource Management for Fisheries Stock Assessment

Robert N. O'Boyle

Abstract A paradigm shift in oceans management is underway, which will influence the future of stock assessment. On one hand, fisheries are increasingly being seen as one ocean sector amongst many, with many new objectives under an ecosystem approach to management (EAM) being sought. On the other hand, there is growing acceptance that stock assessment needs to be considered as only one element of a more comprehensive management strategy evaluation (MSE). Both these trends have implications for science support of future MSEs and the process whereby this science is managed and delivered. There will be a move towards ecosystem models that propose plausible hypotheses of ecosystem functioning. To support these, there will need to be theoretical developments on ecosystem control, spatial dynamics, and the habitat – productivity linkage. Growth in monitoring technology will provide unprecedented opportunities to enhance our knowledge and will fundamentally impact MSE and EAM. Stock assessment will initially be similar to current versions but will increasingly be required to report on ecosystem-related impacts. Decision making will involve evaluation of the merits of competing management strategies in relation to achievement of EAM objectives in the face of uncertainty. The science delivery process will evolve towards review and agreement of management frameworks separate from their scheduled application in assessment. Framework reviews will become quite elaborate, having to deal with all elements of an MSE. Experience with MSE has been limited but positive thus far with experiments in the approach occurring globally. The transition to the new paradigm will however require both a cultural shift in the scientific, ocean management and stakeholder communities and new financial mechanisms. Notwithstanding this, the new paradigm promises to provide a more effective basis for the management of the world's oceans resources.

Keywords Ecosystem approach to management · paradigm shift · management strategy evaluation · stock assessment · science delivery · science institutions

R.N. O'Boyle
Bedford Institute of Oceanography, P.O. Box 1006, Dartmouth, Nova Scotia, Canada B2Y 4A2
e-mail: betasci@eastlink.ca

R.J. Beamish and B.J. Rothschild (eds.), *The Future of Fisheries Science in North America*, 49
Fish & Fisheries Series, © Springer Science + Business Media B.V. 2009

4.1   Introduction

In 1977, many of the world's coastal states extended their jurisdiction out to 200 miles, considering that they could better manage the fisheries resources off their shores than the international management authorities of the time. In the 30 years since then, fisheries have been the dominant human activity impacting the ocean's ecosystems but management efforts to control this impact have met with limited success. Initially, coastal states replaced the displaced foreign fishing fleets with their own, often subsidized, fleets. The overall effect was a doubling in global fishing capacity (Garcia and Newton 1997). Within the new coastal zones, rapid growth in capacity caused numerous problems. Off Canada's east coast, the race to catch fish resulted in first an economic crisis (DFO 1982) and then the collapse of the groundfish stocks in 1992, the latter now recognized as one of the most dramatic in marine history. The reasons for this collapse and the lack of subsequent recovery have been hotly debated (e.g., Myers et al. 1996; Shelton et. al. 2005). What is not often reported is why fisheries management could not contain fishing effort. Again, there are many reasons for this, as exemplified by what occurred off Canada's east coast (Burke et al., 1996). Since the 1970 introduction of the first quota system in the North west Atlantic for Scotian Shelf haddock, there has been an increasing number of regulations imposed to limit the growth in fishing pressure, all to little avail. Quasi-property rights were not introduced until the late 1980s, shortly before the collapse. With continuing overexploitation and the lack of recovery of many resources (Shelton et. al. 2005), further limitations on fishing through both bans on bottom dragging (DFO 2006) and wide spread use of MPAs (Smith et al. 2006) are often suggested.

It is thus useful to ask where we are headed over the next 30 years. Sainsbury (1998) provides a comprehensive overview of potential future stresses on the ocean which I only highlight here. The stress on the oceans and its components are very likely to continue to increase from a variety of competing sources – fisheries, oil and gas, pollution, transport, and so on – due to an increasing human population, economic growth, and the demand for food, which while being likely highly regional, will have knock-down effects throughout the ecosystem. It is also likely that fisheries will not be the predominant impact. Stresses from the other ocean industries may surpass those of fisheries. In addition, the ocean is increasing being used as an alternate source of energy (wind, tidal power, etc.). Finally, climate change promises to produce some of the largest impacts, which are predicted to be felt, at least in the Pacific Ocean, within the next 30–50 years (Overland and Wang 2007).

Along with the growing stress on the ocean ecosystems will come changes to our governance systems. The increase in ocean uses will bring with it more varied and increased sophistication of stakeholders as well as conflict amongst these. Already, some groups are engaging expert "hired guns" to support their positions and failing this, entered into litigation. This trend will no doubt continue and present new challenges to governance systems. Science expertise will be employed throughout the ocean management process – from monitoring through to decision making.

Currently within North America as elsewhere, ocean science is delivered by a range of federal and state governments' institutions, universities, and private foundations. In this future, who is responsible for delivering what will have to be clear to all stakeholders. Given the above, the growth of knowledge may not be able to keep pace with the growth of impacts. This highlights the need for effective processes which allocate limited science resources to the priority requirements.

Given that the ocean will experience increasing uses and that the track record on ocean management is not good, what is the future of ocean management and stock assessment in it?

4.2   A Paradigm Shift in Ocean Management

In his classic, *The Structure of Scientific Revolutions*, Thomas Kuhn (1996) explored the phases of a scientific paradigm shift. He first describes a paradigm as being exemplified by a body of knowledge codified in text books and used as a framework within which the scientific community works. This "normal science" or *paradigm* is constantly challenged by new observations and hypotheses. The accumulation of problems ultimately calls into question the old paradigm, leading to a crisis period, formulation of a new paradigm which is vigorously debated and finally a new paradigm which is codified in new text books. Assuming that this same process applies to management systems and not just science, and this seems reasonable, one can ask "is ocean management in the midst of a paradigm shift?"

Management systems can be thought of as being composed of two main elements – objectives that both broadly and specifically stipulate the ultimate desired state of the unit being managed – an ecosystem or population in a given geographical area – and the suite of controls used to achieve these objectives.

The twentieth century fisheries management paradigm was dominated by objectives strictly associated with optimizing yield extraction from a population. The science was codified in a number of textbooks (e.g., Beverton and Holt 1957; Ricker 1975). More broadly, oceans management has consisted of individual ocean industry sectors working independently to achieve separate objectives. The control of cumulative impacts of all ocean sectors on the ecosystem was either not contemplated or attempted.

Since the beginning of the twenty-first century, there has been a global shift towards an ecosystem approach to management (Gislason et al. 2000; Sinclair et al. 2002) exemplified by objectives based upon broader conservation objectives and the recognition that cumulative impacts across all sectors must be controlled (Rosenberg et al. 2009). Fisheries are increasingly being seen as one sector amongst many. In Canada, suites of ecosystem objectives are been defined to conserve the biodiversity, productivity and habitat of ecosystems (DFO, 2007a; O'Boyle and Jamieson 2006). These objectives are to apply to all ocean sectors impacting the ecosystem. Similar suites of objectives are being developed around the world (Sainsbury and Sumaila 2003) and are being used by managers to address an increasing variety of issues. These include the impact of fisheries on ecosystems (such as bycatch, habitat impacts, and the consequences of genetic changes), the impact of ecosystems on fisheries (such as large-scale fish community changes and climate change) and ecosystem manipulation. The latter might appear to be captured by the first two but it is important to highlight that managers often would like to know what can be done to change an ecosystem component from one undesirable state (e.g., too many apex predators to allow a commercial species to increase in abundance) to another more desirable one. An example of this is the interaction between cod and grey seals on the Scotian Shelf. Managers and industry routinely ask scientists if reducing the seal herd would allow the cod resource to recover. This expansion in the scope of the management objectives is a clear sign of a shift to a new paradigm.

Regarding the control aspect of management, a key consideration is the assessment of how much abundance and/or biomass of a particular population exists. The approach has generally been to fit an assessment model to a set of data and, based upon this, make projections of future yield under a range of harvesting options. Stock assessment operates relatively independently of the other components of the management system. A number of studies conducted since the mid-1990s have highlighted problems with this approach. Burke et al. (1996) indicate that the collapse of the Scotian Shelf groundfish populations was the consequence of problems with a number of components of the management system, not just stock assessment. No one component could be identified as being the cause of the collapse. It is necessary to evaluate the effectiveness of a management system as that – a system. This approach is termed "Management Strategy Evaluation" (MSE). MSE strives to first identify the key uncertainties in each element of a management system and then, through simulation, explore the robustness of different management strategies to these uncertainties (see Stokes et al. 1999 for comprehensive overview of approach). In MSE, the emphasis is not in predicting how many fish will be in the ocean using various management strategies but rather in determining which management strategy is most likely to achieve a stated set of objectives given the uncertainties. It shifts the problem from the absolute to the relative. Here again, one can ask if there is a paradigm shift. There is recognition that the previous approach had significant problems and a change was needed and throughout the world, this change is underway. Here too, then, a paradigm shift is underway.

Overall, both components (objectives and controls) of oceans management are undergoing a paradigm shift. We are moving from stock assessment within the fisheries sector to MSE within an EAM (Fig. 4.1). And this trend will drive the development of new science over the next 3 decades.

EAM will not resolve all problems in the management system. For instance, EAM will not lead to sustainable fishing if there is too much fishing capacity. EAM is about defining overall goals that each sector is to achieve. MSE and EAM are different facets of the new paradigm. Indeed, MSE may be considered a prerequisite to EAM. EAM has its roots in the policy discussion of the early–mid-1990s and only now is its form starting to emerge (Rosenberg et al. 2009). Implementation of an EAM will take time due to the need for new governance structures, new science and management tools and so on. On the other hand, it is possible to implement

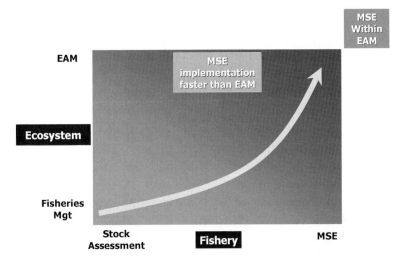

Fig. 4.1 Evolution of an Ecosystem Approach to Management (EAM) and Management Strategy Evaluation (MSE) in the emerging oceans management paradigm

MSE now. Indeed, if MSE had been a standard feature of management systems since 1977, we might not have experienced some of the fishery collapses that have occurred since then. While lessons are still being learnt on how best to conduct an MSE and although it may not be a panacea, it certainly appears to be a step in the right direction.

It is possible that the desire to achieve the EAM agenda will detract from achieving that of MSE. The implementation of MSE has been underway relatively longer than the shift towards EAM but has not captured the attention of many in the oceans community. At conferences and workshops, I have asked participants if they have heard of EAM; many hands go up. When asked about MSE, the response is a lot less emphatic. EAM has captured the imagination of many while MSE is working at the coal face. Putting excessive emphasis on EAM now would be counterproductive; EAM without MSE would exacerbate current problems (Longhurst 2006).

How will trends in the implementation of EAM and MSE influence the future of fisheries science broadly and stock assessment specifically? It is clear that the "stock assessment" of the future will need to evaluate the fishery impacts on achievement of all EAM objectives. This implicates both what science is required and how this science is utilized in the evaluation process.

4.3   Implications for MSE Elements

McAllister et al. (1999) propose that a fisheries management system is composed of an interlinked set of six elements – four (ecosystem/population, observation, implementation, and decision making) in an operational module and two (assessment and

Fig. 4.2 The components of a Management Strategy Evaluation (MSE) (Adapted from McAllister et al. 1999)

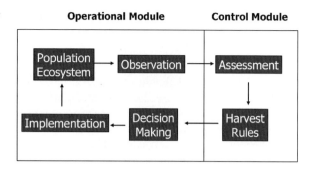

harvest rules) in a control module (Fig. 4.2). Their paper describes in some detail what is involved in each element and how these interact. This framework reflects the key components of a number of specific and general MSE studies (Dichmont et al. 2006; Kell et al. 2006; Sainsbury et al. 2000; Smith et al. 1999) and thus I have used it to organize my comments on the implications of the new ocean management paradigm on each element.

4.3.1 Ecosystem/Population Element

In an MSE, it is important to clearly specify the biological models being used as the underpinning of the analysis. These may not be the same as assumed in the assessment model, a point which we will return to in the following sections. Rather, this element describes, in as quantitative terms as possible, what we know about the biological processes of the ecosystem or population in question. Most importantly, it provides plausible hypotheses of ecosystem and population functioning when (1) there are competing hypotheses of functioning based upon the available data and/or theory and (2) there are gaps in our knowledge of key processes for which we don't even have competing hypotheses. Regarding the latter, Sainsbury (1998) highlights many of the challenges in trying to build plausible models of processes when we don't have alternative hypotheses. What are informative priors when knowledge is lacking? This is where ecosystem modeling can assist management. In an ecosystem model, we can incorporate not only the specifics of the case being studied but also draw upon a broader study of the processes through meta-analyses and ecological theory to hypothesize how ecosystem components may be interacting, and through simulation, evaluate the impact of different assumptions on management performance. This approach may allow us to put constraints upon the priors used. Some assumptions may be theoretically unrealistic. For example, the Eastern Scotian Shelf cod stock collapsed in the early 1990s and since then, the fishery has been closed. Despite this, total mortality, and by inference natural mortality of this and other east coast cod stocks has remained high (Shelton et al. 2005). Why are the cod stocks not recovering? A number of hypotheses are being investigated not least of which is that cod could be in a predator trap (Bundy and

Fanning 2005), i.e., cod abundance is so low that predation at any level by any species prevents cod from increasing in abundance. If one were to study the interaction between cod and seals in isolation of the rest of the ecosystem, these broader processes would be missed and would lead to conclusions that would be unrealistic from an ecosystem functioning point of view. Bundy and Fanning (2005) illustrate how ecosystem models, in this case Ecopath, can assist in constraining some of the possible explanatory hypotheses. This is a good reason why the dynamics of individual populations should be investigated within the broader context of the ecosystem rather than independent of it.

This is only part of the story. An important objective of an EAM is to ensure that an ecosystem doesn't unpredictably change from a desirable to undesirable state. This implies that it has enough ecological resilience (Walker and Salt 2006) to sustain external impacts. This in turn implies that we understand the resilience properties of ecosystems, how they respond to impact and the processes of their recovery. While terrestrial ecological literature is rich on resilience thinking, this body of theory is only starting to enter the marine literature (Duffy and Stachowicz 2006); marine ecologists need to start building these concepts into marine ecosystem models as they will be essential to enhancing our understanding of why ecosystems behave the way they do. If, from a theoretical perspective, marine ecosystems are indeed complex systems (Scheffer and Carpenter 2003), they can be expected to experience the phases of the so-called adaptive cycle (Fig. 4.3). How an ecosystem responds to a stress depends upon its degree of functional and response diversity, the manner in which ecosystem components are linked (their interconnectedness) and the tightness of feedbacks (the rapidity and strength of a response to an impact) (Walter and Salt 2006). These properties need to be built into ecosystem models in order to examine how changes in, say functional biodiversity (e.g., loss of a large component of the filter feeders), can reduce resilience and thus increase the probability of a change to an undesirable state (predator abundance).

So what is the future of ecosystem modeling in support of the new management paradigm? In the longer term, models which are evolutions of EwE and Atlantis

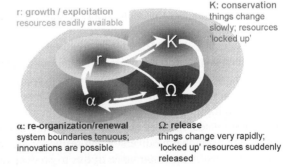

Fig. 4.3 The adaptive cycle of ecosystems (from *Panarchy* by Lance H. Gunderson and C.S. Holling, eds. Copyright © 2002 Island Press. Reproduced by permission of Island Press, Washington, D.C.)

(Fulton et al. 2004; Walters et al. 1997) will no doubt appear. These will not only describe the interactions of the various ecosystem components but also allow examination of how whole subsystems of these ecosystems might react to impacts. They will allow explicit examination of an ecosystem's resilient properties. However, there are very few such models that have so far been developed, never mind being used in management. In the short term, we will likely rely on semiquantitative models of ecosystem functioning which provide an overview of our understanding of the key components and their interconnections. Bayesian belief networks (Pearl 2000) which consist of box models of an ecosystem with linkages expressed in probabilistic terms, and allow expert judgment on the strength of these linkages, are an example of a class of potential models worthy of exploration. On a smaller scale, a number of minimum realistic models have been built to describe subcomponents of the entire ecosystem. The interaction between cod and seals on the Scotian Shelf is one such example (Fu et al. 2001). As subcomponents of an ecosystem are successfully built, they can be linked together into the larger ecosystem models. On an even smaller scale, there are a number of studies (e.g. Phipps 1992) which illustrate how complicated behavior in simulated populations can emerge from using simple rules to govern the interactions between modeled individuals. There are a number of ways to formulate these rules. If these simple rules are based upon some underlying fundamental theory, then one could start by building simple subcomponent models based upon these, which ultimately end in full ecosystem descriptions. So there are a number of possibilities for the future course of development of ecosystem models.

Having said this, we must keep in mind that the intent of MSE is not to develop the one "best model" but rather to identify key uncertainties in our understanding which allow us to develop plausible alternatives of ecosystem functioning. Some models may never reach the large scale due to information gaps. Perhaps we can learn from data-rich situations to apply to these data poor situations. However, this will likely not be possible in many cases as complex systems typically display very nonlinear behavior. This highlights the need for new theory and research on ecosystem functioning. As mentioned above, it might be possible to develop simple rules that govern an organism's behavior based upon some fundamental theory. Typically though, theory will need to be developed for the processes underway at a larger scale. Two main areas of research require attention: ecosystem control and the linkage between productivity and habitat. Regarding ecosystem control, there is a considerable amount of literature describing whether marine ecosystems are top-down (predator) controlled, bottom-up (prey) controlled, or some combination of both (Cury et al. 2000). Frank et al. (2007) are amongst the first to suggest that this control is linked to biodiversity, which can in turn be impacted by overfishing. In their review of ecosystems in the North Atlantic, they suggest that healthy ecosystems are typically bottom-up controlled but overfishing can reduce biodiversity which flips these systems to being top-down controlled. This linkage to biodiversity raises the related issue of how productivity and biodiversity are linked. There is some suggestion that biodiversity and resilience are linked (Duffy and Stachowicz 2006) but what is the importance of this to productivity? Can this be derived from some underlying fundamental principles? How does this influence the recovery process? Hubbell (2001) presented a controversial theory of biodiversity based upon three parameters – community size, speciation rate, and immigration rate – which he claimed produces diversity distributions similar to those observed in nature. While this theory has been challenged, it provides a good example of what can be achieved by examining the consequences of simple assumptions on ecosystem processes. Further, it would be worthwhile investigating how this or competing theories could be extended to incorporate productivity.

The second area requiring research is the habitat – productivity coupling. Much attention has been focused on the need to control the negative impacts of trawling and other fishing gear on the benthic habitat. One has to ask "what is the linkage between benthic habitat and the productivity of the rest of the food chain?" Related to this is how do different benthic habitats respond to impacts? Most habitat classification schemes are based upon some multivariate analysis which classifies habitat according to species composition. While useful in describing habitat distributions, this approach does not easily provide a means of predicting how a habitat will recover from an impact. Kostylev and Hannah (2007) take a quite different approach. Based upon earlier theoretical work of Southwood (1977, 1988), they characterize habitat according to the natural disturbance and physiological stress that the component organisms might experience. They theorize that habitat under high disturbance and low physiological stress would be at low risk to human impact (high recoverability) while the opposite is true in low disturbance, high physiological stress habitats (Fig. 4.4). While this work has yet to be fully tested, it illustrates the value of applying ecological theory to habitat classification. Not only can habitat function be predicted in areas where data are limited but also it allows prediction of habitat behavior when impacted. More such work on this and competing theories and approaches will be required in the future.

Many of the above theories either explicitly or implicitly implicate spatial processes which highlights the need to incorporate these into ecosystem models. This is particularly true when models include habitat-related processes. Networks of marine protected areas are being proposed to control human impacts on the benthos (Smith et al. 2006) although it has been emphasized that they are not a solution to all management problems (Kaiser 2005) and that careful consideration needs to be given when they should be used, how large they should be and their placement. There is very little theory to guide the use of this tool.

Incorporating spatial processes into ecosystem models will also allow evaluation of the impacts of climate change. In the twenty-first century, climate change is likely to have significant impacts on marine ecosystems and their users. Incorporating spatial processes into ecosystem models should be straightforward for the physical oceanographic community. Many of their models are already spatially designed. Such is not the case with population models, which, it could be argued due to sampling limitations, have tended to focus on describing biological processes temporally. This is particularly true of fisheries science. Fishery stock assessments routinely summarize large amounts of data to inform management decisions. Much of the raw data is collected through spatially designed sampling systems and it will be necessary to spatially disaggregate this information for incorporation into future ecosystem models. While the incorporation of spatial processes

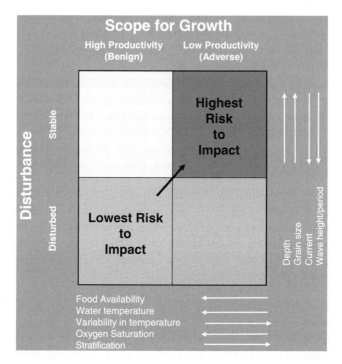

Fig. 4.4 Indicators used to classify benthic habitat based upon the work of Southwood (adapted from O'Boyle et al., 2005b)

into ecosystem models will be a challenge, it provides an opportunity to bring the fields of oceanography and fisheries biology closer together. Space will be the currency common to both.

Lastly, the inclusion of spatial dynamics into ecosystem models will be important to the understanding of fishing fleet dynamics. This is important for a number of reasons. First, it will allow enhanced assessment of the impacts of fishing on the ecosystem and the control of these impacts. It will also allow prediction of fleet behavior and thus provide a valuable means to communicate ecosystem processes to the users of the ocean, who will be able to understand what the models are predicting based upon the information that they are most familiar with. This will be increasingly important as ecosystem models become more complex and seemingly inaccessible to ocean users.

4.3.2 Observation

Over the next 30 years, if current developments are any indication, there will be unprecedented technological advancements, which will greatly enhance our capacity to understand organism, population, and ocean processes. There will no doubt be greater use of existing technologies, e.g., "black boxes" to track the movements of fishing vessels and satellite imagery to provide broad spatial-scale observations on ocean conditions. There will also be refinements of current technologies. For instance, trawl surveys are still the predominant sampling tool to provide long-term trends in abundance and biomass of ecosystem components. There are now available many single, dual, split, and multi-beam acoustic systems that enhance these surveys by providing detail on the spatial distribution of the abundance of biota. This will allow much greater precision on the estimation of abundance and biomass than is now possible. However, it is not yet clear whether or not these technologies will fully replace putting nets in the water and seeing what one catches. While there are ways to infer organism size from acoustic backscatter, species composition is another matter. There have been attempts in the past to determine species from the characteristics of the backscatter signal but in general the net is still the best means to determine what is there as opposed to how much is there. It is likely that sampling tools such as nets will always be needed to supplement observations collected through other more technical means.

One of the largest technological advances will be in our ability, through global initiatives such as the Ocean Tracking Network, to monitor the spatial movements of organisms. Indeed, the availability of these technologies will challenge modellers, especially biological, to incorporate spatial information into their descriptions of populations and ecosystems. This information could have a profound impact on management. More complete description and understanding of spatial processes will lead to more informed use of management tools such as MPAs and a richer dialogue with stakeholders such as fishermen on ocean processes as they generally have a much better grasp of local versus regional scale processes. Overall, this should lead to enhanced buy-in to management.

Multibeam acoustic systems are already opening the door to comprehensive small scale mapping of the ocean bottom and the communities that live there. On the Scotian Shelf, extensive areas have been mapped using multibeam acoustics and, in association with related advances in photographic bottom imaging and sampling technology (Gordon et al. 2006), have led to insights on both habitat descriptions as well as usage. For instance, we have been able to identify associations between specific types of bottom habitat as indicated by multibeam (i.e. gravel & sand) and scallop densities (Fig. 4.5). This has been used to inform the scallop fleet where they should fish to both obtain the highest catch rates as well as keep their operating costs down (Robert 2001). With raising fuel costs, dragging types of gear are starting to meet the upper limits of their profitability. For instance, the beam trawl fishery for nephrops in the North Sea is now becoming uneconomical (M. Dickie Collas, personal communication, 2006). Fishing fleets are investigating numerous means to reduce their costs. These mapping efforts are not only assisting them but leading to a better working relationship between industry and scientists. As well, this is a means to limit the habitat impacts of fishing activity. The bottom mapping is not only identifying habitat to target but also that to avoid. Certainly, maps of bottom type are proving valuable in identifying areas of high and low disturbance from which, as stated above, we can make inferences of the types of organisms that live there (Kostylev and Hannah 2007).

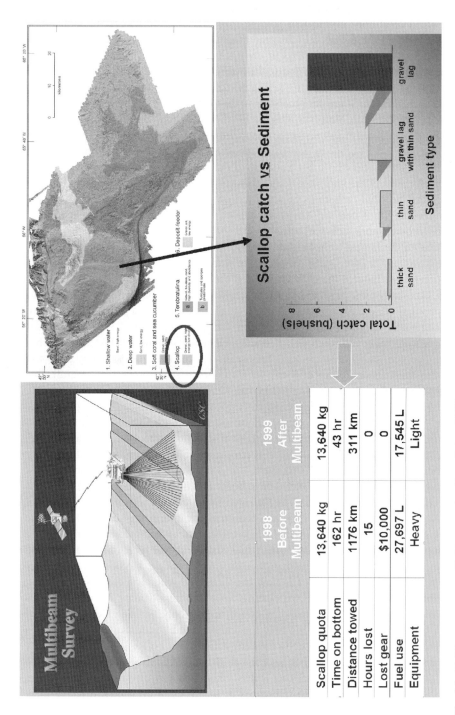

Fig. 4.5 Multibeam surveys and habitat mapping (from Robert 2001)

4.3.3   Assessment and Harvest Rules

Assessment provides the values of the indicators which are used in the harvest rules which in turn are used to decide what management actions – decrease the quota, increase the gear mesh size, etc. – are to be taken to meet a specified objective. They are the key tools to measure the impact of fishing on ecosystem components and thus the performance of management. In the twentieth century paradigm, assessments were primarily restricted to evaluating the impacts of the fishery on the target commercial species, with the main indicators used in the control rules being fishing mortality and spawning biomass. In the twenty-first century ocean management paradigm, there will be a need for an additional type of assessment, one that operates at the ecosystem level. Its purpose will be to measure management performance against a suite of ecosystem objectives and guide management actions to address cumulative impacts on key ecosystem components and processes across all ocean sectors (O'Boyle et al. 2005a). We are starting to see the first versions of these ecosystem level assessments (see for instance the Ecosystem Status Report for the eastern Scotian Shelf; DFO 2003). Because our understanding of ecosystem processes is still rudimentary, these reports generally tend to be descriptive – ocean currents doing this, fish resources doing that – and are not yet at the stage that they can be used to inform management decisions other than in the broadest sense. As ecosystem models and understanding improves, these reports will become an essential communication tool in ocean management.

As the focus of this paper is fisheries, I will not comment further on the ecosystem level assessment other than state that the objectives stated at the ecosystem level will have implications for fisheries stock assessment. If the ecosystem level objective states a need to conserve particular types of benthic habitat, then each sector, including fisheries, must evaluate the share of its impact on that benthic habitat. While the stock assessments will continue to evaluate performance against objectives stated to maintain the productivity of harvested populations, they will also have to evaluate performance of management against the broader set of ecosystem objectives that are relevant to that fishery. This has implications throughout the management system as will be seen below.

In the short-term, stock assessments will likely be an evolution of current variants. Over time, stock assessments have become increasingly complex – from relatively simple surplus production models to fully age-based state-space formulations (Quinn and Deriso 1999). It has been argued that, in general, age-based models tend to perform better than age-aggregated formulations (NAS 1998; Walters and Martell 2004) although there is much to suggest that the specifics of the case in question are important (Punt 1995). Whatever model is used, they generally produce time trends of recruitment, abundance, biomass and fishing mortality, and allow determination of the probability of being greater than or less than specified reference points under different predicted harvest projections (see Fig. 4.6 for an example assessment output based upon Georges Bank haddock).

Some have questioned the need for such complex models (e.g. Hilborn 2002, 2003) saying that simpler, data-based, approaches can not only be effective but also

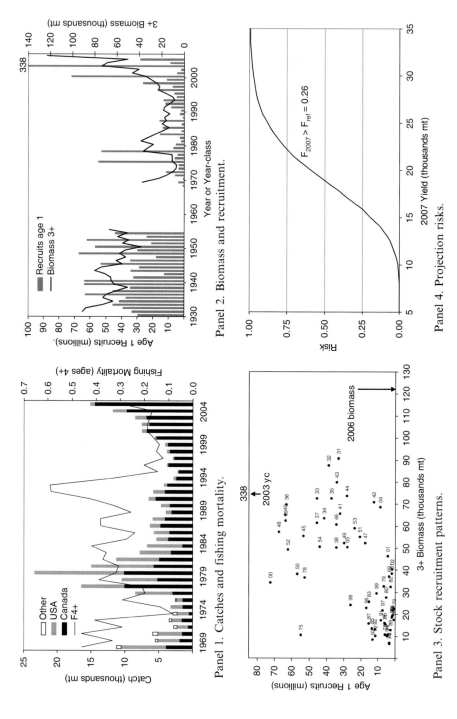

Fig. 4.6 Typical output from twentieth century stock assessment (From TRAC 2006)

lead to greater acceptance in the assessment results by fishermen. Others (Caddy 2002; Halliday et al. 2001; Rose 1997) have proposed that current assessment models do not allow for consideration of the full diversity of information available on a population and in doing so have led to some of the stock collapses (e.g. Northwest Atlantic cod). They have proposed assessment frameworks that include a large suite of indicators with each scaled through expert choice of reference points which divide these into colour zones – typically red, yellow and green. These approaches readily allow integration of local knowledge with the scientific sources, addressing the need to bridge the communications gap between the two groups (Mackinson and Nottestad 1998). The problem with such systems is that they have generally not been based upon a mechanistic understanding of how a particular indicator might respond to some human disturbance. This complicates decision making on what to do about an indicator which is going red, and has led to criticism of this approach (O'Boyle 2003). The issue for stock assessments is neither finding the formulation with the least complexity, or the most indicators, but rather finding that set of indicators that is sufficient to do the job. In other words, what suite of indicators will lead to robust behaviour of the management system in the face of the key uncertainties of the ecosystem? One can develop the most complex model possible but if the level of non-compliance with the regulations is high, assessed fishing mortality may not be the best indicator to use as a basis to limit fishing effort. Size frequencies from sampling and surveys may be better to judge performance towards the objectives. The consequence of an MSE on the assessment element could be simpler models but this needs to be determined on a case by case basis.

Consistent with a shift towards an EAM, there will be an evolution of assessments towards the measurement of performance against a much larger suite of objectives than considered under the old paradigm. One could eventually envision a situation in which there are 20 or so objectives that an assessment must measure performance against. These may relate to different aspects of biodiversity, productivity and habitat as discussed above. For instance, the impact of fisheries on other species, both commercial and non-commercial, will be important to evaluate. There are increasing demands to have the impacts of fisheries on habitats evaluated. Under the twentieth century paradigm, limit and precautionary fishing mortality reference points have been defined generally without regard to collateral habitat impacts. An EAM will require consideration of how fishing mortality and thus fishing effort is related to habitat impact. The fishing mortality reference points may have to be adjusted to take into account this relationship. Consideration of a population's productivity alone may provide one set of fishing mortality reference points which may not be attainable due to the increasing impact on habitat as fishing mortality increases (Fig. 4.7).

The ultimate goal of assessment is to provide the status of a suite of indicators which have been derived from understanding of a comprehensive ecosystem model. One could produce tables looking very much like those of the Traffic Light Method (Caddy 2002) but with understanding of how the indicators relate to the ecosystem dynamics on one side and management actions on the other. Issues that will need to be addressed are similar to those encountered during development of Traffic Light

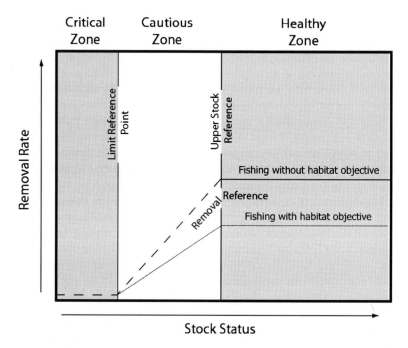

Fig. 4.7 Illustration of how ecosystem objectives related to habitat impacts could modify reference points used for objectives related to stock productivity

Method tables (e.g., redundancy of indicators, multiple use of same data sources, weight given to each indicator and so on, Halliday et al. 2001). It will also be useful to develop indicators, again based upon the ecosystem model, on fleet behavior (catch rates, catch size composition, etc.). This information is often considered in assessments now but generally as an input to the analyses, e.g., catch rates as an indicator of abundance. Such usage has been criticized (Walters and Martell 2004) due to non-linearities in the relationship between catch rate and abundance. However, it would be useful for assessments of the future to produce predictions of catch rates and other fishery-related indicators (an output from the models) to allow stakeholders to "see" the resource through information that they are most familiar with. In other words, assessments of the future will provide views of the ecosystem and its health that both scientists and stakeholders can relate to and in terms that they both understand.

4.3.4 Decision Making

It is possible to have a management system in which the assessment provides the harvest rule with an indicator which immediately leads to a pre-agreed management action. My experience, and perhaps elsewhere as well (Holt 2002) is that resource managers and stakeholders don't like hard and fast rules about what actions to take once a particular reference point is met. They want "wiggle" room. To capture the processes involved, McAllister et al. (1999) added a decision-making element to their description of a management system. For example, the risk plot in Fig. 4.6 (panel 4) shows how the probability of exceeding a fishing mortality reference point increases with yield in the coming year. Managers and the stakeholder community would consider this and weigh the risk to the resource against a suite of other relevant considerations (including socio-economic issues). For instance, many fisheries around the world are essential to the viability of coastal communities. Elsewhere, there may be other externalities that influence the decisions of managers. While EAM efforts are starting to define socioeconomic objectives along with the ecosystem ones, they are far from fully integrating these into the management system and it is unclear how the relative weight given to achievement of the conservation, economic, and social will be determined. There has been little systematic study on how managers use assessment information, including the estimates of uncertainty, in decision-making. This is paradoxical as the scientific community is expending substantial effort in estimating posterior distributions of projected population parameters under different harvest scenarios.

There is thus a critical need to develop understanding on how resource managers make decisions in the face of competing objectives and uncertainty. Walters (1994) provides one possible means on how this might be undertaken. He employed "gaming" or scenario exploration in which participants of a management system, faced with different strategies and challenges, made decisions. Studies such as these would be very worthwhile to pursue. On a related note, there is a need to develop user-friendly software systems that allow managers and stakeholders to explore possible management scenarios. Most current software systems are designed by scientists to meet scientific requirements and are not designed to be user friendly. An early example of user-friendly software to facilitate communication amongst scientists, managers, and stakeholders is ABASIM, developed by Jeremy Prince, Philip Sluczanowski, and John Tonkin in 1992. ABASIM is a computer game which simulates the dynamics of a hypothetical abalone fishery and allows graphical demonstration of how a stock responds to exploitation and how managers can optimize harvesting. During its development, attention was paid to the user interface, with artists employed in its design to ensure that it was user friendly (P. Sluczanowski, personal communication, 1992). During public open houses at the Bedford Institute of Oceanography, we have used ABASIM to illustrate the performance of management systems and have found it to be very popular.

4.4 Implications for Process

4.4.1 The Need for Change

In the twentieth century ocean management paradigm, teams of scientists typically compiled data on a resource on some predefined schedule, and then met to peer review their analyses, the result of which was advice to management on the appropriate harvest levels. The interaction between scientists and managers was relatively restricted to the development of the terms of reference of reviews and communication of assessment results after the review has been completed. There are a number of problems with this approach. First, there is limited examination of alternative management strategies and the overall performance of the management system. As a consequence, the assessment models have grown in complexity as scientists have endeavored to perfect "their part" of the management system. Particular emphasis has been placed upon quantifying the uncertainty in assessments with, as I stated above, little appreciation on how that information is used. Second, during peer review meetings, invited external reviewers have been encouraged to suggest improvements to the formulations. While this is laudable, this can lead to "tinkering" with the formulation to improve the overall fit of a particular model to the available data. There are many possible improvements that could be made and the ones that are adopted may more likely be a function of who the reviewer is rather than the result of improved fundamental understanding of relevant ecosystem processes. One might just be tracking noise in the data. It is more appropriate to develop hypotheses of why the models are not fitting the data and then conduct research to test these. This research obviously takes time and can only be conducted between assessments.

Finally, the peer review process itself is difficult to manage. To understand why, we need to keep in mind the purpose of the peer review. In the twentieth century paradigm, the general purpose of review has been to ensure that the assessment model employed is the most appropriate for the task at hand – the provision of management advice. This can, as stated above, lead to small short-term improvements to the formulation which may or may not lead to a better scientific basis for advice in the longer term. Certainly, inviting international experts to these types of reviews at which there is limited time to reformulate the models, can be a waste of talent and be frustrating to participants. It is better to employ this expertise during the model development stage over a longer period, a recommendation made by a number of international reviewers at a Canada/US assessment review for Georges Bank groundfish stocks (O'Boyle 1998). Another issue is stakeholder engagement in the peer review process. Before the collapse of the east coast groundfish stocks, the scientific advice was developed and peer reviewed by the Canadian Atlantic Fisheries Scientific Advisory Committee (CAFSAC). Only scientists participated in the peer review. Neither managers nor stakeholders were invited to participate in the reviews. When the groundfish stocks collapsed in the early 1990s, CAFSAC was disbanded and the scientific peer review was delegated to the Atlantic regions of the Department of Fisheries and Oceans (DFO). Termed the Regional Advisory Process (RAP), the intent was to both allow regional flexibility in the peer review and allow more stakeholder and manager participation in the process. In 2000, the Canadian government introduced guidelines on Science Advice for Government Effectiveness (SAGE) which made it clear that transparency and openness in the science process was national government policy. This trend towards openness in the peer review process has been underway in several other countries around the world (e.g., USA,

ICES) and in general can be considered a positive development. However, opening the doors of the science peer review process, in eastern Canada at least, has not been without its problems. Initially, the experience was generally positive, with the development of understanding between scientists and stakeholders on the limits of the information that each could bring to the reviews. This resulted in the initiation of numerous DFO Science – industry survey projects which have led to genuine improvements in understanding of ecosystem processes (O'Boyle et al. 1995). However, over time, problems have arisen that may endanger the integrity of the peer review. Perceived conflicts of interest are concerning some scientists who feel that industry is having undue influence on the process.

Some of these issues can be resolved by having a strong chairperson. However, it is also essential to ensure that participants have a clear understanding of why they have been engaged in the review. Some may be participating to obtain the latest information on stock status rather than wishing to contribute to the peer review. Further, if the other elements of the management system are not functioning as they should (e.g., the fisheries management advisory committees), there is a danger that the stakeholders may focus on achieving their preferred ends at the science review rather than on improvements in the scientific basis for advice. With assessments moving towards the consideration of larger suites of indicators to measure performance towards a larger suite of objectives, it is imperative that the peer review is as effective and unbiased as possible.

Separation of the review of the MSE elements, herein termed the "framework review," from its application to inform on-going ocean management should largely address many of the issues presented above and will be a fundamental feature of the twenty-first century ocean management paradigm. The ultimate goal of the framework review will be the development of the most robust strategy to manage the fishery. The framework review will be an extended process, first working through the elements of an MSE and then undertaking the MSE itself to define the most appropriate management strategy for the resource. There will be a shift in emphasis of the scientific input from primarily the assessment to the other elements, with much of this shifted to the ecosystem/population element. The whole process will require extensive teamwork amongst scientists, managers, and stakeholders. The assessment review itself will evolve into a straightforward scheduled application of the MSE framework. This new organizational structure is similar to that proposed by Hilborn (2003) except that here I forecast changes to all elements of the management system, not just assessment.

These changes will have a number of beneficial effects. Focus on review of the framework separate from the assessment will lead to long-term improvements in the former and stability of the scientific basis of the assessment until genuine scientific improvements can be made to it. Scientific expertise will be applied where it is most needed and not applied to assessment by default. As the process will be extended, it will be possible to engage the most appropriate expertise, including that of stakeholders, for the particular element under consideration. From my experience, fishermen have good insight on local processes but less so on the larger-scale ones. Engaging stakeholders on these local issues can greatly assist interpretation of localized trends in the data. Also, it is more appropriate to engage stakeholders in identifying issues in particular time series of data (e.g., gear changes in the fishery) than on the reliability of the last point in that time series (e.g., the level of spawning biomass). The first is more likely associated with their knowledge of the fishery and its interaction with the resource while the latter is more open to a conflict of interest due to short term socioeconomic issues that they may have.

4.4.2 Frameworks and System Performance

In the twenty-first century ocean management paradigm, during the MSE framework review, considerable attention will be given to the ecosystem/population element. Careful consideration of the validity of the key assumptions and processes will be required. More emphasis than is currently the case will be placed on simulation rather than on model fitting. Where the latter is employed, it will be to discern key parameters that could not be obtained from either theory or meta-analyses of similar situations. As noted above, considerable attention will be given to spatial processes and how these might influence the management strategies.

Review of the "observation" element will include issues similar to those currently routinely encountered. Is the sampling appropriate? Is the age determination valid? Is the survey design defensible? However, there will be additional consideration of the benefits of these observational activities in relation to their costs. By under taking a management system wide review, it is more likely that important issues will be uncovered that will require a redirection of resources (e.g., Chen et al. 2003). Typically, sampling variability declines as some exponential function of sampling intensity. For instance, with about 250 stations occupied, the Scotian Shelf surf clam survey is likely oversampling the resource (Fig. 4.8). There is an opportunity to decrease the level of sampling without much loss of information and divert scientific resources to other priority needs (DFO, 2007b). Indeed, it was during an MSE, the first to be held on a Scotian Shelf resource, that this observation was made (DFO, 2007b). It is evident that an MSE is a valuable means to determine funding priorities.

Regarding the "assessment" element of an MSE, the focus of the review will be the identification of the indicators to be used in management decision making. As stated above, an EAM will require expansion of the suite of indicators that need to be considered and will be a product of understanding of the ecosystem and population processes gained through modeling. Thus, assessment will need to focus on the ways and means to produce these indicators that lead to robust and cost effective management strategies.

This also applies to the choice of the harvest rules. Kesteven (1999) provides a case study of a small coastal fishery of NSW Australia which proposes that the times series of catch is a relatively straightforward means to manage the fishery. This approach would obviously need to be confirmed through simulation, but points

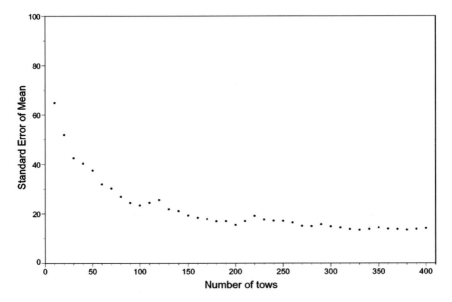

Fig. 4.8 Estimated standard error from the 1996–97 Arctic surf clam survey on Banquereau Bank. Points are standard errors from 30 replicates sampling with replacement from survey tows for the specified number of tows; the current survey has approximately 250 tows (DFO, 2007b)

out that during an MSE, one should be open to the use of "non-model" indicators' which have been validated by the ecosystem/population models.

The new ocean management paradigm will need to be open to different implementation methodologies. Problems in the enforcement of catch and effort reporting have been key factors in many resource failures (e.g., Burke et al. 1996). During a MSE, one will need to evaluate the effectiveness of the implementation of the chosen management strategy and how robust it is to potential problems. Walters (1998) goes as far as to say that the use of large space–time refuges can achieve the stated management objectives without the need of traditional assessment. Again, this approach would have to be evaluated in the MSE but if this can be demonstrated, then it is a viable option.

Once each element of the MSE have been reviewed and the assumptions and uncertainties identified, then the overall system performance for a suite of defined management strategies can be evaluated. Considerable thought will need to go into the definition of the management strategies themselves – catch versus effort regulation, the use of refugia, what management tools to employ when certain indicators are met, and so on. Given the wide variety of options available, as stated earlier, this could be conducted in a "gaming environment," as proposed by Walters (1994). The product of the workshop or workshops would be the details of the management strategy to use, the schedule of future reviews and the process to follow if problems arise. All of this is new ground that will need to be explored. Indeed, the definition of the suite of management strategies is itself an area of required research.

4.4.3 Assessment

Assuming that the framework has been adequately reviewed and the appropriate management strategy has been defined and agreed to, then the assessment will be a straightforward application of the framework to the most up-to-date information to inform management decision making. The emphasis in future assessment meetings will be on ensuring that the framework has been applied appropriately and that it is still valid (e.g., no serious model fitting issues). In relation to the latter, it will be important during framework discussions to outline the diagnostics that assessments would use to determine assessment validity. In essence, assessment review will be for quality assurance.

One can predict a number of side benefits from this approach. First, there will be an overall reduction in the level of resources applied to the assessment element with a shift, as stated earlier, to the other elements, most particularly the ecosystem/population element. This will lead to long-term improvements in the management system as resources will be applied in the elements where they are most needed. Second, stakeholder engagement will be primarily focused on interpretation of the assessment results rather than focused on the details of the assessment itself. This will largely address possible conflict of interest issues. Halliday et al. (2001) have further suggested that with an agreed suite of indicators and reference points, the process of assessment could be automated to the point that the databases could be automatically accessed, the appropriate analyses conducted, the indicators estimated, and the products of the harvest rules produced. These would then be placed on the Internet for all to see. I can see a time when this level of automation might arise but some form of review will always be required, the form of which is open to discussion within an MSE.

4.4.4 Experience with Approach

The MSE approach has been discussed in the scientific literature since the early 1990s (Stokes et al. 1999; Rice and Connolly, 2007). Most analyses have focused on the observation, assessment and harvest rule elements for fisheries, these being mostly in single and sometimes multispecies applications. It is fair to say though that we are very early in our experience and experimentation with the MSE approach and there is still much to learn. Indeed, there are few MSEs for multispecies fisheries, with that of the southern shark and eastern scalefish and shark fishery being a recent example (Smith et al., 2007). Dichmont et al. (2006) provide an illustration of MSE for a short-lived species (northern prawn) and there have been a number of studies exploring the approach in the North Atlantic (Kell et al. 2006). ICES has initiated a project termed FLR (Fisheries Library in R; Kell et al. 2007) to encourage the modular development of software tools to be used in MSEs. In the USA, the National Marine Fisheries Service has, in its review of the stock assessment process (NMFS 2001), determined that MSE within the context of an

EAM is an important initiative to pursue. Overall, MSE is gaining momentum and is starting to make its way into management systems around the world.

In DFO's Maritimes Region on Canada's east coast, there has been a separation of the review of the framework from assessments since 2001. A number of framework reviews have been conducted (see Maritimes RAP website at http://www.mar. dfo-mpo.gc.ca/science/rap/internet/Home.htm). These framework reviews have focused on stock specific issues such as the definition of the management unit, estimation of current state, characterization of the productivity to determine the harvest strategy and the projection procedure and are a step towards an MSE. The experience gained in these framework reviews has led to genuine improvements in the assessment process. Regarding an MSE itself, one for the Scotian Shelf surf clam had been initiated but it soon became apparent that considerably more modeling was required to investigate the spatial dynamics of the resource (DFO, 2007b). This trial MSE uncovered issues and problems to resolve, which is one of the reasons to adopt the approach in the first place.

Likely closest in philosophy to an MSE within DFO is the recovery potential assessment (RPA) required for species that are being considered for listing under the 2003 *Species at Risk Act* (DFO 2004). In these, the potential of a species to recover from a set of impacts is to be assessed with consideration given to the current status, productive potential, recovery targets, and time to recovery under different impact assumptions. These analyses can be quite extensive (see DFO 2005 for an example), considering both the key assumptions underlying the analysis and determining what mitigation is available to address potential impacts on the species.

An ecosystem approach to management as it applies to fisheries is in its early stages of implementation in Canada. Thus far, the bycatch and discard implications of stock specific fisheries are being evaluated and modifications to management plans being considered. The habitat impacts of fishing have been discussed nationally (DFO 2006) and various regions are developing approaches to deal with these. In Maritimes Region, the fisheries management plans are being evaluated against the suite of ecosystem objectives defined through the Eastern Scotian Shelf Integrated Management initiative (Rutherford et al. 2005) to judge their compliance with an EAM. These modest steps in the transition towards a new ocean management paradigm have been very fruitful in not only highlighting what changes need to be made to the management process, but also indicate the need for a shift in attitudes within the ocean community. Indeed, one of the biggest challenges facing the new paradigm is the need for broad-based cultural change.

4.4.5 Need for Cultural Change and Financial Considerations

Paradigm shifts don't occur without a change with in the scientific community itself (Kuhn 1996). Supporters of the old paradigm either convert to the new paradigm or more likely, they pass on with their strongly held beliefs replaced by a new generation enthusiastic to pursue the new paradigm. Extended jurisdiction of coastal zones around the world in 1977 was an important milestone in the management of fisheries in the twentieth century paradigm. In response to the extended jurisdiction, many nations embarked on hiring young scientists to meet the new demands. Thirty years later, many of these scientists are reaching retirement age and will be leaving the profession over the next 5–10 years. The new scientists that will replace them will ultimately define many of the details of the new paradigm.

There are a number of other cultural changes that I see necessary to make the new paradigm a reality. It is the role of scientists to focus on the details of a problem. This is why, when given an assessment to do, they will create the best assessment engine that they can, notwithstanding the fact that problems in other parts of the management system, say enforcement, may nullify the benefits of the "perfect" assessment. For this reason, while science managers need to create EAM and MSE teams that will work on improving their specific element, they also need to actively manage these teams towards the broader goal of ensuring that science resources are applied to priority needs of the management system. Change too must come to the resource management and stakeholder community. Both must be willing to take a broader in decision making to acknowledge the cumulative of fishing and forego perceived short term gains. Overall, there is a need for cultural change in all components of the ocean community.

Managing the new paradigm will not be cheap. The expansion of the objectives under an EAM and the management system elements to consider under an MSE imply more costs than are currently the case. Without some guiding principles, it is possible for the system's administration to become greater than the value of the resource. O'Boyle and Zwanenburg (1997) undertook an analysis of the costs for managing a number of the resources on the Scotian Shelf. They found that the administrative overhead ranged from 2% to 18% depending on the fishery, with shellfish fisheries typically less expensive to manage than groundfish fisheries. One way to proceed is to valuate the services that the ecosystem under management provides and then establish some percent of this as a management overhead, for example 15% annually. Then, each ecosystem management plan, including the sector specific plans, would have to be implemented within this funding.

The source of this funding will be an issue. Currently, most fisheries management systems are publicly funded. In Canada, under the 1992 *Canadian Environmental Assessment Act*, proponents of projects that may have environmental consequences must undertake an impact assessment to their own expense to show that their project would not harm the ecosystem. Under the new ocean management paradigm, a user pay system may be required in which all impacting industries, including fisheries, pay for the incremental costs associated with ensuring that the ecosystem is not irreversibly damaged by their activity. This would include monitoring, such as sampling and surveys, as well as assessment, which in the new paradigm is application of the agreed framework to the current set of data. Public funding would be used to support public good activities which would be of benefit to all ocean industries. This would include ecosystem studies and monitoring and most importantly, management of the peer review process which will receive growing prominence under the new ocean management paradigm.

4.5   Concluding Remarks

We are in the midst of a transition to a new ocean management paradigm and while it is difficult to make out its full shape, an ecosystem approach to management within which is embedded management strategy evaluation will be two of its key components. There are many details to be worked out and the transition will be an evolutionary process with alternate ways and means tried and abandoned as dictated by experience. However, one thing is clear. Getting the new paradigm right will have to be priority for our society. This may be our last chance to prove that the management of the wild resources of the ocean can be conducted in a sustainable manner.

References

Beverton, R.J.H. and S.J. Holt. 1957. On the Dynamics of Exploited Fish Populations. Fishery Investigations Series II, Vol. XIX. Her Majesty's Stationery Office, UK. 533 pp.

Bundy, A. and L. Paul Fanning. 2005. Can Atlantic cod (Gadus morhua) recover? Exploring trophic explanations for the non-recovery of the cod stock on the eastern Scotian Shelf, Canada. Can. J. Fish. Aquat. Sci. 62: 1474–1489.

Burke, D.L., R.N. O'Boyle, P. Partington and M. Sinclair. 1996. Report of the second workshop on Scotia-Fundy groundfish management. Can. Tech. Rep. Fish. Aquat. Sci. 2100: 247.

Caddy, J. 2002. Limit reference points, traffic lights, and holistic approaches to fisheries management with minimal stock assessment input. Fish. Res. 56: 133–137.

Chen, Y., L. Chen and K.L. Stergiou. 2003. Impacts of data quantity on fisheries stock assessment. Aquat. Sci. 65: 92–98.

Cury, P., A. Bakun, R.J.M. Crawford, A. Jarre, R.A. Quinones, L.J. Shannon and H.M. Verheye. 2000. Small pelagics in upwelling systems: patterns of interaction and structural changes in "wasp – waist" ecosystems. ICES J. Mar. Sci. 57: 603–618.

Dichmont, C.M., A. Deng, A.E. Punt, W. Venables and M. Haddon. 2006. Management strategies for short-lived species: the case of Australia's northern prawn fishery. Fish. Res. 82: 204–245.

DFO. 1982. Task Force on Atlantic Fisheries, Navigating Troubled Waters: A New Policy for the Atlantic Fisheries, Supply and Services Canada, December 1982.

DFO. 2003. State of the Eastern Scotian Shelf ecosystem. Canadian Science Advisory Secretariat. Ecosystem Status Report. 2003/004.

DFO. 2004. Revised Framework for Evaluation of Scope for Harm under Section 73 of the Species at Risk Act. DFO Canadian Science Advisory Secretariat. Stock Status Report 2004/048.

DFO. 2005. Recovery Potential Assessment for Winter Skate on the Eastern Scotian Shelf (NAFO Division 4VW). DFO Canadian Science Advisory Secretariat. Science Advisory Report 2005/062.

DFO. 2006. Impacts of Trawl Gears and Scallop Dredges on Benthic Habitats, Populations and Communities. DFO Canadian Science Advisory Secretariat. Science Advisory Report 2006/025.

DFO, 2007a. Guidance Document on Identifying Conservation Priorities and Phrasing Conservation Objectives for Large Ocean Management Areas. DFO Can. Sci. Advis. Sec. Sci. Advis. Rep. 2007/010.

DFO, 2007b. Proceedings of the Maritime Provinces Regional Advisory Process on Assessment and Management Strategy Framework for Banquereau Arctic Surfclam and Ocean Quahogs on Sable Bank and in St. Mary's Bay; 17–18 January 2007 and 4–5 April 2007. DFO Can. Sci. Advis. Sec. Proceed. Ser. 2007/008.

Duffy J.E. and J.J. Stachowicz. 2006. Why biodiversity is important to oceanography: potential roles of genetic, species, and trophic diversity in pelagic ecosystem processes. Mar. Ecol. Progr. Ser. 311: 179–189.

Frank, T.K., B. Petrie and N.L. Shackell. 2007. The ups and downs of trophic control in continental shelf ecosystems. Trends Ecol. Evol. 22: 236–242.

Fu, C., R. Mohn and L.P. Fanning. 2001. Why the Atlantic cod (Gadus morhua) stock off eastern Nova Scotia has not recovered. Can. J. Fish. Aquat. Sci. 58: 1613–1623.

Fulton, E.A., A.D.M. Smith and A.E. Punt. 2004. Ecological Indicators of the Ecosystem Effects of Fishing: Final Report. Report No. R99/1546. Australian Fisheries Management Authority, Canberra.

Garcia, S.M. and C. Newton. 1997. Current situation, trends and prospects in world capture fisheries. In E.L. Pikitch, D.D. Huppert and M.P. Sissenwine (Eds). Global Trends: Fisheries Management. American Fisheries Society Symposium 20. Bethesda, MD. pp. 3–27.

Gislason, H., M. Sinclair, K. Sainsbury and R. O'Boyle. 2000. Symposium overview: incorporating ecosystem objectives within fisheries management. ICES J. Mar. Sci. 57: 468–475.

Gordon, D.C., McKeown, D.L., Steeves, G., Vass, W.P., Bentham, K., Chin-Yee, M. 2006. Canadian imaging and sampling technology for studying benthic habitat and biological communities. In: Todd BJ, Greene HG (eds) Characterization and mapping of sea floor conditions for benthic habitat delineation. Geol Assoc Canada Spec Publ 47: 29–37.

Halliday, R.G., L.P. Fanning and R.K. Mohn. 2001. Use of the Traffic Light Method in Fishery Management Planning. Canadian Science Advisory Secretariat. Research Document. 2001/108.

Hilborn, R. 2002. The dark side of reference points. Bull. Mar. Sci. 70: 403–408.

Hilborn, R. 2003. The state of the art in stock assessment: where we are and where we are going. Sci. Mar. 67 (Suppl 1): 15–20.

Holling, C.S, and L.H. Gunderson. 2002. Resilience and adaptive cycles. In: Panarchy: Understanding transformations in human and natural systems. Eds. Gunderson L.H. and C.S. Holling. Island Press. Washington, DC. 450 pp.

Holt, S. 2002. The whaling controversy. Fish. Res. 54: 145–151.

Hubbell, S. 2001. The Unified Neutral Theory of Biodiversity and Biogeography. Princeton University Press, Princeton, NJ. 375 pp.

Kaiser, M. 2005. Are marine protected areas a red herring or fisheries panacea? Can. J. Fish. Aquat. Sci. 62: 1194–1199.

Kell, L.T., G.M. Pilling, G.P. Kirkwood, M.A. Pastoors, B. Mesnil, K. Korsbrekke, P. Abaunza, R. Aps, A. Biseau, P. Kinzlik, C.L. Needle, B.A. Roel and C. Ulrich. 2006. An evaluation of multi-annual management strategies for ICES roundfish stocks. ICES J. Mar. Sci. 63: 12–24.

Kell, L.T., I. Mosqueira, P. Grosjean, J-M. Fromentin, D. Garcia, R. Hillary, E. Jardim, S. Martle, M.A. Pastoors, J.J. Poos, F. Scott and R.D. Scott. 2007. FLR: An open source framework for the evaluation and development of management strategies. ICES Journal of Marine Science. 64: 640–646.

Kesteven, G.L. 1999. Stock assessments and the management of fishing activities. Fish. Res. 44: 105–112.

Kostylev, V.E., and Hannah, C.G., 2007, Process-driven characterization and mapping of seabed habitats, in Todd, B.J. and Greene, H.G., eds., Mapping the Seafloor for Habitat Characterization: Geological Association of Canada, Special Paper 47, pp. 171–184.

Kuhn, T. 1996. The Structure of Scientific Revolutions. Third Edition. University of Chicago Press, Chicago, IL. 212 pp.

Longhurst, A. 2006. The sustainability myth. Fish. Res. 81: 107–112.

Mackinson, S. and L. Nottestad. 1998. Combining local and scientific knowledge. Rev. Fish Biol. Fisheries. 8: 481–490.

McAllister, M.K., P.J. Starr, V.R. Restrepo and G.P. Kirkwood. 1999. Formulating quantitative methods to evaluate fishery-management systems: what fishery process should be modelled and what tradeoffs should be made? ICES J. Mar. Sci. 56: 900– 916.

Myers, R.A., J.A. Hutchings and N.J. Barrowman. 1996. Hypotheses for the decline of cod in the North Atlantic. Mar. Ecol. Progr. Ser. 138: 293–308.

NAS. 1998. Improving Fish Stock Assessments. National Academy Press, 177 pp.

NMFS 2001. Marine Fisheries Stock Assessment Improvement Plan. Report of the National Marine Fisheries Service National Task Force for Improving Fish Stock Assessments. U.S. Dep. Commerce, NOAA Tech. Memo. NMFS-F/SPO-56, 69 p., 25 appendices. Available at http://www.st.nmfs.gov/st2/index.html

O'Boyle, R. 1998. Proceedings of the Transboundary Resources Assessment Committee. 20–24 April 1998. Canadian Science Advisory Secretariat. Proceedings 1998/10.

O'Boyle, R. 2003. Proceedings of the RAP Meeting on Assessment Frameworks and Decision Rules for 4X-5Y Cod and Haddock. Canadian Science Advisory Secretariat. Proceedings 2003/008.

O'Boyle, R. and G. Jamieson. 2006. Observations on the implementation of ecosystem-based management: experiences on Canada's east and west coasts. Fish. Res. 79: 1–12.

O'Boyle, R. and K.C.T. Zwanenburg. 1997. A comparison of the benefits and costs of quota versus effort-based fisheries management. In D.A. Hancock, D.C. Smith, A. Grant and J.P. Beumer (Eds). Developing and Sustaining World Fisheries Resources. The State of Science and Management, 2nd World Fisheries Congress. CSIRO Publishing, Collingwood, VIC, Australia. pp. 283–290.

O'Boyle, R., D. Beanlands, P. Fanning, J. Hunt, P. Hurley, T. Lambert, J. Simon and K. Zwanenburg. 1995. An overview of joint science/industry surveys on the Scotian Shelf, Bay of Fundy, and Georges Bank. DFO Atlantic Fisheries Research Document 95/133.

O'Boyle, R., M. Sinclair, P. Keizer, K. Lee, D. Ricard and P. Yeats. 2005a. Indicators for ecosystem-based management on the Scotian Shelf: bridging the gap between theory and practice. ICES J. Mar. Sci. 62: 598–605.

O'Boyle, R., V. Kostylev, H. Breeze, T. Hall, G. Herbert, T. Worcester, D. Ricard and M. Sinclair. 2005b. Developing an Ecosystem – Based Management Framework for Benthic Communities: A Case Study of the Scotian Shelf. ICES CM 2005/ BB:18.

Overland, J.E. and M. Wang. 2007. Future climate of the North Pacific Ocean. Trans. Am. Geophys. Union 88: 178–182.

Pearl, J. 2000. Causality. Models, Reasoning and Inference. Cambridge University Press, Cambridge, UK. 379 pp.

Phipps, M.J. 1992. From local to global: the lesson of cellular automata. In D.L. DeAngelis and L.J. Gross (Eds). Individual-Based Models and Approaches in Ecology. Chapman & Hall, New York. pp. 165–187.

Punt, A.E. 1995. The performance of a production – model management procedure. Fish. Res. 21: 349–374.

Quinn, T.J. and R.B. Desiro. 1999. Quantitative Fish Dynamics. Oxford University Press, Oxford, UK. 542 pp.

Rice, J.C. and P.L. Connolly. 2007. Fisheries Management Strategies: an introduction by the Conveners. ICES Journal of Marine Science 64: 577–579.

Ricker, W.A. 1975. Computation and interpretation of biological statistics of fish populations. Bull. Fish. Res. Board Can. 191: 1–382.

Robert, G. 2001. Impact of multibeam surveys on scallop fishing efficiency and stock assessment. Bedford Institute of Oceanography 2001 in Review. Fisheries and Oceans Canada and Natural Resources Canada. 84 pp.

Rose, G. 1997. The trouble with fisheries science. Rev. Fish Biol. Fish. 7: 365–370.

Rosenberg, A.A., M.L. Mooney-Seus, I. Kiessling, C.B. Morgensen, R. O'Boyle, and J. Peacey. 2009. Lessons from national-level implementation across the world. In McLeod, K.L. and

H.M. Leslie [eds] Ecosystem-Based Management for the Oceans. Island Press. Washington, D.C. In Press.

Rutherford, R.J., G.J. Herbert and S.S. Coffen-Smout. 2005. Integrated ocean management and the collaborative planning process: the Eastern Scotian Shelf Integrated Management (ESSIM) initiative. Mar. Policy 29: 75–83.

Sainsbury, K. 1998. Living marine resources – assessment for the 21st century: what will be needed and how will it be provided? Fishery Stock Assessment Models. Alaska Sea Grant College Program, AK. SG. 98-01. pp. 1–40.

Sainsbury, K. and U.R. Sumaila. 2003. Incorporating ecosystem objectives into management of sustainable marine fisheries, including 'best practice' reference points and use of marine protected areas. In M. Sinclair and G. Valdimarsson (Eds). Responsible Fisheries in the Marine Ecosystem. FAO, Rome/CABI, Wallingford, UK. pp. 343–361.

Sainsbury, K., A.E. Punt and A.D.M. Smith. 2000. Design of operational management strategies for achieving fishery ecosystem objectives. ICES J. Mar. Sci. 57: 731–741.

Scheffer, M. and St.R. Carpenter. 2003. Catastrophic regime shifts in ecosystems: linking theory to observation. Trends Ecol. Evol. 18: 648–656.

Shelton, P.A., A.F. Sinclair, G.A. Chouinard, R. Mohn and D.E. Duplisea. 2005. Fishing under low productivity conditions is further delaying recovery of Northwest Atlantic cod (Gadus morhua). Can. J. Fish. Aquat. Sci. 63: 235–238.

Sinclair, M., R. Arnason, J. Csirke, Z. Karnicki, J. Sigurjonsson, H.R. Skjoldal and G. Valdimarsson. 2002. Responsible fisheries in the marine ecosystem. Fish. Res. 58: 255–265.

Smith, A.D.M., K.J. Sainsbury and R.A. Stevens. 1999. Implementing effective fisheries-management systems – management strategy evaluation and the Australian partnership approach. ICES. J. Mar. Sci. 56: 967–979.

Smith, J.L., K. Lewis and J. Langhren. 2006. A policy and planning framework for marine protected area networks in Canada's Oceans. WWF – Canada, Halifax. 105 pp.

Smith, A.D.M, E.J. Fulton, A.J. Hobday, D.C. Smith and P. Shoulder. 2007. Scientific tools to support the practical implementation of ecosystem-based fisheries management. ICES Journal of Marine Science. 64: 633–639.

Stokes, T.K., D.S. Butterworth and R.L. Stephenson. 1999. Introduction. Confronting uncertainty in the evaluation and implementation of fisheries-management systems. ICES J. Mar. Sci. 56: 795–796.

Southwood, T.R.E. 1977. Habitat, the templet for ecological strategies? J. Anim. Ecol., 46: 337–365.

Southwood, T.R.E. 1988. Tactics, strategies and templets. Oikos 52: 3–18.

TRAC. 2006. Eastern Georges Bank Haddock. TRAC Status Report 2006/02.

Walker, B. and D. Salt. 2006. Resilience Thinking. Sustaining Ecosystems and People in a Changing World. Island Press, Washington, DC. 174 pp.

Walters, C. 1994. Use of gaming procedures in evaluation of management experiments. Can. J. Fish. Aquat. Sci. 51: 2705–2714.

Walters, C. 1998. Designing fisheries management systems that do not depend upon accurate stock assessment. In T.J. Pitcher, P.J.B. Hart and D. Pauly's (Eds). Reinventing Fisheries Management. Kluwer, London, UK. 468 pp.

Walters, C.J. and S.J.D. Martell. 2004. Fisheries Ecology and Management. Princeton University Press, Princeton, NJ. 448 pp.

Walters, C.J., V. Christensen and D. Pauly. 1997. Structuring dynamic models of exploited ecosystems from trophic mass–balance assessments. Rev. Fish Biol. Fish. 7: 1–34.

Chapter 5
Ecosystem-Based Management of What? An Emerging Approach for Balancing Conflicting Objectives in Marine Resource Management

Isaac C. Kaplan and Phillip Levin

Abstract Managing marine resources has always been challenging, but this task looms ever larger as society demands more seafood while also requiring that we act as careful stewards of marine ecosystems. Evaluating management strategies in light of the diverse and changing demands of society for the goods and services the oceans provide requires that we clearly expose trade-offs among conflicting objectives. In this paper, we describe an approach using an Atlantis ecosystem model to evaluate management strategies and potential trade-offs between economic and conservation goals in the California Current ecosystem. We simulate a range of fishing intensities, and evaluate potential trade-offs between harvest maximization and the structure of the food web. Our results reveal that fishing combined with life history traits will alter the composition of the community such that short-lived, productive species replace longer-lived, lower productivity species. From an economic perspective, sustainably fishing productive high value species (Dungeness crab, hake, and squid) while overfishing less valuable, low productivity species (some rockfish) may seem like a wise choice; however, from a conservation perspective such a strategy would be completely unacceptable. We use the ecosystem model to visualize these trade-offs between economic and conservation concerns. We measure conservation and ecosystem structure by evaluating a suite of ecosystem indicators, such as ratios of the abundance of functional groups, and mean trophic level. The ratios of piscivore to planktivore, benthic to pelagic fish, and scavenger to piscivore all showed substantial shifts in community structure as levels of harvest increased. The mean trophic level of biological groups in the model was not sensitive to fishing intensity, and did not capture the associated shifts in the structure of the food web. Overall, we illustrate a simulation approach that can examine trade-offs between harvest and community-level indices of ecosystem structure. An ecosystem approach to management requires that we synthesize diverse physical, biological, and socioeconomic data and think critically about the ways in which our decisions affect the ecosystem services we value.

I.C. Kaplan and P. Levin
NOAA Fisheries, Fishery Resource Analysis and Monitoring Division,
2725 Montlake Blvd. E., Seattle WA 98112
Isaac.Kaplan@noaa.gov

R.J. Beamish and B.J. Rothschild (eds.), *The Future of Fisheries Science in North America*, 77
Fish & Fisheries Series, © Springer Science + Business Media B.V. 2009

The whole trend of research and education is toward specialization on particular objects or particular organisms. These are stressed while the assemblage to which they belong is ignored or forgotten, together with the fact that they are to be regarded as integral parts of the system of nature. Shelford (1933)

Keywords California Current · fisheries · ecosystem model · trade-offs · ecosystem indicators · Atlantis

5.1   Introduction

Ecosystem approaches to marine resource management are a response to today's perception of deepening troubles in the world's oceans. There is now widespread agreement among scientists and policy makers that an integrated approach to management of marine resources holds the greatest promise for the long-term delivery of ecosystem services. However, the move to an ecosystem approach to management (EAM) is a byproduct of decades of discourse. Nearly 75 years ago, the Ecological Society of America concluded that a single species focus was undermining conservation efforts in terrestrial systems (Shelford 1933). Rather than focus on single species, the Ecological Society of America argued that it is the "entire series of plants and animals which live together in any community which is of primary interest" (Shelford 1933). However, it was not until the 1990s that this sentiment became firmly established as an ecosystem approach to the management of public lands in the United States (Thomas 1996). More recently, in 1999, a blue ribbon report to Congress (NOAA 1999) ushered in an era of EAM of marine resources in the United States. This report foreshadowed the conclusions of two additional high-level commissions that explored marine resource management (Pew Oceans Commission 2003; U.S. Commission on Ocean Policy 2004), and EAM is now an overriding goal for marine systems in the United States.

   While ecosystem approaches to management are often viewed as a solution to diverse problems in the oceans of the world, the objectives of such an approach can differ substantially among groups with varying interests (Hilborn 2007). For instance, resource agencies typically seek to foster multiple human uses, subject to some environmental constraints. In this case, an EAM will attempt to produce a sustained ecosystem service (i.e., fisheries yield) while minimizing environmental impact. As a consequence, stated objectives promote commercial and recreational uses rather than the status of the ecosystem (Arkema et al. 2006). On the other hand, conservation organizations frequently view the goal of EAM as maintaining biodiversity and/ or ecosystem function (Pikitch et al. 2004). In this case the objective of EAM is to promote ecological integrity while allowing human use. Thus, objectives formulated under this vision focus on building resilience in the ecosystem.

   These different visions of marine EAM are plainly highlighted by Hilborn (2007) in his comments about a recent "consensus statement" about Ecosystem-based Management (McLeod et al. 2005). Hilborn, a preeminent fisheries biologist, notes that the 221 signatories of the consensus statement are mostly academic ecologists who have traditionally supported habitat and ecosystem preservation, and this objective is clearly in conflict with maximum biological utilization (Hilborn 2007). Similarly, the government mandated U.S. Commission on Ocean Policy strongly emphasized EAM and its utility for the sustainable use of ocean resources while the privately funded Pew Commission emphasized biodiversity and ecosystem function (Granek et al. 2005). Failure to recognize these fundamentally different aims can result in misplaced criticism. For instance, Arkema and colleagues (2006) note that EAM actions by resource management agencies do not account for ecological factors emphasized in the academic literature; however, if the goal of a resource agency is to maximize yield while minimizing damage to the environment, it may not be necessary to consider such ecological factors. Thus, as we move forward with EAM we must consider the question – do ecosystem approaches to management protect and serve human needs, or is it a means that humans can use to protect nature (cf. Simberloff 1999)?

5.1.1 "Multiple Use" in Marine Ecosystems

Resource agencies in the U.S. typically have mandates that require consideration of both the needs of humans and the ecosystem. The U.S. Forest Service, for instance, uses "an ecological approach to achieve the multiple-use management of the national forests and grasslands" (Thomas 1996). Similarly, NOAA Fisheries is charged with managing natural marine resources in US waters to maintain economically viable harvest while also conserving "healthy" ecosystems (NMFS 2004). The philosophy of "multiple use" is a commendable objective, but when one of the uses is predator removal, conflicts with conservation objectives seem likely (Zabel et al. 2003; Hilborn 2007). Such conflicts arise because the goal of fisheries is simply to provide a sustainable harvest of a target species. Such an approach usually ignores a broad suite of interactions among exploited species and between exploited species and other members of the community. For example, in an analysis of the Baltic Sea ecosystem, Zabel and colleagues (2003) show that the prosecution of the Baltic Sea fishery, even at a limited level, results in fundamental shifts in the structure of the community. They show that levels of harvest consistent with sustainable fishing render target species ecologically unimportant. In this case, the goal of sustainably fishing a few target species is incompatible with conservation goals. In the Baltic, this occurs because of strong top-down forcing (Harvey et al. 2003). Similar trophic cascades may occur in a number of other systems (Jackson et al. 2001; Frank et al. 2007), perhaps making the result from the Baltic widespread.

Managing marine resources has always been challenging, but this task looms ever larger as society demands more seafood while also requiring that we act as careful stewards of marine ecosystems. In an era when conflicting social expectations are forcing a reassessment of management policies (e.g., U.S. Oceans Commission 2004), it is not surprising to see a range of new management proposals come forward. Evaluating management strategies in light of the diverse and changing demands of society for the goods and services the oceans provide requires that we clearly expose trade-offs among conflicting objectives (Walters and Martell 2004), and develop effective means to operate along these trade-offs (Mangel 2000b).

In this paper, we describe an approach using simulation models that we are under-taking to evaluate management strategies and potential trade-offs between economic and conservation goals in the California Current ecosystem.

5.1.2 Ecosystem-Level Simulation Models as Tools to Evaluate Management Strategies

In most cases, we do not know how alterations of community composition by fishing or other human activities will affect ecosystem structure and function. Marine systems are immense, variable, and affected by many physical and biological forces; thus they are inherently difficult to study. For the most part, our views of marine eco-systems are static: we do not have the contrasts in time or space necessary to monitor the effects of fishing or other disturbances. This perspective has severely limited our ability to predict how communities will respond to anthropogenic pressures.

Because management advice often stems from predictive models, and because of the irreducible uncertainties inherent in marine ecosystems (Mangel 2000a), moving to a framework that allows us to address ecosystem considerations requires the development of complex multispecies or ecosystem models. While the added complexity of ecosystem models allows increased realism, this comes at the cost of increasing the uncertainty of predictions. Ecosystems are filled with uncertainty, and, even if we fully understand the processes governing ecosystem structure and function, observations of the state of the ecosystem also involve sampling error. Indeed, Ludwig and Walters (1985) have argued that simple models tend to be more useful and typically outperform more realistic, complex models. Importantly, however, ecosystem models do not necessarily have to be "right" to provide accu-rate policy predictions. For instance Essington (2004) showed that multispecies models correctly predicted the optimal allocation of fishing effort even though they incorrectly specified details of the predator–prey interaction. Thus, while complex models that simulate ecosystem dynamics are an emerging, and, at times, impre-cise, tool (Hollowed et al. 2000), they have proven useful for testing ecosystem indicators, generating hypotheses about past and future impacts of altered fishing and predation rates, and for screening potential management policies (Fulton et al. 2005; Field 2004; Cox et al. 2002; Hinke et al. 2004).

In this paper we use Atlantis, a modeling approach developed by Fulton and colleagues (Fulton 2001; Fulton et al. 2004), to evaluate a range of management strategies, and to predict their outcome in terms of both economic and conser-vation goals. Atlantis achieves the crucial goal of integrating physical, chemi-cal, ecological, and fisheries dynamics in a three-dimensional, spatially explicit domain. In Atlantis, marine ecosystem dynamics are represented by spatially explicit sub-models that simulate oceanographic processes, biogeochemical factors driving primary production, food web relations among functional groups, habitat interactions, and fishing. Here we apply the Atlantis framework to the California Current Ecosystem of the US West Coast. We simulate a range of fishing intensities, and evaluate potential trade-offs between harvest maximization and structure of the food web. We consider the direct effects on harvested species of economic interest, and the indirect effects on other components of the food web. The results can be interpreted on the level of individual species abundance, or through the use of ecosystem indicators.

5.2 Methods

5.2.1 Study Region

The California Current Atlantis model is fully detailed in Brand et al. (2007). Our model domain extends from Cape Flattery, Washington to Point Conception, California. To allow explicit representation of fish migrations and movement, we divided the model area into eight coastal regions, each with six zones defined by bathymetric contours ranging from the coast to 1,200 m depth. These 48 boxes are flanked by 14 non-dynamic boundary boxes on the seaward, northern, and southern edges. The model also divides the water column into depth layers, ranging from one depth layer for near-shore boxes to seven depth layers for the offshore boundary boxes. We define habitat per area, including sediment type (hard or soft) and kelp and seagrass coverage.

5.2.2 Model Structure

The Atlantis model is fully described in Fulton (2001, 2004) and Fulton et al. (2005). Here we give a brief description of the generic model structure, and specific attributes of the model for the California Current.

The ecological module of the California Current model simulates the dynamics of 54 functional groups in the food web, using nitrogen as a common currency between groups. Concentrations of nitrogen (ammonia and nitrate) in the water column are governed by uptake by autotrophs, excretion by consumers, nitrification, and denitrification. Biological functional groups include habitat-forming species like kelp, corals, and sponges, as well as additional benthic invertebrates, vertebrates, phytoplankton, zooplankton, refractory and labile detritus, and carrion. Primary producer and invertebrate abundances are modeled as aggregated biomass pools (Table 5.1). Vertebrates groups are comprised of ten age classes (Table 5.2). We track the abundance of each group in each box and depth zone through time, using a 12 h time step.

Growth of vertebrates is based on von Bertalanffy growth parameters, but varies with consumption. For simplicity, for this paper we chose a Holling type II functional response for predation. Predation is also constrained by gape limitation, such that predators can only consume prey below a size threshold (e.g., 40% for predators that are fish, and 25% for birds). Atlantis tracks the average size of individuals

Table 5.1 Invertebrate functional groups, and species contained within each group

Group	Species
Carnivorous Infauna	Polychaetes, Nematodes, Burrowing Crustacea, Peanut Worms, Flatworms
Deposit Feeders	Amphipods, Isopods, Small Crustacea, Snails, Ghost Shrimp, Sea Cucumber, Worms, Sea Mouse, Sea Slug, Barnacles, Solanogaster, Hermit Crabs
Deep Benthic Filter Feeders	Anemones, Deep Corals, Lampshells, Reticulate Sea Anemone, Rough Purple Sea Anemone, Swimming Sea Anemone, Gigantic Sea Anemone, Corals, Sponges
Shallow Benthic Filter Feeders	Barnacles, Seafan, Soft Corals, Gorgonian Corals, Black Coral, Green Colonial Tunicate, Sea Pens, Sea Whips, Sea Potato, Vase Sponge, Mussels, Scallops
Other Benthic Filter Feeders	Geoducks, Barnacles, Razor Clams, Littleneck, Manila Clams, Misc. Bivalves, Vancouver Scallop, Glass Scallop, Green Urchin, Red Urchin
Benthic Herbivorous Grazers	Snails, Abalone, Nudibranchs, Sand Dollars, Nake Solarelle, Dorid Nudibranchs, Limpets, Heart Sea Urchin, Spot Prawns, Pandalid shrimp
Prawns	Crangon and Mysid Shrimp
	Meroplankton
Deep Macro-Zoobenthos	Sea Stars, Moonsnail, Whelk, Leather Sea Star, Bat Star, Sunflower Sea Star, Common Mud Star, Crinoids, Brittle Sea Star, Basketstar
Shallow Macro-Zoobenthos	Giant, Bigeye, Yellowring, and Smoothskin Octopus, and Flapjack Devilfish
Mega-Zoobenthos	Dungeness Crab, Tanner Crab, Spiny Lobster, Pinchbug Crab, Red Rock Crab, Graceful Rock Crab, Spider Crab, Grooved Tanner Crab, Bairid, Scarlet King Crab, California King Crab
Meiobenthos	Flagellates, Cilliates, Nematodes
Cephalopods	Market Squid, Japetella, Gonatus, Chiroteuthis, Abraliopsis, Robust Clubhook, Rhomboid Squid, Sandpaper Squid, Vampire Squid
Gelatinous Zooplankton	Salps, Jellyfish, Ctenophores, Comb Jellies
Large Zooplankton	Euphausiids, Chaetognaths, Pelagic Shrimp, Pelagic Polychaetes, Crimson Pasiphaeid
Mesozooplankton	Copepods, Cladocera
Microzooplankton	Ciliates, Dinoflagellates, Nanoflagellates, Gymnodinoids, Protozoa
Large Phytoplankton	Diatoms
Small Phytoplankton	Microphytoplankton
Seagrass	
Macroalgae	Kelp
Benthic Bacteria	
Pelagic Bacteria	
Meiobenth	
Carrion	
Labile Detritus	
Refractory Detritus	

Table 5.2 Vertebrate functional groups, species, and scientific names. "Proportion" refers to the relative abundance of each species in the functional group. Relative abundance was taken from the 1998–2003 NWFSC FRAM shelf/slope trawl survey. Life history parameters were weighted by these proportions

Group	Species	Scientific name	Proportion
Deep Vertical Migrators	Lampfish	Myctophidae	0.34
	Pacific Viperfish	*Chauliodus macouni*	0.22
	Lanternfish	Myctophidae	0.17
	Longfin dragonfish	*Tactostoma macropus*	0.15
Shallow Small Rockfish	Shortbelly	*Sebastes jordani*	0.43
	Stripetail	*Sebastes saxicola*	0.39
	Greenstriped	*Sebastes elongates*	0.12
Deep Small Rockfish	Longspine Thornyhead	*Sebastolobus altivelis*	0.59
	Sharpchin	*Sebastes zacentrus*	0.19
	Splitnose	*Sebastes diploproa*	0.18
	Aurora	*Sebastes aurora*	0.03
Deep Misc. Fish	Pacific Grenadiers	Coryphaenoides *acrolepis*	0.41
	Giant Grenadiers	*Albatrossia pectoralis*	0.31
	Grenadier	Macrouridae	0.08
	Bigfin Eelpout	*Lycodes cortezianus*	0.05
Misc. Nearshore Demersal	Croaker, White	*Genyonemus lineatus*	NA
	Sculpin	Cottidae	NA
	Midshipman	*Porichthys notatus*	NA
Small Flatfish	Dover Sole	*Microstomus pacificus*	0.72
	Rex Sole	*Glyptocephalus zachirus*	0.13
	Pacific Sanddab	*Citharichthys sordidus*	0.08
Deep Large Rockfish	Shortspine Thornyhead	*Sebastolobus alascanus*	0.72
	Darkblotched	*Sebastes crameri*	0.23
Shallow Large Rockfish	Redstriped	*Sebastes proriger*	0.88
	Yelloweye	*Sebastes ruberrimus*	0.05
Mid-water Rockfish	Chilipepper	*Sebastes goodie*	0.53
	Widow	*Sebastes entomelas*	0.18
	Pacific Ocean Perch	*Sebastes alutus*	0.14
	Yellowtail	*Sebastes flavidus*	0.1
Hake	Pacific Hake	*Merluccius productus*	0.97
Sablefish	Sablefish	*Anopoploma fimbria*	1
Large Pelagic Fish	Jack Mackerel	*Trachurus symetricus*	0.92
	Pacific Mackerel	*Scomber japonicus*	0.08
Small Planktivores	Anchovies	*Engraulis mordax*	0.59
	Sardines	*Sardinops sagax*	0.39
	Herring	*Clupea harengus pallasi*	0.02
Salmon	Chinook	*Oncorhynchus tshawytscha*	0.95
	Coho	*Oncorhynchus kisutch*	0.05
Large Flatfish	Arrowtooth Flndr	*Atheresthes stomias*	0.71
	Halibut, Pacific	*Hippoglossus stenolepis*	0.15
	Petrale Sole	*Eopsetta jordani*	0.14
Large Demersal Predators	Lingcod	*Ophiodon elongatus*	1

(continued)

Table 5.2 (continued)

Group	Species	Scientific name	Proportion
Large Pelagic Predators	Albacore	*Thunnus alalunga*	1
Small Demersal Sharks	Dogfish	*Squalus acanthias*	0.73
	Spotted Ratfish	*Hydrolagus colliei*	0.17
	Brown Catshark	*Apristurus brunneus*	0.07
	Pacific Angel	*Squatina californica*	0.06
Large Demersal Sharks	Sleeper	*Somniosus pacificus*	0.97
	Sixgills	*Hexanchus griseus*	0.03
Misc. Pelagic Sharks	Soupfin	*Galeorhinus galeus*	0.88
Skates and Rays	Longnose Skate	*Raja rhina*	0.63
	Bering Skate	*Bathyraja interrupta*	0.13
	Skate	Rajidae	0.11
	Roughtail Skate	*Bathyraja trachura*	0.07
Migrating Birds	Sooty Shearwaters	*Puffinus griseus*	0.9
Diving Seabirds	Common Murre	*Uria aalge*	0.59
	Rhinoceros Auklet	*Cerorhnca monocerata*	0.19
	Cormorants and Shags	*Phalacrocoraxidae*	0.16
Surface Seabirds	Gulls	*Larus glaucescens*	0.81
	Brown Pelican	*Pelecanus occidentalis*	0.09
	Storm Petrels	*Oceanites* spp.	0.05
Sea Otters	Sea Otter	*Enhydra lutris*	1
Pinnipeds	N. Fur Seals	*Callorhinus ursinus*	0.25
	N. Elephant Seals	*Mirounga angustirostris*	0.42
	Ca Sea Lions	*Zalophus californianus*	0.35
	Harbor Seals	*Phoca vitulina*	0.2
Toothed Whales	Sperm	*Physeter macrocephalus*	0.67
	Cuvier's Beaked Whale	*Ziphius cavirostris*	0.17
	Baird's Beaked Whale	*Berardius bairdii*	0.07
	Beaked Whale	*Mesoplodon* spp.	0.05
	Resident, Offshore Orca	*Orcinus orca*	0.03
Baleen Whales	Gray	*Eschrichtius robustus*	0.62
	Fin	*balaenoptera physalus*	0.18
	Blue	*Balaenoptera musculus*	0.15
Transient Orcas	Transient Orcas	*Orcinus orca*	1

of each vertebrate age class in each box. This is apportioned into structural weight (bones, scales) and reserve weight (muscle and gonads). We based recruitment on Beverton-Holt parameters, with fish recruits entering the population at the age at which they settle to the bottom and begin feeding. We allowed density-dependent movement of fish, sea birds, and marine mammals between boxes, as well as advection of plankton. Movement transfers abundance towards neighboring cells with higher potential growth rates. For Pacific hake, albacore, small pelagic fish, marine mammals and birds, we also forced the model with seasonal migrations.

The model is forced with circulation, salinity, and temperature outputs from a Regional Ocean Modeling System (ROMS) of the Northeast Pacific. Temperature influences the respiration rates of each biological group, and each group also has a defined thermal tolerance and a narrower thermal range for spawning. Current velocities across each box face advect nutrients and plankton groups. Salinity does not affect the biological model. We repeated an 8 year Northeast Pacific ROMS output (1996–2003) four times, to span the 25-year duration of our simulations.

5.2.3 Data Sources

Model parameters are necessarily derived from a wide variety of sources, which are detailed in Brand et al. (2007). Briefly, fish life history parameters were primarily drawn from Love (1996), Love et al. (2002), Cailliet et al. (2000), or Fish Base (Froese and Pauly 2005). Marine mammal life history parameters came from many sources, notably Carretta et al. (2005) and Perrin et al. (2002), and seabird life history parameters were drawn largely from Schreiber and Burger (2002), and also from Russell (1999). Species-level parameters were weighted by the relative biomass of those species to form life history parameters for the functional group. Species within each functional group have similar life histories, diets, and distributions. Invertebrate functional groups were fairly coarse, in some cases resolving phyla into one or two functional groups (Table 5.1). Fish, mammal, and bird functional groups generally contain 3–12 species (Table 5.2).

Estimates of fish abundance rely upon published stock assessments (primarily from 2003), or on two NMFS trawl surveys: the 1998–2002 NWFSC slope survey (Builder Ramsey et al. 2002; Keller et al. 2005, 2006a, b) and the 2003 NWFSC "extended" shelf-slope survey. We adjusted survey catch by catchabilities recommended in Millar and Methot (2002) and Rogers et al. (1996). In cases where neither stock assessments nor reliable trawl survey data were available, we used estimates from Field (2004). Estimates of marine mammal abundance were taken from published stock assessments (Angliss and Lodge 2004; Carretta et al. 2005). Seabird abundances were derived from Parrish and Logerwell (2001).

5.2.4 Model Scenario and Application

The results presented here are from 25-year model runs subject to a range of fishing intensities. The initial conditions for the biological model include abundance and weight-at-age of each vertebrate group in each area, and biomass per area for all other groups. These initial conditions are based on data from approximately 1995–2005.

Fishing was parameterized based on initial abundance of each group. We identified all functional groups that are landed by US West Coast fisheries, using the PacFIN fish ticket landings database. We then simulated the harvest of a constant amount (metric tons) of these groups per year, ranging from 0 × initial abundance to 1 × initial abundance. The increments for harvest were [0 0.01 0.03 0.05 0.075 0.1 0.15 0.2 0.25 0.3 0.5 0.7 1.0] × initial abundance.

We also tracked ecosystem indicators using a subset of indices explored by Fulton and colleagues (2005). Our ecosystem indicators included the ratio of pelagic to benthic fish, the ratio of planktivorous to piscivorous fish, the ratio of piscivores to scavengers, and average trophic level. Pelagic fish included small planktivores (e.g., sardines and anchovies), large pelagic fish (mackerel), pelagic sharks, and deep vertical migrators (myctophids). Benthic fish included all flatfish and rockfish, large demersal predators (cabezon and lingcod), skates and rays, miscellaneous nearshore fish (e.g., sculpin), large demersal fish, and deep miscellaneous fish (hagfish, grenadiers, and eelpouts). Planktivores included small planktivorous fish (e.g., anchovy and sardine) and deep vertical migrators (myctophids), and piscivores included all other shark and fish groups. Scavengers included megazoobenthos (e.g., large crabs), deposit feeders (e.g., amphipods and isopods), carnivorous infauna (e.g., polychaetes), and large demersal sharks (e.g., sixgill sharks). Trophic level was derived from Fish Base (Froese and Pauly 2005) and Field (2004). We calculated trophic level of all biological groups in the system (excluding primary producers), envisioning that they could be sampled by fishery independent surveys, rather than in commercial catch. We also calculated the average trophic level of harvested species only.

5.3   Results

As expected, fishing led to depletion of harvested species, at a rate related to the life history characteristics of the species. For example, slow-growing, unproductive species such as midwater rockfish were substantially depleted at harvest levels of 0.1 × initial abundance, and were driven to extinction by harvest levels of 0.3 × initial abundance (Fig. 5.1). In contrast, more productive stocks such as planktivores and squid sustained harvest levels of 0.3 and higher (Fig. 5.1). Unlike exploited species, unharvested upper trophic level species such as pinnipeds showed little or no response to fishing. However, some indirect effects of fishing were evident in lower trophic levels. For instance, gelatinous zooplankton are not harvested, but declined due to indirect effects of fishing, such as depletion of prey groups, including juvenile fish (Fig. 5.1). Some effects of fishing may have cascaded all the way to primary producers. Large phytoplankton (diatoms), for example, showed slight increases in abundance as fishing removed consumer groups; though, this effect was quite modest. Overall, however, we saw few striking examples of strong indirect effects of fishing.

Our exploration of the response of a very limited suite of ecosystem indicators to our fishing scenarios revealed important nonlinearities. The ratios of piscivore to planktivore, benthic to pelagic fish, and scavenger to piscivore all showed substantial shifts in community structure as levels of harvest increased. These shifts were driven by the depletion of less productive stocks at increasing levels of harvest

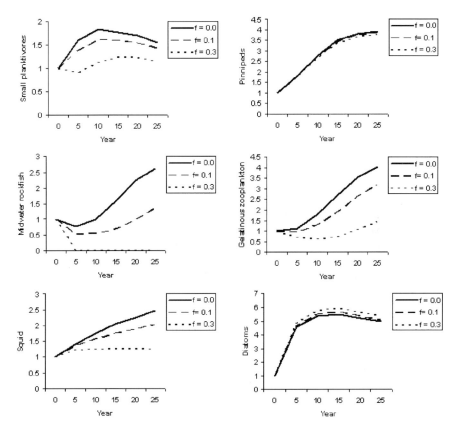

Fig. 5.1 Abundance of four functional groups over the course of 25 year simulations, relative to initial abundance. The three lines indicate harvest levels of 0, 0.1, and 0.3 × initial abundance

(Figs. 5.2–5.4). All three ratios shifted sharply at harvest levels of 0.1–0.15, the level at which most rockfish and flatfish groups were heavily depleted. At a level of harvest equal to 0.1 × initial abundance, all rockfish fell to below 25% of initial conditions (i.e., 2003 abundance) within 25 years, except for midwater rockfish that sustained fishing levels of 0.1 but were driven to extinction by fishing levels of 0.15. Large demersal predators (cabezon and lingcod) and skates and rays were also driven to near extinction by fishing levels of 0.1. Small demersal sharks and large and small flatfish were depleted to 30–50% at harvest levels of 0.1, and to near 0 by harvest levels of 0.15. At fishing levels of 0.2–0.3, fish biomass was dominated by large pelagic fish (mackerel) and small planktivorous fish (anchovies and sardines). Fishing levels greater than 0.3 × initial abundance led to complete depletion of all harvested species.

Scavenger abundance was relatively insensitive to fishing scenarios. This, in concert with the decline of unproductive piscivores, resulted in an increase in the ratio of scavengers to piscivores (Fig. 5.4). These scavengers did not consume offal

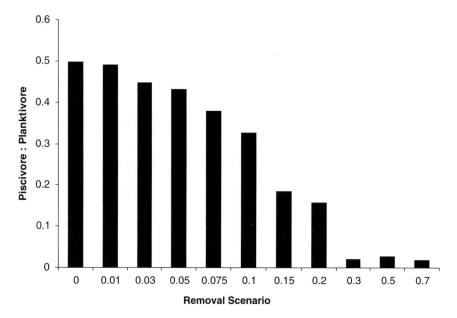

Fig. 5.2 Ratio of piscivore biomass to planktivore biomass, at the end of 25-year simulations, with annual harvest of target species ranging from 0 to 0.7 × initial abundance

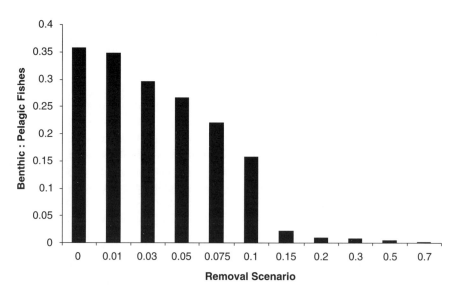

Fig. 5.3 Ratio of benthic to pelagic fish biomass, at the end of 25-year simulations, with annual harvest of target species ranging from 0 to 0.7 × initial abundance

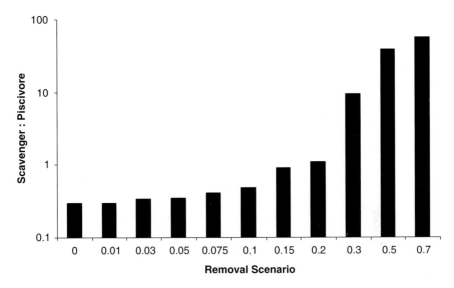

Fig. 5.4 Ratio of scavenger to piscivore biomass, at the end of 25-year simulations, with annual harvest of target species ranging from 0 to 0.7 × initial abundance

or discards from the fishing fleet; if they had, the ratio of scavenger to piscivore may have shown a temporary spike during the early years of the simulations. Instead, we saw fairly constant abundance of these groups for the 25 year runs shown here. Additional runs (not shown) show these groups persisting beyond 50 years, even with complete depletion of harvested fish species.

The mean trophic level of biological groups in the model was not sensitive to fishing intensity or to the associated shifts in the structure of the food web (Fig. 5.5). Total trophic level of all groups excluding primary producers did not decline, largely because most of the biomass was in unfished species that showed little response to harvest. The trophic level of target species showed only a minor decline. Although the ecosystem structure switched from being dominated by groundfish to one dominated by pelagic species as fishing increased, the relatively high trophic level of large pelagic fish (mackerel) and small planktivorous fish (3.5 and 3.08), similar to many rockfish and flatfish, led to little change in average trophic level.

5.4 Discussion

Ecosystem approaches to management attempt to move forward by explicitly considering multiple ecological and socioeconomic objectives. As resource managers adopt the principles of EAM, many decisions represent a junction where a potentially overwhelming assortment of choices face policy makers. Because human decisions affect the structure and dynamics of marine ecosystems, these decisions

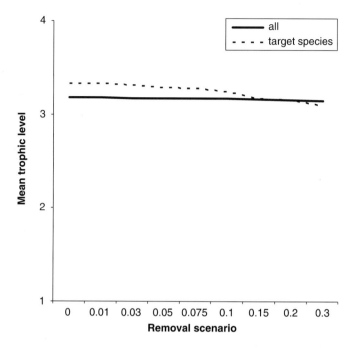

Fig. 5.5 Mean trophic level of all consumers (solid line) and all harvested species (dashed line), with annual harvest of target groups ranging from 0 to 0.3 × initial abundance. These indices are based on abundance in the system at the end of 25 year simulations, not on mean trophic level of the catch

have important implications for a diverse set of ecosystem services. While science cannot solve such conflicts, it can expose trade-offs between different goals, and thus clarify for managers the consequences of decisions. Fisheries management has long dealt with trade-offs such as those occurring among different fisheries targets, current versus future harvest, and abundance of target species versus fishing effort (Walters and Martell 2004); however, because ecological and fisheries metrics are often drawn from different analytical frameworks, examination of trade-offs between fisheries yield and conservation has been complex (Helvey 2004). In this paper, we have illustrated a simulation approach that can examine both single species outputs and community-level indices as a function of fundamental fisheries measure (the fraction of a stock that is harvested).

Models such as the Atlantis model we employed here are useful frameworks because they allow us to combine management input (i.e., what can we do) with a biophysical ecosystem model that captures key uncertainties (i.e., what do we know). Placed in a policy context, both management input and the ecosystem model are then subject to the values of the public (i.e., what do we care about). If consensus can be achieved regarding the general parameterization of an ecosystem model, then decision makers can focus on the values and beliefs that underlie potential disputes among stakeholders advocating for conflicting ecosystem services. Even the parameterization of Atlantis (and other such models) forces researchers and managers to acknowledge that our knowledge of the ecosystem is not simply discovered, but is constructed based on concepts available to individuals (Harms 2004). Thus, parameterizing and employing ecosystem models may most fruitfully be thought of as a process by which groups uncover useful understanding. This common understanding can then highlight areas where consensus may be achieved or topics that require clarification or negotiation.

We view the Atlantis model of the northern California Current as an important strategic tool to aid policy development. Management policies are ultimately an integration of the human dimension of fisheries with the biological and oceanographic realm. Atlantis allows us to simulate these components, to consider the effect of alternate assumptions and scenarios, and to evaluate the outcome from a suite of management strategies. Such management strategies can include levels of harvest, restrictions on bycatch or gear, or improving decision making tools such as stock assessments or ecological indicators. Atlantis serves well as a platform for this sort of management strategy evaluation (MSE; Kirkwood 1997; Sainsbury et al. 2000). Atlantis is not intended for short-term estimation of quotas; instead, it can be used to qualitatively rank alternative management strategies, and to evaluate trade-offs or conflicts between goals such as harvest maximization and ecosystem structure.

The results from the simulations shown here illustrate how differences among the life history of species affect the feasibility of pursuing economic gain at the expense of conservation concerns. Our results, similar to the empirical results of Levin and colleagues (2006), reveal that fishing combined with life history traits will alter the composition of the community such that short-lived, productive species replace longer-lived, lower productivity species. Thus, from an economic perspective, sustainably fishing productive high value species while overfishing less valuable, low productivity species may seem like a wise choice. In the California Current, for instance, much of the value of the fisheries comes from highly productive stocks of Dungeness crabs, hake, and squid. From a solely economic point of view, if the harvest of these species directly or indirectly resulted in the demise of unproductive lower value species (e.g., some rockfish), then this might be acceptable. Obviously, however, from a conservation perspective such a strategy would be completely unacceptable. A model such as Atlantis is useful for visualizing this trade-off and for negotiating management strategies that consider both economic and conservation concerns.

In our simulations, we primarily saw direct effects of fishing, with only modest impacts on unharvested species. This result is different from a similar food web (Ecopath with Ecosim) model for the California Current (Field 2004; Field et al. 2006), which found trade-offs in abundance of species such as hake and crustaceans. Differences between the two models could be caused by several factors, including (1) higher rates of primary production in the Atlantis model, (2) the full age structure in Atlantis (not present in Field (2004)), which means that overfishing predators reduces top-down control on mid-trophic level species, but also reduces the availability of juvenile predators as forage, or (3) other differences in bottom-up forcing and predator–prey relationships. It is possible that under other parameterizations of the model, fishing could have strong indirect effects on the biology of nontarget species, even at sustainable levels of fishing mortality. Further work to fit the model to data and to investigate sensitivity to model parameterization can be used to consider how likely indirect effects are in the California Current.

5.4.1 A Look to the Future

> We are drowning in information, while starving for wisdom. The world henceforth will be run by synthesizers, people able to put together the right information at the right time, think critically about it, and make important choices wisely. E. O. Wilson (1998)

Fisheries scientists have long known that the ecosystem is important to the dynamics of fish stocks, and ecologists have also recognized that fisheries impact the ecosystem in a number of important ways. Thus, constructing models to reveal the obvious is, perhaps, of limited value. On the other hand, the reams of ecological and fisheries data have not told us how to incorporate ecosystem considerations in marine resource management. As Wilson (1998) expressed, information by itself does not constitute knowledge. EAM requires that we synthesize diverse physical, biological, and socioeconomic data and think critically about the ways in which our decisions affect the ecosystem services we value. Of course, because people differ in their values, the diverse services society wishes to acquire from ecosystems are often at odds. Our goal in this paper was to briefly illustrate an ecosystem simulator that we believe will allow us to move the state of marine resource management from one where we generically understand that the ecosystem and fisheries are important and interconnected to one where we know how to strategically consider different ecosystem futures under different management scenarios. The approach we develop here can begin to help frame the questions to be posed and evaluate the likely consequences of different management options.

The time is right for a paradigm shift in how we manage our marine resources (Mangel and Levin 2005). We have accumulated data, expertise, and new tools. The future of marine ecosystems lies in the hands of scientists and managers who can take this collection of knowledge and experiences and synthesize it and truly implement ecosystem-based management. We have the right information, and now is the right time.

References

Angliss, R. P., and K. L. Lodge. 2004. Alaska Marine Mammal Stock Assessments, 2003. U.S. Department of Commerce. NOAA-AFSC, 230 p.

Arkema, K. K., S. C. Abramson, and B. M. Dewsbury. 2006. Marine ecosystem-based management: from characterization to implementation. Frontiers in Ecology and the Environment 4:525–532.

Brand, E.J., I.C. Kaplan. C.J. Harvey, P.S. Levin, E.A. Fulton, A.J. Hermann, and J.C. Field. 2007. A spatially explicit ecosystem model of the California Current's food web and oceanography. U.S. Dept. Commer., NOAA Tech. Memo. NMFS-NWFSC-84, 145 p.

Builder Ramsey, T., T. A. Turk, E. L. Fruh, J. R. Wallace, B. H. Horness, A. J. Cook, K. L. Bosley, D. J. Kamikawa, L. C. Hufnagle Jr., and K. Piner. 2002. The 1999 Northwest Fisheries Science Center Pacific West Coast Upper Continental Slope Trawl Survey of Groundfish Resources off Washington, Oregon, and California: Estimates of Distribution, Abundance, and Length Composition. U.S. Department of Commerce, NOAA Technical Memorandum. NMFS-NWFSC-55.

Cailliet, G. M., E. J. Burton, J. M. Cope, L. A. Kerr, R. J. Larson, R. N. Lea, D. VenTresca, and E. Knaggs. 2000. Biological characteristics of nearshore fishes of California: a review of existing knowledge and proposed additional studies. For the Pacific ocean interjurisdictional fisheries management plan coordination and development project.

Carretta, J., K. Forney, M. Muto, J. Barlow, J. Baker, B. Hanson, and M. Lowry. 2005. U.S. Pacific marine mammal stock assessments: 2004. NOAA-TM-NMFS-SWFSC-375. NMFS.

Cox, S. P., T. E. Essington, J. F. Kitchell, S. J. D. Martell, C. J. Walters, C. Boggs, and I. Kaplan. 2002. Reconstructing ecosystem dynamics in the central Pacific Ocean, 1952–1998. II. A preliminary assessment of the trophic impacts of fishing and effects on tuna dynamics. Canadian Journal of Fisheries and Aquatic Sciences 59:1736–1747.

Essington, T. E. 2004. Getting the right answer from the wrong model: Evaluating the senstivity of multispecies fisheries advice to uncertain species interactions. Bull. Mar. Sci. 74:563–581.

Field, J. 2004. Application of Ecosystem-Based Fishery Management Approaches in the Northern California Current. University of Washington, Seattle, WA.

Field, J. C., R. C. Francis, and K. Aydin. 2006. Top-down modeling and bottom-up dynamics: linking a fisheries-based ecosystem model with climate hypotheses in the Northern California Current. Progress in Oceanography 68:238–270.

Frank, K. T., B. Petrie, and N. L. Shackell. 2007. The ups and downs of trophic control in continental shelf ecosystems. Trends in Ecology and Evolution 22:236–242.

Froese, R., and D. Pauly. 2005. FishBase. In R. Froese and D. Pauly, editors. World Wide Web electronic publication.

Fulton, E. A. 2001. The effects of model structure and complexity on the behavior and performance of marine ecosystem models. Doctoral thesis. University of Tasmania, Hobart, Tasmania, Australia.

Fulton, E. A. 2004. Biogeochemical marine ecosystem models II: the effect of physiological detail on model performance. Ecological Modelling 173:371–406.

Fulton, E. A., A. D. M. Smith, and A. E. Punt. 2005. Which ecological indicators can robustly detect effects of fishing? ICES Journal of Marine Science 62:540–551.

Granek, E. F., D. R. Brumbaugh, S. A. Heppell, S. S. Heppell, and D. Secord. 2005. A blueprint for the oceans: implications of two national commission reports for conservation practitioners. Conservation Biology 19:1008–1018.

Harms, W. F. 2004. Information and Meaning in Evolutionary Processes. Cambridge University Press, Cambridge.

Harvey, C. J., S. P. Cox, T. E. Essington, S. Hansson, and J. F. Kitchell. 2003. An ecosystem model of food web and fisheries interactions in the Baltic Sea. ICES Journal of Marine Science 60:939–950.

Helvey, M. 2004. Seeking consensus on designing Marine Protected Areas: keeping the fishing community engaged. Coastal Management 32:173–190.

Hilborn, R. 2007. Defining success in fisheries and conflicts in objectives. Marine Policy 31(2):153–158.

Hinke, J. T., I. C. Kaplan, K. Aydin, G. M. Watters, R. J. Olson, and J. F. Kitchell. 2004. Visualizing the food-web effects of fishing for tunas in the Pacific Ocean. Ecology and Society 9(1):10.

Hollowed, A. B., N. Bax, R. Beamish, J. Collie, M. Fogarty, P. Livingston, J. Pope, and J. C. Rice. 2000. Are multispecies models an improvement on single-species models for measuring fishing impacts on marine ecosystems? ICES Journal of Marine Science 57:707–719.

Jackson, J. B. C., M. X. Kirby, W. H. Berger, K. A. Bjorndal, L. W. Botsford, B. J. Bourque, R. H. Bradbury, R. Cooke, J. Erlandson, J. A. Estes, T. P. Hughes, S. Kidwell, C. B. Lange, H. S. Lenihan, J. M. Pandolfi, C. H. Peterson, R. S. Steneck, M. J. Tegner, and R. R. Warner. 2001. Historical overfishing and the recent collapse of coastal ecosystems. Science 293:629–637.

Keller, A. A., T. L. Wick, E. L. Fruh, K. L. Bosley, D. J. Kamikawa, J. R. Wallace, and B. H. Horness. 2005. The 2000 U.S. West Coast Upper Continental Slope Trawl Survey of Groundfish Resources off Washington, Oregon, and California: Estimates of Distribution, Abundance, and Length Composition. U.S. Department of Commerce, NOAA Technical Memorandum. NMFS-NWFSC-70.

Keller, A. A., E. L. Fruh, K. L. Bosley, D. J. Kamikawa, J. R. Wallace, B. H. Horness, V. H. Simon, and V. J. Tuttle. 2006a. The 2001 U.S. West Coast Upper Continental Slope Trawl Survey of Groundfish Resources off Washington, Oregon, and California: Estimates of Distribution, Abundance, and Length Composition. U.S. Department of Commerce, NOAA Technical Memorandum. NMFS-NWFSC-72.

Keller, A. A., B. H. Horness, V. J. Tuttle, J. R. Wallace, V. H. Simon, E. L. Fruh, K. L. Bosley, and D. J. Kamikawa. 2006b. The 2002 U.S. West Coast Upper Continental Slope Trawl Survey of Groundfish Resources off Washington, Oregon, and California: Estimates of Distribution, Abundance, and Length Composition. U.S. Department of Commerce, NOAA Technical Memorandum. NMFS-NWFSC-75, 189 p.

Kirkwood, G. P. 1997. The revised management procedure of the international whaling commission. In Global Trends: Fishery Management. E. K. Pikitch, D. D. Huppert, and M. P. Sissenwine, editors. American Fisheries Society Symposium 20, Bethesda, MD, pp. 91–99.

Levin, P. S., E. E. Holmes, K. R. Piner, and C. J. Harvey. 2006. Shifts in a Pacific Ocean fish assemblage: the potential influence of exploitation. Conservation Biology 20:1181–1190.

Love, M. 1996. Probably More Than You Wanted to Know About the Fishes of the Pacific Coast. Really Big Press, Santa Barbara, CA, 381 pp.

Love, M. S., M. Yoklavich, and L. Thorsteinson. 2002. The Rockfishes of the Northeast Pacific. University of California Press, Berkeley, CA, 404 pp.

Ludwig, D., and C. J. Walters. 1985. Are age-structured models appropriate for catch-effort data? Canadian Journal of Fisheries and Aquatic Sciences 42:1066–1072.

Mangel, M. 2000a. Irreducible uncertainties, sustainable fisheries and marine reserves. Evolutionary Ecology Research 2:547–557.

Mangel, M. 2000b. On the fraction of habitat allocated to marine reserves. Ecology Letters 3:15–12.

Mangel, M. and P.S. Levin 2005. Regime, phase and paradigm shifts: making community ecology the basic science for fisheries. Philosophical Transactions of the Royal Society B. 360:95–105.

McLeod, K. L., J. Lubchenco, S. R. Palumbi, and A. A. Rosenberg. 2005. Scientific Consensus Statement on Marine Ecosystem-Based Management. Signed by 221 academic scientists and policy experts with relevant expertise Communication Partnership for Science and the Sea at http://compassonline.org/?q = EBM.

Millar, R. B., and R. D. Methot. 2002. Age-structured meta-analysis of U.S. West Coast rockfish (Scorpaenidae) populations and hierarchical modeling of trawl survey catchabilities. Can. J. Fish. Aquat. Sci. 59:383–392.

NMFS. 2004. NMFS strategic plan for fisheries research. U.S. Department of Commerce, NOAA Technical Memorandum. NMFS F/SPO-61, Silverspring, MD.

NOAA. 1999. Ecosystem-based fishery management. A report to congress by the Ecosystem Principles Advisory Panel. National Marine Fisheries Service, Silver Springs, MD.

Parrish, J., and E. Loggerwell. 2001. Seabirds as indicators, seabirds as predators. In J. Parrish and K. Litle, editors. PNCERS 2000 Annual Report. Coastal Ocean Programs. NOAA, pp. 87–92.

Perrin, W., B. Wursig, and H. Thewissen, editors. 2002. Encyclopedia of Marine Mammals. Academic, San Diego, CA.

Pew Oceans Commission. 2003. America's living oceans: charting a course for sea change. A report to the nation. Pew Oceans Commission, Arlington, VA.

Pikitch, E. K., C. Santora, E. A. Babcock, A. Bakun, R. Bonfil, D. O. Conover, P. Dayton, P. Doukakis, D. Fluharty, B. Heneman, E. D. Houde, J. Link, P. A. Livingston, M. Mangel, M. K. McAllister, J. Pope, and K. J. Sainsbury. 2004. Ecosystem-based fishery management. Science 305:346–347.

Rogers, J. B., M. E. Wilkins, D. Kamikawa, F. Wallace, T. Builder, M. Zimmerman, M. Kander, and B. Culver. 1996. Status of the remaining rockfish in the Sebastes complex in 1996 and recommendations for management in 1997. Appendix E: Status of the Pacific Coast ground-fish fishery through 1996 and recommended acceptable biological catches for 1997. Pacific Fishery Management Council, Portland, OR.

Russell, R. W. 1999. Comparative Demography and Life History Tactics of Seabirds: Implications for Conservation and Marine Monitoring. American Fisheries Society, Bethesda, MD.

Sainsbury, K. J., A. E. Punt, and A. D. M. Smith. 2000. Design of operational management strategies for achieving fishery ecosystem objectives. ICES Journal of Marine Science 57:731–741.

Schreiber, E., and J. Burger. 2002. Biology of Marine Birds. CRC Press, Boca Raton, FL.

Shelford, V. E. 1933. The preservation of natural biotic communities. Ecology 14:240–245.

U. S. Commission on Ocean Policy. 2004. An ocean blueprint for the 21st century. Final report to the Present and Congress.

Thomas, J.W. 1996. Forest Service perspective on ecosystem management. Ecological Applications 6(3):703–705.

Walters, C.J., and S.J.D. Martell. 2004. Fisheries Ecology and Management. Princeton University Press, Princetion N.J. 448 pp.

Wilson, E. O. 1998. Consilience: The Unity of Knowledge. Alfred A. Knopf, New York, 332 pp.

Zabel, R. W., C. J. Harvey, S. L. G. T. P. Katz, and P. S. Levin. 2003. Ecologically sustainable yield. American Scientist 91:150–157.

Chapter 6
The Changing Nature of Fisheries Management and Implications for Science*

Kevin Stringer, Marc Clemens, and Denis Rivard

Abstract The relationship between fisheries management and fisheries science has become more complex and more challenging over recent years as we move from a fish stock-focused approach to the management of fisheries with the objective of maximum sustainable yield, to an approach with multiple objectives encompassing the precautionary approach, ecosystem-based management, and industry economic viability. At its core, the precautionary approach is about taking more cautious measures in the face of uncertainty. Linked to this is the growing recognition of the need to take on ecosystem approach to fisheries management. The increasing lack of stability in ocean conditions and uncertainty around the effect of changing ocean conditions has enhanced the need to be more comprehensive in our approach. All this serves to make fisheries management more complex than it was in the past. Whereas fisheries science advice in the past was focused largely on stock biomass and productivity, fisheries science is now being asked to provide advice, information, and analysis on stock interactions and predation, on spawning seasons and locations, on sensitive areas of significance to the species, on the effect of various gear technologies on benthic communities, on the effect of the increasing number of invasive species in the ecosystem, and on changing ocean conditions and their potential effect on stock dynamics now and in the future. With static funding resources over the past decade, scientists have struggled to find ways to respond to these new queries and to still provide basic stock status advice that continues to be and will continue to be the core scientific requirement for making fisheries management decisions. This state of affairs has stretched the capacity

K. Stringer and M. Clemens
Fisheries and Aquaculture Management, Department of Fisheries and Oceans,
Government of Canada, 200 Kent Street, Ottawa, Ontario, Canada K1A 0E6
e-mail: Kevin.Stringer@NRCan.gc.ca, marc.clemens@dfo-mpo.gc.ca

D. Rivard
Ecosystem Science Directorate, Department of Fisheries and Oceans,
Government of Canada, 200 Kent street, Ottawa, Ontario, Canada K1A 0E6
e-mail: rivardd@gmail.com

*The contents of this paper reflect the views of the authors and do not necessarily reflect the views of the Department of Fisheries and Oceans

R.J. Beamish and B.J. Rothschild (eds.), *The Future of Fisheries Science in North America*,   97
Fish & Fisheries Series, © Springer Science + Business Media B.V. 2009

of fisheries science to respond to the growing array of information requests that are now considered necessary to make responsible decisions. Indeed, the growing complexity and challenges for fisheries management, for fisheries science and for the fishing industry as a whole, has significantly stretched static resources but has been addressed by unique responses, depending on the circumstances, and the development of new partnerships and working arrangements between fisheries managers, scientists, and the fishing industry.

Keywords Ecosystem-based management · fisheries management · fisheries science · framework · policy · sustainability

6.1   Introduction

This paper is about the changing nature of the relationship between fisheries managers and scientists; in particular, as we see it in Canada in our day-to-day involvement in fisheries management. Fisheries science has always been the primary partner of fisheries management. Because of shifts in policy and program emphasis, a more complex legal framework and, most significantly, because of a developing set of processes in which fisheries management decisions are made, the relationship between fisheries management and fisheries science has become more complex, more muddied or less certain, and, in some cases, has become less effective. At the same time, given the policy, legal, and governance frameworks that are developing, it is more important than ever that fisheries managers and scientists be working closely together.

While the changes in policy and program emphasis, legal frameworks, and decision processes have been frustrating for all, it is probably fisheries science that has had the most difficulty adjusting. It is becoming progressively more difficult for scientists to understand how the science advice was taken into account in the final management decision of a particular stock. The relationship between fisheries managers and scientists has always invoked some level of natural tension, largely due to their different perspective in the management decision-making process. While managers tend to think scientists are stuck in their models, scientists have a tendency to feel that managers may disregard not only the model but also the principles in an effort to do whatever is necessary to get through the fishing season without due consideration to long-term consequences. While there has always been this tension between fisheries management and fisheries science, the challenges associated with this natural response are growing as the policy, legal, and governance frameworks become more complex. The future of the fisheries resource is dependent on our ability to take steps together to reestablish or better establish the fisheries science and fisheries management relationship, accepting that a certain level of tension is likely both natural and beneficial for the resource.

The relationship between fisheries science and fisheries management was based on a conservation or sustainable development framework and was quite simple. Fisheries managers would seek and receive science advice on stock biomass, on productivity, and specific advice related to allowable catch levels based on maximum sustainable yield (MSY) or its proxies (Rivard and Maguire 1993). Fisheries management would follow the advice when it could, but no matter what the decision, it was a fairly straightforward approach, based mainly on a single stock and MSY-based objectives. The process was simple, almost linear, with information flowing from the scientists to the consultation process led by fisheries management. In the Canadian management system, the process culminated in a decision by the Minister of the Department of Fisheries and Oceans (DFO) and the implementation of fishing plans that established Total Allowable Catches (TACs) for each stock.

6.2 New Realities

Over the last decade or two, this relatively straightforward approach has gradually been replaced by more complex frameworks that challenge the traditional approach. The current challenges are significant and come in three areas: policy, legal, and governance (Fig. 6.1). The consequence is that fisheries management has become both more complex and more crowded with new concepts.

6.2.1 Changes in Conservation Frameworks

Over the past 10 or 15 years, new objectives, new goals, and new management frameworks have effectively changed the nature of fisheries management. Some years ago, the precautionary approach was introduced as "the" encompassing objective and framework for fisheries management (FAO 1995). This was mostly driven by worldwide concerns that there were a growing number of stocks that were not maintaining their desired productivity. The approach that focused on precaution in the face of uncertainty and pre-agreed management actions to avoid serious or irreversible harm complicated the management (Richards and Schnute 1999; Rice et al. 1999; Rice and Rivard 2002; Rivard and Rice 2003; Shelton et al. 2003; DFO 2006). For fisheries managers, it meant adopting a new set of reference points with decision rules in decision frameworks to avoid serious or irreversible harm to stocks.

More recently, an ecosystem approach to fisheries management was introduced as the new principle for fisheries management and perhaps for the management of everything in the oceans (Mace 2000; Link 2002; O'Boyle and Keizer 2003; ICES 2003). Whether this meant managing an ecosystem, biodiversity, or managing fisheries as part of overall integrated management, the ecosystem approach changed the knowledge required for management and complicated the decision making. It was no longer sufficient to manage solely on the basis of stock productivity. It became necessary to conserve biodiversity and to ensure that the impacts of harvesting technologies on marine and freshwater habitats were mitigated. It is fair to say that

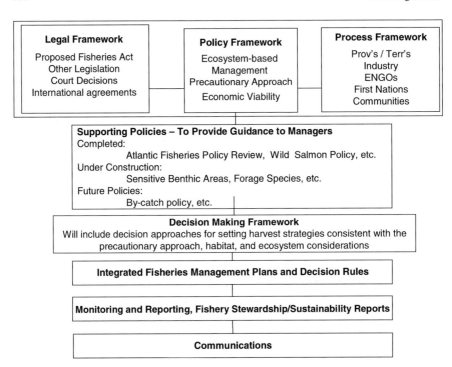

Fig. 6.1 The resource management sustainable development framework integrates the relevant elements of the legal framework, the policy framework, and the process framework. It is supported by a comprehensive set of policies to provide guidance to managers on issues related to licensing, ecosystem management, specific species conservation issues, and other issues related to conservation, socioeconomic aspects or regional considerations. The fishery decision making framework incorporates the precautionary approach (PA) and is implemented through Integrated Fisheries Management Plans and Decision Rules. The framework includes monitoring and reporting on program activities, including Fishery Stewardship and Sustainability Reports (EBM: Ecosystem-Based Management; OBFM: Objective-Based Fisheries Management; ENGOs: Environmental Non-Governmental Organizations)

ecosystem-based management further crowded the conservation and sustainable development frameworks.

In Canada and elsewhere (e.g., New Zealand, USA, etc.), managers and policy makers developed their own hybrid brand of approaches in objectives-based fisheries management in an attempt to include conservation and ecosystem considerations and socioeconomic factors (Clark and O'Boyle 2001). Objectives-based fisheries management frameworks invited managers to be clear about the management objectives and more transparent in the way competing objectives were taken into account in decision making. It brought in socioeconomic factors and industry viability objectives to be addressed through fisheries management. Canada has recently publicly stated that it adopts an "ocean to plate" approach to the management of fisheries that links industry viability to long-term resource sustainability. This approach focuses on understanding the effect of fisheries management policies and programs on all aspects of the use of seafood. The "ocean to plate" is emerging as a national initiative aiming at a better integration of fish harvesting, processing, as well as distribution and marketing to maximize economic value and help lead to an economically viable industry.

At the same time, worldwide, eco-certification and related consumer-driven stewardship initiatives are making resource conservation and sustainability an access to markets issue. Increasingly, fish distributors, buyers, and restaurants in different countries are seeking confirmation and demonstration that Canadian fish are caught in a sustainable way.

The growing demand from markets to demonstrate sustainability is requiring that fishery agencies consider a broader set of conservation objectives in managing fisheries. Canada's interest in an "ocean to plate" approach will help fisheries to adapt and prosper in the changing global marketplace for seafood.

With the precautionary approach, ecosystem-based management, objective-based fisheries management and an "ocean to plate" approach all legitimately competing as "the" encompassing concept for the management of fisheries, the sustainability policy environment became even more challenging. The fisheries management environment is complex not only because of competing policy concepts regarding conservation, but also because of a growing, more diversified, and quite restrictive set of legal frameworks. In Canada, major changes are occurring in the federal legislation, international agreements, and Court decisions.

6.2.2   Changes in Federal Legislation

New federal legislation, the Species at Risk Act (Canada 2002) and in a number of other countries requires that no matter what principles and policies we may wish to apply in the management of fisheries, very specific actions need to be taken to manage species protected under the act. This legislation adds new considerations for fisheries managers that include finding ways to avoid catching listed species or to mitigate the impact of fishing activities on them.

In Canada, the Oceans Act (Canada 1996) added the requirement of managing oceans, as well as fishing activities. It brought a new set of tools, such as Marine Protected Areas (ICES 2007) and integrated oceans management, to complement more traditional approaches of fisheries management. Foremost, it broadened the "client-base," from the users of fish resources to "ocean users" and invited community-based approaches.

DFO has been seeking to introduce or pass a new Fisheries Act legislation (Bill C-45) which could provide new tools not only to manage fisheries, but also for cooperation between fisheries management and fisheries science. A new Fisheries Act has the potential for changing the way managers, scientists, and stakeholders interact. The proposed legislation would likely also recognize the new principles for conservation frameworks, such as the precautionary approach, ecosystem-based management, and industry viability.

6.2.3   Changes Related to International Agreements

Another area of jurisprudence that complicates management is the set of international protocols and laws that have developed over time. Canada, like many other
countries, has signed a number of international agreements and protocols over the
last 10 or 15 years that affect how DFO manages fisheries. Some of these are about
adhering to the principles and approaches previously noted, such as the precautionary approach (UNFA 1995) and the ecosystem-based approach (Smith et al.
2007). Others establish specific targets for management, such as the Johannesburg
Declaration on Sustainable Development (WSSD 2002). Still other international
agreements, like the Convention for International Trade of Endangered Species
(CITES), are longstanding and have the force of law.

6.2.4   Challenges Arising from Recent Court Decisions

Court decisions are the third area of jurisprudence that complicates the management.
In Canada, there have been a number of decisions from the Courts – particularly in
the area of Aboriginal rights – that affect decisions and complicate the application
of any specific policy framework. The Sparrow[1] decision on the Aboriginal right
to fish for food, social, and ceremonial purposes, for example, profoundly effected
the ability to manage fisheries in line with any specific sustainability principle and
to seek specific conservation outcomes. There are several other Court decisions –
Marshall, Larocque,[2] and others, that have had a significant effect on the legal environment for the management of fisheries in Canada and that effectively constrain
how policy sustainability frameworks are applied.

6.3   Discussion

6.3.1   Changing the Relationship with Fisheries Science

Fisheries management processes and governance have become more complex in
recent years. In addition to competing concepts or frameworks for conservation
(single species focuses, precautionary approach, ecosystem approach), there is
an increasingly complex domestic legal framework (proposed new Fisheries Act,
Oceans Act, Species at Risk Act), a set of international agreements on fisheries

---

[1]Judgement of the Supreme Court of Canada made in 1990; *R. v. Sparrow (1990) 1 S.C.R. 1075*

[2]The Marshall case refers to a judgement of the Supreme Court of Canada made in 1999 in *R. v. Marshall (1999) 3 S.C.R 533* and the Larocque case refers to a decision of the Federal Court of Canada made in 2005 *(Larocque v. Canada (Minister of Fisheries and Oceans), 2005 FC 694)*

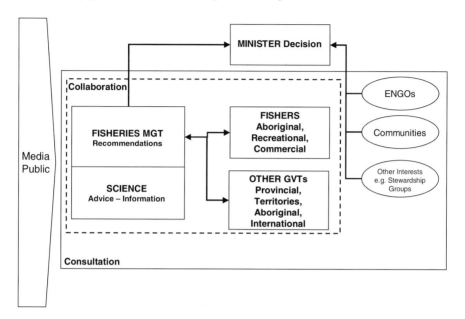

Fig. 6.2 Schematic of the emerging decision process, involving complex interactions for collaboration and consultation from a wide variety of stakeholders. Fisheries management, science, fishers, and other governments are working in a collaborative mode, with consultation engaging a wide variety of stakeholders such as Environmental Non-Governmental Organizations (ENGOs), Communities, Recreational Fishers, and other interest groups

management and sustainable development (UNFA 1995; WSSD 2002), as well as court decisions and land claims with Aboriginal groups that affect the way management decision making is to be made. The new management model (Fig. 6.2) has changed considerably from a process where fisheries science would provide advice and fisheries management would engage with industry and then would provide advice to the Minister. Now, the industry and the provinces have become far more influential and involved in the decision-making process. Industry and the provinces have a place at the decision table where they expect to bring in their views, their principles, and their own policy frameworks. In many cases, industry includes a significant array of interests, from commercial fishers, to Aboriginal fishers, to recreational fishers. In these circumstances, it is difficult to apply any kind of formula approach to fisheries decision making when a multitude of stakeholders expect their views and their principles to be considered equally with departmental advice.

Given these current practices and expectations, any attempt to base decisions solely on a science-based framework is effectively doomed to failure – or at least to inconsistent success unless that science-based framework is accepted by all parties. Traditional stakeholders (industry, provinces) and Aboriginal groups also need to be engaged and, increasingly, DFO feels compelled to engage more than its traditional partners, including community groups, environmental non-governmental organizations, local governments, and the public.

It is largely fisheries management that has contended with the developing myriad of processes. Concerns about scientific objectivity and the need for scientific integrity have always resulted in a tendency for scientists to resist becoming too closely or directly engaged in the decision process. However, both fisheries managers and scientists have come to recognize that they can no longer operate in isolation. The complexity of the processes inherently require a closer relationship between fisheries managers and scientists, one that can be more responsive and timely and one that interacts directly with Industry, Provinces, and others at the decision making table. This is a conclusion largely reached through practice and necessity.

With the new complex conservation policy frameworks, the growing legal framework and the new governance processes in which decisions are made on fisheries, it is ever more difficult for scientists to see their advice reflected in fishery decisions unless they are directly involved in the decision-making process. Moreover, without fisheries science at the table, there is a greater risk that advice from science will not be effectively incorporated into decisions. Scientists need to be more involved alongside fisheries managers, with industry, with provinces, with aboriginal groups, with recreational fishery groups, and with environmental groups. Fisheries managers are generally not as well equipped to understand, communicate and, ultimately, defend the science; and scientists will benefit from the exposure to the new management challenges. Another challenge is how to achieve the right balance without compromising the independence and integrity of the science and of the scientists. The new partners in decision making are not just to be consulted, but more and more to be partners in decision making. They have different interests from managers and scientists – economic viability, business stability, enhancing the property nature of the fishery, social and cultural goals, among others. The current trend to eco-certification (Jacquet and Pauly 2006) brings the common interests of the various groups including DFO's more in line today than previously and should help move everyone to a science-based outcome focused on sustainability. Industry and commercial fishers want to hear and to participate in the message of science and accept the need for sustainability to maintain and grow their markets. In short, industry and commercial fishers need the science to support their case for eco-certification given the increased focus on proving sustainable harvesting practices to the marketplace, through eco-certification.

6.3.2   Implications for Science and Fisheries Management

Because of all of the policy frameworks, all of the legal frameworks, and all of the processes that the management of fisheries requires, fish managers are asking more from science. Fisheries management now requires advice on limit reference points, on harvest strategies, on the effects of climate change, on the effects of changes of in-river temperatures, on the role of species in the ecosystem, on predation practices and spawning habits, on the effect of fishing on the benthos, the ocean floor, and more. At the same time, fisheries managers have not stopped asking for any of the traditional science required within the annual management cycle. They still want an assessment of the status of fish stocks, estimates of mortality and, ultimately, projections to support "the" number (i.e., the Total Allowable Catch) – because these are long used and easily understood. All of this is happening when there are competing demands on limited marine science resources. It is no longer just fisheries management asking questions. It is also ocean managers asking about the effects of other users on ecosystems. The task for fisheries science becomes one of evaluating the cumulative effects of a number of sectors. Aquaculture is also adding to the science agenda as are invasive species, species-at-risk, the impact of seismic activities as of result of oil exploration and activities not traditionally considered to be part of managing commercial fish stocks. The question for the future is how to reposition fisheries management and fisheries science to address the new challenges. For science staff, it is largely about identifying how research and its products could be used for a broader range of clients. It is adopting, where possible, ecosystem-based surveys rather than individual stock surveys and monitoring. It is about selecting research projects strategically and securing them in comprehensive research plans. It is conducting analyses that integrate large data sets across multiple research domains. It is developing new information bases with greater levels of details and at finer scales. It is about giving less advice on tactical questions such as allowable catch levels and more advice on management strategies using robust harvest rules that can be used for tactical (annual) decisions and reduce the demand for ongoing support for routine management decisions. While this may be difficult and controversial in some circles, scientists have to teach fisheries managers, industry leaders and other influential stakeholders in addition to being "doers."

For fisheries managers, the task is to understand and embrace the ecosystem based approach, the precautionary approach, decision rules, etc., and to find ways to incorporate them into decision making. Fisheries managers also have to make a better effort to understand and appreciate the advice of science and not just look for the number to emerge from the model. Managers need to develop a better understanding of the nature of uncertainty and risk associated with management decisions. Fisheries managers need to know the right questions to ask. Because of the changing nature of the advice required, it becomes more important to document the questions formally and to communicate them in a timely fashion for consideration by scientists. Fisheries managers also have to become better at interpreting complex analyses because complex analyses are what are needed to describe a fishery in an aquatic ecosystem context. Fisheries managers need to become better partners and champions of the science. Finally, fisheries managers must ensure that objectives are clearly stated, that decision tools are well documented and used, that measures of efficiency and effectiveness are in place and that corrective measures are taken using established practices. Fisheries management and science need each other more than ever in this complex policy environment and the complex set of processes for taking decisions on conservation. Sound management, sustainable development and the future of fishery resources depends on it.

6.3.3 Key Elements of a Resource Management Sustainable Development Framework

Given the complexity in the policy, legal, and governance frameworks, is it possible to identify one organizing principle or approach that captures sustainable development while respecting the multitude of interests and processes? The fact is that it is not possible. This cannot be boiled down to one formula, to one approach, or to one decision framework. The challenge is not to resolve this question in a simplistic way, but to manage the complexity.

In Canada, this is being pursued through a number of initiatives. First, there is a need for an overall resource management sustainable development framework that reflects all principles of sustainable development. The framework needs to be rigorous, but it also has to reflect reality and what managers and scientists can actually do. This framework needs to recognize all the dimensions of sustainability. Second, there needs to be a set of policies within that framework that provides guidance. These are general guidance documents that offer support for decision makers on the various aspects within the policy frameworks such as managing forage species, dealing with sensitive benthic habitat areas, managing bycatch, and accounting for species interactions. Third, there is a need for a decision-making framework to guide fisheries management decisions. It needs to be consistent with and supportive of the application of the precautionary approach and it has to be practical and reflective of the realities of fisheries management. It has to embrace the emerging approaches based on performance, including evaluating harvest strategies in their ability to meet specific stock characteristics with reasonable chance of success (Kell et al. 2007). Fourth, managers and scientists must engage industry to develop simple decision rules by fishery, and by fleet. Fifth, there is a need to assess progress in the face of this complex set of principles and frameworks. A simple set of measures have to be developed for reporting on progress internally within DFO and externally. We are considering a "checklist" as the basis for Report Cards on fisheries. The checklist could assist in reviewing and measuring the biological and management aspects of resource stewardship and fisheries sustainability including the level of information available to make decisions, the status of the stock, the effectiveness of management measures, and progress towards implementing the PA, ecosystem considerations, and other management measures. The checklist will ask questions for an individual fishery to cover all aspects of resource stewardship and sustainability. Both fisheries science and fisheries management elements are to be covered through the checklist. For both categories, the checklist will cover three general subjects: knowledge available, objectives and reference points, and implementation. For example under "knowledge available," there will be questions about whether there are estimates of abundance and estimates of annual exploitation or harvest rates for the stock. Under "objectives and reference points," someone could be asked to report on limit and target reference points that may have been established for the stock. Under "fisheries management – objectives and reference points," someone may be asked to report on the existence of formal harvest rules consistent with the precautionary approach for the stock. And under "fisheries management – implementation," special attention may be paid to the management process in place to identify sensitive benthic marine areas within the geographic extent of the fishery or to the measures in place to conserve and protect sensitive benthic areas. The point is that we have to expand the traditional approaches of fisheries management, which were mainly addressing questions related to stock productivity, to include the other aspects of resource stewardship and sustainability (Fig. 6.3). For instance, conservation-related axes can include aspects related to habitat, biodiversity, and species interactions, in addition to productivity. Other axes are related to planning, the knowledge base and adequacy of implementation. The "planning" axes would ensure that management objectives are clearly stated and are properly translated into management targets and conservation limits. The "knowledge" axes indicate that the science knowledge necessary for managing the fisheries is available and that the knowledge of the fishery is sufficient for management. The "implementation" axes would ensure that the conservation frameworks are implemented through science programs, fisheries

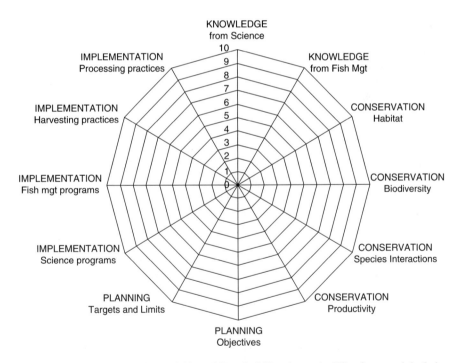

Fig. 6.3 Axes of Resource Stewardship and Sustainability. A sustainability framework includes more axes than the traditional focus on productivity. For instance, conservation can include aspects related to habitat, biodiversity, and species interactions, in addition to productivity. Other axes are related to planning, knowledge, and implementation, each of them having sub-elements that deserve attention.

management programs, using a set of measures governing harvesting practices, as well as processing practices. The challenge is to ensure that a review of a fishery considers all axes and that a low score in any one axis is taken as an indication of some degree of poor stewardship that must be corrected.

International agreements and initiatives, domestic legislation and policies, and growing market pressures for "certified" products are requiring that a broader set of conservation factors and new frameworks be taken into account when managing fisheries. Furthermore, while conceptual frameworks for conservation abound, the difficulty is in implementation. The checklist is one example of a tool that could assist in implementation. It is an enabling tool to not only assist scientists and managers in understanding the key elements of their programs but also to support government and industry leaders in advancing on sustainability. Figure 6.4 provides four examples of how the information collected in a checklist could be used to construct profiles of resource stewardship and sustainability for a fishery or group of fisheries. And finally there is a need for open, transparent, and easily accessible ways to report to clients. It is not sufficient to wait for public campaigns

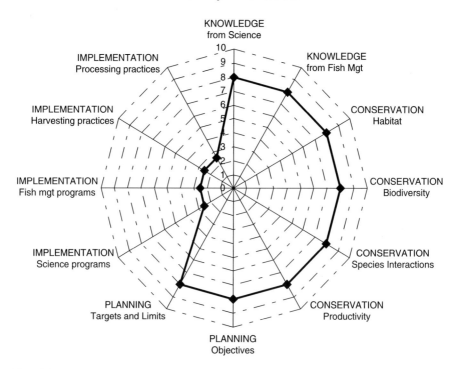

Fig. 6.4 Resource Stewardship and Sustainability Profiles. Measuring progress sustainability along each axis could serve to establish stewardship and sustainability profiles characterizing a given situation. Four profiles are illustrated, each achieving poor or good scores along a particular set of axes thereby characterizing situations of "Poor Implementation," "Poor Planning," "High Stewardship with Good Conservation," or "Poor Stewardship with Poor Conservation"

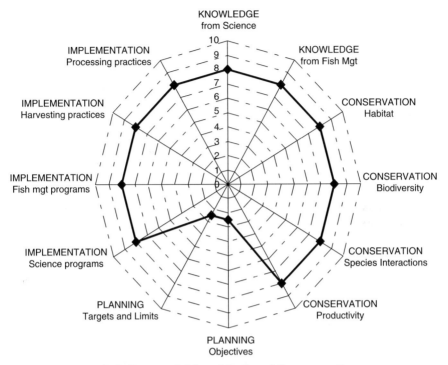

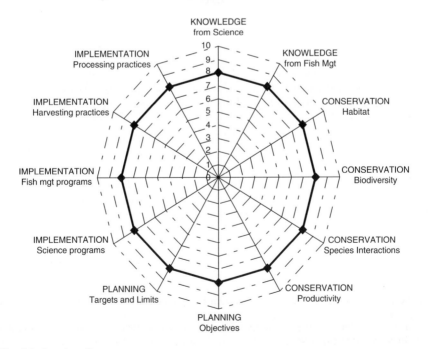

Fig. 6.4 (continued)

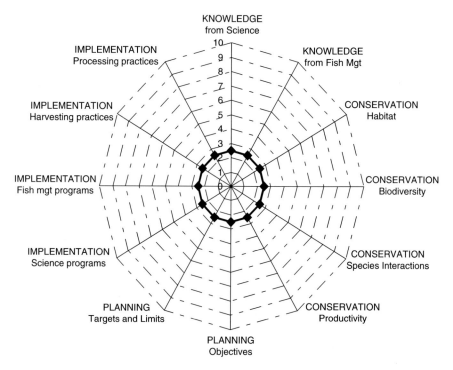

Fig. 6.4 (continued)

such as Seafood Watch in USA (http://www.mbayaq.org/cr/seafoodwatch.asp) or SeaChoice (www.seachoice.org) in Canada before taking action. We must be proactive in addressing sustainability in all its dimensions and complexity.

In conclusion, the path to implementation includes an overall framework for sustainable fisheries, a set of policies to support it, a decision-making framework to guide decisions on harvest levels, a way to assess how we are doing, and tools to communicate all of this to ourselves, industry, the public, and our markets. A framework constructed along these lines constitutes a significant step towards lessening the current uncertain and complex environment and will bring fisheries managers and fisheries science staff closer together.

References

Canada (1996) Oceans Act (1996, c. 31) Can Gov Pub, Communications Canada, Ottawa, Canada K1A 0S9

Canada (2002) Species-At-Risk Act (2002, c.29). Can Gov Pub, Communications Canada, Ottawa, Canada K1A 0S9

Clark S, O'Boyle R (2001) Proceedings of the 14th Canada-USA Scientific Discussions, January 22–25, 2001. MBL Conference Center, Woods Hole, Massachusetts. Northeast Fish Sci Cen Ref Doc 01–07, 53p

Department of Fisheries and Oceans (DFO) (2006) A Harvest Strategy Compliant with the Precautionary Approach. DFO Can Sci Adv Sec Rep 2006/023, 7p

Food and Agriculture Organization (FAO) (1995) Precautionary Approach to Fisheries. Part 1. Guidelines on the Precautionary Approach to Capture Fisheries and Species Introductions. Technical Consultations on the Precautionary Approach to Capture Fisheries. Lysekil, Sweden, June 6–13, 1995. FAO Tech Pap 350, Part 1. FAO, Rome 1995, 52p

ICES (2003) Report of the ICES Advisory Committee on Ecosystems. ICES Coop Res Rep 262, 229p

ICES (2007) Report of the Workshop on Fisheries Management in Marine Protected Areas (WKFMMPA), April 3–5, 2006. ICES Headquarters, ICES CM 2006/MHC:10, 91p

Jacquet JL, Pauly D (2006) The Rise of Seafood Awareness Campaigns in An Era of Collapsing Fisheries. Mar Pol, doi:10.1016/j.marpol.2006.09.003, 6p

Kell LT, Mosqueira I, Grosjean P, Fromentin J-M, Garcia D, Hillary R, Jardim E, Mardle S, Pastoors MA, Poos JJ, Scott F, Scott RD (2007) FLR: An Open-Source Framework for the Evaluation and Development of Management Strategies. ICES J Mar Sci, Advance Access published on March 29, 2007, doi:10.1093/icesjms/fsm012

Link JS (2002) Ecological Considerations in Fisheries Management: When Does It Matter? Fish 27(4):10–17

Mace PM (ed) (2000) Incorporating Ecosystem Considerations into Stock Assessment and Management Advice. Proceedings of the 6th NMFS National Stock Assessment Workshop, Northwest Fisheries Science Center, March 28–30 2000, Seattle, WA. NOAA Tech Memo NMFS-F/SPO-46, December 2000, 78p

O'Boyle R, Keizer P (2003) Proceedings of Three Workshops to Investigate the Unpacking Process in Support of Ecosystem-based Management; February–July 2002. DFO Can Sci Adv Sec Proc Ser 2003/004, 30p

Rice J, Rivard D (2002) Proceedings of the DFO Workshop on Implementing the Precautionary Approach in Assessments and Advice. Can Sci Adv Sec Proc Ser 2002/009, 99p

Rice J, Schnute J, Haigh R (eds) (1999) Proceedings of a Workshop on Implementing the Precautionary Approach in Canada. DFO Can Stock Assess Sec Proc 98/18, 73p

Richards LJ, Schnute JT (1999) Science Strategic Project on the Precautionary Approach in Canada – Proceedings of the Second Workshop, November 1–5, 1999, Nanaimo, B.C. DFO Can Stock Assess Sec Proc 1999/41, 96p

Rivard D, Maguire, JJ (1993) Reference Points for Fisheries Management: The Eastern Canada Experience. In: Smith SJ, Hunt JJ, Rivard D (eds) Risk Evaluation and Biological Reference Points for Fisheries Management. Can Spec Pub Fish Aquat Sci 120: 31–57

Rivard D, Rice J (2003) National Workshop on Reference Points for Gadoids. Can Sci Adv Sec Proc Ser 2002/033, 16p

Shelton PA, Rice JC, Rivard D, Chouinard GA, Fréchet A (2003) Recent Progress on the Implementation of the Precautionary Approach on Canadian Cod Stocks Leading to the Re-introduction of the Moratorium. ICES CM 2002/Y:15, 11p

Smith ADM, Fulton EJ, Hobday AJ, Smith DC, Shoulder P (2007) Scientific Tools to Support the Practical Implementation of Ecosystem-Based Fisheries Management. ICES J Mar Sci, Advance Access published on April 16, 2007, doi:10.1093/icesjms/fsm041

United Nations Fisheries Agreement (UNFA) (1995) United Nations Fisheries Agreement on Straddling Fish Stocks and Highly Migratory Fish Stocks. UN General Assembly Annex 3, A/CONF.164/37, September 8, 1995, pp 45–81

World Summit on Sustainable Development (WSSD) (2002) World Summit on Sustainable Development, Johannesburg, South Africa, September 2–4, 2002. UN Division for Sustainable Development

Chapter 7
Astonishment, Stupefaction, and a Naturalist's Approach to Ecosystem-Based Fisheries Studies

Kevin D. E. Stokesbury, Bradley P. Harris, and Michael C. Marino II

Abstract Fisheries science is empirical and improving the accuracy, precision, and quality of data is paramount. New technologies (e.g., optics and acoustics) may better equip researchers to examine ecosystems from the individual to the community level. However, these new technologies are capable of collecting a great deal of data. How will these data be used in ecosystem modeling? Do we attempt to measure everything and incorporate it into an "omniscient" model? Perhaps not everything that can be measured should be. Will automated data collection and processing kill scientific intuition? We suggest a naturalist's approach to fisheries assessment, which is built on the fundamentals of ecological methodology, and takes full advantage of new technologies to obtain absolute measures and to estimate their associated uncertainty.

Keywords Absolute estimate · ecosystem · fisheries · uncertainty

A naturalist is concerned with levels of organization from the individual organism to the ecosystem, focusing on identification, life history, distribution, abundance, and interrelationships. Ecology includes all things of concern to the naturalist (Mayr 1997). We view a naturalist's approach as a way of conducting scientific research where observations and new classifications of natural phenomena suggest hypotheses which direct new rounds of data gathering (Wilson 1994). E.O. Wilson describes Robert H. MacArthur as a scientist who approached his research with the "strength of an inherent naturalist." E.O. Wilson, E. Mayr and many other great scientists exemplify the naturalist's approach.

In marine science, technology allows the exploration of the world's oceans and as Jules Verne (1870) predicted we are frequently astonished and stupefied by the ocean's

K.D.E. Stokesbury, B.P. Harris, and M.C. Marino II
Department of Fisheries oceanography, School for Marine Science and Technology,
University of Massachusetts Dartmouth, 838 South Rodney French Boulevard,
New Bedford, Massachusetts 02744-1221
e-mails: kstokesbury@umassd.edu,
bharris@umassd.edu, mmarino@umassd.edu

R.J. Beamish and B.J. Rothschild (eds.), *The Future of Fisheries Science in North America*, 113
Fish & Fisheries Series, © Springer Science + Business Media B.V. 2009

beauty and complexity. In less than one century we have gone from never visiting the ocean depths to mapping and observing the physical world including the deep ocean. In one life time we have reduced an unknown wilderness to nature reserves requiring protection due in part to the exploding human population (Wilson 1994).

Clearly, technology is a double-edged sword. In fisheries, technology improves our ability to understand the ocean but increases our exploitation power. The modern fishing vessel makes extensive use of technology with an increasing number of vessels equipped with sophisticated GPS mapping tools and integrated acoustic systems, which allow fishers to record precise catch locations, information on fish distributions, and seabed characteristics. Technology is also changing the way we approach marine science. Remote sensing and computer modeling technologies are growing rapidly and are allowing scientists to investigate the marine environment in new ways. Will this approach to the ocean so remove the scientist that the subject becomes an abstraction? Will we produce a generation of marine scientists who rarely experience the ocean and therefore lack the intuition that develops from direct contact?

The modern fisheries scientist deals in a paradoxical situation. With developing technologies vast amounts of oceanographic data are available, and yet for the practical question of population or stock assessment, data are frequently limited. Fishing vessels may be extremely efficient yet most management strategies use effort controls making the fishery less efficient. The stakes are high as this latent efficiency could rapidly lead to overfishing. The greatest disparity exists between what the public expects fisheries management to accomplish and what can actually be scientifically achieved with present sampling techniques and traditional single-stock data sets. Currently, marine scientists are expected to use an ecosystem approach to fisheries management that is, "geographically specified fisheries management that takes account of knowledge and uncertainties about, and among, biotic, abiotic, and human components of ecosystems, and strives to balance diverse societal objectives" (Sissenwine and Mace 2003). To do this, a change in thinking towards fisheries sampling and modeling is required.

Sissenwine and Mace (2003) suggest that scientists who provide advice to fisheries management must have a desire to apply scientific results to policy decisions and a willingness to give advice in the face of a high degree of uncertainty. The classic scientific attitude, however, is tentative and provisional (Copi 1982; Moore 1993; Mayr 1997). Fisheries scientists then find themselves in a precarious position. To conduct science we must put forth each idea as a hypothesis, based on the available facts, which is accepted only as long as the evidence supports it. However, fisheries management requires scientific "advice," which often contains a high degree of uncertainty, and is frequently used in a dogmatic fashion. How do we navigate this situation and move into an ecosystem approach to fisheries management?

Hypothesis-driven science and an ecosystem approach to management share a common requirement, absolute measures (counts per unit area or volume). To test a hypothesis requires an experimental design where absolute measures are collected that are accurate, precise, and have associated measures of the error. To estimate the vital statistics of a population also requires an absolute measure. These are critical for determining inter- and intraspecific interactions, community structure, energy flow, nutrient cycles, etc. (Krebs 1989). Harvest strategies also require absolute estimates of abundance. These are the same requirements for an ecosystem approach to fisheries management, which strives to measure and predict the interactions between the biotic, abiotic, and human components of the ecosystem. Therefore, the way to move towards an ecosystem approach to fisheries management is to accept two fundamental principles:

1. Absolute measures must be sought in assessments at the individual, population, and ecosystem level.
2. Total uncertainty must be measured.

By examining uncertainty we begin to understand what our absolute measures mean and to see connections within and between ecological levels (e.g., individuals and communities). Measuring total uncertainty involves estimating the accuracy, bias, and precision (variance) of the observation (Krebs 1989).

In fisheries, abundance data bias can be broken down into sampling selectivity and efficiency, each with its own variance (Gunderson 1993). Selectivity is the range of sizes and morphologies of individuals captured by a specific gear, and efficiency is the proportion of individuals caught by the gear compared to the total number of individuals in the gear's path (Fig. 7.1). The majority of scientific data used in fisheries stock assessment today are collected using equipment which produces relative indices of abundance, for example, the catch per unit effort (CPUE). Thus measures of size-specific selectivity and efficiency are required for each species to translate these data into absolute estimates (Gunderson 1993; for an example see Stokesbury et al. 1999a).

New technologies that may enable us to estimate absolute abundance and uncertainty are being developed. Two promising examples with the capability for fishery stock assessments are optic (video and still) and acoustic surveys. Although

Fig. 7.1 Commercial groundfish otter trawl on Georges Bank demonstrating the effect of mesh size selectivity on a catch of Spiny Dogfish (*Squalus acanthias*) (Photo: David Martins, F/V *Heritage*)

both have limitations, their power lays in the ability to directly measure marine fishes and ecosystems in absolute units (individuals m^{-2} or m^{-3}). We provide two examples to demonstrate the use of optics to assess the absolute abundance of a benthic bivalve, and acoustics to estimate the rate of natural mortality for a juvenile clupeid.

7.1 Video Survey of the New England Sea Scallop Resource, 1999 to 2006

To directly assess the USA sea scallop (*Placopecten magellanicus*) population we designed a cooperative survey that was inexpensive, technologically simple, adaptable to spatial management, and provided near real-time data. Further, we needed a non-intrusive sampling tool because approximately 50–80% of the scallop resource is in Marine Protected Areas (MPA), most of which prohibit mobile gear. Our goal was to sample spatially specific scallop density, shell height distribution, recruitment, natural mortality, macrobenthos, and substrate distributions, and to measure the uncertainties associated with these data. Therefore, we selected a centric systematic grid-based video survey design in which four quadrats (replicates) were sampled at each station.

Quadrat techniques are extremely useful in measuring sea scallop abundance, as they yield estimates of density, spatial distribution, and associated habitat on a number of scales (mm^2 to m^2) with well-established statistical procedures (Stokesbury and Himmelman 1993). A downward facing video camera mounted on a solid steel pyramid frame provides a robust and inexpensive quadrat sample.

We determined the appropriate sampling design and frequency by examining a high density of sea scallops in the northeastern corner (504 km^2) of the Nantucket Lightship Closed Area, a 6,167 km^2 MPA in the Great South Channel southeast of Cape Cod, Massachusetts, USA. Fishers identified this area as extremely productive for sea scallops. Mean scallop density was 1.0 m^{-2} (SE = 0.41) at 18 randomly selected stations, where we sampled ten replicate quadrats per station (Stokesbury 2002). Based on this variance to mean ratio, approximately 200 stations resulted in estimates of absolute scallop density with 5–15% precision for the normal and negative binomial distributions, respectively.

The upper limit of quadrat size was restricted by water clarity; 2.8 m^2 was the largest field of view offering consistent visibility given the conditions on Georges Bank. At this camera height, 1.57 m above the sea floor, each image pixel corresponded to $\approx 3 \times 3$ mm (S-VHS), which was sufficiently precise to measure scallops ± 6 mm. Scallops partially visible along the edges of the field of view were also counted so the sample area was increased to 3.2 m^2 (based on half the mean shell height) to correct for edge bias (Krebs 1989; Stokesbury 2002).

Sea scallops aggregate on scales ranging from square centimeters to tens of square kilometers and one scallop in 3.2 m^2 is a high density (0.31 scallops m^{-2}) (Brand 1991; Stokesbury and Himmelman 1993). By increasing the number of quadrats to four per station the observed sample area increased to 12.8 m^2 permitting the detection of scallop densities as low as 0.07 scallops m^{-2}, which corresponds to minimum commercially viable density (Brand 1991). Further, the additional time required to sample four quadrats instead of one was minimal compared to the overall deployment and retrieval of the sampling gear.

Most continental shelf scale fisheries surveys employ a random-stratified design, which assumes that the distribution of the animal is known and can be divided into strata containing a homogeneous distribution. For many marine species this assumption is unfounded in part due to the uncertainties associated with sampling gear selectivity and efficiency. We used a centric systematic design where stations are evenly distributed over the sample area based on a random starting position. The grid-based design is simple, samples evenly across the entire survey area, facilitates mapping and geostatistical analysis (intrinsic and transitive) and has been successfully used to survey scallops on Georges Bank (Thouzeau et al. 1991a, b; Rivoirard et al. 2000; Stokesbury 2002). This design also does not assume previous knowledge of the scallop's distribution within the footprint of the fishery or how that distribution might change over time.

We began annual surveys of areas with high scallop density in three MPA on Georges Bank in 1999 using a 1.57 km grid (Stokesbury 2002) (Fig. 7.2). In 2003,

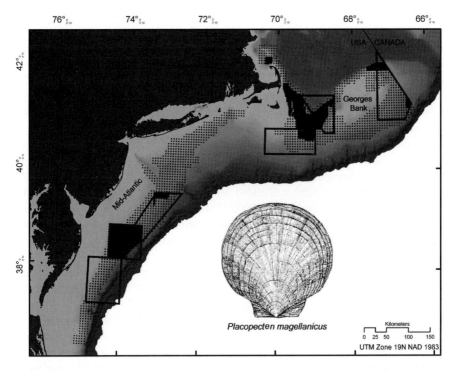

Fig. 7.2 The video survey of the mid-Atlantic and Georges Bank sea scallop resource from 1999 to 2006 (65,000 stations). Each dot is a survey station where four replicate quadrat samples are taken, each using three video cameras. The continental shelf-scale survey is a 5.56 km grid and 2.26, and 1.57 km grids (dark areas) are used for area-specific surveys. The color depth contours show the continental shelf to 2,000 m and the black boundaries are Marine Protected Areas

we expanded the video survey to cover the scallop resource in US waters, surveying 54,800 km² using a 5.56 km grid (Fig. 7.2). The survey area was selected based on the 2000 fishing effort footprint (Rago et al. 2000).

The sampling pyramid was designed for deployment from a scallop fishing vessel (Stokesbury 2002; Stokesbury et al. 2004) (Fig. 7.3). Two scientists, a Captain and Mate were able to survey about 50 stations every 24 h on the 5.56 km grid. For each quadrat, the time, date, depth, and location were recorded, and all visible fish and macrobenthos were counted including those partially visible along the edge. Improvements were made to the sampling methods during the 2005 video survey, including the development and integration of a relational database management system (RDBMS) and the addition of an Ocean Imaging® digital still camera system (6.1 megapixel) (Fig. 7.4).

After each survey the footage was reviewed in the laboratory and a still image of each quadrat was digitized using Image Pro Plus® software. Within each quadrat, macrobenthos, and fish were recounted and the substrate was verified (Stokesbury 2002; Stokesbury et al. 2004). When possible, fish and macrobenthos were identified to species, otherwise animals were grouped into categories based on taxonomic orders (Fig. 7.4).

Undergraduate and graduate students are trained to analyze the footage and images. Manual image analysis is time consuming, but provides invaluable insights to the marine ecosystem and frequently leads to new ideas which often develop into hypotheses and graduate research projects. Other research groups are developing automated imaging and analysis systems, but an important negative by-product of automation at sea and in the laboratory will be the loss of intuition developed from visually examining footage and images of the biotic and abiotic components of the marine ecosystem.

Counts of scallops and other macrobenthos were standardized to individuals per square meter and presence or absence was recorded for the colonial organisms (e.g., sponges, hydrozoans/bryozoans). Mean densities and standard errors of scallops and macrobenthos were calculated using equations for a two-stage sampling design (Cochran 1977; Stokesbury 2002; Stokesbury et al. 2004; Stokesbury and Harris 2006). The absolute number of scallops within a survey area was calculated by multiplying the mean number of scallops per square meters by the total area surveyed (Stokesbury 2002).

Both optical and acoustic systems require calibration. We deployed the video pyramid in the 340,687 l optic-acoustic seawater tank at the University of Massachusetts School for Marine Science and Technology in New Bedford, MA. We conducted calibration experiments using live scallops and shells on sand and granule/pebble substrate, and used grid patterns to examine camera lens curvature. These experiments revealed both the bias and precision of our measurements. The mean measurement of a grid of tiles was 46.4 mm (SD = 2.63) while the true value was true width 48.5 mm (N = 180) indicating the bias and precision associated with distance measures for the 3.2 m quadrat.

Sea stars (Asteroidea) and sea scallops (*P. magellanicus*) were the most abundant macrobenthic organisms in the survey area (Stokesbury et al. 2004). Sea scallop

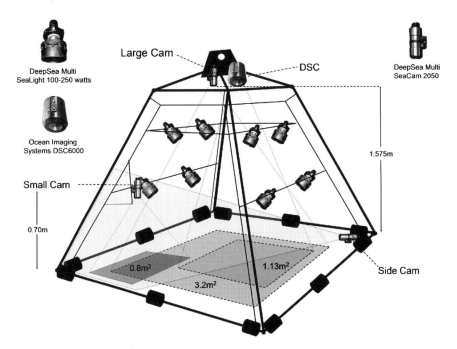

Fig. 7.3 The video survey sampling pyramid (2006 configuration). Three live-feed S-VHS camera, and one 6.1 megapixel digital still camera (DSC) sample the sea floor

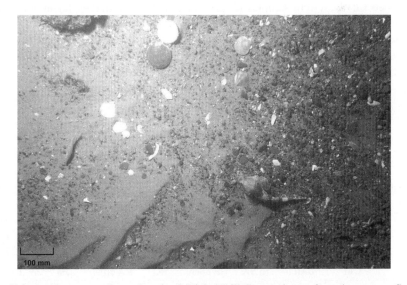

Fig. 7.4 A video survey Ocean Imaging® Digital Still Camera image from the western Great South Channel showing Sea scallops (*Placopecten magellanicus*), Longhorned sculpin (*Myoxocephalus octodecimspinosus*), juvenile gadoid, possibly Atlantic cod (*Gadus morhua*), Sea star (*Asterias* sp.), Hydrozoans and Bryozoans (on cobble) and shell debris, sand ripple, granule/pebble, and a cobble substratum

density varied between 2003 and 2006 (Fig. 7.5). In 2003, we observed a very large settlement of juvenile scallops in the mid-Atlantic (densities ≈ 40 scallops m^{-2}) and the resource reached the highest abundance ever recorded at 16.25 (95% CI ± 3.97) billion scallops, approximately 179,000 t of meats (Stokesbury et al. 2004). By 2004, overall scallop abundance had declined by 47% largely due to heavy fishing on the recruits in the mid-Atlantic (Fig. 7.5). This area became the Elephant Trunk Closed Area in 2005. From 2004 to 2006 scallop density remained constant at 9.3 (95% CI ± 1.38), 8.8 (95% CI ± 1.47), and 8.4 (95% CI ± 1.21) billion individuals in 2004, 2005, and 2006, respectively.

The proportion of sea scallop biomass within closed areas has increased from 61% in 2003 to 82% in 2006. Adjusting to variations in scallop density, size distribution, and fisheries management decisions (area closures/openings) requires redrawing assessment boundaries and calculating biomass for different spatial areas. The grid-based survey design facilitates estimates of density and size distribution for any part of the survey area. Further, the design and absolute density estimates allows the testing of hypotheses using established statistical designs, for example the examination of fishing impact on the marine benthic habitat in a Before-After-Control-Impact experiment (Stokesbury and Harris 2006).

These video survey data have been used in sea scallop fisheries management plan since 1999. Although useful in single species assessments, the true power of optics-based survey is the ability to sample habitats. "Ecosystem-based approach" implies that fisheries-relevant ecological processes, and fisheries themselves, need to be documented in the form of maps (Pauly et al. 2003). In its present configuration the video survey system samples all the visible biotic and abiotic features in each quadrat, and our database now has over four million georeferenced observations of macrobenthos and substrate of the Northwest Atlantic continental shelf. In-depth analysis of these data is only beginning and we believe the ramifications of these surveys for assessments from individuals to ecosystems will be profound.

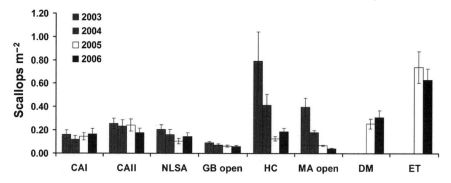

Fig. 7.5 The mean number of sea scallops m^2 (± SE) observed in Closed Area I (CAI), Closed Area II (CAII), Nantucket Lightship Area (NLCA), Georges Bank open area (GB open), Hudson Canyon closed Area (HC), mid-Atlantic open (MA open), Delmarva suggested closed area (DM), and the Elephant Trunk closed Area (ET), during the 2003 to 2006 video surveys

7.2   Acoustic Sampling Pacific Herring in Prince William Sound, Alaska

The Pacific herring (*Clupea pallasii*) population crashed in 1993, 4 years after the *Exxon Valdez* spilled 36,000 t of North Slope crude oil into Prince William Sound. The Sound Ecosystem Assessment (SEA) program began in 1994 as a 5-year integrated study of the physics and biology of Prince William Sound affecting the survival of the juvenile life-stages of pink salmon and Pacific herring. SEA employed simultaneous bottom-up and top-down investigations of parts of the ecosystem during seasons identified as critical to the survival of juveniles (see review in Pearcy 2001). Estimates of absolute density are required to directly measure vital statistics, such as natural mortality (Krebs 1989). We used acoustic surveys to assess the natural mortality for juvenile Pacific herring. These acoustic samples are a small component of the larger ecosystem study.

Clupeid population size is highly variable and fluctuations may result from environmental change or overfishing. These characteristics coupled with the lack of local life history information confounded attempts to determine sources of mortality and the recovery time of the Prince William Sound Pacific herring population (Norcross et al. 2001).

Four bays, Eaglek, Whale, Zaikof, and Simpson, were selected as study sites because they are in Northern, Southern, Eastern, and Western Prince William Sound, respectively. These four bays were acoustically surveyed nine times between June 1996 and March 1998. During each 7-day cruise, each bay was surveyed three times over 24 h beginning at 0000 h, 0800 h, and 1600 h.

Five vessels were used during each survey; three commercial fishing vessels (≈ 16.8 m) deployed the acoustic and oceanographic equipment and fished the seines, a trawler (Alaska Department of Fish and Game R/V *Pandalus*, ≈ 20 m), and a crew vessel (≈ 25 m) where the samples were processed. The acoustic vessel followed a set transect pattern along the shore (MacLennan and Simmonds 1992; Gunderson 1993). Fish schools were measured using a 120 kHz BioSonics 101 echosounder with a preamplifier dual-beam transducer (Stokesbury et al. 2000; 2002).

Once the acoustic vessel surveyed a fish school it was sampled to determine species composition and size structure. Fish were captured using a modified bottom trawl in deep water (1.52 × 2.13 m Nor'Eastern Astoria V trawl doors, head rope 21.3 m, foot rope 29.0 m, estimated 3 × 20.0 m mouth, 10.2 cm mesh wings, 8.9 cm middle, and a 32.0 mm cod end liner), an anchovy seine in surface water (250.0 × 34.0 m and 20.0 m, 25.0 mm stretch mesh), or a small salmon fry seine in shallow water (50.0 × 8.0 m, 3.0 mm stretch mesh deployed from a 6 m skiff equipped with a 70 horsepower engine). One of the main drawbacks of acoustic surveys is that size frequencies and species compositions typically rely on net samples with their associated assumptions of efficiency and selectivity. Each collection was speciated and 1,000 herring, along with the other dominant fish species, were randomly subsampled. Fork length (mm) and wet weight (g) were recorded from 450 herring and the remaining fish were measured to fork length. Once the subsample was collected the remaining fish were released unharmed from the seine as collections were often large.

A length-dependent scaling constant was used to convert the reflected acoustic energy into a biomass estimate (Stokesbury et al. 2000; 2002). The number of fish observed (a_x) during one survey was compared to the subsequent survey (a_{x+1}) approximately 1 year later for each bay. Instantaneous natural mortality rates (M_x d^{-1}) were calculated using the equation:

$$M_x = \log_e (a_{x+1}/a_x)/(t_{x+1}-t_x) \qquad (7.1)$$

where $t_{x+1} - t_x$ equals the time interval between surveys in days (Ricker 1975; Krebs 1989; Stokesbury et al. 2002). We were able to calculate these estimates only because the acoustic surveys provided an absolute measure of the number of juvenile herring per unit volume. Although this equation can work with relative estimates of density, the catchability coefficient must be the same for both the numerator and the denominator so that they cancel each other out. With juvenile fish (particularly clupeids) this is seldom the case as their growth rates are very fast over the summer and vary both spatially and temporally (Stokesbury et al. 1999b). Estimates of instantaneous mortality are extremely sensitive to these effects (Taggart and Frank 1990).

The average instantaneous natural mortality rates for young-of-the-year Pacific herring were 0.009 (SD = 0.002) and 0.016 (SD = 0.012)d^{-1} for the 1996 and 1997 cohorts, respectively. The average instantaneous natural mortality rates for 1-year-old Pacific herring were 0.003 (SD =0.007) and 0.008 (SD = 0.005)d^{-1} for the 1995 and 1996 cohorts, respectively. The highest natural mortality occurred directly after the young of the year metamorphosed from larvae in late July until the beginning of oceanic winter in late October (Stokesbury et al. 2002).

Acoustic techniques are improving at a rapid rate (refer to Chapters 17, 18, and 21). The survey we have presented used a basic acoustic system compared to the more recent technologies. However, by measuring absolute densities we were able to estimate a previously unobtainable natural mortality rate for Pacific herring. The link between this single species estimate and the ecosystem is clear as natural mortality of Pacific herring, a key prey species in the Prince William Sound ecosystem, is directly related to predation rates and primary productivity through growth and winter survival, as well as being critical for recruitment into fisheries.

7.3 Conclusion

Marine ecology, of which fisheries science is a subset, deals with the interactions between organisms and their living and nonliving environment at three fundamental levels: the individual, the species/population, and the community/ecosystem (Mayr 1997). Our ability to examine these different levels is undergoing tremendous growth with new technologies, particularly optics and acoustics. These new technologies also enable us to collect extremely large quantities of data, but it is critical to remember that, "Not everything that can be measured should be" (Krebs 1989).

We suggest a naturalist's approach to fisheries assessment that is built on ecological methodology laid down by Green (1979) and Krebs (1989), which takes full advantage of new technologies to obtain absolute measures and estimates associated total uncertainty. Further, we suggest the following criteria in designing surveys and experiments:

1. Hypothesis-driven approach with experimental design based on observations to determine the appropriate sampling design and scale; the temporal and spatial scales of the sampling design must match that of the hypotheses.
2. Use as much information as possible in collecting initial observations; include historic literature, perspective of non-scientists, and especially observations and perceptions from fishers and other resource users.
3. Start as simply as possible with a scalable sampling design and build a mosaic as knowledge increases.
4. Make your experimental design as adaptive to new technologies as possible; absolute measures are essential.
5. Incorporate spatial and temporal variability in your experimental design (strongly consider systematic sampling designs).
6. Use collection and analysis procedures that allow for the development of your intuition and understanding of the ecosystem (automation can kill intuition).

By applying a naturalist's approach it is possible to determine the abundance of individuals, measure intra- and interspecific relationships, and delineate the community structure in which they live. Further, by measuring uncertainty we can have enough confidence in our understanding and interpretation of the marine ecosystem to provide scientific ecosystem-based fisheries management advice.

Acknowledgments We thank our colleagues who worked on these projects and helped us develop these ideas. We thank V Holliday, K Foote, B Rothschild, and R Beamish for encouragement, guidance, and support. M Stokesbury reviewed a draft of the MS providing helpful comments. These projects were funded by the *Exxon Valdez* Oil Spill Trustee Council through the Sound Ecosystem Assessment (SEA) project to the University of Alaska, Fairbanks, and the Prince William Sound Science Center and by SMAST, the Massachusetts Division of Marine Fisheries, NOAA, and the sea scallop fishery and supporting industries. The views expressed herein are those of the authors and do not necessarily reflect the views of NOAA or any other agencies.

References

Brand AR (1991) Scallop ecology: distributions and behaviour. In: Shumway SE (ed) Scallops: Biology, Ecology and Aquaculture. Elsevier, New York, pp 517–584
Cochran WG (1977) Sampling Techniques, Wiley, New York
Copi IM (1982) Introduction to logic, Macmillan, New York
Green RH (1979) Sampling design and statistical methods for environmental biologists. Wiley, New York

Gunderson RD (1993) Surveys of Fisheries Resources. Wiley, New York

Krebs CJ (1989) Ecological Methodology, Harper & Row, New York

MacLennan DN, Simmonds EJ (1992) Fisheries Acoustics, Chapman & Hall, London

Mayr E (1997) This is Biology: The Science of the Living World, Harvard University Press, Cambridge

Moore JA (1993) Science as a Way of Knowing, Harvard University Press, Cambridge

Norcross BL, Brown ED, Foy RJ, Frandsen M, Gay S, Kline Jr. TC, Mason DM, Patrick EV, Paul AJ, Stokesbury K (2001) A synthesis of the life history and ecology of juvenile Pacific herring in Prince William Sound, Alaska. Fish Oceanog (Suppl) 10:42–57

Pauly D, Watson R, Christensen V (2003) Ecological geography as a framework for a transition toward responsible fishing. In: Sinclair M, Valdimarsson G (eds) Responsible Fisheries in the Marine Ecosystem. CAB International, Cambridge, pp 87–102

Pearcy WG (2001) (Guest Editor) Ecosystem-level studies of juvenile herring and juvenile pink salmon in Prince William Sound, Alaska. Fish Oceanog (Suppl) 10:1–193

Rago P, Murawski S, Stokesbury K, DuPaul W, McSherry M (2000) Integrated management of the sea scallop fishery in the northeast USA: research and commercial vessel surveys, observers, and vessel monitoring systems ICES Mar Sci Symp CM2000/W:13, pp. 18

Ricker WE (1975) Computation and interpretation of biological statistics of fish populations. Bull Fish Res Board Can 191:382

Rivoirard J, Simmonds J, Foote KG, Fernandes P, Bez N (2000) Geostastics for estimating fish abundance. Blackwell Science, Oxford

Sissenwine M, Mace PM (2003) Governance for responsible fisheries: an ecosystem approach. In: Sinclair M, Valdimarsson G (eds) Responsible Fisheries in the Marine Ecosystem. CAB International, Cambridge, pp 363–392

Stokesbury KDE (2002) Estimation of sea scallop abundance in closed areas of Georges Bank, USA. Trans Am Fish Soc 131:1081–1092

Stokesbury KDE, Harris BP (2006) Impact of limited short-term sea scallop fishery on epibenthic community of Georges Bank closed areas. Mar Ecol Prog Ser 307:85–100

Stokesbury KDE, Himmelman JH (1993) Spatial distribution of the giant scallop *Placopecten magellanicus* in unharvested beds in the Baie des Chaleurs, Québec. Mar Ecol Prog Ser 96:159–168

Stokesbury KDE, Bichy J, Ross SW (1999a) Selectivity and efficiency of two otter trawls used to assess estuarine fish and macroinvertebrate populations in North Carolina. Estuaries 22: 882–888

Stokesbury KDE, Foy RJ, Norcross BL (1999b) Spatial and temporal variability in juvenile Pacific herring, *Clupea pallasi*, growth in Prince William Sound, Alaska. Environ Biol Fish 56:409–418

Stokesbury KDE, Kirsch J, Brown ED, Thomas GL, Norcross BL (2000) Spatial distributions of Pacific herring, *Clupea pallasi*, and walleye pollock, *Theragra chalcogramma*, in Prince William Sound, Alaska. Fish Bull 98:400–409

Stokesbury KDE, Kirsch J, Patrick EV, Norcross BL (2002) Natural mortality of juvenile Pacific herring (*Clupea pallasi*) in Prince William Sound, Alaska. Can J Fish Aquat Sci 59:416–423

Stokesbury KDE, Harris BP, Marino II MC, Nogueira JI (2004) Estimation of sea scallop abundance using a video survey in off-shore USA waters. J Shell Res 23:33–44

Taggart CT, Frank KF (1990) Perspectives on larval fish ecology and recruitment processes: probing the scales of relationships. In: Sherman K, Alexander LM (eds) Patterns, processes, and yields of large marine ecosystems. AAAS Selected Symposium Series, p 151–164

Thouzeau G, Robert G, Smith SJ (1991a) Spatial variability in distribution and growth of juvenile and adult sea scallops *Placopecten magellanicus* (Gmelin) on eastern Georges Bank (Northwest Atlantic). Mar Ecol Prog Ser 74:205–218

Thouzeau G, Robert G, Ugarte R (1991b) Faunal assemblages of benthic megainvertebrates inhabiting sea scallop grounds from eastern Georges Bank, in relation to environmental factors. Mar Ecol Prog Ser 74:61–82

Verne J (1870) Twenty Thousand Leagues Under the Seas, translated by W Butcher (1998) Oxford University Press, New York

Wilson EO (1994) Naturalist, Island Press, Washington, DC

Chapter 8
The High-Dimensional Future
of Fishery Science

Jon T. Schnute and Laura J. Richards

Abstract Classical physics describes the universe in four dimensions: three for space and one for time. Modern theories propose additional hidden dimensions, too tightly curled up to be measured with current technology. Similarly, fisheries ecology considers dimensions associated with fish species, their genetic structure, prey organisms, the physical environment, and other factors. Many ecological dimensions effectively remain hidden because they are too numerous to explore systematically. For example, fishery models often apply simple biological assumptions to summary population data. As the planet undergoes regime changes, partly due to human activity, familiar patterns from the past may fail to recur. New regimes move ecological systems into unfamiliar regions within the high-dimensional space of possibilities. Similarly, human perceptions extend to new dimensions with modern technologies, such as satellite cameras, robotic floats, genetic microarrays, and high-performance computers. We look to a future with serious environmental issues and advanced technologies for addressing them. Scientific progress requires a deeper understanding of high-dimensional ecosystems, along with communication that stimulates human populations to take appropriate actions.

Keywords Climate change · data visualization · ecological simulation model · high performance computing · regime change

8.1 Introduction

According to classical physics, we live in a universe of three spatial dimensions and one of time. Relativity shows that these combine seamlessly into a four-dimensional *space–time continuum*. Modern string theories suggest that the universe also contains hidden dimensions, so tightly curled up that we don't notice them

J.T. Schnute and L.J. Richards
Fisheries and Oceans Canada, Pacific Biological Station
3190 Hammond Bay Road, Nanaimo, BC, Canada V9T 6N7

in ordinary experience (Randall 2005). Perhaps our familiar world needs to be extended to 10, 11, or 26 dimensions, where quantum particles consist of tiny strings or membranes vibrating within unseen geometries.

Like physicists, ecologists also encounter hidden dimensions. An ecosystem might include 100 fish species, but these interact with other organisms too numerous to list completely. Aside from plankton species, the hidden dimensions include bacteria and viruses that might be pathogenic or beneficial. The problem of dimensionality appears as complex in ecology as in basic physics, although for different reasons. Someday a unified theory of physics might explain the universe with much fewer than 100 dimensions.

Statistics and other sciences embrace the concept of dimensionality in the phrase *degrees of freedom*. Following an analogy by Randall (op. cit., pp. 1–2), we may have first learned about this kind of freedom in the crib. We had two dimensions for crawling about, but a third for climbing over the rails and escaping. (Randall calls this the beginning of a 'disinformation campaign' that our world has only three spatial dimensions.) Extra degrees of freedom can be liberating, but they also quickly become daunting. Maybe physicists have some hope in a unifying theory of everything, but it seems obvious that no such model of a complex ecosystem will ever exist. The number of dimensions is too huge. Even physicists had to resort to *statistical* mechanics when describing the behavior of 6×10^{23} molecules (Avogadro's number) in a fluid or gas. No computer has a memory large enough to keep track of every one. Similarly, an ecologist can't follow the detailed behavior of every organism. The best available model of an ecosystem is the ecosystem itself!

Partly for these reasons, quantitative fishery science began by examining reality at the higher level of fish populations and summary statistics. Schnute (2006) describes some of this history in the context of pioneering work by Ricker (1975). Since that 1975 publication, dramatic changes have taken place in technology and theory. In 2008, the science community now has much more extensive data, communications media (such as the Internet), computing resources, and statistical techniques. Both authors of this paper have witnessed major changes in fishery science due to these remarkable developments.

While fishery science may have progressed during the last 3 decades, many scientists argue that the fisheries themselves are in crisis (Pauly et al. 1998; Buckworth 1998). Furthermore, on a larger scale, the entire planet almost certainly is experiencing climate change (IPCC 2007a). Coincident with new data, technology, and theory, we find ourselves facing a world that may be operating by rules we have never encountered before.

Regime changes often force scientists to investigate new dimensions in the space of possibilities. We gain new degrees of freedom, but like a baby leaving its crib, we don't know what lies beyond the rails. Different relationships might link the past to the future. How can we best use our current resources to address problems in a changing world? Answers, if they exist, lie inside a space with many dimensions. We have become pretty good spectators of unfolding events, but can we alter our own behavior to avoid potential disasters? To what extent can we formulate reasonable predictions that would guide sound policy decisions? In the sections that follow, we examine these questions in the context of changing regimes for the planet, data collection, and scientific technology.

8.2   Planetary Regime Changes

In 2007, the Intergovernmental Panel on Climate Change (IPCC) and former US Vice President Al Gore, Jr., received the Nobel Peace Prize 'for their efforts to build up and disseminate greater knowledge about man-made climate change, and to lay the foundations for the measures that are needed to counteract such change' (http://nobelprize.org/nobel_prizes/peace/laureates/2007/). The IPCC had just released its Fourth Assessment Report, where the first of three parts (IPCC 2007a) describes the physical basis for climate change induced by greenhouse gas emissions. The authors provide a summary (IPCC 2007b) that we recommend to our readers. In particular, we cite four of their key conclusions:

- 'Global atmospheric concentrations of carbon dioxide, methane, and nitrous oxide have increased markedly as a result of human activities since 1750 and now far exceed preindustrial values determined from ice cores spanning many thousands of years.'
- 'Warming of the climate system is unequivocal, as is now evident from observations of increases in global average air and ocean temperatures, widespread melting of snow and ice, and rising global average sea level.'
- 'Most of the observed increase in global average temperatures since the mid-twentieth century is *very likely* due to the observed increase in anthropogenic greenhouse gas concentrations.' (The authors define *very likely* as a probability of occurrence greater than 90%.)
- 'Anthropogenic warming and sea level rise would continue for centuries due to the time scales associated with climate processes and feedbacks, even if greenhouse gas concentrations were to be stabilized.'

In short, the world's climate is changing. We're entering a period of global warming attributable (with 90% probability) to human activity. Warming will continue for centuries even if we alter our activities now. As we emerge from the cradle of the twentieth century, we find some uncomfortable new degrees of freedom in the world around us.

OK, but so what? Physics still works; in fact, it has been used to reach the conclusions above. How have the rules changed? Here's an example. Increased levels of carbon dioxide, part of the climate change scenario, cause the oceans to become more acidic. (Water H_2O and carbon dioxide CO_2 combine to make carbonic acid H_2CO_3.) A more acidic ocean can cause degradation to corals in tropical zones and calcareous phytoplankton in temperate zones (Feely et al. 2004; http://www.ocean-acidification.net/). By impacting the bottom of the food chain, acidification potentially could have negative effects on species at higher levels, including commercial fish. For example, Pacific salmon (*Oncorhyncus* spp.) eat pteropods, small winged snails among the calcareous phytoplankton.

Some recent ecosystem changes illustrate the potential for large-scale shifts that may relate to climate change, fishing, or both. For example, the species community on the Eastern Scotian Shelf of Canada, once dominated by large-bodied demersal fish, now consists primarily of small demersal and pelagic fish, along with benthic macroinvertebrates (Choi et al. 2005). Elsewhere, particularly in systems impacted by human activities, jellyfish have started replacing commercially fished species (Xian et al. 2005). Nonindigenous species have seriously impacted terrestrial and aquatic systems and exacerbated concerns about biodiversity (Carlton 1996).

Over the last decade, such changes have fostered a move towards a precautionary approach and increased our uncertainty in future projections. We have personal experience with the return of sockeye salmon (*Oncorhynchus nerka*) to the Fraser River, British Columbia, where detailed forecasts of return abundance are conducted each year (Schnute et al. 2000). These forecasts use historical data on numbers of returning salmon and corresponding counts of spawning fish from the previous generation. In 2007, less than one quarter of the forecast number actually returned, leaving a larger discrepancy than anticipated from past experience. While such forecasts always have inherent variability, 2007 was unprecedented in that the returning salmon had to survive 3 consecutive years of above-average sea surface temperatures. Global warming projections from climate models cast doubt on the long-term viability of this stock (Morrison et al. 2002), and the events of 2007 could be an early warning signal.

8.3 Data Regime Changes

As industrial development released greenhouse gases and altered the earth's climate, it also produced remarkable abilities to capture information. Modern authors tell engaging stories about important historical developments. For example, an accurate eighteenth century clock (Sobel 1995), invented by John Harrison, enabled sailors to measure longitude reliably. Similarly, the nineteenth century transatlantic telegraph cable (Gordon 2002), promoted by Cyrus Field, dramatically altered communication between North America and Europe. Messages that previously required a 2-week boat trip could now be sent almost instantly. Both developments, a clock and a cable, changed the practical human perception of space and time.

We now use satellite technology as a centerpiece of global communication, with audio and visual content rather than telegraphic messages. Satellite sensors reveal detailed characteristics of terrestrial and aquatic ecosystems, such as temperature, altimetry, winds, and color. Data that previously would have required expensive, time-consuming surveys now come from orbiting cameras. This information provides indicators of ocean productivity, including its seasonal cycles and spatial variation (Polovina and Howell 2005). In particular, satellite imagery shows oceanographic features associated with 'hot spots' of fish concentration (Zainuddin et al. 2006).

The US Apollo 8 mission, the first manned flight to circumnavigate the moon, captured some of the earliest photos of the earth from space. One of these, taken on Christmas Eve in 1968, shows our planet rising above the moon's horizon. The photo, dubbed *Earth Rise*, appears on a NASA web site (http://lunar.gsfc.nasa.gov/images/gallery/apollo08_earthrise.jpg). Forty years later in 2008, satellite photos give an almost casual familiary to the global earth. You can leave your gravity-imposed cradle by logging on to the Internet and viewing potentially millions of earth images recorded by satellites. For example, you can use the 'Google Earth' program (http://earth.google.com/) to view almost any location on the planet. This remarkable application illustrates the potential for providing massive amounts of information on a human scale. From a fishery perspective, browse the satellite images available at http://oceancolor.gsfc.nasa.gov/FEATURE/gallery.html. These show weather events, plankton blooms, dust storms, sedimentation, and other important features of the ecosystem.

Satellites primarily sense the earth's surface features, not the three-dimensional oceanic world occupied by marine fish. The ARGO project addresses this know-ledge gap with probes that profile the ocean to depths of 2,000 m (Freeland and Cummins 2005; http://www.argo.net/). Currently, more than 3,000 robotic floats scattered across the world's oceans relay temperature and salinity data every 10 days via satellite to central repositories. From here, files can be accessed via the Internet. Although the program is too new to cite explicit benefits for fishery science, the potential gains from ARGO data appear enormous. Once again, we perceive a new dimension, in this case the water column inhabited by marine species.

Although satellites and robotic floats inform us about the properties of the water column, they can't report on how fish use the available habitat. New electronic tags provide sensors attached to individual fish, with an archival ability to record depth, temperature, and other data. To date, most studies have focussed on the movements of large migratory species, such as tunas (Block et al. 2005). Archival tags require that fish be captured to attach the tag and later recaptured to collect the stored data. Although this procedure limits sample size, the available data indicate that fish can be highly selective in their use of ocean space. For example, Walker et al. (2007) reported that Pacific salmon in offshore waters tend to inhabit the top 40–60 m, above the thermocline, but occasionally can be found at depths of 80–120 m. Movements are diurnal, such that salmon are usually near the surface at night, and deeper during the day. Alternatively, returning salmon can be marked with simpler acoustic tags tracked by receiver arrays. This method has allowed measurement of migration and survival rates (English et al. 2005).

Genetic data add further dimensions to our understanding of fish biology. For example, genomics gives insights into the physiological reactions of fish to their environment. Microarrays can monitor thousands of genes simultaneously, making it possible to sift through a large number of gene expressions to identify those that distinguish distinct groups. A medical application might try to distinguish normal from cancerous cells. In fishery science, gene expression profiling of 16,000 salmon genes (von Schalburg et al. 2005) allows screening for genetic markers associated with particular conditions, such as the physiological change expressed by Pacific salmon returning from the ocean to fresh water for spawning (Miller et al. 2007). The voluminous high dimensional data used in this technique often get compressed into a two-dimensional image called a *heat map*, in which colors represent degrees of gene expression for visual pattern recognition (http://en.wikipedia.org/wiki/Expression_profiling).

8.4 Technological Regime Changes

When we began our careers, fishery scientists had cumbersome tools available for quantitative analysis. Programs could be submitted to mainframe computers via decks of punched cards or interactive terminals. Programmable calculators offered a small measure of independence from mainframes, but with very limited capability. So-called 'graphics' could be produced on a line printer with individual data points represented as asterisks. The general public collected music on vinyl records or audio cassettes, as well as (eventually) movies on video cassettes.

The audio change from monaural to stereo represents a dimensional change recognizable to the human ear, as do further changes to technologies that support more than two speakers. Similarly, the changes in film and television from black and white to color provide dimensional changes for the human eye. In each case, the information content goes up substantially. The concepts of information and dimension are closely related. For example, some mobile devices use '12-bit' color, with 4 bits assigned to each of the primary colours red, green, and blue. Four bits gives $2^4 = 16$ possible levels, so that 12 bits can be used to encode about $16 \times 16 \times 16 = 2^{12} = 4{,}096$ colors. Similarly, '24-bit' color assigns 8 bits to each primary color, giving about 16.8 million (2^{24}) colors. Admittedly, we have to be somewhat careful here because colors appear to be three-dimensional on a red-green-blue scale, but 12 or 24 dimensional on a binary scale. No matter how you evaluate it, however, a detailed color image can convey much more information to the human eye than asterisks printed by a line printer.

Combinatorial possibilities play a major role in modern technology. Currently, a computer projector typically displays $1{,}024 \times 768 = 786{,}432$ pixels. If each pixel has a 24-bit color (that's 3 'bytes' in computer terminology), it will require more than 2 MB to store that image. Although compression algorithms can reduce storage requirements, the number mounts quickly if we consider satellite images taken frequently. If we want to build an ecosystem model with ten spatial cells, in which each cell can interact with all others, we need to consider $10 \times 10 = 100$ possible interactions. That would have been impossible on early programmable calculators, but entirely feasible on a modern personal computer. The current generation of models, however, sometimes aims to incorporate millions of cells. This gives trillions of potential interactions, and changes the problem qualitatively. If the calculation works at all on a personal computer, it might require weeks or even years to complete. A quick simulation to see how things might work becomes impossible.

The trend toward bigger and faster computer chips in personal computers has reached a technical barrier. To maintain momentum, chip technology itself has evolved a new dimension associated with the number of core processors. Dual-core, quad-core, and higher level processors are now routinely available in the marketplace. The technique of *parallel processing* allows calculations to be divided among multiple processors in a single computer or a network of computers. A new generation of software is required for distributing computational tasks among the available computer resources.

Computers have two key dimensions: memory and speed. A *byte* (8 bits) is a common unit of memory. Generations of desktop computers have taught us about *kilobytes* (10^3 bytes), *megabytes* (10^6 bytes), *gigabytes* (10^9 bytes), and perhaps *terabytes* (10^{12} bytes) corresponding to thousands, millions, billions, trillions of bytes. The list goes on, with the prefixes *peta, exa, zetta*, and *yotta*, corresponding to 10^{15}, 10^{18}, 10^{21}, and 10^{24}. (We're using a decimal scale here, rather than an alternative binary scale with $2^{10} = 1,024 \approx 10^3$.) Similarly, the acronym *flops* denotes '*fl*oating point *o*perations *p*er *s*econd', an approximate measure of the work a computer can do (http://en.wikipedia.org/wiki/FLOPS). Like measures of memory, *kiloflops, megaflops, gigaflops*, etc., refer to operational speeds. Snazzy computers can store lots of data bytes and perform a large number of flops. The speed measurement gets slightly confusing because chips are normally rated in cycles per second, where one cycle per second is called a *hertz*. For example, this paper is being typed on a 3 gigahertz machine that performs 3 billion cycles per second. Chips can usually do one or more floating point operation per cycle, and typical desktop computers in 2008 perform gigaflops of work on gigabytes of memory with hard disk storage in the order of a terrabyte.

To understand the current state of high-performance computing, shift your scale of thinking upward to teraflops of speed, terabytes of memory, and petabytes of archival data. For example, the US Library of Congress (http://www.loc.gov/webcapture/faq.html) now collects and preserves web data 'because an ever-increasing amount of the world's cultural and intellectual output is created in digital formats and does not exist in any physical form.' In May, 2007, the Library had collected more than 70 terabytes of data. Google Earth (Chang et al. 2006, Table 2) uses a similar amount of storage for its raw image files. The Internet has made it so easy to access such information that it gains a status similar to that of files stored on a (very slow) local hard drive.

The concept of remote resources distributed among sites and organizations leads to a technology called *grid computing*. Berman (2007) compares a computational grid with the more familiar electrical power grid. Just as we connect an appliance to the power grid without regard for the actual source of power, we might similarly connect a computing device to a grid that supplies information and computing power. However, these two grids do have important differences. A power grid delivers a single, easily quantified commodity. A computing grid would provide a wider spectrum of products and services with seamless integration, yet without a simple measure of performance. Unlike a supercomputer with a large number of parallel processors in direct communication, a grid uses separate computers running in parallel and communicating across a network (like the Internet). For more information, see http://en.wikipedia.org/wiki/Grid_computing.

California's San Diego Supercomputer Center (SDSC; http://www.sdsc.edu/) illustrates the potential for high-performance computing on the scale of teraflops and petabytes. Funded by the US National Science Foundation, SDSC has a broad mission of using technology to advance science by providing very large-scale computing resources to US scientists. For example, they have facilitated the development of 'TeraShake', a simulation model of earthquakes in southern California that generates some dramatic visual images (http://visservices.sdsc.edu/projects/scec/terashake/).

Similarly, the College of Engineering at the University of California Santa Barbara (UCSB) uses high-performance computers for computational ecology (Helly et al. 1995; http://www.engineering.ucsb.edu/articles/what_is_computational_ecology). In a video about this program, Prof. John Gilbert explains the importance of a common programming language. Ecologists and other applied scientists need to communicate with specialists in parallel programming and high-performance computing. They can have meaningful discussions 'from day one' because the ecologist writes his code in a high-level language already available for desktop computers. A program called *Star-P* made by Interactive Supercomputing (http://www.interactivesupercomputing. com/) then implements this code on a supercomputer.

8.5   Fishery Science in the Future

In the sections above, we've outlined recent global changes in climate, data, and technology. Circumstances have evolved to produce many new degrees of freedom: bigger problems and bigger tools for resolving them. The 'future', as we might have imagined it fairly recently, is already here. To us, some of the developments described above still seem more like science fiction than accomplished science. This admission may shed doubt on our skills at prognostication, but we're skeptical that anyone can make reliable predictions about such large-scale phenomena over the next few decades. We invite our readers to join us in speculating about directions for the future.

To a considerable extent, human behavior is governed by practical feasibility. Science fiction movies often assume a 'beaming' technology that enables people to be transported instantly from place to place. Given this astonishing (but now almost mundane) capability, plot lines often involve characters beamed out of desperate situations at the last possible moment. Or else, the desperate moment arrives, and the beamer has suddenly gone on the fritz. Unfortunately, however handy it might seem, a real world ecologist doesn't have the prerogative of requesting 'Beam me back to base camp.' Similarly, in the fifteenth century, when Columbus arrived in the Bahamas, he couldn't send a telegraph to Queen Isabella in Spain. Electricity hadn't been discovered yet, and Cryus Field didn't lay a cable across the Atlantic until the nineteenth century. If an ecologist in 1978 (30 years ago) wanted an ecosystem model with several million spatial cells, he couldn't build it because computers lacked the requisite bytes and flops.

In our current world's technology, many things once impossible have now become possible. Simulation models can run at highly detailed scales of space and time, linked to massive databases. This new potential stimulates a great deal of scientific activity. A naïve extrapolation into the future suggests that elaborate simulations with powerful visualizations will become more common in the future. Over the last 3 decades, we've watched expensive, cumbersome mainframe computers and line printers evolve into cheap, powerful desktop computers with high-resolution graphics and printers that can render family photographs. In the next 30 years, will we find petaflop and petabyte supercomputers on every desktop, perhaps with scientific graphics replaced by three-dimensional movies that purport to represent reality? Perhaps computer-generated movie technologies, now used by directors like Stephen Spielberg, will become standard home equipment. In another scenario, personal computers might become simple boxes that connect to a grid of information and computation servers, like terminals on an old mainframe. Perhaps it will be cheaper for the planet to have a few massive systems that serve computation and information on demand, like ice cream from an Italian gelateria.

Imagine a scenario in which a resource manager says, 'I need to understand that ecosystem.' Is this an achievable goal, or is it as unrealistic as 'Beam me back to base camp.'? Do we have a tool for deciphering ecosystems? Clearly, we have many potential components, but are they enough? What are realistic limits? More to the point, what do we mean by the words 'understand', 'ecosystem', and 'I'?

The words 'I' and 'understand' play key roles here, because they remind us that the intended audience is a resource manager, a human with an amazing mental ability to form concepts (Hofstadter 2007). Berman (2006) estimates that a human brain at the micron level is equivalent to about 1 petabyte, although the world accumulates about 2 exabytes of information annually. Our brains have major structures linked to our eyes and ears, so we find visual images and audio content particularly engaging. At a practical level of understanding, we need to assimilate information this way. Whatever else may be going on, our approach has elements of *artificial intelligence* (AI) in which information gets condensed into understandable concepts.

As Hofstadter (2007, p. 205) points out, 'sometimes being mired down in gobs of detailed knowledge is the exact thing that blocks deep understanding.' We live in an era of information overload. How do we sift through massive quantities of data to find important nuggets of insight? Fishery and ecosystem models of the past used a much thinner veneer of information, with many fewer dimensions than the data available now. This made the (apparent) information content easier to grasp, and simple biologically-motivated models provided a reasonable basis for analysis. With higher dimensional data, an appropriate framework for analysis becomes less obvious, and generic techniques of pattern recognition (Ripley 1996) play a larger role. Our earlier comment on genomic heat maps illustrates this point.

On October 8, 2005, a robot vehicle named Stanley drove itself without human intervention across 132 miles of desert from Barstow, California, to Primm, Nevada (Orenstein 2005; Thrun et al. 2006). For 'his' makers at the Stanford Artificial Intelligence Laboratory (SAIL), Stanley won a $2 million dollar prize. Commenting on this achievement, Hofstadter (2007, p. 80) points out that:

'Aboard any such vehicle are one or more television cameras (and laser rangefinders and other kinds of sensors) equipped with extra processors that allow the vehicle to make sense of its environment. No amount of simplistic analysis of just the colors or the raw shapes on the screen is going to provide good advice as to how to get around obstacles without toppling or getting stuck. Such a system, in order to drive itself successfully, has to have a nontrivial storehouse of prepackaged knowledge structures that can be selectively triggered by the scene outside. Thus, some knowledge of such abstractions as "road", "hill", "gulley", "mud" "rock", "tree", "sand", and many others will be needed if the vehicle is going to avoid getting stuck in mud, trapped in a gulley, or wedged between two boulders. The television cameras and the rangefinders (etc.) provide only the simplest *initial* stages of the vehicle's "perceptual process", and the triggering of various knowledge structures of the sort that were just mentioned corresponds to the far end, the *symbolic* end, of the process.'

It seems almost as if Stanley could 'think', but Hofstadter (op. cit., p. 190) disputes that claim. For example, Stanley had no personal satisfaction in passing H1, a rival from Carnegie-Mellon University.

Compared to Stanley crossing the desert, ecosystem management during climate change poses much greater challenges. The underlying concepts, analogs of 'hill', 'gulley', etc., remain partially undefined, as do the goals. What is the 'Primm, Nevada' of ecosystem management? We're charting a course through unknown territory with limited guidelines. Simulation models can provide a solid means for articulating these issues, but even if they run on supercomputers with teraflops and petabytes, they still rely ultimately on mathematical assumptions that definitely fail to capture a much more complex reality, with still more hidden dimensions.

As the computational environment gets more elaborate, the underlying assumptions get progressively hidden within user interfaces and visual output. When users explore the complex world of the simulation itself, they tend to lose sight of its hypothetical character. Simulations can become more like a video game than a scientific exercise. It sometimes happens that an amazing feature of the model turns out to be a bug in the code. When simulations on high-performance systems start to produce output that rivals modern movies, they potentially become even more deceptive by portraying a convincing world that misrepresents reality. Like the mythical sculptor Pygmalion, who fell in love with his own ivory statue, the ecosystem modeler can become enamored with a beautiful simulation that required a great deal of work (Schnute and Richards 2001). Perhaps this effect is enhanced when the program runs on a very expensive system operated by many talented people.

Despite our words of caution, we certainly acknowledge the value of computer models for managing fisheries and ecosystems. One big advantage of supercomputers is that they allow us to conduct rapid tests of complicated scenarios, so that we can explore the possibilities much more deeply than in the past. Fishery science has formalized this concept in a technique called 'Management Strategy Evaluation' (MSE) that tests potential courses of action against a broad range of possible versions of reality (Schnute et al. 2007). After a computer-intensive testing process, stakeholders try to choose actions that produce favorable outcomes robust to various rules of ecosystem behavior.

People particularly want models to make predictions. What will happen if we follow plan A or plan B? Which plan is better? Not surprisingly, scientists tend to shy away from forecasting future events. It's easy to be wrong, and scientific arguments usually demonstrate that the current data are inadequate. However, prediction offers a constructive challenge to understanding. When a prediction fails, we have an explicit context in which to evaluate our current knowledge base. Refinements force us to identify issues that really matter for practical purposes.

The target 'audience' for Stanley's sensors is a navigation program that operates automatically. The target audience for ecosystem sensors and the resulting concepts is potentially the global human population, or at least communities that share environmental resources. Unlike Stanley, people don't automatically follow an action plan, even if it's a pretty good one. Usually they disagree among themselves about the virtues or failures of any proposed plan. We look to a future with serious environmental issues and advanced technology for addressing them. Scientific progress requires a deeper understanding of high-dimensional ecosystems, with communication that stimulates human populations to take appropriate actions. Nothing in evolutionary history quite matches this collective challenge for humanity.

References

Berman F (2006) One hundred years of data. Presentation at the 2006 International Conference on Digital Government Research, May 2006, San Diego, CA. Available at http://www.sdsc.edu/about/director/talks.html

Berman F (2007) Grids in context. Presentation to the Open Science Grid (OSG) All Hands Meeting, March 2007, San Diego, CA. Available at http://www.sdsc.edu/about/director/talks.html

Block BA, Teo SLH, Walli A, Boustany A, Stokesbury MJW, Farwell CJ, Weng KC, Dewar H, Williams TD (2005) Electronic tagging and population structure of Atlantic bluefin tuna. Nature 434: 1121–1127.

Buckworth R (1998) World fisheries are in crisis? We must respond! In: Reinventing fisheries management, edited by Pitcher. Kluwer, Dordrecht, The Netherlands, pp. 3–17.

Carlton JT (1996) Marine bioinvasions: the alteration of marine ecosystems by nonindigenous species. Oceanography 9: 36–43.

Chang F, Dean J, Ghemawat S, Hsieh WC, Wallach DA, Burrows M, Chandra T, Fikes A, Gruber RE (2006) Bigtable: a distributed storage system for structured data. Available as a PDF file at http://labs.google.com/papers/bigtable.html

Choi JS, Frank KT, Petrie BD, Leggett WC (2005) Integrated assessment of a large marine ecosystem: a case study of the devolution of the eastern scotian shelf, Canada. Oceanography and Marine Biology Annual Review 43: 47–67.

English KK, Koski WR, Sliwinski C, Blakley A, Cass A, Woodey JC (2005) Migration timing and river survival of late-run Fraser River sockeye salmon estimated using radiotelemetry techniques. Transactions of the American Fisheries Society 134: 1342–1365.

Feely RA, Sabine CL, Lee K, Berelson W, Kleypas J, Fabry V, Millero FJ (2004) Impact of anthropogenic CO_2 on the $CaCO_3$ system in the oceans. Science 305: 362–366.

Freeland HJ, Cummins PF (2005) Argo: a new tool for environmental monitoring and assessment of the world's oceans, an example from the NE Pacific. Progress in Oceanography 64: 31–44.

Gordon JS (2002) A thread across the ocean: the heroic story of the transatlantic cable. Walker, New York, 272 p.

Helly J, Case T, Davis F, Levin S, Michener W (1995) The State of Computational Ecology. National Center for Ecological Analysis and Synthesis, Santa Barbara, CA. Research Paper No. 1. Available at http://www.nceas.ucsb.edu/nceas-web/projects/2295/nceas-paper1

Hofstadter D (2007) I am a strange loop. Basic Books, New York, 412 p.

IPCC (2007a) Climate Change 2007: The Physical Science Basis. Contribution of Working Group I to the Fourth Assessment Report of the Intergovernmental Panel on Climate

Change, edited by Solomon S, Qin D, Manning M, Chen Z, Marquis M, Averyt KB, Tignor M, Miller HL. Cambridge University Press, Cambridge, 996 p. Available at http://www.ipcc.ch/ipccreports/ar4-wg1.htm

IPCC (2007b) Summary for Policymakers. In: Climate change 2007: The Physical Science Basis. Contribution of Working Group I to the Fourth Assessment Report of the Intergovernmental Panel on Climate Change, edited by Solomon S, Qin D, Manning M, Chen Z, Marquis M, Averyt KB, Tignor M, Miller HL. Cambridge University Press, Cambridge/New York, 18 p. Available at http://www.ipcc.ch/pdf/assessment-report/ar4/wg1/ar4-wg1-spm.pdf

Miller KM, Kaukinen K, Li S, Farrell AP, Patterson DA (2007) Expression profiling of Fraser River late-run sockeye salmon: migration physiology uncovered using cDNA microarray technology. American Fisheries Society Symposium 54: 101–104.

Morrison J, Quick MC, Foreman MGG (2002) Climate change in the Fraser River watershed: flow and temperature projections. Journal of Hydrology 263: 230–244.

Orenstein D (2005) Stanford team's win in robot car race nets $2 million prize. Stanford Report, October 11, 2005. Available at http://news-service.stanford.edu/news/2005/october12/stanleyfinish-100905.html

Pauly D, Christensen V, Dalsgaard J, Froese R, Torres Jr F (1998) Fishing down marine food webs. Science 279: 860–863.

Polovina JJ, Howell EA (2005) Ecosystem indicators derived from satellite remotely sensed oceanographic data for the North Pacific. ICES Journal of Marine Science 62: 319–327.

Randall L (2005) Warped passages: unraveling the mysteries of the universe's hidden dimensions. HarperCollins, New York, 500 p.

Ricker WE (1975) Computation and interpretation of biological statistics of fish populations. Bulletin of the Fisheries Research Board of Canada, No. 191. 382 p.

Ripley BD (1996) Pattern recognition and neural networks. Cambridge University Press, Cambridge, 403 p.

Schnute JT (2006) Curiosity, recruitment, and chaos: a tribute to Bill Ricker's inquiring mind. Environmental Biology of Fishes 75: 95–110.

Schnute JT, Richards LJ (2001) Use and abuse of fishery models. Canadian Journal of Fisheries and Aquatic Sciences 58: 10–17.

Schnute JT, Cass A, Richards LJ (2000) A Bayesian decision analysis to set escapement goals for Fraser River sockeye salmon (*Oncorhynchus nerka*). Canadian Journal of Fisheries and Aquatic Sciences 57: 962–979.

Schnute JT, Maunder MN, Ianelli JN (2007) Designing tools to evaluate fishery management strategies: can the scientific community deliver? ICES Journal of Marine Science 64: 1077–1084.

Sobel D (1995) Longitude: the true story of a lone genius who solved the greatest scientific problem of his time. Walker, New York, 224 p.

Thrun S, Montemerlo M, Dahlkamp H, Staven D, Aron A, Diebel J, Fong P, Gale J, Halpenny M, Hoffmann G, Lau K, Oakley C, Palatucci M, Pratt V, Stang P, Strohband S, Dupont C, Jendrossek LE, Koelen C, Markey C, Rummel C, van Niekerk J, Jensen E, Alessandrini P, Bradski G, Davies B, Ettinger S, Kaehler A, Nefian A, Mahoney P (2006) Stanley, the robot that won the DARPA grand challenge. Journal of Field Robotics 23: 661–692. Available at http://robots.stanford.edu/papers/thrun.stanley05.html

von Schalburg KR, Rise ML, Cooper GA, Brown GD, Gibbs AR, Nelson CC, Davidson WS, Koop BF (2005) Fish and chips: Various methodologies demonstrate utility of a 16,006-gene salmonid microarray. BMC Genomics 6: 126. Available at http://www.biomedcentral.com/1471-2164/6/126

Walker RV, Sviridov VV, Urawa S, Azumaya T (2007) Spatio-temporal variation in vertical distributions of Pacific salmon in the ocean. North Pacific Anadromous Fish Commission Bulletin 4: 193–201. Available at http://www.npafc.org/new/pub_bulletin4.html

Xian W, Kang B, Liu R (2005) Jellyfish blooms in the Yangtze Estuary. Science 307: 41.

Zainuddin M, Kiyofuji H, Saitoh K, Saitoh S (2006) Using multi-sensor satellite remote sensing and catch data to detect ocean hot spots for albacore (*Thunnus alalunga*) in the northwestern North Pacific. Deep Sea Research Part II 53: 419–431.

Chapter 9
Stock Assessment: Operational Models in Support of Fisheries Management

Richard D. Methot Jr.

Abstract Fishery stock assessment models connect ecosystem data to quantitative fishery management. Control rules that calculate annual catch limits and targets from stock assessment results are a common component of US Fishery Management Plans. Ideally, the outcome of such control rules are updated annually on the basis of stock assessment forecasts to track fluctuations in stock abundance. When the stock assessment – fishery management enterprise achieves this level of throughput, they truly are operational models, much as the complex physical models used to routinely update climate forecasts. In reality, many contemporary assessments are closer to an individual scientific investigation than to an operational model. As a result, the review of each stock assessment is extensive and the lag between data acquisition and quota adjustment may extend to several years. If the future stock assessment process is to move towards an operational status, there will need to be changes in three aspects of the process. First, key data streams will themselves need to be made more operational and corporate so that relevant data are immediately available and trusted. Second, stock assessment models need to be made more capable of including diverse relevant data and comprehensively calculating levels of uncertainty, while also being more completely tested, documented, and standardized. The class of models called integrated analysis has these characteristics and is described here, with emphasis on the features of the Stock Synthesis model. Areas of future model development, especially to include more ecosystem and environmental factors, are explored. Third, increased throughput of assessment updates will require streamlining of the extensive review process now routinely required before stock assessment results can serve as the scientific basis for fishery management. Emphasizing review of broadly applicable assessment data and methods, rather than each final result, is a logical step in this streamlining, while maintaining public trust in the final results.

Keywords Fish stock assessment · population dynamics · fishery management

R.D. Methot Jr.
NOAA Fisheries Service, Seattle, WA 98112, USA

R.J. Beamish and B.J. Rothschild (eds.), *The Future of Fisheries Science in North America*,   137
Fish & Fisheries Series, © Springer Science + Business Media B.V. 2009

9.1   Introduction

Modern fisheries stock assessment models (Quinn 2003) are the nexus between our growing scientific capability to understand factors affecting the population dynamics of harvested fish stocks, and the expanding demands for quantitative management of their fisheries. These models assimilate a diverse collection of data, produce forecasts linked to historical estimates, and provide a framework for comprehensive evaluation of model uncertainty and risk assessment for proposed management actions (Maunder et al. 2009; Schnute et al. 2007). Assessment models are increasingly able to incorporate spatial structure (Punt et al. 2000) and the influence of environmental and ecosystem factors (Maunder and Watters 2003). Multispecies stock assessment models are beginning to appear. This paper describes the role of fish stock assessment models in providing ongoing quantitative advice for fishery management and provides an overview of the rapidly evolving capabilities of a class of assessment models termed statistical catch-at-age analysis or integrated analysis, with particular emphasis on the Stock Synthesis model (Methot 1989, 2000). Areas of future model development, including increased linkage to environmental and ecosystem data, will be explored.

Quantitative management of US marine fisheries received a large impetus with the 1976 Magnuson Fishery Conservation and Management Act. The 1996 reauthorization of the Act required fishery management plans to prevent overfishing, rebuild previously overfished stocks, and obtain optimum yield from the fishery (NOAA 1996). Quantitative criteria related to the reproductive potential of the stock were required to: gauge the occurrence of overfishing, determine whether a stock was overfished (depleted) and in need of rebuilding, forecast potential rates of rebuilding for previously overfished stocks, and guide management towards a level of catch that will produce optimum yield but is no greater than the level that would produce maximum sustainable yield. NOAA responded with an update to guidelines for implementation of the Act and subsequent technical guidance (Restrepo et al. 1998). The 2006 reauthorization of the MSA upped the ante further by requiring establishment of annual catch limits in each fishery such that overfishing does not occur, and that these annual catch limits be based upon the scientific recommendations of the Fishery Management Council's Scientific and Statistical Committees or an established peer review process. In addition, as more fisheries are managed with individual fishery quotas, the demand for more precision in total quota determinations will only go up.

The need for expansion and improvement of the stock assessment enterprise (NRC 1998) led to development of the Marine Fisheries Stock Assessment Improvement Plan (Mace et al. 2001) by NOAA's National Marine Fisheries Service. The plan identified three tiers of improvement: (1) mining existing data to extend basic assessments to as many stocks as possible; (2) expanding data collection and assessments to provide adequate assessments for major fish stocks and at least baseline monitoring of minor stocks; and (3) reaching to an ecosystem level of assessment for representative stocks in each region. In addition to direct investment in data collection and assessments, NMFS initiated programs such as the Sea Grant Fellowship in Population Dynamics to train new assessment scientists, the Stock Assessment Toolbox to provide a standardized interface for many assessment models, and the Center for Independent Experts (Brown et al. 2006) to increase the rigor of assessment reviews.

The fact that fish are components of marine ecosystems and are influenced by ecosystem and environmental factors is not news (Hjort 1914). Yet, over the mid–late twentieth century, the fisheries assessment community evolved methods that analyzed data often solely collected from the fishery itself and which incorporated simplifying assumptions that left little room for direct incorporation of environmental and ecosystem factors. The focus on empirical description of the state of the stock was a logical outcome of the need to provide quantitative criteria to enable science-based fishery management decisions. Although such models have provided useful short-term guidance regarding adjustments in fishery regulations and the sustainability of a general level of fishery catch, their black-box nature made them poor candidates to serve as direct tools to understand and investigate the non-fishery factors that also influence the abundance of fish stocks. In parallel, fisheries science continued a strong emphasis on studying factors that affect the growth, mortality, and reproduction of fish stocks in an ecosystem context, but opportunities to directly incorporate this growing body of knowledge lagged.

Fish assessment models do not ignore the fact that fish stocks respond to environmental and ecosystem conditions; they just treat it as a reaction to be measured but not predicted. For example, empirical measurements of annual body weight-at-age (i.e., growth) are directly incorporated as detailed data into many age-structured assessment models. Fishery and survey age composition data are used by the models to estimate the annual level of recruitment of young fish into the population. But these methods for dealing with environment-caused variations in growth and recruitment are entirely empirical. The results, particularly for recruitment time series, provide input for subsequent investigation of possible environmental causes of the fluctuations, but it is only recently that stock assessment models have begun to include the environmental information directly as an additional source of information about the fluctuations (Maunder and Watters 2003; Schirripa and Colbert 2006). Some ways in which environmental and ecosystem information can improve fish stock assessments will be explored later in this paper.

9.2   Stock Assessment Overview

A stock assessment is the collection, analysis and reporting of demographic information to determine the effects of fishing on fish stocks (Mace et al. 2001). There are three basic categories of information that must be provided in order to produce accurate assessments: total catch, abundance trend, and life history characteristics. Deficiencies in one cannot be overcome by excessive data in another.

First, there must be accounting of the total catch. Historically, assessments would use the landed commercial catch. For fisheries of interest, this was usually the dominant component of the total catch and was the component that was most completely available for the entire time series. While simple assessment models require only catch biomass as a sufficient description of the fishery impact, more detailed models require that the catch be broken down into catch numbers-at-age to more precisely assign age-specific mortality and to calculate the time series of annual recruitments that must have occurred in order to have supported this catch (Pope 1972). As fisheries have evolved and scrutiny of their total impact has increased, so has the requirement for complete catch accounting. Today's assessments use total catch by fleet, including commercial and recreational landed and discarded catch in target fisheries and bycatch in other fisheries. Further, studies of discard mortality are used to calculate the total mortal catch. Biological samples characterize the age/size/gender of the catch. Fleets are separated to provide the ability to calculate the differential demographic impact of fleets that principally harvest younger/smaller versus older/larger fish. Where this demographic sampling level is high, the resultant time series of fishery catch-at-age is an influential source of information on fishery removal patterns and the time series of recruitment to the fished stock.

Second, there must be some measure of abundance. Assessments that lack a measure of the level, or at least trend, in stock abundance will not be able to achieve confident results (NRC 1998). Ideally, there will be a time series of survey observations, each calibrated to provide an absolute estimate of stock abundance. Absolute calibration is difficult to attain and the typical goal is to have a time series of observations that track the relative trend in stock abundance. From a statistical perspective, this relative abundance trend is best obtained from a fishery-independent survey so that the sampling protocols can be highly standardized and applied over the range of the stock in a statistically-based sampling strategy. Some surveys have been conducted by a single vessel, or its directly calibrated replacement. Others have relied upon use of multiple chartered vessels and have absorbed the added variability of between vessel variability into the total variability of the survey results (Helser et al. 2004). In some cases, basic fishing technology such as longlines or bottom trawls have been adapted and standardized for use as a survey sampling tool. In other cases, specialized technologies have been developed such as hydroacoustics deployed from Fisheries Survey Vessels, egg and larval surveys, or underwater imaging systems deployed from Remotely Operated Vessels or even Autonomous Underwater Vehicles (ref this book). As with the fishery catch, there should be sufficient demographic sampling of the survey catch to describe the life history segment of the total population that is being monitored and to provide more detailed information on the trends in abundance by age and size. Fishery-independent resource surveys conducted from larger, multiple capability vessels are also a valuable platform on which to piggyback ecosystem observations.

Standardized, fishery-independent surveys are not available for many stocks and the fallback is to use a proxy measure of average fish density per unit area calculated by statistical processing of fishery catch rate (CPUE) data obtained from logbooks or observers (Maunder and Punt 2004). While such CPUE data can appear highly precise due to the thousands of fishery logbook observations included in some analyses, the shortcoming is the inability to confidently assert that the units of fishing effort can be standardized over each year of the time series to the degree that fishery-independent survey methods are standardized. Increasingly, assessment methods are able to relax the assumption of constant catchability to make best advantage of fishery CPUE data (Bence and Wilberg 2006). The topic of time-varying catchability will be explored further in the model section.

Third, there should be sufficient information regarding the stock's life history characteristics. Although biomass-dynamics models operate with just a time series of catch and catch per unit effort, basic age-structured models require the capability to determine fish age from some biological structure such as otoliths, fin rays, or scales; and a measure of body weight-at-age and natural mortality (usually assumed constant across the age range available to the fishery). Because the goal is to analyze the impact of the fishery on the reproductive potential of the fish stock, a normal additional requirement is information on percentage mature at age. Spawning biomass so calculated is still a crude measure of reproductive potential and a more accurate measure will take into account fecundity-at-age and even larval quality if it differs by age (Bobko and Berkeley 2004). Some species such as hermaphrodites and nest-breeders require information on the male's contribution to reproductive potential. A major challenge is maintaining sufficient sampling over time to track changes in growth and maturity. This is especially important if these changes are density-dependent or have long-term trends due to environmental or other factors.

One highly influential factor, natural mortality, is more ecological than biological. Biological measurements of individuals may indicate a fish's relative predisposition to predation, parasitism, disease, and other causes of natural mortality or may provide measurements of growth and reproductive factors that appear correlated with natural mortality. However, natural mortality itself is the average probability of death from non-fishery causes, so is not directly observable from individual fish. Most direct estimates of natural mortality have been obtained by sampling the age composition from pre-fishery periods or lightly fished components of the stock, but such estimates do not directly measure the natural mortality occurring in the current, fished component of the stock.

From the catch, relative abundance, and life history information, assessment models can infer the abundance of the population that must have existed in order to exhibit the observed trend in the abundance indicator while producing the observed level of catch. An adequate assessment should provide an estimate of the time series of stock abundance and fishing mortality and an analysis to determine sustainable levels of fishing mortality and the resultant expected level of catch and stock abundance. The accuracy and precision of the results depends on the quality and quantity of the data, and also on the characteristics of the stock's history. If there is little contrast in the time series of catch and relative abundance, then a wide range of combinations of average fishing mortality and average stock abundance may be consistent with the available data. But if the stock has been monitored through at least one major cycle of lowered abundance and subsequent rebuilding, then more precise estimates of stock abundance and productivity can be obtained from stock trend and absolute catch data. There are two corollaries to this situation: the maximum productivity of a newly fished stock cannot be well determined until it has been fished at a moderate level for a sufficiently long time, nor can the rebuilding target of an overfished stock be well estimated if significant monitoring does not begin until after the stock has already been depleted to a low level of abundance. However, if fishery-independent surveys can employ technologies that are directly calibrated to provide measures of absolute stock abundance, rather than a relative trend, then just a few years of surveys provides immediate stock assessment information regardless of the level of fishery exploitation.

9.3   Scientific Advice for Fishery Management

The results of stock assessments serve as the basis for long-term and short-term fishery management decisions. First, the assessment provides the basis for status determinations. These status determination criteria are specified in regional fishery management plans guided by the National Standard 1 Guidelines of the Magnuson-Stevens Sustainable Fisheries Act and technical guidance. Loosely they entail: (1) determining whether overfishing is occurring by comparing the current level of fishing mortality to a limit level that is based upon the level that would produce maximum sustainable yield; and (2) comparison of current reproductive potential (usually measured just as spawning biomass) to a limit level (usually set to approximately half the level that would produce maximum sustainable yield) as a measure of stock depletion and a trigger for development of a rebuilding plan. Second, assessments provide forecasts of the expected future catch and stock abundance associated with proposed harvest policies. Thus they provide the basis for calculation of the expected time period for rebuilding of previously overfished stocks and for implementation of the harvest policy that will produce optimum yield from the fishery. Finally, the time series of abundance, mortality, and productivity produced by single-species stock assessments provide input to ecosystem food web models. Indeed, the multi-decadal stock assessment results are among the most quantitative and well-documented results available for such ecosystem models.

A single stock assessment can provide sufficient information to serve as the basis for a one-time status determination and for setting fishery management targets. However, stock assessments are also expected to serve as a core component of an ongoing fishery management system. Status determinations need to be updated, rebuilding of overfished stocks needs to be tracked, and catch levels need to be adjusted to maintain fishing mortality targets. Achievement of these additional goals means that assessments must be updated frequently to track changes in stock conditions due to natural and fishery factors. In effect, they become part of the operational model used to provide fishery management advice. Harvest control rules serve to translate stock assessment forecasts into target and limit levels of fishery catch (Fig. 9.1). The term, operational model, distinguishes such assess-

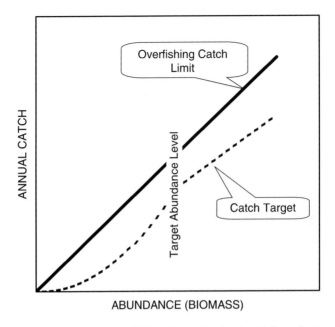

Fig. 9.1 Harvest control rules calculate limit and target levels of catch from short-term forecasts of stock abundance. In this hypothetical example, the target is a smaller fraction of the limit at lower levels of stock abundance in order to guard against further depletion

ments from one-time, stand-alone scientific investigations. An operational model provides timely updates to inform a set of clients of rapidly changing conditions. For example, the frequently updated forecast of the path and intensity of a hurricane is calculated using a complex model of the system calibrated with data collected on the time scale of hours. The results, including probability distributions, are rapidly produced and disseminated to the public in an easily understood graphical format. The fishery assessment operational model is identical in concept. The major differences being that the fishery biological model is less well-understood than the physical model used for hurricane forecasting; the time scale is months–years, not hours–days; and the output of the fishery model feeds directly into a regulatory framework rather than a public information framework.

The requirement for fishery assessments to serve as part of an operational model for management of US marine fisheries has increased with passage of the Magnuson-Stevens Fishery Conservation and Management Reauthorization Act of 2006. By 2010, Fishery Management Plans must specify annual catch limits for each fishery, based on scientific advice and at a level such that overfishing does not occur. Clearly, quantitative stock assessments are key to implementation of these requirements. Stock assessments can provide estimates of the level of catch that would be considered overfishing, and can provide a probability distribution for the chance of overfishing relative to a range of possible annual catch limit levels (Prager et al. 2003).

It is feasible for a simple model that only tracks recent trends in stock abundance to serve as the basis for adjustment of fishery target levels of catch. However, such a simple model is not well suited to assure that the target level is itself correct, nor is it well suited to integrating multiple current data sources into a forecast of future stock conditions. Further, when affected constituents see a simple model's inability to track some observed trends in the data, the importance of environmental and ecosystem factors are usually invoked, and the demand for more complex analysis begins. Although a model should not be more complex than necessary to assimilate the available data, the best solution is not necessarily to start with a simple model and to move to a more complex model as data allow. Instead, a smoothly scalable approach is to build a fully detailed model and to collapse its details down to the level that is estimable with the available data. Further, the complex model provides the framework for more comprehensively calculating the uncertainty in model results due to factors for which there are insufficient data.

9.4 Integrated Analysis Assessment Models

A class of models that has evolved over the past 25 years to meet the growing assessment challenge is termed integrated analysis or statistical catch-at-age analysis. A recent review can be found in Schnute et al. (2007). Such models were first developed in the 1980s (Fournier and Archibald 1982; Methot 1989) and began to see widespread use and rapid evolution by the late 1990s (McAllister et al. 1994; Quinn 2003). Integrated analysis models work as a simulation of the underlying population dynamics calibrated with the available data. They tend to cast the goodness-of-fit to the model in terms of data elements that retain the statistical characteristics of the raw data. This distinguishes integrated analysis from models such as Virtual Population Analysis that work more as a transformation of a particular type of preprocessed data, in this case fishery catch-at-age. Most integrated analysis models have an age-structured population dynamics sub-model, but the class itself is more general and includes strictly length-structured population models (Chen et al., 2005). The current generation of integrated analysis models has broad and flexible capabilities: age and/or length structured; spatial structure; environmental inputs; and other features that have them evolving towards multispecies models (Livingston and Methot 1998; Sitar et al. 1999). The general characteristics of integrated analysis models are described here, with particular emphasis on the features incorporated in the Stock Synthesis model.[1]

Integrated analysis models incorporate a linked set of sub-models (Fig. 9.2). The core sub-model contains the population dynamics. This is where the processes of birth, death, and growth create the time series of estimated population abundance

---

[1]NOAA Fisheries Toolbox Version 2.10, 2006. Stock Synthesis 2 Program, Version 2.00c. [Internet address: http://nft.nefsc.noaa.gov].

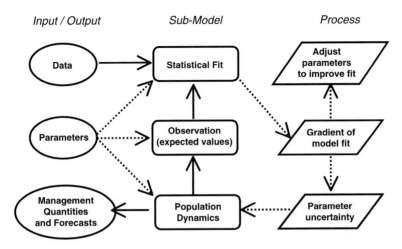

Input / Output   Sub-Model   Process

Fig. 9.2 Integrated analysis models consist of a linked set of sub-models: population dynamics, observation, and statistical. The statistical model compares expected values from the observation sub-model to the data and calculates the gradient of the goodness-of-fit with regard to the parameters in order to iteratively adjust the parameter values to achieve the best fit

and mortality. Some of these processes are represented by fixed input quantities and others are calculated from parameters being estimated by the model. For example, annual recruitment of young fish into the modeled population and annual fishing mortality by each fleet are usually calculated from model parameters. On the other hand, natural mortality is usually based on a fixed input value because data available to include in the model are not informative about the exact levels. Growth (e.g., body weight-at-age) is fixed in some models and estimated within others. As integrated analysis models evolve, there has been a move towards estimation of more parameters within the models and to utilize Bayesian methods to provide additional information in the form of a prior probability distribution for the value of the parameters. The estimation of more parameters has also evolved towards development of approaches that allow some parameters to have values that vary over time either as freely fluctuating quantities or as linkages to additional model inputs such as ecosystem or environmental factors.

Next is the observation sub-model where the processes of catchability, selectivity, aging imprecision, and other factors are modeled to create expected values for the types of available data. Like the population dynamics sub-model, the observation sub-model represents some processes with fixed inputs and others as relationships that incorporate estimated model parameters. This observation sub-model is where integrated analysis obtains much of its strength. Rather than preprocess and adjust the data so that it is in terms of the underlying population dynamics, integrated analysis models build knowledge of the observation process into the creation of expected values for the data. For example, the virtual population analysis model assumes that fish ages are determined without any error, so if the otolith reading process is known to have some variability between readers, the inverse of this reader variance should be applied to the age data before feeding the adjusted data into the VPA assessment model. When doing so, the variance associated with reader imprecision is not carried through to final management quantities. Besides, it is difficult to sharpen data that have already been blurred by the observation process. In integrated analysis, the opposite approach is taken. Information about reader imprecision is used to blur the expected value of model estimates of age composition so the model's estimates are blurred to the same degree that it is believed that the data have been blurred. Because the blurring process is built into the model, its effect on variance of model outputs is fully incorporated.

Third is the statistical sub-model where the goodness-of-fit to the data is calculated in terms of the negative of the logarithm of the probability of the observations, termed the negative log-likelihood or NLL for short. The NLL for each diverse type of data is basically calculated by scaling each observation's deviation from the predicted value according to the statistical form and magnitude of the error distribution for that data source.

This NLL basis means that the degree to which the model doesn't exactly fit each type of data is scaled in comparable terms. So, even though the model may contain tens of NLL components (age composition from fishery A, length composition from fishery B, % discard from fishery A, abundance from survey C, catch per unit effort from fishery B, recruitment index from survey D, etc.), the NLL components can be added together into a total NLL that is a meaningful measure of the total goodness-of-fit to all the data. Key to a successful model is inclusion of all relevant processes that have contributed to the observed data so that all the deviations are due to measurement error, and using the correct level of variance for these measurement errors. Of course, determining when the model is at the "sweet-spot" of complexity such that hidden processes are not misinterpreted as high measurement error is part of the art of model building.

Many integrated analysis models are written in the C++ computer language using ADMB, which was developed in the private sector by Otter Research (http://otter-rsch.com/) to facilitate the development of complex models. It employs automatic differentiation so that the gradient of the NLL with respect to each parameter's value can be calculated analytically, thus greatly speeding the iterative search for the set of parameters that maximizes the goodness-of-fit. When the gradient is large, this means that the parameter is influential and is relatively far from the value that maximizes the NLL. As the model searches for the set of parameter values that maximizes the NLL, it is searching for values at which the gradient goes to zero, hence where no further improvements can be made (Fig. 9.2). At that point, the model also calculates the curvature of the NLL surface with respect to the parameters. Where the curvature is strong, this means that small movements of the parameters away from the best fitting point will have large degradation in the NLL, thus meaning that the best value of the parameter is precisely determined. Where the curvature is weak, this means that the parameter's value has little effect on the NLL, which means that the data included in the model do not have much information about the best value of that particular parameter. It is not uncommon for the NLL surface to form a ridge with respect to a pair of parameters.

The strength of this ridge represents the correlation between these two parameters. The model may have tens to hundreds of parameters being estimated, so the multidimensional shape of the NLL surface is complex indeed. Because the curvature of the NLL surface is not necessarily parabolic and symmetric, as assumed by the normal distribution theory for obtaining confidence intervals, integrated analysis methods also use nonparametric approaches to calculate the shape of the NLL surface. In the Monte Carlo Markov Chain approach, after finding the best-fitting set of parameter values, the model then semi-randomly moves around the parameter space, each time calculating the NLL and building up an empirical representation of the multidimensional NLL surface.

The present state-of-the-art for assessment modeling routinely incorporates two forms of uncertainty in forecasts of stock abundance and potential fishery yield: (1) uncertainty in model parameters and current stock conditions based on goodness-of-fit between the model and the available data; and (2) expected future year-to-year fluctuations in productivity (recruitment) (Prager et al. 2003; Brodziak et al. 1998). These two components of variability may capture most of the total uncertainty, but there are other factors to consider: model structure, management implementation, and ecosystem factors.

Every model's structure is an approximation of the myriad of processes actually affecting fish stocks and creating the set of available data. Alternative models will make different assumptions and process the available data in different ways. The use of alternative models is important for understanding the basis for and robustness of any model's results. Considering a range of complex and simple model structures can clarify the additional insight that the more complex model provides as it incorporates a richer set of data. Where data are highly informative about stock conditions, good alternative models should produce similar results. But as the quantity and quality of data weakens, alternative model assumptions will have more influence on the results. Model-averaging and decision tables (Patterson et al. 2001) are two principal approaches to dealing with model structure.

Imperfect implementation of forecast catch levels is an additional factor to consider when conducting medium- or long-term forecasts of the stock's response to fishing. When the fishery is managed principally through input controls such as number of licenses or number of days at sea, the implementation error occurs because such measures are imperfect at holding fishing mortality to exactly the prescribed level. When the fishery is managed principally through the output control of quotas, then there is implementation error in controlling the fishery catch to that level and implementation error in forecasting the correct level of a future quota based on imperfect knowledge of stock abundance. The potential impact of these implementation errors are principally investigated through Management Strategy Evaluations that simulate the entire system of stock dynamics, imperfect assessment, imperfect management implementation, and feedback of actual catch to stock dynamics (Butterworth and Bergh 1993; Smith et al. 1999; Patterson et al. 2001).

An additional aspect of uncertainty is with regard to ecosystem factors. In particular, adult natural mortality and juvenile natural mortality (e.g., the mean spawner–recruitment relationship) are plausibly related to whole ecosystem conditions and it is conceivable that a multispecies ecosystem model will someday be able to estimate how they change and include this information in each stock's assessment. However, today's single-species assessment models have none of these data so these factors are assumed to be constant at some average level. A more complete characterization of the uncertainty in single-species assessments should seek a means to acknowledge the suppressed uncertainty caused by keeping these factors constant.

9.5   Generalized Model: Stock Synthesis

One of the highly generalized integrated analysis models is termed Stock Synthesis (Methot 2000), now implemented as Stock Synthesis 2 in the NOAA Fisheries Assessment Toolbox and popularly known as SS2. SS2 is a third-generation inte-grated analysis model. The first was developed in the mid-1980s specifically for assessment of anchovy off the coast of California (Methot 1989). The second was a generalized model developed principally for assessment of groundfish off the US west coast and Alaska (Methot 2000). It existed in two versions: one was a length-age-structured model developed for situations with predominately length data, and the other was an age-structured model with capability for movement between geographic regions. This third-generation model merges the length-age and age-area second-generation models, adds additional features, and is coded in ADMB to gain speed and powerful methods for evaluating uncertainty. There are three major aspects of SS2's adaptability that have contributed to is widespread use: (1) it is highly flexible in its ability to have multiple fisheries and surveys with diverse characteristics; (2) its parameters have a rich set of controls to allow prior constraints, time-varying flexibility, and linkages to environmental data; and (3) its structure allows it to be scaled down to simple, data-limited cases using only two estimated parameters, and up to complex data-rich situations requiring hundreds of parameters.

In the population sub-model of SS2, annual total recruitment is calculated as a deviation from an estimated spawner–recruitment curve, which in turn describes the central tendency of the time series of recruitments (Fig. 9.3). The magnitude of each recruitment deviation is informed by the data, including environmental data, in the model yet constrained by an overall distribution function so that estimates of historical, data-limited, and forecast recruitment deviations will have the same distribution properties as the recruitment deviations that are well informed by the data (Fig. 9.4) (Maunder et al. 2006). In this regard, SS2 performs similar to sto-chastic stock reduction analysis (Walters and Martell 2004), but SS2 also includes a full observation sub-model to make advantage of more complex data where it is available.

Growth of individuals is defined to follow a von Bertalanffy function, the para-meters of which are estimable in the model when sufficient size and age data are included. In fact, SS2 could be configured to estimate only the growth parameters

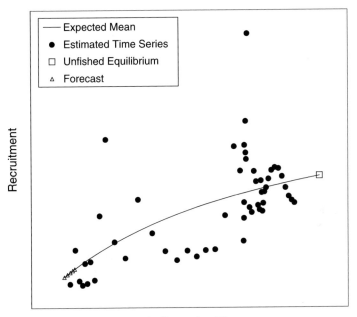

Fig. 9.3 Annual recruitment is defined as a deviation from the estimated spawner–recruitment relationship. The relationship defines the historical unfished equilibrium and provides a basis for calculating maximum sustainable yield and forecasting future levels of recruitment. In this example, during early years (high biomass) there are no data to inform the model about recruitment fluctuations, so estimated recruitments remain close to the mean relationship

and then calculate a yield per recruit analysis from a data set that had no catch and just a single size composition observation that had informative size modes. Of course, the full model capability is realized when it is allowed to estimate growth parameters from a time series of observations while taking into account the influence of size-selectivity sampling, aging error, and other processes that can otherwise bias estimates of growth. As with all SS2 parameters, the growth parameters can vary over time using a variety of methods including random annual deviations, separate parameter values for specified time blocks, and functional linkage to environmental data. For growth, an additional feature is the calculation of a year-class-specific growth deviation for situations such as abundant year classes having density-dependent suppression of growth.

SS2 tracks population numbers-at-age within each of several possible sub-divisions, termed platoons (Goodyear 1984). A platoon in SS2 is a collection of individuals that share the same biological characteristics and probability of being captured by a fishery or observed by a survey. Each platoon is defined to have a normal distribution of size-at-age that interacts with the size selectivity of each fishery and survey to create the unique observed size-at-age for that fishery/survey (Fig. 9.5). While a single platoon model is feasible to configure, it is more common

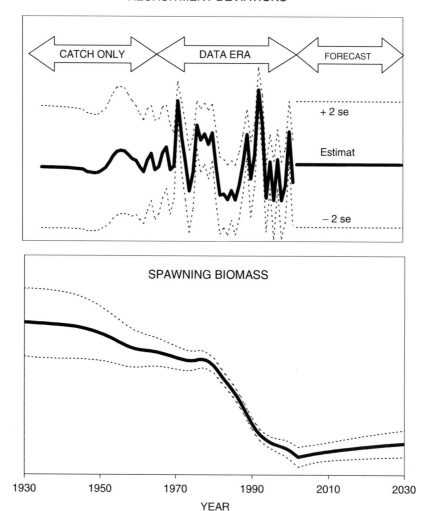

Fig. 9.4 The underlying spawner–recruitment relationship in Fig. 9.3 allows the model to produce population estimates, with variance, during far historical periods with no data other than catch, a data-rich era, and a forecast period. The transition between these periods is mostly transparent with the data phasing in and then back out

to partition the population into male and female platoons so that their dimorphic growth, mortality, and fishery selectivity characteristics can be calculated. When recruitment occurs in multiple seasons of the year, each such birth season adds platoons. In addition, it is possible to define multiple growth patterns, each with unique growth characteristics and receiving a fraction of the total number of recruits. A configuration with multiple areas and multiple growth patterns allows

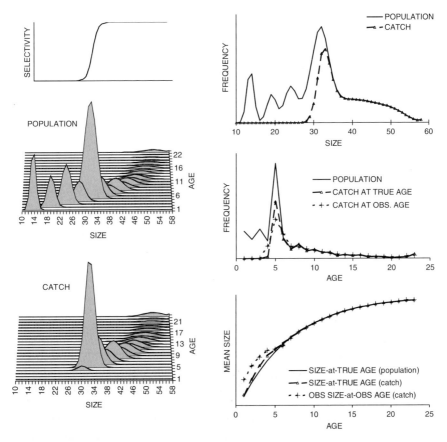

Fig. 9.5 Size-selectivity acts upon the population's total size composition to create the size composition of catch. It also acts upon the normal distribution of size-at-age to create a unique observed size-at-age for that fishery that is used to calculate that fishery's body weight at age. In the observation sub-model, aging error will blur the occurrence of a strong year class into adjacent ages, thus affecting the observed size-at-age for those weaker year classes

investigation of geographic clines in growth. Finally, each gender, birth-season, growth pattern platoon can be further subdivided into up to five sub-platoons to better track the consequences of size-selective mortality. We cannot distinguish such sub-platoons in the data, but we can assert that an underlying growth process that is built up from multiple platoons is a more accurate representation of the natural range of individual growth trajectories (Kristensen et al. 2005) than what can be provided by a single, homogeneous platoon (Fig. 9.6). Fast-growing sub-platoons will enter a size-selective fishery at a younger age, and thus experience greater cumulative fishing mortality and resultant reduced survival to older ages.

Each fishery or survey included in a SS2 configuration has a pattern of selectivity that can be in terms of age, size, or both and to include gender differences. Selectivity defines the fraction of a particular size (or age) that is captured relative

Fig. 9.6 Each platoon of
fish in the model can be
divided into 1, 3, or 5 independent sub-platoons with a
specified fraction of the
total variability in size-at-
age between sub-platoons
versus within sub-platoon.
Here five sub-platoons have
the amount of variability
within sub-platoon equal to
70% of the variability
between sub-platoons

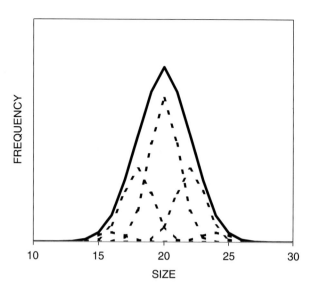

to capture rate for the size (or age) that has a selectivity of 1.0. By providing both
age and size options, the model can, for example, be configured to estimate small
fish selectivity as a function of size if it is believed that it is mostly a function of
gear technical characteristics (e.g., mesh size), while also estimating older fish
selectivity as a function of age if it is believed that it is mostly due to an age-based
diffusion into microhabitats that are relatively inaccessible to the fishery or survey
sampling gear. Various parameterizations of selectivity are available. These can
be as simple as specification of knife-edge selectivity occurring at a particular age
(no estimated parameters). Functional forms include a two-parameter logistic function, a six-parameter double normal (Fig. 9.7), nonparametric forms with a separate
parameter for each age, and others.

SS2 incorporates two options for modeling the fishery catch. The first option
employs Pope's (1972) approximation to calculation of fishery mortality. Here, the
population's numbers-at-age are decayed to the middle of the season according to
natural mortality alone. Then the catch-at-age for each fishery is calculated as a
harvest rate times the selectivity at age. Each fisheries harvest rate is calculated such
that the total catch (either in numbers or biomass) matches the observed catch. The
survivors after all fishery removals are then decayed to the end of season. In this first
approach, the harvest rates are simply an array of values to match the observed catch
and do not enter the model as explicit parameters. The second option treats fishing
mortality as a continuous process simultaneous with natural mortality. Here the
fishing mortality rates are estimated as model parameters. The first option is faster,
especially in models with long time series and large number of fisheries, and the second option performs more robustly when fishing mortalities are high. When fishing
mortalities are low or multiple seasons are used to reduce the cumulative mortality
within any season, the two approaches produce equivalent results. Fishery catch can
be in terms of numbers or biomass and different fisheries can be in different units.

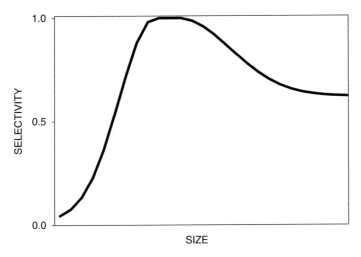

Fig. 9.7 The double normal selectivity function is commonly employed in SS2 when a dome-shaped selectivity pattern is needed. Six parameters control the function's shape

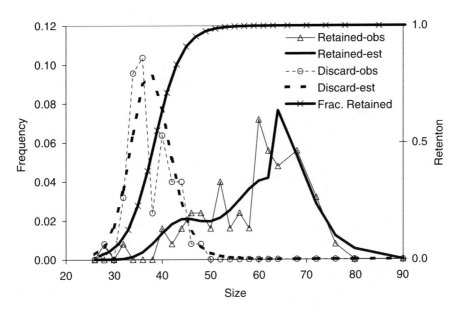

Fig. 9.8 SS2 allows for partitioning of the total catch into discarded and retained components and the calculation of expected values (est) for data from each component. Inclusion of discard and retained size composition data (obs) in this example allowed estimation of the retention function within SS2

Fishery catch can be partitioned into discarded and retained components (Punt et al. 2006) (Fig. 9.8). Further, the discarded partition can have a length-specific survival function defined. Thus the total fishing mortality for a given fleet is the retained catch plus the mortal fraction of the discarded fish. By partitioning the total catch into discard and retained components, SS2 is then able to develop expected values for discard only, retained only, or total catch samples of size and age composition.

With the above description of discarding, we have partly transitioned from the population sub-model to the observation sub-model. SS2 can produce expected values for several kinds of common fishery and survey data. All of these expected values start from a size/age/gender array of selected fish calculated by applying the size/age/gender selectivity for each gear to the size/age/gender population array at that point in time (Fig. 9.5). When summed across ages, the result is the expected size composition, which can then be compared to an observation of the size composition for that fishery or survey. When summed across sizes, the result is the expected value of the sampled age composition. But stopping at this stage would omit the effect of aging imprecision. The process of determining age from otoliths or scales or other structures involves some uncertainty that is expected to blur the observed age composition (Tyler et al. 1989; Kimura and Lyons 1991). Rather than try to remove this blurring effect from the data before providing the data to the model, integrated analysis models like SS2 build the blurring process into the model so that the expected values are blurred to the same degree that the data are blurred. While it is feasible to provide the model with both size and age composition data, this will tend to double weight the information from some fish. An alternative approach available in SS2 is to examine the age composition data conditional on being within a subset of the size/gender range (Methot 2000; Punt et al. 2000). Thus, the model considers the additional information provided by age data over and above the information already provided by the fish size.

The size and age composition expected values can be compared to data for each gender, summed across genders, or treating the size/gender information as a joint distribution that preserves information on sex ratio. When multiple platoons and sub-platoons are used, the selectivity and resultant mortality is applied to each and the results are summed across platoons because the difference between platoons is invisible to the observation process. This is analogous to creating a two-gender model to deal with a known difference in growth between males and females, then summing the expected values across the two genders to provide a combined gender expected value where data have not been partitioned into males in females.

SS2 can also include information from surveys of stock abundance. The sum of the age/size/gender "selected" fish from a particular fishery or survey is the estimated total abundance of this selected component of the population. The sum can be in terms of numbers of fish, or in terms of biomass by incorporating the model's estimate of body weight at age. This sum times a scaling factor is the expected value for the survey observations. This scaling factor is also termed a catchability coefficient, q. Ideally, q would be independently measured and calibrated, as is possible for some acoustic and visual surveys. Some bottom trawl surveys where the area-swept, herding, and escapement has been measured may also be analyzed as a fully calibrated survey (Somerton et al. 2007).

Most surveys, however, have intangible factors in the catchability that have defied direct calibration to date. Where standardization of methods has allowed assertion that q has remained constant at some unknown value, then it is feasible to treat the survey as a measure of the stock's relative trend and to allow the assessment model to estimate q internally as a scaling factor between the units of the survey measurement and the units of estimated, selected population abundance. In this case, it takes accumulation of a time series of observations before any meaningful assessment information can be obtained from the survey. In the worst case, the intangible components of q cannot be comfortably asserted to be constant over time. This is most common when fishery catch per unit of effort is used as an index of population abundance. However, even in this situation it is not uncommon for the assessment configuration to maintain the assumption of constant q or a prespecified drift in q over time.

As an example of the kind of analysis that can be conducted with SS2, consider three survey scenarios. In the first scenario, a single research vessel conducts an annual relative index survey for 20 years and is then replaced by a similar vessel that is calibrated to the first vessel so that a continuous time series of survey observations can be analyzed with the assessment model under a constant q assumption. In the second scenario, several (2–4) chartered vessels conduct replicate surveys each year and the combined results of these surveys are analyzed as a relative index under the constant q assumption. Differences in q between vessels is not directly calibrated and becomes part of the system noise. In the third scenario, thousands of logbook observations from hundreds of vessels are processed using a statistical model to develop an annual CPUE index that is analyzed by the assessment model under the constant q assumption. All three scenarios have assumed a constant q in the assessment analysis and the fishery CPUE scenario may produce a more precise model result because of the large number of observations. What's wrong with this picture and why is there value in single-vessel standardized surveys over statistically standardized fishery data?

A more holistic approach would acknowledge that there is always some fluctuation in q (Millar and Methot 2002; Francis et al. 2003; Bence and Wilberg 2006) and some survey methods are better than others at keeping these fluctuations small. We expect a more constant q from scenario 1 in the preceding paragraph than from scenario 3, so the model should be configured to use this knowledge. In SS2, the q parameter can be specified as having an annual random deviation or an annual random walk, each with prior value with variance. Thus it is feasible to directly incorporate information on the degree of confidence in the constancy of q. In the first scenario above, the annual q parameter could be specified to have only a small random walk from the previous year's q value except in the year of vessel transition in which case the value of the vessel inter-calibration and its variance would provide information on the size of a larger change in q that year, and this degree of change would be updated each subsequent year as more data are collected using the new vessel. In the second scenario, the degree of variability in estimated vessel effect among the chartered survey vessels could be used to constrain the degree of random drift in q for their combined survey result. In the third scenario, the lack of direct standardization of the fishery CPUE data could be used to assign a larger variability to the possible random walk in catchability (Bence and Wilberg 2006).

Of course, allowing too much variability in q means that the population signal possibly found in the survey trend will be lost to the estimation of time-varying q. The result of this more holistic treatment of variability in q would be demonstration of the improved overall precision in assessment model results that can be attained when good standardization and vessel inter-calibration has occurred.

The statistical sub-model of SS2 is generally as described above for integrated analysis models. All data going into the model have an associated level of variance that determines the scaling of the deviations between the data and the model's estimates of expected values. Because these estimates of data variance may themselves be inaccurate and because the structure of the model may not be flexible enough to adequately represent all processes that created the actual data, the SS2 approach provides opportunity for adjustment of the data variance. Model outputs include statistics that compare the average goodness of fit to the level of input data variance. This can then guide changes in the level of model flexibility (more or less parameters) and adjustment of the input variance levels to better represent the subsequent model capability to match these data. With the input and output variances so tuned, the final estimates of variance in model outputs better represent the relative contribution of all sources of data. As the model evolves towards more use of random effects for factors such as annual survey catchability, it will be necessary to develop better protocols for balancing the magnitude of these random effects versus adjustment of the variance terms.

Although model fitting is in terms of the total NLL, it is prudent and necessary to examine the model's fit to each data component individually and graphically (Richards et al. 1997). The best-fitting set of parameters will be a compromise. They will not provide the best possible fit to any one component, nor should they produce an unreasonably poor fit to any influential data. Visualizing and quantifying the residuals in the fit to each type of included data helps the modeler identify places where adjustments need to be made in the model structure. Too tight a fit means that some aspect of the model is too flexible and has too many free parameters. Unreasonable patterns in residuals usually mean that some process affecting the real data has not yet been included in the model. The art of model building is largely about the selection of the best degree of model complexity and best balance of fit among the various data components.

Management quantities and forecasts are an important feature of SS2. Following estimation of the model's parameters, the values of various management quantities are calculated and a forecast is conducted using a selected management policy or specified catch level. By integrating this management layer into the overall model, the variance of the estimated parameters can be propagated to the management quantities and forecast, thus facilitating a description of the risk of various possible management scenarios. Because the entire model works as a simulation, it is feasible for a single model configuration to start in a pre-data, lightly fished era; extend through a data-rich era to the present day; and continue into a forecast era (Fig. 9.4). The transition from the estimation era to the forecast era is transparent and denoted only by the phasing out of the data availability, but even that transition is blurred as very recent environmental data are included in today's models.

The management quantities include the fishing intensity level that would produce a specified level of spawning biomass per recruit, a specified spawning biomass level as a fraction of unfished biomass (taking into account the spawner–recruitment relationship), and the fishing intensity level that would produce maximum sustainable yield. The search for these quantities is across a range of fishing intensity levels conditioned upon a particular allocation of the intensity among fishing fleets and the size/age pattern of selectivity for each fleet. This means that the level of fishing mortality varies between ages in a complex way and a single value representation of fishing mortality is ambiguous. Because of this ambiguity, SS2 reports the level of fishing intensity in terms of spawning potential ratio (Prager et al. 1987). The forecast can use the current fishing intensity or any one of the calculated management quantities. The forecast incorporates biomass-based adjustments to future fishing intensity levels according to the harvest policies in the west coast groundfish Fishery Management Plan (Ralston 2002).

Environmental data can be incorporated into SS2 is two ways. First, any SS2 parameter can be defined to be a function of an input environmental data time series. For example, this could be used to set the annual expected recruitment deviation to be a function of sea surface temperature (Schirripa and Colbert 2006) according to the method described in Maunders and Watters (2003). It could also be used to set a fishery selectivity parameter to be a function of an environmental variable such as wind speed or any other predictor variable such as mean depth of fishing. Survey catchability could be similarly linked to an input variable. Growth could be linked to environmental variables such as temperature or ecosystem productivity. Natural mortality could be linked to predator abundance (Methot 1989; Livingston and Methot 1998).

The first approach to environmental linkage described above is based upon the intuitive concept that the environmental factor has caused changes in the population process being modeled. But this intuition has a degree of naivety. The environmental variables we measure are, hopefully, a good indicator of the myriad and complex factors that actually cause the changes in population processes. But they are only indicators; for example, they are data that may be informative about the process. The second method of including environmental data in SS2 exploits this alternative logic. Currently, SS2 only implements this alternative for recruitment by having the environmental data enter the model as if they are a survey of the age zero recruitment deviations. These data then compete and/or reinforce other data in the model to produce the best-fitting estimates of recruitment. From a statistical perspective, the model does not care if the direct recruitment data come from a fishery-independent survey of 5-month-old juveniles that indexes the numbers of recruits, or from an environmental measurement that indexes the deviation of recruitment relative to the level predicted from the spawner–recruitment relationship. Thus, method two uses environmental data as a correlate to help explain recruitment, whereas method one uses environmental data to cause the recruitment deviations. In the future, SS2 and other integrated analysis models are expected to evolve to more use of such random effects procedures. In this approach, any parameter could be defined as having an annual change, the random effect, and the magnitude of these changes can be estimated through inclusion of conventional as well as environmental data.

9.6 Getting to Ecosystem

The NMFS Stock Assessment Improvement Plan described Tier II as elevating stock assessments to new national standards of excellence and Tier III assessments as reaching to an ecosystem level. What does it mean to achieve an ecosystem level for fishery stock assessments? Characterizing Tier II assessments helps provide a foundation for this discussion. Tier II assessments are empirical descriptions of a stock's status. They measure stock abundance and fishing mortality, compare these to reference levels, calculate fishing mortality levels that are sustainable given current conditions, and translate abundance and fishing mortality targets into a forecast of short-term catch levels. In doing so, the assessment model defines the system as containing the stock of fish and its fishery. Outside influences are recognized to cause random perturbations to the system, but these perturbations are considered measurable within the system and not to require understanding about how the outside influences cause the perturbations. Tier II assessments treat the fishery reference levels as entirely derivable from information inside the "system" and treat the outside influences as providing only random noise without directional trends. Such Tier II stock assessments do have a one-directional link to ecosystem analyses because the time series output of Tier II assessments is valuable input and validation to holistic ecosystem food web models.

Getting to Tier III means expanding the scope of the assessment system so that more of the outside influences become part of the analyzed system. As a first step, fishery assessment models can include more environmental and ecosystem inputs so that factors in the assessment system are explicitly linked to these inputs. Integrated analysis models have already begun evolving in this direction as described in this paper, and some examples of linked multispecies models have appeared. The next step will essentially be a merger of the expanding scope of these integrated analysis models and the increasing detail and data assimilation capabilities of holistic ecosystem food web models. Before reaching that stage, and perhaps even at today's stage of model evolution, it seems relevant to ask whether it would be beneficial to explicitly develop two linked scales of model complexity (Hilborn 2003). The more complex, ecosystem-linked model would be the strategic model used to determine target harvest rates that achieve optimum yield from fisheries while explicitly accounting for ecosystem effects. The less complex model would be the tactical model that uses simple data inputs to guide tweaking the fishery up and down to implement the policies determined from the complex model.

The following section identifies some ways in which Tier II assessments can evolve towards Tier III. In general these fall into two categories: (1) allowing change in factors that now must be assumed to be constant, and (2) improving predictions for factors that are currently allowed to fluctuate in a random manner.

The first category includes natural mortality and the shape and scale of the spawner–recruitment relationship. The second category principally includes annual recruitment deviations and body growth. Environmentally caused fluctuations in survey catchability could be included here also.

Natural mortality (M) is the 900 lb gorilla of stock assessment parameters. It is arguably the parameter that is least estimable from conventional assessment data and the parameter that is most dependent on shifts in the ecosystem predator–prey relationships. Where fishing mortality (F) is much greater than natural mortality, then the exact value of M has little effect on estimates of the trend in abundance of the stock, but well-controlled fisheries are not likely to have F much, if at all, greater than M. While small error in M is unlikely to cause an assessment to misestimate the trend in stock abundance, the absolute level of stock abundance is directly related to the level of M. Natural mortality can be an estimable parameter in an assessment model, but robust performance of such a model generally requires informative and precise survey and age composition data with verifiable selectivity and catchability characteristics. Without data that is truly and unambiguously informative about M, the model will adjust M to attempt to explain other patterns in the data. Consequently, M is usually held fixed in assessment models at a level estimated from the age composition from pre-fishery periods, or lightly fished components of the stock, or from empirical relationships between the few direct estimates of M and more easily obtained life history parameters. Validating the accuracy of these M estimates for today's fully exploited ecosystem and obtaining time and age-varying values is an extreme challenge. Getting contemporary information on M is one of the greatest contributions that ecosystem studies could give to stock assessment.

The estimated spawner–recruitment relationship (S–R) represents the average level of recruitment expected from a specified level of reproductive output. Walters and Martell (2004) contains a broad examination of the various factors that go into estimation and interpretation of this relationship. They note that R is not produced by a single S–R relationship; rather it is the result of a myriad of sequential life history stages, each with various potentials for density-dependent factors. So, under what conditions can a simple two-parameter S–R relationship adequately describe the historical pattern in the data, serve as the basis for estimation of the long-term productive capacity of the stock, and forecast the expected level of future recruitment?

First is the measurement of spawning biomass. It is not uncommon for fish stocks to exhibit changes in age-specific maturity over time and it is possible that the shift to earlier maturation is part of the stock's compensatory response to the additional adult mortality imposed by the fishery. Yet such changes in maturity are usually not measured as a time series and it is more common for contemporary, "better" estimates of maturity and fecundity to be used to calculate the spawning output throughout the time series. Thus, the estimated curvature in the S–R can be confounded with the degree to which maturation has shifted and whether these temporal shifts have been taken into account in the calculation of spawning biomass. Investigation of the degree to which density-dependent shifts in maturation occur in harvested fish stocks could lead to improved standard practices for dealing with such shifts in the face of inadequate information.

Second is the assertion that the S-R is constant in the face of ecosystem shifts. The S–R basically represents the cumulative mortality through the egg–larval–juvenile stages and how this mortality changes with stock abundance. However, juvenile fish can be prey to many species of fish and many of these species may have exhibited changes in abundance over the same decades that are being analyzed for the S–R of the subject species. In many systems, the S–R data have been collected over a period of time in which the abundance of many species has been reduced due to fishing. Frankly, the coastal ecosystems are coming into a new state with human fishers as an introduced predator. Fishing mortality's primary effect is in reducing the abundance of older fish, which have the greatest tendency to be piscivorous. So it is possible that while fishing has reduced the abundance of spawners that produce juveniles of species A, fishing has also changed the abundance of species B and C that are predators on the juveniles of species A. A major contribution of ecosystem food web studies could be identification of the stocks that are most in need of including ecosystem interactions in their S–R relationships.

A third issue is the sequencing of environmentally caused and density-dependent mortality. The S–R relationship is routinely written such that environmentally caused variability occurs after the action of density-dependent survival. The ecological arguments that would support such a relationship make sense for a species like salmon. For example, when salmon spawners are super-abundant they may spawn in marginally suitable reaches of streams that do not support high egg survival. Subsequent to this density-dependent stage is the estuarine and early ocean period in which it is recognized that highly variable environmental conditions will cause variation in survival of juveniles. However, what makes sense for the predominant life history of marine fish? Isn't it generally accepted that high variability in survival occurs in the early larval stage and that density-dependence is most likely during the subsequent juvenile stage as they settle into limited habitats or otherwise behaviorally interact? If so, shouldn't the S–R relationship be formulated such that most variability occurs before density-dependence acts, in which case the density-dependence should dampen the environmentally caused variability in larval survival? Further consideration of this alternative S–R formulation could perhaps make more sense of the uninformative scatter found in many sets of spawner–recruitment data.

Fourth is the effect of environmental factors, which has both long-term and short-term consequences. Long-term environmental patterns can bias estimates of S–R parameters. As fishing has moved average spawning biomass from a moderately high level to a moderately low level over a period of decades, the observed change in average recruitment is the basis for the estimated S–R relationship. However, if the decadal time scale of some environmental shifts also affects recruitment, then the S–R estimate is confounded with the environmental changes. Unambiguously disentangling the spawner effect from the long-term environmental effect seems nearly impossible until many decades of monitoring are available or until process studies are able to estimate either the spawner effect or the environmental effect without resorting to simple correlation studies.

   The short-term effect of the environment on annual recruitment is more a matter of improved forecasting rather than disentangling historical relationships. Conventionally available fishery and survey data are sufficient to estimate annual fluctuations in recruitment. However, these data cannot provide direct information on recruitment until the fish are old enough to appear in these sources of data. They provide precise, accurate estimates but they are fundamentally retrospective estimates. This may be sufficient for long-lived species in which the young recruiting year classes are only a small fraction of the population and fishery, but more timely recruitment estimates are needed for short lived species, species with extremely high recruitment fluctuations, or species with management plans that seek to closely track maximum potential yield. More timely estimates of recent recruitment fluctuations can be made by conducting a survey that measures abundance of pre-recruits at a young age, and/or by determining and measuring environmental covariates that provide good predictions of recruitment. Survey cost, timeliness of estimates, and precision are factors that influence the relative merits of initiating a pre-recruit survey versus initiating a research program to determine a relevant environmental predictor. In practice, a program desiring a better recruitment predictor will probably need to do both in order to provide necessary cross-validation.

   Growth and reproduction, like recruitment, are empirically measurable from available data, thus can be allowed to change over time in assessment models. Tier III models will include mechanisms that link growth and reproduction to population abundance and ecosystem/environmental factors.

   A final aspect of more realistic Tier III assessment models is spatial structure. The need for spatially explicit models is growing as we consider the dynamics of populations that have a significant fraction of their abundance within marine reserves (Holland 2002; Punt and Methot 2004; Field et al. 2006). Conventional models that treat the stock as if diffusion was infinitely high can produce biased results when applied to populations that have low rates of mixing between areas. Spatial population models may be needed to combine information from multimodal survey programs in which direct observation methods measure the absolute fish abundance on rocky habitats while conventional trawl surveys measure relative trends on adjacent smoother habitats. In principle, it is straightforward to code the model to include multiple geographic zones and to allow fish movement between zones. The data requirements for such a model are feasible in some of our highly monitored fisheries today. However, spatially disaggregating historical data and obtaining information on rates of fish movement are daunting steps.

   Some of the above suggestions have substantial new data requirements. In particular, tagging studies to obtain movement rates and predation monitoring programs to measure natural mortality rates are expensive field programs, the value of which should be evaluated against the potential gains in assessment accuracy and precision. The suggestions regarding temporal shifts in maturation, the form of the spawner–recruitment relationship, and long-term climate effects on the spawner–recruitment relationship could, in some instances, be addressed through additional investigation using currently available data. Finally, suggestions for more flexible model capabilities can be implemented as next generation assessment models are developed. In the short term, expanded model capabilities to admit time-varying factors may demonstrate decreased precision in model results, but this will establish a framework for better identification of the data needed to truly improve the precision of model results.

9.7 Operational Model

Stock assessments provide operational support for quantitative management of fisheries. The integrated analysis models described in this paper have the capability to link the estimation of the population's historical abundance, to the inference of biological reference points and forecasts of future population trends and potential catch. In order to provide more timely updates for more stocks, efficiencies are needed at each step in the sequence from collection of data through delivery of results.

The first obvious step is the need for timely access to accurate, precise, and comprehensive fishery and survey data. A great fraction of the total assessment time and energy goes into discovering, quality-checking, and calibrating historical data that don't quite meet the standards of today's data collection systems. Greater efficiency in this process can be obtained by taking a horizontally integrated approach rather than a vertically integrated approach. Most data collection systems collect data on multiple species, so analysis and review of these data is best accomplished at the same time across all relevant species, rather than species by species as they are assessed. Likewise, life history analysis methods are likely to be relevant for many related species, so also can be clustered into a methods-oriented review rather than fully opening the topic for each species as its assessment is reviewed. Timely availability of contemporary data can be improved through better data systems: more electronic recording of data in the field and more sophistication in the databases to quickly accomplish quality checks and delivery of data to end-users.

The second step is the set of models. These must be at the right level of complexity: simple enough to be rapidly updated without extensive diagnostics, and complex enough to adjust for confounding effects of non-fishery factors. Because of the great diversity of data situations and fish life history patterns, standardized models must have a flexible structure that is scalable in complexity to the particular situation being analyzed. Quick tactical models may need to be paired with more complex strategic models to achieve the right mix of overall capabilities. Once developed, such standardized models allow less experienced users to fully participate in assessment modeling, they facilitate improved communication as results are being reviewed, and they enable development of a more comprehensive suite of tools to disseminate model results to a wide range of clients and the interested public. A downside of increased standardization among the models used for management purposes is a stifling of research and creativity. Recognizing this as a possibility can perhaps be turned into greater emphasis on explicit research on model development. We need to decide what technical and ecosystem processes can be included within the modeled system of the operational models, and which must be fire-walled off into exploratory analyses designed to improve the next generation of operational models.

Third is the process in which the model is used to develop technical advice for the regulation of fisheries. This regulatory link creates a high level of controversy for all aspects of the stock assessment enterprise. Fishery data and assessment models receive an extraordinary degree of public scrutiny and formal review. Increased throughput of assessment updates will require streamlining of the extensive review process now routinely required before stock assessment results can serve as the scientific basis for fishery management. Emphasizing review of broadly applicable assessment data and methods, rather than each final result, is a logical step in this streamlining, while maintaining public trust in the final results.

References

Bence, James R. and M. J. Wilberg. 2006. Performance of time-varying catchability estimators in statistical catch-at-age analysis. Canadian Journal of Fisheries and Aquatic Sciences 63: 2275–2285.

Bobko, S. J. and S. A. Berkeley. 2004. Maturity, ovarian cycle, fecundity, and age-specific parturition of black rockfish (Sebastes melanops). Fishery Bulletin 102: 418–429.

Brodziak, J., P. Rago, and R. Conser. 1998. A general approach for making short-term stochastic projections from an age-structured fisheries assessment model. In F. Funk, T. Quinn II, J. Heifetz, J. Ianelli, J. Powers, J. Schweigert, P. Sullivan, and C.-I. Zhang (Eds.), Proceedings of the International Symposium on Fishery Stock Assessment Models for the 21st century. Alaska Sea Grant College Program, University of Alaska, Fairbanks, AK.

Brown, S. K., M. Shivlani, D. Die, D. B. Sampson, and T. A. Ting. 2006. The Center for Independent Experts: The National External Peer Review Program of NOAA's National Marine Fisheries Service. Fisheries 31 (12): 590–600.

Butterworth, D. S. and M. O. Bergh. 1993. The development of a management procedure for the South African anchovy resource, pp. 83–99. In S. J. Smith, J. J. Hunt, and D. Rivard (Eds.), Risk Evaluation and Biological Reference Points for Fisheries Management. Canadian Special Publications in Fisheries and Aquatic Science 120.

Chen, Y., M. Kanaiwa, and C. Wilson. 2005. Developing a Bayesian stock assessment framework for the American lobster fishery in the Gulf of Maine. New Zealand Journal of Freshwater and Marine Sciences 39: 645–660.

Field, J. C., A. E. Punt, R. D. Methot, and C. J. Thomson. 2006. Does MPA mean 'Major Problem for Assessments'? Considering the consequences of place-based management systems. Fish and Fisheries 7: 284–302.

Fournier, D. A. and C. P. Archibald. 1982. A general theory for analyzing catch at age data. Canadian Journal of Fisheries and Aquatic Sciences 39: 1195–1207.

Francis, R. I. C. C., R. J. Hurst, and J. A. Renwick. 2003. Quantifying annual variation in catchability for commercial and research fishing. Fishery Bulletin 101: 293–304.

Goodyear, C. P. 1984. Analysis of potential yield per recruit for striped bass produced in Chesapeake Bay. North American Journal of Fisheries Management 4: 488–496.

Helser, T., A. E. Punt and R. D. Methot. 2004. A generalized linear mixed model analysis of a multi-vessel fishery resource survey. Fisheries Research 70: 251–264.

Hilborn, R. 2003. The state of the art in stock assessment: where we are and where we are going. Scientia Marina 67 (Suppl 1): 15–20.

Hjort, J. 1914. Fluctuations in the great fisheries of northern Europe viewed in the light of biological research. Rapports et Proces-Verbaux des Réunions du Conseil Permanent International pour l'Exploration de la Mer 20: 1–228.

Holland, D. S. 2002. Integrating marine protected areas into models for fishery assessment and management. Natural Resource Modeling 15: 369–386.

Kimura, D. K. and J. J. Lyons. 1991. Between-reader bias and variability in the age-determination process. Fishery Bulletin 89: 3–60.

Kristensen, K., P. Lewy, and J. E. Beyer. 2005. Maximum likelihood estimation in a size-spectra model. Canadian Journal of Fisheries and Aquatic Sciences 99: 1–10.

Livingston, P. A. and R. D. Methot. 1998. Incorporation of predation into a population assessment model of eastern Bering Sea walleye pollock. In F. Funk, T. J. Quinn II, J. Heifetz, J. N. Ianelli, J. E. Powers, J. F. Schweigert, P. J. Sullivan, and C.-I. Zhang (Eds.), Fishery Stock Assessment Models.. Alaska Sea Grant AK-SG-98-01. University of Alaska Fairbanks 1037 p.

Mace, P. M., N. W. Bartoo, A. B. Hollowed, P. Kleiber, R. D. Methot, S. A. Murawski, J. E. Powers, and G. P. Scott. 2001. National Marine Fisheries Service Stock Assessment Improvement Plan. Report of the NMFS National Task Force for Improving Fish Stock Assessments. NOAA Technical Memorandum NMFS-F/SPO-56. 76 pp.

Maunder, M. N. and A. E. Punt. 2004. Standardizing catch and effort data: a review of recent approaches. Fisheries Research 70: 141–159.

Maunder, M. N. and G. M. Watters. 2003. A general framework for integrating environmental time series into stock assessment models: model description, simulation testing, and example. Fishery Bulletin 101: 89–99.

Maunder, M. N., S. J. Harley, and J. Hampton. 2006. Including parameter uncertainty in forward projections of computationally intensive statistical population dynamic models. ICES Journal of Marine Science 63: 969–979.

Maunder M. N., J. T. Schnute, and J. Ianelli. 2009. Computers in Fisheries Population Dynamics. In B. Megrey and E. Moksness (eds.), Computers in Fisheries Research. Kluwer Academic Publishers. p 337–372. ISBN: 978-1-4020-8635-9.

McAllister, M. K., E. K. Pikitch, A. E. Punt, and R. Hilborn. 1994. A Bayesian approach to stock assessment and harvest decisions using the sampling/importance resampling algorithm. Canadian Journal of Fisheries and Aquatic Sciences 51: 2673–2687.

Methot, R. D. 1989. Synthetic estimates of historical abundance and mortality for northern anchovy. American Fisheries Society Symposium 6: 66–82.

Methot, R. D. 2000. Technical Description of the Stock Synthesis Assessment Program. National Marine Fisheries Service, Seattle, WA. NOAA Tech Memo. NMFS-NWFSC-43: 46 p.

Millar, R. B. and R. D. Methot. 2002. Age-structured meta-analysis of the U.S. West Coast rockfish (Scorpaenidae) populations and hierarchical modeling of trawl survey catchabilities. Canadian Journal of Fisheries and Aquatic Sciences 59: 383–392.

National Research Council (NRC). 1998. Improving Fish Stock Assessments. National Academy Press, Washington, DC. 177 pp.

NOAA. 1996. Magnuson-Stevens Fishery Conservation and Management Act. NOAA. Technical Memorandum NMFS-F/SPO-23. Silver Spring, MD.

Otter Research Ltd. 2005. An introduction to AD Model Builder Version 7.1.1 for use in nonlinear modeling and statistics. Sidney, BC, Canada. (http://admb-project.org/)

Patterson, K., R. Cook, C. Darby, S. Gavaris, L. Kell, P. Lewy, B. Mesnil, A. E. Punt, V. Restrepo, D. W. Skagen, and G. Stefansson. 2001. Estimating uncertainty in fish stock assessment and forecasting. Fish and Fisheries 2: 125–157.

Pope, J. 1972. An investigation of the accuracy of Virtual Population Analysis using cohort analysis. ICNAF Research Bulletin 9: 65–74. Also available in D. H. Cushing (Ed.). 1983. Key Papers on Fish Populations. IRL Press, Oxford. 405 p.

Prager, M. H., J. F. O'Brien, and S. B. Saila. 1987. Using lifetime fecundity to compare management strategies: a case history for striped bass. North American Journal of Fisheries Management 7: 403–409.

Prager, M. H., C. E. Porch, K. W. Shertzer, and J. F. Caddy. 2003. Targets and limits for manage-ment of fisheries: a simple probability-based approach. North American Journal of Fisheries Management 23: 349–361.

Punt, A. E. and R. D. Methot. 2004. Effects of marine protected areas on the assessment of marine fisheries. American Fisheries Society Symposium 42: 133–154.

Punt, A. E., F. Pribac, T. I. Walker, B. L. Taylor, and J. D. Prince. 2000. Stock assessment of school shark, *Galeorhinus galeus*, based on a spatially explicit population dynamics model. Marine and Freshwater Research 51: 205–220.

Punt, A. E., D. C. Smith, G. N. Tuck, and R. D. Methot. 2006. Including discard data in fisher-ies stock assessments: Two case studies from south-eastern Australia. Fisheries Research 79: 239–250.

Quinn, T. J. I. 2003. Ruminations on the development and future of population dynamics models in fisheries. Natural Resource Modeling 16: 341–392.

Richards, L. J., J. T. Schnute, and N. Olsen. 1997. Visualizing catch-age analysis: a case study. Canadian Journal of Fisheries and Aquatic Sciences 54: 1646–1658.

Ralston, S. 2002. West coast groundfish harvest policy. North American Journal of Fisheries Management 22: 249–250.

Restrepo, V. R., G. G. Thompson, P. M. Mace, W. L. Gabriel, L. L. Low, A. D. MacCall, R. D. Methot, J. E. Powers, B. L. Taylor, P. R. Wade, and J. F. Witzig. 1998. Technical Guidance on the Use of Precautionary Approaches to Implementing National Standard 1 of the Magnuson-Stevens Fishery Conservation and Management Act. NOAA Technical Memorandum. NMFS–F/SPO–31.

Schirripa, M. J. and J. J. Colbert. 2006. Interannual changes in sablefish (*Anoplopoma fimbria*) recruitment in relation to oceanographic conditions within the California Current System. Fisheries Oceanography 15 (1): 25–36.

Schnute, J. T., M. N. Maunder, and J. N. Ianelli. 2007. Designing tools to evaluate fishery manage-ment strategies: can the scientific community deliver? ICES Journal of Marine Science 64: 1077–1084.

Sitar, S. P., J. R. Bence, J. E. Johnson, M. P. Ebener, and W. W. Taylor. 1999. Lake Trout mortality and abundance in Southern Lake Huron. North American Journal of Fisheries Management 19: 881–900.

Smith, A. D. M., K. J. Sainsbury, and R. A. Stevens. 1999. Implementing effective fisheries man-agement systems–management strategy evaluation and the Australian partnership approach. ICES Journal of Marine Science 56: 967–979.

Somerton, D. A., P. T. Munro, and K. L. Weinberg. 2007. Whole-gear efficiency of a benthic survey trawl for flatfish. Fishery Bulletin 105: 278–291.

Tyler, A. V., R. J. Beamish, and G. A. McFarlane. 1989. Implications of age determination errors to yield estimates. 27–35. In R. J. Beamish and G. A. McFarlane (Eds.), Effects of Ocean Variability on Recruitment and an Evaluation of Parameters Used in Stock Assessment Models. Canadian Special Publications in Fisheries and Aquatic Science 108: 27–35.

Walters, C. J. and S. J. D. Martell. 2004. Fisheries Ecology and Management. Princeton University Press, Princeton, NJ. 399 p.

Chapter 10
Fisheries Science in the Future

Randall M. Peterman

Abstract Fisheries scientists face an exciting but demanding future. Decision makers, stakeholders, and the public will be seeking advice from scientists on a wide range of topics, including the management implications of climatic change and other changes in fish habitats, complex trophic interactions, altered relationships among environmental and biological variables, and a broad range of ecosystem and other indicators that reflect multifaceted objectives of various interest groups. These and other sources of complexity will be routinely incorporated into stock assessment models. These models will also be expected to take several types of uncertainty explicitly into account in order to provide risk assessments for risk-based management decisions. To thrive in such an environment, young fisheries scientists will ideally need to have not only a solid ecological background, but also a high level of quantitative skills (statistics and simulation modeling), interdisciplinary training, and an ability to work collaboratively with, and communicate effectively with, conservation groups, fishing industry organizations, university researchers, government scientists, and decision makers.

Keywords Stock assessment · nonstationarity · climate change · human dynamics · young fisheries scientists · management advice

10.1 Introduction

This paper is a personal view of topics related to the nature of fisheries science in the future, with an emphasis on fish stock assessment. By 'fisheries science' I mean the monitoring, scientific research, and analyses that are conducted to produce advice that is useful for decision makers, stakeholders, and members of the public who

R.M. Peterman
School of Resource and Environmental Management,
Simon Fraser University, Burnaby, British Columbia, Canada, V5A 1S6
e-mail: *peterman@sfu.ca*

R.J. Beamish and B.J. Rothschild (eds.), *The Future of Fisheries Science in North America,*   167
Fish & Fisheries Series, © Springer Science + Business Media B.V. 2009

manage or otherwise affect aquatic ecosystems. I will not discuss fisheries *management per se*, but rather the scientific advice that can lead to well-informed decisions. My views on fisheries science inevitably reflect the biases of my own experience in research and teaching graduate students, so these views should be interpreted in that context. The intended audience for this paper includes not only fisheries biologists and scientists in management agencies, environmental consulting firms, the fishing industry, and non-governmental organizations, but also young scientists who are trying to prepare themselves for a future career that is filled with as many uncertainties as the fisheries data they analyze. In addition, fisheries managers might glean useful ideas about what they should expect from future scientific advisors.

Overall, I believe that the future of fisheries science is bright, although there are significant challenges ahead. For example, not only are there increasing demands on fisheries scientists owing to the diverse interests in ecosystem-based management, risk management, biological conservation, and other major policy initiatives, but natural system dynamics are being altered as climatic and other human-induced changes unfold. Fortunately, data sets collected with improved sampling designs, along with new statistical and simulation modeling methods, will likely facilitate the work of fisheries scientists and lead to more comprehensive scientific advice for decision makers and stakeholders. Although this paper discusses many topics related to fish stock assessment, several of the ideas are also directly relevant to research on the natural dynamics of aquatic systems.

10.2   Stock Assessment

Fisheries stock assessment, which is the process of quantitatively estimating potential outcomes of various contemplated management actions (Hilborn and Walters 1992), is likely to continue getting more sophisticated and complex, just as it has in past decades (Quinn and Deriso 1999). Three major factors push in this direction. (1) The trend toward ecosystem-based management, although still somewhat ill-defined, is forcing many stock assessment scientists to consider a wider range of indicators of an aquatic system's status than the single-species emphasis in the past. (2) Similarly, management objectives and their associated indicators increasingly reflect the diversity of stakeholder groups that have input to decision makers. (3) As scientists learn more about the complexities of aquatic systems and use more types of data, they usually attempt to add details to their stock assessment models. For instance, many models now include trophic and competitive interactions, the effects of which are often nonlinear, lagged, and/or cumulative. Computer programming skills are high among fisheries researchers in universities and management agencies, creating few barriers to adding such interactions to stock assessment models. However, these multifaceted interactions also increase requirements for the data to parameterize them.

Such data needs raise the issue that our limited *knowledge* will tend to counteract the movement toward ever-increasing complexity in fish stock assessment models. The limited quality and quantity of both data and knowledge will tend to keep models (at least peer-reviewed ones) from going too far into details when not justified by the data. This adage comes to mind: 'The analysis should be as simple as possible, but no simpler than necessary' to capture essential features of systems that are required to answer management questions (Morgan and Henrion 1990).

Some readers might point out that there will probably continue to be rapid growth in *data* about the ocean and its inhabitants through the growing number and types of remote sensing satellites, buoys, gliders, drifters, 'smart tags', remotely operated underwater vehicles (ROVs) and video cameras, and internet-cable-linked seafloor observatories such as NEPTUNE off the west coast of North America (http://www. neptunecanada.ca/index.html). The Internet has already greatly broadened access to such data, and it is difficult to forecast what web-based tools will be available in the future to facilitate use of such growing databases. However, I would counter the view that we will have an abundance of data by noting that vast quantities of *data* are not the same as large amounts of *knowledge* about how the aquatic system works and how it will respond to future natural disturbances and human activities. The large data sets resulting from our new data-gathering systems must be analyzed in ways that can generate such knowledge. For example, the tens of thousands of data points that make up time series of oceanographic conditions at many spatial locations must first be analyzed with sophisticated statistical techniques to succinctly describe patterns in space and time. Only then can scientists determine whether those conditions are related to changes in fish survival rates.

However, if the experience in the space sciences over the last 4 decades is any indication, we may be entering a period when the sheer amount of raw data from electronic sensors ('a fire hose of data streams' according to one anonymous sage) will overwhelm our ability to analyze them in a timely manner. For instance, there are computer tapes and hard disks filled with many terabytes of unprocessed data downloaded from terrestrial mapping satellites. There is no doubt that, in collaboration with climate scientists and oceanographers, fisheries scientists will become increasingly sophisticated in applying new methods to analyze similar large data sets to generate the knowledge that they need (e.g., Myers 2001). However, I am skeptical that limited budgets will be used correctly to balance the high costs of advanced and exciting new electronic sensing systems with the relatively modest costs of skilled people needed to analyze the resulting data. As evidence for this viewpoint, just think about how many data sets are already sitting unanalyzed in files of fisheries scientists, oceanographers, and climatologists.

Modern-day stock assessment is fundamentally a process of estimating biological risks owing to our limited understanding of the structure and functioning of aquatic systems and uncertainties inherent in data. Bayesian statistical methods for describing and estimating those uncertainties are becoming commonplace and will likely play a greater role in future fish stock assessments as more scientists learn the methods and software. As well, expectations will continue to rise for taking uncertainties into account. Good analyses of stochastic models generate probability distributions of potential outcomes for each possible management action. Ideally, those distributions should also include economic and social indicators (see section below on 'Advice for managers'). Scientists would then provide outputs from such models to decision makers (i.e., risk managers), who must then make difficult choices among management options. They must take into account not only the quantified information provided by stock assessment scientists, but also other factors that are difficult to quantify, such as social and economic benefits/costs of each action, and long-run intangible benefits to society.

Scientists can help managers make difficult decisions by conducting extensive sensitivity analyses to extend their initial, 'best-fit' analysis. To be of greatest value to decision makers, sensitivity analyses should not just focus on describing probability distributions of outcomes. Instead, they should also show how the rank order of alternative management actions changes under a variety of assumptions about uncertainties in (1) parameter values, (2) structural forms of model components, (3) management objectives (including component indicators and degrees of risk aversion), and (4) other components of the analysis. Such sensitivity analyses will help identify those management actions that are most robust to uncertainties, i.e., that remain best across a wide range of assumptions. One example of this approach is the International Whaling Commission's identification of a management procedure that dealt with uncertainty about sub-stock structure of certain whale populations (e.g., de la Mare 1996). Another reason for conducting sensitivity analyses is to identify the highest priorities for future research, where additional knowledge would sufficiently reduce uncertainty to where it might affect the rank order of management options.

Stock assessment scientists can also help managers by explicitly providing quantitative estimates of tradeoffs that are likely to occur between indicators for any given set of alternative management actions. Uncertainties in estimates of those tradeoffs would also be useful to indicate the confidence placed on the mean or median values. For instance, analysts might estimate that management option A will likely produce \$500,000 (±\$100,000) in catch and a 50% (±10%) chance of depleting the population below some acceptable level, whereas option B will produce only \$300,000 (±\$100,000) in catch but will reduce the chance of depletion to 20% (±10%). Managers armed with such explicit information about tradeoffs are more likely to reach a well-informed decision than without such numbers.

This last example raises another point that is broadly debated and where practices vary widely. How much information about uncertainties should stock assessment scientists convey to decision makers? Of course, there is no definitive answer because it depends on the situation and the people involved in giving scientific advice and in making decisions. Too much information about uncertainties may lead to confusion or misunderstanding by managers, most of whom are not quantitatively trained. In the increasingly common situations in which stakeholders are involved in decision making, presentation of uncertainties around results may lead some to choose whichever end of the probability distribution or confidence interval is most suitable for their interests and to press managers to act accordingly. For this latter reason, a former high-level fisheries manager once told me that he routinely asked his scientific analysts to avoid mentioning such uncertainties. I have great respect for decision makers who must make difficult tradeoff decisions, and I have never been in such a position. Even so, I believe that more information about the implications of management options is better than less. Scientific advisors can reduce the abuse of information about uncertainties by emphasizing results about which we are quite certain over those about which we are less certain (Rosenberg 2007). As well, sensitivity analyses provide an opportunity to educate everyone involved in decision making about the potential detrimental effects of choosing just one end or the other of an uncertainty range.

Fisheries scientists face a broader strategic question related to how to present uncertainties. In situations where economic and social considerations have histori-cally tended to dominate ecological ones (possibly because ecologists are so thor-ough at describing the uncertainties in their analyses), should scientists inflate the estimated biological concerns to hedge the resulting management decisions toward a more biologically conservative decision? I would emphatically answer 'no'. Instead, scientists should 'tell it like it is' based on the data and without any delib-erate hedging assumptions, and should provide decision makers with extremely thorough sensitivity analyses that indicate potential long- and short-term outcomes across a wide range of assumptions and management actions. Furthermore, if scientists get caught in the game of making biologically conservative assumptions in order to hedge results in that direction, they will at the very least lose credibil-ity with decision makers and at the very best, they will create confusion because decision makers won't know how much the results shown to them have already been biased. A more proactive approach is for fisheries managers to ask where the uncertainties are in the analyses of the *economic and social* impacts of proposed management options (Peterman 2004). Usually there are few or no such analyses. One can then ask, how much weight should a decision maker put on estimates of outcomes that do not reflect uncertainties that we know must exist in the system? In such cases, the thorough evaluation of *ecological* uncertainties should at least give some comfort to decision makers about the possible ecological outcomes.

Effective communication is a topic that runs through several of the examples above. Stock assessment scientists need to further improve the quality and crea-tivity of their presentations of complex quantitative analyses to decision makers and stakeholders, even though some innovative approaches are already emerging (e.g. AMOEBA plots in Collie et al. 2003). Scientists should participate in meetings with decision makers and stakeholders to ensure that everyone thoroughly under-stands the stock assessment results, their implications, and especially the rationale behind any counterintuitive or controversial outcomes. Good documentation and communication packages take considerable time to prepare and refine, so sufficient time and technical support should be given to scientists to achieve this goal. One exciting prospect for the future comes from researchers in computing science and cognitive science/psychology who have developed virtual-reality hardware and software to interactively visualize extremely complex data sets (e.g., van Dam et al. 2002). Not only can such visualization systems help scientists understand complex relationships among multiple components, but they can also help decision makers to see implications of complex tradeoffs among multiple indicators. I hope to see fisheries scientists actively involved in this type of work in the future.

A key challenge for stock assessment scientists is to determine 'Which improve-ments in stock assessment models (for instance, by including more detail such as spatial structure and movement among sub-populations, temporally changing productivity, etc.) will actually lead to improved performance in terms of management objectives?' This question must be asked because changes to stock assessment models are common, yet some of them do little but add unnecessary minor variation to predicted yields and abundances. Changes to stock assessment models must be viewed as only one part of a much larger fisheries management system, which includes steps for data collection, data analysis, decision-making, and regulation-setting, interpretation of those regulations by local managers, enforcement of regulations, and actual effects on the natural system by fish harvesting and other human activities (Fig. 10.1). Furthermore, each of these steps has uncertainties (Fig. 10.1). For instance, environmental variability and sampling variation lead to bias and imprecision in estimates from observed data, and there is communication error among scientists, managers, and stakeholders. As well, there is 'outcome uncertainty', which is the deviation between management targets and actual outcomes. Outcome uncertainty is affected by the degree of compliance with regulations by harvesters and by physical and biological sources of variation in catchability, for instance (Rosenberg and Brault 1993; Rice and Richards 1996; Holt and Peterman 2006). This important source of variability has often been ignored in stock assessments but is increasingly being recognized as important.

The question posed at the start of the previous paragraph is currently being answered by using computer models that simulate the entire set of system components just described (Fig. 10.1). They include dynamic simulations of (1) the natural physical and biological system as well as the sampling of that system with observation error (the 'operating model' component, Punt 1992), (2) scientific assessments

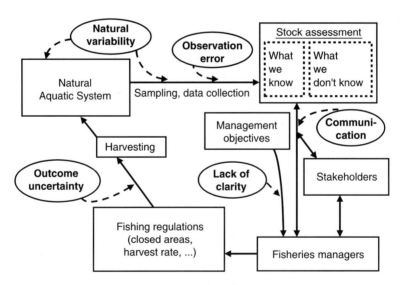

Fig. 10.1 A conceptual diagram of the flow of information and actions in a typical fishery system. Rectangles represent components of the system, solid arrows indicate flows of information and actions among components, and ellipses represent major sources of uncertainty (Adapted from Peterman 2004. With permission of the publisher)

(estimation of parameters and state variables), (3) subsequent management choices of regulations based on the simulated assessments, (4) responses of simulated harvesters to those regulations, and (5) effects of that harvesting (or other human activities) on fish population dynamics. Walters and Martell (2004) point out that Hilborn (1979) first conducted such 'closed-loop simulations' (a term drawn from adaptive control theory), which have subsequently been called 'management-procedure evaluations' (Butterworth and Punt 1999) and 'management-strategy evaluations' (MSEs) (Sainsbury et al. 2000). The latter two names reflect an essential purpose of this modeling approach, namely to determine how well management objectives can be met by a given management procedure (composed of the processes of data collection, data analysis/parameter estimation, state-dependent decision-making rules, and implementation of regulations), especially given uncertainty in all of these components as well as natural variability (Walters 1986; de la Mare 1996). Such fisheries models are already in wide use. For instance, see the entire issue of volume 56(6) of the ICES Journal of Marine Science in 1999. These MSEs can also be used to determine the relative importance of different sources of uncertainty for meeting objectives. For example, outcome uncertainty might swamp improvements in sampling or stock assessment methods, suggesting that much tighter enforcement might be needed rather than a more detailed assessment model. Such MSE models of fisheries systems have the same benefits and limitations as other simulation models, but they attempt to take into account as many sources of major uncertainties as are feasible. I agree with Sainsbury (1998), who says that despite the challenges of building models to conduct management-strategy evaluations of fisheries, such models are going to play increasingly important roles in the future.

10.3   Changing Physical and Biological Conditions

Changes in climate and their effects on freshwater and marine systems are at the top of research agendas of many fisheries scientists, but such changes are only one example of a much broader problem that must be addressed more effectively in the future. That problem is non-stationarity, which reflects a changing mean and/ or variance over time in factors such as oceanographic currents or productivity of fish populations. Non-stationary processes include phenomena variously called low-frequency variability, regime shifts, decadal-scale variability, autocorrelation, and time trends. For decades, most methods of analysis in fisheries science have assumed stationarity (constant mean and variance) for the processes being studied. However, there is now strong evidence that non-stationarity is very common and that whatever form it takes, such changes in productivity have important implications for scientists, managers, conservation goals, and the fishing industry because they may lead to incorrect conclusions about the status of fish populations or appropriate management actions (Walters 1987). Fisheries scientists have been applying various types of models and statistical methods to deal with this situation. These include explicitly modeling process variation as well as observation error through state-space models and errors-in-variables models, but there is still much more to do in the future to capture the many types of underlying sources of variation.

Changes in physical and biological conditions at both short and long spatial and temporal scales also create another problem, confounding, which will continue to plague fisheries scientists and managers. Confounding of interpretation occurs when observed changes cannot be uniquely attributed to causal mechanisms because two or more natural processes change simultaneously, or even at the same time as human activities. Such confounding usually leads to lack of clarity about which actions managers should take. The fisheries literature is filled with cases like this, where debates raged for years about the causes of collapses of fish populations because environmental changes occurred at about the same time as major changes in management regulations or harvesting effects (e.g., Pacific sardine in the 1950s and 1960s, northern cod in eastern Canada in the 1990s). No amount of sophisticated statistical analysis will remove substantial confounding. To reduce uncertainty about causal mechanisms, we can implement active adaptive management, whereby experimental designs of management actions can help separate causal mechanisms of future changes (Walters 1986). In the presence of time-trended climatic changes, such experiments would need to use a "staircase design", in which replicates of actions are begun at different times in different places (Walters et al. 1988). Even if a carefully designed experiment is not possible, scientists, harvesters, and managers should be looking for opportunities to create comparison groups. However, the barriers to taking this experimental approach are still large, despite some prominent successes in fisheries systems (Walters 2007), so counting on active adaptive management to reduce the problems of confounding may be overly optimistic in many situations.

Much has been written about potential effects of climatic change on fish populations, so I will not focus on such forecasts here. Instead, I will discuss five examples of how such changes, however they unfold in the future, affect how fisheries science should be conducted. First is the general problem of non-stationarity. As explained above, to the extent that climatic change alters underlying average values or variances of parameters and processes, historical data will become less relevant for estimating current, let alone future, status of fish stocks and their responses to management options. Thus, stock assessment scientists will need to mainly use recent or more representative subsets of data chosen from the long historical data series. For example, scientists can use informal data-discounting schemes or more rigorous state-space and time-series models that attempt to track underlying climate-induced changes.

A second, related problem, is that historical data may have insufficient contrast or range to cover forecasted future conditions. Climate models indicate that, at least in temperate regions, there will be changes in both freshwater habitats (such as magnitude and seasonal timing of snowmelt, summer low-water levels, summer high temperatures, and flash-flood levels), and marine habitats (through altered mixed-layer depths, timing, and magnitude of spring blooms, surface, and sub-surface currents). There will also likely be changes in abundances of prey, predators, and competitors. Undesirable invasive species, blooms of toxic algae, and appearances of masses of coccolithophores may also become more common as habitats are disrupted (we have seen the latter in the Bering Sea). Without historical data to reflect extremes of such climate-driven conditions, forecasts of effects on fish population dynamics are going to be highly uncertain because they will be extrapolations beyond the range of observed values of the independent variables. Nonlinearities are so common in natural systems that such extrapolations should be interpreted very cautiously. This challenging situation suggests an important point, that more extensive, statistically powerful, and efficient sampling and monitoring designs will be needed in the future to provide rapid feedback for stock assessment and decision making because unexpected changes will occur due to imperfect forecasts from models. If the time lag is too great between monitoring and providing that feedback, management objectives are less likely to be met.

A data-related issue provides the third example of how climatic change will affect how scientists conduct fisheries research. Widely used indicators of ocean conditions such as sea-surface temperature (SST) may no longer be good representations of characteristics of water masses encountered by fish. For instance, during global warming, nearshore SST may no longer be as correlated with Pacific salmon survival rates as in the past (Mueter et al. 2002) because the relationships among component processes that contribute to SST (winds, surface currents, upwelling, solar radiation) may change, as might their relationships with prey and predator abundances. So not only will non-stationarity result in changing values of variables, but relationships among variables could also change. There are already many cases in which relationships between environmental variables and fish variables have broken down (Drinkwater and Myers 1987; Walters and Collie 1988), and climatic change may increase the number of such cases. These potential problems again add caution to interpretation of analyses based on existing relationships between environmental and fish variables.

Fourth, many scientists believe that the most reliable forecasts of fish-related variables in the face of climatic change will be about spatial distributions at the edges of their ranges, rather than for overall survival or growth rates. This view, in combination with the basin-model concept of fish distribution (MacCall 1990) and the observation that recruitment of fish populations tends to be most variable at the edges (Myers 1991), suggests that early-warning signs of climate-driven changes in growth, reproduction, and survival rates, are most likely to appear at those edges, rather than in the most ideal habitats. Such changes have already been observed for various fish species at the southern or northern ends of their ranges in the North Atlantic and North Pacific Oceans (e.g., Atlantic cod, *Gadus morhua*, Drinkwater 2005 and Rothschild 2007; and Pacific hake, *Merluccius productus*, Benson et al. 2002). Increased concerns about conservation of such populations will lead to scientific dilemmas and policy conflicts. Specifically, hard questions will be asked about how much should be spent on research on those populations of concern that may appear doomed to extirpation and how, for instance, the U.S. Endangered Species Act and Canada's Species-at-Risk Act will be applied in such cases, given the effect of such rare populations on reducing harvest rates on other, more abundant populations.

Finally, in the future, we will need two parallel research streams in fisheries science, mechanistic, and non-mechanistic. We should continue to focus on understanding mechanisms linking changes in climatic forcing to growth, reproduction, survival, and distribution of fish populations. However, the chain of such linkages between physical variables and fish population responses is long, and each link contains numerous uncertainties, some of which are very large. It may therefore be a long time before we can make reliable, mechanistically based forecasts of effects of climatic change on most fish species at local or even regional levels. Therefore, to provide useful advice to decision makers in the meantime, we also need to conduct research on purely descriptive statistical methods that do _not_ rely on a detailed knowledge of such mechanisms. Such methods include neural networks and other techniques in the newly emerging field of 'ecological informatics' (Chon and Park 2006). Much may also be gained by conducting more time series modeling (Chatfield 1989), including use of Kalman filtering and other such 'noise-reduction' methods, to more clearly estimate the underlying 'signal' (Walters 1986; Peterman et al. 2000). In order for non-mechanistic statistical models to be useful for managers, they should be linked with the efforts to design the more rigorous and rapid-detection sampling programs that were noted above. Stock assessment scientists could then use models for management-strategy evaluations to identify appropriate management responses to changes observed in the aquatic system (e.g., the evaluation of rotational harvest strategies on sea cucumber in British Columbia, Humble et al. 2007). Of course, non-mechanistic models are also vulnerable to being wrong due to not taking into account changing relationships among variables, as noted previously. It is not clear, though, how much better forecasts from mechanistically based research will be, especially given the points above about the range of historical data and the questionable future usefulness of today's best environmental indicators of changes in fish populations.

10.4 Human Dynamics

Stock assessment models are incomplete in their evaluations of potential effects of management actions unless they dynamically consider the influence of human activities. A well-known example of this is the economically driven 'high-grading' of catches (to retain higher-valued fish and discard others) that has frequently occurred after certain regulations were implemented. Lack of full compliance by fish harvesters with regulations is another familiar example. Therefore, data analyses and stock assessment models of the future should incorporate humans as dynamic, not static, elements, while also representing uncertainties in those components. There are already examples of this approach, but they are the exception, rather than the norm (e.g., Hilborn 1985; Gillis et al. 1995; Dorn et al. 2001; Ulrich et al. 2002 to name but a few). The paucity of examples is largely due to a lack of data on how human actions can be dynamically linked to the ecological system. One way to improve this situation is for fisheries scientists to conduct more research in conjunction with social scientists, perhaps posing hypotheses about humans being analogous to natural predators and attempting to understand their dynamics in the context of numerical responses, functional responses, and nonlinear responses to various incentives or disincentives created by management policies. As well, stock assessment scientists should explicitly parameterize outcome uncertainty and include it in their models to reflect possible non-compliance with regulations and unexpected selectivity of fishing gear. This need for modeling outcome uncertainty will become much more important in the future, in part because of limited government enforcement budgets.

10.5 Shifting Roles of Institutions and Organizations

To this point, I have discussed fisheries science without reference to where it is conducted. Obviously, the main places have been government research laboratories and universities. However, changes are coming. Governments and universities will probably continue to be the main locations for scientific research (including stock assessment), but non-governmental organizations will probably grow in importance. This is in part due to constrained government budgets, which create an incentive to move more research from government budgets to the private sector. One well-known example is New Zealand's crown corporation, the National Center for Fisheries and Aquaculture in the National Institute of Water and Atmospheric Research (NIWA). I have not yet seen rigorous evaluations of whether such new arrangements are more successful in terms of maintaining productive aquatic systems that also generate economic and social benefits over the long term.

In addition, conservation groups are now either directly doing research in aquatic science or are providing large sources of funding to university and government researchers. The Monterey Bay Aquarium Research Institute (MBARI), founded by entrepreneur David Packard, is a prime example. With over 200 scientists and other staff, MBARI is larger than some government research laboratories in aquatic sciences (www.mbari.org). The Gordon and Betty Moore Foundation of San Francisco is another major supporter of research in marine environments, with about $44 million contributed to such work in 2006 alone (www.moore.org). I expect that we will see increasing influence of such non-governmental organizations on the direction of fisheries science in the future, in both absolute and relative terms due to decreasing government financial flexibility. In addition, several large conservation organizations related to the marine and freshwater environments are hiring more highly trained and experienced scientists to conduct their own independent analyses.

Next, 'eco-certification' groups such as the Marine Stewardship Council are coordinating evaluations of the quality of scientific information that supports management of particular fisheries. We have already seen examples, for instance, where scientists in the Alaska Department of Fish and Game were tasked with quantitatively determining spawning (escapement) goals for some Alaskan salmon populations when a certification team noted a deficiency in justifications for those goals (Marine Stewardship Council 2000). In the future, the lure of becoming certified may create incentives for more thorough research and stock assessments than might otherwise have been conducted. To the extent that this feedback improves knowledge from research, transnational certifying organizations such as the Marine Stewardship Council can be viewed as altering global standards for fisheries science, as well as management (Constance and Bonanno 2000).

Fishing industry groups are another type of non-governmental organization that is likely to have more influence on fisheries science in the future than at present. As governments transfer more costs to the fishing industry, the latter justifiably wants more involvement in stock assessments. Many non-profit research arms of the industry have been created around the world from royalties on catches. They often hire their own scientists who conduct independent stock assessments or, in the best arrangements, who work closely with government scientists to conduct collaborative stock assessments (e.g., Canadian Sablefish Association, Herring Conservation and Research Society, and the Canadian Groundfish Research and Conservation Society).

Governments and these fishing industry groups are both increasingly funding stock assessment research at universities as well, in part due to the shortage of quantitatively trained fisheries scientists. Such university-based research includes development of operating models and management-strategy evaluations, better ways to deal with non-stationarity, and new approaches to assessments in extremely data-limited situations. These initiatives provide excellent applied examples for educating graduate students, post-doctoral fellows, and research assistants in fisheries science. In university graduate schools, such as my own, where students obtain an interdisciplinary education as well as training in quantitative methods of fisheries science, novel stock assessment models that include human dynamics are more likely to be developed than in standard natural science units or government fisheries science laboratories which do not traditionally employ social scientists.

Rapid advances in fisheries science may come from future collaborations among all of these groups. There are already many successful examples of close functional linkages between scientists in different institutions, for instance, in cases where government laboratories are located on or near university campuses or high-profile non-profit research institutes. Well-known examples include the (1) Woods Hole Oceanographic Institution and the U.S. National Marine Fisheries Service's (NMFS) Biological Laboratory, (2) the Scripps Institution of Oceanography and the NMFS Southwest Fisheries Center, and (3) the U.S. Fish and Wildlife Service's Co-operative Fish and Wildlife Units at university campuses. I hope to see such collaborative arrangements expand in the future to include non-governmental conservation organizations and fishing industry groups, all working together to improve scientific knowledge that would provide new long-term solutions to difficult management issues.

The reduction in government-based research in fisheries science and the resulting increase in research by other organizations create two important problems that will need to be addressed. First, it will be a challenge to coordinate and balance the research knowledge produced by different groups that may have different objectives. Second, if governments reduce their commitment to long-term monitoring and surveys, it may be difficult to maintain the quality of data sets, many of which will be essential for detecting climate-driven changes, for instance. Reduced government monitoring will therefore either lead to weaker scientific advice to decision makers or to a greater responsibility on non-governmental organizations to conduct monitoring and make data readily available through on-line databases (with appropriate oversight from government staff to maintain quality control). However, to date, most environmental NGOs have not wanted to pay for such basic government operations as ongoing monitoring. This dilemma is yet to be resolved.

10.6   Synopsis

Previous sections describe many substantial challenges for fisheries scientists in the future, most of which have important management implications. For example, analyses of potential relationships among biological variables and/or environmental variables will be plagued by nonstationarity, confounding caused by simultaneously changing factors, and (potentially) altered relationships among variables as a result of climate change. Compared to the present, fish stock assessment models will likely become more complex biologically, with a greater diversity of indicators of objectives, and more explicit consideration of uncertainties (both process variation and observation errors). There will be an increasing need to simulate human activities besides fishing as dynamic components of fisheries systems. More extensive data and advanced statistical analyses will be required to estimate parameters for such models. There will be greater challenges in communicating such complex and quantitative material to managers, as well as stakeholders, particularly if the latter become more involved in decision making. Finally, fisheries scientists will find themselves amid changing roles for existing research-intensive institutions and new collaborations with emerging ones like conservation organizations and fishing industry groups. Below, I provide advice for how managers and young scientists can cope with these challenges while perhaps also improving fisheries situations in the future.

10.7   Advice for Managers

Fisheries managers should expect certain procedures from stock assessment scientists. For instance, early in a stock assessment process, scientists should ask for management objectives so that they can quantitatively evaluate the relative merits of different management actions. The objectives will include broad categories (e.g., economic, biological, and social) representing various interest groups, indicators for those categories, their relative weightings, and targets as well as unacceptable constraint levels for the indicators. In addition, the time frame for achieving the objectives should be specified, along with acceptable limits on the uncertainty of achieving them. The nature and extent of managers' risk aversion should also be identified. Such questions are rarely easy to answer because there are usually many possibilities and no clear definitive answers. One way to handle this situation is for managers to provide scientists with ranges of answers that can then be run through stock assessment models using extensive sensitivity analyses, as described above. By iteratively repeating the process of analysis and reporting results to managers, scientists might help bring to light new relevant indicators. Ideally, some management options will also be ruled out due to their rankings being too sensitive to uncertainties.

Second, scientists should explicitly incorporate key uncertainties into their analyses and describe for managers the resulting uncertainties in outcomes of contemplated actions. There are several types of uncertainties, including those in: (1) the original data (imprecision and bias), (2) forms of relationships among components of analyses, (3) estimated parameter values, and (4) components of management objectives. As described above, scientists should conduct extensive sensitivity analyses on these uncertainties to directly help managers by showing how the rank order of alternative management actions changes under a variety of assumptions about these uncertain components. Scientists should also quantify how tradeoffs between indicators vary across alternative management actions so as to make the implications of managers' choices clear. Such tradeoffs among social, economic, and biological objectives are common, so it is critical that fisheries managers expect the same high level of standards for taking uncertainties into account in the social and economic advice as they do in the biologically-based advice they receive (Peterman 2004). Currently, this does not appear to be the case in most situations. Typically, stock assessment scientists are expected to explicitly take many ecological uncertainties into account, yet advisors/stakeholders who provide input to managers regarding economic and social implications do not usually provide such information on uncertainties, let alone take them into account in their estimates of economic and social impacts of proposed management actions. Many managers then unjustifiably conclude that the ecologists know less about what is going to happen than those who do not even mention uncertainties! This is illogical and may lead to decisions that do not reflect serious ecological risks, which could detrimentally influence economic and social risks as well.

10.8   Top Eight Points of Advice for Young Fisheries Scientists

Forecasting the future of fisheries science and what scientists will need to know is fraught with as many pitfalls as forecasting next year's abundance of a fish stock. However, it is feasible to give advice about generic skills and experience for preparing for future work in this field. Young scientists, including those in graduate school, should ideally seek to build the following portfolio of eight types of skills and experience.

1. There is no substitute for a solid ecological background. This does not mean that young scientists need to have specific, taxonomically detailed knowledge about the full range of aquatic organisms. Instead, they need extensive exposure to theoretical and empirical examples of the types of processes that generate dynamics of aquatic systems at various spatial and temporal scales, such as linear and nonlinear interactions among organisms and with their environments, lag effects, thresholds, cumulative effects. Such processes affect spatial distributions of populations and interactions among predators, prey, parasites, and competitors. Ideally, learning about these processes from an applied perspective helps to link their importance to environmental management questions.

2. Tied to the first point is general knowledge about atmospheric and physical/biological oceanographic processes and their effects on conditions in oceans, rivers, and terrestrial habitats. Again, the focus should be on key processes that create the dynamics of these systems.

3. A thorough knowledge of the literature is essential, including research that came before electronic databases! We must build on what we have already learned and not re-invent the wheel. Ideally, young scientists should aim to become a "walking library" to the extent that they know the literature well enough to engage in high-level discussions.

4. Excellent quantitative skills will become even more essential than they already are for fisheries scientists. People with these quantitative skills will be clearly distinguishable from the many others who only have a pure ecological background. Two types of quantitative skills will be most beneficial. First, statistical methods currently used by fisheries scientists go far beyond simple classical hypothesis tests and instead emphasize estimation of parameters and their properties (Hilborn and Mangel 1997). Some widely used advanced methods are nonlinear regression, hierarchical models (including mixed-effects models), time-series models, Generalized Linear Models (GLM), Generalized Additive Models (GAM), bootstrapping, and spatial statistics/GIS (Geographic Information Systems) methods, to name just a few. Newer methods will likely emerge in the future as researchers in the discipline of statistics become more actively engaged in applied fields like fisheries and witness the problems that we face. Bayesian statistical methods will likely be commonplace, including analyses with multiple uncertain parameters that use importance resampling, Monte Carlo Markov Chain (MCMC) methods, or other techniques that are yet to emerge for such complex problems.

   The second type of quantitative skill needed by future fisheries scientists is the ability to write their own simulation models using common programming languages. Currently popular ones are Visual Basic, C++, R, or S-Plus. These models will continue to be largely stochastic and will iterate across many uncertainties in a decision analysis framework (Walters 1986).

   Statistical models and simulation models are being increasingly combined. For example, before implementing new statistical estimation algorithms, we can determine their statistical properties such as bias and precision using operating models (Hilborn and Walters 1992). Such models create 'real' (simulated) data sets with known, user-specified properties and then simulate the sampling and parameter estimation steps. Estimates are then compared with the 'real' values to determine bias.

5. Most fisheries scientists will benefit from knowing how to design statistically rigorous sampling programs and from taking part in field surveys to see first-hand what sources of bias and observation error can be embedded in data. Cooperative education programs and internships can provide young scientists with such experience. Ideally, students' thesis topics should be chosen to facilitate development of the first five types of skills above while at the same time applying them to real-world fisheries problems.

6. Fisheries scientists of the future will also benefit from adding some exposure to other disciplines to their solid disciplinary training. Such interdisciplinary work should include not only other areas in natural sciences such as oceanography, hydrology, and climate science, but also some social sciences such as resource economics, environmental law, and policy analysis. Such scientists will then be more likely to conduct analyses needed by decision makers to make difficult tradeoff decisions.

7. Fisheries scientists often work in collaborative teams. However, the most effective contributors to such team efforts are those who not only have experience with such groups but who also have strong ecological and quantitative backgrounds, as described above, and have independent problem-solving skills based on good creative thinking as well as critical thinking.

8. Excellent written and oral communication skills are essential. These are required to communicate complex technical material clearly to non-technical audiences (managers, stakeholders). Such communications are commonly hindered because some recipients of scientific advice may not properly interpret apparently simple concepts such as 'probability', which has been shown to have five other interpretations besides 'chance' (Teigen 1994).

If young fisheries scientists can obtain all of these skills, and if some of them move on to become decision makers, future fisheries will indeed be in good hands!

Acknowledgments I am grateful to the people who provided critical comments on this manuscript: Ashleen Benson, Sean Cox, Jeremy Collie, an anonymous reviewer, Richard Beamish, and Brian Rothschild.

References

Benson AJ, McFarlane GA, Allen SE, Dower JF (2002) Changes in Pacific hake (*Merluccius productus*) migration patterns and juvenile growth related to the 1989 regime shift. Can J Fish Aquat Sci **59**:1969–1979

Butterworth DS, Punt AE (1999) Experiences in the evaluation and implementation of management procedures. ICES J Mar Sci **56**:985–998

Chatfield C (1989) The Analysis of Time Series: An Introduction. 4th edition. Chapman & Hall, London, 241 p

Chon T-S, Park Y-S (2006) Ecological informatics as an advanced interdisciplinary interpretation of ecosystems. Ecol Inform **1(3)**:213–217

Collie JS, Gislason H, Vinther M (2003) Using AMOEBAs to display multispecies, multifleet fisheries advice. ICES J Mar Sci **60**:709–720

Constance DH, Bonanno A (2000) Regulating the global fisheries: The world wildlife fund, unilever, and the marine stewardship council. Agric Human Values **17**:125–139

de la Mare WK (1996) Some recent developments in the management of marine living resources. In: Floyd RB, Sheppard AW, De Barro PJ (eds) Frontiers in Population Ecology. CSIRO, Melbourne, pp 599–616

Dorn MW (2001) Fishing behaviour of factory trawlers: A hierarchical model of information processing and decision-making. ICES J Mar Sci **58**:238–252

Drinkwater KF (2005) The response of Atlantic cod (*Gadus morhua*) to future climate change. ICES J Mar Sci **62**:1327–1337

Drinkwater KF, Myers RA (1987) Testing predictions of marine fish and shellfish landings from environmental variables. Can J Fish Aquat Sci **44**:1568–1573

Gillis DM, Peterman RM, Pikitch EK (1995) Implications of trip regulations for high-grading: A model of the behavior of fishermen. Can J Fish Aquat Sci **52**:402–415

Hilborn R (1979) Comparison of fisheries control systems that utilize catch and effort data. J Fish Res Board Can **36(12)**:1477–1489

Hilborn R (1985) Fleet dynamics and individual variation: Why some people catch more fish than others. Can J Fish Aquat Sci **42**:2–13

Hilborn R, Mangel M (1997) The Ecological Detective: Confronting Models with Data. Princeton Monographs in Population Biology. Vol. **28**. Princeton University Press, Princeton, NJ, 315 p

Hilborn R, Walters CJ (1992) Quantitative Fisheries Stock Assessment: Choice, Dynamics, and Uncertainty. Chapman & Hall, New York, 570 p

Holt CA, Peterman RM (2006) Missing the target: Uncertainties in achieving management goals in fisheries on Fraser River, British Columbia, sockeye salmon. Can J Fish Aquat Sci **63**: 2722–2733

Humble SR, Hand CM, de la Mare WK (2007) Review of data collected during the annual sea cucumber (*Parastichopus californicus*) fishery in British Columbia and recommendations for a rotational harvest strategy based on simulation modelling. DFO Can Sci Advis Sec Res Doc **2007/054**

MacCall AD (1990) Dynamic Geography of Marine Fish Populations. University of Washington Press, Seattle, WA, 153 p

Marine Stewardship Council (2000) The summary report on certification of commercial salmon fisheries in Alaska, 33 p (accessed on line at www.msc.org, September 2000)

Morgan MG, Henrion M (1990) Uncertainty: A Guide to Dealing with Uncertainty in Quantitative Risk and Policy Analysis. Cambridge University Press, Cambridge, 332 p

Mueter FJ, Peterman RM, Pyper BJ (2002) Opposite effects of ocean temperature on survival rates of 120 stocks of Pacific salmon (*Oncorhynchus* spp.) in northern and southern areas. Can J Fish Aquat Sci **59**:456–463, plus the erratum printed in Can J Fish Aquat Sci **60**:757

Myers RA (1991) Recruitment variability and range of three fish species. NAFO Sci Coun Stud **16**:21–24

Myers RA (2001) Stock and recruitment: Generalizations about maximum reproductive rate, density dependence, and variability using meta-analytic approaches. ICES J Mar Sci **58**:937–951

Peterman RM (2004) Possible solutions to some challenges facing fisheries scientists and managers. ICES J Mar Sci **61**:1331–1343

Peterman RM, Pyper BJ, Grout JA (2000) Comparison of parameter estimation methods for detecting climate-induced changes in productivity of Pacific salmon (*Onchorhynchus* spp.). Can J Fish Aquat Sci **57**:181–191

Punt AE (1992) Selecting management methodologies for marine resources, with an illustration for southern African hake. In: Payne AIL, Brink KH, Mann KH, Hilborn R (eds) Benguela Trophic Functioning. S Afr J Mar Sci **12**:943–958

Quinn TJ, Deriso RB (1999) Quantitative Fish Dynamics. Oxford University Press, Oxford

Rice JC, Richards LJ (1996) A framework for reducing implementation uncertainty in fisheries management. N Am J Fish Man **16**:488–494

Rosenberg AA (2007) Fishing for certainty. Nature **449**:989

Rosenberg AA, Brault S (1993) Choosing a management strategy for stock rebuilding when control is uncertain. In: Smith SJ, Hunt JJ, Rivard D (eds) Risk Evaluation and Biological Reference Points for Fisheries Management. Can Spec Pub Fish Aquat Sci **120**:243–249

Rothschild BJ (2007) Coherence of Atlantic cod stock dynamics in the Northwest Atlantic Ocean. Trans Am Fish Soc **136(3)**:858–874

Sainsbury K (1998) Living marine resources assessment for the 21st century: What will be needed and how will it be provided? In: Funk F, Quinn, II TJ, Heifetz JN, Ianelli JE, Powers JF, Schweigert JF, Sullivan PJ, Zhang C-I (eds) Fishery Stock Assessment Models. Alaska Sea Grant College Program AK-SG-98-01, University of Alaska, Fairbanks, AK, pp 1–40

Sainsbury KJ, Punt AE, Smith ADM (2000) Design of operational management strategies for achieving fishery ecosystem objectives. ICES J Mar Sci **57**:731–741

Teigen KH (1994) Variants of subjective probabilities: Concepts, norms, and biases. In: Wright G, Ayton P (eds) Subjective Probability. Wiley, New York, pp 211–238

Ulrich C, Pascoe S, Sparre PJ, De Wilde J-W, Marchal P (2002) Influence of trends in fishing power on bioeconomics in the North Sea flatfish fishery regulated by catches or by effort quotas. Can J Fish Aquat Sci **59**:829–843

van Dam A, Laidlaw DH, Simpson RM (2002) Experiments in immersive virtual reality for scientific visualization. Comput Graph **26**:535–555

Walters CJ (1986) Adaptive Management of Renewable Resources. MacMillan, New York, 374 pp

Walters CJ (1987) Nonstationarity of production relationships in exploited populations. Can J Fish Aquat Sci **44(Suppl 2)**:156–165

Walters CJ (2007) Is adaptive management helping to solve fisheries problems? Ambio **36(4)**:304–307

Walters CJ, Collie JS (1988) Is research on environmental factors useful to fisheries management? Can J Fish Aquat Sci **45**:1848–1854

Walters CJ, Martell SJD (2004) Fisheries Ecology and Management. Princeton University Press, Princeton, NJ, 399 p

Walters CJ, Collie JS, Webb T (1988) Experimental designs for estimating transient responses to management disturbances. Can J Fish Aquat Sci **45**:530–538

Chapter 11
Some Observations on the Role of Trophodynamic Models for Ecosystem Approaches to Fisheries

Mariano Koen-Alonso

Abstract Useful trophodynamic models within an ecosystem approach to fisheries need to be simple enough that we can actually learn from them, but complex enough that we can believe that their results are reliable. Reliability is commonly assessed by comparing model outputs with data, but many food web models grow in complexity very quickly while the sources of information to parameterize and evaluate them do not grow as fast. Within the diversity of modeling frameworks currently available, their adequacy for addressing management issues largely depends on the scope and specific goals of the modeling exercise (e.g., exploration of competing hypotheses, strategic advice, tactical advice). Nonetheless, any management-oriented implementation needs to attain a balance between uncertainty and realism, and optimal model performance is expected to be achieved at some intermediate level of complexity. Furthermore, trophodynamic models can capture key aspects of ecosystem dynamics, but implementing ecosystem approaches to fisheries will also require integrating this dynamics and its uncertainty with all other aspects of the management process. One natural way of doing this is by considering trophodynamic models as operating models within a management strategy evaluation framework. Highly complex ecosystem models can be used to fulfill this role, but simpler models (e.g., "minimum realistic models [MRM]") are likely to be a better starting point. Recent advances in food web theory can provide useful building blocks and guidance for developing these "minimum realistic models." Although these models can be good candidates for capturing the joint dynamics of core components of the system, we can never be that we are not missing something important. Due to this fundamental uncertainty, embedding these simple trophodynamic models into management strategy evaluation frameworks and pursuing multiple modeling approaches can provide a sensible venue for assessing the robustness of management strategies to our alternative ways of representing trophodynamic effects. Also, because "minimum realistic models" are focused on a relatively small number of

M. Koen-Alonso
Northwest Atlantic Fisheries Centre, Fisheries and Oceans Canada,
80 East White Hills Rd., P.O. Box 5667, St. John's, NL, Canada
e-mail: mariano.koen-alonso@dfo-mpo.gc.ca

R.J. Beamish and B.J. Rothschild (eds.), *The Future of Fisheries Science in North America*,   185
Fish & Fisheries Series, © Springer Science + Business Media B.V. 2009

ecosystem components, they are well suited for matching their data requirements with operationally feasible monitoring programs. Despite the specific modeling approaches we choose, developing these operating models in conjunction with field programs that can provide the appropriate data to parameterize them and to assess their performance is fundamental to achieve reliable trophodynamic models for ecosystem approaches to fisheries.

Keywords Ecosystem-based management · food web · management strategy evaluation · maximum feasible model · minimum realistic model · operating model

11.1 Introduction

Trophodynamic models describe the joint dynamics of populations linked by trophic interactions. The origin of modern predator–prey theory can be traced back to the original Lotka–Volterra predator–prey model (Berryman 1992), and interestingly enough, one issue that Volterra was trying to explain with his predator–prey equations was the observed fluctuations in fisheries resources (Volterra 1928). Despite this early connection, fishery science actually developed within the realm of population ecology (Quinn II 2003; Mangel and Levin 2005), using single-species population dynamics models as core tools in the stock-assessment process (e.g., Quinn II and Deriso 1999).

Today, there is global consensus that fisheries management needs to move from its traditional single-species focus towards more holistic approaches (Clark et al. 2001; Garcia et al. 2003; Pikitch et al. 2004; Marasco et al. 2007). These new frameworks are aimed to integrate multiple uses of aquatic resources while maintaining fully functional ecosystems. This way of thinking emerges from the realization that (a) human activities impact much more than their intended targets, (b) these impacts can be of sufficient extent and magnitude to compromise ecosystem structure and function, (c) different human activities can affect ecosystem components in ways that may hinder each others resource base and/or affect ecosystem functioning due to their cumulative impacts, and (d) physical forcing (e.g., seasonality, environmental stochasticity) can be a critical driver of population dynamics, but environmental conditions can also show multiannual trends and patterns that significantly impact ecosystem productivity and organization (e.g., regime shifts).

Regardless, if only a handful of commercial stocks actually generate direct economic and/or social benefits, it is now clear that we are dealing with ecosystems altered by exploitation not just exploited populations. Hence, achieving a sustainable use of fisheries resources requires management scoped at the scale of the actual object subject to the exploitation impacts, not just at the scale of the components being directly targeted by human activities.

With this shift in perspective, community ecologists have been called upon to take a more central role in the provision of science support for management

(Mangel and Levin 2005), and with it, trophodynamic models are coming full circle to address, as in the days of Volterra, the joint dynamics of fisheries resources.

Along these lines, the goal of this chapter is essentially exploratory; it is an attempt to sketch how I see trophodynamic models fitting within an ecosystem approach to fisheries. To do so, I will start by outlining core elements of the overall approach, the type of applications trophodynamic models have been used for in that context, and some of the general issues we face when dealing with trophodynamic models. Then, I will go on to briefly describe the management strategy evaluation framework and argue that operating models within this framework will require the consideration of trophodynamic effects. Next, I will suggest that currently minimum realistic models are probably our best option to balance model complexity and reliability, making them a promising avenue for developing trophodynamic operating models. Consequently, I will provide a brief characterization of these models and their main caveats, and will summarize elements of food web theory that can provide an envelope for thinking about them. Based on these elements I will put forward some ideas and speculations that could be helpful for building trophodynamic minimum realistic models that can be used as operating models for management strategy evaluation. Finally, I will close with a succinct, but hopefully compelling, advocacy about the tremendous importance of gathering field data which are consistent with, and supportive of, the modeling approach being pursued.

11.2   Ecosystem Approach to Fisheries

The term "ecosystem approach to fisheries" refers to the holistic approach to management indicated above, but many other names have also been coined (e.g., "ecosystem-based fisheries management") (Link 2002b). These terms often broaden to include activities other than fishing (e.g., "ecosystem approach to management," "ecosystem-based management") (Garcia et al. 2003; Arkema et al. 2006).

Notwithstanding this diversity of names, and granting that each description may put the emphasis on different aspects, activities, and scales, a recent analysis did not find any significant difference among 18 proposed definitions for ecosystem-based management (Arkema et al. 2006). Some key ideas at the core of many of these definitions are (Christensen et al. 1996; Garcia et al. 2003; Arkema et al. 2006):

- The use of all resources should be sustainable in the long term (both in the ecological and socioeconomical senses).
- Management should be driven by explicit goals (i.e., objective-oriented) and must integrate, at some point, all human activities which share the ecosystem.
- Ecosystem structure and function must be preserved or restored (if feasible) in those cases where they have been significantly altered by human impacts. This idea implies the necessity of some operational understanding of ecosystem functioning.

- Uncertainties, weaknesses, and knowledge gaps need to be explicitly recognized and addressed (i.e., the approach should be explicitly designed as a permanent work in progress, always challenging itself), and management practices need to include provisions for unexpected changes in system conditions.

In the same way that the ecosystem approach to fisheries broadens the scope of management goals, the science supporting it is expected to require a wider set of tools and activities. For example, these activities would likely include expanding monitoring programs to incorporate variables beyond those of commercially important stocks (Link et al. 2002), selecting (i.e., identifying and formally evaluating their performance) and tracking a variety of community/ecosystem indicators (Fulton et al. 2005; Jennings 2005; Rice and Rochet 2005), and developing models which can capture the interactions among both ecosystem components and human activities (Yodzis 1994; Christensen and Walters 2003; Butterworth and Plaganyi 2004; Plaganyi 2007).

Although this discussion will revolve around modeling, it is important to highlight that the science required for ecosystem approaches to fisheries involves much more than just modeling. The task at hand is too big for one methodology to do it all. Modeling and simulation exercises are well suited for addressing issues for which we can reasonably describe/approximate the ecological mechanisms involved at the appropriate temporal and spatial scales. Trying to get at mechanistic explanations is essential in order to move from purely statistical descriptions towards actual understanding of what drives the dynamics of exploited ecosystems. It is at this point where models can show their full potential because they allow very precise representations of our ideas about ecological mechanisms and processes. Hence, we can use them for testing these ideas, to explore hypotheses of how the system may behave in the future, and if proven robust, to serve as tools for providing advice.

In the same way that modeling is only one component of the science required in support of ecosystem approaches to fisheries, this modeling work should not be constrained to multispecies or whole ecosystem models either. The advocacy for the use of these type of models in management is sometimes viewed with skepticism (Quinn II 2003), and some of these criticisms are valid; models in support of ecosystem approaches to fisheries do not necessarily require to be fully resolved neither up to the minimum biological details nor to capture the full extent of ecological complexity. The ecosystem approach to fisheries defines a conceptual domain within which management frameworks are developed, and like within any management framework, models need to be good enough to reliably inform the decision-making process. If this can be done with a simple model, then there is no reason not to do it. Still, multispecies and ecosystem models will play an increasingly important role as ecosystem approaches to fisheries are implemented and applied as some of the management questions we are facing today simply cannot be answered with single-species models (Lilly et al. 2000; Link 2002a; Butterworth and Punt 2003; Stefansson 2003b; Aydin et al. 2005). In this context, it is important to be explicit about which roles we expect these multispecies and ecosystem models to occupy, and what the main areas of concern are expected to be.

11.3   Trophic Models for Ecosystem Approaches to Fisheries

We can essentially distinguish three broad classes of applications for trophic models in a management context: conceptual, strategic, and tactical. The first one is essentially exploratory and research-driven; these applications are aimed to investigate an alternative hypothesis about system functionality in qualitative or semi-quantitative terms; such models are developed to probe our ideas about system structure and function (Yodzis 1998; Aydin et al. 2005; Guénette et al. 2006; Pope et al. 2007). These explorations are often very useful to set general boundaries for system behavior, to characterize major system features, and to promote integration and standardization of data sources (Mohn and Bowen 1996; Bundy 2001; Ciannelli et al. 2004; Trzcinski et al. 2006). This class of application has been the most common so far and probably will continue to be in the near future. Exploratory models often develop into the second class of applications, providing insights towards strategic management advice (Punt and Butterworth 1995; Yodzis 1998; Gislason 1999; Link 2003). These applications do not necessarily attempt to provide precise quantitative outputs; rather, their goal is to provide information to assess and compare medium- to long-term management paths and policies. The use of trophodynamic models as a tool for providing strategic advice is the class of applications that will likely show the highest growth in the years to come as ecosystem approaches to fisheries are more widely implemented. The third class of applications involves using trophic models to provide quantitative evaluations directly relevant to the short-term management processes (i.e., tactical advice, e.g., quota setting). Although work has been done with this class of application in mind (Magnússon 1995; Vinther 2001; Stefansson 2003a), this role is still being played for the most part by single-species stock-assessment models (e.g., Patterson et al. 2001).

When dealing with trophodynamic modeling, we also need to keep in mind that there is no single theory to describe predator–prey dynamics. For example, models derived from the basic Lotka–Volterra predator–prey model are different from those derived from Leslie's predator–prey model. In the former, consumption becomes the source from which predator growth occurs (i.e., prey consumed through a conversion factor becomes predator), while in the later consumption is used to set the limits for predator growth (i.e., the consumption is used to scale the predator's carrying capacity and does not necessarily affect its intrinsic growth rate) (Turchin 2003). This difference persists today embodied in the current Lotka–Volterra compliant predator–prey models (Turchin 2003) and those based in the logistic theory of food web dynamics (Berryman et al. 1995). One important consequence of these alternative perspectives is that, for models using biomass as currency, only Lotka–Volterra compliant ones respect the conservation of biomass principle (Ginzburg 1998). Nonetheless, the appropriateness of these mathematical representations of trophodynamics is still open to debate (Ginzburg 1998; Berryman 1999) and this discussion is particularly relevant for fisheries applications. Multispecies fisheries models commonly use consumption estimates by predators to estimate mortality rates for the prey, but many of them (most typically, as expected, those using numbers of individuals as model currency) do not use this very same consumption to generate growth rates for predator populations, though some of them can use prey availability to modulate growth rates (e.g., Gislason 1999; Begley and Howell 2004).

Another issue of debate is which equation should be used to model the functional response. The mathematical form of this function can have a large impact on model dynamics and predictions (Yodzis 1994; Mackinson et al. 2003; Fulton et al. 2003b; Koen-Alonso and Yodzis 2005) and there is a quite diverse range of alternative formulations available (Gentleman et al. 2003; Koen-Alonso 2007; Walters and Christensen 2007).

The type and level of model detail (e.g., age-structured, size-structured, biomass-based models) and resolution (e.g., how many species, spatial considerations) also pose issues that need to be dealt with (Fulton et al. 2003a, b; Sabo et al. 2005; Pinnegar et al. 2005; De Roos et al. 2007). Among them, the question of the entities being modeled deserves particular attention. Although using taxonomical species as the nodes in food web models is probably the most common approach, functional groups that capture trophically similar species is also a common practice. However, specific size-classes or stages of different species are often the ones that constitute a functional group; the relative body size of predators and prey is an important component in determining trophic interactions (e.g., Cohen et al. 1993, 2003). In many cases, food webs defined in terms of body-size entities can provide a more accurate depiction of the actual trophic structure (Raffaelli 2007), and the body of research focused on size-structured modeling of food webs is growing (Benoît and Rochet 2004; Andersen and Beyer 2006; Hall et al. 2006; Pope et al. 2007; Maury et al. 2007a, b).

From this diversity of theoretical options a fair set of alternative approaches have been applied to fisheries issues and are currently available for modeling aquatic food webs (Plaganyi 2007). These options include, but are not restricted to, well-known modeling frameworks like Ecopath with Ecosim (Walters et al. 1997; Christensen and Walters 2003), Atlantis (Fulton et al. 2004b; Smith et al. 2007), and Gadget (Stefansson 2003a; Begley and Howell 2004). As highlighted by Plaganyi (2007)'s thorough review, each modeling approach has its own virtues and caveats and its adequacy will depend on the specific objectives being sought. Given the uncertainties associated with model structure and parameterization, together with the often large number of assumptions and approximations required, it is widely recognized that robust answers will only be achieved if different modeling approaches actually agree (Fulton et al. 2003a). Furthermore, the use of multiple, structurally different modeling approaches to properly estimate uncertainty in stock assessments and forecasting is also being advocated for classical single-species models (Patterson et al. 2001). Thus, if we need more than one model to assess robustness of results, the issue is not really about which modeling approach is better (we already know that all of them are wrong in one way or another). The real issue is how, and in which context, different trophodynamic models could be employed to become useful for ecosystem approaches to fisheries.

Although all models have a place in research-driven applications, as we move towards practical implementations aimed to inform strategic and/or tactical management decisions, I believe useful trophodynamic models need to be simple enough that we can take them apart and fully understand why they behave the way they do, but complex enough that we can take their results seriously for the specific level at which we are aiming to provide the advice. Furthermore, we need to convince ourselves – and others – that model results are where reliable, reliability is essentially assessed by comparing model outputs with data and by providing some estimate about how uncertain the model outputs are expected to be.

Many food web models grow in complexity very quickly while the sources of information to parameterize and evaluate them do not grow as fast. Furthermore, adding complexity to a model increases uncertainty around its predictions (Hakanson 1995), and too much complexity can lead to too much uncertainty, problems with interpretation of model's parameters and predictions, and high data requirements (Wilson and Pascoe 2006). Optimal model performance is expected to be achieved at some intermediate level of complexity, which provides an acceptable compromise between uncertainty and realism (Costanza and Sklar 1985; Hakanson 1995; Wilson and Pascoe 2006).

Furthermore, trophodynamics represents only a fraction of the uncertainty that any modeling exercise within an ecosystem approach to fisheries be faced with will. We have to integrate this uncertainty with all other sources in order to provide a balanced assessment of alternative management actions. One obvious way of doing this is by embedding trophodynamic models within management strategy evaluation (MSE) or similar frameworks (Butterworth and Punt 2003; Butterworth and Plaganyi 2004).

11.4 Management Strategy Evaluation

Management strategy evaluation (MSE), also known as management procedures (Smith et al. 1999; Butterworth and Punt 1999), is essentially a closed-loop analysis of the full management cycle (Walters and Martell 2004). Conceptually, it consists of three basic compartments: an operating model, an assessment model, and a management model (Fig. 11.1) (Kell et al. 2006). The operating model represents the exploited system in all aspects that are relevant to management as close to reality as possible, including uncertainties related to our lack of knowledge and natural variability. This operational model is used to simulate the dynamics of the system, generating data streams which resemble those typically available from monitoring programs. These data are then analyzed using the assessment model to produce an evaluation of the status of the exploited system. Based on this evaluation, the management model, which ideally simulates both the harvest control mechanisms and the efficacy of their implementations, is applied to the operating model. Finally, the operating model is used to simulate the dynamics after management, and the cycle starts again.

Basic loop for Management Strategy Evaluation

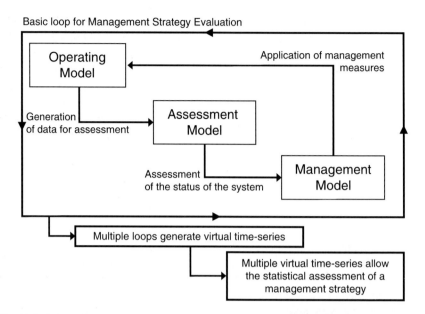

Fig. 11.1 Schematic representation of the Management Strategy Evaluation framework

The implementation of MSE allows the comparison of multiple management options while considering a comprehensive range of uncertainties in both the biological and management processes. Key to the strength of MSE approaches is the capacity of the operational models to replicate all or at least the most contrasting but still plausible system behaviors. Crucial to the task of defining plausibility boundaries is the "conditioning" of the operating model (Kell et al. 2006; Rademeyer et al. 2007). This is often achieved by fitting the operating model to the data, so that this model reasonably describes available information (Kell et al. 2006).

Even if we are unable to implement a fully complete MSE framework, it is still extremely useful to conceptualize the modeling work aimed to support ecosystem approaches to fisheries as a component of MSE. Keeping this perspective helps to identify where major gaps could be (e.g., maybe our biggest problems are not necessarily on the ecology side), and to develop an integrated vision of how trophodynamic models can contribute in practice to the overall process of providing management advice.

11.5   Trophodynamic Models as Operating Models

If model variability (i.e., parameter and structural uncertainty) and, more importantly, its trend over time can be properly described with a single-species model, then such a model can be used as an operating model in a reliable way. However, most fisheries at a regional level are a multispecies activity, not only due to the use of non-selective gears, but also because multiple fisheries are developed within the same ecosystem and their individual management plans affect the same ecologically functional unit. Thus, achieving sustainability at the ecosystem level requires trade-offs among different fisheries. These trade-offs need to be based on some of how understanding the different target species are actually linked. Therefore, if a reliable operating model should consider the effects of other fisheries, this will likely lead to the need for trophodynamic operating models. It is through food web interactions that most of these effects are transmitted and modulated (Bax 1998). Of course, this does not preclude that after considering trophodynamic effects, the conclusion can be that management strategies are robust to them and their variability and they can be safely ignored, but such a result still requires their evaluation to begin with.

Another important reason to pursue trophodynamic models as operating models is that the final outcome of indirect effects often contradicts our initial expectations (Yodzis 1988, 1996, 1998). Early system responses to a given perturbation can go in exactly the opposite direction of the final state of the system, a state which is achieved when all the effects of the perturbation have worked themselves out through the food web. Direct effects will manifest themselves first, while indirect effects will show their impact later in time and in waves, first those coming from short pathways and last those from the longest pathways in the food web (Yodzis 1996). Because of these features of complex systems it is virtually impossible to know a priori how parameters in a single-species model need to be modified to mimic multispecies effects, and adding white noise to model parameters will not always do the job (Lawton 1988; Vasseur and Yodzis 2004).

Trophodynamic models are in principle well suited to play the role of operating models (Butterworth and Punt 2003; Butterworth and Plaganyi 2004) and the research is underway to implement a whole ecosystem model as an operating model in a MSE framework (Smith et al. 2007). However, most MSE implementations still use single-species models as operating models (Butterworth and Punt 2003; Kell et al. 2006), with the reason commonly put forward as the lack of reliability of current multispecies and ecosystem models (Butterworth and Punt 2003; Rademeyer et al. 2007). This appears to be a "Catch 22" situation. Trophodynamic models containing many species and/or lots of details will likely capture a wider range of plausible behaviors, but the information to "condition" them will be increasingly scarce. This poor "conditioning capability" increases the uncertainty about model predictions, and hinders any potential forecasting ability that the model may have. Practical implementations require reliable trophodynamic operating models that can balance uncertainty and resolution.

In most cases this has been attempted by building minimum realistic models (Butterworth and Punt 2003; Stefansson 2003b; Kell et al. 2006). These models are designed with specific questions in mind and contain only a limited number of species. The goal is only to include those species that are most likely to have important interactions with the species of interest (Butterworth and Harwood 1991; Punt and Butterworth 1995). Although this is, in principle, a reasonable proposition, determining which set of species is the minimum necessary to mimic the dynamics of the real system is not straightforward (Yodzis 1998, 2000; Fulton et al. 2003a; Pinnegar et al. 2005). Nonetheless, I believe that this is a powerful concept for developing practical operational models within MSE which deserves closer examination.

11.6   Minimum Realistic Models

The concept of minimum realistic model (MRM) emerged as a tool to explore the impact of seals on fisheries resources (more specifically hake) in the Benguela ecosystem (Butterworth and Harwood 1991). The idea was to develop "relevant, decoupled portions of a much larger potential multispecies model with a reduced number of parameters for each of the most commercially important fish species preyed upon seals" (Butterworth and Harwood 1991). In the Benguela case, the model was developed by including species and interactions in order to capture in the model most of the known predation mortalities of medium to large hake (Punt and Butterworth 1995).

The main advantage of MRMs is that they simplify the full food web into a more manageable problem by reducing the complexity of the model, and hence, the amount of information required to parameterize it, while keeping at the same time the full dynamic description of the reduced system. Their biggest weakness is our inability for testing if the interactions and species included suffice to approximate the real system. We have no way of knowing beforehand if we have missed critical components that can drastically change the picture we use for providing advice.

Let us define a true MRM as a simple food web model that can reliably mimic the dynamics that its constituent species/functional groups will show in the real system when exposed to specific perturbations (e.g., fishing). Then, the safer way of producing a true MRM implies starting from a highly complex model that we believe properly captures the system dynamics and compare its behavior with simplified versions of itself (Yodzis 1998, 2000; Fulton et al. 2003a, b, 2004a, c; Pinnegar et al. 2005). Of course, the initial, highly complex, baseline models are themselves simplifications of reality (i.e., they are just "realistic models" instead of "minimum realistic" ones), and hence, also subject to the preconceptions, biases, and interests of their developers (Pinnegar et al. 2005). Nonetheless, these models still provide a place to start.

This "experimental" approach is certainly useful for developing MRMs. Fulton et al. (2003a) summarized many of the lessons learned. In relation with trophic complexity, they indicated that food webs simplified by aggregation or omission of groups to less than 25% of their original size were not able to represent the original model, but the adequate level of simplification is system-dependent. Aggregating species with similar prey and predators into functional groups, provided that their dynamics are not dramatically different, can also be useful, while pooling functional groups is less successful than omitting the least important groups entirely. Most remarkably, the identity of the components and links included in the model can be the most important determinant of model performance, and there is no reliable algorithm to make these decisions for us.

Building MRMs using this "experimental" approach has one major disadvantage; we need to have a highly detailed model of the system in the first place. In many cases this may not be even possible, and candidate MRMs will likely be built using more fragmentary information. Under these circumstances, current food web theory could help us in developing a more conceptual approach to MRMs and give us some guidance about how to put them together.

11.7   Food Web Theory and MRMs

Although there is no general theory of MRMs, there are several elements of theory that can be seen as providing the basic underpinnings, and together with recent empirical analyses, they could constitute a useful straw-man framework for MRMs.

If MRM is a theoretically sound concept we need to identify a set of focal species and show that the dynamics of the subweb model formed by them is not significantly altered by the omission of the rest of the community. Schaffer (1981) showed that if the dynamics of the omitted species are much faster than the focal ones, then we can "abstract" from the full system of equations-which describe the complete food web-a smaller subset which reasonably describes the dynamics of the focal species. The accuracy of this approximation depends on the relative difference in the timescales between focal and nonfocal species dynamics (Schaffer 1981). One good example is the derivation of the classical competition model from a full trophodynamic model with explicit representation of the resources dynamics (Yodzis 1989). In this case, the resources being competed for become implicit in the final competition equations by assuming that their dynamics are fast enough that they are always at equilibrium (Yodzis 1989).

In the context of press perturbation experiments, Bender et al. (1984) demonstrated that when the interactions of the focal species on the nonfocal ones **or** the interactions of nonfocal species on focal ones are all very small, then the error incurred by dismissing the nonfocal species is also small (Bender et al. 1984). This result is particularly insightful because it shows that even if the impact of the nonfocal species onto the focal ones is large, as long as the reverse is not true, the study of the subweb containing only focal species is still possible and meaningful (Yodzis 1989).

These results suggest that a subset of species could exist such that the modeling of their dynamics will reasonably mimic at least some of the features of the full food web. Hence, MRMs appear as a theoretically plausible concept, but the key aspects that can make them possible are the relative speeds of the species dynamics and the strength of their interactions.

Another concept relevant to this discussion is the idea of modularity in ecosystems. Compartments or modules in food webs are subgroups of species in which many strong interactions occur within the subgroups and few weak interactions occur between the subgroups. If they exist in real food webs, then these compartments are natural candidates for MRMs.

If we just state the obvious that every food web is actually a subweb embedded within an even larger system (Winemiller and Layman 2005), then the issue of modularity is trivially true at the global scale; the earth's biosphere is a functional unit, but the food web dynamics of the Newfoundland Shelf system has little impact on the Benguela marine ecosystem. Both systems are compartments within the biosphere, but the links among species across these compartments are far weaker than the links within compartments. Hence, studying the dynamics in one without considering the other will pose no danger. This rationale holds true for major habitat divisions (Pimm and Lawton 1980), but answering this question at scales that could matter for management applications is not that simple nor general (Pimm and Lawton 1980; Moore and Hunt 1988; Raffaelli and Hall 1992; Dunne et al. 2005; Montoya et al. 2006).

One particularly relevant aspect of this debate is the connection between food web structure and system stability (May 1972; McCann 2000; Dunne et al. 2005). On purely topological grounds (i.e., without including species dynamics), the pattern of connections observed in real food webs has in itself important stability implications (Solé and Montoya 2001; Montoya and Solé 2002; Dunne et al. 2002b; Montoya et al. 2006). Although real food webs show a high level of connectance (Williams et al. 2002), some species are more connected than others (Solé and Montoya 2001; Montoya and Solé 2002). Although the precise form of the degree distribution can vary among ecosystems (Montoya and Solé 2002; Dunne et al. 2002a), it has been shown that real ecological networks are more robust to the random elimination of species than to the directed removal of the highly connected ones (Solé and Montoya 2001; Dunne et al. 2002b). If species are successively extirpated starting with the most connected one the food web quickly breaks down into several disconnected subwebs and the relative amount of secondary extinctions is much higher (Dunne et al. 2002b; Montoya et al. 2006).

When near-equilibrium dynamics are considered, Yodzis (1981) showed that real food webs are more stable than their randomly assembled counterparts, and Moore and Hunt (1988) found that these same food webs were blocked in smaller subwebs. A recent analysis of highly resolved food webs also found evidence of modularity, and this compartmentalization increased food web stability by retaining the impact of the perturbations within a single compartment (Krause et al. 2003). Although the existence of modules/clusters in real food webs is still an open question (Dunne et al. 2005; Montoya et al. 2006), it is clear that the stability of real systems is highly dependent on the structural features of the network and the strength of the links among components.

Interaction strengths in real food webs are known to be biased towards weak links (Paine 1992; Berlow et al. 2004), and strong interactions tend to be associated with short loops in the network (Neutel et al. 2002). This type of distribution of interaction strengths has a significant impact on system stability (de Ruiter et al. 1995; McCann et al. 1998; McCann 2000; Neutel et al. 2002). A trophodynamic system exclusively composed by strong links is highly unstable and often renders oscil-latory or even chaotic dynamics (McCann et al. 1998). Theoretical studies have shown that if the system is characterized by only a few strong interactions embed-ded in a matrix of weak ones, these weak links act as dampeners of these oscilla-tions and enhance the stability of the entire system (McCann et al. 1998). The weak link effect among species can also be scaled up to stabilize subsystems or modules (the weak subsystem effect), where top predators can act as couplers in space of weak and strong pathways (McCann et al. 2005a, b).

Although strong links may channel most of the energy flow in real ecosystems, a weak link may not be weak all the time. Berlow (1999) has clearly shown that some weak links can have a high variance, meaning that they can become strong links under certain circumstances. Hence, focusing only on links that are always strong may not necessarily be a good idea. Links among species are adaptive, they change over time, and adaptive food webs are far more stable than nonadaptive ones (Kondoh 2003; Kondoh 2005). The critical concept here is that stability may be generated by the combination of a few strong and many weak links, but which links are the strong ones is something that changes over time. The question is: do all links become strong at one point or another?

A recent study combining theoretical modeling with empirical evidence indi-cates that real food webs appear to be organized in such a way that top predators act as couplers of energy channels which differ in productivity and turnover rate (Rooney et al. 2006). This study also showed that stability is maximized when the energy flow between these channels is asymmetrical and where top predators can rapidly switch among them.

The very existence of these clearly detectable energy channels among a wide variety of real ecosystems (Rooney et al. 2006) strongly advocates, in my opinion, for some general internal modularity or clustering in the way energy is transferred up in food webs. Considering the high connectance observed in real food webs (Williams et al. 2002), this discovery also suggests that even though interaction strengths can change over time, the main pathways that channel the energy would likely be a relatively narrower set when compared with all possible pathways in a food web.

11.8 From Theory to Practice: Some Ideas to Consider When Developing MRMs

Some of the elements summarized in Section 11.7 have received closer attention in the context of their potential implications for developing MRMs. Here I will build upon few examples to advance some ideas and speculations that may be useful when we face the challenge of developing MRMs.

Yodzis (1998) evaluated the impact of eliminating weak links using the "experimental" approach based on his 29-species model of the Benguela system.

He concluded that, when the link strength was estimated by proportions in the diet of the predator, eliminating all links under 5% generated a reasonable approximation to the full model predictions. This cutoff level eliminated 52% of the original links, but did not reduce the number of species in the model. Eliminating links up to 10% significantly altered the quantitative results, but still kept the qualitative output while retaining all species in the model.

Yodzis (2000) also used his Benguela model to explore the issue of modularity. He found that if the interaction between predator and prey species did not consider "other-prey effects" in the functional response (i.e., each predator–prey interaction was represented by a single-species functional response), then a subset of species can be identified that functions as a reliable module for the full system response for the question he was posing (the effect of culling fur seals on hake yields). However, when multispecies functional responses were considered, the effect became diffused in the food web and no sensible module could be found (Yodzis 2000).

At first, these results appear to be quite discouraging for any attempt of building MRMs. Although we can dismiss half of the links based on their strength, we still need to consider essentially all food web components. However, a closer examination of the analysis can give us some hope. In Yodzis' (1998) analysis, each time that a link in the original food web was eliminated, the remaining links for a given predator were rescaled as a function of the simplified diet. This means that the strength of each remaining interaction was inflated instead of keeping its original magnitude.

With respect to modularity, Yodzis's (2000) results indicate that the existence of a trustworthy subweb module is dependent on an assumption (absence of "other-prey effects") that can hardly hold for real ecosystems. However, this analysis also provides one useful insight; dismissing the "other-prey effect" in the functional responses implies assuming that each link in the food web acts as an independent channel of energy within the network (i.e., the interaction strength between two species – *sensu* Rooney et al. (2006) – only depends on the state of those two interacting species). If we consider that most links are weak, it is a fair starting point to consider that only the few strong links (or those weak ones but with high variance) will likely be the ones responsible for any major change in the channeling of energy. If we focus on these links we should, at least in principle, be capable of addressing most of the variability in the flow of energy through a given node in the web.

It has been shown that the highly connected species in a food web can be critical to ensure food web stability (Solé and Montoya 2001; Dunne et al. 2002b), and McCann et al. (1998, 2005a, b) have made the case for the importance of weak links as the stabilizing mechanism that allows complex systems to persist. Rooney et al. (2006) have shown that distinct pathways of energy exist in real systems, where top predators act as couplers of these channels.

So far, most MRMs have been developed with very precise questions in mind (e.g., Punt and Butterworth 1995). However, if the goal is to implement them as operating models for MSE then the set of questions will be broader necessarily. How can we approach the task of building MRMs in this broader setting? Here are some thoughts.

Despite being trivially obvious, we should start from the beginning. Real ecosystems are thermodynamic systems, and hence constrained by the laws of physics. Hence, even though we may use many currencies in our models (numbers, biomass, carbon units, etc.), energy cannot be created from nothing, it flows among components in the ecosystem and gets stored in their mass. Furthermore, the transfer of energy is never perfect and the efficiency of these transformations is always less than unity. These conditions always apply and should be used to constrain model behavior and aid in model design (Yodzis and Innes 1992).

In this context, candidate MRM must try to capture the main flows of energy in the system. These pathways will most likely involve the most dominant species in the system and the most connected ones (which can or cannot be the same), and unlike many simple multispecies fisheries models, which are mainly focused on commercial species, they would also necessarily include key species at different trophic levels from basal species to top predators. Basal species in these operating models will essentially provide the main source of the energy that percolates through the model. On the other extreme of the food web, top predators are emerging as having a particularly important role in linking different energy channels, hence the dynamic of these predators will need to be fully represented as well, most critically its potential ability for switching among different energy pathways.

Models that try to capture the main energy pathways do not necessarily need to be overly resolved. For example, demersal fish communities of many marine systems have k-dominance curves where 80% or more of the abundance is encompassed by the ten most dominant species (Bianchi et al. 2000). This suggest that two-species models would likely be too small, but moderate size models with just a few species or functional groups (maybe 10–15 components) are possibly adequate to reasonably capture the main flows of energy in the system that would likely have the highest impacts in terms of comparing alternative management options. Although this last statement is just speculation, I believe it is an idea worth exploring. Nonetheless, on the practical side of things, these moderate size models are possibly in the upper range of complexity for which we can reasonably implement well-resolved long-term ecosystem survey programs to gather the necessary data for parameterization and model testing.

In terms of describing the dynamics of the links among components, a reasonable starting point is to implement multispecies functional responses, but only considering strong interactions or weak ones with high variance. The number of links explicitly modeled will likely be small in comparison with the full spectrum of trophic linkages that any given species may have (Williams et al. 2002), and since the links will be strong ones, the potential for model instability would be high. However, the stabilizing effect provided by the network of weak links can be mimicked by considering "other food" components in the functional response in the same way that allochtonous inputs have been modeled (Huxel and McCann 1998).

Including "other food" effects when modeling functional responses is certainly not a new idea (Sparre 1991; Stefansson 2003a; Koen-Alonso and Yodzis 2005). However, the reasoning behind it here is essentially different than in previous applications. Most times other food effects are justified under the premise of allowing a more realistic representation of the prey field (e.g., it is unlikely that a predator will feed only on the species included in the model). The rationale I am suggesting here is explicitly directed to approximate the stability effects associated with non-modeled weak links. This keeps the explicitly modeled strong links within realistic magnitudes by avoiding the inflation effect associated with a narrower representation of the diet.

If candidate MRMs are developed for capturing the major pathways of energy in the system, the "other food" components in the functional response must necessarily be relatively small; most of the energy within the model would be expected to be generated by the basal species while the input of energy from "other food" would be expected to be low. This perspective is consistent with the finding that simple food web models are stabilized by low levels of allochtonous inputs (Huxel and McCann 1998).

In terms of functional response formulations, many models use Type 2 functions as a default. However, we should keep in mind that the functional responses in these candidate MRMs represent the overall population response, and not the individual foraging behavior. The trophic relationships represented in these models are population-level functional responses which typically integrate these individual foraging behavior in space and time, and often across population structure (Ives et al. 1999). In this context, Type 3 formulations have proven to confer more stability to both simple and complex food web models (e.g., Williams and Martinez 2004), and are also a frequent phenomenological emergent from many ecologically plausible mechanisms like prey switching, predator learning, and the presence of prey refuges (Koen-Alonso 2007). Therefore, I believe using multispecies Type 3 formulations is a better starting point than customary Type 2 functions.

In a nutshell, I think the most promising avenue for trophodynamic models to have a useful practical role within ecosystem approaches to fisheries is to develop them to fulfill the role of operating models within MSE. In such a role, I also believe we should aim to develop candidate MRMs as operating models, but where these candidate MRMs are bounded by bioenergetic considerations we should aim to capture the main pathways of energy within the exploited community. In this context, it is particularly important to pay attention to top predators linking different energy pathways, and dominant and highly connected species, regardless of their trophic level or commercial value. Finally, I suggest we make full use of current ideas about the role of weak and strong links in food webs by implementing functional responses that can mimic trophic switching and improve model stability.

While this possible way forward may not necessarily render operating models fully functional in terms of providing complete tactical advice, I believe they can capture the main signals associated with trophic interactions while keeping model complexity within reasonable boundaries. The strategic-level MSEs emerging from implementing these operating models can then be used to provide a reliable envelope and boundary conditions for more focused analyses aimed to generate the specifics needed for tactical decisions.

11.9 Final Thoughts: Minimum Realistic or Maximum Feasible Models?

Although some of the ideas sketched in the previous sections may indeed render true MRMs, we will never know for certain whether or not we are missing something critical. Despite this fundamental uncertainty management decisions need to be made, and MSE approaches provide a compelling framework to test how our alternative conceptions about the real world, and our interactions with it, may turn out. In this context, using MRMs as operating models is an attractive proposition. These models are comparatively simple, fully dynamic, and we can take advantage of allometric theory, predator–prey size relationships (i.e., size-structured predation), and basic bioenergetics to help us parameterize the models (Yodzis and Innes 1992; Brown et al. 2004; Pope et al. 2007) and to mechanistically include variables like temperature (Gillooly et al. 2001; Savage et al. 2004; Vasseur and McCann 2005), providing a venue for exploring some of the potential impacts of climate change.

However, in order to trust any model enough to use it as an operating model, the model needs to properly reproduce available data. It is only this comparison with the real world that will make it believable. Furthermore, we have displayed a remarkable creativity in developing alternative mathematical formulations to represent ideas about how the real world functions, and hence, only the contrast with reality can allow us to choose among (or amalgamate) equally reasonable theoretical options.

This need for model validation highlights the importance of developing monitoring programs and surveys that match both the data requirements and the outputs of the models we intend to use. Whole ecosystem models are most definitively useful to explore ideas and develop concepts; however, fully validating them is beyond reach for the most part but is essential in a decision-making application. Financial and logistical constraints will always restrict the scope of long-term ecological monitoring programs (Caughlan and Oakley 2001), and hence, the availability of data for validation will always be limited in one way or another. Considering this, MRMs could provide a better platform to achieve a reasonable balance between the amount of data actually available for parameterization and validation and the number of assumptions hardwired in the models. Nonetheless, even if MRMs are simple by comparison with full ecosystem models, it is likely that there will never be enough information to truly consider any model as a true MRM. Instead of "minimum realistic models" we should think about these models as "maximum feasible models" because even though the possibility of missing important elements exists, they might still be the best simple trophodynamic models we can put together without seriously hindering our capacity for model validation and testing. Hence, if minimum realistic models actually are maximum feasible models, then approaching the operating model role with multiple modeling approaches is not just a desired feature, it is a must have one.

We can safely say that even the best models are only as good as the data we feed into them, and without data, they are just formalized representations of our ideas about the world. Despite how interesting and insightful these formalizations are to us, quite often the real world mischievously ignores them. This is something that we should not forget.

Acknowledgments I wish to thank the American Institute of Fishery Research Biologists for the opportunity to participate in its Fiftieth Anniversary Symposium and discuss the ideas explored in this chapter; especially Jake Rice and Mike Fogarty who were instrumental for this to happen. Despite the rush I put them through, Pierre Pepin, Alejandro Buren, Beth Fulton, and Andre Punt provided useful comments on preliminary versions of this paper. An anonymous reviewer also provided good suggestions to improve the manuscript. In the last few years, many people have had significant influences on my views and opinions about food web modeling and its role in fisheries management. At risk of forgetting many, I want to give special thanks to Kevin McCann, Neil Rooney, Dave Vasseur, Geoff Evans, Jake Rice, Pierre Pepin, Alida Bundy, Jason Link, John Harwood, Beth Fulton, and André Punt. With some of them I have had many discussions, with others, just a few, but in all cases they have been insightful and thought-provoking. Of course, all errors are mine. Thanks to Dick Beamish for his incredible patience with me. Finally, I want to dedicate this work to the memory of Peter Yodzis. I am as certain he would have deeply enjoyed discussing and dissecting these ideas as I am sure many of them would have not survived his sharp scrutiny.

References

Andersen, K.H. and Beyer, J.E. 2006. Asymptotic size determines species abundance in the marine size spectrum. Am Nat **168**: 54–61.

Arkema, K.K., Abramson, S.C., and Dewsbury, B.M. 2006. Marine ecosystem-based management: from characterization to implementation. Front Ecol Environ **4**: 525–532.

Aydin, K.Y., McFarlane, G.A., King, J.R., Megrey, B.A., and Myers, K.W. 2005. Linking oceanic food webs to coastal production and growth rates of Pacific salmon (Oncorhynchus spp.), using models on three scales. Deep-Sea Res II **52**: 757–780.

Bax, N.J. 1998. The significance and prediction of predation in marine fisheries. ICES J Mar Sci **55**: 997–1030.

Begley, J. and Howell, D. 2004. An overview of Gadget, the Globally applicable Area-Disaggregated General Ecosystem Toolbox. ICES CM **2004/FF:13**: 1–15.

Bender, E.A., Case, T.J., and Gilpin, M.E. 1984. Perturbation experiments in community ecology: theory and practice. Ecology **65**: 1–13.

Benoît, E. and Rochet, M.-J. 2004. A continuous model of biomass size spectra governed by predation and the effects of fishing on them. J Theor Biol **226**: 9–21.

Berlow, E.L. 1999. Strong effects of weak interactions in ecological communities. Nature **398**: 330–334.

Berlow, E.L., Neutel, A.-M., Cohen, J.E., de Ruiter, P.C., Ebenman, B., Emmerson, M., Fox, J.W., Jansen, V.A.A., Jones, J.I., Kokkoris, G.D., Logofet, D.O., McKane, A.J., Montoya, J.M., and Petchey, O. 2004. Interaction strengths in food webs: issues and opportunities. J Anim Ecol **73**: 585–598.

Berryman, A.A. 1992. The origins and evolution of predator-prey theory. Ecology **73**: 1530–1535.

Berryman, A.A. 1999. Alternative perspectives on consumer-resource dynamics: a reply to Ginzburg. J Anim Ecol **68**: 1263–1266.

Berryman, A.A., Michalski, J., Gutierrez, A.P., and Arditi, R. 1995. Logistic theory of food web dynamics. Ecology **76**: 336–343.

Bianchi, G., Gislason, H., Graham, K., Hill, L., Jin, X., Koranteng, K., Sanchez, F., and Zwanenburg, K. 2000. Impact of fishing on size composition and diversity of demersal fish communities. ICES J Mar Sci **57**: 558–571.

Brown, J.H., Allen, A.P., Savage, V.M., and West, G.B. 2004. Toward a metabolic theory of ecology. Ecology **85**: 1771–1789.

Bundy, A. 2001. Fishing on ecosystems: the interplay of fishing and predation in Newfoundland-Labrador. Can J Fish Aquat Sci **58**: 1153–1167.

Butterworth, D.S. and Harwood, J. 1991. Report on the Benguela Ecology Programme workshop on seal-fishery biological interactions. Rep Benguela Ecol Prog S Afr **22**: 1–22.

Butterworth, D.S. and Plaganyi, E.E. 2004. A brief introduction to some approaches to multispecies/ ecosystem modelling in the context of their possible application in the management of South African fisheries. Afr J Mar Sci 53–61.

Butterworth, D.S. and Punt, A.E. 1999. Experiences in the evaluation and implementation of management procedures. ICES J Mar Sci **56**: 985–998.

Butterworth, D.S. and Punt, A.E. 2003. The role of harvest control laws, risk and uncertainty and the precautionary approach in ecosystem-based management. *In* Responsible fisheries in the marine ecosystem. FAO and CABI, pp. 311–319.

Caughlan, L. and Oakley, K.L. 2001. Cost considerations for long-term ecological monitoring. Ecol Indicat **1**: 123–134.

Christensen, N.L., Bartuska, A.M., Brown, J.H., Carpenter, S., D'Antonio, C., Francis, R., Franklin, J.F., MacMahon, J.A., Noss, R.F., Parsons, D.J., Peterson, C.H., Turner, M.G., and Woodmansee, R.G. 1996. The report of the Ecological Society of America Committee on the scientific basis for ecosystem management. Ecol Appl **6**: 665–691.

Christensen, V. and Walters, C.J. 2003. Ecopath with Ecosim: methods, capabilities and limitations. Ecol Model **172**: 109–139.

Ciannelli, L., Robson, B.W., Francis, R.C., Aydin, K., and Brodeur, R.D. 2004. Boundaries of open marine ecosystems: an application to the Pribilof Archipelago, Southeast Bering Sea. Ecol Appl **14**: 942–953.

Clark, J.S., Carpenter, S.R., Barber, M., Collins, S., Dobson, A., Foley, J.A., Lodge, D.M., Pascual, M., Pielke Jr, R., Pizer, W., Pringle, C., Reid, W.V., Rose, K.A., Sala, O., Schlesinger, W.H., Wall, D.H., and Wear, D. 2001. Ecological forecasts: an emerging imperative. Science **293**: 657–660.

Cohen, J.E., Pimm, S.L., Yodzis, P., and Saldana, J. 1993. Body sizes of animal predators and animal prey in food webs. J Anim Ecol **62**: 67–78.

Cohen, J.E., Jonsson, T., and Carpenter, S.R. 2003. Ecological community description using the food web, species abundance and body size. Proc Natl Acad Sci USA **100**: 1781–1786.

Costanza, R. and Sklar, F.H. 1985. Articulation, accuracy and effectiveness of mathematical models: a review of freshwater wetland applications. Ecol Model **27**: 45–68.

De Roos, A.M., Schellekens, T., van Kooten, T., van de Wolfshaar, K., Claessen, D., and Persson, L. 2007. Food-dependent growth leads to overcompensation in stage-specific biomass when mortality increases: the influence of maturation versus reproduction regulation. Am Nat **170**: E59–E76.

de Ruiter, P.C., Neutel, A.-M., and Moore, J.C. 1995. Energetics, patterns of interaction strengths, and stability in real ecosystems. Science **269**: 1257–1260.

Dunne, J., Brose, U., Williams, R.J., and Martinez, N. 2005. Modeling food-web dynamics: complexity-stability implications. *In* Aquatic food webs: an ecosystem approach. Oxford University Press, Oxford. pp. 117–129.

Dunne, J.A., Williams, R.J., and Martinez, N.D. 2002a. Food-web structure and network theory: The role of connectance and size. Proc Natl Acad Sci USA **99**: 12917–12922.

Dunne, J.A., Williams, R.J., and Martinez, N.D. 2002b. Network structure and biodiversity loss in food webs: robustness increases with connectance. Ecol Lett **5**: 558–567.

Fulton, E.A., Smith, A.D.M., and Johnson, C.R. 2003a. Effect of complexity on marine ecosystem models. Mar Ecol Prog Ser **253**: 1–16.

Fulton, E.A., Smith, A.D.M., and Johnson, C.R. 2003b. Mortality and predation in ecosystem models: is it important how these are expressed? Ecol Model **169**: 157–178.

Fulton, E.A., Parslow, J.S., Smith, A.D.M., and Johnson, C.R. 2004a. Biogeochemical marine ecosystem models II: the effect of physiological detail on model performance. Ecol Model **173:** 371–406.

Fulton, E.A., Smith, A.D.M., and Johnson, C.R. 2004b. Biogeochemical marine ecosystem models I: IGBEM – a model of marine bay ecosystems. Ecol Model **174:** 267–307.

Fulton, E.A., Smith, A.D.M., and Johnson, C.R. 2004c. Effects of spatial resolution on the performance and interpretation of marine ecosystem models. Ecol Model **176:** 27–42.

Fulton, E.A., Smith, A.D.M., and Punt, A.E. 2005. Which ecological indicators can robustly detect effects of fishing? ICES J Mar Sci **62:** 540–551.

Garcia, S.M., Zerbi, A., Aliaume, C., Do Chi, T., and Lasserre, G. 2003. The ecosystem approach to fisheries. Issues, terminology, principles institutional foundations, implementation and outlook. FAO Fisheries Technical Paper **443:** 1–71.

Gentleman, W., Leising, A., Frost, B., Strom, S., and Murray, J. 2003. Functional responses for zooplankton feeding on multiple resources: a review of assumptions and biological dynamics. Deep-Sea Res II **50:** 2847–2875.

Gillooly, J.F., Brown, J.H., West, G.B., Savage, V.M., and Charnov, E.L. 2001. Effects of size and temperature on metabolic rate. Science **293:** 2248–2251.

Ginzburg, L.R. 1998. Assuming reproduction to be a function of consumption raises doubts about some popular predator-prey models. J Anim Ecol **67:** 325–327.

Gislason, H. 1999. Single and multispecies reference points for Baltic fish stocks. ICES J Mar Sci **56:** 571–583.

Guénette, S., Heymans, S.J.J., Christensen, V., and Trites, A.W. 2006. Ecosystem models show combined effects of fishing, predation, competition, and ocean productivity on Steller sea lions (*Eumetopias jubatus*) in Alaska. Can J Fish Aquat Sci **63:** 2495–2517.

Hakanson, L. 1995. Optimal size of predictive models. Ecol Model **78:** 195–204.

Hall, S.J., Collie, J.S., Duplisea, D.E., Jennings, S., Bravington, M., and Link, J.S. 2006. A length-based multispecies model for evaluating community responses to fishing. Can J Fish Aquat Sci **63:** 1344–1359.

Huxel, G.R. and McCann, K.S. 1998. Food web stability: the influence of trophic flows across habitats. Am Nat **152:** 460–469.

Ives, A.R., Schooler, S.S., Jagar, V.J., Knuteson, S.E., Grbic, M., and Settle, W.H. 1999. Variability and parasitoid foraging efficiency: a case study of pea aphids and *Aphidius ervi*. Am Nat **154:** 652–673.

Jennings, S. 2005. Indicators to support an ecosystem approach to fisheries. Fish Fish. **6:** 212–232.

Kell, L., De Olivera, J.A.A., Punt, A.E., McAllister, M.K., and Kuikka, S. 2006. Operational Management Procedures: An introduction to the use of evaluation frameworks. *In* The knowledge base for fisheries management. Elsevier, Amsterdam, pp. 379–407.

Koen-Alonso, M. 2007. A process-oriented approach to the multispecies functional response. *In* From energetics to ecosystems: the dynamics and structure of ecological systems. Springer, Heidelberg/Dordrecht, pp. 1–36.

Koen-Alonso, M. and Yodzis, P. 2005. Multispecies modelling of some components of the northern and central Patagonia marine community, Argentina. Can J Fish Aquat Sci **62:** 1490–1512.

Kondoh, M. 2003. Foraging adaptation and the relationship between food-web complexity and stability. Science **299:** 1388–1391.

Kondoh, M. 2005. Is biodiversity maintained by food-web complexity? The adaptive food-web hypothesis. *In* Aquatic food webs: an ecosystem approach. Oxford University Press, Oxford, pp. 130–142.

Krause, A.E., Frank, K.A., Mason, D.M., Ulanowicz, R.E., and Taylor, W.W. 2003. Compartments revealed in food-web structure. Nature **426:** 282–285.

Lawton, J.H. 1988. More time means more variation. Nature **334:** 563.

Lilly, G.R., Parsons, D.G., and Kulka, D.W. 2000. Was the increase in shrimp biomass on the northeast Newfoundland shelf a consequence of a release in predation pressure from cod? J Northw Atl Fish Sci **27:** 45–61.

Link, J.S. 2002a. Ecological considerations in fisheries management: when does it matter? Fisheries **27**: 10–17.

Link, J.S. 2002b. What does ecosystem-based fisheries management mean? Fisheries **27**: 18–21.

Link, J.S. 2003. A model of aggregate biomass tradeoffs. ICES CM **2003/Y:08**: 1–28.

Link, J.S., Brodziak, J.K.T., Edwards, S.F., Overholtz, W.J., Mountain, D., Jossi, J.W., Smith, T.D., and Fogarty, M.J. 2002. Marine ecosystem assessment in a fisheries management context. Can J Fish Aquat Sci **59**: 1429–1440.

Mackinson, S., Blanchard, J.L., Pinnegar, J.K., and Scott, R. 2003. Consequences of alternative functional response formulations in models exploring whale-fishery interactions. Mar Mamm Sci **19**: 661–681.

Magnússon, K.G. 1995. An overview of the multispecies VPA – theory and applications. Rev Fish Biol Fisher **5**: 1573–5184.

Mangel, M. and Levin, P.S. 2005. Regime, phase and paradigm shift: making community ecology the basic science for fisheries. Phil Trans R Soc B **360**: 95–105.

Marasco, R.J., Goodman, D., Grimes, C.B., Lawson, P.W., Punt, A.E., and Quinn II, T.J. 2007. Ecosystem-based fisheries management: some practical suggestions. Can J Fish Aquat Sci **64**: 928–939.

Maury, O., Faugeras, B., Shin, Y.J., Poggiale, J.C., Ben Ari, T., and Marsac, F. 2007a. Modeling environmental effects on the size-structured energy flow through marine ecosystems. Part 1: The model. Prog Oceanogr **74**: 479–499.

Maury, O., Shin, Y.J., Faugeras, B., Ben Ari, T., and Marsac, F. 2007b. Modeling environmental effects on the size-structured energy flow through marine ecosystems. Part 2: Simulations. Prog Oceanogr **74**: 500–514.

May, R.M. 1972. Will a large complex system be stable? Nature **238**: 413–414.

McCann, K.S. 2000. The diversity-stability debate. Nature **405**: 228–233.

McCann, K.S., Hastings, A., and Huxel, G.R. 1998. Weak trophic interactions and the balance of nature. Nature **395**: 794–798.

McCann, K.S., Rasmussen, J.B., and Umbanhowar, J. 2005a. The dynamics of spatially coupled food webs. Ecol Lett **8**: 513–523.

McCann, K.S., Rasmussen, J.B., Umbanhowar, J., and Humphries, M. 2005b. The role of space, time, and variability in food web dynamics. *In* Dynamic food webs: multispecies assemblages, ecosystem development and environmental change. Academic, San Diego, CA, pp. 56–70.

Mohn, R. and Bowen, W.D. 1996. Grey seal predation on the eastern Scotian Shelf: modelling the impact on Atlantic cod. Can J Fish Aquat Sci **53**: 2722–2738.

Montoya, J.M. and Solé, R.V. 2002. Small world patterns in food webs. J Theor Biol **214**: 405–412.

Montoya, J.M., Pimm, S.L., and Solé, R.V. 2006. Ecological networks and their fragility. Nature **442**: 259–264.

Moore, J.C. and Hunt, H.W. 1988. Resource compartmentation and the stability of real ecosystems. Nature **333**: 261–263.

Neutel, A.M., Heesterbeek, J.A.P., and De Ruiter, P.C. 2002. Stability in real food webs: weak links in long loops. Science **296**: 1120–1123.

Paine, R.T. 1992. Food-web analysis through field measurements of per capita interaction strength. Nature **355**: 73–75.

Patterson, K., Cook, R., Darby, C., Gavaris, S., Kell, L., Lewy, P., Mesnil, B., Punt, A., Restrepo, V., Skagen, D.W., and Stefansson, G. 2001. Estimating uncertainty in fish stock assessment and forecasting. Fish Fish. **2**: 125–157.

Pikitch, E.K., Santora, C., Babcock, E.A., Bakun, A., Bonfil, R., Conover, D.O., Dayton, P., Doukakis, P., Fluharty, D., Heneman, B., Houde, E.D., Link, J., Livingston, P.A., Mangel, M., McAllister, M.K., Pope, J., and Sainsbury, K.J. 2004. Ecosystem-based fishery management. Science **305**: 346–347.

Pimm, S.L. and Lawton, J.H. 1980. Are food web divided into compartments? J Anim Ecol **49**: 879–898.

Pinnegar, J.K., Blanchard, J.L., Mackinson, S., Scott, R.D., and Duplisea, D.E. 2005. Aggregation and removal of weak-links in food-web models: system stability and recovery from disturbance. Ecol Model **2–4:** 229–248.

Plaganyi, E.E. 2007. Models for an ecosystem approach to fisheries. FAO Fisheries Technical Paper **477:** 1–108.

Pope, J.G., Rice, J.C., Daan, N., Jennings, S., and Gislason, H. 2007. Modelling an exploited marine fish community with 15 parameters – results from simple size-based models. ICES J Mar Sci **63:** 1029–1044.

Punt, A.E. and Butterworth, D.S. 1995. The effects of future consumption by the Cape fur seal on catches and catch rates of the Cape hakes. 4. Modelling the biological interactions between Cape fur seals *Arctocephalus pusillus* and the Cape hakes *Merluccius capensis* and *M. paradoxus*. S Afr J Mar Sci **16:** 255–285.

Quinn II, T.J. 2003. Ruminations on the development and future of population dynamics models in fisheries. Nat Res Model **16:** 341–392.

Quinn II, T.J. and Deriso, R.B. 1999. Quantitative fish dynamics. Oxford University Press, New York.

Rademeyer, R.A., Plaganyi, E.E., and Butterworth, D.S. 2007. Tips and tricks in designing management procedures. ICES J Mar Sci **64:** 618–625.

Raffaelli, D. 2007. Food webs, body size and the curse of the latin binomial. *In* From energetics to ecosystems: the dynamics and structure of ecological systems. Springer, Heidelberg/Dordrecht, pp. 53–64.

Raffaelli, D. and Hall, S.J. 1992. Compartments and predation in an estuarine food web. J Anim Ecol **61:** 551–560.

Rice, J.C. and Rochet, M.J. 2005. A framework for selecting a suite of indicators for fisheries management. ICES J Mar Sci **62:** 516–527.

Rooney, N., McCann, K.S., Gellner, G., and Moore, J.C. 2006. Structural asymmetry and the stability of diverse food webs. Nature **442:** 265–269.

Sabo, J.L., Beisner, B.E., Berlow, E.L., Cuddington, K., Hastings, A., Koen-Alonso, M., McCann, K.S., Melian, C., and Moore, J. 2005. Population dynamics and food web structure: predicting measurable food web properties with minimal detail and resolution. *In* Dynamic food webs: Multispecies assemblages, ecosystem development, and environmental change. Academic, San Diego, CA, pp. 437–450.

Savage, V.M., Gillooly, J.F., Brown, J.H., West, G.B., and Charnov, E.L. 2004. Effects of body size and temperature on population growth. Am Nat **163:** 429–441.

Schaffer, W.M.i. 1981. Ecological abstraction: the consequences of reduced dimensionality in ecological models. Ecol Monogr **51:** 383–401.

Smith, A.D.M., Sainsbury, K.J., and Stevens, R.A. 1999. Implementing effective fisheries-management systems – management strategy evaluation and the Australian partnership approach. ICES J Mar Sci **56:** 967–979.

Smith, A.D.M., Fulton, E.J., Hobday, A.J., Smith, D.C., and Shoulder, P. 2007. Scientific tools to support the practical implementation of ecosystem-based fisheries management. ICES J Mar Sci **64:** 633–639.

Solé, R.V. and Montoya, J.M. 2001. Complexity and fragility in ecological networks. Proc Roy Soc Lond Series B **268:** 2039–2045.

Sparre, P.J. 1991. Introduction to multispecies virtual population analysis. ICES Mar Sci Symp **193:** 12–21.

Stefansson, G. 2003a. Issues in multispecies models. Nat Resour Model **16:** 415–437.

Stefansson, G. 2003b. Multi-species and ecosystem models in a management context. *In* Responsible fisheries in the marine ecosystem. FAO and CABI, pp. 171–188.

Trzcinski, K.M., Mohn, R., and Bowen, W.D. 2006. Continued decline of an Atlantic cod population: how important is gray seal predation? Ecol Appl **16:** 2276–2292.

Turchin, P. 2003. Complex population dynamics: a theoretical/empirical synthesis. Princeton University Press, Princeton, NJ.

Vasseur, D.A. and McCann, K.S. 2005. A mechanistic approach for modeling temperature-dependent consumer-resource dynamics. Am Nat **166**: 184–198.

Vasseur, D.A. and Yodzis, P. 2004. The color of environmental noise. Ecology **85**: 1146–1152.

Vinther, M. 2001. Ad hoc multispecies VPA tuning applied for the Baltic and North Sea fish stocks. ICES J Mar Sci **58**: 311–320.

Volterra, V. 1928. Variazioni e fluttuazioni del numero d'individui in specie animali conviventi (Reprinted in English). *In* Chapman, R.N. (1931) Animal ecology. McGraw-Hill, New York, pp. 409–448.

Walters, C.J. and Christensen, V. 2007. Adding realism to foraging arena predictions of trophic flow rates in Ecosim ecosystem models: shared foraging arenas and bout feeding. Ecol Model **209**: 342–350.

Walters, C.J. and Martell, S.J.D. 2004. Fisheries ecology and management. Princeton University Press, Princeton, NJ.

Walters, C.J., Christensen, V., and Pauly, D. 1997. Structuring dynamic models of exploited ecosystems from trophic mass-balance assessments. Rev Fish Biol Fish **7**: 139–172.

Williams, R.J. and Martinez, N.D. 2004. Stabilization of chaotic and non-permanent food web dynamics. Eur Phys J B **38**: 297–303.

Williams, R.J., Berlow, E.L., Dunne, J.A., Barabási, A.-L., and Martinez, N.D. 2002. Two degrees of separation in complex food webs. Proc Natl Acad Sci USA **99**: 12913–12916.

Wilson, D.C. and Pascoe, S. 2006. Delivering complex scientific advice to multiple stakeholders. *In* The knowledge base for fisheries management. Elsevier, Amsterdam, pp. 329–353.

Winemiller, K.O. and Layman, C.A. 2005. Food web science: moving on the path from abstraction to prediction. *In* Dynamic food webs: multispecies assemblages, ecosystem development and environmental change. Academic, San Diego, CA, pp. 10–23.

Yodzis, P. 1981. The stability of real ecosystems. Nature **289**: 674–676.

Yodzis, P. 1988. The indeterminacy of ecological interactions as perceived through perturbation experiments. Ecology **69**: 508–515.

Yodzis, P. 1989. Introduction to theoretical ecology. Harper & Row, New York.

Yodzis, P. 1994. Predator-prey theory and management of multispecies fisheries. Ecol Appl **4**: 51–58.

Yodzis, P. 1996. Food webs and perturbation experiments: theory and practice. *In* Food webs: integration of patterns and dynamics. Chapman & Hall, New York, pp. 192–200.

Yodzis, P. 1998. Local trophodynamics and the interaction of marine mammals and fisheries in the Benguela ecosystem. Ecology **67**: 635–658.

Yodzis, P. 2000. Diffuse effects in food webs. Ecology **81**: 261–266.

Yodzis, P. and Innes, S. 1992. Body size and consumer-resource dynamics. Am Nat **139**: 1151–1175.

Chapter 12
Why and How Could Indicators
Be Used in an Ecosystem Approach
to Fisheries Management?

Marie-Joëlle Rochet and Verena M. Trenkel

Abstract Fishery papers on ecosystem indicators, or ecological indicators, have flourished over the last 10 years, and many were justified by referring to the ecosystem approach to fisheries (EAF). However, the reason(s) why indicators are relevant to an EAF are not always clear. Still less clear is the way(s) indicators might be used to give management advice in the context of an EAF.

In this chapter we recount the emergence of the indicator concept in the EAF context. The concept being overloaded with too many roles and interpretations, we propose to split it into three separate categories fulfilling the functions of control, audit, and communication. The research needed to make these indicator variants operational is largely interdisciplinary.

Keywords Ecological indicators · Marine ecosystem · Fisheries management

12.1 Introduction

Indicators are needed to evaluate how well a fishery is managed, in relation to specified objectives (Hilborn and Walters 1992). If that is the sole use of indicators, where does it come from that so much of today's fisheries science is devoted to indicators, especially in the context of an ecosystem approach to fisheries (EAF)? In recent years fisheries research has focused primarily on developing indicators; much less effort has gone into developing approaches for actual fisheries management in an ecosystem context, based on indicators. Clearly, if appropriate management performance indicators are monitored, they will reveal if something is wrong, but do not necessarily tell what is wrong, or what management action should be implemented to mitigate effects. In the context of single

M.-J. Rochet and V.M. Trenkel
Département Ecologie et Modèles pour l'Halieutique, IFREMER, B.P. 21105,
44311 Nantes Cedex 03, France
e-mail mjrochet@ifremer.fr

R.J. Beamish and B.J. Rothschild (eds.), *The Future of Fisheries Science in North America*, 209
Fish & Fisheries Series, © Springer Science + Business Media B.V. 2009

stock assessments, the dual role of indicators as triggers for management measures ("control" function) and elements of management performance reporting ("audit" function) has emerged for various practical and historical reasons (Rice and Rivard 2007), where the second role corresponds to the function mentioned by Hilborn and Walters cited above. The dual role of stock indicators seems to have pervaded to ecosystem indicators. For an ecosystem approach for forest management, it has even been suggested that using the same indicators for both purposes might be misleading. For example, if an indicator species is used for a given habitat or community, managing the indicator species instead of (and potentially at the detriment of) the habitat or community of interest might be rewarding in terms of management performance (Simberloff 1999). In this chapter, we examine whether and how indicators could be used for giving ecosystem-based fishery management advice, i.e., control function, in addition to management performance evaluation, i.e., audit function. We first briefly recount the emergence of the term indicator in the EAF context. Then, we review how indicators have been proposed to be used for management advice, and what might be the alternative approaches. Finally, we suggest further research directions for indicators to become usable for fisheries management in an ecosystem context.

12.2 A Brief History

Fisheries publications on ecosystem indicators, or ecological indicators, have flourished over the last 10 years. Searching in the Aquatic Sciences and Fisheries Abstracts database for publications with simultaneously the key words *ecosystem*, *indicators*, and *fishery management*, the first four results are found for 1985, three of which refer to the same study; the next entry is for 1991. Then, an exponential-type increase of the number of publications began in the 1990s, with 38 publications on these topics in 2005 (Fig. 12.1). By contrast, publications mentioning both *ecosystem* and *fishery management* are found back to the earliest years in the database (1971) and probably much earlier. On the other hand, there were already 15 publications referencing both *indicators* and *fishery management* prior to 1980.

So the ideas of an ecosystem approach to fisheries, and of indicators as tools for fisheries management, were already discussed during the 1970s, whereas their close association emerged during the 1990s. This association is now widely accepted, justifying, for example, the organization of a whole Symposium on "Quantitative ecosystem indicators for fisheries management" in 2004 (Daan et al. 2005). Many studies dealing with indicators begin with a reference, and sometimes some text, about why an ecosystem approach is needed, and then go on introducing their indicator work, without specifying the link between the ecosystem approach and indicators (see, e.g., most papers in Daan et al. 2005). Explicit reasons for why indicators are useful in the ecosystem approach are not presented until the early twenty-first century (Table 12.1). The indicator boom in the 1990s may rather be ascribed to the call by Agenda 21 adopted at the Rio Earth Summit in 1992 for the development of "the concept of indicators of sustainable development in order to

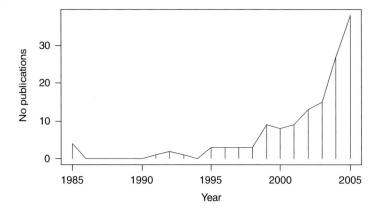

Fig. 12.1 Number of publications in the Aquatic Sciences and Fisheries Abstracts database referring together to *ecosystem*, *indicators*, and *fishery management* in their title, abstract, or keywords (NB the ASFA database gathers references since 1971)

Table 12.1 Quotations from some publications which justified why indicators are needed for an ecosystem approach to fisheries (EAF)

Definition of an indicator	Why are indicators needed for an EAF?	Purpose of indicators	Reference
A variable, pointer, or index related to a criterion. Its fluctuations reveal the variations in those key elements of sustainability in the ecosystem, the fishery resource or the sector and social and economic well-being.	"Information on the contribution of most human activity to sustainable development will be difficult to obtain"	"[T]o enhance communication, transparency, effectiveness and accountability in natural resource management"	FAO (1999)
"[P]ointers that can be used to reveal and monitor the conditions and trends in the fishery sector."	Required by Agenda 21	"[T]o assess development and management performance in relation to the various components of the fishery system"	Garcia et al. (2000)
"[T]wo defining characteristics: (1) quantify information so its significance is more readily apparent; and (2) simplify information about complex phenomena to improve communication"	"[T]o help simplify the volumes of complex primary data often demanded by more ecologically and socially inclusive approaches"	"[T]o assess current conditions, simplify and communicate information, and monitor progress toward [...] sustainability goals"	Pajak (2000)

(continued)

Table 12.1 (continued)

Definition of an indicator	Why are indicators needed for an EAF?	Purpose of indicators	Reference
	"[M]ultispecies models fully describing species interactions have proven expensive to parameterize and are subject to many uncertainties that cast doubt on their predictions. In addition, their output is difficult to portray in an easily comprehensible way"	"[T]o report both simple and quantitative information about complex systems"	Bellail et al. (2003)
	"[I]n view of the limited resources available for developing ever more complex predictive models ", […] switch from "hard predictability" to "soft predictability, which does not require detailed understanding of processes and capability of quantitative predictions of outcomes of specific policies"	"[P]roviding information on the state relative to the specific objectives for management and the direction to move to achieve those objectives"	Degnbol and Jarre (2004)
"[V]ariables, pointers or indices of a phenomenon"	"[A]s [ecosystem] attributes may not be directly measurable, indicators can act as proxies for them"	"i) [D]escribing the pressures […], state […] and response […], ii) tracking progress towards management objectives iii) communicating complex trends […] to a non-specialist audience"	Jennings (2005)
	"[E]cosystems are so complex and unpredictable that suites of indicators are needed to give an adequate picture of their state"		Rice and Rochet (2005)

identify such indicators" (Dahl 2000; Garcia and Staples 2000). Indeed, EAF is largely the application of sustainable development to fisheries (Garcia and Grainger 1997; Scandol et al. 2005). In the fisheries, obviously the second requirement of Agenda 21 has been met, and indicators have been identified, but whether the concept of an indicator is clearly developed is less sure. Niels Daan, for example, at the end of the ICES/SCOR Symposium on Quantitative Ecosystem Indicators for Fisheries Management, complained about the lack of a rigorous definition of an indicator, and the purpose it is supposed to serve (Daan 2005).

Indicators and reference points have been increasingly used in the late twentieth century to advise on stock management (e.g., Caddy and Mahon 1995), especially when the precautionary approach to fisheries management developed (Garcia 2000; Garcia 1996). This allowed advice to be more formal and standardized, and pre-agreed decision rules were expected to avoid faltering and conflicts at the time tough decisions would be needed. For single stocks, two main indicators have been widely used: fishing mortality and spawning stock biomass, relating to the more or less explicit objectives of keeping fishing pressure at a sustained level and maintaining stock reproductive capacity. Both indicators have the dual function of control and audit (Rice and Rivard 2007). Moving to an ecosystem approach implied a broadened view of the components to be considered and a multiplicity of objectives. Implicitly assuming that the two-dimensional stock set could be generalized to a multidimensional ecosystem set, indicators have been sought in vague relationship with both the ecosystem components and the management objectives. Recognizing the complexity of ecosystems, the agreement settled that several indicators would be needed in relationship with each management objective (FAO 1999; Methratta and Link 2006; Rice 2001). In addition, ecosystem management objectives are only broadly defined leaving much room for their translation into operational objectives (ICES 2005b; O'Boyle et al. 2005). As a consequence indicators proliferated; long lists of indicators have been proposed by scientists (Rice 2000; ICES 2005a; Daan et al. 2005) and agreed by agencies (FAO 1999; EEA 2003). Both scientists and agencies then felt the need to shrink these lists, and criteria and frameworks to evaluate and select indicators flourished on their turn (e.g., Rice and Rochet 2005; ICES 2002; FAO 1999; UNCSD 2001). The confusion about indicator roles and the ambiguity of the concept probably explain that these lists of criteria were often impressively long.

The indicator concept was borrowed from environmental management, which has relied on pollution indicators, bio-indicators, and indicator species for a long time. In this field, which is now joining forces with fishery management for an integrated approach, indicators "are designed to inform us quickly and easily about something of interest" because "it is not possible to measure everything" (National Research Council 2000). The multidimensionality and complexity of natural ecosystems and of human impacts implies that all environmental variables cannot be monitored and assimilated, and that indicators have to be used to summarize the information of interest; cost-effectiveness concerns were included in the concept as well (National Research Council 2000; Cairns et al. 1993). In terrestrial ecology or pollution monitoring, indicator species were expected to fulfill this need, although additional indicators at the ecosystem level are increasingly needed as well (Landres 1992). However, already 15 years ago a critical examination cast doubt about the ability of any figure to meet the reliability, cost-effectiveness, and precision requirements implied by this expectation (Landres 1992). Indeed indicators are asked so much (Table 12.1) that it is hard to imagine how any figure could serve all purposes they are intended for. As outlined by the FAO guidelines, "indicators are not an end in themselves" (FAO 1999), they have to be incorporated in broader approaches or "frameworks." We have come up with lists of indicators, but a working operational framework for their use in decision making is still lacking to a great extent. The most developed frameworks to date are the hierarchical structure of the Australian Ecologically Sustainable Development (ESD) Reporting Framework, which divides well-being into ecological, human, and economic components and then further subdivides these components (Chesson and Clayton 1998); and the Pressure-State-Response (PSR) or Driving force-Pressures-State-Impacts-Response (DPSIR) frameworks promoted by FAO (1999).

A last but key point in this review of the background of ecosystem indicators development is that it is generally agreed that an EAF will not be workable without a strong and active participation of stakeholders (FAO 2003; Garcia and Cochrane 2005). Deciding about multiple uses of marine ecosystems at various scales will require negotiations, decentralization, and processes to solve conflicts (Garcia and Cochrane 2005). All stakeholders involved in theses various steps will need to be informed and get some common understanding of the issues at stake, in addition to their own views and perceptions (Degnbol 2005). Indicators might be easier to understand by this wide audience than, e.g., model outputs, and could provide the transparency required to promote dialogue (Degnbol and Jarre 2004).

12.3   Approaches to Giving Advice for an EAF

Stock management has relied mainly on stock assessments. This consists in interpreting catch history, and possibly additional information, with the help of a population dynamics model. In many TAC-based management systems, the model is then used to make projections of future stock states under possible management options and advice is given based on the outcomes. Indicators and reference points have only recently been added, not substituted, to this process. Logically, the first attempts to develop tools for an EAF (long before indicators became unavoidable) were dynamic models (e.g., May et al. 1979; Larkin 1966; Laevastu and Larkins 1981). If ecosystems are viewed as inherently dynamic systems, dynamic models are needed to understand their behavior and provide predictive management advice in the spirit of what was done for stocks. The effort to develop multispecies and ecosystem models is ongoing. However, there was early warning that their results would be difficult to use directly for management purposes, and especially for quantitative prediction, because the numerous parameters of such models would be difficult to estimate even with intensive data collection, and complex nonlinear models may have unexpected behavior (May 1984; Clark 1985). These arguments still hold and have been rewritten at length by many authors since then. The complexity of model output was already perceived as a problem at this time when transparency to stakeholder was not yet a concern (Clark 1985). This does not mean that multispecies and ecosystem models are not useful, however, they can hardly be used to give transparent management advice, and this has been a justification to develop indicators (Table 12.1). Though, multispecies models are not necessarily a prerequisite for dynamic model-based ecosystem advice and very simple dynamics can be used when trophic interactions are not elucidated enough to be modeled quantitatively (e.g., Sainsbury et al. 2000; Constable et al. 2000).

Science would play a different role in an adaptive management framework. In adaptive management, short-term management policies are developed in an experimental design and the outcomes are analyzed for further development and implementation of management policy (Walters 1986). This could be implemented to maintain a fish assemblage with the aim of achieving both multispecies persistence and economically viable catch levels by designing areas with different effort levels (Tyler 1999). Of course catch and survey indicators would be needed to monitor the outcomes, but they are not proposed as the core of the approach. Adaptive management has not been widely adopted in fisheries, ocean, and coastal management because it requires sophisticated modeling, fishers might not be willing to be in the area with low effort, and it is costly (Bundy 1998). However, complex modeling at the basis of the experimental design is not always necessary (Degnbol 2002); experimental design could be made equitable to the involved stakeholders; and poorly managed fisheries are costly too. There are increasing proponents of passive adaptive management, aiming at learning from the past even in the absence of a true experimental design (Degnbol 2002). Here, the role of science is to monitor and provide interpretation of management results and propose changes in management measures.

Other suggested approaches consist in analyzing systems to identify which are the major problems to be solved that are not addressed by single stock management, and where management actions are likely to be efficient. Fletcher (2005) used qualitative risk assessment based on stakeholder involvement to identify and prioritize the issues to be addressed by fishery management. The ecosystem-based management of the Alaska groundfish fisheries seems largely to be based on limiting damaging fishing practices (Witherell et al. 2000). The purpose of sampling and observing here is not to parameterize a dynamic model, but to build a conceptual model of the exploited ecosystem and seek manageable control variables. Nicholas Bax and colleagues (1999) called these people's interactions with the ecosystem components that influence production "leverage points"; in the southeast Australian continental shelf discarding practices, location of fishing effort and transferable ecological stock rights were found to be such leverage points. In the same vein, Beamish and McFarlane (1999), based on a diversity of environmental, plankton, abundance, and catch data, drew species ecosystem management charts showing the key factors affecting species abundance for some species in the Strait of Georgia. This kind of approach addresses the key question of what can be done that is not examined in most indicator works and not always in dynamic modeling, and should be further developed (Jennings 2005).

12.4 How Could Indicators Be Used for Giving EAF Management Advice?

12.4.1 State of the Art

For stock management, indicators are often used in control rules based on reference points that trigger management action. When the threshold value is reached a predefined action should be taken. Inheriting from this habit reference points have been sought for ecosystem indicators, and their existence or at least potentiality is one of the criteria generally put forward for selecting indicators (e.g., FAO 1999). Recognizing that the choice of a reference point is ultimately arbitrary, Link (2005) suggested limit reference points and warning thresholds for a selection of 14 ecosystem indicators, providing as many decision criteria. This approach was adopted for the Convention on the Conservation of Antarctic Marine Living Resources (Constable et al. 2000) and precautionary reference points were set for both prey and predator species, ensuring there would be enough prey to sustain predator populations. In that case stocks are managed taking account of ecosystem considerations, but in many instances the interactions might be more complex and intervention required at a higher level. The next question is then, how to integrate several (or many) criteria to come to a decision? The most developed approach is the traffic light method. This method consists of given time series of indicators by coloring each data point according to the position relative to the reference points: green if the indicator is in the acceptable domain, yellow if in the warning region, and red if beyond the limit (Caddy 2002; Halliday 2001). From there to decision making, weightings, and combination rules still have to be determined, and this is increasingly difficult as the number of indicators rises. Alternatively, giving up identifying reference points for each indicator, a multidimensional reference region can be sought based on a multivariate analysis of indicators (Link et al. 2002): reference regions will appear if the time series is long enough and covers contrasted states of the system. When this is not available, artificial extreme states combining extreme indicator values can be forced into the analysis (Pitcher and Preikhost 2001).

On the other hand, it has been recognized that it would be difficult to define reference points for community metrics, on both technical and theoretical grounds (Jennings and Dulvy 2005; Hall and Mainprize 2004), and for nontarget species, owing to data scarcity (Hall and Mainprize 2004). Thus, it has been suggested to use reference directions instead, that is, desirable or undesirable trends in indicators (Jennings et al. 2005; Jennings 2005) or in the multivariate indicator space (Link et al. 2002). However, determining which trends are desirable or not still requires to assess the state of the system of the beginning of the time series, or a reference state (Rochet et al. 2005). Once this is done, conceptual models can be used to interpret the observed combinations of indicators trends. For example, if both average length decreased and abundance increased in a population over a period of time, we infer that there has been one or several strong year-classes, whereas decreases in both metrics would suggest an increased mortality, attributable to fishing (Rochet et al. 2005). When appropriate metrics are available, this can be further used to suggest the appropriate management actions: if population size has increased while fishing pressure remained stable, some increase in catch (TAC) might be allowed (Trenkel et al. 2007).

Early use of indicators in an ecosystem management context has been made for evaluating the performance of various management strategies via simulations (management strategy evaluation) for which indicators serve both for calculating performance measures and directly in decision rules (Sainsbury et al. 2000).

12.4.2   What Is an Indicator?

Based on the review above it seems essential to keep well separated the various functions of indicators for an EAFM. Table 12.1 suggests that the concept is overloaded and should be split into several concepts, thus several terms. For example, the term "metrics" was previously used when dealing with EAF (e.g., Rice 2000). In the perspective below we will use three terms, isolating pieces of the definitions in Table 12.1:

A metric is a variable which summarizes a process or pattern of interest in an exploited ecosystem. A structured suite of metrics will reveal important changes or differences to decision makers. Metrics are control tools used for giving science-based advice to management bodies.

An index summarizes complex phenomena to reveal important changes or differences to stakeholders. Indices are tools dedicated to the communication with a wide audience.

An indicator is a variable which quantifies how well a fishery is managed, in relation to specified objectives. An indicator typically has an audit function.

So defined, each candidate for one of these categories can be evaluated against a restricted list of criteria (Table 12.2) instead of, e.g., the nine criteria formerly identified by Rice and Rochet (2005). Some properties do not have to be met by individual variables, but rather by a suite of these variables (Table 12.2). For example, cost does not have to be considered separately for each metric or indicator; the cost of the whole suite matters. An additional criterion to be considered might be that each suite of figures should be comprehensive for the purpose it is intended to serve: metrics should provide a good coverage of the ecosystem components and attributes (Jennings 2005); indices should show how stakeholder concerns evolve; indicators should cover the broad management objectives.

Table 12.2 Criteria (From Rice and Rochet 2005) relevant to evaluate metrics, indices, and indicators as defined in the text. i is for criteria to be considered for each individual variable, s for criteria relevant for a suite of variables

Criterion	Metrics	Indices	Indicators
Concreteness		i	
Theoretical basis	i + s	i	
Public awareness		i	
Cost	s	s	s
Accurate measurement	i		
Availability of historic data	i		
Sensitivity to fishing impacts			i
Responsiveness to management actions			i
Specificity to fishing impacts			i

12.4.3   Perspectives

An ecosystem approach to fisheries means a broad scope encompassing nontarget species, habitats, and the whole ecosystem; this implies more diverse and potentially conflicting uses of a complex shared resource, thus a diversity of stakeholders; the overall complexity in turn implies ignorance, indeterminacy, and unpredictability (Garcia 2000; Degnbol 2002; Scandol et al. 2005). Complexity defies the direct use of dynamic models. Metrics and indicators have proliferated with the expectation they would give a picture of complexity without the need of strong assumptions on system drivers, processes, and dynamics. In the process we should not leave aside links, meaning, and the knowledge accumulated for more than a century of fisheries science. Complexity has to be organized and we have tools for this. The recent calls for linking pressure and state metrics are a first step in this direction. But we must remember that one of the main reasons why an EAF has been called for was the interactions between ecosystem components and uses. Then will it be possible to manage specific considerations based on indicators structured in single cause–effect relationships by negotiation, as advocated by Degnbol (2002)? We propose that this process can be helped by a conceptual model linking the problems to be managed and the possible management actions *via* the known interactions in the ecosystem (Fig. 12.2). This model could vary according to specific situations from simple verbal models or PSR frameworks to parameterized dynamic models. It would help to first identify the appropriate metrics, indices, and indicators needed in the process. It would make the relationships between these more transparent to the users. It could also help identify gaps where no solution is available to solve the problems. The model and monitoring system could evolve as knowledge improves and the management is evaluated by the indicators.

The various stakeholders probably have different conceptual models to be, to the extent possible, reconciled and accommodated in a common conceptual model. Both social approaches and negotiation processes, and science as a transcultural and testable knowledge can play a role here. This might require methods to formalize

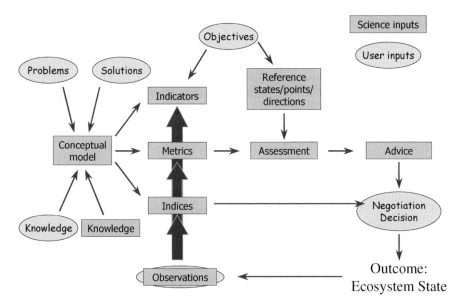

Fig. 12.2 A schematic of possible uses of indicators, metrics, and indices for an EAF management. Bubbles refer to user inputs and rectangles to science inputs. Observations can be both user and science input. Indices would be required for inclusive governance where decisions would be the results of stakeholder negotiation. Problems here are the main issues to be addressed in the ecosystem or fishery to be managed, while solutions are the "leverage points," or manageable control variables

and structure user inputs. The empirical comparative studies for identifying the qualitative links between directions of changes in ecosystem indicators and fishing pressure or management measures as proposed by several authors (Rice 2001; Link et al. 2002) have to be carried out. Qualitative models that would make the relationships between metrics explicit without accurate assumptions about processes and parameterization could be developed, and their properties explored by the developing qualitative methods (Eisenack and Kropp 2001; Puccia and Levins 1985; Ramsey and Veltman 2005; Bernard and Gouzé 2002)

Figure 12.2 helps identifying research needs by discipline (Table 12.3). Many cells in Table 12.3 are merged because interdisciplinary work is urgently required for an EAF (Rice 2005; Degnbol et al. 2006) and this is identified at all points where user inputs (bubbles) and science inputs (rectangles) are linked in Fig. 12.2. We stress the need for improved communication at all stages: conceptual models and indices have to be constructed interactively with constant stakeholder input; similarly metrics and indicators selection needs input from the decision bodies.

Finally, we wish to outline that there is no need to wait for all these components to be developed for undertaking an EAF management. Big issues are already known and some very efficient measures that could be implemented soon are known as well (Degnbol 2002; Witherell et al. 2000), e.g., classification of fishing gears and practices (Garcia 1996). Common sense is still the best approach to complex problems.

Table 12.3 Research needs by discipline to use metrics, indices, and indicators in an EAF. Merged cells indicate needs for interdisciplinary work

Component (from Fig. 12.2)	Social sciences	Natural sciences	Maths
Problems & solutions Stakeholder knowledge	Develop methods and tools to help formulating and structuring required user inputs		
Stakeholder knowledge	Develop methods to merge various sources of knowledge		
Knowledge		Improve, especially about ecosystem dynamics and cumulative impacts (Rosenberg and McLeod 2005)	
Conceptual model	Develop conceptual models to integrate current knowledge and the problems to be solved with the available solutions		
Indicators, metrics, and indices		Derive coherent metrics, indices, and indicators suitesmaking sense from conceptual models	
Reference states/ points/directions	Establish reference states, directions, or points based on specified objectives, data analysis and modeling work		
Assessment		Develop integrated assessment methods based on multiple metrics and a conceptual model	
Advice	Go from assessment to advice, that is, identify among the possible management actions the most suitable ones for the current situation		
Negotiation & decision	Analyze stakeholder decision making and interactions		

12.4.4 Two Examples

Here we illustrate some of the developments identified above by two examples, both at an early stage: identification of fishers' problems in the eastern English Channel; and building a conceptual model for the hake and *Nephrops* fishery in the Bay of Biscay.

To formalize the major issues perceived by the fishers in the eastern English Channel, a survey was conducted among French fishers in June 2006 (Prigent et al., 2008). Twenty-nine semidirective interviews were carried out among fishers of different métiers. Cognitive maps were employed to formalize their experience and knowledge: interviewees were asked to draw a bubble diagram of the Channel ecosystem, with arrows indicating the main determining factors including their direction and influence strength. Cognitive maps can be aggregated to represent the view of a group of persons. For example, Fig. 12.3 shows the average cognitive map of ten trawlers in the eastern English Channel. The largest negative influences on

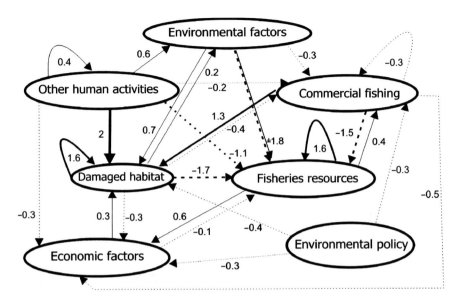

Fig. 12.3 Average cognitive map of ten trawlers in the eastern English Channel (see Prigent et al., 2008). The width of arrows is proportional to the link strength (also indicated by numbers). Continuous lines: positive effects; dotted lines: negative effects. Loops are created during the aggregation process when interviewees identified subcomponents of the bubbles shown here

fisheries resources and commercial fishing can be interpreted as the main concerns of these fishers. Environmental factors (and especially climate change) are perceived to have a strong influence on resources, however, the effect can be negative or positive depending on species. By contrast, a damaged biotope is felt by these fishers to have a strong impact on the resources. They fear water pollution especially by macro-waste (plastic bags, old shoes, pieces of old machines, lost fishing gears, etc.), gravel extraction, underwater cables and offshore wind farms, maritime traffic, and agriculture runoffs might impair resource growth and reproduction. A second concern is overexploitation of resources, they have to share with too many other fishers (especially from other European Union countries). They ascribe the poor state of many stocks in the region to too large catches and discards, threatening fish reproduction. The rapid pace of technological developments in fisheries is perceived as a key problem. This kind of work is essential to outline the problems to be addressed by ecosystem management, and to identify appropriate metrics to be communicated to the fishers.

In the Bay of Biscay there is a large hake fishery exploited by many fleets from several countries using various gears. There is also a valuable *Nephrops* fishery, which is mainly exploited by French trawlers, and happens to take place in a major hake nursery. This fishery generates huge amounts of both hake and *Nephrops* discards (Talidec et al. 2005; Rochet et al. 2006). We apply here the approach outlined above to the specific problem of these discards. Possible solutions include the

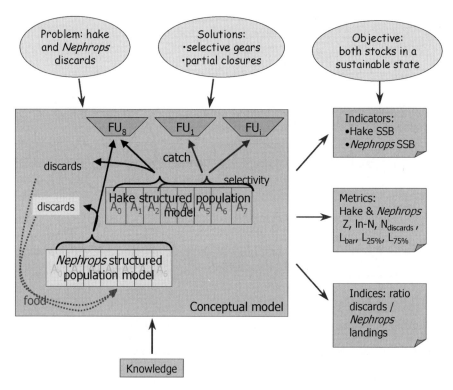

Fig. 12.4 A conceptual model for the *Nephrops* and hake fishery in the Bay of Biscay. FU_j: Fishery units. A_j: age- or size-classes. SSB: Spawning stock biomass, Z: total mortality rate, ln-N: log-transformed population abundance, L_{bar}: average length, $L_{q\%}$: q-percentile of the population length distribution (all estimated from trawl-survey data), $N_{discards}$: number discarded (estimated from onboard observer programs)

use of more selective gears and seasonal closures at hake recruitment peak. Based on the available scientific knowledge (Drouineau et al. 2006), we designed a conceptual model of this system (Fig. 12.4). Hake and *Nephrops* both have structured population dynamics in this model; this does not mean full parameterized models would be required; rather the results of similar models would be used to infer that fishing should impact population abundance and length distribution, which justifies our choice of metrics for each population (Trenkel et al. 2007). Numbers of discarded animals are added to account for the specific problem we address here. The ratios of discarded hake and *Nephrops* per *Nephrops* caught would be self-explanatory indices for the fishers. And if the overall objective is to have both stocks in a sustainable state, hake and *Nephrops* spawning stock biomass (SSB) should be sufficient indicators.

Acknowledgments This work was partially funded by the EU project IMAGE (contract FP6–044227). We thank Serge Garcia for useful comments on an earlier version of this manuscript.

References

Bax, N., Williams, A., Davenport, S. & Bulman, C. (1999) Managing the ecosystem by leverage points: a model for a multispecies fishery. *Ecosystem Approaches for Fisheries Management* (ed University of Alaska Sea Grant), pp. 283–303. Fairbanks, Alaska.

Beamish, R.J. & Mcfarlane, G.A. (1999) Applying ecosystem management to fisheries in the Strait of Georgia. *Ecosystem Approaches for Fisheries Management* (ed University of Alaska Sea Grant), pp. 637–664. Fairbanks, Alaska.

Bellail, R., Bertrand, J., Le Pape, O., Mahé, J.-C., Morin, J., Poulard, J.-C., Rochet, M.-J., Schlaich, I., Souplet, A. & Trenkel, V. (2003) A multispecies dynamic indicator-based approach to the assessment of the impact of fishing on fish communities, pp. 12. ICES Document, CM 2003/V: 02.

Bernard, O. & Gouzé, J.-L. (2002) Global qualitative description of a class of nonlinear dynamical systems. *Artificial Intelligence* **136**, 29–59.

Bundy, A. (1998) The red light and adaptive management. *Reinventing Fisheries Management* (eds T.J. Pitcher, P.J.B. Hart & D. Pauly), pp. 361–368. Kluwer, Dordrecht.

Caddy, J.F. (2002) Limit reference points, traffic lights, and holistic approaches to fisheries management with minimal stock assessment input. *Fisheries Research* **56**, 133–137.

Caddy, J.F. & Mahon, R. (1995) *Reference Points for Fisheries Management*. FAO, Rome.

Cairns, J., Mccormick, P.V. & Niederlehner, B.R. (1993) A proposed framework for developing indicators of ecosystem health. *Hydrobiologia* **263**, 1–44.

Chesson, J. & Clayton, H. (1998) A framework for assessing fisheries with respect to ecologically sustainable development, pp. 60. Bureau of Rural Sciences, Canberra.

Clark, C.W. (1985) *Bioeconomic Modelling and Fisheries Management*. Wiley, New York.

Constable, A.J., De La Mare, W.K., Agnew, D.J., Everson, I. & Millar, D. (2000) Managing fisheries to conserve the Antarctic marine ecosystem: practical implementation of the Convention on the Conservation of Antarctic Marine Living Resources (CCAMLR). *ICES Journal of marine Science* **57**, 778–791.

Daan, N. (2005) An afterthought: ecosystem metrics and pressure indicators. *ICES Journal of Marine Science* **62**, 612–613.

Daan, N., Christensen, V. & Cury, P.M. (eds) (2005) Quantitative ecosystem indicators for fisheries management. *ICES Journal of Marine Science* **62,** 307–614.

Dahl, A.L. (2000) Using indicators to measure sustainability: recent methodological and conceptual developments. *Marine Freshwater Research* **51**, 427–433.

Degnbol, P. (2002) The ecosystem approach and fisheries management institutions: the noble art of addressing complexity and uncertainty with all onboard and on a budget. *Fisheries in the Global Economy. Eleventh Biennial Conference of the International Institute of Fisheries Economics and Trade (IIFET)* (ed B. Shallard), pp. 11. Bruce Shallard and Associates, Wellington, New Zealand.

Degnbol, P. (2005) Indicators as a means of communicating knowledge. *ICES Journal of Marine Science* **62**, 606–611.

Degnbol, P. & Jarre, A. (2004) Review of indicators in fisheries management – a development perspective. *African Journal of Marine Science* **26**, 303–326.

Degnbol, P., Gislason, H., Hanna, S., Jentoft, S., Raakjær Nielsen, J., Sverdrup-Jensen, S. & Wilson, D.C. (2006) Painting the floor with a hammer: technical fixes in fisheries management. *Marine Policy* **30**, 534–543.

Drouineau, H., Mahévas, S., Pelletier, D. & Beliaeff, B. (2006) Assessing the impact of different management options using ISIS-Fish: the French *Merluccius merluccius – Nephrops norvegicus* mixed fishery of the Bay of Biscay. *Aquatic Living Resources* **19**, 15–29.

EEA (2003) Environmental performance indicators for the European Union. http://themes.eea.europa.eu/indicators/

Eisenack, K. & Kropp, J. (2001) Assessment of management options in marine fisheries by qualitative modelling techniques. *Marine Pollution Bulletin* **43**, 215–224.

FAO (1999) Indicators for sustainable development of marine capture fisheries, FAO Technical Guidelines for Responsible Fisheries, No. 8, pp. 68. FAO, Rome.

FAO (2003) The ecosystem approach to fisheries. Issues, terminology, principles, institutional foundations, implementation and outlook. FAO Fisheries Technical Paper. No. 443. FAO, Rome.

Fletcher, W.J. (2005) The application of qualitative risk assessment methodology to prioritize issues for fisheries management. *ICES Journal of Marine Science* **62**, 1576–1587.

Garcia, S.M. (1996) The precautionary approach to fisheries and its implications for fishery research, technology and management: an updated review. *Precautionary Approach to Fisheries. Part 2. Scientific Papers*, pp. 1–65. FAO, Rome.

Garcia, S.M. (2000) The precautionary approach to fisheries: progress review and main issues: 1995–2000. *Report of the CWP Inter-sessional WG on Precautionary Approach Terminology and CWP Sub-group on Publication of Integrated Catch Statistics for the Atlantic*, pp. 21–48. ICES, Copenhagen, Denmark.

Garcia, S.M. & Cochrane, K.L. (2005) Ecosystem approach to fisheries: a review of implementation guidelines. *ICES Journal of Marine Science* **62**, 311–318.

Garcia, S.M. & Grainger, R. (1997) Fisheries management and sustainability: a new perspective of an old problem? *Developing and Sustaining World Fisheries Resources. The State of Science and Management. 2nd World Fisheries Congress* (ed D.A. Hancock, D.C. Smith, A. Grant & J.P. Beumer), pp. 631–654. CSIRO, Australia.

Garcia, S.M. & Staples, D.J. (2000) Sustainability reference systems and indicators for responsible marine capture fisheries: a review of concepts and elements for a set of guidelines. *Marine Freshwater Research* **51**, 385–426.

Hall, S.J. & Mainprize, B. (2004) Towards ecosystem-based fisheries management. *Fish and Fisheries* **5**, 1–20.

Halliday, R.G., Fanning, L.P. & Mohn, R.K. (2001) Use of the traffic light method in fishery management planning, pp. 41. Canadian Science Advisory Secretariat, Ottawa.

Hilborn, R. & Walters, C.F. (1992) *Quantitative Fisheries Stock Assessment. Choice, Dynamics and Uncertainty*. Chapman & Hall, New York.

ICES (2002) Report of the ICES Advisory Committee on Ecosystems, 2002, pp. 131. ICES, Copenhagen.

ICES (2005a) Ecosystem effects of fishing: impacts, metrics, and management strategies, ICES Coop. Res. Rep. No. 272, pp. 177. ICES, Copenhagen.

ICES (2005b) Guidance on the application of the ecosystem approach to management of human activities in the European marine environment, Cooperative Research Report no. 273, pp. 22. ICES, Copenhagen.

Jennings, S. (2005) Indicators to support an ecosystem approach to fisheries. *Fish and Fisheries* **6**, 212–232.

Jennings, S. & Dulvy, N.K. (2005) Reference points and reference directions for size-based indicators of community structure. *ICES Journal of Marine Science* **62**, 397–404.

Laevastu, T. & Larkins, H.A. (1981) *Marine Fisheries Ecosystem. Its Quantitative Evaluation and Management*. Fishing News Books, Farnham.

Landres, P.B. (1992) Ecological indicators: panacea or liability? *Ecological Indicators* (eds D.H. Mckenzie, D.E. Hyatt & V.J. Mcdonalds), pp. 1295–1318. Elsevier Applied Science, London/New York.

Larkin, P.A. (1966) Exploitation in a type of predator-prey relationship. *Journal of the Fisheries Research Board of Canada* **23**, 349–356.

Link, J.S. (2005) Translating ecosystem indicators into decision criteria. *ICES Journal of Marine Science* **62**, 569–576.

Link, J.S., Brodziak, J.K.T., Edwards, S.F., Overholtz, W.J., Mountain, D., Jossi, J.W., Smith, T.D. & Fogarty, M.J. (2002) Marine ecosystem assessment in a fisheries management context. *Canadian Journal of Fisheries and Aquatic Sciences* **59**, 1429–1440.

May, R.M., ed. (1984) Exploitation of Marine Communities: Report of the Dahlem Workshop on Exploitation of Marine Communities, Berlin, April 1–6, 1984. Springer-Verlag, Berlin.

May, R.M., Beddington, J.R., Clark, C.W., Holt, S.J. & Laws, R.M. (1979) Management of multispecies fisheries. *Science* **205**, 267–277.

Methratta, E.T. & Link, J.S. (2006) Evaluation of quantitative indicators for marine fish communities. *Ecological Indicators* **6**, 575–588.

National Research Council (2000) *Ecological Indicators for the Nation*. National Academy Press, Washington, DC.

O'Boyle, R., Sinclair, M., Keizer, P., Lee, K., Ricard, D. & Yeats, P. (2005) Indicators for ecosystem-based management on the Scotian Shelf: bridging the gap between theory and practice. *ICES Journal of Marine Science* **62**, 598–605.

Pajak, P. (2000) Sustainability, ecosystem management, and indicators: thinking globally and acting locally in the 21st century. *Fisheries* **25**, 16–30.

Pitcher, T. & Preikhost, D. (2001) RAPFISH: a rapid appraisal technique to evaluate the sustainability status of fisheries. *Fisheries Research* **49**, 255–270.

Prigent, M., Fontenelle, G., Rochet, M.-J. & Trenkel, V.M. (2008) Using cognitive maps to investigate fishers' ecosystem objectives and knowledge. Ocean & Coastal Management, 51, 450–462

Puccia, C.J. & Levins, R. (1985) *Qualitative Modeling of Complex Systems: An Introduction to Loop Analysis and Time Averaging*. Harvard University Press, Cambridge, MA.

Ramsey, D. & Veltman, C. (2005) Predicting the effects of perturbation on ecological communities: what can qualitative models offer? *Journal of Animal Ecology* **74**, 905–916.

Rice, J. (2001) From science to advice – how to find ecosystem metrics to support management. ICES CM 2001/T: 12, 15.

Rice, J. & Rochet, M.J. (2005) A framework for selecting a suite of indicators for fisheries management. *ICES Journal of Marine Science* **62**, 516–527.

Rice, J.C. (2000) Evaluating fishery impacts using metrics of community structure. *ICES Journal of Marine Science* **57**, 682–688.

Rice, J.C. (2005) Implementation of the ecosystem approach to fisheries management – asynchronous co-evolution at the interface between science and policy. *Marine Ecology Progress Series* **300**, 265–270.

Rice, J.C. & Rivard, D. (2007) The dual role of indicators in optimal fisheries management strategies. *ICES Journal of Marine Science* **64**(4), 775–778.

Rochet, M.-J., Trenkel, V., Bellail, R., Coppin, F., Le Pape, O., Mahé, J.-C., Morin, J., Poulard, J.C., Schlaich, I., Souplet, A., Vérin, Y. & Bertrand, J.A. (2005) Combining indicator trends to assess ongoing changes in exploited fish communities: diagnostic of communities off the coasts of France. *ICES Journal of Marine Science* **62**, 1647–1664.

Rochet, M.-J., Bertignac, M., Fifas, S., Gaudou, O. & Talidec, C. (2006) Estimating discards in the French *Nephrops* fishery in the Bay of Biscay. ICES 2006/K: 24.

Rosenberg, A.A. & McLeod, K.L. (2005) Implementing ecosystem-based approaches to management for the conservation of ecosystem services. *Marine Ecology Progress Series* **300**, 270–274.

Sainsbury, K., Punt, A.E. & Smith, A.D.M. (2000) Design of operational management strategies for achieving fishery ecosystem objectives. *ICES Journal of Marine Science* **57**, 731–741.

Scandol, J.P., Holloway, M.G., Gibbs, P.J. & Astles, K.L. (2005) Ecosystem-based fisheries management: an Australian perspective. *Aquatic Living Resources* **18**, 261–273.

Simberloff, D. (1999) The role of science in the preservation of forest biodiversity. *Forest Ecology and Management* **115**, 101–111.

Talidec, C., Rochet, M.-J., Bertignac, M. & Macher, C. (2005) Discards estimates of nephrops and hake in the Nephrops trawl fishery of the Bay of Biscay: methodology and preliminary results for 2003 and 2004. *Working Document for the ICES Working Group on the Assessment of Southern Shelf Stocks of Hake, Monk and Megrim, WGHMM. Lisbon, Portugal, 10–19 May 2005*.

Trenkel, V.M., Rochet, M.-J. & Mesnil, B. (2007) From model-based prescriptive advice to indicator-based interactive advice. *ICES Journal of Marine Science* **64**(4), 768–774.

Tyler, A.V. (1999) A solution to the conflict between maximizing groundfish yield and maintaining biodiversity. *Ecosystem Approaches for Fisheries Management*, pp. 367–385. Alaska Sea Grant College Program, Fairbanks, Alaska.

UNCSD (2001) *Indicators of Sustainable Development: Guidelines and Methodologies*. United Nations Commission on Sustainable Development, Washington, DC.

Walters, C. (1986) *Adaptive Management of Renewable Resources*. Macmillan, New York.

Witherell, D., Pautzke, C. & Fluharty, D. (2000) An ecosystem-based approach for Alaska groundfish fisheries. *ICES Journal of Marine Science* **57**, 771–777.

Chapter 13
Fisheries Abundance Cycles in Ecosystem and Economic Management of California Fish and Invertebrate Resources

Jerrold G. Norton, Samuel F. Herrick, and Janet E. Mason

Abstract It is important for fishery scientists and ecosystem-based fishery managers to recognize that there may be apparent persistence in an ecosystem followed by ecosystem changes corresponding to different ecological states and different levels of fisheries output; revenues paid to California fishers have varied more than fivefold in inflation adjusted dollars during the 75-year period of our study. Empirical orthogonal function (EOF) analysis of California commercial fish landings from 1928 to 2002 defines a two-dimensional ecological space and the position of indicator species within it. This ecological space appears related to more easily monitored physical environmental processes that can be used to determine the persistence and probable temporal variability in species and ecological states. Observation of ecological changes in limited take marine reserves, the first step in ecosystem-based fishery management, is needed to reveal the proportion of ecosystem cycling that is dependent on exploitation. Future ecosystem-based fishery management may use Individual Transferable Quota shares that confer to fishers a user right to a percentage share of the Total Allowable Catch of fish. This management structure within the ecosystem-based management framework will require a high level of knowledge about ecosystem changes, fishery changes, and the resulting changes in the economic and social environment of the fishery. It is likely that the concepts of ecological space and ecological change presented here will be useful in assessing the population size and Total Allowable Catch of particular species and the importance of these species in sustaining a productive ecosystem for consumptive use and increasingly important nonconsumptive users.

J.G. Norton (✉)
Environmental Research Division, Southwest Fisheries Science Center, 1352 Lighthouse Avenue, Pacific Grove, CA 93950, USA Tel.: 831 - 648 - 9031 (8515);
e-mail: Jerrold.G.Norton@noaa.gov

S.F. Herrick
Fisheries Research Division, Southwest Fisheries Science Center, 8604 La Jolla Shores Drive, La Jolla CA 92037, USA

J.E. Mason
Environmental Research Division, Southwest Fisheries Science Center, 1352 Lighthouse Avenue, Pacific Grove, CA 93950, USA

R.J. Beamish and B.J. Rothschild (eds.), *The Future of Fisheries Science in North America*, 227
Fish & Fisheries Series, © Springer Science + Business Media B.V. 2009

Keywords Assessment · ecosystem · fisheries · management · revenue

13.1 Introduction

There have been difficulties in understanding, regulating, and anticipating fluctuations in the commercial harvest of fish and invertebrate species from the Exclusive Economic Zone (EEZ) of the United States (US) (e.g., Botsford et al. 1997; Longhurst 2006). These difficulties arise, in part, from problems both in establishing historical patterns (e.g., Baumgartner et al. 1992; Rothschild 1995; Klyashtorin 2001; Rosenberg et al. 2005) and in understanding the magnitude and possible consequences of climate changes, particularly with regard to the capacity of the marine ecosystem to provide fishery and other ecosystem resources.

It is well known that the life history and reproductive success of fishes are influenced by environmental fluctuations (e.g., Hollowed and Bailey 1989; McFarlane and Beamish 1992; Beamish 1993; Jacobson and MacCall 1995; Rothschild 1995). However, the lack of a long-term, historical perspective on oceanic ecosystems does not allow year-to-year variations to be viewed within the context of lower frequency, decade-scale ecosystem variations. For instance, the rapid decline of the Pacific sardine fishery in the 1940s and 1950s was the result of heavy fishing during the reproductively unfavorable ocean climate conditions that began in the mid-1940s (Herrick et al. 2007). This led to a collapse of the fishery and local economic hardship as canneries and reduction facilities closed. In accord with the ecological cycle presented below, the sardine fishery reemerged in the early 1990s after ecological conditions had returned to those similar to the conditions in the 1930s and 1940s.

In the following sections we give details on how the cycle of abundance found in the sardine population is also evident in the fisheries of more than a dozen other species. Then we suggest how analysis of these cycles of abundance might be used in ecosystem-based fishery management.

13.2 Background

California State laws require that all sales of fish and invertebrates from primary harvesters (fishers) to fish dealers be recorded and that a species (or species group), weight landed and price per pound be recorded on sales receipts (California Fish and Game Code section 8043). These records of landing weights and ex-vessel revenue have been collected from 1917 to the present. Sales receipts summarized and tabulated by the California Department of Fish and Game (CDF&G) from 1928 to 2002 have been put into Internet-accessible form by the Environmental Research Division of the Southwest Fisheries Science Center (http://www.pfeg.noaa.gov/products/las.html). Mason (2004) edited these data to provide consistency in unit measure and species names. Records collected prior to 1928 are less complete (California Bureau of Commercial Fisheries 1937).

We analyze these California commercial (CACom) landings data to find ecological climate changes that have occurred in the California fishers harvest environment (CFHE), which is that part of the California Current System (CCS) that extends from borders of California (32.5°N–42°N) westward 400–600 km. The coastal CFHE includes the major coastal upwelling areas of the west coast of North America and that part of the CFHE with greatest variability, associated with coastally generated Rossby Waves. The area 200–600 km from the coast includes the persistent, meandering, southerly flow of the California Current. Much of the variability in this offshore area appears associated with northeastern Pacific atmospheric forcing (Enfield and Allen 1980; McGowan et al. 1998; King et al. 1998, Fu and Qiu 2002).

For our analysis we selected landings series that were: (1) identified by species throughout; and (2) were recorded in the CACom landings regularly over the 75-year period from 1928 to 2002. Twenty-nine species were recorded in the landings in more than 55 of the years studied; 25 of the 29 species were in the records and in the CFHE in every year. These species represent diverse habitats and life histories (Table 13.1). Changes in the relative proportion of these species in the commercial catch appear to indicate changes in the CFHE ecological state; so they were considered ecological indicator species. We hypothesize that changes in dominant commercial species reflect changes in the flow of energy through the ecosystem that, in turn, influence the life history of many additional species throughout the CCS (McFarlane and Beamish 1992; Rothschild 1995; Norton and Mason 2003, 2004, 2005).

The near constant presence of the 29 indicator species, harvested by California commercial fisheries since the late 1800s, indicates that markets remained available and expanded as landings increased. Decreased availability, on the other hand, is usually reflected in increased capture costs, which may result in decreased landings if profitability decreases. Population increases and increases in per capita consumption in California, in particular, and the US, as a whole, ensured that the demand for the 29 indicator species has generally increased over time. Foreign demand for California seafood also became increasingly important to California fishers as their catch was incorporated into global markets (Wolf et al. 2001; O'Bannon 2001; Dietz et al. 2003). It is unlikely that large, harvestable concentrations of the 29 indicator species existed off California unnoticed and unharvested by California fishers. Possible exceptions to this assumption are market squid and northern anchovy (Jacobson and Thomson 1993; Yaremko 2001).

Technological advances such as more powerful engines, power winches, sonar, synthetic fiber lines, and methods of oceanic gill netting and long lining led to increased landings of some of the 29 indicator species. Combined landings of the 29 indicator species made up 80–90% of the total landings weight from the CFHE in each year from 1928 to 2002. After 1976, a number of previously lightly exploited species, such as deep water rockfish (*Sebastes* sp. and *Sebastolobus* sp.), urchins (*Strongylocentrotus* sp.), and prawns (*Pandalus platyceros*), added to the resurgence of California fisheries (Mason 2004). At times the CACom landings of a particular species will be proportional to its abundance in the CFHE and at times it may be less proportional (see below). However, if multispecies, ecosystem-based

Table 13.1 Type of habitat, scientific names, and four-letter code for the 29 ecosystem indicator species discussed in the text

Demersal species	Code[a]	Scientific name
Cabezon	Cbzn	*Scorpaenichthys marmoratus*
California halibut	Chlb	*Paralicthys californicus*
California scorpionfish	Scor	*Scorpaena guttata*
California sheephead	Shpd	*Semicossyphus pulcher*
Dungeness (market) crab	Dcrb	*Cancer magister*
Giant sea bass	Gbas	*Stereolepis gigas*
Lingcod	Lcod	*Ophiodon elongatus*
Ocean whitefish	Owfs	*Caulolatilus princeps*
Pacific hake (whiting)	Pwht	*Merluccius productus*
Pacific halibut	Phlb	*Hippoglossus stenolepis*
Spiny lobster	Slbs	*Panulirus interruptus*
Sablefish	Sabl	*Anoplopoma fimbria*
White croaker	Wcrk	*Genyonemus lineatus*
Migratory species		
Albacore	Albc	*Thunnus alalunga*
Pacific barracuda	Cuda	*Sphyraena argentea*
Pacific bluefin tuna	Btna	*Thunnus orientalis*
Pacific bonito	Pbnt	*Sarda chiliensis*
Skipjack tuna	Stna	*Katsuwonus pelamis*
Swordfish	Swrd	*Xiphias gladius*
White seabass	Wbas	*Atractoscion noblis*
Yellowfin tuna	Ytna	*Thunnus albacares*
Yellowtail	Yltl	*Seriola dorsalis*
Pelagic species		
Chub (Pacific) mackerel	Cmck	*Scomber japonicus*
Jack mackerel	Jmck	*Trachurus symmetricus*
Market squid	Msqd	*Loligo opalescens*
Northern anchovy	Nanc	*Engraulis mordax*
Pacific herring	Phrg	*Clupea pallasii*
Pacific pompano (butterfish)	Pbtr	*Peprilus simillimus*
Pacific sardine (pilchard)	Psdn	*Sardinops sagax*

[a]Pacific States Marine Fisheries Commission code used in Fig. 13.1

retrospective analyses are to be made for the 1928–2002 period with consistent and objective information, they will have to be based on the landings data.

13.3 The Cyclic Nature of Changes in the California Marine Ecosystem

When landings time series for the 29 indicator species were examined by empirical orthogonal function (EOF) analysis we found that EOF1 and EOF2 explain 30% and 19%, respectively, of the overall variance of annual CFHE landings for those species (Norton and Mason 2003, 2004). The time variations in the standardized, $\log_e$ transform of the annual landing weight for each of the 29 species were arranged as columns, with rows representing years in the input matrix. Each EOF was stable when various species were removed from the input (Norton and Mason 2004, 2005). Each EOF provides the relationship among species along an orthogonal axis as a "loading value". If a species has a high loading value it will have variation through time that is similar to the EOF itself. Negative loading values indicate inverse correlation to the EOF. When loading values for each species within EOF1 and EOF2 are plotted on orthogonal axes (Fig. 13.1),

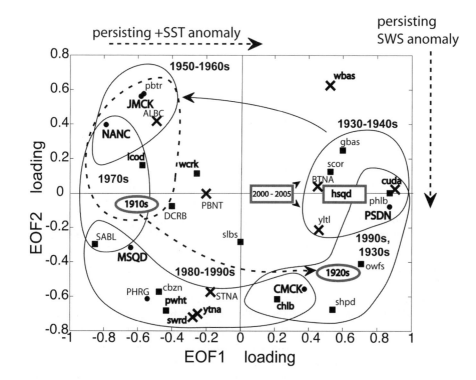

Fig. 13.1 Loading values for each of the 29 indicator species are plotted for EOF1 on the horizontal and for EOF2 on the vertical axes. The species grouped together by closed curves indicate species groups that have maxima during the decades indicated. Filled squares show the location of demersal species, crosses indicate migratory species and filled circles indicate pelagic species in the ecological space defined by the EOF axes. The dotted arrow above the top shows the shift in species composition as positive SST anomalies accumulate (see text for details). The dotted arrow beginning at the top right shows the shifts in species composition as southward wind stress (negative by convention) continues to be anomalously strong. The 29 species names are given as four-letter codes and indicate total metric tons (t) landed during the 1928–2002 period. Species that have contributed more than 100,000 t are shown in uppercase letters, with bold uppercase letters indicating species landings of > a million metric tons. Other species contributing >10,000 t are shown in lower case bold letters. A list of the species plotted and complete species names are given in Table 13.1. The dashed line and arrow from upper left to right within the EOF space indicates inferred changes occurring from 1917 to the late 1920s. A rectangle on the right hand side of the diagram indicates the presence of humboldt squid (hsqd) off California since 1997 and during 1934–1935

they can be grouped into clusters of species (enclosed curves) that have maxima in landings in the same decades. These groupings represent commercial fisheries indicators of ecological states within the "ecological space" defined by EOF1 and EOF2 (Fig. 13.1).

Counterclockwise progression of ecological states is shown in Fig. 13.1, from the 1930s–1940s (middle, right), when the Pacific sardine (PSDN) dominated the landings, to the 1950s–1960s when sardine landings had decreased and landings of northern anchovy (NANC), jack mackerel (JMAC), and albacore (ALBC) increased. These three species remained important in California fisheries through the 1970s along with lingcod (lcod) and sablefish (SABL) that reached their maxima in the 1970s. After the physical climate shift in the mid-1970s (Norton et al. 1985; Ebbesmeyer et al. 1991), a larger group of relatively high-value fish became available in the 1980s (lower, middle in Fig. 13.1). Other species, such as urchins and prawns that had not been part of the earlier California fishery also became commercially important during this period (Kalvass and Hendrix 1997, Mason 2004). These latter species with shorter landings time series were not included in the analysis.

The moratorium on the directed sardine fishery of 1967 allowed the sardine population to rebuild when conditions were again favorable in the 1980s and 1990s. These favorable conditions combined with conservative management resulted in the reestablishment of a viable sardine fishery that has continued to the present. Along with the sardines, came a resurgence in the commercial species that were associated with them 50 years before (middle, right in Fig. 13.1). These included chub mackerel (CMCK), California halibut (chlb), scorpionfish (scor), ocean white fish (owfs), and increasing commercial and recreational landings of barracuda (cuda) and yellowtail (yltl). It is likely that many unmeasured ecosystem components throughout the CCS also underwent cyclic changes from 1930 to 2002.

Although commercial landings data collected prior to 1928 are less complete, they indicate that the species enclosed by the broken line on the left side of Fig. 13.1 were more numerous in the landings before 1920 than they were in the late 1920s. The apparent movement through the ecological space from 1917 to the later 1920s is shown by the broken arrow, with the decade designations in ovals. Early fishery records (California Bureau of Commercial Fisheries 1937) show that more Humboldt (or jumbo) squid (*Dosidicus gigas*) (hsqd) than market squid (MSQD) were landed during 1934–1935. Since that time Humboldt squid have seldom been reported in the CFHE. However, Humboldt squid have become a notable component of the California Current ecosystem since the 1997–1998 El Niño period (Field et al. 2007), as indicated at the middle right in Fig. 13.1. These observations suggest similarity between the 1997–2005 period and the 1930s and early 1940s, as indicated by the EOF analysis. Ongoing investigations suggest that the California Current ecosystem remains in the state indicated by species maxima on the right side of Fig. 13.1, but we cannot assume that the CFHE will remain in this state indefinitely, or that future ecosystem changes will follow the exact paths shown in Fig. 13.1.

13.4   Economic Results of Ecosystem Climate Shifts

Many factors may have affected variation in the total revenue paid to California fishers on landing their catch (total ex-vessel revenue) and the number of boats making landings over the 1928–2002 period (Fig. 13.2). However, it appears that ecosystem change, reflected in the relative abundance of the 29 indicator species, is important in the variation of total California ex-vessel revenue and total boats making landings in California ports. During the first growth period in the 1930s and early 1940s (Fig. 13.2), sardines were the largest catch by both weight and ex-vessel revenue, accounting for as much as 70% of the weight landed and 50% of ex-vessel revenue (Marr 1960; Mason 2004). The first boat maxima (Fig. 13.2) occurred as sardine abundance was decreasing in the late 1940s, and fisheries for albacore, ocean caught salmon (*Oncorhynchus tshawytscha* and *O. kisutch*) and Dungeness crab (dcrb) became relatively more important. Both boat investment (reflected in the number of boats actively involved in the fishery) and total revenue declined to minima in the 1960s. These declines correspond to movement through ecological space from top to middle on the left side of Fig. 13.1. Anchovies were important in the catch in the late 1960s, 1970s, and early 1980s, but generally contributed 20% or less to the total ex-vessel revenue. The second maxima in boats and revenue in the late 1970s and early 1980s had a different dynamic than the first with the increase in boats reporting landings starting before the ex-vessel revenue increased (Fig. 13.2). The growth period into this maximum is represented in eco-logical space by the shift down in the lower left quadrant (Fig. 13.1). Dungeness crab, salmon, yellowfin tuna, herring (PHRG), albacore, skipjack tuna (STNA), swordfish (swrd), and groundfish all contributed more than 10% to annual ex-vessel revenue at times during the late 1970s and early 1980s. This was a period of target

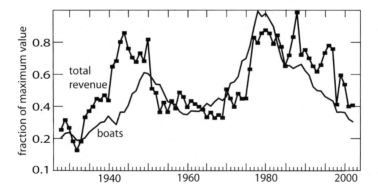

Fig. 13.2 The total ex-vessel revenue and total number of boats reporting landings (Norton and Mason 2003) are shown as percent of maximum values for the 1928–2002 period. All CFHE landings, except offshore factory reduction ship landings of sardines are included. The number of boats fluctuated between 1,446 and 7,353. The total ex-vessel revenue (squares) varied between $36,625,209 and $225,261,590 adjusted to year 2000 dollars

species diversification with more of the 29 indicator species available for harvest (Fig. 13.1). In addition, newly exploited species such as urchins contributed 5–25% of total California ex-vessel revenue.

The period since the late 1980s is characterized by increasing fisheries regulation (Dewees and Weber 2001; Leet et al. 2001, Appendix A) obscuring the relationship between the changing ecosystem and ex-vessel revenue and investment in boats. The decreasing number of boats making landings and the more variable ex-vessel revenue (Fig. 13.2) may indicate that California commercial fishing had become a more professional enterprise, with greatest economic rewards going to more adaptable and persistent operations. However, the return of chub mackerel in the 1980s followed by an increase in sardine biomass to more than a million metric tons (Hill et al. 2006), coupled with the reappearance of yellowtail and barracuda as frequently captured sport species strongly suggests a return in 1990–2005 to ecological conditions similar to those in the 1930s and 1940s. These changes indicate progression across the bottom of Fig. 13.1. Many of these temporal changes in species availability and the ex-vessel revenue to California fishers are discussed for discrete fisheries in Leet et al. (2001). The point is that as the ecosystem changes through "temporarily stable" ecological states, both the economic output of the ecosystem and the character of capital investment (indexed by the number of boats making landings) changes. The 29 indicator species were contributing more than 90% of the total ex-vessel revenue in the 1930s and less than 60% of the total ex-vessel revenue in the 1990s (Mason 2004) showing that the return of the ecosystem to the state of the 1930s did not return the California fishers to this earlier economic environment.

13.5 Ecosystem and Physical Ocean Climate Changes

If the physical environment determines the ecosystem state, then it would be useful to find indices of the physical processes involved, because these physical processes may be more easily monitored than the ecosystem, particularly under conservative harvesting strategies. The possibility of using physical environmental indices to monitor and predict changes in the CFHE and CCS ecosystems is illustrated first for the 29 ecosystem indicator species and second for sardines, which have been one of the most important forage and commercial species harvested during the 1928–2002 period.

Time variable coefficients, $C1$ and $C2$ of EOF1 and EOF2, respectively (also know as principal components) give numerical values for EOF expression in the input matrix through time. When $C1$ and $C2$ are compared to indices representing large-scale physical changes in the CFHE, simple linear relationships are found that may be expressed as,

$$C1 = f(\Sigma A) + t \qquad (13.1)$$

where f may be a scaling function or a constant, ΣA is the integrated physical anomaly or summation of the annual physical anomalies (A) from the record overall mean, and t is a term that will account for noise and other unexplained variation. This relationship and the identity, $A = \log_e e^A$ will be discussed in greater detail below.

Accumulation or summation of anomalies reveals climatic rather than interannual changes (Klyashtorin 2001; Hanley et al. 2002; Norton and Mason 2004).

Changes in landings composition from 1928 to 2002, indicated by $C1$, are closely related ($r^2 = 0.85$) to persisting conditions in the CFHE as indicated by integrated anomalies (ΣA) of sea surface temperature (SST) measured at the Scripps Institution of Oceanography Pier in La Jolla, California (La Jolla A-SST). The relation of La Jolla A-SST to a wide range of SST variance over the California Current region has been discussed by McGowan et al. (1998). La Jolla A-SST and $C1$ are also closely related to equatorial atmosphere–ocean processes (Chelton et al. 1982, Meyers et al. 1998; Norton and Mason 2004, 2005). These observations strongly suggest that the important features of the La Jolla A-SST index represent large-scale phenomena that change the CFHE physical characteristics by surface and subsurface forcing (Enfield and Allen 1980; Chelton et al. 1982; Norton et al. 1985; Meyers et al. 1998; Fu and Qiu 2002). The high correlation between the physical environmental index and EOF1, suggests that the physical index and its trends have potential use in understanding and predicting changes in commercial fish reproductive success (e.g., Herrick et al. 2007).

Changes in landings indicated by $C2$ are closely related to accumulated anomalies in the California southward wind stress (SWS) index of Parrish et al. (2000), August–October accumulated sea level anomaly at San Francisco, California (A-SFSL), the accumulated shore temperature anomaly at Pacific Grove, California (PG A-SST), and the Pacific Circulation Index (PCI), described by King et al. (1998). The r^2 values for all these relationships are >0.65 with inverse correlations for A-SFSL, PG A-SST, and the PCI. Based on the correlations, the physical indices explain 25–35% of the variance in the EOF input matrix described above (Norton and Mason 2003, 2004, 2005).

The relationship of $C2$ to the PCI suggests that forcing of $C2$ is related to large-scale atmospheric processes occurring over the northeastern Pacific Ocean (Enfield and Allen 1980; Chelton et al. 1982; King et al. 1998; Parrish et al. 2000; Norton and Mason 2004) rather than to unique California Current processes. $C2$ temporal patterns may also reflect processes related to the abrupt increase in surface temperatures observed after 1976 over many areas of the earth (Norton et al. 1985; Ebbesmeyer et al. 1991; Cane et al. 1997; Norton 1999; Cobb et al. 2003). The $C2$ temporal pattern also corresponds to variation in the concentration of fish larvae, other zooplankton, and their predators off California and northern Mexico during 1954–1998 (Ainley et al. 1995; McGowan et al. 1998; Norton and Mason 2003).

13.6 Sardine Availability and Physical Processes

Because the sardine landings series is highly loaded in EOF1 and has nearly neutral loading in EOF2 (Fig. 13.1), variation in sardine availability will be similar to the time variation of EOF1. As noted above the accumulated SST anomaly, La Jolla

A-SST, has decadal-scale variability that is similar to the time variation in EOF1 (r^2 = 0.85). The variability of the natural log of the sardine landings series has a similar relationship to La Jolla A-SST (r^2 = 0.81). The relationship between La Jolla A-SST and sardine landings from the CFHE has been discussed by Norton and Mason (2005) and Herrick et al. (2007). In the following we focus on the relationship between La Jolla A-SST and a series of sardine biomass estimates (Conser et al. 1998; Hill et al. 2006), which covers 1934–2004, excluding the 1964–1981 interval when low abundance and landings resulted in less accurate estimates of sardine biomass (Barnes et al. 1992).

An r^2 = 0.70 is found for the linear fit between the natural log of sardine biomass and the La Jolla A-SST. The climate signal is partially obscured by interannual variability. The relationship between La Jolla A-SST and sardine biomass may be expressed as

$$\log_e (V) = f(\Sigma A) + t \tag{13.2}$$

where $\log_e (V)$ is the natural log of the annual biomass estimate (V). f is a scaling function depending partially on the measure of abundance used, ΣA is the sum of the La Jolla A-SST anomalies from the beginning of the record and ending in the the most recent analysis year and t is noise and factors not well represented by ΣA. This relationship defines climate-scale variability in sardine biomass.

Because $A = \log_e e^A$, Eq. (13.2) may be rewritten as,

$$V = F(\prod e^A) + t \tag{13.3}$$

where V is the biomass, F is a scaling function, and $\prod(e^A)$ is the progressive product of e raised to the power of the annual environmental anomaly (A). The value of e^A is computed for every year from the anomaly value for that year. It is seen that the influence of the environment on decadal scales can be given by a multiplier (climate factor) of the existing estimated biomass, or $V_{year} = (V_{year-1})e^A$. If biomass estimates are not available for ecosystems or species known to be strongly influenced by the environment, then the trajectory of their population size may be estimated from physical environmental indices as illustrated in Fig. 13.3, but it should be kept in mind that the $(e^A)_{year}$ climate factors may produce noisy series on interannual scales.

The relationship between the biomass series and the 1-year climate factor projected biomass is shown by the closed and open symbols, respectively, in Fig. 13.3, with r^2 = 0.85 for the linear comparison. When the climate factors are used to project the biomass for 2 and 4 years, the comparison r^2 values drop to 0.62 and 0.50, respectively. When only the 1934 biomass estimate is used to start a series of progressive products (Eq. 13.3), the comparison of the biomass and the projected series is r^2 = 0.68, which demonstrates the climate-following nature of sardine abundance. We estimated sardine biomass for the low sardine abundance years using the 1959 biomass value as an initial scaling point (Fig. 13.3, crosses) and found the lowest biomass estimates to be about half the size of those calculated by Barnes et al. (1992). However, the minimum in 1974 was within a year of the Barnes et al. (1992)

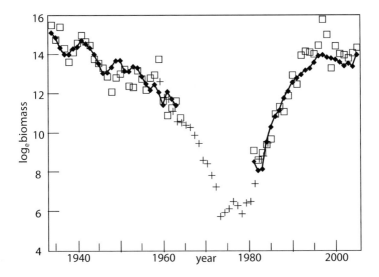

Fig. 13.3 Comparison of biomass estimates, filled diamonds, from Hill et al. (2006) and biomass estimated using climate factors. Open squares show biomass estimates using climate factors to project the next year's biomass. The crosses show estimates of biomass using the published biomass estimate for 1959 and projected biomass estimates using the climate factors from 1960 to 1981

minimum. The cycle of abundance that has occurred with sardines and several other species (e.g., Maunder et al. 2006) on the right side of Fig. 13.1 is clearly shown in Fig. 13.3 with maxima in the 1930s and 1990s and a minimum in the mid-1970s.

These comparisons are presented to demonstrate the utility of the physical environmental indices (climate factors) in interpreting the effects of the ocean climate on Pacific sardines, which is one of the 29 ecosystem indicator species. In principle this can be done with any species that has sufficiently high loading values in the ecosystem space shown in Fig. 13.1. However, species that are not on or near an axis in Fig. 13.1 will be best analyzed by physical indices strongly associated with each axis; three or more terms may be necessary.

A difficulty with fisheries/climate studies is that time series are always too short and it is difficult to know the validity of our conclusions (McGowan 1990). We would be more confident if we knew how the physical environmental changes led to changes in primary and secondary production and had plausible relationships explaining how changes at the bottom of the food chain were transmitted to exploitable fish stocks (e.g., McFarlane and Beamish 1992; Rothschild 1998; Ware and Thomson 2005). However, the apparent widespread nature of the EOF signals and their physical counterparts (Norton and Mason 2004, 2005) suggest that we have found basic relationships in the physical environment that change the ecosystem off California. In addition, cycles of 40–70 years, such as the one we have found in California fisheries statistics, occur in many places throughout the world (Klyashtorin 2001) and in pre-fishery proxy sardine abundance records from sediments off California (Baumgartner et al. 1992).

We would like to improve the confidence in our results by extending the series to three or more cycles, or back to the mid-1700s. This will require proxy records. We are currently investigating the use of fish scales from hypoxic sediments off the California coast (Soutar and Isaacs 1974; Baumgartner et al. 1992) and tropical coral that appear to record, in their chemical composition, the large-scale oceanic events that affect the California Current ecosystem (Cobb et al. 2003; Norton and Mason 2005).

To summarize, we note that: (1) Successful ecosystem-based fishery management (EBFM) will depend on an understanding of the frequency, magnitude, and type of variations that can occur in the ecosystem. (2) The relative abundance of the dominant organisms in the ecosystem will change as the state of the ecosystem changes. (3) The economic output of the ecosystem will probably change with changing species composition. Long-term ex-vessel revenue will fluctuate as ecosystem states change. (4) It will be possible to track and anticipate many ecosystem and species abundance changes through changes in the physical environment. Applicable physical environmental indices suggest basin-wide control of the interdecadal modes of ecosystem variation in the CFHE and throughout the CCS.

13.7 Prospects

Although natural production of wild or unconfined animals is of limited importance for overall human sustenance, consumption of wild marine species remains an important contributor to human health, nutrition, and well-being. With the growth of human populations, the exploitation of marine resources in a shared resource common property setting has led to extensive overfishing and economic inefficiency (Hardin 1968; Arnason 1998; Dietz et al. 2003). Consequently, it has fallen to government agencies to regulate harvests of wild species from within their EEZs in an effort to sustain fisheries production. Ecosystem-based fisheries management has recently been defined by society as the best approach to maintaining ecosystem processes and resources for fishers and other environmental users (National Research Council 1999). The EBFM view acknowledges that there are numerous other values ascribed to the organisms populating an ecosystem, in addition to human nutrition, and that there may be trade-offs between different ecosystem processes and resources (Herrick et al. 2007). The US government enacted the 2007 reauthorization (Public Law 109–479, 2007) of the Magnuson-Stevens Fishery Conservation and Management Act (Public Law 94–265) that mandates an ecosystem-based approach to fisheries management and conservation. Management plans developed by regional fisheries councils, based on scientific analysis from state and federal scientists, will be required to consider both direct and indirect effects of fishing interactions on both targeted and nontargeted species, habitat protection, and other highly valued ecosystem resources (Townsend et al. 2004, Grafton et al. 2006).

Achievement of effective EBFM will be a lengthy, iterative process – one that will require diverse and currently developing scientific methodologies (e.g., Chapters 7,

18, and 21) and interdisciplinary efforts to identify and describe important linkages between varying natural and socioeconomic systems (Norgaard 1989). The key here is to broaden the focus of traditional fisheries conservation management science from a relationship between a targeted species and a fishery, to a more comprehensive outlook that embraces all important species in terms of their ecological, habitat, and fishery interactions, as well as their value to society. Present fishery conservation and management institutions will be progressively reshaped to assess the economic value of an increasing number of important (high energy flux) ecological interactions that occur in the ecosystem. This will allow the ecosystem space outlined in Fig. 13.1 to be filled with many more ecosystem component species such as Humboldt squid, salmon, and perhaps planktonic species.

The California Cooperative Oceanic Fisheries Investigations (CalCOFI) program, of the State of California and the National Marine Fisheries Service (NMFS), has conducted extensive biological and physical oceanographic surveys covering a time period that includes the ecosystem states of the 1950s (upper left in Fig. 13.1) to the current ecological states that appear similar to those observed in the 1930s and 1940s. Much of the CalCOFI biological data can be integrated into an ecological space similar to the one we have presented (Fig. 13.1). The food of long-lived upper trophic level species will undoubtedly shift over time to utilize available nutrient resources. NMFS fish surveys could provide trophic linkage information by analyzing stomach contents using automated molecular genetic tag technology to identify species, increasing our understanding of trophic interactions and changing ecological relationships by several orders of magnitude.

Large-scale physical factors such as decade-scale ocean climate shifts can cause significant changes in marine species communities, reflected in changes in the ecological state (Fig. 13.1) of the ecosystem (MacCall 1996; Rothschild 1995; Klyashtorin 2001; Norton and Mason 2003, 2004, 2005). While the new state may be as diverse and as ecologically valuable as that which it replaced, individual species may no longer be available in desired quantities for either commercial or nonconsumptive users. Although aggregate biomass may not be greatly affected by a shift in ecological state, shifts in proportions of available species may cause significant changes in the fisheries operations and their economic value (Fig. 13.2).

In the two following sections we examine briefly how knowledge of ecosystem states and changes might inform and be informed by two methods of EBFM. Marine protected areas are designed to protect entire ecosystems by area restriction of use and individual transferable quota management systems are designed to engage the fisher as well as the scientist and manager in ecosystem analysis.

13.8   Marine Protected Areas

The development of a network of Marine Protected Areas (MPA), where commercial and recreational activity is either limited or excluded may be an important first step in EBFM and in the preservation, management, and understanding of intact ecosystems. Marine Protected Areas may range in protection level from "no take" and "no disturbance" marine reserves to managed areas with controlled take. The CFHE off California is highly invested in marine protected areas by the Marine Sanctuary Program of NOAA/NOS, the Marine Life Protection Program of California and PFMC groundfish essential fish habitat areas (Helvey 2004).

As the ecosystems of fully protected MPAs revert to states similar to those that existed before major societal disturbance, they will provide scientific information and insight to the fishery scientist regarding the output and stability of unexploited ecosystems in different climate states. It might be argued that some of the cycling through the ecological space presented in Fig. 13.1 was the result of selective and unintended harvest perturbations that fed back positively through the ecosystem to amplify changes in ecological state. If "no take" MPAs are large enough, they may allow this question to be answered for nonmigratory ecosystem components.

13.9   Individual Transferable Quotas

When a fishery is managed under an individual transferable quota (ITQ) system, society maintains ownership of the resource stock, but user rights to the renewable flow from the stock are allocated to fishers in the form of a tenable, secure, and transferable share of an adjustable total allowable catch (TAC), usually as a percentage (Hannesson 2004, Townsend et al. 2004; Bromley 2005; Grafton et al. 2006; Beddington et al. 2007). ITQs provide strong incentives to reduce the race to catch fish and resulting over capitalization, so that greater net economic benefits can be realized from harvesting the TAC, which will be determined within the EBFM framework. Allocating shares of the TAC to individual fishers and then allowing them to trade their shares will result in the TAC being harvested in a more cost-effective manner; fishers themselves will determine the most efficient way to harvest the TAC. In addition, ITQs lessen the need for administratively costly and economically inefficient management measures that attempt to limit the overall harvest to the TAC level by controlling fishing effort.

Within the EBFM-ITQ management system the fisher possesses long-term economic assets that will induce him/her to invest in the ecosystem so that commercially harvested fish stocks become environmental capital capable of generating long-term returns. This gives the fisher a strong incentive to understand ecosystem processes that change the value of the ITQs that he owns. In this system both the fishery scientist and the fisher are concerned with climate shifts within the ecosystem and the species that will be available for harvest. TACs must be adjusted as ecological climate states change. The TAC for some species will have to be reduced to prevent over fishing and preserve "seed" stocks for times when conditions become favorable for life-cycle success, while other populations will become relatively more abundant, allowing higher TAC (Herrick et al. 2007). The fishery scientist puts this knowledge into estimating the TACs for a group of species and the fisher assesses the value of his/her resulting ITQ holdings in terms of potential profits from using his ITQs, as well as the potential economic gains from trading the ITQs in his possession. Collectively, these decisions lead to the economically optimum use of commercial fishery resources for society as a whole (Arnason 1998). However, if anticipated ecosystem output is not realized or cannot be maintained, then there will be social pressure on managers to increase the TAC at the risk of curtailing ecosystem processes and reducing ecosystem resources.

Within the EBFM-ITQ approach the task of the fishery manager is to set the TAC for each important species in the ecosystem and determine initial allocation of quota shares. The initial informational requirements for this management system might seem overwhelming. Managers will need detailed information concerning the economics of the fisheries and the species linkages within the ecosystem, particularly if ecosystem manipulation is introduced (Arnason 1998). The costs of accumulating this knowledge may seem prohibitive, but a complete stock assessment for every managed species may not be needed in years when an analysis of physical environmental influences, such as the climate factors developed for sardines, are sufficient to establish a sufficiently accurate TAC for the following year. We have shown that there are easily detectable climate signals in the variations in abundance of several of the 29 California indicator species. Relationships of this kind and those that utilize other effects of physical environmental forcing (e.g., Chapter 31) will aid greatly in establishing adequate TACs at a reasonable cost.

13.10 Summary and Conclusions

Characteristics of ecosystem shifts, observed over 75 years, can be used to anticipate and manage future changes in ecosystem productivity. Easily applied techniques of monitoring California ecosystem states have been outlined and prediction may be possible, based on persistence and relationships between the ecosystem and the physical environment. Effective EBFM must analyze ecosystem transitions and adjust management strategies appropriately. Recent empirical studies suggest that about 7 years will be required to distinguish an ecosystem shift from more frequent interannual transient shifts (Herrick et al. 2007). The following points are added for consideration.

1. Variable levels of fish and invertebrate production by marine ecosystems will continue to challenge fishery management and society to act proactively to optimize ecosystem processes and resources for a variety of users.
2. Ecosystem-based fishery management will evolve to work with changing ecosystem properties to decrease levels of resource destruction and prevent loss of biological and economic productivity.
3. Fish and invertebrate harvesters will be able to use the knowledge of ecological climate cycles to anticipate changes in the value of the ITQs they own or would like to acquire, allowing greater economic efficiency.
4. Assessment requirements of prospective ecosystem managers will be similar to current requirements, but magnified because the population health of many

ecologically important (high energy flux) species must be understood whether or not they are commercially or recreationally important.

5. Marine protected areas that have a full level of biodiversity may provide baseline studies that indicate the frequency and magnitude of ecological change that is independent of extraction processes.

6. New techniques and methodologies, such as molecular genetic tagging and analysis, will be of increasing importance in ecosystem research, the analysis of fishery sustainability and ecosystem-based management.

References

Ainley DG, Sydeman WJ, Norton J (1995) Upper trophic level predators indicate interannual negative and positive anomalies in the California current food web. Mar Ecol Prog Ser 118: 70–79

Arnason R (1998) Ecological fisheries management using individual transferable share quotas. Ecol Appl 8:S151–S159

Barnes JT, Jacobson LD, MacCall AD, Wolf P (1992) Recent population trends and abundance estimates for the Pacific sardine (*Sardinops sagax*). Calif Coop Ocean Fish Invest Rep 33:60–75

Baumgartner TR, Soutar A, Ferreira-Bartrina V (1992) Reconstruction of the history of Pacific sardine and northern anchovy populations over the past two millennia from sediments of the Santa Barbara Basin, California. Calif Coop Ocean Fish Invest Rep 33:24–40

Beamish RJ (1993) Climate and exceptional fish production off the west coast of North America. Can J Fish Aquat Sci 50:1002–1016

Beddington JR, Agnew DJ, Clark CW (2007) Current problems in the management of marine fisheries. Science 316:1713–1716

Botsford LW, Castilla JC, Peterson CH (1997) The management of fisheries and marine ecosystems. Science 277:509–515

Bromley DW (2005) Purging the frontier from our mind: Crafting a new fisheries policy. Rev Fish Biol Fish 15:217–229

California Bureau of Commercial Fisheries (1937) The commercial fish catch of California for the year 1935. Calif Dep Fish Game Fish Bull 49:170p

Cane MA, Clement AC, Kaplan A, Kushnir Y, Pozdnyakov D, Seager R, Zebiak SE, Murtugudde R (1997) Twentieth-century sea surface temperature trends. Science 275:957–960

Chelton DB, Bernal PA, McGowan JA (1982) Large-scale interannual physical and biological interaction in the California current. Journal of Marine Research 40:1095–1125

Cobb KM, Charles CD, Cheng H, Edwards RL (2003) El Niño/Southern Oscillation and tropical Pacific climate during the last millennium. Nature 424:271–276

Conser R, Hill K, Crone P, Lo N, Felix-Uraga R (1998) The coastal pelagic species fishery management plan. Pacific Fishery Management Council, Portland, OR, 167p

Dewees CM, Weber ML (2001) A review of restricted access fisheries. In: Leet WS, Dewees CM, Klingbeil R, Larsen EJ (eds) Califonia's living marine resources: A status report. University of California – Agricultural and Natural Resources SG01-11, pp 73–76

Dietz T, Ostrom E, Stren PC (2003) The struggle to govern the commons. Science 302: 1907–1912

Ebbesmeyer CC, Cayan DR, McLain DR, Nichols FH, Peterson DH, Redmond KT (1991) 1976 Step in the Pacific Climate: Forty environmental changes between 1968–1975 and 1977–1984. In: Betancourt JL, Tharp VL (eds) Proceedings of Seventh Annual Pacific Climate (PACLIM) Workshop, April 1990. Calif Dept Water Resources Interagency Ecol Studies Prog Tech Rep 26:115–126

Enfield DB, Allen, JS (1980) On the structure and dynamics of monthly mean sea level anomalies along the Pacific coast of North and South America. J Phys Oceanogr 10:557–578

Field JC, Baltz K, Phillips AJ, Walker WA (2007) Range expansion and trophic interactions of the jumbo squid, *Dosidicus gigas*, in the California current. Calif Coop Ocean Fish Invest 48:131–146

Fu L-L, Qiu B (2002) Low frequency variability of the North Pacific Ocean: The roles of boundary-and wind-driven baroclinic Rossby waves. J Geophys Res Oceans 107:3220–3235

Grafton Q, Arnason R, Bjorndal T, Campbell D, Campbell HF, Clark CW, Conner R, Dupont DP, Hannesson R, Hilborn R, Kirkley JE, Kompas T, Lane DE, Munro GR, Pascoe S, Squires D, Stienshamn SI, Turris BR, Weninger Q (2006) Incentive-based approaches to sustainable fisheries. Can J Fish Aquat Sci 63:699–710

Hanley DE, Bourassa ME, O'Brien JJ, Smith SR, Spade ER (2002) A quantitative evaluation of ENSO indexes. J Clim 16:1249–1258

Hannesson R (2004) The privatization of the oceans. MIT Press, Cambridge, MA, 202p

Hardin G (1968) The tragedy of the commons. Science 162:1243–1248

Helvey M (2004) Seeking consensus on designing marine protected areas: Keeping the fishing community engaged. Coast Manage 32:173–190

Herrick SF, Jr, Norton JG, Mason JE, Bessey C (2007) Management application of an empirical model of sardine-climate regime shifts. Mar Policy 31:71–80

Hill KT, Lo NCH, Macewicz BJ, Felix-Uraga R (2006) Assesment of the Pacific sardine (*Sardinops sagax caerulea*) population for U.S. management in 2006. NOAA Technical Memorandum, NOAA-TM-NMFS-SWFSC-386, 75p

Hollowed AB, Bailey KM (1989) New perspectives on the relationship between recruitment of Pacific hake, *Merluccius productus*, and the ocean environment. In: Beamish RJ, McFarlane GA (eds) Effects of ocean variability on recruitment and an evaluation of parameters used in stock assessment models. Can Spec Pub Fish Aquat Sci 108:207–220

Jacobson LD, MacCall AD (1995) Stock-recruitment models for Pacific sardine (*Sardinops sagax*). Can J Fish Aquat Sci 52:566–577

Jacobson LD, Thomson CJ (1993) Opportunity costs and the decision to fish for northern anchovy. N Am J Fish Manage 13:27–34

Kalvass PE, Hendrix JM (1997) The California red sea urchin, *Strongylocentrotus franciscanus*, fishery: Catch, effort, and management trends. Mar Fish Rev 59:1–17

King JR, Ivanov VV, Kurashnov V, Beamish RJ, McFarlane GA (1998) General circulation of the atmosphere over the North Pacific and its relationship to the Aleutian Low. NPAFC Doc 318, 18p (available at http://www.npafc.org)

Klyashtorin LB (2001) Climate change and long-term fluctuations of commercial catches. FAO Fish Tech Pap 410, 86p

Leet WS, Dewees CM, Klingbeil R, Larson EJ (eds) (2001) California's living marine resources: A status report. University of California – Agricultural and Natural Resources SG01-11, 492p

Longhurst A (2006) The sustainability myth. Fish Res 81:107–112

MacCall AD (1996) Patterns of low-frequency variability in the California Current. Calif Coop Ocean Fish Invest Rep 37:100–110

Marr JC (1960) The causes of major variations in the catch of Pacific sardine. In: Rosa H, Murphy G (eds) Proceedings of the world scientific meeting on the biology of sardines and related species. FAO United Nations, Rome, Italy, pp 667–791

Mason JE (2004) Historical patterns from 74 years of commercial landings from California waters. Calif Coop Ocean Fish Invest Rep 45:180–190

Maunder MN, Barnes JT, Aseltine-Nielson D, MacCall AD (2006) The status of California scorpionfish (Scorpaena guttata) off southern California in 2004. Status of the Pacific coast groundfish fishery through 2005: Stock assessment and fishery evaluation, Vol I. Pacific Fishery Management Council, Portland, OR, 125p

McFarlane GA, Beamish RJ (1992) Climatic influence linking copepod production with strong year-classes in sablefish, *Anoplopoma fimbria*. Can J Fish Aquat Sci 49:743–753

McGowan JA (1990) Climate and change in oceanic ecosystems: The value of time-series data. Trends Ecol Evol 5:293–299

McGowan JA, Cayan DR, Dorman LM (1998) Climate-ocean variability and ecosystem response in the northeast Pacific. Science 281:210–217

Meyers SD, Melsom A, Mitchum GT, O'Brien JJ (1998) Detection of the fast Kelvin wave tele-connection due to ENSO. J Geophys Res Oceans 103:27655–27663

National Research Council (1999) Sustaining marine fisheries. National Academy Press, Washington, DC, 164p

Norgaard RB (1989) The case for methodological pluralism. Ecol Econ 1:37–67

Norton JG (1999) Apparent habitat extensions of dolphinfish (*Coryphaena hippurus*) in response to climate transients in the California Current. Sci Mar 63:239–260

Norton JG, Mason JE (2003) Environmental influences on species composition of the commercial harvest of finfish and invertebrates off California. Calif Coop Ocean Fish Invest Rep 44:123–133

Norton JG, Mason JE (2004) Locally and remotely forced environmental influences on California commercial fish and invertebrate landings. Calif Coop Ocean Fish Invest Rep 45:136–145

Norton JG, Mason JE (2005) Relationship of California sardine (*Sardinops sagax*) abundance to climate-scale ecological changes in the California Current system. Calif Coop Ocean Fish Invest Rep 46:83–92

Norton JG, McLain DR, Brainard RE, Husby DM (1985). El Niño event off Baja and Alta California and its ocean climate context. In: Wooster WS, Fluharty DL (eds) Niño effects in the eastern subarctic Pacific Ocean. University of Washington, Seattle, WA, pp 44–72

O'Bannon BK (ed) (2001) Current fishery Statistics 2000, Fisheries of the United States, 2000. U.S. Department of Commerce/NOAA/National Marine Fisheries Service, Silver Spring, MD, 126p

Parrish RH, Schwing FB, Mendelssohn R (2000) Mid-latitude wind stress: The energy source for climatic shifts in the North Pacific Ocean. Fish Oceanogr 9:224–238

Rosenberg AA, Bolster WJ, Alexander KE, Leavenworth WH, Cooper AB, McKenzie MG (2005) The history of ocean resources: Modeling cod biomass using historical records. Front Ecol Environ 3:84–90

Rothschild BJ (1995) Fish stock fluctuations as indicators of multidecadal fluctuations in biological productivity of the ocean. In: Beamish RJ (ed) Climate change and northern fish populations. Can Spec Pub Fish Aquat Sci 121:201–209

Rothschild BJ (1998) Year class strengths of zooplankton in the North Sea and their relation to cod and herring abundance. J Plankton Res 20:1721–1741

Soutar A, Isaacs JD (1974) Abundance of pelagic fish during the 19th and 20th centuries as recorded in anaerobic sediment off the Californias. Fish Bull 72:257–273

Townsend RE, McColl J, Young MD (2004) Design principles for individual transferable quotas. Mar Policy 30:131–141

Ware DM, Thomson RE (2005) Bottom-up ecosystem trophic dynamics determine fish production in the northeast Pacific. Science 308:1280–1284

Wolf P, Smith PE, Bergen DR (2001) Pacific sardine. In: Leet WS, Dewees CM, Klingbeil R, Larsen EJ (eds) Califonia's living marine resources: A status report. University of California – Agricultural and Natural Resources SG01-11, pp 99–302

Yaremko M (2001) California market squid. In: Leet WS, Dewees CM, Klingbeil R, Larson EJ (eds) California's living marine resources: A status report. University of California – Agricultural and Natural Resources SG01-11, pp 295–298

Chapter 14
Ecosystem Models of Fishing Effects: Present Status and a Suggested Future Paradigm

Saul B. Saila

Abstract Quantitative modeling methods applied to anthropomorphic effects of harvesting on aquatic ecosystems have become increasingly utilized tools in the management of fisheries. However, to date traditional modeling approaches have not been found to be very useful as "surrogate experimental systems" in applied ecology, such as fishing effects on entire ecosystems. A review is made of several dynamic ecosystem models, and an effort is made to assess their utility to fisheries management. Both theoretical and simulation models, as well as individual-based models are shown to be of limited utility for various reasons. A new paradigm based on the premise that dynamic behavior of models, which includes fishing effects on the entire aquatic ecosystem, emerges from low-level interactions of independent agents. This concept is illustrated with simple artificial life models of fish schooling and predator–prey relations. These models describe each individual fish, including its interaction with others and the environment. There is no overall controlling program. Thus, the overall behavior of the school and predator–prey relations emerge from local interactions among many individuals. A brief description of the premise on which artificial life is based is included. Examples of other artificial life models at ecological scales are provided and some specifics on artificial life models in a fisheries context are suggested.

Keywords Artificial life · ecosystem-level dynamic model

S.B. Saila
Graduate School of Oceanography, University of Rhode Island,
Narragansett, RI 02881, USA
e-mail: saul.saila@att.net

R.J. Beamish and B.J. Rothschild (eds.), *The Future of Fisheries Science in North America*, 245
Fish & Fisheries Series, © Springer Science + Business Media B.V. 2009

14.1   Introduction

The overall goal of this chapter is an attempt to introduce to fishery scientists interested in ecosystem-level modeling a paradigm, which has been around for some time, but which has not been thoroughly explored for its utility in fisheries-related ecosystem studies. This goal is approached herein by:

(a) Providing a brief introduction and background for a new paradigm
(b) Demonstrating by simple examples some fishery-related applications of artificial life
(c) Speculating on the future of ecosystem-level applications of artificial life models in fisheries management.

Algorithmic ecosystem models are being increasingly advocated and utilized as tools for evaluating effects, such as fishing, on aquatic ecosystems. However, the extent to which these models are able to make meaningful predictions has not been adequately investigated to date, nor have some alternative approaches been seriously considered as yet. A symposium dealing with many aspects of the mathematical analysis of fish stock dynamics has been published by the American Fisheries Society (Edwards and Megrey 1989). Robinson and Frid (2003) provided an assessment of 24 marine ecosystem models and concluded that the ECOPATH with Ecosim model (Walters et al. 1997) and the Andersen and Ursin multispecies extensions to the Beverton and Holt model (Andersen and Ursin 1977) seemed most likely to yield good insights according to the criteria which were applied for evaluation. They also concluded that future consideration should be given to explicitly incorporate spatial factors and extrinsic forcing functions, such as climate, into these models. More recently, Walters and Martell (2004) provided a wealth of important information on the derivation, use, and abuse of various mathematical models which have been utilized for decisions regarding the management of harvested aquatic ecosystems. Many of these models are complex and they include structural diversity as well as dynamic complexity with feedbacks. A general conclusion reached by Walters and Martell was that it is impossible to fully capture the rich behavior of ecosystems in mathematical models. I was impressed by statements on page 286 of their book, which are utilized in the following material. They find that enormous numbers of assumptions are needed whenever one derives a proposed scheme for making predictions about any dynamic system whether or not one makes these assumptions explicit. Assumptions are used to define parameters and forms of functional relations that might provide reasonable approximations to the dynamics of the system. The primary issue is not the quality of the assumptions, but rather the quality of the approximations.

I suggest that aquatic and other ecosystems may also be considered according to a new paradigm as complex adaptive systems of interacting autonomous agents. These systems exhibit behavior, which emerges from the interaction of a large number of elements from lower levels. Complex adaptive systems are those in which there can be oscillations and limit cycles in the populations of individuals. These oscillations and limit cycles are considered as an emergent property of the ecosystems. I believe that the application of conventional mathematics and biology may be of limited utility in understanding the dynamics of such complex ecosystems. The typical reductionist approach to aquatic ecosystem analysis is the top-down approach. This approach is believed to be inadequate for reasons which follow. A new approach to aquatic ecosystem analysis is suggested in which this top-down reductionist approach is replaced by a bottom-up approach using autonomous agents (programs to represent individual animals).

This approach is exemplified by RAM, an artificial life system designed at UCLA (Taylor et al. 1988) for modeling population behavior and evolution. It is, among other things, a versatile tool for expressing in program form theories of population behavior, ecological interaction, and adaptation. I believe artificial life offers a new vocabulary and new approach (paradigm) for describing ecological phenomena, as well as ideas and theories that are otherwise intractable mathematically and experimentally impossible. RAM, as well as the fish schooling system and the predator–prey model described herein are based on the observation that the life of an organism is in many ways similar to the execution of a program, and that the global emergent behavior of a population of interacting organisms is best emulated by the behavior of a corresponding population of autonomous agents.

Ermentrout and Edelstein-Keshet (1993) have reviewed a number of biologically motivated artificial life models. They state that complex biological systems which cannot be precisely quantified by mimicking physical laws can be solved by a series of simple rules that are easy to compute quickly.

Olson and Sequeira (1995) have nicely summarized traditional modeling approaches in ecology and introduced an alternative paradigm based on artificial life. The material which follows is an attempt to introduce this paradigm to fishery scientists interested in measuring aquatic ecosystem effects from harvesting activities by fishers. Adami (2002) has recently introduced a digital life system termed Avida to aid in the study of emergence and evolution of simple ecosystems in real time.

My justification for suggesting artificial life models of aquatic ecosystems is based on the following statements. Model behavior for the study of ecosystem organization have traditionally been expressed formally as systems of algebraic or differential equations. Equational models (top-down attempts at approximating nature) are subject to many limitations. Examples include the following:

1. It would take hundreds (perhaps thousands) of lines of equations to express even a simple model of a fish's behavior as a function of many genetic, memory, and environmental variables that affect its behavior. There seem to be no generally available mathematical tools for dealing with equational systems of such complexity.
2. In general, equational models are not good at dealing with highly nonlinear effects, such as thresholding or if-then-else conditionals, which occur in behavioral and ecological studies. In conventional models coding an organism's behavior implicitly requires a solution to equations that must be integrated.

3. Most mathematical tools, from simple arithmetic through differential calculus depend on the assumption of linearity. Roughly, this means that we get to a value for the whole by adding the values of the parts. More formally, a function is linear if the value of the function, for any set of values assigned to its arguments, is simply a weighted sum of these values.

4. Zadeh (1973), a prominent scientist, has made the following statement regarding model complexity stated as the "principle of incompatibility." This principle says that: "As complexity of a system increases, our ability to make precise and yet significant statements about its behavior diminishes until we reach a threshold beyond which precision and significance (or relevance) become almost mutually exclusive characteristics."

Recent developments in fisheries science include increased use of simulation models, individual-based models, and the AD model builder for solving differential equation-based models. However, these developments all seem to have some shortcomings. Generally speaking, in simulation modeling the system under consideration is broken down to subsystems. These are connected by rate equations, and the levels of these connections (links) are treated as black boxes. Because interactions between sub-models occur only through these links, they limit the dynamics of the entire system. Both individual-based as well as simulation models are subject to this type of high-level limitation. Huston et al. (1988) have introduced and advocated individual-based models of communities and ecosystems to better address some principles of individuality and locality. Grimm (1999) has provided an excellent review of individual-based model uses. These models are sometimes based on the global consequences of local interactions of members of a population. It has been suggested that the major difference between individual-based and artificial life models is whether the simulated inner loop proceeds cell by cell or individual by individual. However, the main point here is that although individual-based models may keep track of individual fish, their interactions are specified (hard coded) at a higher level. In summary, because ecological systems are inherently nonlinear (Allen and Starr 1998) the behavior of the whole system is different from the sum of its parts. Although biological systems seem to be inherently nonlinear, models are often built as though they were linear. This is done by modeling subcomponents and assigning them into a rigid and probably incorrect structure. In some fishery management applications the modeler and user must "tune" the model using system-level data sets. The result of this learning process is to increase the specificity of the model for one species and/or one specific region.

Simulation models of complex aquatic systems seem to be limited by the way they are constructed. Horn et al. (1989) state that simulation model behavior tends to be dominated by structure rather than detail. That is, model behavior may therefore depend more on the way model components are linked together than on the detailed form of the state equations. It should be recalled that interaction among model components can occur only through these links, which may constrain the dynamics of the model system.

In summary, emergent behavior (self-organization) as demonstrated by artificial life models may be a viable alternative to conventional models of aquatic ecosystems. This emergent behavior is a process in which pattern or behavior emerges solely from numerous interactions among lower-level components of the system. The rules specifying interactions among the system components are executed using only local information.

In the following sections simple artificial life models, namely a fish predator–prey model and schooling by fishes will be demonstrated and briefly explained.

14.2 Wa-Tor

The artificial life game Wa-Tor is based largely on an idea first published by Dewdney (1984) and further developed and distributed by Kovach Computing Services (http://www.koycomp.co.wk/wator/index.html). The game Wa-Tor provides a simulation model of predatory–prey relationships.

The conventional Lotka–Volterra system for predatory–prey dynamics is given by the following differential equations:

$$dx/dt = ax-bxy \tag{1}$$

$$dy/dt = -cy+dxy \tag{2}$$

The first term in each equation is the population tendency to increase or decrease if left alone. The second term in each equation is the effect of interaction between the two populations. It should be noted that this system is unstable.

Wa-Tor is a population ecology simulation game based on a random model. The sampled ocean is a toroidal space populated by two species. Predators in this case are termed sharks and the prey are termed fish. The user of Wa-Tor controls the initial number of predators (sharks) and prey (fish), their breeding rates, their life times, and the starvation time for the sharks. Sharks and fish then breed and move about in a random manner. Sharks eat the fish near them. The populations of both species can be watched on a map as well as on three types of graphs (predator–prey graphs, population curves, and age structure graphs). These results can also be saved and imported to spreadsheets for further processing.

The population curve shows the fluctuations through time of both the prey (fish) and the predator (shark) populations. The horizontal axis represents time in chronons (an arbitrary time unit). Both predator and prey move once during a chronon. A maximum of 100 chronons are indicated on the graph. The population of predators and prey each range from 0 to 961.

Generally speaking, the population graph shows smoothly rising and falling populations with the maximum shark population trailing behind the prey fish population. This pattern appears to emulate very closely the results derived from the Lotka–Volterra equations, which have often been used to model changes in populations of prey and predators such as the Hudson Bay hares and lynxes.

The pattern derived from the Wa-Tor game is not the solution of the differential equations of Lotka–Volterra. Instead, a similar pattern is produced simply by the following. As the prey fish population increases the sharks find an abundance of food. Therefore, the chances of a shark starving are lowered and they increase in numbers. Some time soon after, fish are eaten faster than they can breed and the fish population begins to decline. When this happens, the shark population will also begin to decrease after a few chronons. This dynamic system is the result of interactions between the predatory and prey species and does not require the solution of the Lotka–Volterra equations. However, these results clearly emulate a basic model of predator–prey population ecology.

The message from this brief description is that the interaction of relatively simple agents (sharks and fish) in this case provides an emergent system of behavior clearly resembling actual predator–prey data, and it does not require a top-down differential equation model to provide very realistic results.

14.3   Fish Schooling

The original model which was adapted for our fish schooling model was created by Reynolds (1987) as a simulation of bird flocking behavior. In the model, a collection of autonomous but interacting objects referred to as "Boids" inhabit a simulated environment. The modeler specifies the manner in which individual objects (fish in our case) respond to local conditions or events. The global behavior of the aggregate of objects is completely an emergent phenomenon. That is, none of the rules for individual objects depend upon global information. The only updating of the global condition is done on the basis of individual objects responding to local conditions.

Each object (fish) in the aggregate shares the same behavioral tendencies as follows:

(a) Maintaining a minimum distance from other objects in the environment including other fish
(b) Matching velocities with fish in its vicinity
(c) Moving toward the perceived center of mass of the fish in its vicinity.

The above are the only rules, which govern the behavior of the aggregate. These rules simply determine the behavior of a set of interacting objects. This behavior results in a very natural motion of schooling when visualized over time.

With suitable settings for the parameters of the system, a group of fish released at random positions within some volume will collect into a dynamic school, which swims around environmental obstructions in a very natural way. Sometimes the school divides into sub-schools as the entire school swims around both sides of an obstacle. These sub-schools reorganize about their own isolated centers of mass but later remerge into a single school.

An important point to recall here is that nowhere in this model are there rules for behavior of the fish school at the global level. Results at a global level emerge from the local activity taken collectively. I suggest that this "bottom-up" approach to the generation of complex behavior is more powerful and simple than that generated by conventional system models, which are constructed in top-down fashion.

The behavior of the fish school as a whole does not depend on the internal details of the entities from which it is constructed. It depends only on the details of the way in which these entities behave in each other's presence. This model does not capture all the nuances upon which schooling behavior depends, such as lateral line sensitivity and visual acuity. However, a major point of interest is that it is possible to capture within an aggregate of artificial entities a very lifelike behavior. Furthermore, this behavior emerges in the artificial system in the same way as it emerges in the natural system, namely from interaction among individuals. Terzopoulos et al. (1995) describe in detail a virtual marine environment inhabited by realistic artificial fishes. These clearly illustrate emergent group behavior.

In order to better understand the schooling behavior of fishes it is necessary to find out how an individual fish coordinates its behavior with that of other members of the school. Partridge (1982) has clearly indicated that fish rely on both their vision and lateral line systems. These are utilized by individual fish to adjust its speed and heading according to several of its neighbors. There seems to be no evidence of a leader and each individual seems to base its behavior on nearby fish. A likely explanation for schooling is a self-organizing mechanism where individual fish apply a few basic behavioral rules in response to local information from nearby fish. Huth and Wissel (1992) develop a complex mathematical model describing the movement of fish schools in two dimensions. This is a conventional model, which can be compared with the artificial life model mentioned herein.

14.4   Summary and Conclusions

I have included two examples of simple fishery-related models of artificial life, namely Wa-Tor and Boids. The first describes a simple predator–prey model. The second includes a simple marine world inhabited by somewhat lifelike artificial life forms that emulate the aggregate behavior of fishes in a simplified natural habitat. Each fish is an autonomous agent, which responds to a few rules leading to a realistic predator–prey relationship in the first case and simulated schooling in the other case.

I believe that emergent (artificial life) models of ecosystems seem to avoid some of the pitfalls of dominance of structure and brittleness found in more conventional models. They also incorporate fascinating subsets of real world complexity. Therefore, I believe that artificial life models of high-level aquatic ecosystems will become important contributors to a better understanding of ecosystem-level dynamics and useful for testing alternative fishery management strategies, which are virtually impossible to test in a real system.

The following is an example of an emergent artificial life system, which is thought to be of interest to fishery scientists. Taylor et al. (1988) have used an artificial life ecosystem to study population dynamics. They have developed a generalized programming environment termed RAM for building artificial ecosystems.

This example illustrates two principles governing artificial ecosystems. They are:

1. Designers of artificial life systems work toward the goal of minimizing, if not eliminating global control. This is done by encoding behavior of agents only at the agent level.
2. Artificial life systems have a strong spatial component. Therefore, interactions between organisms (agents) represented occur locally, and they depend on their location in the environment.

Olson and Sequeria (1995) have developed an artificial ecosystem shell termed LAGER (Large-scale Automated Generic Ecosystem Replication). It is designed to provide a platform for testing ideas about model ecosystems using an artificial life approach. They describe the LAGER shell in detail.

PARE (Parasitoid ARtificial Ecosystem) is an example of an ecosystem-level application using the LAGER framework. In this case it is a host/parasitoid model based on natural systems involving parasitic wasps and their insect hosts. A detailed description of PARE is provided by the authors.

Adami (2002) states that artificial life models provide opportunities to study the emergence and evolution of simple ecosystems in real time, and contrasts this to individual-based techniques. He also introduces Avida (a digital life system) and discusses its utility for obtaining a better understanding of ecosystems.

The above are only a small sample of artificial life systems with emphasis on ecological applications. However, even these provide an interesting beginning to this subject area.

I do not wish to imply that artificial life systems should be viewed as a replacement for rigorous mathematical models. Modeling complex ecological systems through computer simulations has become an important part of current fishery science. These models often involve large numbers of nonlinear differential equations, which are costly in computer time and difficult in parametric exploration.

It seems to me that the "bottom-up" approach can provide a unique complement to more conventional approaches by being able to reproduce in quantitative detail the results of an experimental procedure. This may result in an opportunity for accelerated development of more realistic ecosystem-level model systems in the foreseeable future. Artificial life models are expected to become more useful in the next decade to address ecosystem-level issues in the aquatic environment. These pertain to the emergence, evolution, maintenance, and stability of such systems when subjected to various forms of perturbation. Artificial life parameters can be set at will, and measurements can be recorded and compared to natural systems. There is no doubt in my mind that artificial life and other computer-related developments will be increasingly utilized by fishery scientists in the next decade.

References

Adami, G. 2002. Ab initio modeling of ecosystems and artificial life. Natural Resource Modeling 19(1): 139–145.

Allen, TFH and TB Starr. 1998. Hierarchy: Perspectives for Ecological Complexity. University of Chicago Press, Chicago, IL.

Andersen, KA and E Ursin. 1977. A multispecies extension of the Beverton and Holt theory of fishing, with accounts of phosphorus circulation and primary production. Fran Danmarks Fiskeri-Og Havundersogelser N.S. 7: 319–435.

Dewdney, AK. 1984. Computer recreations. Scientific American. December: 14–22.

Edwards, EF and BA Megrey. 1989. Mathematical Analysis of Fish Stock Dynamics. American Fisheries Society Symposium 6. Bethesda, MD.

Ermentrout, GB and L Edelstein-Keshet. 1993. Cellular automata approaches to biological modeling. Journal of Theoretical Biology 160: 97–133.

Grimm, V. 1999. Ten years of individual-based modeling in ecology: what have we learned and what could we learn in the future? Ecological Modelling 115: 129–148.

Horn, HS, HH Shugart, and DL Urban. 1989. Simulations as models of forest dynamics. In: J Roughgarden, RM May, and SA Levin (eds) Perspectives in Ecological Theory. Princeton University Press, Princeton, NJ.

Huston, M, D De Angelis, and W Post. 1988. New computer models unifying ecological theory. Bioscience 38: 682–691.

Huth, A and C Wissel. 1992. The simulaton of the movement of fish schools. Journal of Theoretical Biology 156: 365–385.

Olson, RL and RA Sequeira. 1995. An emergent computational approach to the study of ecosystem dynamics. Ecological Modelling 79: 99–120.

Partridge, BI. 1982. The structure and function of fish schools. Scientific American. June: 114–123.

Reynolds, CW. 1987. Flocks, birds and schools: a distributed behavioral model. Computer Graphics 21: 25–34.

Robinson, LA and CLJ Frid. 2003. Dynamic ecosystem models and the evolution of ecosystem effects of fishing: can we make meaningful predictions? Aquatic Conservation: Marine and Freshwater Ecosystems 13: 5–20.

Taylor, CE, DR Jefferson, SR Turner, and SR Goldman. 1988. RAM: artificial life for the exploration of complex systems. pp. 275–295. In: CE Langton (ed) Artificial Life. Addison-Wesley, Redwood City, CA.

Terzopoulos, D, X Tu, and R Grzeszczak. 1995. Artificial fishes: autonomous locomotion, perception, behavior and learning in a simulated physical world. Artificial Life 1(4): 327–351.

Walters, C and SJD Martell. 2004. Fisheries Ecology and Management. Princeton University Press, Princeton, NJ.

Walters, C, V Christensen, and D Pauly. 1997. Structuring dynamic models of exploited ecosystems from trophic mass-balance assessments. Reviews in Fish Biology and Fisheries 7: 1–34.

Zadeh, LA. 1973. Outline of a new approach to the analysis of complex systems and decision processes. IEEE Transactions SMC-3 1: 28–44.

Chapter 15
The Impacts of Environmental Change and Ecosystem Structure on the Early Life Stages of Fish: A Perspective on Establishing Predictive Capacity

Pierre Pepin

Abstract Since Hjort's seminal work in the early twentieth century, researchers studying the early life stages of fish have attempted to understand and clarify the roles of growth and loss to help predict patterns of recruitment variability. Estimating the abundance of eggs and larvae has provided a fishery-independent measure of stock spawning potential. We also have significant skills in predicting the drift and dispersal of fish eggs and larvae, and consequently the role of wind-driven circulation on potential losses. However, there are few instances where the role of trophic dynamics on growth, starvation or mortality has been demonstrated without ambiguity. Here, I discuss what research needs to be carried out to establish a predictive capacity about the roles of food availability and predator abundance in determining development, growth and mortality of fish eggs and larvae. Revisiting laboratory approaches to determine feeding capacity and growth potential, as well as the application of advanced sampling methods and analytical approaches needed to develop an understanding of how larvae are affected by prey and predator, are just two issues that represent key challenges that we must address in order to develop rigorous frameworks within which we can make and test predictions about early life-history dynamics. Each year-class represents the outcome of a series of stochastic processes involving many factors. Although the objective of forecasting year-to-year fluctuations in recruitment is unrealistic, the development of a bio-physical stochastic framework can serve to forecast probabilities of reproductive success in response to physical forcing and food web structure.

Keywords Larval fish · recruitment · stochastic processes · prey-predator interaction · integrative modelling · growth · mortality

P. Pepin
Fisheries and Oceans Canada, Northwest Atlantic Fisheries Centre, P.O. Box 5667,
Street John's, NL A1C 5X1, Canada
e-mail: pierre.pepin@dfo-mpo.gc.ca

R.J. Beamish and B.J. Rothschild (eds.), *The Future of Fisheries Science in North America*, 255
Fish & Fisheries Series, © Springer Science + Business Media B.V. 2009

15.1   Introduction

Since Hjort's (1914) seminal paper dealing with the causes of variations in fishery production, most research efforts directed at understanding the causes of recruitment variability in marine and freshwater fish populations has focussed on the early life-history period. Losses from egg production to metamorphosis (i.e. transformation to the juvenile stage when habitat and/or behaviour change significantly) are substantial (>99%), and subtle variations in either development or mortality rates can result in large differences in the overall number of survivors (Beyer 1989; Houde 1989). Consequently, it would seem logical to concentrate on identifying the key events or processes during this period in the life cycle that might result in the 10- to 100-fold variations in recruitment that are typically observed in most commercial fish populations.

Throughout the 1960s and continuing into the 1990s, a considerable portion of the research effort on early life stages focussed on variants of Hjort's (1914) original hypotheses that (1) food availability plays a major role in determining survival, and (2) advective losses may represent a key element in the life history of fish. Hjort had argued that broadcasting large numbers of poorly developed larvae with limited energy reserves could lead to significant losses due to starvation if individuals were unable to find suitable food resources following yolk absorption. However, the lack of consistent evidence that starvation losses were frequent led many researchers to broaden their investigations to look at the role of food availability on larval growth and development as a determinant of stage survival (Leggett and DeBlois 1994). A variation on this theme, Cushing's (1975) match/mismatch hypothesis, proposed that variations in the relative timing of secondary and larval production could be a source of variation in early life-history survival. The underlying theme in many early concepts dealing with egg and larval survival focussed on identifying the impact of short-term events that could be significant because of the ephemeral nature of the animals under study. In addition to limited energy reserves, the early life history represents a short period of time relative to the life cycle of most fish species. From hatch to metamorphosis, animals undergo a 100- to 1,000-fold increase in body weight, and growth and mortality rates are typically high (up to tens of percents per day). The significant changes in body size imply that larvae are likely to interact with a broad range of prey and predators. Thus any change in the timing, spatial overlap or strength of interactions could have important consequences to the overall survival, and consequently year-class strength. However, the very nature of the animals under study made the task of identifying such events nearly impossible. In general, fish eggs and larvae are minor components of the overall plankton community, which results in a large underlying error in sampling variability (Power and Moser 1999) and requires that a larger proportion of a sample be processed (Smith and Richardson 1977). The rapid development and mortality rates also make it difficult to achieve a synoptic view on temporal and spatial scales that are relevant to identifying the causes of variations in these vital characteristics. Eggs and larvae are also patchily distributed such that most broad-scale sampling programmes are often only suitable for resolving large-scale features of the spatial distribution. In fact, the resolution of most sampling programmes is often at the limit required to address questions relating to large-scale biophysical interactions that may affect the early life stages of fish at the population level.

In an attempt to better identify critical events, some research efforts (e.g. Frank and Leggett 1982; Baumann et al. 2003; Pepin et al. 2003) focussed on studying larval dynamics in smaller regions where sampling intensity could better match the time and spatial scale appropriate to identifying the impact of short-term events on development or survival. If critical events, factors, mechanisms or interactions could be identified with greater accuracy and precision than could be achieved through large-scale surveys, the knowledge could then be applied more broadly, given appropriate data. However, both process and model error limit the capacity to effectively scale up the results of such programmes (Schneider et al. 1999).

The lack of correspondence between early life-history survival or the recruitment with single processes or the occurrence of specific events gradually led research efforts toward multidisciplinary programmes aimed at identifying the combined effects of several factors that regulate reproductive success in marine fish populations, essentially an ecosystem approach to population dynamics. Numerous correlative studies provided evidence that large-scale processes showed varying degrees of association with recruitment processes (Koslow 1984; Myers et al. 1997; Planque and Frédou 1999) but identifying the underlying processes was generally difficult. Because the variability in both biological and physical variables in most aquatic ecosystems is characterised by long-term trends, many of which are correlated, most meta-analytical studies can at best identify a suite of possible mechanisms and interactions. These are often limited by the time series that were included in the analyses because there is a preponderance of physical indices of environmental status relative to biological ones.

In the 1980s and 1990s, a number of regional or large-scale programmes were developed around the vision that by simultaneously sampling and modelling physical processes and prey–predator interactions it should be possible to achieve realistic insight into factors affecting population regulation. By its very nature, this vision required multiple investigation teams, each focusing on one element of the puzzle, but linked to achieve an overall perspective of population regulation. Such research programmes were logistically and financially demanding, and expectations tended to be high. However, the development of these programmes also provided recognition that population regulation during the planktonic phase of the life cycle reflected a continuum of interactions, and that complex interactions could at times balance each other off while stochastic events could result in significant population losses that could elude prediction.

The challenge faced in developing such large-scale programmes did result in significant advances in the development and application of regional ocean circulation models to predict the drift and dispersal of fish eggs and larvae. When coupled with biological models of prey–predator interactions (Werner et al. 2001), these could serve to focus fieldwork on addressing complex processes, thereby increasing effectiveness. Including inter-annual variations in the physical forcing of a region could also provide insight into the expected variations in distribution and biological interactions. The high spatial and temporal resolution of the predictions can also be contrasted with observations of patterns of variation in individual condition derived using biochemical or otolith-based indices that better reflect the short-term variations that may characterise larval development. Together, input from a diversity of elements and factors can provide some predictive capacity of population regulation.

The research on the Baltic Sea cod stock summarised by Köster et al. (2005) provides an example of a successful large-scale multidisciplinary programme that has yielded some retrospective capacity, which identified key biological and physical factors affecting population regulation. That programme identified the direct and indirect roles of environmental change on recruitment variability based on knowledge of regional circulation, larval prey types and availability, and patterns of variation in predator abundance. Elements that contributed to the development of some predictive capacity included a well-developed and calibrated regional circulation model and good life-history knowledge that allowed the identification of areas of population loss (i.e. poor nursery or juvenile habitats). In addition, the region is characterised by strong environmental driving forces that result in significant changes in physical and biological conditions over time. Researchers were also able to identify clear interactions among larvae, their prey and their predators (Hinrichsen et al. 2002; Köster et al. 2005). But what provided the capacity to develop an accurate retrospective representation of the system's dynamics was the high level of regional monitoring that provided data to forecast and hindcast in order to verify that the various interactions had been identified correctly through the research programme. Without a strong and accurate observation system with which to test understanding and develop predictive capacity, much of the research into the dynamics of early life-history stages is difficult to validate and apply.

Despite decades of effort, research into early life-history stages has yielded very limited predictive capacity – most conclusions are largely qualitative. Most of the research findings are heavily reliant on providing a statistical description of the relationships among variables rather than in the development of a robust framework from which processes are described and quantified. Furthermore, most analyses of field collections do not consider differences in the precision with which variables are measured. For example, the vast majority of field studies into the causes of variations in growth rates in larval fish have found a strong influence of temperature but few have found strong evidence that variations in food availability have a statistically significant effect. However, in no instance is there consideration that the underlying uncertainty in measuring temperature and prey availability at one site is inherently different, with temperature measurements being highly repeatable, whereas single plankton samples are highly imprecise. Comparison among cohorts seldom has an underlying expectation of how large a difference can be expected or detected – uncertainty in representation of environmental variability or model projections rarely forms part of the analysis. As a result, empirical relationships are rarely effective predictors if underlying processes are not properly understood.

If we are to look toward the future of research into early life stages, we have to identify topics and approaches in which we *must* develop predictive capability of the impacts of climate change and ecosystem structure on the early life stages of fish. In the following perspective I will argue that progress in early life research requires significant advances in five major subject areas. First, we must challenge drift projections from regional circulation models using a quantitative approach rather than the qualitative approach that has generally been used in the past. Second, we should revisit sampling approaches, how they should differ among variables and redefine how we represent the stochastic nature of the environment. Third, reaction norms for the feeding and growth of larval fish are essential in order to provide a base for comparison with field observations. Fourth, there is an urgent need to quantify the impact of nekton on the dynamics of all elements of the planktonic foodweb, including fish eggs and larvae, if we are to apply an ecosystem approach to management. Finally, we must identify the link between dispersal of early life stages and the spatial structure of spawning and recruitment among sub-stock components if we are to avoid population collapse.

15.2   Biophysical Models

Increases in the need to understand the relationship between marine organisms and their physical environment, coupled with the better availability of computing power, has resulted in the use of coupled biophysical models becoming a common practise in most studies performed at population scales. The degree of complexity and interaction of the biological and physical components varies greatly. In the simplest form, projections from regional ocean circulation models are used to forecast patterns of drift and dispersal, with the biological elements of the study being represented simply as passive particles, which neither change their state nor respond to changes in the currents. Individual-based models (IBMs), where the dynamics of growth and loss of egg and/or larvae are affected by factors such as local temperature or prey availability, are increasingly being coupled with regional circulation models as a means of assessing how variations in a suite of variables contribute to general life-history cycles and how they might contribute to variations in survival among cohorts. More sophisticated models attempt to include a model component that is used to predict the dynamics of lower trophic levels (nutrient, phytoplankton and zooplankton – NPZ), which can in turn serve as prey for larvae. Such models can have greater scope for understanding the effect(s) of environmental variability on regional dynamics, but the complexity of the models themselves makes them accessible to few researchers.

When regional circulation models are driven with realistic wind fields, it then becomes possible to forecast variations in the dispersal of fish eggs and larvae from the spawning grounds, and if data or NPZ predictions are available, then the combined effect of variations in physical and biological factors on distribution and year-class strength can be studied. In many instances, the movement of eggs and larvae is forecast using a climatological average current field, one that reflects the regional density (kg m^{-3}) field measured over several decades. The projections from these simulations provide the equivalent to a life-history model for the population under study, and usually the general knowledge for the stock matches model expectations reasonably well. Ådlandsvik and Sundby (1994) forced a regional circulation model of the Norwegian Shelf and the Barents Sea with winds for the period 1977–1986 to forecast variations in the drift and dispersal of cod eggs and larvae from the Lofoten Islands spawning grounds. Predictions from their simulations provided a reasonable match to observations of the distribution of 0-group cod collected during fall surveys, but the comparison was largely qualitative and the authors could only speculate as to the causes of discrepancies in cases where predictions and observations did not match. In fact, developments in coupled biophysical models of marine populations have been limited by our understanding of their predictive capacity. Further, the degree of sophistication in physical models has progressed much more rapidly than the detail or understanding of the biological components. The comparison of predictions from biophysical models remains largely qualitative and if we are to move forward in our understanding of how changes in environmental forcing affect the organisms or populations that are of interest to scientists and resource managers, it is time that research into biological or fisheries oceanography move toward a quantitative framework for assessing model accuracy. With increasing demands and developments for oceanographic circulation models that are operational in their ability to forecast and adjust their predictions based on the input of new and more recent data (Johannessen et al. 2006), the expectations of managers of the scientific advice derived from biophysical models has moved in a similar direction. However, as with any forecasting, the effect of uncertainty in the various input elements must be considered if we are to develop confidence in our predictive capacity.

At the most basic level, uncertainty in biophysical models comes from uncertainty in regional circulation models as well as uncertainty in the knowledge used to represent the source of the biological elements. Regional ocean circulation models, even those with data assimilation schemes, have a degree of uncertainty in predicted current strength and direction, which is often determined in comparison with one or several observation sites. The effect of this uncertainty is often considered as being part of the random walk component (i.e. diffusion) of particle (drifter) tracking algorithms. However, the reality is that the uncertainty in current strength and direction should have consequences for the forecasted current field (water being incompressible), which will in turn affect the uncertainty in the predicted transport of particles (i.e. organisms). Errors in the current field can be viewed as differences in the realization of the current fields or as introducing autocorrelation in the random walk component of drifter projections, which can, in turn, have important consequences to projections of probable drift and dispersal (Pepin and Helbig 1997). How large the uncertainty in the projected particle field is and how it depends on the accuracy of the circulation model remain two questions that are seldom considered in the interpretation of predictions from biophysical models.

Panteleev et al. (2004) developed a formal framework that could serve to quantify the uncertainty in the reconstruction of passive tracer fields in an ocean with open boundaries and known current fields. Observations, the spatial smoothing terms and the passive tracer conservation equations (rules) served as a way of weakly constraining the predicted changes in the distribution of fish eggs and larvae. In the example they considered, eggs and larvae were tracked over a 7-day period. Although the general features in the projected distribution of organisms remained the same under different parameter constraints, the degree of uncertainty varied considerably from one realisation to the next, which in turn affected the uncertainty with which mortality rates could be estimated. However, in addition to uncertainty in the physical elements of the projection scheme there is also uncertainty in the observations. In most oceanographic sampling programmes, logistic constraints limit sample resolution and replicate samples from each station are seldom collected. In addition, the sequence of stations relative to the regional currents can have a significant impact on the uncertainty with which vital rates are estimated (Helbig and Pepin 1998; Panteleev et al. 2004).

The degree to which uncertainty in circulation parameters used to make drift projections, sampler precision and cruise track will affect the capacity to forecast changes in distribution and estimate the impact of variations in atmospheric and environmental forcing on population dynamics will depend on the specific circumstances under study. For example, in areas with strong circulation features, the uncertainty in drift projections may be limited relative to the uncertainty in biological field observations whereas both elements may be important in areas where the vagaries in ocean currents may be greater than the mean circulation field. The important issue is that the underlying uncertainty in our knowledge needs to be considered in interpreting predictions from coupled models as is the impact on the accuracy of retrospective analyses or in establishing long-term expectations (Helbig and Pepin 2002). Because of the limited observational base on which oceanographers are being asked to develop an understanding of the processes affecting population dynamics in an increasingly variable physical environment (Lehodey et al. 2006), it is essential that both the accuracy and precision of inferences be known. The significance of this knowledge is twofold. First, it allows us to establish quantitatively the magnitude of changes in population abundance we can (or could) predict based on our current knowledge of the ecosystems under study, and thereby allows the identification of key assumptions or processes that are not adequately represented in our description of biological and physical interactions. Second, it allows assessment of the observational programme on which projections are based. The coupled models can be used to determine the optimal spatial and temporal resolution needed to achieve objective measures of accuracy and precision using large numbers of stochastic simulations. The implications are important to inform managers regarding what can be achieved with the current state of knowledge and what level of monitoring is required to apply that knowledge to meet operational objectives.

Because societal needs and expectations define many of the governmental policies about the use and conservations of marine resources, it is becoming critical for the scientific community to provide a clear quantitative measure of their forecasting capability.

15.3   Environmental Sampling and Uncertainty

Multidisciplinary programmes aimed at the study of early life-history dynamics try to characterise variability in physical and biological elements of the ecosystem that would have a direct impact on fish eggs and larvae. The balance between scientific objectives and logistic constraints (e.g. availability of ship time, funds for sample processing, topical expertise) invariably results in a compromise in the collection of data on a regional scale and the concentration on process-oriented efforts, both of which differ in their spatial and temporal resolution.

The survey design needed to accurately describe environmental conditions within a region depends on the variable being sampled. Mackas et al. (1985) and Seuront et al. (1996) showed that as one moves from physical variables, which are dominated by greater variability at large space and time scales, to phytoplankton and zooplankton, a greater proportion of the variability in overall abundance occurs at smaller scales. This does not imply that physical discontinuities do not occur on relative short scales (e.g. fronts, internal waves) but that, in general, the balance between prey–predator interactions, population or somatic growth and behaviour leads to greater patchiness as we move up the food chain. Differences in the scales of variability become important when we consider the repeatability of the observations, particularly if we attempt to contrast observations of local conditions for different variables and attempt to relate our estimates of environmental status to the state of a focal organism (e.g. fish larvae).

Consider the case of trying to relate the patterns of variation in growth of larval fish to the biotic and physical conditions of the water masses in which they are captured. We know from laboratory studies that the metabolic demands of larval fish are greatly affected by temperature (Houde 1989; Pepin 1991) and that variations in food availability are likely to result in differences in growth and survival (e.g. Werner and Blaxter 1980). In a previous analysis (Pepin 2004), I presented evidence that (1) the volume sampled by most plankton nets used to characterise the feeding environment for larval fish is two to five orders of magnitude larger than the daily search ambit of an individual, and (2) the regional environmental volume represented by each sample of the feeding environment is 8–12 orders of magnitude larger than the individual search ambit. If the variable being measured to characterise the larva's environment can be measured with reasonable precision at the sampling site (i.e. repeated samples have a high degree of repeatability) and the decorrelation scales of the variable is large (i.e. it varies predictably over large spatial scales), then the sample can provide an accurate representation of local conditions. If, on the other hand, precision is low and decorrelation scales are short, the sample may not be an accurate representation of the variability among sampling sites because aliasing (i.e. high frequency variability will be masked as low frequency patterns as a result of sampling at an insufficient resolution) can lead to misrepresentation of regional conditions. Thus, if we attempt to contrast the effect of two variables with different precisions and decorrelation scales, any proper statistical treatment should take the inherent uncertainty in each observation into account in order to ensure that the potential impact of each variable included in the comparison is adequately evaluated. Unfortunately, this is seldom the case. Invariably, studies of early life stages of fish often reach the conclusion that the patterns of variation in growth rates are more strongly associated with environmental temperature than any of the other variables sampled, including prey availability (e.g. Campana and Hurley 1989; Buckley et al. 2006). The implications could be that (1) prey availability is sufficient under most circumstances and development is limited by metabolic scope rather than food limitation, or that (2) environmental structure or patchiness in the prey field is inadequately represented by the survey strategy, or (3) a combination of (1) and (2).

Establishing the functional response of larval feeding or growth to environmental conditions should, under idealised conditions, be based on a representation of the environment on scales relevant to the process under consideration, which in the case of larval fish is of the order of tens to hundreds of centimetres. In reality, sampling of the environment at such scales using conventional tools is prohibitive in terms of operational and processing time. A variety of remote sensing methods can provide high-resolution observations (high-frequency acoustics (e.g. Holliday and Pieper 1995; Donaghay 2004), the Optical Plankton Counter (OPC: Herman et al. 2004), and the Video Plankton Recorder (VPR: Davis et al. 1996; Qiao and Davis 2006)) either at one location in a moored configuration, or towed by a ship. Such tools have revealed spatial structure and patchiness in phytoplankton and zooplankton communities that demonstrate a high degree of organization that is not represented effectively by gear types (e.g. towed nets) that are more commonly used in studying population processes at a regional scale. It is unlikely that such remote sensing tools would be used consistently in regional studies because the technical requirements to operate some of these systems and the analytical requirements to process the large amounts of data collected present some challenges in interpreting the observations. Furthermore, taxonomic identification, often required when dealing with selective predators such as larval fish (e.g. Last 1978a, b; Pepin and Penney 1997), is not always possible using high-frequency acoustics or the OPC, although the VPR does allow species identification (Qiao and Davis 2006). However, to effectively model population processes or prey–predator interactions, the occasional high-frequency characterization of a region could serve to define the appropriate underlying statistical error structure (probability distribution) and decorrelation scales that are relevant to the comparison of stomach contents or growth rates with the more limited net-derived estimates of plankton abundance.

Several studies have characterised the underlying variability of sampling devices (Downing et al. 1987; Pinel-Alloul et al. 1988; Cyr et al. 1992; Power and Moser 1999), with a variety of distributions (Poisson, log-normal, gamma, negative binomial) providing an appropriate fit to the data under varying circumstances. In their analysis of data collated from a variety of sources, Downing et al. (1987) and Pinel-Alloul et al. (1988) found that the sample volume (or the distance or duration of tows) had a significant effect on the variance-to-mean relationship, which implies that the scale over which a sample is collected will impact on the underlying distribution used to characterise the 'error' of the observations. Thus the appropriate error to consider in relating the uncertainty of the *independent* variables should depend on the scale over which the interactions are expected to occur. The implication is that a purely statistical assessment of prey–predator relationships without consideration that independent variables have differing levels of precision is likely to increase the probability of not rejecting the null hypothesis when it is false (Type II error). In the case of prey–predator interactions related to the feeding and growth of larval fish, this could lead to an underestimation of the role of biotic interactions or factors in determining survival.

Although a variety of statistical tools allows the use of diverse error structure (e.g. generalised linear models), an underlying assumption is that the independent variables are known accurately. In reality, the independent variables used to represent environmental conditions are measured with considerable error. With advances in resampling techniques, non-parametric methods, and Bayesian statistics, it may be possible to incorporate the underlying uncertainty in our representation of the environment that affects fish eggs and larvae. Only with such techniques may it be possible to provide a statistically reliable assessment of multivariate functional relationships between physical and biological conditions with larval feeding, growth, condition or mortality.

15.4   Reaction Norms

In response to Hjort's (1914) hypothesis that starvation caused by food deprivation at yolk absorption could play a key role in establishing year-class strength, numerous laboratory studies were performed on a variety of species to identify the sensitivity and recovery capacity of larval fish to food availability (Dabrowski 1975; Yin and Blaxter 1987). It is clear that most vital rates (e.g. growth, time to yolk absorption, the period of no return [i.e. the period when a fish must find sufficient food resources], mortality) are strongly affected by environmental temperature but that the effect of larval size and phylogeny are somewhat more debatable (Houde 1989, 1997; Pepin 1991). However, in laboratory settings the concentrations of prey required to achieve realistic growth and survival far exceeded concentrations normally measured in the field, which leads to the hypothesis that prey aggregation and the role that physical forcing may have on patchiness were likely to play an important role in larval dynamics (Lasker et al. 1978). Husbandry of the early stages of marine fish has proven difficult and the combination of rearing conditions that can yield optimal growth and survival for the aquaculture industry has required significant research efforts (e.g. Rosenlund and Skretting 2006). However, beyond the most basic of understanding, the linkage between laboratory and field studies has yielded limited predictive capacity, despite early efforts to identify standard experimental protocols that could allow the comparison of results among studies and species. In a sense, the difficulties of rearing marine fish under laboratory settings may have been a distraction because there have been no attempts to apply broadly based review or meta-analytical approaches to data from many studies and species to determine if the relationship between the sensitivity of measured growth and survival under culture could be generalised or quantified. How much of a change in prey consumption (note that this is not prey availability) is required to change growth rates by a certain percentage, what is the resulting change in survival under the "ideal" laboratory conditions, and what are the maximum ingestion and growth rates that we are likely to observe, are largely unknown for most species.

For most species of marine fish larvae, there is little or no knowledge of the scope for either feeding or growth rates. We know that both feeding and growth will be affected by the functional feeding response in relation to prey availability, the temperature of the environment in which the larvae live and their body size. The scope for metabolic activity is determined by temperature, and most metabolic processes are size-dependent (Peters 1983). If we can establish what the temperature- and size-dependent maximum feeding or growth rate is under *ad libitum* conditions, we can provide a yardstick against which field observations could be contrasted. Folkvord (2005) took this approach in his study of growth in cod (*Gadus morhua*). After developing the necessary skills in husbandry to ensure that laboratory conditions provided growth rates comparable to those observed in mesocosms where food was plentiful, Folkvord (2005) proceeded to raise larval cod over the range of environmental temperatures to which the stock was likely to be exposed. From the non-linear functional relationships he was able to develop, he could then contrast a range of field studies. Through this approach, Folkvord (2005) was able to establish that under most circumstances surviving cod larvae in the sea typically grow at rates close to their size- and temperature-dependent capacity. This does not imply that food limitation will not occur under some circumstances but without the basic knowledge of the temperature- and size-dependent growth potential, such an inference would not have been possible, or at least uncertain. Unfortunately, this is the only species for which such information is currently available. Although numerous other studies of the growth of larval fish in the field have been carried out (see Houde 1989, 1997; Pepin 1991 for reviews), in no other situation is it possible to determine the proximity of the observations to the maximum potential. Because most studies often include only a few cohorts of larvae with a limited degree of variation in environmental conditions (e.g. food, temperature, mortality), establishing a functional empirical relationship that would relate feeding or growth to key environmental factors is likely to be difficult, particularly given different degrees of uncertainty in the independent variables (see previous section).

Most models of larval fish development and growth apply some very basic and fundamental rules concerning ingestion and metabolic requirements but the empirical validation remains limited (Werner et al. 2001). Although the development of functional relationships for maximum temperature- and size-dependence would provide a basis for comparison of field observations, most ecological models of larval fish dynamics recognise that there will be variability among individuals and that the degree of variation will change under different environmental settings. Prey patchiness is likely to increase variability in apparent feeding and/or growth rates, at least within a realistic physiological range, while selective mortality, such as that caused by predators, is likely to decrease the variability among individuals (Pepin 1989). It is therefore essential that the interpretation of the role of variations in prey availability on the observed feeding or growth rates (of survivors), even relative to reaction norms, should also take into account the rate of loss of cohorts because strong selective loss could mask the influence of other sources of variation in overall survival. It is equally important to consider the variability among individuals in order to assess the various factors influencing the survivors collected during field programmes.

15.5   Quantifying the Impact of Nekton

Based on the regularity of the size distribution of organisms in the ocean, Peterson and Wroblewski (1984) predicted that mortality rates would decrease with increasing larval weight (W) according to a relationship $\sim W^{-0.25}$, the general form of which has been confirmed in a number of analyses of field data (Houde 1989; Morse 1989; Pepin 1991) with the caveat that mortality rates increase with temperature. A key assumption in Peterson and Wroblewski's (1984) theory is that mortality is primarily the result of losses to predators. However, none of the confirmatory analyses was able to validate the assumption because none provided information on predator abundance in the various ecosystems included in the reviews.

Bailey and Houde (1989) provided the first overview of the potential for predators to prey on fish eggs and larvae, based principally on laboratory data. Their review provided clear evidence of general size-dependent functional relationships which depended on the relative sizes of prey and predator, as well as the predators' feeding strategy (e.g. ambush, cruising, filter-feeding and raptorial). Small eggs and larvae were likely to be subject to a greater diversity and number of predators than larger ones, and invertebrate predators would likely be more important during the early stages of larval development, with planktivorous fish and large nekton having a greater impact on later developmental stages. Paradis et al. (1996) was able to refine Bailey and Houde's (1989) general conclusions, through a meta-analysis of laboratory and mesocom studies of predation on larval fish, and concluded that peak vulnerability to most predator types occurred when the larva was 10% of the predator's body length. They were also able to establish that individual planktivorous fish had higher feeding rates on fish eggs or larvae than most gelatinous or crustacean zooplankton. How this translates into a potential for population regulation in the field by individual predator taxa depends very much on the specific circumstances of a stock (Paradis et al. 1999; Hansson et al. 2005; Köster et al. 2005; Lynam et al. 2005). Despite clear recognition that predation plays a key role in the survival of the early life stages of fish (Bailey and Houde 1989; Leggett and DeBlois 1994; van der Veer and Leggett 2005), field studies that quantified the impact of predators within a region or followed the progression of a population over time to assess the potential losses to predators remain limited (e.g. Garrison et al. 2000; Baumann et al. 2003; Pepin et al. 2003; Hansson et al. 2005), and there are no instances where predicted and observed losses are compared.

The inability of the research community to develop a quantitative framework to understand the potential impact of nekton on the survival of fish eggs and larvae is probably one of the greatest factors limiting the predictive capability of our current knowledge of early life-history dynamics and their relation to recruitment variability. With well-developed applications of coupled regional circulation models with individual-based depictions of the distribution of and feeding on eggs and larval fish, the description of mortality is often limited to the application of a simple size-dependent term added to the model. Losses from populations of fish eggs and larvae will be dominated by the effects of transport and predation. For the latter, the key is to develop an understanding of encounter rates because vulnerability, once an encounter has occurred, can be derived from laboratory observation. To predict the encounter rates of predators requires knowledge of their aggregation patterns and potential overlap with larvae, the migration or movement patterns of the predators (on a scale of weeks to months), knowledge of their metabolic requirements and estimates of the 'effective' encounter radius (Pepin 2006). Because most nekton can be observed using acoustic sensors and sounders, it is essential that programmes studying early life stages of fish begin to undertake regional observation experiments using moored and mobile devices that could help construct a view of the time and spatial scales of the movement of potential predators. In most instances, our knowledge of the movement, migration and aggregation patterns of large gelatinous zooplankton and pelagic fish (e.g. anchovy, sprat, sardines, herring and mackerel) is often limited to seasonal cycles. There have been few attempts to quantify the causes of variation in the timing of migration cycles in pelagic fish, and most models of the migratory process are at best qualitative.

The importance of ecosystem-level zooplankton–fish interactions has recently gained recognition in the efforts associated with NUMERO (North Pacific Ecosystem Model for Understanding Regional Oceanography) (Werner et al. 2007). However, current modelling skills remain limited to the standard application of bioenergetic principles. Work by Megrey et al. (2007) and Ito et al. (2007) relied on generating spatial variability in prey fields that were not matched to the spatial scale of foraging fish. Patch frequency, distribution and size will undoubtedly affect the foraging patterns of predators and consequently affect their impact on prey. To accurately predict the impact of pelagic fish on early life stages requires a representation of the fine-scale prey field that is relevant to the predator (Letcher et al. 1996) (i.e. the probability distribution of encounters) and behavioural responses to environmental change, typically important for schooling fish (Pitchford and Brindley 2005), will have to be factored in (Werner et al. 2007).

The cumulative survival of early life stages is dependent on the probability of dying during a stage, and the duration of that stage. The latter is dependent on the growth rate whereas the former is dependent on the sources of loss. Research efforts directed at vital rates have been uneven. Growth rates can be measured with relative ease using otolith microstructure, which may also allow the reconstruction of cohort histories with reasonable accuracy if not precision. On the other hand, mortality rates generally require repeated sampling of a population, and the precision of estimates tends to be poor, but without greater effort to link estimates of loss with a causal factor (e.g. Smith et al. 1989), development of predictive capability from early life-history research will be incomplete and inaccurate.

15.6 Population Structure and Connectivity

There is growing recognition that many commercial stocks behave as meta-populations, a complex of units among which some degree of intermingling and exchange takes place. The units may represent functionally distinct elements in some respect (e.g. spawning migrations but not feeding migrations). Some elements may be sources while others act more as sinks, and the viability of each unit is likely to be variable with a certain degree of regional coherence reflecting the spatial decorrelation scale of environmental drivers. The rates, scale and spatial structure of successful exchange, or connectivity, among units drive replenishment (Cowen et al. 2006). The importance of metapopulation structure and connectivity is increasingly being recognised by managers because there is evidence that many collapsed stocks have failed to recover despite compensatory population dynamics. Complex spatial structure along with Allee effects can have important consequences for conventional management forecasts (Frank and Brickman 2000). The early life-history stages represent a key determinant of this aspect of population dynamics that requires the development of a challenging approach to understanding the nature of space in population regulation.

To date, most research dealing with the drift and dispersal of larval fish has dealt with the contribution of key spawning elements to the overall dynamics of a stock. Equally important is the identification of suitable settlement or juvenile habitats in order to properly describe the major source and sink elements of the life-history cycle. However, it will not be sufficient to simply apply circulation models with the spatial resolution needed to characterise most of the important oceanographic features of a region. Within most metapopulations, there exist many small units that can, under the appropriate circumstances, contribute significantly to the overall dynamics. These units may be difficult to identify quantitatively because their contribution may be stochastic in time or there may be many small units that contribute consistently but their importance may increase when pressures on the major stock units increase. Thus the study of connectivity within a region should include a twofold approach. Forward drift projections provide a perspective of potential sources, sinks and loss caused by drift and dispersal. In a quantitative sense the understanding provided can be measured as: What is the probability that animals from a source will reach a sink? On the other hand, backward drift projections can offer a different perspective. It is not possible to perform drift simulations in reverse because dispersal cannot be represented accurately. However, by seeding the entire range of a regional circulation model with 'particles', it is possible to query the output to determine all the possible sources of particles that reach a sink in the time frame relevant to the species of interest. In contrast to forward projections, backward simulations ask what are the likely sources of particles for a sink, which does not yield a probability weighted by the concentration of particles from a source but rather yields a likelihood of possible sources. Ideally, one would like to solve the optimization problem of determining source-sink distributions that, when used to initialise a forward drift projection, lead to the best fit between observed and predicted distributions. However, this is a difficult problem that may have more than one solution and which may depend on the goodness-of-fit criteria. This twofold approach to understanding the drift and dispersal dynamics of stocks that may represent a metapopulation should provide a realistic way of identifying all possible relationships between sources and sinks and help to quantify the variability in connectivity, which will also depend on the distribution of adult fish and the availability of suitable conditions or habitats for young fish.

Verification of a model's projections could be difficult. The use of otolith microchemistry or genetic markers could assist in assessing the accuracy of the models but the resolution capacity of either tool would have to be considered carefully. The inherent variability in such markers, for both sources and sinks, would have to be applied in repeated drift simulations (both forward and backward, as ensemble approaches) to assess if the differentiation in field characteristics is the result of identifiable relationships or simply one of many realizations of a stochastic process.

15.7   Discussion

The range of life-history strategies adopted by marine fish populations are designed to deal with the stochastic uncertainty in environmental conditions that will be faced through the life cycle. Life cycles have evolved to follow the seasonal progression of environmental conditions that an animal is likely to encounter during various stages of development, whether these factors are physical or biological in nature. In the case of most teleostei, reproduction involves the production of hundreds to millions of eggs during a normal life time, with the implication that the majority will not survive. Although there is a deterministic element to the life cycle of most marine fish, the process their life histories have evolved to deal with is in fact the stochastic nature of the environment in which the young are released. Beyer (1989) and Houde (1989) represented the progression from egg to metamorphosis (or recruitment) as a series of stages governed by the probability of survival and the duration of each stage. They argued correctly that the variations in recruitment we observe in commercial stocks represent subtle variations in the overall survival rate, and in the elements of developmental times and loss rates. As a result of our increasing understanding of the patterns of variations in marine ecosystems and the populations that inhabit them, which involve long-term and large-scale processes as the dominant modes, there may now be less emphasis on prediction of outstanding year-classes (whether strong or weak), which may depend on stochastic events with low predictability, and more on a need to determine which factors most affect the production potential of a population. However, throughout most early life-history research efforts, scientists have attempted to characterise and quantify the vital rates affecting cohorts in the hope that we could identify functional relationships with environmental conditions that would allow us to predict substantial changes in survival. Quantification has been directed toward the deterministic nature of both early life stages and their environment whereas the nature of the problem requires that we attempt to describe vital rates and their relationship with biological and physical processes as probability distributions that reflect the stochastic nature of the interactions. Physical processes affect planktonic fish eggs and larvae by altering their overlap and interactions with potential prey and predators, or by moving them into areas where they will not be able to complete their life cycle, thereby affecting the probability distribution of interactions. The amount of food in the stomachs of larvae, their growth rate and their interactions with predators can be described in two ways: the average functional relationship for a given set of conditions and/or as the distribution of likely states around this average relationship. To date, research into the early life-history stages of fish has concentrated on the first half of this description by providing a statistical representation of the functional relationships rather than a mechanistic one, which has had limited success. To increase the predictive capacity of early life-history research, efforts have to be directed toward providing a mechanistic framework that can be used to characterise the stochastic nature of the interactions between fish eggs and larvae with their environment. This will require changing the approach to describing the environment and quantifying interactions among organisms, which will require a shift in the statistical approaches commonly applied (e.g. generalised linear models, generalised additive models) to ones that accurately describe the variability in the observations (e.g. non-parametric density estimators (Davison and Hinkley 1997)). Mechanistic models that predict the variability among individuals (e.g. IBMs) should be challenged with observations of the variability, not only with comparisons based on central moments. The need to provide a description of the probabilistic nature of the interactions between fish larvae and their environment will require that the density of information collected as part of process studies be increased substantially, which will likely require the use of new sensors because many of the current approaches based on nets simply do not have the capacity to collect enough data.

Model resolution and complexity needed to link observations and predictions will have to vary depending on the nature of the forecasting skills sought by scientists and managers (deYoung et al. 2004). The development of predictive skills will require a diversity of approaches, methods and models, each of which involves different compromises. Equally important is the need for models to be a sufficiently thorough representation of the key interactions, and data has to be of sufficient quality to constrain the model, its parameters or boundary conditions. The use of variational or ensemble methods to model validation as applied to early life-history research should be explored. The emphasis has to be on quantitative testing that can potentially lead to model rejection. Key to their validation and application are the decisions concerning how data are used to update predictions and what criteria are used to measure the misfit (i.e. the penalty function) for a multivariate setting.

Research is aimed at understanding the processes that affect a population. The density of information that I believe we need in order to develop predictive capacity from research on the early life-history stages of fish could not be collected routinely. For research programmes to be successful in the provision of advice, they have to identify monitoring approaches that can be applied with reasonable resource requirements and that will provide the essential elements needed to make the findings of the research programmes applicable. It is then the responsibility of management agencies to sustain the necessary observation programmes to make the results of research applicable routinely so that forecasting can be sustained, reliable and with a quantifiable degree of uncertainty.

References

Ådlandsvik B, Sundby S (1994) Modelling the transport of cod larvae from the Lofoten area. ICES Mar Sci Symp 198:379–392

Bailey KM, Houde ED (1989) Predation on eggs and larvae of marine fishes and the recruitment problem. Adv Mar Biol 25:1–83

Baumann H, Pepin P, Davidson FJM, et al. (2003) Reconstruction of environmental histories to investigate patterns of larval radiated (*Ulvaria subbifurcata*) growth and selective survival in a large bay of Newfoundland. ICES J Mar Sci 60:243–258

Beyer JE (1989) Recruitment stability and survival: simple size-specific theory with examples from the early life dynamics of marine fish. Dana 7:45–147

Buckley LJ, Calderone EM, Lough RG, et al. (2006) Ontogenetic and seasonal trends in recent growth rates of Atlantic cod and haddock larvae on Georges Bank: effects of photoperiod and temperature. Mar Ecol Prog Ser 325:205–226

Campana SE, Hurley PCF (1989) An age- and temperature-mediated growth model for cod (*Gadus morhua*) and haddock (*Melanogrammus aeglefinus*) larvae in the Gulf of Maine. Can J Fish Aquat Sci 46:603–613

Cowen RK, Paris CB, Srinivasan A (2006) Scaling of connectivity in marine populations. Science 311:522–527

Cushing DH (1975) Marine ecology and fisheries. Cambridge University Press, Cambridge

Cyr H, Downing JA, Lalonde S, et al. (1992) Sampling larval fish populations: choice of sample number and size. Trans Am Fish Soc 121:356–368

Dabrowski K (1975) Point of no return in the early life of fishes: an energetic attempt to define the food minimum. Wiad Ekol 21:277–293

Davis CS, Gallager SM, Marta M, et al. (1996) Rapid visualization of plankton abundance and taxonomic composition using the video plankton recorder. Deep-Sea Res II: 43:1947–1970

Davison AC, Hinkley DV (1997) Bootstrap methods and their application. Cambridge series in statistical and probabilistic mathematics. Cambridge University Press, Cambridge

deYoung B, Meath M, Werner FE, et al. (2004) Challenges in modelling ocean basin ecosystems. Science 304:1463–1466

Donaghay P (2004) Critical scales for understanding the structure, dynamics, and impacts of zooplankton patches. J Acoust Soc Am 115:2520

Downing JA, Perusse M, Frenette Y (1987) Effect of interreplicate variance on zooplankton sampling design and data analysis. Limnol Oceanogr 32:673–680

Folkvord A (2005) Comparison of size-at-age of larval Atlantic cod (*Gadus morhua*) from different populations based on size- and temperature-dependent growth models. Can J Fish Aquat Sci 62:1037–1052

Frank KT, Brickman D (2000) Allee effects and compensatory population dynamics within a stock complex. Can J Fish Aquat Sci 57:513–517

Frank KT, Leggett WC (1982) Coastal water mass replacement: its effect on zooplankton dynamics and predator-prey complex associated with larval capelin (*Mallotus villosus*). Can J Fish Aquat Sci 39:991–1003

Garrison LP, Michaels W, Link JS, et al. (2000) Predation risk on larval gadids by pelagic fish in the Georges Bank ecosystem. I. Spatial overlap with associated hydrographic features. Can J Fish Aquat Sci 57:2455–2469

Hansson LJ, Moeslund O, Kiørboe T, et al. (2005) Clearance rates of jellyfish and their potential predation impact on zooplankton and fish larvae in a neritic ecosystem (Limfjorden, Denmark). Mar Ecol Prog Ser 304:117–131

Helbig JA, Pepin P (1998) Partitioning the influence of physical processes on the estimation of ichthyoplankton mortality rates I. Theory. Can J Fish Aquat Sci 55:2189–2205

Helbig JA, Pepin P (2002) Modelling the evolution of fish egg distributions using High Frequency Radar. Fish Oceanogr 11:175–188

Herman AW, Beanlands B, Phillips EF (2004) The next generation of optical plankton counter: the laser-OPC. J Plankton Res 26:1135–1145

Hinrichsen HH, Møllman C, Voss R, et al. (2002) Biophysical modelling of larval Baltic cod (*Gadus morhua*) growth and survival. Can J Fish Aquat Sci 59:1858–1873

Hjort J (1914) Fluctuations in the great fisheries of northern Europe viewed in the light of biological research. Rapp P-v Réun Cons Int Explor Mer 20:1–228

Holliday DV, Pieper RE (1995) Bioacoustic oceanography at high frequencies. ICES J Mar Sci 52:279–296

Houde ED (1989) Comparative growth, mortality and energetics of marine fish larvae: temperature and implied latitudinal effects. Fish Bull 87:471–496

Houde ED (1997) Patterns and trends in larval-stage growth and mortality of teleost fish. J Fish Biol 51:52–83

Ito S, Megrey BA, Kishi MJ, et al. (2007) On the interannual variability of the growth of Pacific saury (*Cololabis saira*): a simple 3-box model using NEMURO.FISH. Ecol Model 202:174–183

Johannessen JA, Le Traon P, Robinson I, et al. (2006). Marine environment and security for the European area: toward operational oceanography. Bull Am Meteorol Soc 87:1081–1090

Koslow JA (1984) Recruitment patterns in Northwest Atlantic fish stocks. Can J Fish Aquat Sci 41:1722–1729

Köster FW, Møllman C, Hinrichsen, et al. (2005) Baltic cod recruitment – the impact of climate change on key processes. Can J Fish Aquat Sci 62:1408–1425

Lasker R, Parsons TR, Jansson BO, et al. (1978) The relation between oceanographic conditions and larval anchovy food in the California Current: identification of factors contributing to recruitment failure. Rapp P-v Réun Cons Int Explor Mer 173:212–230

Last JM (1978a) The food of four species of pleuronectiform larvae in the eastern English Channel and southern North Sea. Mar Biol 45:359–368

Last JM (1978b) The food of three species of gadoid larvae in the eastern English Channel and southern North Sea. Mar Biol 48:377–386

Leggett WC, Deblois E (1994) Recruitment in marine fishes: is it regulated by starvation and predation in the egg and larval stages? Neth J Sea Res 2:19–134

Lehodey P, Alheit J, Barange M, et al. (2006). Climate variability, fish and fisheries. J Climate 19:5009–5030

Letcher BH, Rice JA, Crowder LB, et al. (1996) Variability in survival of larval fish: disentangling components with a generalized individual-based model. Can J Fish Aquat Sci 53:787–801

Lynam CP, Heath MR, Hay SJ, et al. (2005). Evidence for impacts by jellyfish on North Sea herring recruitment. Mar Ecol Prog Ser 298:157–167

Mackas DL, Denman KL, Abbott M (1985) Plankton patchiness: biology in the physical vernacular. Bull Mar Sci 37:652–674

Megrey BA, Rose KA, Klumb RA, et al. (2007). A bioenergetics-based population dynamics model of Pacific herring (*Clupea harengus pallasi*) coupled to a lower trophic level nutrient–phytoplankton–zooplankton model: description, calibration, and sensitivity analysis. Ecol Model 202:144–164

Morse WW (1989) Catchability, growth and mortality of larval fish. Fish Bull 87:417–446

Myers RA, Mertz G, Bridson J (1997). Spatial scales of interannual recruitment variations in various marine, anadromous and freshwater fish stocks. Can J Fish Aquat Sci 54:1400–1407

Panteleev GG, deYoung B, Reiss CS, et al. (2004). Passive tracer reconstruction as a least-squares problem with a semi-Lagrangian constraint: an application to fish eggs and larvae. J Mar Res 62:787–814

Paradis AR, Pepin P, Brown JA (1996). Vulnerability to predation of fish eggs and larvae: a review of the influence of the relative size of prey and predator. Can J Fish Aquat Sci 53:1226–1235

Paradis AR, Pépin M, Pepin P (1999) An individual-based model of the size-selective vulnerability of fish larvae to four major predator types: disentangling the effect of size-dependent encounter rates and susceptibility to predation. Can J Fish Aquat Sci 56:1562–1575

Pepin P (1989) Predation and starvation of larval fish: a numerical experiment of size- and growth-dependent survival. Biol Oceanogr 6:23–44

Pepin P (1991) Effect of temperature and size on development, mortality and survival rates of the pelagic early life history stages of marine fish. Can J Fish Aquat Sci 48:503–518

Pepin P (2004). Early life history studies of prey-predator interactions: a perspective from the larva's point of view. Can J Fish Aquat Sci 61:656–671

Pepin P (2006) Estimating the encounter rate of fish eggs by Atlantic capelin (*Mallotus villosus*) based on stomach content analysis. Fish Bull 104:204–214

Pepin P, Helbig JA (1997) The distribution and drift of cod eggs and larvae on the Northeast Newfoundland Shelf. Can J Fish Aquat Sci 54:670–685

Pepin P, Penney RW (1997) Patterns of prey size and taxonomic composition in larval fish: are there general size-dependent models? J Fish Biol 51(Suppl A):84–100

Pepin P, Dower JF, Davidson F (2003) A spatially-explicit study of prey-predator interactions in larval fish: assessing the influence of food and predator abundance on growth and survival. Fish Oceanogr 12:19–33

Peters RH (1983) The ecological significance of body size. Cambridge University Press, Cambridge

Peterson I, Wroblewski JS (1984) Mortality rates of fishes in the pelagic ecosystem. Can J Fish Aquat Sci 41:1117–1120

Pinel-Alloul B, Downing JA, Perusse M, et al. (1988) Spatial heterogeneity in freshwater zooplankton: variation with body size, depth, and scale. Ecology 69:1393–1400

Pitchford JW, Brindley JA (2005) Quantifying the effects of individual and environmental variability in fish recruitment. Fish Oceanogr 14:156–160

Planque B, Frédou T (1999) Temperature and recruitment of Atlantic cod (*Gadus morhua*). Can J Fish Aquat Sci 56:2069–2077

Power JH, Moser BE (1999) Linear model analysis of net catch data using the negative binomial distribution. Can J Fish Aquat Sci 56:191–200

Qiao H, Davis CS (2006). Accurate automatic quantification of taxa-specific *plankton* abundance using dual classification with correction. Mar Ecol Prog Ser 306:51–61

Rosenlund G, Skretting M (2006) Worldwide status and perspective on gadoid culture. ICES J Mar Sci 63:194–197

Schneider DC, Bult T, Gregory RS, et al. (1999) Mortality, movement, and body size: critical scales for Atlantic cod (*Gadus morhua*) in the Northwest Atlantic. Can J Fish Aquat Sci 56(Suppl 1):180–187

Seuront L, Schmitt F, Schertzer D, et al. (1996) Multifractal analysis of Eulerian and Lagrangian variability of physical and biological fields in the ocean. J Nonlin. Processes Geophys 3:236–246

Smith PE, Richardson SL (1977) Standard techniques for pelagic fish egg and larva surveys. FAO Fish Tech Pap 175:100

Smith PE, Santander H, Alheit J (1989) Comparison of the mortality rates of Pacific sardine, *Sardinops sagax*, and Peruvian anchovy, *Engraulis ringens*, eggs off Peru. Fish Bull 87:497–508

van der Veer HW, Leggett WC (2005) Recruitment. In R. Gibson (ed) Flatfish biology and exploitation. Blackwell, Oxford.

Werner FE, Quinlan JA, Lough RG, et al. (2001). Spatially-explicit individual based modeling of marine populations: a review of the advances in the 1990s. Sarsia 86:411–421

Werner FE, Ito SI, Megrey BA, et al. (2007) Synthesis of the NEMURO model studies and future directions of marine ecosystem modelling. Ecol Model 202:211–223

Werner RG, Blaxter JHS (1980) Growth and survival of larval herring (*Clupea harengus*) in relation to prey density. Can J Fish Aquat Sci 37:1063–1069

Yin C, Blaxter JHS (1987) Escape speeds of marine fish larvae during early development and starvation. Mar Biol 96:459–468

Chapter 16
The Promise of an Ecosystem Approach: Lessons from the Past – Hopes for the Future

Wendy Watson-Wright, Jake Rice, Henry Lear, and Barbara Adams

Prepared for the 50th Anniversary Symposium of the American Institute of Fishery Research Biologists

There are few fisheries scientists today who have known the state of the fishery – be it local, national or global – to be other than in a state of crisis of one form or another. Of course we tend to live and observe such matters through the lens and context of our own times. Yet in the scheme of things we are really the new kids on the block. After all, the west coast fishery dates back over a century and the east coast half a millennium.[1] Undoubtedly the first observations about the state of the oceans and the health of the fish were made by the fishermen themselves, and later through the work of the first scientists.

As the AIFRB celebrates its first half-century, we would be remiss if we didn't at least acknowledge our collective history and consider how it might inform our future.

This paper reflects the Canadian experience.

In Canada, the federal Department of Fisheries and Oceans is responsible for developing and implementing strategies that support the nation's economic and ecological interests in the oceans and inland waters. In a country bounded by three

W. Watson-Wright
Assistant Deputy Minister, Science, Fisheries and Oceans Canada,
200 Kent St., Ottawa, ON K1A 0E6 (613) 990-5123

J. Rice
Director, Assessment and Peer Review, Canadian Science Advisory Secretariat,
Fisheries and Oceans Canada

H. Lear
Senior Advisor, Environment and Biodiversity Science,
Eco-system Directorate, Fisheries and Oceans Canada

B. Adams
Executive Director Strategic Science Outreach, Fisheries and Oceans Canada

[1] "*The Aquatic Explorers*", Johnstone, Kenneth, University of Toronto Press, in cooperation with the Fisheries Research Board of Canada, the Scientific Information and Publications Branch, Fisheries and Marine Service, Department of Fisheries and the Environment, 1977, pp. 16.

R.J. Beamish and B.J. Rothschild (eds.), *The Future of Fisheries Science in North America*, 275
Fish & Fisheries Series, © Springer Science + Business Media B.V. 2009

oceans and with the world's largest freshwater system, the challenges to scientists to help policy makers make sound decisions – in the face of constant threats to the sustainability of the fishery – are formidable.

A century ago the fisheries scientists of the day faced the same geographic realities. Although the resource challenges were different in scope and scale, the demands for scientific proposals that would help sustain the livelihoods of so many of the country's citizens were no less challenging.

At the turn of the nineteenth century, although the Atlantic fishing grounds had been fished by humans for so many hundreds of years, little was known about the huge array of aquatic animals and plants that lived below. It was too alien a world for humans to explore.

As the nineteenth century gave way to the twentieth century, the concerns for fishery began to deepen. Throughout the eastern North Atlantic the fish began to disappear and no one knew why. In this day and age it is hard to credit the degree to which the disappearance of the fish constituted an international crisis for people on both sides of the North Atlantic. A huge food supply, a major item of commerce and a large source of employment was endangered. Fishing nations responded to the crisis by initiating ambitious hatchery programs in a vain effort to replace missing fish. They also built new biological stations dedicated to developing a dynamic new aquatic science that could lead them to an understanding of what was happening to an environment they were desperate to understand. Although *the problem* in the western Atlantic at that time was more perceived than real, *the anxiety* was just as great as it was in the eastern Atlantic ("Canada's Original Ocean Scientists").[2]

Canada was a young nation with a population of about five million citizens. Scientists were operating at the beginning of a long learning curve when it came to understanding the environmental realities of the Canadian fishery.

It was around that time that a federal Board of Management (to become the Biological Board of Canada in 1912 and the Fisheries Research Board in 1937)[3] was established in Canada so that scientific observation and discovery could begin in earnest – mainly through the work of scientists at marine research stations on our east and west coasts.

The Pacific Biological Station was established in 1908, in Nanaimo on Vancouver Island. Its first curator was the Reverend George Taylor, a self-taught field naturalist who lobbied the Canadian government to adopt a scientific approach to the study and regulation of fisheries. His insistence on scientific rigor, applied problem solving, and peer-reviewed publications helped the station establish a world-class reputation in fisheries research. He was joined by the *formidable* "Bill" Ricker, best known for the "Ricker Curve," a mathematical model of fish population dynamics, which is still used the world over to determine average catches for regional fisheries.

---

[2] From a series of documentaries that depicts the lives and scientific contributions of pioneering scientists in marine biology and oceanography in Canada. The series was originally produced between 1994 and 2004 by the First Fisheries Science Documentary Society under the guidance of Department of Fisheries and Oceans.

[3] Op. cit. The Aquatic Explorers, pp. 77 and 163.

Over the decades the Pacific Station, like its Atlantic counterpart, developed a stellar reputation for its groundbreaking research ("Canada's Original Ocean Scientists").[4]

In the east, the St. Andrews Biological Station in the province of New Brunswick was established in 1899. Still in operation today, St. Andrews is the oldest marine station in Canada. It was chosen primarily because of the diverse biological and physical environment in the Bay of Fundy and its proximity to important commercial fisheries for herring, groundfish, and invertebrate species.

Like their west coast colleagues the pioneering scientists in marine biology and oceanography who staffed St. Andrews were quite remarkable in the lives they led and the contributions they made to the field.

Perhaps the most famous was Dr. A.G. Huntsman known by many as the Fisherman's Friend. He was a pioneer Canadian oceanographer and fisheries biologist best known for his research on Atlantic salmon. He also made important contributions to the study of aquatic invertebrates, aquatic ecology, growth and fatigue in fishes, fish migration, the economics of fishing, and fish technology. In short, a sort of Renaissance man for the fish!

Huntsman challenged prevailing trends in research, academia, and the bureaucracies of the day. In the process he would transform the little station into an internationally recognized marine science center.

It was on the fishing docks and boatyards of New Brunswick and Nova Scotia that he came into direct contact with commercial fishermen and their families. He was disturbed by the plight and the uncertainty of their lives and was determined to devote himself to improving their lot by applying science to increase their catches and find better ways to market their fish.

Huntsman had a very strong loyalty to fishermen for two reasons. First, he honestly was in support of them and wanted to improve their conditions. Second, he thought that they were good observers, that what they saw was often unbiased and very useful to science.

He believed that more might be achieved by going offshore to investigate the environmental conditions out where the fish lived. His attitude was that the fish are a product of where they are.

He benefited substantially from his collaboration with the Norwegian scientist Johan Hjort. Hjort's was the first systematic survey of the physical oceanography of Atlantic Canada, and it marked the beginning of efforts, particularly at St. Andrews, to study how the changes in oceanographic conditions can affect fisheries.

Working with Hjort taught Huntsman the value of more comprehensive studies in fisheries science and broadened our horizons to see the whole ocean as our laboratory ("Canada's Original Ocean Scientists").[5]

Such was the fisheries snapshot of the 1900s in Canada and the beginning of looking at the broader interactions of ecology and the environment.

As we fast-forward to the 1970s and the early 1980s our view and our research began to narrow. The cod in the Northwest Atlantic and the salmon on our Pacific Coast

---

[4] Op. cit. From the documentary series.

[5] Ibid. The description of the span of Huntsman's career from the documentary series.

were disappearing quickly. There was a clamor for politicians to do something. To do *anything*. Whole communities saw their livelihoods threatened and the pressure was enormous for someone, not the least of all the scientists, to come up with solutions.

It was a time when "mission-oriented research" became the order of the day and there was a shift away from independent research.

We would learn to our dismay that such an approach was short-sighted and did not produce the results we were looking for.

It has become all too clear that the singular focus on individual *population-based* fisheries science was not and has not been working satisfactorily. When the management of the fishery became single species-based, the single-species population dictated the relevant scale for research and management. And of course such an approach by itself *was doomed* to failure since ecosystem relationships are played out and applied in space. It was time to go back to the drawing board.

And so we find ourselves today returning to an approach reminiscent of those early years – an approach that can grapple with the complex and multiple interactions that are at play in and on our oceans and freshwater systems.

If the imperative to move to place-based fisheries science is clear, we can say that we have already made a start. Spatial management tools such as the designation of Marine-Protected Areas in recent years are a concrete example of this changing foundation for fisheries science. But they are just a first evidence of the change, not the final product. Nevertheless, these first overtures from population-based approaches to space-based ones represent a huge step for fisheries science.

Of course most of the fishing nations of the world have already made policy commitments to an ecosystem approach (EA) and to integrated management (IM) frameworks – perhaps no more so than in Canada and the United States. The question is – in a world of environmental uncertainty – how will these policy commitments play out in the decades to come?

Much of what we need to do differently from the past would at first blush, seem well beyond the purview of the research scientist. How so?

Well, first and foremost, we need to get better buy-in to the *operational implications* of the concepts from all the stakeholders whose cooperation is needed. And although we have given perhaps a more than passing nod to that requirement in the decades past, it is now simply overwhelming.

It is probably fair to say that in principle at least most resource users endorse the concepts inherent in EA and IM. But it is far from clear that they have the same understanding of what these policy concepts mean for them or indeed if they even apply to them. Because of course, they and we all have different agendas – agendas that are often at odds with each other.

It is clear that scientists cannot save the fishery on their own. It would be incredibly arrogant to think they could. It seems logical that science has a key role in two principal tasks: helping put EA and IM into practice, and communicating how to do that to all affected parties. It is not entirely clear which is the more challenging task.

Of all the dilemmas that both politicians and scientists must deal with, perhaps the most confounding – and perhaps the most interesting from a scientific point of view – is the *matter of uncertainty*. Being scientists, we of course want to codify and categorize uncertainty in all its guises. The task might be a little akin to herding cats; but nonetheless that is where we are.

But if one thing is clear, it is that "uncertainty" will no longer be tolerated as an excuse for our past errors, or as a scare tactic to try to motivate or deter managers and policy makers from acting. As scientists we are now called on to make "uncertainty" an explicit component of formal management systems.

So what will the formal systems that incorporate uncertainty directly look like? There are a number of models at play.

First there is "model uncertainty." It encompasses what we do not know about the functional relationships linking parts of the ecosystem dynamically – including the human and nonhuman parts and the living and nonliving parts.

Then there is "parameter uncertainty." Even when we know the shape of the functional relationships, we rarely know exactly how strong they are – exactly how much of X causes how great a response in Y.

These two aspects of uncertainty have been central to fisheries science for the past 50 years.

Now the proposition gets trickier. There is the uncertainty about future states of nature. Managers do not need advice on what they should have done last year – they want advice on what to do next year. To do that, we have to forecast what that future state of nature will be.

We do not forecast perfectly yet, and undoubtedly never will. But there are now many success stories in using our accumulating knowledge of climate, weather, physical, and biological oceanography. They have allowed us to say "something" useful to managers about some specific risks and dangers lurking in the coming years and forewarning them when and where special precautionary measures might be appropriate.

In this regard technology has been a great boon to fisheries biologists. We need to keep our eye out for and take advantage of new technologies that come our way. But conversely we have to remember to a large extent the crises in world fisheries have also been a *product* of technology. Unfortunately, the resort to technology *may also* have driven us away from ecological studies to the collection, storage, and manipulation of data with perhaps not enough thought as to whether that data is relevant to ecological pressures or processes, or indeed, if the data will ever be used at all!

New gadgets and gear can be very seductive and they certainly have their proper place in our endeavors. On balance we should use them to our advantage, but they should not be the main driver or the end-all of what we do.

Because we will never completely eliminate uncertainty about future states of nature, we have interesting research challenges in finding the most useful and effective ways to structure and package that remaining uncertainty in ways that allow managers and policy makers to consider the uncertainty correctly in *their* tasks.

And that brings us to the matter of *"implementation uncertainty."* It is the "elephant in the room" uncertainty that not everyone wants to acknowledge because it confirms our worst fears that there are some things we can never completely control or fix.

"*Implementation uncertainty*" is the uncertainty we have about how the fishery will really unfold, when policy X or regulation Y is enacted. Managers and their science advisors may have quite specific expectations for what a new policy will do, but fishers – not to mention the fish – have their own expectations about what they are going to do on or in the water. Those expectations may not lead them to react to a management measure in the way that the science advisors or managers intended.

Consider again what we know about the first three types of uncertainty. We know that strides have been made in addressing all three through good fisheries science. We know that we are on the verge of reaping greater benefits from past investments in observing systems and operational oceanography, and in research linking ocean dynamics, ecosystem dynamics, and fish population dynamics.

The next generation of models will have even less model and parameter uncertainty, as they are refined with the data provided by the great increase in ocean observing systems. They will support better forecasts, reducing uncertainty about future states of nature as well. It will take more investment in all these types of research to realize these benefits, but we know how to do the research and to apply it to management and policy needs. We just need the resources to get on with the job.

But what about the Implementation Uncertainty? Can we make the same claims there? The matter is not yet clear.

Whether they are fishing at sea or in our lakes and rivers – there is no getting around the fact that resource users are going to see a far more complex management system under EA and IM. Under EA they will see complicated models and rules being the basis for decisions that affect their lives, and under IM they will see their interests being "negotiated" against the interests of other users of "*their* water*," most of whom they would not consider legitimate players in determining *their* futures.

Do we expect them to comply more readily when such regimes are in place? Do we expect each management measure to have a greater chance of doing exactly what we intended in an IM system than in a single-species management plan? When we have had, at best, limited success at making the social and economic objectives for just the fisheries explicit enough to fit into management strategies, how will we extract objectives for the whole ecosystem?

The truth is we just do not know the answers to these questions.

It is understandable if research scientists might want to throw up their hands, if not in disgust then perhaps in frustration and ask why they are being dragged into this messy affair. Such is life in the twenty-first century. But at a more elemental level perhaps, one could argue, that is what Huntsman recognized when he went out and talked to the fishermen about their experience and observations, when he went out to sea to gain an understanding of the environment of where the fish lived and the fishermen worked, and in his collaboration with Johan Hjort to better understand physical oceanography.

Well, just as Huntsman recognized the need for partners, so do we. We need them badly and we need them now. With sociologists, anthropologists, and political scientists. If for no other reason than to help us understand issues of motivation and behavior, and why society is so reluctant to go slow on exploitation and is ready to risk all for short-term profit or political expediency.

Over the longer term these partnerships will close the loop on placing fisheries science effectively in an EA and IM framework. The research will inform managers and policy makers about the consequences of choices on all four dimensions of sustainability: economic, social, institutional, andecological.

Moreover, much of the work of the social sciences is inherently place-based, so the partnerships would naturally help us move our ecological sciences in a direction we would otherwise find challenging to pursue on our own.

Finally, those whose activities are being "managed" (if that is the right word) – the participants in the fisheries and the communities dependent on them – need to have meaningful roles in those partnerships. In the context of complex partnerships and in integrated rather than single-use management, life for them is just going to get *much more complex*. Only if they can see their concerns considered as viable voices can we expect to make the progress on reducing Implementation Uncertainty.

Science needs to remain committed to the broader issues of ecosystem change and how that translates into fish health and productivity or lack of it. Focusing on short-term goals will not advance our knowledge, wisdom, or ability to deal with the larger issues.

We need more research on the ecological, biological, physical, and chemical processes that contribute to ecosystem dynamics in periods of regime shift especially during the impending climate changes on the horizon.

Further, with so many competing interests and industries all carving out their piece of the pie off our coasts, and in and on our freshwater systems, there is often a disconnect between what scientists do and say and who is listening and what they choose to hear.

It is clear that even in our esoteric specialty we cannot continue the way we have been. Perhaps because fisheries biology is so esoteric we need integrators and scientists who will help us all get a better handle on "the big picture." For our part, we as research scientists need to be better communicators. We not only have to be right, we have to be seen to be right!

It is a big challenge for us all. It may and probably will be a frustrating exercise in the years to come. We know that the decisions we make now will pay off in the future yet there is often little incentive for those on the ground to comply. In the fishery everything is local and the immediacy of livelihood can be an understandable mitigating factor to changing behavior. So we may not win any popularity contests with our recommendations.

Even though technologies and human behavior have changed over time, the nature of nature has not – except to the extent that humans have negatively impacted it. We know what does not or has not worked.

The biologist John Culliney has said, *"The oceans are the planet's last great living wilderness, man's only remaining frontier on earth and perhaps his last chance to prove himself a rational species."*

It is a sobering thought.

However, such an outlook need not be so bleak, irredeemable, or inevitable.

As we look to the ecosystem approach and to integrated management we can take some comfort in the wisdom of those "old guys" and how their experience might indeed inform our future. In this context we could say that we are completing the circle and returning to the place from whence we came after several decades in the wilderness.

That bodes well for our future endeavors.

Chapter 17
Technology for Evaluating Marine Ecosystems in the Early Twenty-First Century

Dale Vance Holliday

Abbreviations ADCP: Acoustic Doppler Current Profiler; AUV: Autonomous Underwater Vehicle; CDOM: Colored Dissolved Organic Material; CTD: Conductivity, Temperature, Depth; LOPC: Laser Optical Plankton Counter; LOCO: Layered Organization in the Coastal Ocean; LIDAR: LIght Detection and RAnging; LEO: Long-term Ecosystem Observatory; NOPP: National Oceanographic Partnership Program; ORCAS: Ocean Response Coastal Analysis System; OGO: Ocean Groundfish Observatory; ONR: Office of Naval Research; ROV: Remotely Operated Vehicle; SAS: Synthetic Aperture Sonar; TAPS: Tracor Acoustical Profiling System; UMass: University of Massachusetts; URI: University of Rhode Island; VPR: Video Plankton Recorder; ZOOVIS: ZOOplankton Visualization and Imaging System

Abstract Measuring, monitoring, and predicting oceanic and coastal conditions are widely acknowledged as essential activities in support of long-term ecosystem-based fishery management efforts. Efforts are underway to build new administrative and technical infrastructures to support collecting oceanographic data, assimilate it into models, and ensure its availability to the public, managers, and scientists in a timely fashion. In large part, however, the success of coastal and ocean observing systems will depend on what kinds of measurements are made and their relevance to the success or failure of recruitment in an exploited population. From an ecosystem perspective, habitat characterization, measurements of the abundance and distribution of a target species, its predators, competitors, and food resources should be made on scales similar to those experienced by individuals of the target species. The ultimate value of fixed observatories, mobile platforms, and state-of-the-art data distribution infrastructures critically depends on the availability and use of appropriate sensors. Sensor technology may be the weakest link in evolving plans for a transition to ecosystem-based management. Although the distribution of sensing capabilities for the various ocean parameters, and plant and

D.V. Holliday
School for Marine Science and Technology, University of Massachusetts, Dartmouth,
5034 Roscrea Avenue, San Diego, CA 92117, USA
e-mail: vholliday@umassd.edu

R.J. Beamish and B.J. Rothschild (eds.), *The Future of Fisheries Science in North America*,   283
Fish & Fisheries Series, © Springer Science + Business Media B.V. 2009

animal populations that make up an ecosystem is uneven, a few promising sensor developments are highlighted, perceived roadblocks to developing new sensors are noted, and speculations are made on probable future developments in sensor technology for ecosystem assessment.

Keywords Acoustical sensors · bioacoustics · ecosystems · optical sensors · sensor technology

17.1 Introduction

A famous American philosopher, Yogi Berra, is alleged to have said "Prediction is hard, especially when it is about the future!" Keeping this in mind, in this contribution we speculate about how technology for assessing aquatic ecosystems will evolve over roughly the next 20 years. We focus on sensor technology for making measurements intended to provide data for generating and validating models of ecosystems that might eventually become predictive and worthy of use in the management of living resources in the sea, and in those waterways and aquatic coastal zones that border our coasts. While many of the sensors we may need for use in the sea will also be useful in limnology, other than making note of that here, our focus will be mostly on the ocean.

We will make some predictions about the evolution of sensor technology between now and this seemingly distant horizon. Given that the will exists to dedicate resources to their development, most of the sensor technologies we identify already have sufficient scientific underpinnings to give us some assurance that they can become practical, useful tools for studying the sea and its inhabitants.

For moderately complex sensors, the time from concept to maturity is about a decade. Thus, a 20-year horizon only allows time to bring the development of two generations of *really new* technology, methods and sensors for describing an ecosystem to completion. One of the most important messages you should take from this is that we must start *now*, or they will not be operational and available for use by the next generation of marine scientists.

17.2 Looking Forward: The Next 20 Years

17.2.1 Challenges to Be Met

Before proceeding to a discussion of specific technologies that may become available over the next 2 decades, it is worth mentioning some factors that have limited progress in developing new technologies over the past 20 years. Perhaps we can learn from our experiences.

Funding (or the lack thereof) is often given as a reason for progress (or *vice versa*). Indeed, there has been a significant international investment in acquiring relatively mature technologies for use by agencies charged with assessing fish stocks, but there has *not* been a recent major investment in *developing* really new technology for application in either fisheries assessment or basic research. On the positive side, an increasing number of fisheries research vessels are now outfitted with multifrequency echosounders and a few even have multi-beam sonars. On the negative side, numerous acoustical methods proposed as early as the 1960s have not yet been tried, even at a basic research level. Theoretically sound concepts, e.g., transducers that adjust their active aperture with time to maintain a constant beamwidth with depth (or range) have been proven but not implemented on a wide scale in fisheries acoustics, even though maintaining the same sampling volume and having an increasing directivity index with time from the transmission has some clear advantages (Zakharia 1999). The use of advanced signal processing methods in fisheries has also received far too little attention in fisheries acoustics (Holliday et al. 1980; Zakharia 2001). The use of multiple acoustical propagation paths in fish biomass assessment work is likewise underutilized, or not used at all.

Critically needed funding for research and development of new sensors that can be used for ecosystem characterization has been limited and intermittent. This is in spite of decades of talk about moving toward multispecies and ecosystem-based management. Relatively rare exceptions can be found in a few promising sensor developments that are based on optical scattering and absorption, optical imaging, and in the autonomous measurement of nutrients and water chemistry. However, in both fisheries and ecosystem sensor research and development, the available funding has too often been applied to making incremental improvements in old sensors rather than to the invention of brand new tools. Improvements for existing sensors are unquestionably needed, but in the author's opinion there is a serious imbalance, with the result that the pipeline of really new ideas and sensor technologies is nearly empty. Funding limits aside, there are other important limits on progress, and many of them may not be so obvious.

In recent years there has often been an assumption that the individuals who use high-tech sensors are best equipped to develop new ones. This is only rarely the case. *There are important differences in the backgrounds and motivations of people who are good at **developing** new sensors for use in the sea, as opposed to those who have the capability of **making effective use** of such sensors to solve science problems or manage marine resources.* To achieve a goal of being able to predict the consequences of our actions as a part of a sustainable resource management program, we need to support both cultures at sustainable levels. To be most effective, they need to work closely together, while respecting each other's views and strengths. It is critical, however, that it is clearly understood that the *innovative **developer*** of new technology is *not interchangeable* with the *intelligent **user*** of such technologies, and *vice versa*. There is a very small, shrinking, worldwide pool of innovative **developers** and a rapidly growing pool of intelligent **users** of technology. It is critical that the ocean sciences community begins to understand this imbalance and recognize that there is a distinction between these two groups. Existing tools are largely for measuring standing stocks – mostly stocks of commercially exploited nekton. Such tools have so far proven inadequate for achieving goals involving estimating the derivatives of standing stocks, i.e., prediction. We need new modeling approaches and measurement methods, and most likely new kinds of sensors with which we can measure ecosystem descriptors that we probably do not currently measure (e.g., rates of growth and predation at all trophic levels).

We are now *critically* short of successful technology *developers*. The question is "why"? There are several reasons. First, academia, where many good ideas for new sensors originate, usually does not reward a *developer mentality*. The time frame for *idea-to-practical-tool* is about 10 years. This is a significant fraction of one's academic career. If a young assistant professor starts to work on a new technology early in his/her career, pursuit of only one relatively complex development project can take most of the time during which tenure will be either gained or lost. Producing large numbers of science publications is a dominant metric for advancement. Taking, say, 10 years to develop a single new tool is not the route to a half-dozen high-quality first author publications each year. In addition, funding rarely is guaranteed for the whole development cycle, even if progress can be demonstrated. One must *feed* an experienced *staff* (e.g., graduate students, postdocs, and technicians) through the whole cycle, or lose significant time to retraining. There is no magic formula for success in increasing the pool of *innovative **developers***, but the factors mentioned above discourage the growth of that pool. Recognizing this is a first step toward changing the situation. The imbalance between *users* and *developers*, and the balance of funding that is available to each community should be addressed sooner, rather than later.

In industry there is little motivation to hire and let the brightest young people enter into decade-long development programs that will have a small consumer base when a new product is ready to enter the marketplace. At the very most, tens to hundreds of complex, high-tech sensors might be sold each year to support fisheries management or research efforts, worldwide. Further, acceptance of a new tool by a very conservative fisheries and biological oceanography research community might take 5 years or more, beyond the time that a new sensor is ready to market. Combined with the fact that scientists and institutions usually only replace their instruments when they are either damaged beyond repair or are lost, recovering the cost of a decade-long trip from a good idea to an off-the-shelf product is simply impractical in most cases, even when part of the development cost is sponsored by governmental agencies. Unlike car manufacturers and watchmakers, purveyors of scientific instruments do not build in obsolescence. Most oceanographic instruments are expected to last for one's career, or a significant fraction thereof. Whether in a large or a small business, success is measured in terms of return on investment, at least annually. In small businesses, there is a further constant threat of a shortfall in cash flow. Many such businesses are perpetually only weeks or months from bankruptcy. Survival demands an absolute focus on the narrow marketplace for a successful oceanographic sensor. There is often a necessary resistance to broadening a successful product line if it requires a capital investment or will divert limited engineering or marketing resources.

There are exceptions in industry, but only a few. One worthy of mention is a commercial manufacturer of echosounders and sonars, many of whose instruments find application in fishing fleets as well as on research and stock assessment vessels. In this case, at least some of the cost of technology needed to produce state-of-the-art sensors for use in research and fisheries assessment was leveraged by virtue of its position in the defense industry as well as from a national investment. Unfortunately, dual-use sensors with which both fisheries and ecosystems can be characterized are rare. Attempts to use sensors developed for one of these markets in the other usually leads to nonoptimal compromises in approach and performance.

Scientists at government laboratories are generally intelligent users of technology, rather than developers. In the fisheries community, charters for these institutions often start and stop with annual stock assessments. Improving the precision of measurements is a primary goal, but changing the way one makes the measurements, or their accuracy, often simply leads to questions about how data from the new method will fit together with historical data, and in the end, change tends to be viewed as bad. In most industrial nations there are a limited number of important instances of individuals who fit into the *innovative developer* category. These are people with exceptional imaginations and whose character will not let them rest when the *status quo* is not good enough. These exceptions seem to succeed in spite of their institutional systems.

While three or four exceptions come to mind, space limits us to a mention of only one such instance, the development of ship- or air-borne LIDAR for detecting fish, fish schools and thin layers of plankton. The development of this technology for use in fisheries and ecosystem assessment is far from complete, but at present it seems likely to find an important place in our toolbox of sensors before the end of the first decade of development. Further, LIDAR has the potential for considerably more development and probable widespread use during the second 10-year period. It is not hard to see a parallel between the development of LIDAR and the evolution of fisheries acoustics. The development of LIDAR is about where fisheries acoustics was in the late 1970s or early 1980s. Although its use may be limited to examining the upper ocean, the euphotic zone is an important part of the marine ecosystem. In an analogy between multifrequency and multi-beam acoustics, it does not take much imagination to see a future in the simultaneous use of calibrated multi-wavelength and multi-beam LIDAR systems. Adding passive hyperspectral imaging to a LIDAR system could also be of considerable value, especially if data streams from each could be quantitatively combined.

Unfortunately, there seems to be no magic bullet that assures us of a future plethora of new sensors for use in fisheries and ecosystem science. It is clear that academia, industry, and government labs working alone have not filled the technology pipeline with new sensors. An innovative program was started a few years ago by the US Congress to partially address this issue. The National Ocean Partnership Program (NOPP) combines the resources of academia, industry, and government labs in a unique way, but it also suffers from some of the same problems that have persisted in the past. On the funding side, contributions to annual NOPP budgets are at the discretion of individual participating agencies and can vary with other budget demands within each federal agency. Dedicated multiyear (line item) funding would be an improvement and allow better continuity and planning for long-term technology development projects. The NOPP remains a relatively new experiment, but it is an innovative tool for the management of basic and applied research in technology development and it does have some potential. At the very least, it is not "business as usual." It remains to be seen if, in the long term the NOPP program funds a good balance of *intelligent users* and *innovative developers*. Basic research is the business of taking reasonable risks in the hope that new ideas can be developed into practical tools. Unfortunately, many funding agencies have become very risk averse. This is just one more challenge that must be faced if the pipeline is to be recharged.

If sensors that are already on the drawing boards are to have a chance for further development and be in use by the beginning of 2017, we must address three major issues. First, it is critical to understand that while innovation and freedom to pursue ideas is critical in basic research, where most ideas for new technological approaches to sensing fish or ecosystems descriptors begin, the time to routine deployment often depends as much on the *consistency of effort* applied throughout the development cycle as on the absolute amount of the funding.

Second, there are several easily identifiable factors that will *limit success* in the development and use of a new instrument, or an old one that is being modified for use in the sea. Although not a comprehensive list, such factors include the difficulty of repackaging a prototype for *immersion*, often at significant depths; redesign for use in *long-term deployments* (e.g., to use extremely small amounts of power); and attention to stringent design requirements involving resistance to *corrosion* and *fouling*. Any oceanographic instrument that is expected to make reliable, high-quality, long-term observations, whether tethered, deployed on the seabed, or drifting at the surface or in the water column must meet a higher design standard than does a sensor or system whose main use is in a lab, or during a typical cruise on an oceanographic ship for a few weeks. If you have no power plug, you will likely need to design for extremely low power consumption (milliamperes at 12 V operating; microamperes when sleeping; a typical PC draws several amperes at 120 V AC). Dealing with fouling is generally more difficult for sensors based on the use of optics than acoustics, but it must be considered for any sensor that is to be left in seawater for very long. One must also recognize that it is absolutely essential to design not only for performance and cost, but also for *maintainability*. A simple mechanical repackaging of an existing sensor is rarely sufficient and often leads to failure. Although trial and error may eventually work to overcome these challenges, it is usually more efficient if someone with a background in physics or electronics engineering can address them early in a development program.

Third, *uncertain continuity of funding* is a deadly enemy of innovation. Innovation only rarely comes from large institutions that are capable of, or interested in, carrying highly skilled teams across funding gaps. Once a key person in a team who is developing new technology, whether the principal investigator or a technician with special skills, is reassigned or leaves because he/she is unsure of his/her future and job security, the project is almost certainly going to be significantly delayed, if not doomed. This happens much more frequently than is generally recognized by funding agencies where responses to annual budget cycles often create slowdowns or gaps in support. To be fair, many of the delays and gaps are often the result of uneven workloads in grants and contracts offices, but whatever the cause, any deviation from a planned funding profile can have serious negative consequences for individuals and research programs. Sometimes, rather than salary gaps, discontinuities are as simple as major delays in obtaining ship time to test a prototype or to collect ground truth data at the right time in the development cycle. Broken promises between agencies, and between agencies and their principal investigators have devastated or delayed numerous programs that might have otherwise been carried out either more efficiently or to a successful conclusion in a timelier manner. Other promising programs have simply disappeared without being given a chance to develop to maturity. These are all management problems, and can be successfully addressed if the will to do so is there.

A comprehensive look at the technology we will need to support an ecosystems-based management of fisheries would fill a large book. Having already mentioned LIDAR, we can only touch on a few other examples of what we think can be accomplished in the next 2 decades. Some of the ecosystems sensors listed below will mature soon, others will take the entire 20 years, and perhaps a few more, but our intent is to highlight a few specific sensors and innovative modes of deployment that will be representative of the technology we will be using in the future. Examples have been chosen to give the reader a sense of the complexity that can be anticipated, and the promise of what can be achieved.

17.2.2 The First Decade: 2007–2017

The first 10 years are relatively easy. Most of the sensors that will reach maturity in the next decade are already on drawing boards, or exist as prototypes. The primary challenge is to select a few development programs to highlight from a group of very modest size. The pool to choose from is relatively small because, as mentioned earlier, funding and other factors have limited new starts and thereby the number of *innovative developers* who have been able to consistently hold a program together for several years.

In spite of these very real challenges it still seems reasonable to predict that there will be substantial advances in several aspects of sensor technology with applications to ecosystem-based fisheries management. Some advances will come in the form of new sensor types, while others will involve repackaging of sensors that now work in laboratory environments. A variety of optical sensors that make use of multiple wavelengths and multiple angles of scattering or absorption are becoming widely available. Other sensors make use of spectral fluorescence. Making a high-quality measurement of scattering or absorption is only part of the problem. Algorithms that make them useful for separating such parameters as CDOM, phycocyanin, phycoerythrin, chlorophyll-*a,* and inorganic particulates are required before one can interpret the basic measurement. Such algorithms are now being developed and validated (Sullivan et al. 2005). Data from optical sensors are rapidly being integrated into suites of standard oceanographic measurements, along with ADCP measurements of currents, estimates of water-column turbulence, and simple CTD profiles. Assuring that a sensor is, and remains, calibrated is also critical for both optical and acoustical sensors. Stable, very high-resolution spectral sensors for determining water-column properties are also approaching off-the-shelf levels of maturity, as are sensors for measuring acoustical backscattering and bioluminescence.

In part because acoustical scattering from the seabed has not attracted much attention until the last few years, sensors for the quantitative acoustical assessment of the seabed are not yet as mature as those that address water-column properties. Present-day acoustical descriptions of seafloor habitats are largely non-quantitative, but the possibility of real advances toward physics-based interpretation of the acoustical reflectivity of the seabed and its inhabitants is now a realistic goal. While useful, important advances will almost certainly be made toward this goal during the next decade; there will also likely be some significant challenges and opportunities in the following decade for the application of acoustical tools to the characterization of the benthic habitat.

The commercial availability of multi-beam sonars is a good example of an important technology change that is now underway. While there is room for new developments in this field during the next decade, this technology is quickly changing the way we view fish schools and observe fish behavior. Even now, practical challenges involving the calibration of many sonar beams are being addressed in order to make this new tool useful in assessing fish biomass and determining patterns of fish distribution in relation to the physics and chemistry of the water column. Only about a decade ago, calibrating one beam was a major challenge for fisheries acousticians. Engineers and scientists are working hard to calibrate sonars with dozens of beams today, and they will have to deal with hundreds in the next decade. Much improved spatial resolutions that derive from having access to multiple overlapping beams and the advent of systems that operate at higher acoustical frequencies are offering the potential for describing both fish and plankton distributions with spatial resolutions of ca. 10 cm in three spatial dimensions. Such systems will be deployed on new platforms that will unobtrusively place a sonar or multi-beam sounder near a school to make up for higher absorption losses that come with using higher frequency acoustics. Larger transducer bandwidths and better efficiencies for transducers also contribute to improved sensor resolutions in time and space.

Originally proposed in the 1960s and early 1970s, Doppler measurements have been recently employed to describe scatterer motions by, and within, fish schools and aggregations. One can envision a day when Doppler will be available as a parameter in 3-D visualizations of schools in situ. Faster capabilities in processing signals should enable rapidly repeated sequential measurements with displays of the details of acceleration within schools as fish search, encounter patches of food, and begin to forage. If multifrequency capabilities are included in such a system, it should be possible to visualize predator–prey interactions at previously impossible scales and with detail that will allow us to begin to quantify behavior in an environment in which our eyes are inadequate for direct observations. Combinations of multi-beam technology and multifrequency methods are nearly within our grasp. Combinations of multi-static and multifrequency methods are also just beyond today's horizon. Additional aspects of the synergy to be gained by combining these technologies include remote classification and identification. These are briefly addressed later in this chapter, as are combinations of methods being developed separately in underwater optics and underwater acoustics.

The use of low-frequency sound beams traveling via several different propagation modes to detect and study fish schools at distances of tens to hundreds of kilometers was suggested in the 1960s. The feasibility of using low-frequency sound for fisheries work has recently been demonstrated in different configurations, including both one-way absorption and backscattering. For the absorption method, when applied to fish with swimbladders, a capability for extracting fish size has been established (Diachok and Wales 2005; Diachok et al. 2005). Parametric acoustical sources remain undeveloped for use in fisheries, but still offer some advantages in array size for the generation of narrow low-frequency beams. Monitoring and surveying fish schools and school groups at ranges of hundreds of kilometers are now achievable. While there is work to be done before these methods can be used to obtain quantitative biomass estimates, at present they could be used to study schooling behavior and large-scale migrations, resolving details that cannot be obtained in any other way (Makris et al. 2006). The advancement of these methods lacks only dedicated advocates within the fisheries community.

Sometimes a change in the mode of deployment of an existing sensor is at least as critical as the performance of the sensor itself. Optical and multifrequency acoustical sensors that have been available for several decades have recently been deployed on water-column profilers that allow one to resolve thin horizontal plankton layers that sometimes contain nearly 80% of the plankton biomass in the water column within a few layers that are each less than a meter thick. When present, these layers are thought to be an important food resource, strongly impacting the survival of larval fish. Multifrequency acoustical scattering sensors have been used to demonstrate that thin zooplankton layers are both remarkably persistent and robust even in the presence of strong physical forcing and mixing. They can cover tens of horizontal kilometers and can persist for weeks. Multiple mechanisms have been identified as associated with thin layer formation and dissolution. Some of these are physical, while others involve phytoplankton and zooplankton behavior. The availability of increased temporal and spatial resolution in new optical and acoustical sensors, combined with new methods for their deployment are directly responsible for the discovery of these previously unknown structures in the marine food web. We envision an increased use of these tools and others like them.

17.2.3   The Second Decade: 2017–2027

Beyond the first 10 years we can anticipate that smaller and smarter sensors with faster response times, larger dynamic ranges, and more built-in decision-making capability will be conceived, engineered, tested, and produced. The development of novel power sources and new battery chemistries should allow sensors to work longer in remote locations, with the consequence that they could be used autonomously in deployment modes we have not yet considered. Some sensing devices may "mine" the sea or the seabed for much of their power. Better battery chemistries should also lead to smaller sensors, an important factor for "smart" tags of several kinds, where the battery is now responsible for much of a tag's volume and weight.

In addition to conventional modes of deployment from ships and on moorings, marine observatories will begin to be populated with an increasingly diverse set of sensors that will measure and report the parameters that modelers will need for prediction. The first generation of observatories, e.g., the Rutgers University Long-term Ecosystem Observatory (LEO), has proven that the concept is viable. For reasons of cost and maintenance, a second generation of marine observatories may not grow in number or scope as fast as the ocean sciences community may wish, but one should consider the first few of these as prototypes that can be expanded when resources come available. Meanwhile, they can be viewed as opportunities for testing numerous technologies for a third, more extensive generation of observatories. Although the visionaries who see the potential for a worldwide infrastructure may not find progress toward their implementation moving as fast as they might desire, they should continue to recognize that even a limited number of regional facilities can change the way we view the sea. It is also important to realize, however, that infrastructure alone is insufficient. Without the development and use of new, robust sensors that will be useful in marine observatories, especially those for studying life in the sea, the full potential of these expensive facilities to support fisheries and ecosystem management objectives will not be reached. At present, most of the sensors conceptually intended for use in the first generation of observatories seem to be the same ones that we routinely deploy from other platforms. A renewed focus on how they can be used for water-column measurements of relevant ecosystem parameters is needed.

The most serious of the present limits on sensor performance involves inadequate spatial and temporal resolution. The amount of data we can telemeter from remote places is also a problem. These limits are largely dictated by embedded computing power, battery technology, and the bandwidth available to the science community for real-time communications at reasonable costs. Over the next 20 years, we can anticipate that these constraints could be largely removed, or at least greatly diminished. In the second decade, data from many environmental and ecosystem sensors, including those that measure biomass, will be accessible in near-real time. It is not entirely outside reason that both the rates at which the data are accessed and the positioning of the sensors spatially will become adaptive. It will be possible to deploy and maintain autonomous sensors in formations (spatial arrays) that allow one to resolve unaliased spatial patterns and fine-scale details of the phenomena being monitored or studied.

Synthetic aperture sonar (SAS) is a technology that is now in its infancy. It is too expensive for most nonmilitary applications at present. However, it should come of age during the second decade and can be expected to become a useful, practical tool for seabed classification and detailed mapping as well as for studying the animals that live within the water column and just below the sea surface.

The performance of new sensors should begin to approach the limits of spatial and temporal resolution imposed by the physics of the propagation of sound and light under water and the ever-changing undersea environment. Removing the environmental limits on resolution and the quantity of data we can telemeter is very important because we ultimately must observe individual organisms in situ, at scales that describe the things they do in their daily ambit. This ability is one of the most significant advantages that terrestrial ecologists have over their peers who work in the diverse habitats of the sea. When we can resolve, observe, and track individuals, we may be able to understand the many facets of their behavior. We might be able to understand what characteristics of their environment are important and why they choose specific behaviors over others. At that point, we may have a chance to build realistic, adaptive ecosystem models that will assimilate a sufficient amount of data to have real predictive value.

It should not be a major problem to assure that measurement arrays, e.g., fields of drifters or autonomous underwater vehicles (AUVs) become *self-healing* when a sensor fails, and *self-adjusting* in response to the measured ocean environment with minimal interference from shore- or ship-based personnel assuming that is desirable. This is currently a routine procedure with some autonomous underwater vehicle operators, although it remains to be done automatically, without operator intervention. As our understanding of the relationships between the ocean's chemistry and physics, and each component of the food web improves we can expect a day to come when shore-bound modelers can purposefully rearrange arrays of sampling sensors to optimize the quality, quantity, and even the kind of data being fed into their predictive models.

Advances in computing power should encourage the development of sophisticated acoustical and optical cameras for deployment on autonomous underwater vehicle (AUV) and remotely operated vehicle (ROV) platforms. Those sensors can be self-focusing, complementary, if not synergistic, and would be able to correct for statistical propagation (scintillation) and scattering effects in the environment that are now beyond the computing power available outside a laboratory. Platform motions could also be removed in situ and synthetic aperture sonars should allow much improved spatial resolutions over those that can currently be achieved.

Advanced optical and acoustical instrumentation for trawls and nets can be anticipated. It is possible that today's advances in underwater acoustic telemetry in the surf zone will lead to a capability for higher data rates between a ship and nets or trawls operating in the vicinity of a ship's wake. This should not only allow better control of the net's position, but should allow one to get better acoustical or optical images of the dynamic responses of the net while being towed and of the organisms one is attempting to collect. Hopefully, this will allow a better understanding of fish behavior under the stress of imminent capture. That, in turn, should have a positive impact on reducing bycatch, an improved size-selectivity for fishing gear, and lead to better species-specific fishing gear designs.

Autonomous underwater vehicles currently have the capability to detect and track very low levels of current-borne chemical traces, e.g., oil or gas seeps, to their sources. It is entirely plausible that this technology could be improved to the point that autonomous vehicles could be used to track natural biological exudates and chemical trails to their sources. For example, if specific chemicals are associated with corals, it might be possible to use autonomous vehicles to map distributions of sparse coral colonies or even individual organisms, enumerating them in the deep ocean as well as in shelf areas. This technology might be applied to a variety of sessile organisms, e.g., oyster beds or even small, isolated clumps of seagrass. Life might be detected and mapped in places that are difficult to access in ships or boats, e.g., on the seafloor under the Arctic ice. Early detection and mapping of chemicals and water conditions associated with the advent of harmful algal blooms might also be quite useful and practical within this time frame.

Inexpensive tags that report data acoustically to arrays of listening posts are in place today. These methods will only get less costly and more widely used in the future. Tags, that when recovered might contain data from a half-dozen or more sensors are not unreasonable to consider. Increases in raw computing power, and the likelihood of huge increases in digital memories with large decreases in power consumption, make it likely that future generations of tags for fish will go well beyond what we have today. Costs will likely decrease and they will become smaller so that they can be used on more species and sizes of organisms. When combined with the advances in chemical sensors that we see happening today, it is not unreasonable to think about tags that measure different facets of the physiological state of a tagged animal.

Finally, it is easy to see how more data can advance our progress toward models that predict the future state of ecosystems. These data will only lead to better management if they can be displayed and synthesized in ways that have yet to be addressed. While it is likely that computing power will increase, without a dedicated effort it seems unlikely that the data visualization and modeling needed for fisheries and ecosystem management will be automatically developed in parallel with the technological advances in sensor technology that are now planned, much less those that have yet to be conceived in the next decade. We must integrate, synthesize, and display data from hundreds of platforms, carrying dozens of sensors, resolving the physical and chemical environment in 4-D. This must be done in multiple, nested domains ranging in size from large marine ecosystems, down to scales that resolve the ambit of individual fish larvae. This is a daunting task. Unless we plan and budget for the distribution, processing, visualization, and interpretation of the data, we will not obtain the end result we desire.

17.3   Looking into the Technology Pipeline: A Few Examples

We now turn to a few specific sensors and deployment methods that should evolve during the next decade. Our frequent use of acoustics in these examples is simply recognition that at medium and large distances from a sensor, sound is less dramatically attenuated than is electromagnetic energy, and is usually a better probe to be used for looking at greater distances into the sea than is optics. Optical methods, e.g., multi-wavelength scattering sensors, on the other hand, are often used when sensing very small particles in close proximity to the sensor (e.g., phytoplankton). They are also often useful for determining levels of CDOM, detritus, or sediment load. LIDAR, deployed on aircraft, is developing to be an exceptionally good way to observe the near-surface water column very quickly over hundreds, or perhaps even a thousand kilometers, in a few hours. Several specific ocean environments that should benefit from the availability and deployment of new technology are also discussed briefly.

We admit to a bias in the selection of the six projects we will discuss below in some detail, as the author is involved with each of these projects at levels ranging from being the principal investigator (PI), to that of a co-PI, or simply as a consultant. The first three projects discussed below involve research begun by Charles Greenlaw and the author under ONR sponsorship while they were at BAE Systems. The last three projects involve several of our co-principal investigators at the University of Massachusetts at Dartmouth, the University of Rhode Island's (URI) Graduate School of Oceanography, and NOAA's National Marine Fisheries Service and Pacific Marine Environmental Laboratory.

17.3.1 Combining Multi-static and Multifrequency Acoustical Methods

Acoustical scattering from an organism depends on the sound's frequency spectrum and the angles at which the sound reaches and then leaves the scatterer. Measurements of the sound scattered at many angles and frequencies around an organism can be made simultaneously. This greatly expands the number of independent measurements available for use when one is attempting to extract the characteristics of an organism, i.e., its size, shape, and physical properties.

Measurements were made of "echo" levels *versus* time for broadband acoustical scattering at multiple angles around a molded RTV silicone cylinder with hemispherical end caps. We use "echo" in a generic sense here, including backscattering, but also meaning signal arrivals from any angle with respect to the incident wave after it interacted with the target (i.e., backscattering, forward scattering, and scattering at all other angles). The cylinder had a diameter of 7.14 mm, and including the two hemispherical end caps, had an overall length of 35.72 mm. The principal axis of the cylinder was positioned in the horizontal plane, rotated to an aspect of $-15°$ from the plane of the incident, broadband pulse (ca. 0.300–0.800 MHz). Measurements included those between $10°$ and $350°$, but near-backscattering angles were excluded because the receiver physically masked the source at $0°$.

The direct path arrivals at the receiver appear as a hyperbolic arc of high-level scattering extending in angle between ca. $145°$ and $215°$ (Fig. 17.1). The hyperbolic arcs with maximum values near $195\,\mu s$ are reflections from the water surface. The specular arrival from the cylinder segment is near $30°$ and consists of multiple arrivals at ca. 25, 40, 58, and $80\,\mu s$. The remainder of the echo structure in the time delay – the multi-static angle plane shown is due to refraction around the cylinder,

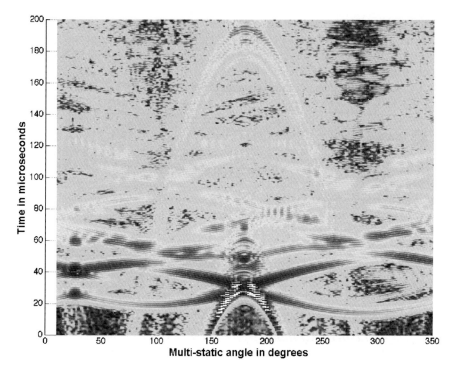

Fig. 17.1 A broadband acoustic pulse produces a complex pattern of echo levels *versus* time for forward- and multi-static angles. In this figure, the abscissa represents the angle between the source transducer and the receiving transducer, measured in a plane with the target at the origin. The ordinate is time after the pulse trigger, delayed to center the target echoes on the plot. Color indicates echo level on a decibel scale. The target is a molded RTV cylinder with hemispherical ends, with its principal axis set at $-15°$ from the normal to the incident sound wave. The complex interaction of the incident broadband sound pulse with the target results in numerous distinct echoes at each receiver angle. The pattern of this echo structure *versus* time and multi-static angle is unique for this target. The arcs with peaks at $180°$ are artifacts of the experimental setup arising from the direct pulse and a surface reflection

multiple internal reflections, possible absorption phenomena, creeping waves and coherent interactions between all of the acoustic waves generated during the echo formation process.

The target strength of the cylinder and the geometry described above can also be displayed as a function of a nondimensional frequency (*ke*) and the multi-static scattering angle (Fig. 17.2). Taken together the data displayed in Figs. 17.1 and 17.2 forms a *fingerprint* of the target, and includes information about its shape, size, orientation, and physical properties. Although the *fingerprints* are quite complex, each feature contains information about the target. This kind of detailed data is also required if today's mathematical models of scattering are to be improved. If they are to be universally useful, such models must be based on the first principles of acoustical scattering, i.e., they must be physics-based, rather than empirical. Historically, there has been, and will likely continue to be, an iterative interaction

Scattered intensity vs multistatic angle and non-dimensional frequency for A=–15°

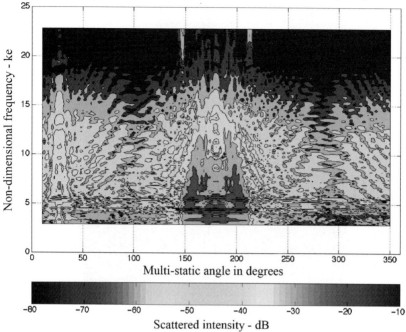

Multi-static angle in degrees

Scattered intensity - dB

Fig. 17.2 The multi-static "target strength" of the cylinder described in the text is displayed in a plane defined by the nondimensional frequency (product of the acoustic wavenumber, k, and the cylinder's diameter, e) and the multi-static angle. The angle between an incident sound wave and the direction to multiple receivers is shown on the abscissa. The intensity of the sound at the receiver is coded in color. Red represents high scattering levels, blue is low. Higher target strengths occur near the specular angle of 30° and in the forward scattering direction. Useful target strengths occur at many angles and nondimensional frequencies, however, suggesting that a sampled set of these data could be used to extract information about this target

between those who model scattering from marine organisms and those who measure scattering, either in a laboratory tank or in situ.

A prototype sensor has been built to make measurements on individual organisms during a vertical cast from a ship (Fig. 17.3). The purpose of making these measurements will be to take a step toward reconstructing scatterer shapes, sizes, and compressibility from multifrequency, multi-static acoustical scattering data. The prototype was designed for 2-D measurements, but it can easily be expanded to 3-D by adding sensors at locations with viewing angles outside of a single horizontal plane when it is understood how to the extract information desired from the 2-D images.

The challenge is to extract information from acoustical *echoes* and to interpret it in biophysical terms that can be used to characterize the object from which the scattering originates. Hopefully, this can lead to new approaches to the remote classification and identification of targets to genera or species. Acoustics, unlike optics, always penetrates the interior of a marine organism, thus in principle at least, there is a chance that even details of internal structures can be resolved.

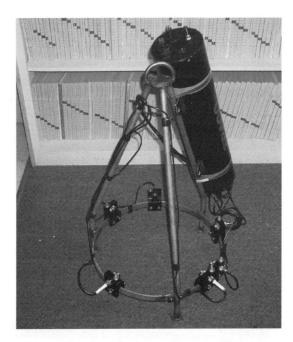

Fig. 17.3 A circular frame holds several broadband transducers aimed at the center of the circular plane area. The long black cylinder is a pressure case for electronics. This configuration uses two transmitting transducers in order to cover a wide frequency band (*bottom right*). Pulses are directed across the diameter of the measurement plane. Multiple receiving transducers are located at known, adjustable, angles in the horizontal plane and are used to intercept echoes when one (or more) scatterer is detected in a very small volume in the center of the circle formed by the frame

With a library of acoustical *fingerprints* such as the one shown, or better yet with an appropriate set of theoretical signatures based on first principles of the relevant scattering physics, one might use well-established inverse methods to derive robust estimates of size, abundance, and in some cases, the physical properties of the organisms, perhaps even for their internal parts.

17.3.2   The Seabed as a Marine Habitat

Characterizing the benthic habitat is the subject of exceptional interest at present. In many countries, a statutory requirement has been established that requires mapping and characterization of those habitats by agencies whose charter involves managing the coastal zone ecosystem.

For now, it seems likely that charts made with commercially available echosounders and side-scan sonars, supplemented and interpreted with directed ground truth measurements will be the toolset of choice for the characterization of coastal zone habitats. In part, that such charts can be prepared with relatively little expense and with sensors that are readily available will continue to drive scientists and managers to use these tools.

Acoustical methods and imaging optics are both being used for this purpose, but with today's technology there are significant opportunities for error in interpretation of the scattering data. Commercial systems and software are currently being used to convert echoes into bottom *type*. They often use empirical relationships that may be nonunique, even though local ground truth is also used in the processing. Rigorous, fully validated quantitative methods must be developed if acoustical classification of the seabed is to reach its full potential.

The seabed is not a static environment, a fact that is presently largely ignored. Numerous processes, both physical and biological are constantly acting to change those characteristics that control the amount of scattering one gets from the ocean floor. The temporal spectrum of processes that can change the seabed ranges from seconds to eons. Relevant physical processes include waves, tides, and currents that cause resuspension and redeposition of sediments. Biological processes include burrowing, sediment cementation by exudates from biological organisms, and the formation of supersaturated oxygen in the pore waters of sandy sediments, where subsequent formation of bubbles is mediated by both physical and biological activity. The physical, chemical, and biological parameters that describe this special ocean ecosystem are constantly changing, and the algorithms that allow one to interpret acoustical scattering from the water–sediment interface and below, must reveal those changes. Charts of acoustical properties made today, may well be quite different later today, tomorrow, in a week, or a year from the time the data are collected.

All is not lost however. It seems likely that in the future, acoustical measurements and methods alike could be based on the physics of the sound-scattering process at, and within, the seabed. With an understanding of that process, and if users insist on having better tools, then increasingly reliable and more capable sensors and algorithms for this purpose should become available over the next 2 decades. A recent ICES Cooperative Research Report (Anderson et al. 2007) addresses today's technology and the potential for tomorrow's technology as it relates to the characterization of the marine habitat and what lives there. This publication addresses the subject in considerable detail. If one is interested in seafloor habitat characterization, the report should be mandatory reading.

17.3.3 Primary Production in Shallow, Sandy Benthic Environments

At one time, ecology textbooks suggested that shallow, sandy littoral zones were nearly devoid of life. This is proving to be wrong. If one would like to measure primary production in these environments, there may be a relatively simple way to address the problem acoustically.

A simple, broadband acoustical sensor has been used in a laboratory environment to demonstrate that primary production can be detected in marine beach sands. At the present early stage in the development of this method, the concept is only being tested in situ on sandy sediments in shallow water.

In the laboratory, a glass aquarium was placed in a window. The window, on which normal sunlight was incident, was masked with a neutral density filter to simulate ambient light intensity at ca. 20 m. The scattered sound intensity from a sand bed collected from a local beach changed by a factor of 100 over each diel period (Fig. 17.4).

In the lab, acoustic scattering levels were strongly correlated with diel changes in sunlight, but with a lag (Holliday et al. 2003a, 2004). The root cause of the observed variability in sound scattering was determined to be primary production by resident episammic flora and the epipelon. These are different kinds of micro-algae that live on and between individual sand grains. With enough light, they produce O_2 supersaturation of the pore water in the sand. In the Gulf of Mexico dissolved O_2 levels in the top few millimeters of sand have been measured at levels six times those in the atmosphere. Several mechanisms can induce bubbles to form in supersaturated pore waters. Sometimes the bubbles coalesce, rise through a tortuous route to the sediment – water interface and either get trapped on the sediment surface or break loose and rise to through the water column to the sea surface. The presence of bubbles in the interstitial pores between the sand grains, and on the sand's surface is what changes the sound scattering during a daily light cycle.

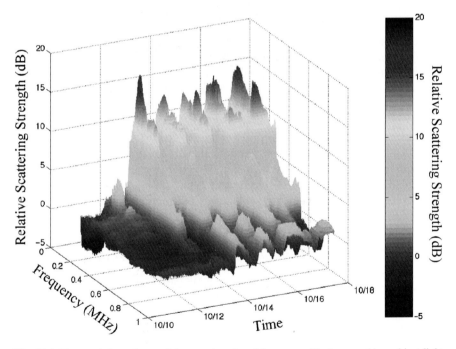

Fig. 17.4 Temporal dependence of the sound scattered from a sand bed exposed to ambient light over a period of ca. 6 days in October 2003. The spectra, collected at 1-min intervals (32 ping coherent averages) were normalized to a spectrum collected after the sand's surface was smoothed at the beginning of the experiment. Thus, the ordinate represents the evolution of scattering relative to the scattering at the time the sand bed was smoothed

A lightweight frame, designed for deployment on the seabed from a small boat serves as a mount for a small wideband acoustical transducer on a short arm (Fig. 17.5). The arm can be rotated to direct a wideband acoustical pulse toward the seabed at a known angle. A mount on which a camera can be placed to view the ensonified area can be seen at the opposite end of the arm on which the transducer is mounted.

We have observed large scattering changes in several shallow sandy locations on the shelf of northern and southern California, and those data are being analyzed in an attempt to make sure that the changes were from the production of O_2, and not to wave or current-related sediment resuspension, bioturbation, the migration of benthic fauna through the ensonified area, or uneven settling of the instrument frame. It is hoped that visual monitoring of the seabed will be helpful in determining whether primary production can be a factor in changing seabed sound scattering in the sea, as was the case in the laboratory.

This experiment has been included as an example of how acoustical technology can be used to gain insight into very small ecosystems – i.e., one that supports what is now known to be a complex community of tiny organisms that live in the pore

Fig. 17.5 A small, wideband acoustical transducer (*arrow*) is shown mounted on a simple frame that can be set on the seabed from a small boat. The white case contains electronics to transmit pulses at a preset angle to ensonify the water – sediment interface. The echoes are detected and recorded in a digital format in a solid-state memory. The black cylindrical case contains a battery. A mount on the right end of the arm at the top is designed to hold a camera

water between, and on sand grains at a beach. This project is in its infancy, but the hope is that above some threshold, if there is ever a scientific need, one might be able to monitor primary production remotely in this way. This brief discussion was also included here to demonstrate that not all new technologies and sensors must be complex or expensive.

17.3.4 Integration of Acoustics and Optics

At the University of Massachusetts at Dartmouth, under the guidance of Brian Rothschild and Kevin Stokesbury, a bottom-moored system is being developed to complement traditional methods of assessing groundfish in the "dead zone" (Miksis-Olds et al. 2006). The "dead zone" is the first few meters above the seabed where traditional, hull-mounted echosounders on ships fail to operate properly because of interference from bottom scattering. The Ocean Groundfish Observatory (OGO) is a bottom-mounted sensor array that combines the strengths of both acoustics and optics, using a multi-beam 200 kHz acoustical system in up-looking and side-looking modes to measure the ventral or side-aspect target strengths of individual fish. The size of the fish, and often the species, are determined with a low-light-level imaging camera and parallel laser beams as seen in the left panel of Fig. 17.6. The laser beams give one the ability to establish a spatial scale in the vicinity of the fish. With this system, target strength data, along with fish size, orientation, tilt, and target aspect angle can be accumulated in situ for local species over long periods of time.

The echo from the fish (Fig. 17.6) appears at a distance of about 1.25 m (4 ft) from the transducer. Because the system is calibrated, the fish's target strength, near

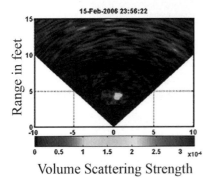
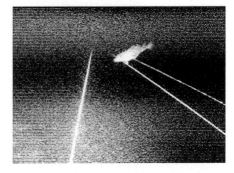

Fig. 17.6 An echo collected with the multi-beam 200 kHz sonar in the UMass Dartmouth's Ocean Groundfish Observatory (OGO) system is shown in the left panel. One of four low-light cameras on the bottom-mounted platform revealed that the acoustical target was a cod. Four parallel laser beams, three of which are in the photo of the fish in the right panel, are visible as light is scattered from plankton and detritus in the water. Their geometry allows one to estimate the fish's length to be 30.5 cm. These data were collected by Jen Miksis-Olds, Ernest Bernard, and Christopher Jakubiak (School of Marine Science and Technology, UMass Dartmouth) during a deployment in February 2006

side-aspect, can easily be calculated. What appears to be either multiple fish or a patch of plankton can be seen at longer ranges from the transducer. Those scatterers are outside the detection range of the low-light-level camera. If aggregations of fish were present, it should be possible to use traditional echo integration methods to assess their abundance, given the histograms being obtained for target strength at different angles and fish sizes. This is not often done in traditional survey work at target aspects other than dorsal. As statistics on target strength versus the angle of ensonification are accumulated during long deployments, along with size and species, data from this sensor system may eventually allow one to work with schools or aggregations from many angles of observation. Some measurement capability has already been implemented for examining the physical descriptors of the ecosystem, and the OGO system may also eventually be outfitted with a variety of sensors that will more completely characterize the ecosystem, including the plankton.

17.3.5 Resolving Fine-Scale Vertical Structure in an Ecosystem

Heterogeneity is one of the most common characteristics of aquatic ecosystems. Variability is present in time and space for virtually any parameter one wishes to use in describing the physics and chemistry of a marine ecosystem and the distribution of life therein. When data are collected to describe how some property, e.g., water density, temperature, or phosphate, varies with time, depth, or distance in the horizontal, the intervals at which the measurements are made must be sufficiently short to resolve the fastest changes. If this requirement is not met, then any pattern derived from the sampled data may deviate significantly from the real pattern. Once under-sampling occurs, no amount of averaging, filtering, or complicated processing of the aliased data will allow one to recover the real pattern.

The rules for how close independent samples must be in time or space are described by theorems originally ascribed to Nyquist (1924) and Shannon (1948). Platt and Denman (1975) provide a scholarly discussion of these criteria in relation to ecosystem modeling. Those principles apply generally to any measurement wherein the desired result is a true representation of the spatial pattern of a measured attribute, how it varies with time, or both.

Percy Donaghay's Ocean Response Coastal Analysis System (ORCAS) profiler developed at the University of Rhode Island (URI) was specifically designed to ensure that one is able to measure simultaneously a wide range of ecosystem attributes to see how they depend on depth while having an assurance that aliasing is not a problem. Arrays of such profilers are used to attack the problem of aliasing in the horizontal. A product of Percy Donaghay's research team at the University of Rhode Island, the Ocean Response Coastal Analysis System (ORCAS) profiler is moored to the seafloor (Donaghay et al. 1992; Dekshenieks et al. 2001; Donaghay 2004; Babin et al. 2005). It uses positive buoyancy to un-spool wire on the "up-cast." During a "down-cast" the wire is wound onto the winch spool. The result is a carefully controlled profiling rate that can be as slow as 0.5 cm/s. The water column is

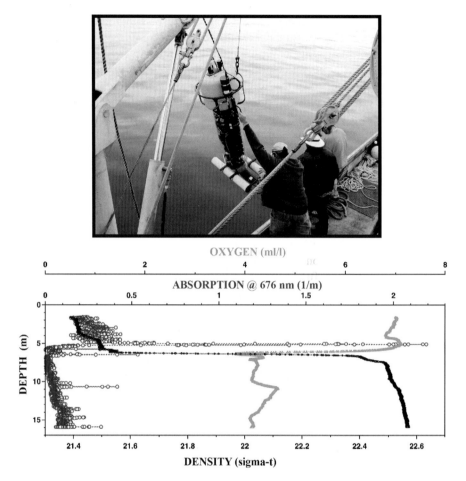

Fig. 17.7 The ORCAS profiling instrument platform carries a suite of high-technology sensors selected to study primary and secondary production in relation to fine-scale vertical structure in the physics and chemistry of the coastal ocean. Sub-meter scale sampling is done during water-column profiles whose extents and durations can be programmed remotely by telemetry. Data profiles can be requested and sent to a shore station on command, or at programmed intervals. The top panel shows the deployment of an ORCAS profiler package. The bottom panel illustrates vertical profiles of dissolved O_2, light absorption at 676 nm, and water density collected at a station in East Sound, Orcas Island, Washington. These data are from an ONR multi-investigator study titled Layered Organization in the Coastal Ocean (LOCO) and were collected by Percy Donaghay and Jim Sullivan, Graduate School of Oceanography, URI

sampled by whatever sensor suite the ORCAS carries. It is capable of autonomous operation and some adaptive behavior is built into the onboard controller. When the ORCAS platform (Fig. 17.7, top panel), instrumented with a wide variety of optical, acoustical, and physical sensors, nears the surface as it executes a profile, an antenna sticks out of the water and transmits the accumulated data to shore. At that point new instructions can be transmitted to the buoy about its next task. If there are no new instructions, the buoy returns to a preset depth near the bottom and waits for the time of next programmed sampling event. Additional rules that would allow the ORCAS to make autonomous, data-dependent decisions could easily be implemented. The system can also be controlled with interactive commands being forwarded by operators on shore after they examine the data it transmits via a telemetry link. In either mode, one can achieve unaliased centimeter-scale vertical sampling resolution with multiple ecosystem sensors in the water column.

Depending on the problem at hand, a typical sensor suite deployed with an ORCAS profiler includes a CTD and sensors for O_2, spectral absorption, spectral attenuation, spectral scattering, backscattering at 532 nm, chlorophyll-*a* fluorescence, and CDOM fluorescence. Sensors are also available for other wavelengths of light, and for such parameters as Eh and pH.

Optical spectra are used to characterize the particulate material (e.g., provide estimates of particle size distribution and pigment composition) as well as to estimate the quantity of particulate material (e.g., phytoplankton chlorophyll concentration). The use of spectral information to sort out various water properties that contribute to the absorption and scattering of light is a relatively new development in optical oceanography. To date, much of the emphasis in optics has been on using ratios of scattering at a limited number of discrete wavelengths. Although the underlying physics is different, this approach parallels the development of inverse algorithms in acoustical oceanography, where spectral levels and shapes are used to determine swimbladder sizes in fish, and size-abundance distributions for zooplankton. It seems likely that, as the use of stable, calibrated, underwater hyperspectral sensors becomes routine, more sophisticated inverse algorithms that use quasi-continuous spectra will naturally follow. Also, in parallel with developments in acoustics, one can predict that multi-wavelength, multi-static sensors will greatly increase the amount of information that can be derived from optical sensors.

A chemistry package, designed by Al Hanson's team at SubChem Systems has been successfully deployed by the ORCAS. It has been used to measure the vertical structure of nutrients in situ (e.g., NO_3, NO_2, PO_4, NH_4, silicate, and iron [II]) with exceptionally high resolution. Other sensors sometimes used on the ORCAS measure current shear, turbulence, and the spectrum of downwelling light. Wideband acoustical zooplankton sensors have also been deployed on these profilers, and water bottles have been used to collect phytoplankton and small zooplankton from sub-meter thick layers.

Profiles made with the ORCAS in East Sound, Washington, revealed a complex water-column vertical structure (Fig. 17.7, bottom panel). The reader is left the exercise of determining how well the environment would have been characterized with bottle or net samples from, say, 5- or 10-m discrete depth intervals. Such was the state of the art not so many years ago! Thin layers in optical absorption were observed at several distinct depths and each was correlated with features in the profiles of dissolved oxygen and water density. Data for the absorption profile at 440 nm and a discussion of the vertical distribution of zooplankton at different sizes can be found in Holliday et al. (2003b). An ability to simultaneously resolve the fine-scale features illustrated in Fig. 17.7, and to study how zooplankton and fish respond to phytoplankton, the physics and chemistry of the ocean, and distributions of nutrients at these same scales is likely to change our perception of how life really is in the sea. The next step, already being addressed at elementary levels, is to understand how plankton behavior affects survival and reproduction in fish larvae, and ultimately in populations across the marine food web in general.

Working with oceanographers from University of Rhode Island and UC Santa Cruz over several years, we examined thin layers in about a dozen different kinds of coastal ecosystems around the United States. Layer structures and the processes that led to their formation and dissipation differ from site to site, but thin layers are very common features in coastal ecosystems. There is also good evidence that they exist offshore. Based on the ORCAS optical data and data from a Tracor Acoustical Profiling System (TAPS) that was deployed in an upward-looking multifrequency sounder mode nearby, it is clear that vertical resolutions for studying both phytoplankton and zooplankton had to be ca. 10 cm or better if we were not to have totally missed detecting most of the plankton biomass when this kind of layer exists. With traditional methods, vertical structure in both water density and O_2 would very likely have gone unnoticed as well. Thin plankton layers form a food resource comparable to the patches and the broad chlorophyll maximum near the pycnocline. Cassie (1963), Lasker (1975), and Mullin and Brooks (1976) once determined from trophic arguments that structures containing biomass that was higher than the average background were critical for larval fish survival and recruitment. For thin zooplankton layers, over 78% of the zooplankton biomass has been observed in a complex of three such sub-meter thick layers at a location off southern California, where the water column was 100 m deep (Holliday et al. 1998).

17.3.6   TAPS-8 in the Coastal Gulf of Alaska
and the Bering Sea

Having seen numerous examples of the zooplankton response to phytoplankton thin layers, we were reluctant to undertake the project discussed next. Only one depth could be instrumented. There was a significant chance that a single sensor would not be at the right depth to be detected if the biomass was distributed in a thin horizontal layer. We were, and are still, convinced of the essential value of having enough vertical resolution to detect all of the zooplankton biomass in a profile of the water column. We were finally convinced to proceed with the project because doing so would introduce multiple frequency zooplankton sensors into an environment where the assimilation of data from this kind of sensor by predictive models is going to be critical. It was also an opportunity to once again prove that communication of relatively large acoustical data sets from very remote locations to shore-based scientists is actually practical, and that near real-time data analysis and distribution of the results are also possible with today's technology.

With those caveats, we now describe an ongoing project involving the author, Charles Greenlaw, Jeff Napp (NOAA/NMFS/AFSC); and Phyllis Stabeno, Bill Floering, and Bill Parker (NOAA/PMEL); and Tony Jenkins (University of Washington). High-frequency acoustical zooplankton sensing technology was used in the Coastal Gulf of Alaska to monitor zooplankton and micronekton at a single depth for several months in each of 2002, 2003, and 2004. The duration of the deployment each year was driven by ship schedules. For programmatic reasons the TAPS-8 system (Fig. 17.8) was moved to the Alaskan shelf in the Bering Sea for the 2006 field season. In 2007, the TAPS-8 sensor was successfully deployed and retrieved from the difficult-to-access Bering Sea.

When deployed, the beams for this system are directed horizontally from the vertically suspended in-line mooring cage. There are several operating modes, but in one, samples are taken at intervals extending about 16 m horizontally from the cage. This mode was designed to test a hypothesis that algae gradually accumulates on the cage and that it may provide an attractive food source for grazing by herbivores and omnivores that migrate to the depth of the cage each evening. Given our results to date, on average the effect seems to extend to a distance of ca. 1.5 m from the cage. Unfortunately, the cage undoubtedly rotates, and it is not instrumented to distinguish between upstream and downstream orientations of the acoustic beams.

Biovolume variations for zooplankton and micronekton were measured at ca. 20-min intervals with the TAPS-8 in the Gulf of Alaska from mid-April through August in 2004. For practical purposes within the scope of this contribution, biovolume is the same as plankton displacement volume. The biovolume-size distribution is illustrated during a single week in April for organisms shaped like copepods and for krill-like shapes, at this location likely euphausiids (Fig. 17.9, left and right panels). The instrument package autonomously sampled a single depth near 20 m in the upper water column every 24 min in 2002 for >130 days, every 20 min in 2003 for >113 days, and every 20 min in 2004 for >142 days. Data were recorded internally and accessed on recovery. These dates were determined by ship schedules. In addition to the acoustics, we also obtained data from CTDs, thermistors, fluorometers, nitrate sensors, a met station, and current sensors.

In 2006, we moved the sensor to a mooring on the Alaskan shelf in the Bering Sea. It was deployed in mid-April and retrieved at the end of September. The sensor was located on the wire at 17 m on a subsurface mooring that was separate from a large surface buoy equipped with a met station and satellite telemetry. The TAPS reported its data to the surface mooring via a digital acoustic modem instead of a

Fig. 17.8 Mooring cage configuration with TAPS-8 acoustical zooplankton sensor and two battery packs. The array of TAPS transducers is on the flat part of the pressure case nearest the reader. The other two cases contain batteries. When moored, the cage is vertical, or "in-line", with the mooring wire

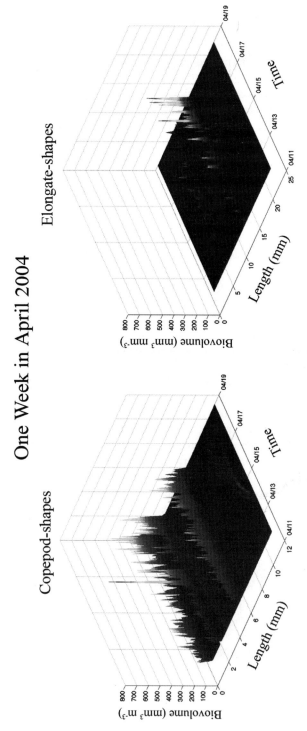

Fig. 17.9 The variability in acoustically estimated zooplankton biomass, expressed as biovolume, is shown for organisms with shapes approximating those of calanoid copepods (*left panel*) and for elongate scatterers that are similar in shape to krill or euphausiids (*right panel*). These data are for a single week (April 12–19) in 2004. The TAPS-8 was located at 20 m depth on a mooring in the middle of the shelf, south of Seward, AK in 180 m of water. This location is on the southern edge of the Alaska Coastal Current

wire. Internal recording was also retained as a backup. From the surface buoy, the data were transmitted to an Iridium satellite and relayed to NOAA's Alaska Fisheries Science Center in Seattle. From there, they were automatically distributed to an FTP site at the BAE Systems lab in San Diego. The data were processed several times a week. Multiple- frequency volume scattering measurements were transformed to biovolume distributions similar to those shown, and that information was e-mailed, along with summary interpretations, to the other principal investigators.

The TAPS-8 system is one of several state-of-the-art systems currently under development for autonomous estimation of zooplankton abundance. The ZOOplankton Visualization and Imaging System, ZOOVIS (Benfield et al. 2003) has its roots in the Video Plankton Recorder or VPR (Davis et al. 1992). The Laser Optical Plankton Counter (LOPC) builds on the original OPC design, uses laser-imaging methods, is in its second generation, and is available from commercial sources (Herman 1992). We consider the acoustical and optical sensors complementary, rather than competitive.

17.4   Conclusions

Whether a TAPS, or some other device, it is anticipated that clusters of sensors appropriate for describing an ecosystem on sub-meter scales must become commonplace. To be useful, such sensors will have to be adapted to a variety of platforms. For example, one might deploy the TAPS-8 on a wave-driven profiler or one of the University of Rhode Island's ORCAS profilers in order to get full water-column zooplankton and micronekton profiles. Advanced sensors should also be put on drifters and gliders. For some kinds of sensors and research programs, it would be both convenient and cost-effective to use methods for deploying high-tech sensors in modes similar to those being used by the University of Rhode Island's Tom Rossby, whose ADCPs are used on ships and ferries of opportunity for extended periods.

Physical parameters such as temperature, salinity, and turbulence must be measured with vertical resolutions similar to those now achievable with the ORCAS and TAPS sensor systems. This is *not* currently common practice in the physical oceanography community.

Optical sensors are being used to acquire high-resolution data on phytoplankton. It would not be a great stretch of the imagination to find a way to trigger the acquisition of bottle samples in interesting phytoplankton, nutrient, or chemical fine structure, on command, or better yet without human intervention. In some cases, even when organisms are detected with advanced sensors, a small sample of the water and the life in it greatly adds to the information we can extract from a profile. At present, there are no good ways to get physical samples of sub-meter thick vertical structures in the water column. This challenge needs to be addressed.

We have not attempted to review much of the progress in making in situ measurements of nutrients and ocean chemistry. We just note that Al Hanson at URI is making rapid, substantial advances in this area, and has deployed his sensors on Percy Donaghay's ORCAS profiler very successfully in support of thin layers research. Similar sensors have also been very successfully used on AUVs.

Insofar as zooplankton acoustics goes, an operational TAPS-16 has been built, as has an operating prototype of a 96-frequency sensor. This type of sensor would achieve an order of magnitude better resolution in animal size than we have with the 8-frequency unit, far surpassing a 21-frequency sensor (MAPS) that we used for a number of years in the 1970s and 1980s. This basic technology is the subject of numerous papers in scientific journals beginning in about 1972, and the basic principles appear in several textbooks. We predict that adding another independent dimension, multi-static measurements, to the multifrequency ones should further enhance our attempts to achieve our long-term goal of better understanding how aquatic organisms not only survive, but also often thrive in the sea.

Finally, considerable care is in order when reviewing crazy outlandish proposals. I cannot resist closing with a quote from a reviewer of my first proposal to the NSF. The subject was the use of high-frequency acoustics to study zooplankton. His comment was terse and to the point – "There is no sound at a megahertz." Fortunately, someone else thought pursuing my outlandish idea was worth the risk. We urge reviewers and panel members alike to have an open mind and to be willing to take a few risks when faced with the need to make funding recommendations. Without risks being taken, few really novel sensors will be developed during the next 2 decades.

Acknowledgments Some of the projects we discussed, especially those in the last part of this contribution have origins that can be traced to incidental discussions that took place over 40 years ago. Over the years, Charles Greenlaw, David Doan, and numerous technicians and summer interns helped turn raw concepts into functioning sensors. Percy Donaghay, Rick Pieper, and Paul Smith have provided invaluable reviews and comments as we have tried to understand how best to use new sensor technologies to discern how the zooplankton and fish are distributed in relation to their environment.

Funding has been from multiple sources, including ONR Contract N00014-00-D-0122 and its predecessors to BAE Systems; NOAA Contracts AB133F055U3288 and 50ABNF100050 to BAE Systems; and ONR Grant N00014-07-1-0639 to the University of Rhode Island.

References

Anderson J, Holliday DV, Kloser R, et al. (2007) Acoustic seabed classification of marine physical and biological landscapes. ICES Coop Res Rep 286:183

Babin M, Cullen JJ, Roesler CS, et al. (2005) New approaches and technologies for observing harmful algal blooms. Oceanography 18:210–227

Benfield MC, Schwehm CJ, Fredericks RG, et al. (2003) ZOOVIS: A high-resolution digital still camera system for measurement of fine-scale zooplankton distributions. In Strutton P and Seuront L (eds) Scales in aquatic ecology: measurement, analysis and simulation. CRC Press, Boca Raton, FL

Cassie RM (1963) Microdistribution of plankton. Oceanogr Mar Biol Annu Rev 1:223–252

Davis CS, Gallager SM, Berman MS, et al. (1992) The video plankton recorder (VPR): Design and initial results. Archiv Hydrobiol – Beih Ergeb Limnol 36:67–81

Dekshenieks MM, Donaghay PL, Sullivan JM, et al. (2001) Temporal and spatial occurrence of thin phytoplankton layers in relation to physical processes. Mar Ecol Prog Ser 223:61–71

Diachok O, Wales S (2005) Concurrent inversion of geo- and bio-acoustic parameters from transmission loss measurements in the Yellow Sea. J Acoust Soc Am 117:1965–1976

Diachok O, Smith P, Wales S, et al. (2005) Bioacoustic absorption spectroscopy: The promise of classification by fish size and species. J Acoust Soc Am 118:1907–1908(A)

Donaghay PL (2004) Profiling systems for understanding the dynamics and impacts of thin layers of harmful algae in stratified coastal waters. Proceedings of the 4th Irish Marine Biotoxin Science Workshop 44–53

Donaghay PL, Rines HM, Sieburth JM (1992) Simultaneous sampling of fine scale biological, chemical and physical structure in stratified waters. Arch Hydrobiol 36:1–14

Holliday DV, Edwards RL, Yayanos AA, et al. (1980) Problems in marine biological measurements. In Diemers FP, Vernberg FJ, Mirkes DZ (eds) Advanced concepts in ocean measurements for marine biology. University of South Carolina Press, Columbia, SC

Holliday DV, Pieper RE, Greenlaw CF, et al. (1998) Acoustical sensing of small scale vertical structures in zooplankton assemblages. Oceanography 11:18–23

Holliday DV, Donaghay PL, Greenlaw CF, et al. (2003a). Advances in defining fine- and microscale pattern in marine plankton. Aquat Living Res 16:131–136

Holliday DV, Greenlaw CF, Thistle D, et al. (2003b). A biological source of bubbles in sandy marine sediments. J Acoust Soc Am 114:2317–2318(A)

Holliday DV, Greenlaw CF, Rines JEB, et al. (2004) Diel variations in acoustical scattering from a sandy seabed. Proc ICES Ann Sci Conf ICES CM 2004/T:01

Herman AW (1992) Design and calibration of a new optical plankton counter capable of sizing small zooplankton. Deep-Sea Res 39:395–415

Lasker R (1975) Field criteria for survival of anchovy larvae: The relation between inshore chlorophyll maximum layers and successful first feeding. Fish Bull 73:453–462

Makris NC, Ratilal P, Symonds DT, et al. (2006) Fish population and behavior revealed by instantaneous continental shelf–scale imaging. Science 311:660–663

Miksis-Olds JL, Bernard E, Jakubiak CJ, et al. (2006) Combining acoustics and video to image fish in the acoustic dead zone. J Acoust Soc Am 120:3059(A)

Mullin MM, Brooks ER (1976) Some consequences of distributional heterogeneity of phytoplankton and zooplankton. Limnol Oceanogr 21:784–796

Nyquist H (1924) Certain factors affecting telegraph speed. Bell Syst Tech J 3:324–346

Platt T, Denman KL (1975) Spectral analysis in ecology. Annu Rev Ecol System 6:189–210

Shannon CE (1948) A mathematical theory of communication. Bell Syst Tech J 27:379–423, 623–656. This work was republished as Weaver W, Shannon CE (1949) The mathematical theory of communication. University of Illinois Press, Urbana, IL

Sullivan JM, Twardowski MS, Donaghay PL, et al. (2005) Use of optical scattering to discriminate particle types in coastal waters. Appl Opt 44:1667–1680

Zakharia ME (1999) Improving echograms' resolution by wideband pulse compression and dynamic focusing. J Acoust Soc Am 105:051(A)

Zakharia ME (2001) Wideband and correlation techniques and their application to fisheries acoustics: Existing prototypes and future trends. J Acoust Soc Am 109:2304(A)

Chapter 18
Acoustic Methods: Brief Review and Prospects for Advancing Fisheries Research

Kenneth G. Foote

Abstract Acoustic methods are widely used in fisheries research, often providing vital information that can be obtained in no other way. In reviewing active methods, phenomena of sound scattering are first described. The means of ensonification and detection, the generic sonar, is described. Examples include the traditional scientific echo sounder and the following six classes of sonar: multibeam, sidescan, acoustic lens-based, parametric, synthetic aperture, and conventional low-frequency sonars. Methods of data processing, quantification, and data interpretation are addressed. In reviewing passive methods, sounds produced by organisms are exemplified. The traditional means of detecting sound, the hydrophone, is then described together with various configurations of hydrophones. Methods of data analysis and classification are outlined. Calibration is addressed separately for sonars and hydrophones. Applications of the various methods are cited. Potential applications of new, improved, or refined acoustic methods to outstanding problems in fisheries and fisheries habitat research are indicated.

Keywords fish acoustics · fishery acoustics · ecosystem acoustics

18.1   Introduction

Acoustic methods are popular in fisheries research because these are fishery-independent, quantitative, noninvasive, remote, rapid, and synoptic (Forbes and Nakken 1972). They are capable of sensing multiple scales, from centimeters to thousands of kilometers in sailed distance. They can sense both organisms and their habitat, including water column, bottom, and sub-bottom.

While the present nominal subject is fish, this is interpreted in the broader sense of aquatic organisms, both individuals and aggregations. Certainly any organism possessing physical properties that differ from those of the ambient immersion

K. G. Foote
Woods Hole Oceanographic Institution

R.J. Beamish and B.J. Rothschild (eds.), *The Future of Fisheries Science in North America*,   313
Fish & Fisheries Series, © Springer Science + Business Media B.V. 2009

medium, both freshwater and seawater, will respond to an incident acoustic wave by generating echoes (Medwin and Clay 1998). Thus, if the mass density, elasticity, compressibility, and/or a sound speed in the organism differ from the respective medium properties, an echo will be formed. The incident sound will be scattered in other directions too.

Because of differences in physical properties between organisms and ambient medium, sonar can be used in the traditional, active mode for detection of organisms. A signal of known type and power level can be transmitted by means of a set of controlling electronics and coupled transducer. The signal will propagate away from the transducer. As it encounters regions of the medium with differing properties, also called heterogeneities, the sound is generally redistributed, or scattered, in all directions. This makes possible detection of the scattered sound with a transducer and suitable receiver electronics. If the detection is accomplished with the same transducer that was used in transmission, the sonar is genuinely monostatic. If the receiving transducer is separate from the transmitting transducer, the sonar is bistatic, although if the two transducers are located near to one another, the sonar will effectively be monostatic. It is assumed in the following that the sonar operation is monostatic.

Three conditions are recognized for being able to detect organisms by active sonar: that they can be ensonified, that their echoes can be distinguished, or resolved, from other echoes, and that the level of scattering is sufficiently strong relative to the background noise. The art of fisheries acoustics is choosing the right sonar, operating parameters, and signal processing to ensure reliable, unambiguous detection, and in knowing what the limits are for error estimation. It is also a powerful impetus for the design and development of new sonars or refinement of existing sonars.

In a number of important cases, active acoustic detection of organisms is not particularly effective or even feasible, for example, when fish are near the bottom or bottom structures such as reefs. In some other cases, active acoustic detection may be highly undesirable, for example, in studies of marine mammals, although occasionally achieved with or without intent (Dunn 1969; Love 1973; Levenson 1974; Lucifredi and Stein 2007). An alternative to the use of active acoustics is passive acoustics. Insofar as marine life makes sound, this can, in principle, be detected.

The previous description of sonar detection applies to passive detection too, but without active ensonification. Rather, a sonar transducer, e.g., hydrophone, only listens, and received signals are processed to detect, quantify, and sometimes even classify the causative organisms. It is noted that the sound produced by organisms may be deliberate, as in the generation of echolocation or communication signals (Tyack 2000; Tyack and Clark 2000), or it may be incidental, as in nest-building on the bottom or in swimming, through production of a hydrodynamic disturbance (Ezerskii and Selivanovskii 1987).

In all instances of active and passive detection of fish and other organisms, noise is present. It is intrinsic to the immersion medium, transduction, and electronics. In active acoustics, it may arise from scattering by entrained air bubbles (NDRC 1946; Leighton 1994), marine snow or suspended sediment (Schaafsma 1992; Thorne et al. 1993), temperature and salinity microstructure (Goodman 1990; Lavery and

Ross 2007), and turbulent microstructure (Seim et al. 1995; Lavery et al. 2003), among other causes. When produced by a large number or aggregation of scatterers, the noise is called reverberation. The noise, as unwanted signal, may also be due to other, uninteresting water column or surface organisms, from which interesting echoes or sounds must be distinguished. This problem is largely, but not completely, ignored in the following. As suggested, choice of the acoustic instrument or method can be decisive for any one application. It almost always involves compromise, optimization, and/or trade-offs to maximize the detection of interesting signals and minimize interference from uninteresting signals, or noise.

The aim here is to describe a number of active and passive acoustic methods that can be used to detect and quantify aquatic life, and in some cases to classify this too. These methods include the traditional tools of the scientific echo sounder and hydrophone because of the continued development, viability, and versatility of associated methods. Other tools will also be described, as these are currently expanding the researcher's repertoire, enabling new knowledge to be acquired about fish in ecosystems, ultimately to foster sound management of both fish resources and their habitat.

18.2   Active Acoustic Methods

18.2.1   Scattering Phenomenology

The basis for active acoustic detection of aquatic organisms is their scattering of incident sound. It is the nature of sound, as well as other wave phenomena, to propagate outwards from a source. When the wave intercepts a region of space whose essential properties differ from those of the ambient medium, assumed to be homogeneous, secondary waves are generated (Rayleigh 1896). The outwards propagation of these secondary waves away from such a heterogeneity is called scattering. In the case of sound, heterogeneities, or scatterers, are defined by the properties of mass density and sound speed. The sound speed can be that of longitudinal and/or transverse waves. Sound speed differences can also be expressed in terms of elastic constants, elastic modulii, or compressibility.

The magnitude of scattering depends roughly on the contrast in properties between the ambient medium and heterogeneities, occasionally principally on the impedance, or product of mass density and longitudinal-wave sound speed. It also depends on the size and shape of the scatterer relative to the wavelength of incident sound.

Rather hard and dense materials will scatter sound, but a material such as steel has a mass density of about 7,800 kg/m^3 and sound speed of 6,000 m/s. Relative to the nominal properties of water, namely 1,000 kg/m^3 and 1,500 m/s, the ratio of impedances is of order 30. In the case of air, however, with mass density 1.3 kg/m^3 and sound speed 330 m/s, the ratio of impedances is of order 0.0003. This contrast is much greater than that of steel and water, suggesting the effectiveness of air in scattering sound (Minnaert 1933). This is borne out by scattering from the surface and by scattering from air bubbles entrained by breaking waves, which can degrade the performance of surface-deployed sonars when sufficiently numerous (Novarini and Bruno 1982).

As active scatterers, fish are conveniently divided into two classes, as they possess or lack a swimbladder. There are subdivisions of the swimbladder-bearing class. Among fish of commercial importance, physoclists have closed swimbladders, with inflation and deflation controlled through the action of the *rete mirabile*. Examples are cod (*Gadus morhua*) and rockfish (*Sebastes* spp.). Physostomes have open swimbladders, with an exterior duct controlled by a sphincter muscle. An example is herring (*Clupea harengus*). Some swimbladders are wax-filled, partially or completely. Examples are some mesopelagic fishes (Kleckner and Gibbs 1972) and orange roughy (*Hoplostethus atlanticus*) (McClatchie and Ye 2000).

In the case of fish lacking a swimbladder, the principal scattering parts may be relatively hard, dense bone or relatively compliant, low-density liquids in the liver or other tissues. Representative examples are mackerel (*Scomber scombrus*), some shark species (Baldridge 1970), and the castor oil fish (*Ruvettus pretiosus*) (Bone 1972).

Zooplankton have been classified as acoustic scatterers as they contain gas pockets or gas-filled organs, hard shells, or are liquid-like (Stanton et al. 1998).

The several classifications are keys to modeling acoustic scattering by a diversity of aquatic organisms exceeding the few taxonomic classes mentioned here. Some models are analytical, depending on the use of certain special shapes as approximations to the shape of the principal scattering organ. Others are numerical, without approximation of shape. Ultimately, the models are used to gain insight into scattering dependences or to quantify the magnitude of scattering for interpreting *in situ* scattering data.

18.2.2   Sonar Systems

In essence, a sonar system consists of a box of electronics and a coupled transducer or a transducer array. The electronics control the excitation of the transducer and reception of echoes. Typically, the electronics, the transducer, and the signal processing have been optimized so that the sonar can be operated reliably over long periods of time. Such sonars typically admit of calibration, enabling them to be used quantitatively to known levels of accuracy and precision.

Received signals are generally boosted at an early stage, in a process called preamplification, to minimize effects of additive electronic and other noise introduced at later stages. Attempts are generally made to compensate for the purely physical effects of geometrical spreading of the sonar beam and echoes and medium absorption through application of range compensation. In older systems this is done electronically through so-called time-varied gain (TVG) (Mitson 1983). In more recent systems, this is done through digital processing. Two particularly useful functions are the following. (i) Single-scatterer compensation. In the intensity domain, the received signal is multiplied by $r^4 10^{\alpha r/5}$, where r is the range in meters and α is the absorption coefficient in decibels per meter. In the logarithmic domain, the received signal is increased by adding $40 \log r + 2\alpha r$. (ii) Scattering-layer compensation. In the intensity domain, the received signal is multiplied by $r^2 10^{\alpha r/5}$. In the logarithmic domain, the received signal is increased by adding $20 \log r + 2\alpha r$. Other functions may also be available. Occasionally no compensation is applied to the received signal other than the mentioned preamplification.

Other forms of signal processing may be applied in the receiver, as when the transducer is composed of multiple elements that are mechanically and electronically distinct. In such cases, it may be possible to shape single beams or form multiple beams. As in the case of range compensation, this is typically done electronically or digitally. In the case of a lens-based sonar, the beams are formed instantaneously.

It may be imagined that transducers are critical elements in sonar operations. Their physical configuration and associated signal processing determine the type of sonar. This may be further distinguished by the acoustic frequency because of differences in transducer composition with frequency due to the need to generate and receive sound efficiently. The mechanism that is used to generate or receive sound at one frequency or in one frequency band is not necessarily useable in another band and often is not. It is recognized that there is a limit to how much a transducer can be scaled in dimensions to access other parts of the frequency spectrum because ultimately it is used in water, with its own properties of mass density and elasticity, or compressibility. A particular transducer is shown in cross section and in an array in Fig. 18.1. This is accompanied by two other examples of transducer arrays.

18.2.2.1   Scientific Echo Sounder

A scientific echo sounder is a relatively simple but especially powerful form of sonar system. The transducer may be integral in the sense that it consists of a single element or an array of elements that are wired together electrically, hence can form only a single beam. The transducer may be compounded of distinct elements in an array, allowing two or four beams to be performed.

In dual-beam processing (Ehrenberg 1974, 1979), the transducer array is typically circular, with a central core of elements and a concentric outer ring of elements. All elements are used in transmission to form a rather narrow beam. In reception the central core of elements forms a broad beam and the entire array forms a narrow beam. This enables the angle of resolved targets relative to the transducer axis to be determined, hence the backscattering cross section too.

In split-beam processing (Ehrenberg 1979), a circular or square array of elements is divided into quadrants. All elements are excited together to transmit a narrow beam. In reception, each quadrant signal is received separately and used to form half beams. The differences in phase of port and starboard beams and fore and aft

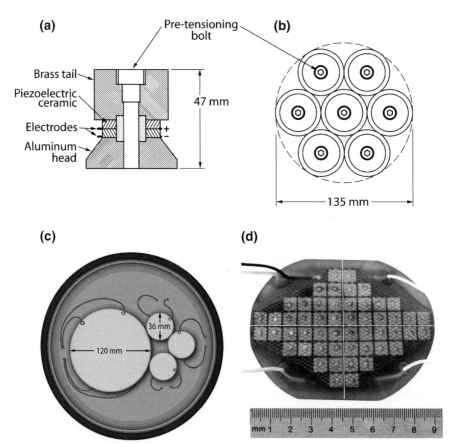

Fig. 18.1 Illustration of a transducer element and three arrays. (**a**) Schematic diagram of the cross section of a broadband transducer element nominally spanning the frequency range 25–40 kHz, developed by RESON, with principal radiation in the downward direction as oriented here (Foote 1998). (**b**) Sketch of array geometry using elements of the type shown in (**a**). (**c**) Diagram of a 120-kHz transducer array, indicating wiring of elements by exaggerated solder joints on the conducting upper surface of each element, before molding with polyurethane (R. McClure, BioSonics, Inc.). (**d**) Photograph of an elliptical 200-kHz transducer array before molding with polyurethane (P. Nealson, Hydroacoustics Technology, Inc.)

beams allows the respective angular positions of a resolved target to be determined. The beam pattern in the direction of the target is thus known and can be subtracted, in the logarithmic domain, from the echo strength to determine the target strength. Over a succession of pings, targets can be tracked (Brede et al. 1990).

The primary data provided by an echo sounder are exemplified by the echogram. This aligns successive received signals in a display, as in Fig. 18.2. Interpretation of this display begins with an understanding of how the coupled transducer is being used. In many applications, it is mounted on or towed from a moving vessel. If oriented downwards, the echogram images the water column along vertical sections,

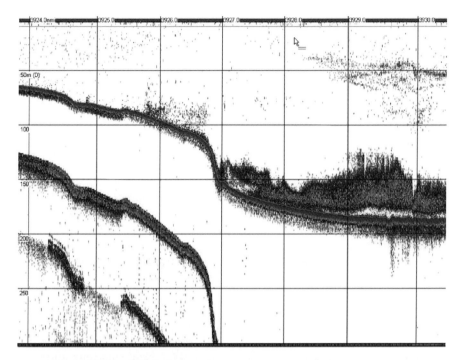

Fig. 18.2 Echogram collected with the EK500/38-kHz scientific echo sounder system, showing a dense, near-bottom aggregation of herring (*Clupea harengus*) in the Gulf of Maine over the approximate depth range 150–180 m. The diffuse scattering layer at 50 m is believed to be plankton. The total displayed depth range is 0–300 m, but without data from the surface to the transducer depth extended by several meters to allow for completion of the transmission and intense close-range reverberation. The total sailed distance is about 8 nautical miles The echogram is displayed by commercial postprocessing software prepared by SonarData Pty. Ltd. (J. M. Jech, NOAA, National Marine Fisheries Service, Northeast Fisheries Science Center)

typically from the transducer depth to the bottom. The transducer may also be deployed from a fixed position, such as a surface mooring or bottom structure, with vertical or horizontal orientation, for example. In these cases, the echogram presents a record of potential movement of targets intersecting the transducer beam.

If the transducer has a single beam, it can measure and display the echo strength at a given range. If the transducer has dual beams, it can also measure and display the target strength and angle of resolved targets from the transducer axis. If the transducer has split beams, it can measure the echo strength, target strength, and two orthogonal angles, hence position of resolved targets relative to the transducer.

Scientific echo sounders generally have automated processing routines that can express the received signal as a volume backscattering strength, given that the echo sounder is calibrated. These can also identify individual, or resolved, targets according to selectable criteria, and extract parameters associated with these. This quantity can be integrated in the intensity domain with respect to range, and averaged over a series of echoes, resulting in measures of area backscattering, discussed further below.

While the automated processing of echo sounder data is very useful in ordinary operations, it has inherent limitations. These are addressed principally in separate, independent postprocessing systems. Design principles for commercial systems have been enunciated in the documentation of the Bergen Echo Integrator (BEI) (Foote et al. 1991). These enable erroneous bottom detections to be corrected, echo features to be classified by the user, and measures of area backscattering to be allocated to scatterer classes. Noise-dominated parts of the echo record can be avoided, and criteria for single-target detection can be revised and reapplied to the primary data. Alternatively, echo integration noise can be removed (Korneliussen 2000). Provision is made for the processing of multiple-transducer and multiple-frequency data (Ona et al. 2004). The method of synthetic echograms has promoted this process by relating multiple-frequency echo data at a given location to the respective echo data at the designated reference frequency (Korneliussen and Ona 2003).

Postprocessing of echo sounder data presupposes availability of the unprocessed, raw data. Scientific echo sounders routinely provide these.

Thus far, transducer frequencies have not been considered explicitly. The foregoing comments are quite general.

The earliest scientific echo sounders were based on resonant, ultrasonic transducers, which are still widely used. Because of their resonant nature, they are quite effective at responding to an electrical excitation and launching a narrowband acoustic signal in water. At 38 kHz, for example, transducer efficiencies are often of order 50–70%, with acoustic power generation of the order of 1–2 kW for transducer areas of about 1,000 cm^2. This frequency is near the optimum when balancing detection criteria for fish of commercial interest (Furusawa 1991).

The same resonant transducers are also very efficient at converting acoustic vibrations to electrical signals. The dynamic range of modern scientific echo sounders based on resonant transducers is of order 120–160 dB. Thus, the ratio of the maximum signal that does not cause receiver saturation to the minimum detectable signal is of order 10^6–10^8 in the amplitude domain. Common operating frequencies of resonant transducers include 18, 30, 38, 50, 70, 105, 120, and 200 kHz, with wavelengths ranging from 8 to 0.8 cm, as well as some higher frequencies extending into the low megahertz range. Nominal pulse durations are 0.1–1 ms, with corresponding range resolution 7.5–75 cm, respectively. Typical beamwidths are in the range 2–12°, with 7° being the median. Use of multiple resonant transducers is enabling classification to be performed (Holliday 1977; Holliday and Pieper 1980; Holliday et al. 1989; Madureira et al. 1993; Mitson et al. 1996; Korneliussen and Ona 2002, 2003; Jech and Michaels 2006).

Some research scientific echo sounders are exploiting the use of broadband transducer technology. In one system, the Broadband Acoustic Scattering Signatures (BASS) system, seven nominally octave-bandwidth transducers span the range from 25 kHz to 3.2 MHz (Foote et al. 2005a). Corresponding processing includes spectral classification, which has made possible identification of copepods and euphausiids *in situ* (Foote et al. 2005a).

In a radical departure from ultrasonic echo sounding, an Edgetech source is being exploited for its relatively low, broad bandwidth, nominally 1.7–20 kHz (Stanton et al. 2007). Its greatest sensitivity is over the octave band 2–4 kHz (Stanton et al. 2006), which embraces common resonance frequencies of fish swimbladders (Løvik and Hovem 1979). The receiver sensitivity is also being enhanced, through so-called pulse compression (Chu and Stanton 1998), which principally involves cross correlation of the transmit signal with the received signal.

Scientific echo sounders have been used to characterize the seafloor by means of a novel method due to Orlowski (1984, 1989). This compares the second bottom echo over the path: transducer–bottom–surface–bottom–transducer, to the ordinary first bottom echo. The ratio of the energy in the respective echoes is indicative of the combination of bottom roughness and substrate, that is, type. While the connection may not be entirely clear, devices based on the principle, such as RoxAnn, are commercially successful (Schiagintweit 1993). Given the high quality of data on targets in the water column, contained in the primary echo, it is entirely feasible to relate such data to information on the seafloor habitat.

18.2.2.2   Multibeam Sonar

The achievement of processing multiple signals simultaneously, as in dual- or split-beam processing, has been considerably extended in multibeam sonar. Many elements are typically assembled in an array. Signals arising from individual elements or small clusters of elements are received and processed essentially simultaneously. Electronic or digital phasing enables directional beams to be formed with various orientations. These can render spatially extended scatterers, such as the seafloor and structures, which represents traditional applications, e.g., bathymetry and observing and/or mapping offshore oil platforms.

In recent years, increases in computing power have enabled the entire water-column signals associated with each channel to be stored and processed, not just that associated with the dominant scatterer or scatterers at a fixed range or within a fixed range interval.

Some examples of narrowband sonar arrays are described. The operating frequency of the RESON SeaBat 8101 is 240 kHz. It forms 101 beams with individual-beam resolution of $1.5 \times 1.5°$, thus spanning a total sector, or swath, of 150° in the equatorial plane. The Simrad SM2000 Multibeam Sonar operates at 90 or 200 kHz, each with 128 beams. Individual beamwidths are $1.5 \times 20°$ and $1.5 \times 1.5°$ in the respective imaging and echo sounding modes. The total spanned swath is typically 120 or 150°, depending on the exact configuration of the transducer elements.

Developments in multibeam sonar are ongoing, commensurate with increases in computing power. The RESON SeaBat 7125 at 400 kHz, for example, can form 256 equi-angle beams of resolution 0.5° or 512 equi-distant beams. The Simrad ME70 scientific multibeam echo sounder (Trenkel et al. 2006) and MS70 scientific multibeam sonar (Ona et al. 2006) were designed specifically for fisheries research (Andersen et al. 2006). The ME70 can form beams of widths 2–20°, covering swaths from 60 to 140°, with up to 45 individual beams in the most directional case. The MS70 forms 500 beams, each of order 3–4°, in a two-dimensional array of dimensions 25 beams in the horizontal plane and 20 beams in the vertical plane. Both the ME70 and MS70 are broadband, with nominal spectral sensitivity over the band 70–120 kHz.

Multibeam sonars that provide the water-column signal have been applied to a number of problems in fisheries research. These include quantification of fish schools (Misund et al. 1992) and observation of effects of vessel passage on fish behavior (Misund and Aglen 1992; Misund 1993; Gerlotto et al. 1999; Soria et al. 1996). Predator–prey interactions have also been observed (Nøttestad and Axelsen 1999; Axelsen et al. 2001; Benoit-Bird and Au 2003). The routine use of multibeam sonar in bathymetry witnesses to its potential in mapping fish in the water column relative to the bottom (Mayer et al. 2002), as well as in more general fisheries habitat studies (Reynolds et al. 2001).

18.2.2.3 Sidescan Sonar

The basis of sidescan sonar is a linear array of acoustic transducer elements (Fish and Carr 1990, 2001). The simplest configuration is that of a single row of identical elements. These may be wired together, producing a beam that is extremely narrow in the plane that includes the array, and quite broad in the plane that is orthogonal to the array. It is also possible to shape the beam by weighting the various elements in a process called shading (Urick 1983). This can be used to reduce sidelobes at the slight costs of broadening the main lobe and reducing sensitivity.

Sidescan sonars are typically mounted on vehicles that are towed from a vessel, usually with identical arrays mounted on the port and starboard sides. The arrays may be tilted downwards, for example, 20° from the horizontal, for better coverage of the bottom. As the transmitted signal propagates outwards, echoes are generated by ensonified objects.

The main application of sidescan sonar is for imaging the bottom. An example is shown in Fig. 18.3. In this, a 100-kHz sidescan sonar has imaged patches on the seafloor off Chatham, Massachusetts. These are dense beds of mussels (*Mytilus edulis*).

In another application of sidescan sonar, but operating at 420 kHz, images of the seafloor in Monterey Bay, California, have shown distinct mottling. Underwater divers and use of video photography have established that this is due to clusters of benthic egg capsules of the squid *Loligo opalescens* (Foote et al. 2006). By measuring the small patches comprising the mottle within a geographical information system it has been possible to quantify the potential recruitment of the squid.

Migrating salmon in shallow water have been counted by sidescan sonar at 100 and 330 kHz (Trevorrow 1998). Near-surface salmonids have also been counted by sidescan sonar at 330 kHz (Trevorrow 2001). Near-bottom groundfish have been counted and in some cases imaged by a very directional sidescan sonar operating at 1.3 and 1.4 MHz, with horizontal beamwidth of 0.1° and vertical beamwidth of about 180° (Barans and Holliday 1983).

Some sidescan sonar arrays consist of multiple rows of elements. These enable formation of multiple beams in the plane orthogonal to the array axis, hence with the

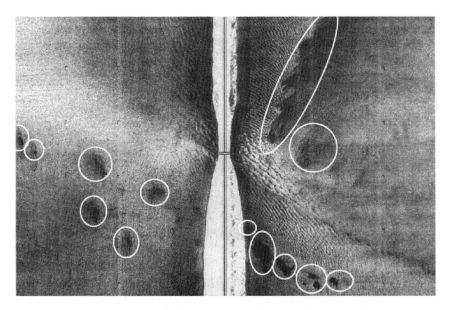

Fig. 18.3 Sidescan sonar image of mussel (*Mytilus edulis*) beds in a sea area off Chatham, Cape Cod, Masssachusetts, measuring 200 m in width and 130 m in length. The nominal operating frequency is 100 kHz. The beds are encircled. The large feature running from the upper left to lower right is geophysical: it is a large sand wave, with wavelength greater than 100 m, which is due to strong tidal currents, exceeding two knots or 1 m/s in this area (Fish and Carr 1990)

potential of interferometry and bathymetry. A major interferometric sidescan sonar was the Geological Long Range Inclined Asdic (GLORIA) (Rusby et al. 1973). Its operating frequency was 6.4 kHz, with beamwidth of 2° in the horizontal plane and 10° in the vertical plane, achieved by configuring 144 transducer elements in a rectangular array 5 m long and 1.2 m high. Near-thermocline scattering observed with GLORIA has been postulated as being due to fish (Weston et al. 1991).

Recently, the interferometric capability has been refined to the extent that the particular sidescan sonars are being called multibeam sidescan sonars. An example is the Teledyne Benthos C3D Side Scan/Bathymetry System operating at 200 kHz, with 1° beamwidth in the horizontal plane and range-independent resolutions of 5.5 cm perpendicular to the sonar track and 5 cm in the vertical plane.

The mentioned applications to shellfish and benthic squid egg clusters illustrate the capacity of sidescan sonar to map biological resources in the context of their seafloor habitat. The integrated use of sidescan sonar and video technology for fisheries habitat studies has been noted (Barans and Holliday 1983).

18.2.2.4   Acoustic Lens-Based Sonar

A radical departure from conventional sonar design is the use of acoustic lens to accomplish beamforming (Clay and Medwin 1977). The principle of operation is analogous to that of an optical lens-based system, e.g., microscope or refracting telescope. In the optical case, the speed of light in the lens is generally less than that in air or *in vacuo*, and the lens is shaped to exploit this difference to focus scattered or other incident light on a plane. Both convex and concave lenses are typically used in conjunction in a compound system to overcome dispersion, or frequency-dependent effects.

In the case of sonar, both convex and concave lenses are used. These are fashioned of materials in which the speed of sound is different from that of the ambient water medium. The range in material sound speeds is greater than that of light speeds in the analogous optical case, with sound speeds being significantly less than that in the ambient medium for some materials, and significantly greater in some other materials. Examples of materials with relatively low sound speed are fluorocarbon liquids (Wade et al. 1975) and silicone rubbers (Folds and Hanlin 1975). Polystyrene and syntactic foams are examples of materials with relatively high sound speed. Dispersion in materials used in acoustic lenses is negligible as a practical matter.

Acoustic lens-based sonars operate by transmitting a signal, then receiving echoes with a lens system that focuses the echoes on a plane, which gives an instantaneous record of the positions of the scatterers in range and angle. The imaging plane consists of a series of transducer elements that convert the acoustic pressure, or displacement, into an electrical signal, which is then recorded.

A prominent commercial acoustic lens-based sonar is the Dual-Frequency Identification Sonar (DIDSON) (Belcher et al. 2002; Belcher 2007). This operates at 1.1 and 1.8 MHz or at 0.7 and 1.2 MHz. At 1.1 MHz it forms 48 beams of width 0.4°. At 1.8 MHz it forms 96 beams of width 0.3°. In both cases, a swath of 29° is spanned, with orthogonal beamwidth of 14°. Operating ranges are nominally 36 m at 1.1 MHz and 9 m at 1.8 MHz.

Experience has demonstrated the effectiveness of the sonar in imaging both water-column scatterers and surfaces, e.g., fish near a hydroelectric power station (Moursund et al. 2003). It has also been used to count salmon in Alaskan rivers (Burwen et al. 2004; Holmes et al. 2006a). It has been used to image salmon redds (Tiffan et al. 2004). Its potential for habitat studies of rockfish may rival or exceed that of very directional sidescan sonar (Barans and Holliday 1983).

18.2.2.5 Parametric Sonar

Parametric sonar is based on the concept of the parametric acoustic array (Westervelt 1963). Two acoustic signals are transmitted collinearly at the same time. Because of the intrinsic nonlinearity of the medium, quantified by the parameter of nonlinearity (Beyer 1974; Everbach 1997), the two waves interact, forming new acoustic waves at the sum and difference frequencies. Significantly, the difference-frequency wave is exceedingly directional, roughly resembling that of the primary waves. If the primary frequencies are only slightly different, the difference frequency will be quite low. Given a strong frequency dependence in absorption, the primary waves, together with the sum-frequency wave, will attenuate very rapidly relative to that of the low difference-frequency wave, which can propagate to much greater ranges.

Two other special properties of this nonlinearly generated difference-frequency wave are that it has no sidelobes and it can be modulated quite easily in frequency.

The conversion efficiency from primary to secondary difference-frequency waves is rather low however, hence leading to trade-off considerations in choosing a sonar for a potential application (Moffett and Konrad 1997). Nonetheless, in one commercial system, the Simrad PS18 Parametric Sub-bottom Profiler, with primary frequencies in the band 15–21 kHz, difference frequencies span the band 1–6 kHz, and the source level at 4 kHz is 204 dB re 1 μPa at 1 m (Dybedal 1993).

Applications in fisheries research that require rather low frequencies, say in the mid-frequency sonar band, 1–10 kHz, but with a rather compact physical source, might very well make use of the parametric sonar. This has been done by Diachok (1999, 2000) in a method for long-range detection of swimbladder-bearing fish by fluctuations in the forward-scattered field. This represents a continuation of early work in which large fluctuations were observed in conventional sonar transmissions at 1 kHz for ranges from 1.9 to 137 km in the Bristol Channel (Weston 1967; Weston et al. 1969). The fluctuations were associated with swimbladder-bearing fish (Weston et al. 1969; Weston 1972).

The parametric sonar may also be useful in characterizing the shallow substrate habitat of some organisms.

18.2.2.6   Synthetic Aperture Sonar

When the track of a moving sonar is relatively stable and/or known to high precision, then echoes derived from multiple pings can be viewed as arising from a virtual array, or aperture, in which each pinging position effectively represents an array element (Gough and Hayes 1989). Thus, the several returns can be processed coherently, as though derived from separate transducer elements or arrays, also called channels. Application of beamforming can then achieve an extraordinary angular resolution far exceeding that intrinsic to the sonar itself, given proper allowance for refraction (Rolt and Schmidt 1994).

Synthetic aperture radar (SAR) is well known, but the sonar analogue, synthetic aperture sonar (SAS), is relatively new, and applications in fisheries await development. One intriguing development is the use of dolphin-like sonar signals for three-dimensional imaging (Altes and Moore 1997). As autonomous underwater vehicles (AUVs) begin to see wider application, their stable track capability will enable SAS technology to be applied to both fish and the bottom habitat, as it is already to the detection of buried targets and sub-bottom geological structure (Lynch et al. 2006).

18.2.2.7   Conventional Low-Frequency Sonar

Rather low-frequency sonars, with operating frequencies of order 1 kHz or less, have been used to detect echoes from swimbladder-bearing fish at considerable ranges.

In a study conducted in 1958, Weston and Revie (1971) used a bottom-mounted 1-kHz sonar to transmit signals outwards in the Bristol Channel. With a transmit beamwidth of 15° and receive beamwidth of 4°, echoes due to moving targets were detected. These were attributed to the Cornish pilchard (*Sardina pilchardus*).

The 6.4-kHz GLORIA sidescan sonar (Rusby et al. 1973), mentioned above, succeeded in detecting herring at ranges up to 15 km in water depth 120–170 m.

Makris et al. (2006) have continued developing this method, using a moored low-frequency sound source, with transmit band 390–440 Hz, and a receiver towed in the vicinity of the source, thus rendering the configuration bistatic. Fish aggregations were detected and imaged at ranges 10–20 km.

18.2.3   Methods of Quantification and Data Interpretation

Scattering data can be quantified in a number of ways. If individual scatterers are resolved, then it is possible count these. By knowing the acoustic sampling volume, the numerical density of scatterers can be computed directly in a process called echo counting (Mitson and Wood 1961).

If individual scatterers cannot be resolved, then it is possible to quantify the acoustic density in a process called echo integration (Forbes and Nakken 1972; MacLennan 1990; Foote and Stanton 2000). The received echo signal is compensated for attention by layer-scattering, applying a range compensation function in the amplitude domain of $r10^{\alpha r/10}$, where r is the range and α is the absorption coefficient in decibels per unit range. The resultant amplified signal is then squared and integrated between two range limits. When divided by the range interval and multiplied by a constant determined by calibration, the result is a value of volume backscattering coefficient (Medwin and Clay 1998). This is typically averaged over a series of successive echoes. In the logarithmic domain this quantity is the mean volume backscattering strength.

Echo integration can also be applied to individually resolved scatterers. Thus, it is a more general method than echo counting.

The mean volume backscattering coefficient s_v is often integrated between two range limits, resulting in a value for the area backscattering coefficient s_A. This is a measure of the cumulative backscattering within the transducer beam over the specified range interval. It is the product of the numerical area density of scatterers, ρ_A, and mean characteristic backscattering cross section σ:

$$s_A = \rho_A \, \sigma$$

In modern terminology, this is the fundamental equation of echo integration (Foote and Knudsen 1994; Foote and Stanton 2000), but it has an older antecedent involving system-specific units of measurement (Forbes and Nakken 1972).

In ordinary operations with a monostatic sonar, s_A is measured and a value for σ is assumed. The equation can be solved for ρ_A, which is a measure of biological density, e.g., number of fish per unit area.

Such measurements of fish density are routinely made according to a systematic sampling regime executed along line transects. When integrated over an area, the abundance of the surveyed fish can be assessed (Foote and Stefánsson 1993). Examples of abundance surveys based on the echo integration method are numerous. They include 0-group cod, saithe, haddock, polar cod, redfish, capelin, and herring in the Barents Sea (Haug and Nakken 1977), Peruvian anchoveta (Johannesson and Robles 1977), northern anchovy along the California coast (Mais 1977), juvenile sockeye salmon in lakes (Mathisen et al. 1977), blue whiting northwest of the British Isles (Midttun and Nakken 1977), Pacific hake and herring in Puget Sound and herring in Carr Inlet in southeast Alaska (Thorne 1977), Icelandic summer spawning herring (Jakobsson 1983), walleye pollock (Traynor and Nelson 1985; Wespestad and Megrey 1990), herring in the North Sea (Bailey and Simmonds 1990), spawning Cape anchovy off South Africa (Hampton et al. 1990), herring in the Gulf of Maine and Georges Bank (Overholtz et al. 2006). Zooplankton stocks have also been assessed by the echo integration method, e.g., the euphausiids *Meganyctiphanes norvegica* off Georges Bank (Greene et al. 1988) and *Euphausia superba* in the Southern Ocean (Hewitt and Demer 2000; Lawson et al. 2004). Shellfish stocks have also been assessed in the same way, e.g., razor clam and surf clam along the coast of central Chile (Tarifeño et al. 1990).

Fish density can be measured and abundance can be assessed with data from other sonars too. Echo integration is being performed with multibeam sonar data, but this is at an early stage of development. Echo integration, as well as echo counting, can be performed with other sonars too, with potential major gains in information.

Scattering data are typically interpreted or judged on the basis of physical capture, other direct observation, or prior knowledge of the local biology. They are then assigned to a particular scatterer class, if possible distinguished by both species and size. Insofar as such information can be known with a degree of certainty, so is the measure of biological density determined by echo counting or echo integration.

The backscattering cross section σ is seen to be a key quantity in the equation of echo integration. It is an important quantity in the equation of echo counting too, since the acoustic sampling volume depends on the echo strength relative to the detection threshold (Foote 1991a).

Much effort has been and will continue to be expended in determining σ or the corresponding logarithmic measure of target strength. Any general listing of methods, e.g., MacLennan (1990) and Foote (1991b), will include purely empirical determinations, theoretical determinations, as by the modeling exercises outlined briefly above, and the in-between, namely theory-guided, quasi-empirical methods.

The dependence of target strength on frequency will be of increasing interest because of the recognized potential of spectral methods for classification (Holliday 1980). Use of the method of synthetic echograms (Korneliussen and Ona 2003), outlined above, is realizing spectral classification for fish.

In addition to the methods of echo counting and echo integration, methods based on echo statistics offer alternative approaches to the problem of quantification. Three seminal papers present concepts and examples for both resolved and unresolved echo fluctuations, including those from fish (Clay and Heist 1984; Stanton 1985; Stanton and Clay 1986). These methods continue to be developed.

18.2.4 Active Sonar Calibration

An advantage of acoustics in fisheries research is its quantitative nature. To realize this most fully, calibration is essential. The standard-target calibration method is a particular method that is straightforward, efficient, and accurate.

The principles of standard-target calibration are well known (Foote et al. 1987). For a sonar system of particular operating frequency and bandwidth, a standard target is chosen or designed on the basis of its scattering characteristics, which can be determined *a priori*, through theoretical modeling. The target is thus a primary standard, which avoids problems associated with the use of secondary or lesser calibration standards.

The standard target is placed at a known position in the beam of the sonar transducer or transducer array. The echo spectrum S_R, for example, can be measured. This can be related to the transmit signal spectrum S_T thus:

$$S_R = S_T FPH,$$

where F is the standard-target far-field form function, P is the two-way acoustic path loss, and H is the overall frequency response function of the combined transmit-receive system (Foote et al. 1999; Foote 2000). The quantity F is related to the backscattering cross section σ by the relation (Foote 1982),

$$\sigma = A\pi \int |S_T FH|^2 dv / \int |S_T H|^2 dv,$$

where the integration is performed with respect to frequency v over the entire spectrum. In the limiting cases of transmission of a monochromatic signal or perfectly narrow receiver bandwidth, $\sigma = 4\pi |F|^2$, which corresponds to the idealized textbook case.

If the transmit signal or receiver frequency response function is relatively narrow, σ may vary relatively significantly over the band of interest. If the bandwidth is substantial, however, and variation in σ with respect to frequency are significant, the measurements may have to be repeated with a second target, which is chosen or designed so that its backscattering cross section is substantial and does not vary rapidly in the same regions as that of the first target.

Typical standard targets are spherical or disklike in shape. Spherical targets may be made of a homogeneous solid elastic material such as an aluminum alloy, electrolytic-grade copper, or tungsten carbide with a cobalt binder. They can also be hollow, as in the case of ceramic-shelled flotation spheres (Stachiw and Peters 2005; Weston et al. 2005).

Some examples are a 60-mm-diameter sphere of electrolytic-grade copper for use with 38-kHz scientific echo sounders (Foote 1982), a 38.1-mm-diameter sphere made of tungsten carbide with 6% cobalt binder for use with multibeam sonars at 200 kHz (Foote et al. 2005b), or 280-mm-diameter sphere of aluminum alloy for use with a parametric sonar sensitive over the band 1–6 kHz (Foote et al. 2007).

Calibration protocols have been devised for use of standard targets with scientific echo sounders (Foote et al. 1987) and multibeam sonars (Foote et al. 2005b). Protocols have been outlined for calibration of the parametric sonar (Foote et al. 2007).

18.3   Passive Acoustic Methods

As noted in Section 18.1, there are many situations in which fish and other organisms are not accessible to active acoustics or in which it is not desirable to attempt to ensonify animals. For organisms that make sound, passive acoustic detection may offer a feasible way to study these remotely, noninvasively, and quantitatively (Rountree et al. 2006).

18.3.1   Vocalizations and Other Sounds

18.3.1.1   Fish

Many fish species produce sound. In an early series of investigations, sound production was examined for 208 species distributed over 53 orders (Fish and Mowbray 1970). The majority of these were soniferous. More than 800 species are now known to produce sound (Kaatz 2002). Their aggregate geographical distribution is global.

Mechanisms for sound production are pulsations of the swimbladder, if present; stridulation; hydrodynamic movements; vibration of substrate, body, or tendon; and release of air. If the sounds are made with intent, they are regarded as being biological; if they are made incidentally, without intent, they are regarded as being mechanical.

Communication sounds are often sonic, or within the audible range, typically less than several kilohertz. However, the gulf menhaden (*Brevoortia patronus*) can detect ultrasound in the band 40–80 kHz (Mann et al. 2001); the American shad (*Alosa sapidissima*), 25–120 kHz (Plachta and Popper 2002); alewife (*A. pseudoharengus*), 117–133 kHz (Dunning et al. 1992); and blueback herring (*A. aestivalis*), 110–140 kHz (Nestler et al. 1992).

Although data on source level are scarce, measurements have been made on several fish. For cod (*Gadus morhua*), it is 120–133 dB re 1 μPa at 1 m for frequencies in the range 20–120 Hz (Midling et al. 2002). For croaking gouramis (*Trichopsis vittata*), it is 81–124 dB re 1 μPa at 1 m for frequencies 1.5–3 kHz (Wysocki and Ladich 2002).

Examples of sounds made by a cod are shown in Fig. 18.4. These include both signals in time, called oscillograms, and time-varying spectra, called spectrograms.

18.3.1.2 Aquatic Mammals

There are many examples of sound production by aquatic mammals. These include cetaceans, or order of aquatic mammals comprising whales, porpoises, and dolphins, and Pinnepedia, or suborder comprising seals, sea lions, and walrus, among others. The variety of produced sounds is enormous. Examples of sounds made by a sperm whale are shown in Fig. 18.5.

Among mysticetes, or baleen whales, vocalizations are often characteristic of the species. Blue whales (*Balaenoptera musculus*) and fin whales (*B. physalus*) both produce powerful low-frequency sounds. Blue whales produce sequences of tones in the frequency band 14–222 Hz, with duration 36 s and source level 188 dB re 1 μPa at 1 m. Fin whales produce frequency-modulated pulses of sound descending

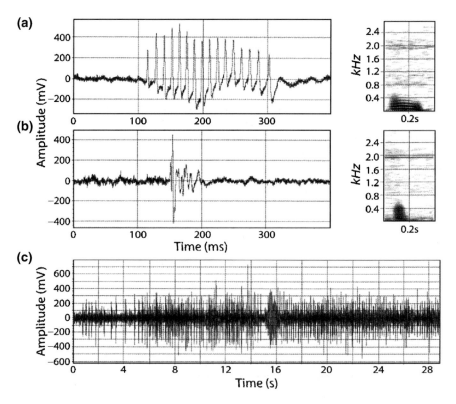

Fig. 18.4 Oscillograms and spectrograms of sounds made by a single cod (*Gadus morhua*): (**a**) grunt, (**b**) single knock, and (**c**) series of knocks (Midling et al. 2002)

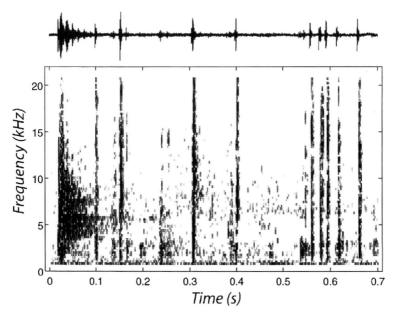

Fig. 18.5 Oscillogram (*upper*) and spectrogram (*lower*) of sperm whale (*Physeter macrocephalus*) vocalization (Tyack 2000)

from 23 to 18 Hz over 1 s, with comparable source levels (Watkins et al. 1987). Bowhead whales (*Balaena mysticetus*) and humpback whales (*Megaptera novaeangliae*) produce complicated sequences of signals, called songs, in the audible range, 100–4,000 Hz (Tyack 2000; Tyack and Clark 2000). Source levels are in the range 129–189 dB re 1 μPa at 1 m. Sounds due to slaps of humpback whale flukes and flippers have been observed with source levels in the range 183–192 dB re 1 μPa at 1 m (Thompson et al. 1986).

Odontocetes, or toothed whales, generally produce much high-frequency vocalizations than do mysticetes, typically in the range 0.5–250 kHz (Popper and Edds-Walton 1997). The repertoire of odontocete sounds includes tones and pulsed calls, which are important in intraspecific communications (Weilgart and Whitehead 1993, 1997; Tyack 2000, 2001; Gordon and Tyack 2002). It also includes clicks, or short-duration broadband signals, used in echolocation. These can be quite powerful. For example, source levels of the bottlenose dolphin (*Tursiops truncatus*), false killer whale (*Pseudorca crassidens*), and narwhal (*Monodon monoceros*) are 218–228 dB re 1 μPa at 1 m (Au 1993).

Many pinniped species produce sound underwater. Vocalizations span the range from pure tones emitted by the harp seal (*Pagophilus groenlandicus*) (Møhl et al. 1975) to the complex, song-like signals of the bearded seal (*Erignathus barbatus*) (Ray et al. 1969). The source level of harp seal vocalizations is about 135–140 dB re 1 μPa at 1 m (Watkins and Schevill 1979; Terhune and Ronald 1986).

18.3.1.3 Non-chordates

A number of non-chordate species are also soniferous (Fish and Mowbray 1970). Snapping shrimps of the family Alpheidae are well known for their contribution to the ambient noise spectrum over the band 0.1–24 kHz in temperate waters shallower than about 55 m. The 425 species of snapping shrimp (Schmitz 2002) have a worldwide geographical distribution that is coastal between 40°N and 40°S (Fish and Mowbray 1970). The mechanism of sound production involves a high rotational speed of the dactyl, which forces a jet of water out of the socket at high speed, causing the pressure to fall below that of the vapor pressure, thus generating cavitation bubbles (Versluis et al. 2000). When the pressure rises, the bubble collapses. The source level of single specimens of *Crangon californiensis*, *C. clamator*, and *Synlpheus locking-toni* are in the range 146–160 dB re 1 μPa at 1 m. Specimens of *Alpheus heterochaelis* generate a broader spectrum of sound, exceeding 200 kHz, with source levels in the range 183–190 dB re 1 μPa at 1 m (Schmitz 2002).

Other crustacean species of Arthropoda also produce sound. These include the subclasses Cirripedia, or barnacles, and Malacostraca, including orders Stomatopoda, Amphipoda, and Decapoda (Schmitz 2002). The order of Decapoda includes shrimp, American lobster (*Homarus americanus*), and fiddler crab (*Uca* spp.), among others. The sound production mechanism is stridulation of antennae over toothed surfaces of the carapace.

In the phylum Mollusca, the black mussel (*M. edulis*) produces a crackling sound over the band 1–4 kHz (Fish and Mowbray 1970). The mechanism of sound production is movement of the mussel, with associated rupture of the byssal threads anchoring it to rocks.

In the phylum Echinodermata, a number of sea urchin species are soniferous. These include the New Zealand species *Evechinus chloroticus*, with emissions in the band 20–5,000 Hz, New England species *Strongylocentrotus dröbachiensis*, and tropical species *Diadema setosum* (Fish and Mowbray 1970).

18.3.2 Detection of Animal Sounds

The basis for observing or measuring animals by passive acoustics is the sounds that they make, including temporal and spectral characteristics and source level. As in the case of application of active sonars, such technical parameters are critical for choosing the means of detection, namely hydrophones or arrays of hydrophones. A hydrophone is basically a microphone that is suitable for use underwater. The mentioned technical parameters are also important for deciding on issues of deployment, such as range to sources of sound, and proximity of hydrophones to boundaries such as bottom, bottom structures, and surface. Exercise of the passive sonar equation (Urick 1983) can guide deployments, as in ensuring a sufficient signal-to-noise ratio.

A variety of hydrophone configurations has been used in systematic investigations of biological sound sources. Some of these are outlined. All have a current applicability, and most are in use. They will serve future applications too, but also from more sophisticated platforms, such as autonomous underwater vehicles and various ocean observatories, and with advanced signal processing methods.

18.3.2.1 Single Hydrophones

Single hydrophones have a long history of use in investigations of biological sound sources. They have been used to record sounds made by fish, marine mammals, and non-chordates. Examples of methods of deployment are suspensions from a float (Cummings et al. 1986) and anchoring on the bottom (Edds 1988). Hydrophones have also been mounted on poles for *in situ* recording of fish sounds (Lobel 2001). Recently a hydrophone was attached to a Webb Slocum glider (Webb et al. 2001) to record low-frequency sound transmissions in shallow water (Dossot et al. 2007). Its potential for recording biological signals is evident.

18.3.2.2 Paired Hydrophones

Paired hydrophones have been used for localization of marine mammals, e.g., humpback whales in the West Indies (Winn et al. 1975) and bowhead whales when migrating past Point Barrow, Alaska, in the spring (Clark and Johnson 1984). In both of these examples the array was mounted horizontally. Vertical orientations have also been used, as in surveying sperm whales and other cetaceans in the southeastern Caribbean (Watkins and Moore 1982).

18.3.2.3 Three-, Four-, and Five-Hydrophone Arrays

An array of three hydrophones at the vertices of a right triangle with equal base lengths 1.5 m has been mounted 0.5 m above the bottom. This has been used to determine the direction of sounds from right whales, as well as track them (Clark 1980). Bowhead whales have been counted following detection with a linear array of three or four hydrophones (Clark and Ellison 1988), with localization to a range 3–4 times the array length, based on 3–5 hydrophones (Clark and Ellison 2000). Three-dimensional positioning of sound sources has been accomplished with a nonrigid three-dimensional array of four hydrophones, with two hydrophones suspended from a surface float and two hydrophones suspended from the ends of a vessel, with separation distance of order 30 m (Watkins and Schevill 1972). Relative positions were determined through an *in situ* calibration effected with two acoustic transmitters, called pingers, with observation of time differences.

18.3.2.4   Sonobuoys

A sonobuoy is a system composed principally of a hydrophone, radio transmitter with antenna, and float. Deployment of a number over a wide area can achieve synopticity in surveying marine mammals. In one study, a total of 82 sonobuoys were deployed in the eastern Caribbean Sea (Levenson and Leapley 1978). Deployment in special areas, such as open leads in ice (Ljungblad et al. 1982) or from ice edges or ridges (Cummings and Holliday 1985), might enable observations to be made, or enable observations to be made with the highest relative sensitivity or accuracy. Attachment of a Global Positioning System (GPS) receiver to sonobuoys or other free-drifting buoys can enable detections and localization of marine mammals, e.g., foraging blue whales, using a three-sonobuoy array with 1.8-km spacing (Hayes et al. 2000). A set of ten sonobuoys and a single hydrophone have been used to define the spawning areas of red drum, weakfish, spotted sea trout, and silver perch (Luczkovich and Sprague 2002).

18.3.2.5   Towed Arrays

Towed arrays are well developed and widely used in geophysical exploration. In one case, cetacean sounds were determined by a 45-m-long array, with spectral sensitivity over the band 20 Hz–15 kHz (Thomas et al. 1986). A three-element array with 9-m length has been used to record dolphin whistles (Thode et al. 2000). Odontocetes in the Southern Ocean have been surveyed by a towed hydrophone array (Leaper et al. 2000). Recently, a six-element hydrophone array has been towed from an autonomous underwater vehicle (Sullivan et al. 2006; Holmes et al. 2006b). While this was used for physical measurements, it has a clear potential for biological measurements too.

18.3.2.6   Passive Sonar

A sonar system can also be used passively as a sophisticated receiver. Sperm whale vocalizations have been recorded with a low-frequency sonar in the Mediterranean Sea (Pavan et al. 2000).

18.3.3   Methods of Analysis and Classification

Sounds produced by fish and other aquatic organisms have been reported in the time domain by so-called oscillograms and in the frequency domain by power spectra. Because of the complexity of some animal sounds, the two presentations are often combined in a spectrogram, as in Figs. 18.4 and 18.5.

These analyses of produced sounds are quantitative. Subjective analyses have also been performed, with characterization of sounds through terms such as chirps, clicks, growls, knocks, mews, moans, shrieks, trumpet-like blasts, and whistles, among others. Association of sounds, whether quantitative or qualitative, accomplishes classification. This is an active area of research, with much potential for surveying soniferous fish and other organisms that are difficult to observe by active acoustics.

Another form of analysis consists of processing of signals received on multiple hydrophones. By measuring arrival-time or phase differences in the respective cases of broadband and narrowband signals, it is possible to determine direction. This process may be aided by cross correlation of temporal waveforms (Clark and Ellison 2000) or by cross correlation of filtered spectrograms (Altes 1980; Clark et al. 1986). Matched filtering, neural networks, and spectrogram correlation (Clark et al. 1987) have been used to determine the presence of vocalizing bowhead whales (Mellinger and Clark 2000). Neural networks have been used to classify vocalizations by false killer whales (Murray et al. 1998), as well as killer whales (Deecke et al. 1999).

18.3.4 Passive Acoustic System Calibration

Much of the analysis performed on animal sounds depends on the time of arrival or the time signal itself. Both of these measurements generally require or would benefit from calibration.

As described earlier, calibration is the process whereby the performance characteristics of a system are determined. In the case of hydrophones, these characteristics may include, among other things, sensitivity, dynamic range, linearity of response, and frequency response system characteristics of the system. The quality of observations with hydrophones depends fundamentally on the performance of these devices and their associated electronics.

Methods for single-hydrophone calibration are described by Bobber (1970) and Urick (1983). A recent source of information on comparison methods is Robinson et al. (2006).

Use of multiple-hydrophone systems often requires combination or comparison of signals from individual hydrophones. This may be aided by transmission of omnidirectional signals to determine hydrophone position and/or establish a consistent time reference.

18.4 Summary

Methods of active and passive acoustics have been reviewed for their current and potential applications in fisheries research, but understood here as referring more broadly to aquatic ecosystem research. Traditional active acoustic methods, such as those based on scientific echo sounders, continue to be developed, as through increasing bandwidth. Other sonars that lack a significant history of application in fisheries possess a huge potential, as for sharpening and/or increasing the effective sampling volume or range of spectral sensitivity. In some cases, the habitat can be imaged at the same time. Such sonars include multibeam sonars that provide the water-column signal, sidescan sonars, acoustic lens-based sonars, and parametric sonars. Each of these has its own set of characteristics that can be highly advantageous in extending the tools used by fisheries researchers independently of fisheries data.

Passive acoustic methods also have a long history of application, but they are under rapid development, for example, to survey soniferous fish in reef environments or to detect marine mammals. The range of aquatic organisms that can be observed is thus being extended.

Signal processing is a *sine qua non* of these acoustical methods. They are fundamental to the visualization of data, as well as to the detection, identification, and quantification of organisms. Methods of sonar and hydrophone signal processing are being steadily developed and refined, for example, to exploit bandwidth and beamwidth, ultimately to increase detection ranges, improve or extend identification, improve quantitative results, and provide contextual information on the habitat.

Calibration is essential to achieving and maintaining high performance in acoustical applications to fish and other organisms. While efficient methods exist for sonar systems, such as scientific echo sounders and multibeam sonars, protocols remain to be developed for other acoustical systems.

Acknowledgments B.J. Rothschild and R. Beamish are thanked for their invitation to participate in the Fiftieth Anniversary Symposiums of the American Institute of Fishery Research Biologists, as well the opportunity to help organize the session on technology. M. Parmenter is thanked for casting the manuscript into the official style.

References

Altes RA (1980) Detection, estimation, and classification with spectrograms. J Acoust Soc Am 67:1232–1246

Altes RA, Moore PWB (1997) Bionic synthetic aperture sonar. J Acoust Soc Am 102:3123

Andersen LN, Berg S, Gammelsæter OB, Lunde EB (2006) New scientific multibeam systems (ME70 and MS70) for fishery research applications. J Acoust Soc Am 120:3017

Au WWL (1993) The sonar of dolphins. Springer, New York

Axelsen BE, Anker-Nilssen T, Fossum P, Kvamme C, Nøttestad L (2001) Pretty patterns but a simple strategy: predator-prey interactions between juvenile herring and Atlantic puffins observed with multibeam sonar. Can J Zool 79:1586–1596

Bailey RS, Simmonds EJ (1990) The use of acoustic surveys in the assessment of the North Sea herring stock and a comparison with other methods. Rapp P-v Réun Cons Int Explor Mer 189:9–17

Baldridge HD (1970) Sinking factors and average densities of Florida sharks as functions of liver buoyancy. Copeia 1970:744–754

Barans CA, Holliday DV (1983) A practical technique for assessing some snapper/grouper stocks. Bull Mar Sci 33:176–181

Belcher EO (2007) Vision in turbid water. In: Proceedings of MTS/ADCI Underwater Intervention 2007 Conference, New Orleans, LA, 5 pp

Belcher EO, Fox WLJ, Hanot WH (2002) Dual-frequency acoustic camera: a candidate for an obstacle avoidance, gap-filler, and identification sensor for untethered underwater vehicles. In: Proceedings of the MTS/IEEE Oceans 2002 Conference, Biloxi, MS, pp 1234–1238

Benoit-Bird K, Au W (2003) Hawaiian spinner dolphins aggregate midwater food resources through cooperative foraging. J Acoust Soc Am 114:2300

Beyer RT (1974) Nonlinear acoustics. Naval Ship Systems Command, Washington, DC

Bobber RJ (1970) Underwater electroacoustic measurements. Naval Research Laboratory, Washington, DC

Bone Q (1972) Buoyancy and hydrodynamic functions of integument in the castor oil fish, *Ruvettus pretiousus* (Pisces: Gempylidae). Copeia 1972:75–87

Brede R, Kristensen FH, Solli H, Ona E (1990) Target tracking with a split-beam echo sounder. Rapp P-v Réun Cons Int Explor Mer 189:254–263

Burwen D, Maxwell S, Pfisterer C (2004) Investigations into the application of a new sonar system for assessing fish passage in Alaskan rivers. J Acoust Soc Am 115:2547

Chu D, Stanton TK (1998) Application of pulse compression techniques to broadband acoustic scattering by live individual zooplankton. J Acoust Soc Am 104:39–55

Clark CW (1980) A real-time direction finding device for determining the bearing to the underwater sounds of southern right whales, *Eubalaena australis*. J Acoust Soc Am 68:508–511

Clark CW, Ellison WT (1988) Numbers and distributions of bowhead whales, Balaena mysticetus, based on the 1985 acoustic study off Pt. Barrow, Alaska. Rep Int Whal Commn 38:365–370

Clark CW, Ellison WT (2000) Calibration and comparison of the acoustic location methods used during the spring migration of the bowhead whale, *Balaena mysticetus*, off Pt. Barrow, Alaska, 1984–1993. J Acoust Soc Am 107:3509–3517

Clark CW, Johnson JH (1984) The sounds of the bowhead whale, *Balaena mysticetus*, during the spring migrations of 1979 and 1980. Can J Zool 62:1436–1441

Clark CW, Ellison WT, Beeman K (1986) Acoustic tracking of migrating bowhead whales. In: Proceedings of the MTS/IEEE Oceans Conference 1986, pp 341–346

Clark CW, Marler P, Beeman K (1987) Quantitative analysis of animal vocal phonology: an application to swamp sparrow song. Ethology 76:101–115

Clay CS, Heist BG (1984) Acoustic scattering by fish - acoustic models and a two-parameter fit. J Acoust Soc Am 75:1077–1083

Clay CS, Medwin H (1977) Acoustical oceanography: principles and applications. Wiley, New York

Cummings WC, Holliday DV (1985) Passive acoustic location of bowhead whales in a population census off Point Barrow, Alaska. J Acoust Soc Am 78:1163–1169

Cummings WC, Thompson PO, Ha SJ (1986) Sounds from Bryde, *Balaenoptera edeni*, and finback, *B. physalus*, whales in the Gulf of California. Fish Bull 84:359–370

Deecke VB, Ford JKB, Spong P (1999) Quantifying complex patterns of bioacoustic variation: use of a neural network to compare killer whale (*Orcinus orca*) dialects. J Acoust Soc Am 105:2499–2507

Diachok O (1999) Effects of absorptivity due to fish on transmission loss in shallow water. J Acoust Soc Am 105:2107–2128

Diachok O (2000) Absorption spectroscopy: a new approach to estimation of biomass. Fish Res 47:231–244

Dossot GA, Miller JH, Potty GR, Morre KA, Holmes JD, Lynch JF (2007) Acoustic measurements in shallow water using an ocean glider. J Acoust Soc Am 121:3108

Dunn JL (1969) Airborne measurements of the acoustic characteristics of a sperm whale. J Acoust Soc Am 46:1052–1054

Dunning DJ, Ross QE, Geoghegan P, Reichle JJ, Menezes JK, Watson JK (1992) Alewives avoid high-frequency sound. N Am J Fish Manage 12:407–416

Dybedal J (1993) TOPAS: parametric end-fire array used in offshore applications. In: Hobaek H (ed) Advances in nonlinear acoustics. World Scientific, Singapore, pp 264–275

Edds PL (1988) Characteristics of finback *Balaenoptera physalus* vocalizations in the St. Lawrence Estuary. Bioacoustics 1:131–149

Ehrenberg JE (1974) Two applications for a dual-beam transducer in hydroacoustic fish assessment systems. Proc IEEE Conf Eng Ocean Environ 1:152–154

Ehrenberg JE (1979) A comparative analysis of in situ methods for directly measuring the acoustic target strength of indivudual fish. IEEE J Ocean Eng 4:141–152

Everbach EC (1997) Parameters of nonlinearity of acoustic media. In: Crocker MJ (ed) Encyclopedia of acoustics. Vol 1. Wiley, New York, pp 219–226

Ezerskii AB, Selivanovskii DA (1987) Backscattering of sound by the hydrodynamic wakes of marine animals. Sov Phys Acoust 33:370–372

Fish JP, Carr HA (1990) Sound underwater images, a guide to the generation and interpretation of side scan sonar data. 2nd edn. Lower Cape Publishing, Orleans, MA

Fish JP, Carr HA (2001) Sound reflections, advanced applications of side scan sonar. Lower Cape Publishing, Orleans, MA

Fish MP, Mowbray WH (1970) Sounds of western north Atlantic fishes: a reference file of biological underwater sounds. Johns Hopkins Press, Baltimore, MD

Folds DL, Hanlin J (1975) Focusing properties of a solid four-element ultrasonic lens. J Acoust Soc Am 58:72–77

Foote KG (1982) Optimizing copper spheres for precision calibration of hydroacoustic equipment. J Acoust Soc Am 71:742–747

Foote KG (1991a) Acoustic sampling volume. J Acoust Soc Am 90:959–964

Foote KG (1991b) Summary of methods for determining fish target strength at ultrasonic frequencies. ICES J Mar Sci 48:211–217

Foote, KG (1998) Broadband acoustic scattering signatures of fish and zooplankton (BASS). In: Proceedings of the Third European Marine Science Technology Conference, Lisbon, Portugal, 23–27 May 1998. Vol 3, pp 1011–1025

Foote KG (2000) Standard-target calibration of broadband sonars. J Acoust Soc Am 108:2484

Foote KG, Knudsen HP (1994) Physical measurement with modern echo integrators. J Acoust Soc Jpn (E) 15:393–395

Foote KG, Stanton TK (2000) Acoustical methods. In: Harris RP, Wiebe PH, Lenz J, Skjoldal HR, Huntley M (eds) ICES Zooplankton Methodology Manual. Academic, London, pp 223–258

Foote KG, Stefánsson G (1993) Definition of the problem of estimating fish abundance over an area from acoustic line-transect measurements of density. ICES J Mar Sci 50:369–381

Foote KG, Knudsen HP, Vestnes G, MacLennan DN, Simmonds EJ (1987) Calibration of acoustic instruments for fish density estimation: a practical guide. ICES Coop Res Rep 144:1–69

Foote KG, Knudsen HP, Korneliussen JR, Nordbø PE, Røang K (1991) Postprocessing system for echo sounder data. J Acoust Soc Am 90:38–47

Foote KG, Atkins PR, Bongiovanni C, Francis DTI, Eriksen PK, Larsen M, Mortensen T (1999) Measuring the frequency response function of a seven-octave-bandwidth echo sounder. Proc Inst Acoust 21(1):88–95

Foote KG, Atkins PR, Francis DTI, Knutsen T (2005a) Measuring echo spectra of marine organisms over a wide bandwidth. In: Papadakis JS, Bjørnø L (eds) Proceedings of International Conference on Underwater Acoustic Measurements: Technologies and Results, Heraklion, Crete, Greece, 28 June–1 July 2005, pp 501–508

Foote KG, Chu D, Hammar TR, Baldwin KC, Mayer LA, Hufnagle LC, Jr, Jech JM (2005b) Protocols for calibrating multibeam sonar. J Acoust Soc Am 117:2013–2027

Foote KG, Hanlon RT, Iampietro PJ, Kvitek RG (2006) Acoustic detection and quantification of benthic egg beds of the squid *Loligo opalescens* in Monterey Bay, California. J Acoust Soc Am 119:844–856

Foote KG, Francis DTI, Atkins PR (2007) Calibration sphere for low-frequency parametric sonars. J Acoust Soc Am 121:1482–1490

Forbes ST, Nakken O (1972) Manual of methods for fisheries resource survey and appraisal. Part 2. The use of acoustical instruments of fish detection and abundance estimation. FAO Man Fish Sci (5):1–138

Furusawa M (1991) Designing quantitative echo sounders. J Acoust Soc Am 90:26–36

Gerlotto F, Soria M, Fréon P (1999) From 2D to 3D: the use of multi-beam sonar for a new approach in fisheries acoustics. Can J Fish Aquat Sci 56:6–12

Goodman L (1990) Acoustic scattering from ocean microstructure. J Geophys Res 95:11557–11573

Gordon J, Tyack P (2002) Sound and cetaceans. In: Evans PGH, Raga JA (eds) Marine mammals: biology and conservation. Kluwer/Plenum, London, pp 139–196

Gough PT, Hayes MP (1989) Test results using a prototype synthetic aperture sonar. J Acoust Soc Am 86:2328–2333

Greene CH, Wiebe PH, Burczynski J, Youngbluth MJ (1988) Acoustical detection of high-density krill demersal layers in the submarine canyons off Georges Bank. Science 241:359–361

Hampton I, Armstrong MJ, Jolly GM, Shelton PA (1990) Assessment of anchovy spawner biomass off South Africa through combined acoustic and egg-production surveys. Rapp P-v Réun Cons Int Explor Mer 189:18–32

Haug A, Nakken O (1977) Echo abundance indices of 0-group fish in the Barents Sea, 1965–1972. Rapp P-v Réun Cons Int Explor Mer 170:259–264

Hayes SA, Mellinger DK, Croll DA, Costa DP, Borsani JF (2000) An inexpensive passive acoustic system for recording and localizing wild animal sounds. J Acoust Soc Am 107:3552–3555

Hewitt RP, Demer DA (2000) The use of acoustic sampling to estimate the dispersion and abundance of euphausiids, with an emphasis on Antarctic krill, *Euphausia superba*. Fish Res 47:215–229

Holliday DV (1977) Extracting bio-physical information from acoustic signatures of marine organisms. In: Anderson NR, Zahuranec BJ (eds) Oceanic sound scattering prediction. Plenum, New York, pp 619–624

Holliday DV (1980) Use of acoustic frequency diversity for marine biological measurements. In: Diemer FP, Vernberg FJ, Mirkes DZ (eds) Advanced concepts in ocean measurements for marine biology. University of South Carolina, Columbia, SC, pp 423–460

Holliday DV, Pieper RE (1980) Volume scattering strengths and zooplankton disributions at acoustic frequencies between 0.5 and 3 MHz. J Acoust Soc Am 67:135–146

Holliday DV, Pieper RE, Kleppel GS (1989) Determination of zooplankton size and distribution with multi-frequency acoustic technology. J Cons Int Explor Mer 46:52–61

Holmes JA, Cronkite GMW, Enzenhofer HJ, Mulligan TJ (2006a) Accuracy and precision of fish-count data from a "dual-frequency identification sonar" (DIDSON) imaging system. ICES J Mar Soc 63:543–555

Holmes JD, Carey WM, Lynch JF (2006b) Results from an autonomous underwater vehicle towed hydrophone array experiment in Nantucket Sound. J Acoust Soc Am 120:EL15–EL21

Jakobsson J (1983) Echo surveying of the Icelandic summer spawning herring 1973–1982. FAO Fish Rep 300:240–248

Jech JM, Michaels WL (2006) A multifrequency method to classify and evaluate fisheries acoustics data. Can J Fish Aquat Sci 63:2225–2235

Johannesson KA, Robles AN (1977) Echo surveys of Peruvian anchoveta. Rapp P-v Réun Cons Int Explor Mer 170:237–244

Kaatz IM (2002) Multiple sound-producing mechanisms in teleost fishes and hypotheses regarding their behavioural significance. Bioacoustics 12:230–233

Kleckner RC, Gibbs RH, Jr (1972) Swimbladder structure of Mediterranean midwater fishes and a method of comparing swimbladder data with acoustic profiles. Mediterranean Biological Studies Final Report. Smithsonian Institution, Washington, DC. Vol 1, Pt 4, pp 230–281

Korneliussen RJ (2000) Measurment and removal of echo integration noise. ICES J Mar Sci 57:1204–1217

Korneliussen RJ, Ona E (2002) An operational system for processing and visualizing multi-frequency acoustic data. ICES J Mar Sci 59:293–313

Korneliussen RJ, Ona E (2003) Synthetic echograms generated from the relative frequency response. ICES J Mar Sci 60:636–640

Lavery AC, Ross T (2007) Acoustic scattering from double-diffusive microstructure. J Acoust Soc Am 122:1449–1462

Lavery AC, Schmitt RW, Stanton TK (2003) High frequency acoustic scattering from turbulent microstructure: the importance of density fluctuations. J Acoust Soc Am 114:2685–2697

Lawson GL, Wiebe PH, Ashjian CJ, Gallager SM, Davis CS, Warren JD (2004) Acoustically-inferred zooplankton distribution in relation to hydrography west of the Antarctic Peninsula. Deep-Sea Res II 51:2041–2072

Leaper R, Gillespie D, Papastavrou (2000) Results of passive acoustic surveys for odontocetes in the Southern Ocean. J Cetacean Res Manage 2:187–196

Leighton TG (1994) The acoustic bubble. Academic, San Diego, CA

Levenson C (1974) Source level and bistatic target strength of the sperm whale (*Physeter catodon*) measured from an oceanographic aircraft. J Acoust Soc Am 55:1100–1103

Levenson C, Leapley WT (1978) Distribution of humpback whales (*Megaptera novaeangliae*) in the Caribbean determined by a rapid acoustic method. J Fish Res Board Can 35:1150–1152

Ljungblad DK, Thompson PO, Moore SE (1982) Underwater sounds recorded from migrating bowhead whales, *Balaena mysticetus*, in 1979. J Acoust Soc Am 71:477–482

Lobel PS (2001) Acoustic behavior of cichlid fishes. J Aquaricult Aquat Sci 9:167–186

Love RH (1973) Target strengths of humpback whales *Megaptera novaeangliae*. J Acoust Soc Am 54:1312–1315

Løvik A, Hovem JM (1979) An experimental investigation of swimbladder resonance in fishes. J Acoust Soc Am 66:850–854

Lucifredi I, Stein PJ (2007) Gray whale target strength measurements and the analysis of the backscattered response. J Acoust Soc Am 121:1383–1391

Luczkovich JJ, Sprague MW (2002) Using passive acoustics to monitor estuarine fish populations. Bioacoustics 12:289–291

Lynch JF, Chu D, Austin T, Carey W, Pierce A, Holmes J (2006) Detection and classification of buried targets and sub-bottom geoacoustic inversion with an AUV carried low frequency acoustic source and a towed array. In: Proceedings of the MTS/IEEE Oceans 2006 Conference, Boston, MA, 5 pp

MacLennan DN (1990) Acoustical measurement of fish abundance. J Acoust Soc Am 87:1–15

Madureira LSP, Everson I, Murphy EJ (1993) Interpretation of acoustic data at two frequencies to discriminate between Antarctic krill (*Euphausia superba* Dana) and other scatterers. J Plankton Res 15:787–802

Mais KF (1977) Acoustic surveys of northern anchovies in the California Current system, 1966–1972. Rapp P-v Réun Cons Int Explor Mer 170:287–295

Makris NC, Ratilal P, Symonds DT, Jagannathan S, Lee S, Nero RW (2006) Fish population and behavior revealed by instantaneous continental shelf-scale imaging. Science 311:660–663

Mann DA, Higgs DM, Tavolga WN, Souza MJ, Popper AN (2001) Ultrasound detection by clupeiform fishes. J Acoust Soc Am 109:3048–3054

Mathisen OA, Croker TR, Nunnallee EP (1977) Acoustic estimation of juvenile sockeye salmon. Rapp P-v Réun Cons Int Explor Mer 170:279–286

Mayer L, Li Y, Melvin G (2002) 3D visualization for pelagic fisheries research and assessment. ICES J Mar Sci 59:216–225

McClatchie S, Ye Z (2000) Target strength of an oily deep-water fish, orange roughy (*Hoplostethus atlanticus*) II. Modeling. J Acoust Soc Am 107:1280–1285

Medwin H, Clay CS (1998) Fundamentals of acoustical oceanography. Academic, Boston, MA

Mellinger DK, Clark CW (2000) Recognizing transient low-frequency whale sounds by spectrogram correlation. J Acoust Soc Am 107:3518–3529

Midling K, Soldal AV, Fosseidengen JE, Øvredal JT (2002) Calls of the Atlantic cod: does captivity restrict their vocal repertoire? Bioacoustics 12:233–235

Midttun L, Nakken O (1977) Some results of abundance estimation studies with echo integrators. Rapp P-v Réun Cons Int Explor Mer 170:253–258

Minnaert M (1933) On musical air bubbles and the sound of running water. Phil Mag 16:235–248

Misund OA (1993) Abundance estimation of fish schools based on a relationship between school area and school biomass. Aquat Living Resour 6:235–241

Misund OA, Aglen A (1992) Swimming behaviour of fish schools in the North Sea during acoustic surveying and pelagic trawl sampling. ICES J Mar Sci 49:325–334

Misund OA, Aglen A, Beltestad AK, Dalen J (1992) Relationships between the geometric dimensions and biomass of schools. ICES J Mar Sci 49:305–315

Mitson RB (1983) Fisheries sonar (incorporating Underwater observation using sonar by DG Tucker). Fishing News Books, Farnham, Surray, England

Mitson RB, Wood RJ (1961) An automatic method of counting fish echoes. J Cons int Explor Mer 26:281–291

Mitson RB, Simard Y, Goss C (1996) Use of a two-frequency algorithm to determine size and abundance of plankton in three widely spaced locations. ICES J Mar Sci 53:209–215

Moffett MB, Konrad WL (1997) Nonlinear sources and receivers. In: Crocker MJ (ed) Encyclopedia of acoustics. Vol. 1. Wiley, New York, 607–617

Møhl B, Terhune JM, Ronald K (1975) Underwater calls of the harp seal, *Pagophilus groenlandicus*. Rapp P-v Réun Cons Int Explor Mer 169:533–543

Moursund RA, Carlson TJ, Peters RD (2003) A fisheries application of a dual-frequency indentification sonar acoustic camera. ICES J Mar Sci 60:678–683

Murray SO, Mercado E, Roitblat HL (1998) The neural network classification of false killer whale (*Pseudorca crassidens*) vocalizations. J Acoust Soc Am 104:3626–3633

NDRC National Defense Research Committee (1946) Physics of sound in the sea. Reprinted 1969 by Department of the Navy Headquarters Naval Material Command, Washington, DC

Nestler JM, Ploskey GR, Pickens J, Menezes J, Schilt C (1992) Responses of blueback herring to high-frequency sound and implications for reducing entrainment at hydropower dams. N Am J Fish Manage 12:667–683

Nøttestad L, Axelsen BE (1999) Herring schooling manoeuvres in response to killer whale attacks. Can J Zool 77:1540–1546

Novarini JC, Bruno DR (1982) Effects of the sub-surface bubble layer on sound propagation. J Acoust Soc Am 72:510–514

Ona E, Korneliussen R, Knudsen HP, Røang K, Eliassen I, Heggelund Y, Patel D (2004) The Bergen multifrequency analyzer (BMA): a new toolbox for acoustic categorization and species identification. J Acoust Soc Am 115:2584

Ona E, Dalen J, Knudsen HP, Patel R, Andersen LN, Berg S (2006) First data from sea trials with the new MS70 multibeam sonar. J Acoust Soc Am 120:3017

Orlowski A (1984) Application of multiple echoes energy measurements for evaluation of sea bottom type. Oceanologia 19:61–78

Orlowski A (1989) Application of acoustic methods to correlation of fish density distribution and the type of sea bottom. Proc Inst Acoust 11:179–185

Overholtz WJ, Jech JM, Michaels WL, Jacobson LD (2006) Empirical comparisons of survey designs in acoustic surveys of Gulf of Maine-Georges Bank Atlantic herring. J Northw Atl Fish Sci 36:127–144

Pavan G, Hayward TJ, Borsani JF, Priano M, Manghi M, Fossati C, Gordon J (2000) Time patterns of sperm whale codas recorded in the Mediterranean Sea 1985–1996. J Acoust Soc Am 107:3487–3495

Plachta DTT, Popper AN (2002) Neuronal and behavioural responses of American shad *Alosa sapidissima* to ultrasound stimuli. Bioacoustics 12:191–193

Popper AN, Edds-Walton PL (1997) Bioacoustics of marine vertebrates. In: Crocker MJ (ed) Encyclopedia of acoustics. Wiley, New York, pp 1831–1836

Ray C, Watkins WA, Burns JJ (1969) The underwater song of *Erignathus* (bearded seal). Zoologica 54:79–83

Rayleigh JWS (1896) The theory of sound. 2nd edn. Revised and enlarged. Reprinted 1945, Dover, New York

Reynolds JR, Highsmith RC, Konar B, Wheat CG, Doudna D (2001) Fisheries and fisheries habitat investigations using undersea technology. In: Proceedings of the MTS/IEEE Oceans 2001 Conference, Honolulu, HI, pp 812–820

Robinson SP, Harris PM, Ablitt J, Hayman G, Thompson A, van Buren AL, Zalesak JF, Enyakov AM, Purcell C, Houqing Z, Yuebing W, Yue Z, Botha P, Krüger D (2006) An international key comparison of free-field hydrophone calibrations in the frequency range 1 to 500 kHz. J Acoust Soc Am 120:1366–1373

Rolt KD, Schmidt H (1994) Effects of refraction on synthetic aperture sonar imaging. J Acoust Soc Am 95:3424–3429

Rountree RA, Gilmore RG, Goudey CA, Hawkins AD, Luczkovich JJ, Mann DA (2006) Listening to fish: applications of passive acoustics to fisheries science. Fisheries 31:433–446

Rusby JSM, Somers ML, Revie J, McCartney BS, Stubbs AR (1973) An experimental survey of a herring fishery by long-range sonar. Mar Biol 22:271–292

Schaafsma AS (1992) In situ acoustic attenuation spectroscopy of sediment suspension. In: Weydert M (ed) European Conference on Underwater Acoustics, pp 177–180

Schiagintweit GEO (1993) Real-time acoustic bottom classification for hydrography: A field evaluation of RoxAnn. In: Proceedings of the MTS/IEEE Oceans 1993 Conference 3:214–219

Schmitz B (2002) Sound production in crustacea with special reference to the Alpheidae. In: Wiese K (ed) The crustacean nervous system. Springer, Berlin, pp 536–547

Seim HE, Gregg MC, Miyamoto RT (1995) Acoustic backscatter from turbulent microstructure. J Atmos Ocean Tech 12:367–380

Soria M, Fréon P, Gerlotto F (1996) Analysis of vessel influence on spatial behaviour of fish schools using a multi-beam sonar and consequences for biomass estimates by echo sounder. ICES J Mar Sci 53:453–458

Stachiw JD, Peters D (2005) Alumina ceramic 10 in flotation spheres for deep submergence ROV/AUV systems. In: Proceedings of the MTS/IEEE Oceans 2005 Conference, Washington, DC, 8 pp

Stanton TK (1985) Density estimates of biological sound scatterers using sonar echo peak PDFs. J Acoust Soc Am 78:1868–1873

Stanton TK, Clay CS (1986) Sonar echo statistics as a remote-sensing tool: volume and seafloor. IEEE J Oceanic Eng 11:79–96

Stanton TK, Chu D, Wiebe PH (1998) Sound scattering by several zooplankton groups. II. Scattering models. J Acoust Soc Am 103:236–253

Stanton TK, Chu D, Jech JM, Irish JD (2006) Statistical behavior of echoes from swimbladder-bearing fish at 2–4 kHz. In: Proceedings of the MTS/IEEE Oceans 2006 Conference, Boston, MA, 3 pp

Stanton TK, Chu D, Jech JM, Irish JD (2007) A broadband echosounder for resonance classification of swimbladder-bearing fish. In: Proceedings of the IEEE Oceans 2007 Conference, Aberdeen, UK, 3 pp

Sullivan EJ, Holmes JD, Carey WM, Lynch JF (2006) Broadband passive synthetic aperture: experimental results. J Acoust Soc Am 120:EL49–EL54

Tarifeño E, Andrade Y, Montesinos J (1990) An echo-acoustic method for assessing clam populations on a sandy bottom. Rapp P-v Réun Cons Int Explor Mer 189:95–100

Terhune JM, Ronald K (1986) Distant and near-range functions of harp seal underwater calls. Can J Zool 64:1065–1070

Thode A, Norris T, Barlow J (2000) Frequency beamforming of dolphin whistles using a sparse three-element towed array. J Acoust Soc Am 107:3581–3584

Thomas JA, Fisher SR, Ferm LM, Holt RS (1986) Acoustic detection of cetaceans using a towed array of hydrophones. Rep Int Whal Commn, Special Issue 8:139–148

Thompson PO, Cummings WC, Ha SJ (1986) Sounds, source levels, and associated behavior of humpback whales, Southeast Alaska. J Acoust Soc Am 80:735–740

Thorne RE (1977) Acoustic assessment of Pacific hake and herring stocks in Puget Sound, Washington and southeastern Alaska. Rapp P-v Réun Cons Int Explor Mer 170:265–278

Thorne PD, Hardcastle PJ, Soulsby RL (1993) Analysis of acoustic measurements of suspended sediments. J Geophys Res 98:899–910

Tiffan KF, Rondorf DW, Skalicky JJ (2004) Imaging fall Chinook salmon redds in the Columbia River with a dual-frequency identification sonar. N Am J Fish Manage 24:1421–1426

Traynor JJ, Nelson MO (1985) Methods of the U.S. hydroacoustic (echo integrator-midwater trawl) survey. Int North Pac Fish Comm Bull 44:30–38

Trenkel V, Mazauric V, Berger L (2006) First results with the new scientific multibeam echo-sounder ME70. J Acoust Soc Am 120:3017

Trevorrow MV (1998) Salmon and herring school detection in shallow waters using sidescan sonars. Fish Res 35:5–14

Trevorrow MV (2001) An evaluation of a steerable sidescan sonar for surveys of near-surface fish. Fish Res 50:221–234

Tyack PL (2000) Functional aspects of cetacean communication. In: Mann J, Connor RC, Tyack PL, Whitehead H (eds) Cetacean societies: field studies of dolphins and whales. University of Chicago Press, Chicago, IL, pp 270–307

Tyack PL (2001) Marine mammal overview. In: Steele JH, Turekian KK, Thorpe SA (eds) Encyclopedia of ocean sciences. Academic, San Diego, CA, pp 1611–1621

Tyack PL, Clark CW (2000) Communication and acoustic behavior of dolphins and whales. In: Au W, Popper AS, Fay R (eds) Hearing by whales and dolphins. Springer, New York, pp 156–224

Urick RJ (1983) Principles of underwater sound. 3rd edn. McGraw-Hill, New York

Versluis M, Schmitz B, von der Heydt A, Lohse D (2000) How snapping shrimp snap: through cavitating bubbles. Science 289:2114–2117

Wade G, Coelle-Vera A, Schlussler L, Pei SC (1975) Acoustic lenses and low-velocity fluids for improving Bragg-diffraction images. Acoust Hologr 6:345–362

Watkins WA, Moore KE (1982) An underwater acoustic survey for sperm whales (*Physeter catodon*) and other cetaceans in the southeast Caribbean. Cetology 46:1–7

Watkins WA, Schevill WE (1972) Sound source location by arrival-times on a non-rigid three-dimensional hydrophone array. Deep-Sea Res 19:691–706

Watkins WA, Schevill WE (1979) Distinctive characteristics of underwater calls of the harp seal, *Phoca groenlandica*, during the breeding season. J Acoust Soc Am 66:983–988

Watkins WA, Tyack P, Moore KE, Bird JE (1987) The 20-Hz signals of finback whales (*Balaenoptera physalus*). J Acoust Soc Am 82:1901–1912

Webb DC, Simonetti PJ, Jones CP (2001) SLOCUM: an underwater glider propelled by environmental energy. IEEE J Oceanic Eng 26:447–452

Weilgart L, Whitehead H (1993) Coda communication by sperm whales (*Physeter macrocephalus*) off the Galápagos Islands. Can J Zool 71:744–752

Weilgart L, Whitehead H (1997) Group-specific dialects and geographical variation in coda repertoire in South Pacific sperm whales. Behav Ecol Sociobiol 40:277–285

Wespestad VG, Megrey BA (1990) Assessment of walleye pollock stocks in the eastern North Pacific Ocean: an integrated analysis using research survey and commercial fisheries data. Rapp P-v Réun Cons Int Explor Mer 189:33–49

Westervelt PJ (1963) Parametric acoustic array. J Acoust Soc Am 35:535–537

Weston DE (1967) Sound propagation in the presence of bladder fish. In: Albers VM (ed) Underwater acoustics. Plenum, New York, pp 55–88

Weston DE (1972) Fisheries significance of the attenuation due to fish. J Cons int Explor Mer 34:306–308

Weston DE, Revie J (1971) Fish echoes on a long range sonar display. J Sound Vib 17:105–112

Weston DE, Horrigan AA, Thomas SJL, Revie J (1969) Studies of sound transmission fluctuations in shallow coastal waters. Phil Trans Roy Soc Lond 265:567–607

Weston DE, Somers ML, Revie J (1991) GLORIA interference patterns with modes akin to surface-duct modes. J Acoust Soc Am 89:2180–2184

Weston S, Stachiw J, Merewether R, Olsson M, Jemmott G (2005) Alumina ceramic 3.6 in flotation spheres for 11 km ROV/AUV systems. In: Proceedings of the MTS/IEEE Oceans 2005 Conference, Washington, DC, 6 pp

Winn HE, Edel RK, Taruski AG (1975) Population estimate of the humpback whale (*Megaptera novae-angliae*) in the West Indies by visual and acoustic techniques. J Fish Res Board Can 32:499–506

Wysocki LE, Ladich F (2002) Ontogeny of hearing and sound production in fishes. Bioacoustics 12:183–189

Chapter 19
Combining Techniques for Remotely Assessing Pelagic Nekton: Getting the Whole Picture

James Churnside, Richard Brodeur, John Horne, Patrick Adam, Kelly Benoit-Bird, Douglas C. Reese, Amanda Kaltenberg, and Evelyn Brown

Abstract A variety of observational techniques either have been developed or are under development for fisheries research. These techniques have greatly increased the quantity and quality of information that can be obtained from a research survey and it is anticipated that this trend will continue. Traditional ship-based surveys will be supplemented by data collected from fixed moorings, autonomous underwater vehicles, aircraft, and satellites. Each of these platforms is limited in the spatial and temporal scales that can be sampled. By combining data from multiple platforms and sensors, we will be able to obtain a more complete picture of the components of a particular ecosystem over a greater range of scales. This is particularly true for pelagic nekton, which can move independent of fluid motion. In many cases, the observational difficulties created by this mobility can be mitigated by the use of aircraft, which can cover large areas with optical instruments such as imagers and Light Detection and Ranging (lidar).

Keywords Fisheries surveys · lidar · pelagic nekton

J. Churnside
NOAA Earth System Research Laboratory, CSD3, 325 Broadway, Boulder, CO 80305, USA
e-mail: james.h.churnside@noaa.gov

R. Brodeur, D. C. Reese
Northwest Fisheries Science Center, National Marine Fisheries Service,
Hatfield Marine Science Center, Newport, OR 97365, USA
e-mails: Rick.Brodeur@noaa.gov, dreese@lifetime.oregonstate.edu

J. Horne, P. Adam
University of Washington, School of Fisheries, Box 355020, Seattle, WA 98195, USA
e-mails: jhorne@u.washington.edu, patrick1@u.washington.edu

K. Benoit-Bird, A. Kaltenberg
College of Oceanic and Atmospheric Sciences, Oregon State University, 104 COAS Admin
Bldg, Corvallis, OR 97331-5503, USA
e-mails: kbenoit@coas.oregonstate.edu, akaltenb@coas.oregonstate.edu

E. Brown
Flying Fish Ltd., P.O. Box 169, Husum, WA 98623, USA
e-mail: husumbandb@earthlink.net

R.J. Beamish and B.J. Rothschild (eds.), *The Future of Fisheries Science in North America*,   345
Fish & Fisheries Series, © Springer Science + Business Media B.V. 2009

19.1   Introduction

Fisheries science has always utilized, and often initiated, technological and analytical developments in sampling and data processing. Sampling technologies for pelagic species have traditionally focused on specialized hardware, with each device designed to sample a single or a suite of conspecifics. As the development of individual instruments matured, additional sensors or samplers were often added to a common platform. Data collection among instruments has remained largely autonomous, with data integration occurring after collection.

Continued increases in computing power and data storage have expanded the potential for the acquisition of large data sets. Expanded acquisition often includes higher sample resolution, extended ranges of sampling, or the addition of data streams from the same or colocated instruments. A significant challenge when collecting data from multiple instruments is to synchronize and integrate sampling over space and time. In the recent past, advances in sampling technologies have shifted from an emphasis on increased data acquisition to the analysis and interpretation of the data collected.

At the present time, much of what we know about the distribution of pelagic fishes and plankton is based on data from trawl or ship-mounted acoustic surveys along transects (Gunderson 1993). Both survey methods are conducted from vessels that are relatively slow-moving and are vulnerable to aliasing, depending on the direction of a survey relative to movements by species of interest. Data from trawl surveys are constrained by unknown species-specific gear selectivity and the movement and aggregation patterns of the target animals. Acoustic data are constrained by the lack of ability to differentiate among species and to detect targets at boundaries such as the water surface and the bottom (Fréon and Misund 1999). Near-surface areas inhabited by many pelagic species are particularly difficult to sample with conventional downward-looking echosounders due to the depth of the transducers, the dead zone right below the transducer face, and potential vessel avoidance by fish dwelling near the surface.

While it is clear from this volume that there is a wide variety of tools available for fisheries research, no single instrument and/or platform is capable of providing a complete picture of a marine ecosystem. One problem is that there is only a limited range of spatial and temporal scales that can be accessed from each of the available platforms; the other is that there is only a limited amount of information that can be obtained by each of the available instruments. One solution to the constraints of single sensors and platforms is to integrate and combine information obtained by various techniques to obtain the "whole picture." The epipelagic ecosystem presents particular challenges because of the mobility of its constituents. It also has the unique property that many of its constituents can be observed using airborne optical techniques such as light detection and ranging (lidar), which provides access to a range of spatial and temporal scales not possible from other platforms.

19.2 Techniques

For the purposes of this chapter, we will define a technique as a piece of equipment and the platform used to deploy it. By definition, a hull-mounted echosounder installed on a ship is a technique that is different from the same echosounder deployed on a fixed mooring. The first example provides spatial information about the distribution of fish, while the second provides temporal information. Similarly, an echosounder installed on a ship is a technique different from a trawl on the same ship. The information obtained from each of the two sampling gears is very different and even though the sampling equipment is on a common platform, the spatial and temporal scales sampled by the two gears are also different. Clearly, some techniques are more appropriate than others to describe the spatial and temporal dynamics of the pelagic ecosystem, but there is no single technique that is capable of providing the complete picture at all spatial and temporal scales. In this chapter, we consider combinations of techniques that can be used to sample the pelagic habitat. Various techniques have been combined in the past, but we see the future of fisheries research as involving increased numbers and synchrony of combinations of techniques.

Nets and other sampling gear will always be a critical component of fisheries research. Recent developments in net technology will increase the range and spatial resolution of net sampling (Godø 2009, this volume). However, these are limited to deployment from surface ships, with a corresponding limit to the area that can be covered.

Vessel-mounted acoustics are used regularly in resource assessment and ecosystem description. Transducers from echosounders or sounders may be permanently attached to the hull of a vessel or mounted on a platform that is towed by the ship. Hull-mounted acoustic systems are typically maintained by large government institutions that conduct assessment and research on commercially important fish and invertebrates. Because all acoustic surveys include a direct sampling component, pelagic and demersal fish trawls, plankton nets, and water sampling instruments such as conductivity–temperature–depth (CTD) recorders are commonly found on acoustic research vessels. Acoustic technologies such as echosounders or multibeam sonars may also be mounted on small platforms, commonly called towed bodies. These instrument packages may be dedicated to acoustic sampling or may contain other biological and physical samplers (e.g., Wiebe et al. 2002). A towed package has the advantages of sampling waters close to the surface and can be used on any vessel. Hull-mounted systems are typically able to sample in rougher weather, are installed on vessels that include all instruments needed to conduct a full survey, and contain the infrastructure that integrates data collection and storage. Recent developments in acoustics are providing better target discrimination and calibration than ever before, and further improvements are predicted (Foote 2009, this volume; Holliday 2009, this volume).

Optical techniques are particularly well suited for aircraft deployment. The advantage of aircraft is that large areas of the ocean can be covered quickly at a small fraction of the cost per kilometer of a surface ship. Infrared radiometers provide sea-surface temperature (SST), which is an important physical characteristic of the epipelagic environment. Visible radiometers provide estimates of the primary productivity of the near-surface layer. Visible imaging systems detect schools of fish near the surface. Lidar detects fish and plankton deeper in the water column.

Aircraft have been used by the fishing industry for many years to efficiently locate fish (e.g., herring, sardine, and menhaden) and to direct fishing efforts by purse seiners. The objective of the fishers is to locate dense concentrations of fish close to a processing plant, a sampling strategy that starkly contrasts from that of a fisheries abundance or research survey. Because schools of epipelagic fish are visible from the air (Fig. 19.1), species identification is generally possible using school morphology, color, and characteristic flashing patterns of individual fish.

The application of lidar to fisheries research is a relatively new technique (Brown et al. 2002; Carrera et al. 2006; Churnside et al. 2001, 2003; Churnside and Thorne 2005; Churnside and Wilson 2004; Lo et al. 2000; Tenningen et al. 2006). The National Oceanic and Atmospheric Administration (NOAA) fish lidar installation is shown in Fig. 19.2. A schematic diagram of the lidar (Fig. 19.3) shows the major components. A short (12 ns) pulse of green (532 nm) light is generated by the laser. The light from the laser is pointed about 15° from nadir by a pair of steering mirrors, which provides alignment with the receiver telescope. The beam is also expanded to be eye-safe for humans and marine mammals (Zorn et al. 2000) at the water surface. Light scattered from the ocean is polarization-filtered and collected by the telescope. Polarization orthogonal to the transmitted polarization is typically used because it provides the highest contrast between fish and small particles in the water (Vasilkov et al. 2001). The telescope is focused using an

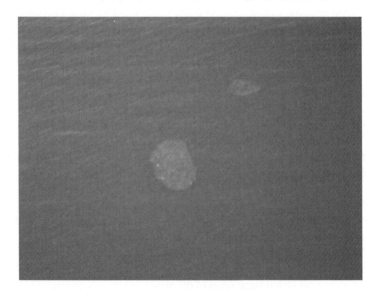

Fig. 19.1 Aerial photo of two schools of menhaden in the Atlantic Ocean about 10 km east of Ocean City, New Jersey. The larger school is about 15 m in the long direction (Photo courtesy of Brian Lubinsky, US Fish and Wildlife Service)

Fig. 19.2 Installation of fish lidar in the NOAA Twin Otter aircraft

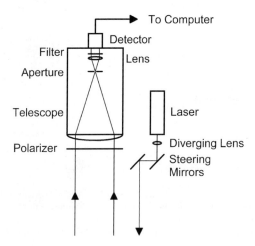

Fig. 19.3 Schematic diagram of the NOAA fish lidar

interference filter, which reduces the amount of background light, onto a pho-
tomultiplier tube, which converts the light into an electronic signal. This signal
is digitized at a rate of 1 GHz, producing a profile of the scattering in the water
with a depth resolution of 11 cm. A second receiver channel collects the scattered
light that is co-polarized with the transmitter to provide more information about
small-particle scattering.

Data from lidar are similar to echosounder data, in that they provide a depth
profile of returned or backscattered energy. It is important to note that the
propagation and scattering physics of light and sound differ. The main consequence
of this is that lidar can only penetrate to about 50 m depth in the clearest waters, and
even less in typical coastal waters. Depending on the carrier frequency, scientific
echosounders can sample the water column over several hundred meters. Figure
19.4 (a–d) presents several examples of scattering profiles from the lidar.

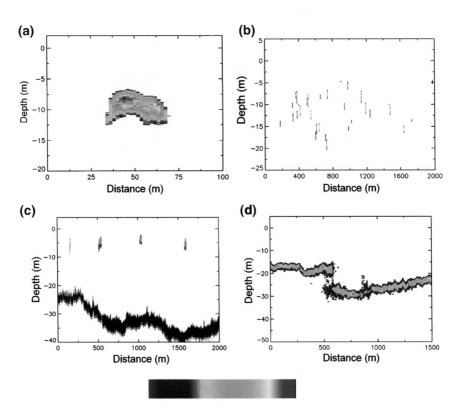

Fig. 19.4 Examples of fish school backscatter profiles from lidar, with relative color scale at
the bottom. (**a**) School of sardines off the Oregon coast; (**b**) aggregation of salmon sharks in the
Gulf of Alaska; (**c**) four schools of herring over the bottom in the southeastern Bering Sea (note
the bottom return is black where it is off the color scale); (**d**) thin zooplankton layer off the
Oregon coast

Data from optical techniques are also acquired from satellites. Sea-surface temperature and primary productivity maps are currently available. Lidar from satellites may be possible in the future, although the ability to resolve small fish schools with this technique is unlikely.

19.3   Combinations of Techniques

Combining sampling equipment on the same platform is the easiest way to combine and integrate techniques. The spatial and temporal range of the data will match while sample resolution will depend on instrument aperture and sample frequency. Matching sample data using a common resolution is easier when samples are synchronized on the same platform. Another relatively straightforward integration of data types is the combination of ship and satellite data. Despite the huge difference in platform speed ($10\,km\ s^{-1}$ for a satellite in low-earth orbit compared with $5\,m\ s^{-1}$ for a surface ship), the spatial and temporal resolutions of the two data sources are compatible. Aircraft provide access to large areas in short time frames, which results in a more synoptic data stream than ships can provide, but the ability to combine aircraft and ship data is more difficult.

A traditional combination of sampling techniques places acoustics and fishing gear on the same ship. A systematic sampling of preplanned transects is occasionally interrupted to sample dense concentrations of all observed acoustic targets. The systematic acoustic survey provides information on fish distribution, while samples from trawls are used to collect data on species composition, length distribution, and physiological or reproductive condition of fish.

Another combination of sampling techniques uses video and acoustic equipment mounted on trawl nets. These gears are used to monitor fish behavior around and in the net. These techniques are discussed elsewhere in this volume (Godø 2009, this volume).

The combination of vessel-mounted and moored acoustics can be used to collect spatially and temporally independent density data. Vessel-mounted acoustics generate spatially indexed data that may be confounded by the passage of time while sampling, but is typically used to map density distributions in an area, estimate abundance, and to size fish or invertebrate populations of interest. Moored acoustics are fixed in space and spatial coverage is restricted by the transducer beam angle and the depth of the water column. Moored acoustics are used to investigate vertical movement patterns over daily cycles or the timing and flux of migrations (Onsrud et al. 2005). An array of moorings can be placed to capture horizontal movement patterns if a *priori* knowledge of migration direction is available. Moored acoustic arrays, typically deployed on a smaller scale (e.g., 3 km in length with 0.5 km spacing) compared to ship-based acoustic surveys, which may cover hundreds or thousands of nautical miles, are planned to expand in both scope and coverage (e.g., see http://www.oceantrackingnetwork.org/index.html).

A limited number of surveys have combined vessel-based and moored acoustics. These studies have typically been limited to a single, downward-looking echo-sounder, such as the Bergen Acoustic Buoy system (Skaret et al. 2006) as the moored component. In this configuration, the acoustic buoy records a time series of nekton density, but the directional component of nekton horizontal flux remains unknown. To detect the horizontal flux of mesopelagic nekton, Benoit-Bird and Au (2004) used a linear array of upward-looking moored echosounders. Even more detailed information about seasonal fish movements and predator–prey interactions has been obtained by a combination of acoustic sensors in a Norwegian fjord (Godø 2009, this volume).

To investigate the four-dimensional flux of nekton, we have combined vessel-mounted and moored acoustics in a small-scale grid (1 nautical square mile) survey. One upward-looking 200 kHz acoustic system was moored at each corner of the grid enabling the vertical, latitudinal, and longitudinal movement of nekton to be resolved over 7 days at a resolution of 4 s. To complement data recorded by the moored acoustics, the spatial distribution of nekton within the survey grid was surveyed along transects using 38, 120, and 200 kHz echosounders. Coincident measurements were made at each of the four corners, providing observations of spatially and temporally indexed densities. The vessel completed sets of eight transects (four north–south, four east–west) to record the fine-scale distribution of nekton within the survey grid and to estimate the horizontal flux of nekton through the survey area.

Radiometers in low-earth orbit provide maps of sea-surface temperature and primary productivity of near-surface waters. These radiometers do not cover the entire earth during each orbit, but build a composite image over a period of approximately 1 week. With consideration of data losses due to cloud cover, a monthly composite may be required to complete an image. This time period is very similar to the timescale of a vessel-based acoustic survey. In both cases, the assumption is made that there are no significant changes in the density distributions of the species or ecosystem components during the survey time period. An example of the combination of satellite data with an acoustic/trawl ship survey (Fig. 19.5) shows how the backscattered acoustic energy is concentrated in the upwelling zone near the coast. The trawl results are used to relate the backscattered energy to the length and density distributions of sardines.

Understanding pelagic ecosystems is difficult due to the sheer volume of these environments, making it problematic and costly to sample. As mentioned above, satellites provide coverage over large spatial scales; however, complete coverage is often disrupted by cloud cover. One way to overcome this dilemma is to combine the satellite data with data obtained from aircraft. Aircraft are capable of covering large areas over short periods of time, thus making them ideal for environments that are typically difficult or costly to sample. By combining satellite and aircraft data, it is possible to obtain data on animal distributions along with important habitat characteristics such as SST and chlorophyll concentrations, over large spatial scales at high temporal resolutions. In addition, for areas which are not sampled by satellites (e.g., due to cloud cover or areas close to shore) information can be obtained by

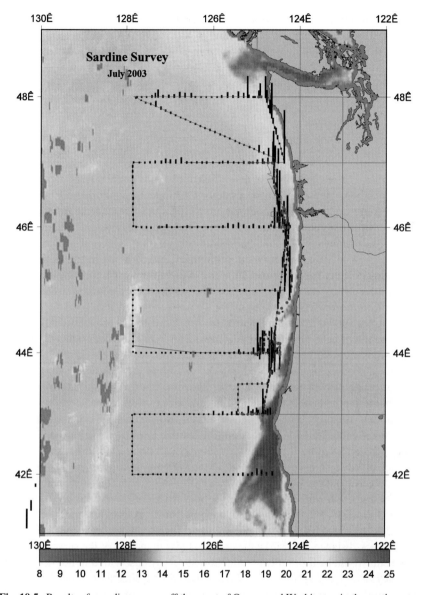

Fig. 19.5 Results of a sardine survey off the coast of Oregon and Washington in the northwestern United States. The background image presents satellite-derived sea-surface temperature (°C) according to the color bar. Vertical bars along the survey tracks present relative acoustic return. Blue (red) circles mark locations of trawls that did (did not) contain sardines (Figure courtesy of Richard Charter, NOAA National Marine Fisheries Service)

data collected from aircraft. To get a more complete picture of what is going on in an ecosystem, trawl data can be included to provide information on species, community dynamics, and size distributions. In addition, trawls provide a means to ground-truth visual observations on species identification from aircraft.

19.4 The Future

Advances in fisheries research will require data collected over a wide range of spatial and temporal scales and effective and efficient methods to meld large data volumes into a coherent picture of an ecosystem. Continuous monitoring of the earth's surface from space will continue with improved spatial resolution, more accurate measurements of primary production in coastal areas, and the addition of surface-salinity measurements. Continuous monitoring of the water column at specific locations will only increase with the deployment of additional instrumented moorings. Episodic surveys using aircraft as sample platforms will increase and should improve the effectiveness of coordinated ship surveys. This increased use of aircraft as sample platforms will be driven by the synoptic nature of aerial surveys and their low cost per survey kilometer relative to ship surveys. The greatest challenge for researchers will be to effectively integrate and analyze these data sources.

The most straightforward way to combine techniques is to use adaptive surveys. Satellite and mooring data would be used to define the area and timing for the most effective aircraft and ship surveys. The area and timing of ship surveys are often based on historical information, but these may shift in the short term due to interannual variability in water temperatures and potentially become unreliable over longer time periods due to shifts in habitat associated with global climate change or local regime shifts. Aircraft surveys can be used to direct ship surveys to efficiently use ship time. To illustrate the latter, consider the case of Fig. 19.5. A lidar instrument package was flown over the ship during the systematic survey. Had the ship been directed by the lidar survey, it could have covered less than the full transects, but still captured most of the backscattered acoustic energy. Specifically, 90% of the acoustic energy could have been obtained from 65% of the survey track distance, or 65% of the energy at less than 30% of the survey distance. The result would be a significant reduction in survey cost with a minor decrease in accuracy. Alternatively, the same ship time could be directed to more productive areas, producing a more accurate survey at the same cost. Caveats to this strategy include the assumption that all species of interest are detected by both sensors, and that the time required by the ship to transit to high-density areas observed during the aerial survey are small relative to the time periods when animal distributions change significantly.

One difficult aspect when studying the interaction between distributions of marine animals and habitat components that affects distributions is obtaining data at high spatial and temporal scales. Historically, researchers studying nekton have relied on ship-based observations. It is now possible to integrate data from a variety of platforms. Increases in computing power and data storage capabilities allow researchers to manipulate and analyze large quantities of data. Initial data products include images or "snapshots" of an ecosystem. The Geographic Information System (GIS) is an important tool for fisheries biologists and is being used more frequently to display relationships between the distributions of pelagic marine life and important habitat components. Increased computing power also enables geostatistical tools to interpolate data values in areas where data were not collected (e.g., Reese and Brodeur 2006). Combining data from multiple platforms within a GIS framework facilitates a synoptic and complete coverage of important fish habitat. The production of images within GIS software can easily be shared with resource managers, scientists, politicians, and the general public.

Another way to combine techniques is to use statistical parameter estimation. For example, if we have statistical models of the probability density functions of lidar-returned energy, acoustic backscattered energy, and the trawl catches of sardine biomass, we can combine those data to predict the value of the sardine biomass. We want to predict sardine biomass S such that the probability p of biomass given observed lidar L, acoustic A, and trawl T is a maximum (i.e., $p(S|L,A,T)$]. Using Bayes' theorem, this quantity will be maximized when $p(L,A,T|S)\,p(S)$ is at a maximum. The first factor describes the probability density of the observed signals for a given biomass and the second describes the probability density of the overall sardine biomass. The first factor includes all uncertainties within surveys such as under-sampling, fish detectability, and sensor noise. It is based on an understanding of the characteristics of the various techniques. The second factor is based on the variability in biomass from historical data.

Finally, different techniques might be combined using data assimilation into bio-physical models (e.g., Besiktepe et al. 2003). Data can be inserted into the models at the temporal and spatial scales available. In a classical view of four-dimensional modeling, one specifies the conditions at all points at time zero and allows the model to develop from there to the time of interest, according to the appropriate differential equations. In a model that includes assimilation, the model predictions will be compared with observations made after the initial time of the model. Model results will be adjusted to more closely conform to the observations, and the model run will proceed. Consistency of the model (i.e., the requirement that it satisfy the underlying equations) can be used to ensure that different data streams are weighted according to their intrinsic reliability. For this approach to be effective, we need better models than those currently available (Methot 2009, this volume), and we will probably also need more computer power than has been available to date.

We have described a vision of the future that includes improvements in sampling, acoustic, and optical technologies. However, we think even greater advances are possible by combining the results from different techniques, each of which has different strengths and weaknesses. This will be particularly effective as results from different platforms are combined to provide a complete picture of an ecosystem over a greater range of spatial and temporal scales. The potential for this is most evident in the epipelagic ecosystem, which is accessible to observation from ships, moorings, autonomous vehicles, aircraft, and satellites.

References

Benoit-Bird KJ, Au WWL (2004) Diel migration dynamics of an island-associated sound-scattering layer. Deep-Sea Res. 51:707–719

Besiktepe ST, Lermusiaux PFJ, Robinson AR (2003) Coupled physical and biogeochemical data-driven simulations of Massachusetts Bay in late summer: real-time and postcruise data assimilation. J Mar Syst 40–41:171–212

Brown E D, Churnside JH, Collins RL, et al. (2002) Remote sensing of capelin and other biological features in the North Pacific using lidar and video technology. ICES J Mar Sci 59: 1120–1130

Carrera P, Churnside JH, Boyra G, et al. (2006) Comparison of airborne lidar with echosounders: a case study in the coastal Atlantic waters of southern Europe. ICES J Mar Sci 63:1736–1750

Churnside JH, Demer DA, Mahmoudi B (2003) A comparison of lidar and echosounder measurements of fish schools in the Gulf of Mexico. ICES J Mar Sci 60:147–154

Churnside JH, Sawada K, Okumura T (2001) A comparison of airborne lidar and echo sounder performance in fisheries. J Mar Acoust Soc Jpn 28:49–61

Churnside JH, Thorne RE (2005) Comparison of airborne lidar measurements with 420 kHz echo-sounder measurements of zooplankton. Appl Opt 44:5504–5511

Churnside JH, Wilson JJ (2004) Airborne lidar imaging of salmon. Appl Opt 43:1416–1424

Churnside JH, Wilson JJ, Tatarskii VV (2001) Airborne lidar for fisheries applications. Opt Eng 40:406–414

Fréon P, Misund OA (1999) Dynamics of pelagic fish distribution and behaviour: effects on fisheries and stock assessments. Fishing News Books, Oxford

Gunderson DR (1993) Surveys of fisheries resources. Wiley, New York

Lo NCH, Hunter JR, Churnside JH (2000) Modeling statistical performance of an airborne lidar survey for anchovy. Fish Bull 98:264–282

Onsrud MSR, Kaartvedt S, Breien MT (2005) In situ swimming speed and swimming behaviour of fish feeding on the krill *Meganyctiphanes norvegica*. Can J Fish Aquat Sci 62:1822–1832

Reese DC, Brodeur RD (2006) Identifying and characterizing biological hotspots in the Northern California Current. Deep-Sea Res II 53:291–314

Skaret G, Slotte A, Handegard NO, et al. (2006) Pre-spawning herring in a protected area showed only moderate reaction to a surveying vessel. Fish Res 78:359–367

Tenningen E, Churnside JH, Slotte A, et al. (2006) Lidar target-strength measurements on north-east Atlantic mackerel (*Scomber scombrus*). ICES J Mar Sci 63:677–682

Vasilkov AP, Goldin YA, Gureev BA, et al. (2001) Airborne polarized lidar detection of scattering layers in the ocean. Appl Opt 40:4353–4364

Wiebe PH, Stanton TK, Greene CH, et al. (2002) BIOMAPER-II: an integrated instrument platform for coupled biological and physical measurements in coastal and oceanic regimes. IEEE J Ocean Eng 27(3):700–716

Zorn HM, Churnside JH, Oliver CW (2000) Laser safety thresholds for cetaceans and pinnipeds. Mar Mammal Sci 16:186–200

Chapter 20
Using the Seabed AUV to Assess Populations of Groundfish in Untrawlable Areas

M. Elizabeth Clarke, Nick Tolimieri, and Hanumant Singh

Abstract Near-bottom groundfish communities were surveyed in the waters off California and Oregon using the Seabed autonomous underwater vehicle (AUV) at depths ranging from 100 to 500 m. These surveys were designed to test the utility of the Seabed AUV in surveying groundfish and their associated habitat. The long-term goal of these tests is to fill the need for cost-effective, non-extractive, fishery-independent surveys in untrawlable areas. During nine dives we collected information on the species composition, abundance, and size of many groundfish species, and over 30,000 images were collected from the optically calibrated camera. Habitat classification on a fine spatial scale was easily accomplished and allowed habitat associations of many species to be determined. An assessment of one of the most easily identifiable species, rosethorn rockfish *Sebastes helvomaculatus*, also showed that the size composition of this species varied over habitat types. These surveys were the first using the Seabed AUV to survey fishes in these habitats and provided insights for sample design and enhancements that would optimize the AUV for future operational surveys.

Keywords Autonomous underwater vehicle, AUV · fishery-independent surveys · groundfish · underwater imaging

20.1 Introduction

Fishery-independent surveys of groundfish populations provide the basic information needed for monitoring and assessing fish stocks. These assessments are a key component of the scientific advice to decisions makers for management of groundfish.

M. E. Clarke(⊠), N. Tolimieri

Northwest Fisheries Science Center, NOAA Fisheries Service, 2725 Montlake Blvd E.,
Seattle, WA 98136, USA
e-mail: elizabeth.clarke@noaa.gov

H. Singh
Department of Applied Ocean Physics and Engineering,
Woods Hole Oceanographic Institution, Woods Hole, MA 02543-1109, USA

In fisheries research, fishery-independent data are becoming increasingly important in the light of growing concerns about the availability and applicability of fishery-dependent data. Fishery-dependent data are of limited utility in monitoring populations when catch limits are very restrictive. It is therefore apparent that there is a need to invest in additional indices of abundance. It is also important in light of the limited catch available for some overfished species that new methods that are non-extractive should also be developed.

Traditionally groundfish have been surveyed primarily from vessels using the same extractive methods similar to those used in the capture fisheries. On the west coast most of the surveys for groundfish have been conducted using bottom trawl gear. This gear cannot be effectively deployed in rocky habitat, and these rugged areas are the primary habitat of many groundfish species. Currently a bottom trawl survey is the primary fishery-independent source of information for stock assessments (Keller et al. 2006). This survey is conducted annually from the Mexican border to the Canadian border in water depths between 50–1200 m. Over 700 stations are randomly selected in designated strata and are occupied in trawlable areas. The survey area is subdivided into approximately 12,000 cells that are 1.5×2.0 nm. Each of four vessels is randomly assigned a set of 180 cells. While this survey provides comprehensive information in areas accessible by bottom trawls much of the area is likely untrawlable (Zimmermann 2003).

In general the assumption is made that bottom trawl surveys represent to some degree untrawlable areas. In some cases the observed trawl densities of fish biomass are expanded to both trawlable and untrawlable grounds. If in untrawlable areas the groundfish species composition, size composition, and biomass are unknown, in the bottom trawl survey estimates may result in inaccurate estimates of fish biomass. Thus, if the densities differ between the trawlable and untrawlable grounds, then the density observed from the trawl samples would not be expected to be representative of the population density. This situation is most problematic if the distribution of fish on trawlable and untrawlable grounds is density-dependent (Fretwell and Lucas 1970) or if the size distribution between untrawlable and trawlable areas varies considerably. Autonomous underwater vehicles (AUVs) are relatively small, untethered, unmanned, self-propelled vehicles able to carry a variety of sensors along preprogrammed mission trajectories. AUVs have already been employed as a survey platform to detect fish in midwater using acoustic methods (Fernandes et al. 2000, 2003). Studies have also shown that these electric powered platforms may have minimal fish avoidance (Fernandes et al. 2000, 2003; Griffiths et al. 2001). A unique bottom tracking AUV, the Seabed, has been used to monitor coral reef habitat (Armstrong et al. 2006). We are attempting to further develop the Seabed as a tool to conduct non-extractive surveys of groundfish. This particular AUV we believe has great potential because its unique bottom hugging capability makes it particularly appropriate for this task (Singh et al. 2004).

In this paper we discuss the capabilities of the AUV, the results of initial tests with the current AUV, and recommend improvements that are needed to use this AUV as an operational tool.

20.2   Methods and Results

In the fall of 2005, the Seabed AUV was deployed off the research vessel *Thomas G. Thompson* to assess the potential for AUVs to conduct surveys in untrawlable areas. The Seabed AUV is a multihull, hover-capable vehicle, which, unlike traditional torpedo-shaped AUVs, is capable of working extremely close to the seafloor while maintaining very precise altitude (3 ± 0.05 m) and navigation (1 m) control (Fig. 20.1). Its small footprint coupled with its 2,000 m depth rating makes it an ideal platform for conducting surveys at the continental shelf and upper slope depths deployed from on ships ranging from standard oceanographic vessels to smaller fishing vessels of opportunity.

The Seabed was utilized at three different locations on the west coast of the United States over the course of 14 days. These areas included Daisy Bank (two dives) and Coquille Bank (one dive) off the coast of Oregon and St Lucia Bank (six dives) off the coast of Central California (Fig. 20.2). During these dives over 30,000 images were collected from the optically calibrated camera. On average each image viewed an area of 3.02 m^2 (± 0.32 S.D.).

The suite of sensors on board the AUV include 1.2 megapixel 12-bit high dynamic-range camera and associated strobe, a 230 kHz Delta-T multibeam imaging system, a 1.2 MHz Acoustic Doppler Current Profiler (ADCP), fluorometers, a pumped CTD, and methane sensor. Typical mission durations for the current vehicle allow it to run with its suite of sensors for 6–8 h covering distances of up to 10–15 km on a single dive.

The AUV, which can run at speeds between 0.3 and 1 m/s, was programmed to run at minimum speed and to maintain a fixed distance from the bottom of 2.5 m. The initial position of the AUV was determined by shipboard GPS. Measurements of velocity over the bottom, heading, altitude, pitch, roll, and integrated position

Fig. 20.1 The Seabed AUV

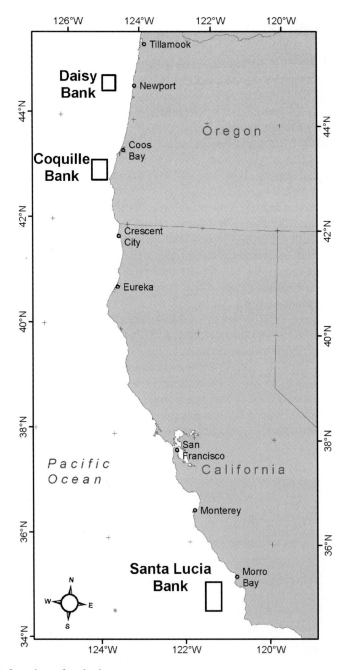

Fig. 20.2 Locations of study sites

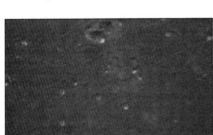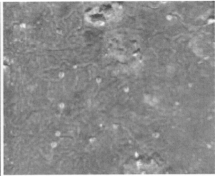

Fig. 20.3 *Left*: Original image of burrowing irregular urchins (*Spatangoida* sp.). *Right*: The same image after color compensation. The high dynamic range camera on board Seabed allows us to compensate for the nonlinear attenuation of light underwater to obtain high-resolution imagery with high color fidelity

are provided by a 1.2 MHz ADCP. More information on the Seabed components, controls, and navigation are listed in Singh et al. (2004).

The sensors, the AUV, and its associated systems are all vertically integrated. Thus the imagery can be easily color-corrected (Fig. 20.3), merged with the navigation and attitude data, photomosaicked, and then analyzed for species counts, sizes, and distributions with an easy to use Graphical User Interface (Ferrini and Singh 2006).

20.2.1   General Sampling Protocol

At Daisy and Coquille Banks the AUV ran similar survey tracks (Figs. 20.4 and 20.5). On each dive, the AUV completed four transects, each approximately 0.5 km in length. The first transect was run within the center of the rocky habitat or within the center of the proposed closed area, as appropriate. The second transect was run within the rocky habitat (or closed area) but along the margin of that habitat. The third and fourth transects were run in the soft sediment (or outside the proposed closed area) with one transect along the margin and one farther from the rocky area (or closed area). This design allowed us to examine the effects of habitat type and edge effects. Along each transect the AUV took approximately 600 photo quadrats (frames, $3.0\,m^2 \pm 0.32$ S.D.). From each transect, we analyzed 200 randomly chosen frames (see below).

For the dives at St Lucia Bank, the AUV tracks differed from those at Daisy and Coquille Bank (Fig. 20.6) because of different overall sampling objectives. At St. Lucia Bank we were primarily concerned with collecting data inside and outside of proposed trawl closure areas. Dive 15 followed a track line similar to those at Daisy and Coquille Bank but with three transects within the proposed closed area and three outside. Dive 10 was in rocky habitat also crossing the boundary of the closed area with one transect inside and one transect outside. Dive 11 was inside

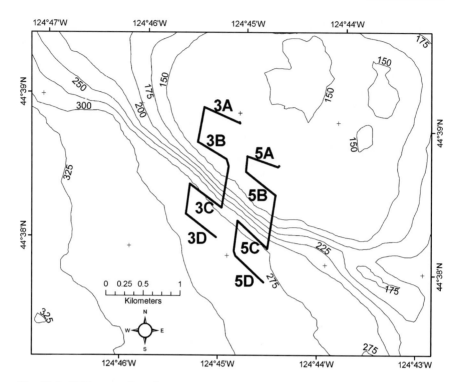

Fig. 20.4 AUV survey lines for two dives at Daisy Bank. Within each dive the four labeled, parallel lines were data collection transects. The three parallel lines at 90° are transit lines where data were not collected. The first two lines (A and B) are within rocky habitat; the second two lines (C and D) are at the base of the bank in sediment

the closed area in rocky habitat and Dive 12 was outside of the closed area in rocky habit. Dive 16 was within the closed area in soft sediment. Dive 14 was an exploratory dive conducted in an area where petrale sole *Eopsetta jordani* were reported to occur at high density.

In many cases, especially at Daisy and Coquille Banks, habitat is confounded by depth because the soft sediment areas were deeper than the adjacent rocky area. In a fisheries sense, this is not a problem. Trawl surveys frequently sample the softer sediments around these rocky outcrops. Clearly we would like to know whether sampling in the trawlable area next to an untrawlable area gives us good information about populations on the rocky areas.

20.2.2 Subsampling Schemes

In order to determine the best method for analysis of the large number of images collected, we evaluated several subsampling schemes by conducting power analyses

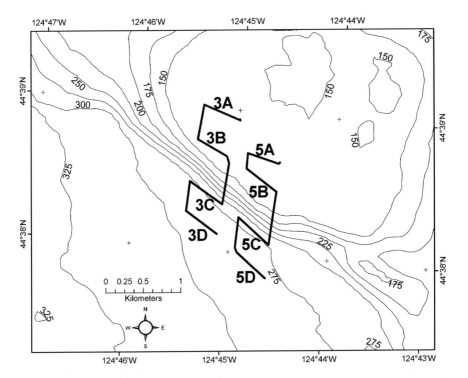

Fig. 20.5 AUV survey lines for one dive at Coquille Bank. Within the dive the four labeled, parallel lines were data collection transects. The one long line between the labeled lines is a transit line where data were not collected. The first two lines (7A and 7B) are within rocky habitat; the second two lines (7C and 7D) are at the base of the bank in sediment

on a subsample of the data from Daisy and St Lucia Banks prior to completing the complete analysis of the frames. We considered three alternatives: (1) divide the main transect into three randomly placed but nonoverlapping subtransects 100 frames long, and analyze all frames to generate a total count for each subtransect; (2) as above, but analyze only 50 alternate frames, and (3) analyze 90 random frames from each full transect treating each frame as a replicate. In this sampling regime, the main transect (with four per dive) was considered a location (e.g., center of the rocky area or margin of the rocky area). For both cases (1) and (2), $n=3$ for each location, while in case (3), $n=90$. In case (1) double counts were deleted since frames overlapped and individual fish could be counted twice.

For the power analysis, we examined differences between Daisy Bank and St Lucia Bank in terms of the number of rockfish *Sebastes* spp. (six 100-frame sections from Daisy Bank and three 100-frame sections from St. Lucia Bank). Since the data were counts, we used log-linear models (generalized linear models with log-link and Poisson distribution).

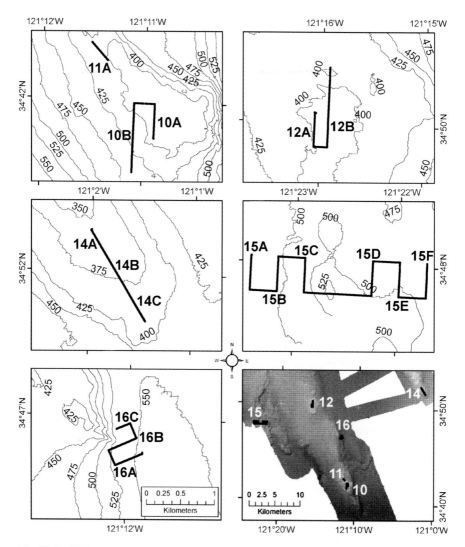

Fig. 20.6 AUV survey lines for dives at St. Lucia Bank. Within the dives 10, 12, 15, and 16, the labeled, parallel lines were data collection transects. The lines between the labeled lines are transit lines where data were not collected. Dives 11 and 14 were exploratory dives in unique habitats and sampling was conducted in one continuous line. The lower right panel shows the locations of all sampling sites on sun-illuminated bathymetry

We estimated the statistical power of the GLM tests following Willis et al. (2003) who provide a conversion from standard power analysis, which assumes homogeneity of variance, to the Poisson situation where variance equals the mean and the data may also be overdispersed such that $\sigma^2 = \phi\mu$, where ϕ is the overdispersion parameter.

Table 20.1 Sample size needed to detect various multiplicative effect sizes. Note that for the two methods, the replicate is a 100-frame section of a 0.5 km long main transect (considered a location). Thus while 46 subtransects would be required to detect a 50% difference (effect size of 1.5) in the number of fish between two sites using the all frames approach, this would require analyzing 4,600 frames

Method	φ (overdispersion parameter)	Desired effect size		
		×1.25	×1.5	×2.0
All frames in a section of a subtransect	21.38	164 subtransects (16,400 frames)	46 subtransects (4,600 frames)	15 subtransects (1,500 frames)
Alternate frames in a section of subtransect	14.74	135 subtransects (6,750 frames)	38 subtransects (1,900 frames)	12 subtransects (600 frames)
Random frames selected from an entire transect (location)	2.4	1,435 frames	403 frames	124 frames

An approximate upper bound on type II error rate is given by the value β obtained as the probability of having standard normal quantile z_β given by:

$$z_\beta = \frac{\log(k)}{\sqrt{\dfrac{\phi}{n\mu_1} \dfrac{k+1}{k}}} - z_{\alpha/2}$$

where k is the ratio of the two means $k=\mu_2/\mu_1$ and μ_1 is the smaller of the two means. The lower bound on power is then $1 - \beta$. As usual, n is the sample size. The standard normal quantile exceeds the value z_β with probability β. The value α is the type I error rate (here 0.05) such that $z_{\alpha/2}=z_{0.025}=1.96$. It is relevant to note that overdispersion and low mean abundance in the smallest of the means being compared reduces power.

Results of the power analysis suggested that analyzing individual, random frames was a more effective approach for two reasons (Table 20.1). First, analyzing individual, random frames results in a higher sample size for the same level of effort ($n=90$ versus $n=3$ per location). Second, the other two methods resulted in highly overdispersed data (indicating clumping). Ideally the overdispersion parameter should be equal to 1.0, but values less that 3.0 are generally considered acceptable.

20.2.3 Species Identification

In many cases the success of identification of species using remote optical methods is dependent on species-specific morphological characteristics and markings being identifiable from a photograph. In the case of the images from the Seabed AUV, the identification of fish species is dependent on the analyst's ability to discern these characteristics from an overhead view of the organism. Flatfish and skates are particularly easy to identify from this view (Fig. 20.7). Certain rockfish such as *Sebastes helvomaculatus*, rosethorn rockfish, and *S. diploproa*, splitnose rockfish, have morphology or markings that along with the known bathymetric and geographic distributions make them easily identifiable from the overhead images at least over portions of their ranges. Members of species groups such as thornyheads, *Sebastolobus* spp. can be identified but shortspine *Sebastolobus alascanus* and longspine thornyheads *S. altivelis* generally cannot be distinguished from each other. For marine invertebrates, where there are specific morphological characteristics and markings that are easily identifiable from the overhead view the identification of these invertebrates to a particular taxonomic or morphologic level and in many cases to species is possible. One benefit of the vertical images is that the spatial relationships between fishes and invertebrates and associated habitats are easily viewed and quantified (Fig. 20.8).

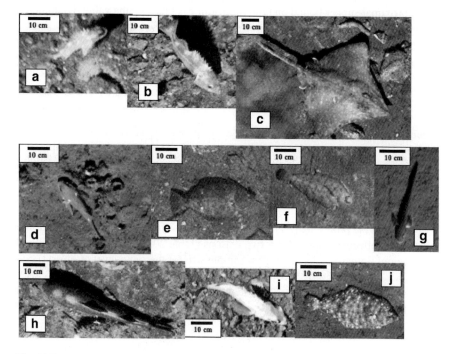

Fig. 20.7 Typical groundfish identifiable from AUV images. (**a**) Rosethorn rockfish off Oregon, *Sebastes helvomaculatus*; (**b**) blackgill rockfish, *S. melanostomas*; (**c**) longnose skate, *Raja rhina*; (**d**) thornyhead, *Sebastolobus spp.*; (**e**) petrale sole, *Eopsetta jordani*; (**f**) rex sole, *Glyptocephalus zachirus*; (**g**) bigfin eelpout, *Lycodes cortezianus*; (**h**) sablefish, *Anoplopoma fimbria*; (**i**) splitnose rockfish, *S. diploproa*; (**j**) Dover sole, *Microstomus pacificus*

Fig. 20.8 Blackgill rockfish *Sebastes melanostomus* and "vase" sponge

20.2.4 Monitoring of Habitat Associations

The habitat was classified and the number of rockfish were counted at each of 8 transects on Daisy Bank, 4 transects on Coquille Bank and 17 transects on St. Lucia Bank. Habitat was classified using a simplified two-letter classification scheme (after Hixon et al. [1991] and Stein et al. [1992]) where the first letter indicates the primary substrate type (50% of substrate or greater) and the second letter indicates the secondary substrate type (greater than 20% but less than 50% of the substrate type). In this scheme the letter R indicates rock ridge, F indicates flat rock, B indicates boulder, C indicates cobble, P indicates pebble, S indicates sand and M indicates mud. Rockfish were counted with the assistance of the Graphical User Interface developed for this purpose (Ferrini and Singh 2006) (Fig. 20.9). This revealed that rockfishes were much more abundant on transects where the percentage of rocky habitat was the highest and lowest where mud and sand predominated (Fig. 20.10).

20.2.5 Size Composition of Rosethorn Rockfish

An analysis of the size composition of one of the most easily identified species, rosethorn rockfish, *S. helvomaculatus*, was conducted on dives that occurred on Coquille and Daisy Bank. This analysis was limited to Daisy and Coquille Banks

Fig. 20.9 An example of the Graphical User Interface (GUI) that was used to easily analyze images collected by the Seabed AUV

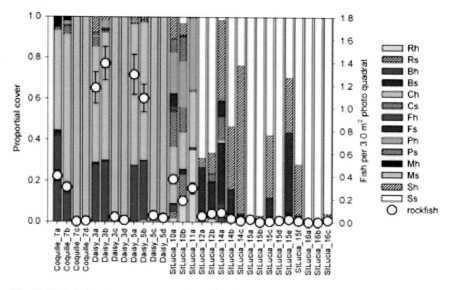

Fig. 20.10 Relationship between rockfish and associated habitat type on 4 transects on Coquille Bank, 8 transects on Daisy Bank and 17 transects on St. Lucia Bank (where R=rock ridge, F=flat rock, B=boulder, C=cobble, P=pebble, S=sand, M=mud, and an upper case letter indicates that greater than 50% was classified as that type and lower case indicates that greater than 20% but less than 50% was classified as that habitat type)

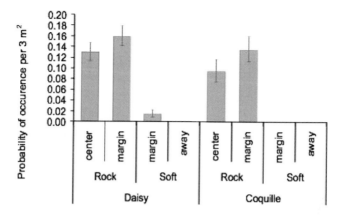

Fig. 20.11 Density of rosethorn rockfish *Sebastes helvomaculatus* in four habitats at Daisy and Coquille Banks. Because rosethorn were either present (one fish) or absent in the photoquadrats, data were analyzed using a logistic regression model. Probability of occurrence is qualitatively similar to density

because in more southerly areas rosethorns could be easily confused with several other species (Yoklavich et al. 2000). Rosethorn are typical of the deep rock habitat (Love and Yoklavich 2006), and adults also use transitional areas between rock and mud (Love et al. 2002). For four transects (200 random quadrats per transect), all of the rosethorn rockfish were counted and their body size in grams estimated from length measurements using published length–weight relationships (Love et al. 2002). The habitat (transect not the individual frame) on which each fish was found was categorized as either rocky or soft and as either on the edge or the center of those habitats. More rosethorn were found on rocky habitat than on adjacent soft sediment (Fig. 20.11). But larger fish and more biomass were found on the reef margin than in the center (Figs. 20.12 and 20.13). Only large fish ever ranged off of the rocky areas.

20.3   Discussion

The primary focus of this research was to determine the utility of a bottom-tracking AUV as a tool for assessing the abundance of groundfish in and around rocky, untrawlable areas. The advantages routinely described in surveys with AUVs versus Remotely Operated Vehicles (ROVs) were apparent in this application (Bingham et al. 2002). The Seabed AUV tether-free maneuvering ability and near-bottom performance were significant assets when surveying in rocky areas. Furthermore, the deploying vessel was freed after deployment to conduct other operations such as high-resolution multibeam sonar mapping of the sea floor and water column oceanography. This led to increased efficiency in the use of research vessel time.

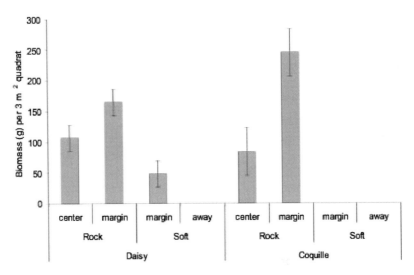

Fig. 20.12 Biomass of rosethorn rockfish *Sebastes helvomaculatus* in various habitats on Daisy and Coquille Bank

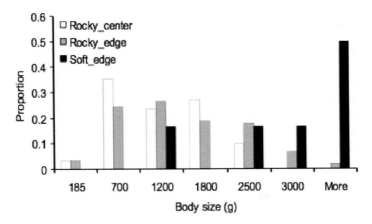

Fig. 20.13 Size of individual rosethorn rockfish *Sebastes helvomaculatus* in rocky versus soft habitat

The lack of tether also avoided problems such as entanglements. The lack of the tether did have the disadvantage that unlike with tethered vehicles (ROVs) no images could be viewed in real time. However, the unique ability of the vehicle to drive at a fixed distance from the bottom was a clear advantage over many other platforms. Not only were observations simplified since the distance from the bottom was not constantly changing but the ability to maintain a fixed altitude from the bottom even in rocky areas allowed the vehicle to avoid collisions. In addition, the difficulty in target identification and verification associated with acoustic surveys from AUVs (Fernandes et al. 2003) was not evident here. However, for some species we have yet to develop clear characters that allow identification from an overhead view. Therefore, one of the next steps needed is to develop keys of local species from an overhead view. For some species the addition of a side or forward-looking oblique camera will give a view similar to that seen from other human-occupied submersibles and ROVs and will aid in identification.

An enormous amount of data is collected during AUV deployment. The image storage and processing can be a bottleneck in the data analysis. However, our analysis of the various subsampling schemes shows that there are efficient ways to subsample our data that retain appropriate power while minimizing the number of frames that must be examined. Nonetheless, the specific subsampling protocol used here should not be considered definitive as we continue to investigate other possibilities. Tools also have been developed to improve the efficiency in the analysis of the images (Ferrini and Singh 2006). Future automation of the analyses will help make the AUV an even more powerful tool. We are currently developing automated target recognition tools to scan frames and separate those that contain fish and subsequently categorize those fish into general groups.

The patterns that we see in regards to the habitat utilization of groundfish are similar to those described for deep water habitats in our region (Love and Yoklavich 2006). It is clear that in addition to information on fish abundance and general habitat utilization, categorizations of fine scale habitat utilization by fish and invertebrates and the spatial relationships between fish and invertebrates are possible. In future analyses these finer scale patterns of habitat utilization will be examined and may provide insights into fish habitat relationships.

Finally, this tool unlike many other traditional survey tools allows easy analysis of size distributions of fish on fine scales. The fish and invertebrates can be measured directly unlike indirect methods necessary when using technologies such as acoustics. Our analysis shows that there are fine scale patterns in size that may be masked when survey tools that integrate over fairly large spatial scales are used. It is evident that understanding these fine scale size distributions may be critical to intercalibrating the information collected with trawls and information collected during surveys in rocky habitat.

The Seabed AUV has great potential as a direct observation tool to estimate the density of benthic fishes in high relief areas that are not accessible by other tools. In addition, differences in fish densities and sizes between trawlable and untrawlable areas can be observed that may have implications for the estimation of abundance of groundfish.

Acknowledgements We thank two anonymous reviewers for comments of the manuscript. Thank you to Curt Whitmire who produced many of the maps and Julia Clemons and Keri York who identified and counted biota in images. Thank you to Waldo Wakefield for advising on data analyses and many other aspects of the project and for reviewing early versions of this work. We also thank R. Mulcaster, C.W. Alcock, A. Quixano, S. Panza and M.C. Saavedra for inspiration.

References

Armstrong RA, Singh H, Torres J, Nemeth RS, Can A, Roman C, Eustice R, Riggs L, Garcia-Moliner G (2006) Characterizing the deep insular shelf coral reef habitat of the Hind Bank marine conservation district (US Virgin Islands) using the Seabed autonomous underwater vehicle. Cont. Shelf Res. 26: 194–205

Fernandes P, Brierley A, Simmonds E, Millard N, McPhail S, Armstrong F, Stevenson P, Squires M (2000) Fish do not avoid survey vessels. Nature 404: 35–36

Fernandes P, Stevenson P, Brierley A, Armstrong F, Simmonds E (2003) Autonomous underwater vehicles: future platforms for fisheries acoustics. ICES J. Mar. Sci. 60: 684–691

Ferrini VL, Singh H (2006) FISH_ROCK: A Tool for Identifying and Counting Organisms in Bottom Photographs. Woods Hole Oceanog. Institute Technical Report. WHOI-2006-1, 28 pp.

Fretwell SD, Lucas HL (1970) On territorial behavior and other factors influencing habitat distribution in birds. I. Theoretical development. Acta Biotheor. 19: 16–19

Griffiths G, Enoch P, Millard NW (2001) On the radiated noise of the Autosub autonomous underwater vehicle. ICES J. Mar. Sci. 58: 1195–1200

Hixon MA, Tissot BN, Pearcy WG (1991) Fish Assemblages of Rocky Banks of the Pacific Northwest (Heceta, Coquille, and Daisy Banks). U.S. Minerals Management Service. OCS Study MMS 91-0052, Camarillo, CA.

Keller AA, Horness BH, Tuttle VJ, Wallace JR, Simon VH, Fruh EL, Bosley KL, Kamikawa DJ (2006) The 2002 U.S. West Coast Upper Continental Slope Trawl Survey of Groundfish Resources off Washington, Oregon, and California: Estimates of Distribution, Abundance, and Length Composition. U.S. Department of Commerce, NOAA Technical Memorandum. NMFS-NWFSC-75, 189 pp.

Love MM, Yoklavich M (2006) Deep rock habitats. In: Allen L, Pondella DJ, Horn MH (eds) The Ecology of Marine Fishes: California and Adjacent Waters. University of California Press, Berkeley, CA, 670 pp.

Love MM, Yoklavich M, Thorsteinson L (2002) The Rockfishes of the Northeast Pacific. University of California Press, Berkeley, CA, 414 pp.

Singh H, Can A, Eustice R, Lerner S, McPhee N, Pizarro O, Roman C (2004) Seabed AUV offers new platform for high-resolution imaging. EOS, Transactions of the AGU 85: 289, 294–295

Stein DL, Tissot BN, Hixon MA (1992) Fish-habitat associations on a deep reef at the edge of the Oregon continental shelf. U.S. Fish. Bull. 90: 540–551

Yoklavich M, Greene HG, Cailliet G, Sullivan D, Lea R, Love M (2000) Habitat associations of deep-water rockfishes in a submarine canyon: an example of a natural refuge. U.S. Fish. Bull. 98: 625–641

Zimmermann M (2003) Calculation of untrawlable areas within the boundaries of a bottom trawl survey. Can. J. Fish. Aquat. Sci. 60: 657–669

Chapter 21
Technology Answers to the Requirements Set by the Ecosystem Approach

Olav Rune Godø

Abstract Time series of abundance indices from scientific surveys are often the backbone in assessing the present state and expected development of exploited fish stocks. However, landing statistics, which have associated uncertainties often set the historic trends, and thus might be misleading with respect to ecosystem dynamics. The extended demand in the ecosystem approach is to consider the welfare of the whole ecosystem. Can this be done adequately with traditional tools? And what solutions can be expected from new technologies?

Future methods must enable quantitative observation of biotic densities with adequate resolution in both time and space. We also need to quantify the dynamics, including inter- and intra-specific competition and interactions between biology and environment. Advanced technology and knowledge have created a new scientific base for the ecosystem approach. Remote sensing techniques based on acoustics and optics offer both detailed and overview pictures, and can be deployed in time and space from innovative platforms and vessels of opportunity. Remote categorisation of information, e.g. species and size identification, is no longer a dream and modern observation techniques give the scientists information about processes with adequate time resolution. In the short term, we need to uncover the actual efficiencies of sampling trawls. The research should aim at establishing sampling tools based on knowledge of behavioural stimuli and responses of the target species rather than traditional ideas in trawl construction. In the long term, the limiting factor is not the technology, but our ability to develop integrated observation-modelling solutions that merge complex data from a multitude of sensors and platforms and extract the essential information.

Keywords Ecosystem survey · technology · acoustics · optics · platforms · sensors · integrated monitoring

O.R. Godø
Institute of Marine Research, P.O. Box 1870 Nordnes 5817 Bergen, Norway
e-mail olavrune@imr.no

R.J. Beamish and B.J. Rothschild (eds.), *The Future of Fisheries Science in North America*, 373
Fish & Fisheries Series, © Springer Science + Business Media B.V. 2009

21.1 Introduction

The ecosystem approach to fisheries (EAF) has been defined by FAO (2001) as "sustainable fisheries management taking into account the impacts of fisheries on the marine ecosystem and the impacts of the marine ecosystem on fisheries". The declarations and agreements on the ecosystem approach reflect a common understanding of the need to extend traditional fisheries management to sector-crossing management with due acknowledgement of the multiple users of the marine ecosystem (FAO 2001; UN 2002). This requires a shift in the information gathered in support of management decisions (FAO 2001; UN 2002). Improved understanding and assessment of the ecosystem is called for, with emphasis on ecosystem structure, functioning and variability (FAO 2001; UN 2002). International fisheries management organisations like the International Council for Exploration of the Sea (ICES, www.ices.dk; Misund and Skjoldal 2005), the Northwest Atlantic Fisheries Organization (NAFO, www.nafo.ca) and the North Pacific Marine Science Organization (PICES, www.pices.int; Livingston 2005) organise international scientific efforts aimed at achieving these ambitious goals.

Traditionally, fisheries have been managed through scientific monitoring of the target stocks. Catch statistics and sampling the catch composition reveal what has been removed, while scientific surveys provide evidence on the state of the stock in the most recent years. Analysis of the time series from this routine monitoring work is normally the main foundation for the management of the most valuable fish stocks. Although the simplicity of the single-species approach is often said to cause inefficiency in current practices to prevent over-exploitation and degradation of ecosystem integrity, the actual causes are often found elsewhere (Sissenwine and Murawski 2004). Another component of the EAF discussion is the use of Marine Protected Areas (MPAs) as an instrument for implementation. The Marine Ecology Progress Series covers the scientific (Browman and Stergiou 2004), the political and socio-economic implications of the EAF (Stergiou and Browman 2005) in two Theme Sections. In both sections the authors give limited or no attention to technology as a driver or instrument to secure responsible implementation of the EAF. Nevertheless, the lack of knowledge and understanding of the ecosystem is stressed by several authors, as are the problems associated with inadequate and erroneous catch information (see, e.g. Sissenwine and Murawski 2004; Tudela and Short 2005) and fear failure of the EAF if the concept is not backed by enough scientific substance. It is interesting to note the very limited attention in the EAF literature to methodology issues related to improving the level of knowledge or ways to expand the degree of detailed monitoring. While the lack of ecosystem understanding is recognised, there are major proposed actions such as the establishment of Management Protected Areas (MPAs) to serve as an "untouched" refuge for ecosystem diversity, and the development of efficient ecosystem indicators that are easy to monitor and provide information about state of the ecosystem or some part of it (Jennings 2005).

So what is the role of technology in the development of the EAF? Technology is a driver for the development of science in general. In particular, the inaccessible marine environment would have remained undiscovered if technology had not enabled sampling at larger depth and in more detail. In this chapter, I first give some perspectives on the impact of technology in fisheries and fisheries science. Next, I demonstrate new approaches and possibilities opened by technology. I will further discuss the application of the described technologies in relation to the demands set by the EAF. For example, how can technology support the quality of indicator development, and how can we establish a scientific basis for MPAs without an adequate methodology to monitor their state and development? Finally, the ecosystem approach demands the integration of information. Thus, in Section 21.4 I discuss how data can be synthesised into usable information, through better integration of model development and field technologies. The overarching goal of this chapter is to elucidate gaps in the performance of current scientific approaches to management and to demonstrate the potential of new technology to solve challenges arising with the EAF.

21.2   Experiences in the Past

During the last 100–150 years, technology has been the major factor driving the expansion of commercial fishing. From a scientific point of view we have moved from the question "how can we harvest more from the unlimited resources of the open ocean" to "how can we harvest sustainably from the limited resources" (see Anderson 2002). This signifies a fundamental change in focus and simultaneously we are reminded of the conflicts and suffering experienced by the commercial fishermen along with the dramatic development. The herring and cod stories from the Northeast Atlantic (Toresen and Østvedt 2000; Godø 2003) may serve as examples of interaction between technological development and fisheries within a 100-year perspective. Just as important, we should note the risk imposed when technological development co-occurs with climate-driven ecosystem shifts.

21.2.1   Technology in the Development of Fisheries

Fishing techniques were traditionally developed locally, and innovative methods gradually expanded to other areas. The development of fishing methods has mostly depended on advances made in other fields, e.g. the motorisation of vessels, improved vessel design and sonar. Notable steps in the time line of commercial fishing technology are often accompanied by steps in the exploitation level of fish stocks. Some of these are summarised in the following table.

Technology steps from 1900	Result
Motorisation of fishing vessels	Expansion of fishermen's home range
	Development of active (towed) fishing gears
	Development of modern oceanic fishing
Fish finding instrumentation	Increased efficiency
	Development of new fishing gears
Mechanisation of fishing	Increased efficiency
	Handling larger gears
	Handling more gears
Selective fishing gears	Regulations to limit unwanted species and sizes
– Mesh selection	(notably juveniles) in the catch
– Selection devices	– Size discrimination
– Gear construction	– Size and species discrimination
	– Size and species discrimination
	– Seabed habitat protection
	Unaccounted mortalities caused by regulation
Gears improving quality of catch	
(under development)	Live catch
	Feeding the catch to optimise economic return

During the first 70 years of the past century, development of commercial fishing technology concentrated on improving the capture efficiency, but in the final years of the century the focus changed dramatically towards means of improving selectivity (Valdemarsen 2001). Also, there is a trend towards more focus on technology that improves the catch quality, even to the extent of maintaining living animals on board. The interest for merging wild fisheries and aquaculture is increasing (Bombeo-Tuburan et al. 2001), although it is important to consider the ecological effects (Mous et al. 2006). The development of fishing gear that brings living krill onboard the vessel is another example demonstrating present development. Thus, the present trend towards technology and methods offering high selectivity and quality gives due attention to the conservation and sustainability of the target species. FAO and ICES have since World War II had a key role in the global development of fisheries technology (see, e.g. Ben-Tuvia and Dickson 1968; Walsh et al. 2002).

21.2.2   Technology in Development of Fisheries Science

Technological development, both in general and that specific to commercial fishing, has been decisive for the implementation of new technology in marine science. The crucial importance of technology is illustrated in Fig. 21.1. Here you see a photo of John Murray and Johan Hjort with colleagues on the deck of R/V *Michael Sars* in 1910. Murray funded an expedition of the modern Norwegian vessel, R/V *Michael Sars*, with Johan Hjort as the scientific leader. Their book from this expedition (Murray and Hjort 1912) became a milestone in marine science. A major technology driver for this achievement was the big winch seen behind the authors

Fig. 21.1 John Murray and Johan Hjort with colleagues onboard new Michael Sars in 1910. Note the big winch behind the scientist. This gave the scientists access to larger depths and unknown species in the northern Mid Atlantic. Source: Institute of Marine Resource, Bergen, Norway.

in Fig. 21.1. This machine provided access to depths never previously explored, and resulted in the description of many species new to science.

During the twentieth century, we see several similar steps in the development of fisheries science. For example, anti-submarine research during the World Wars prepared the ground for modern sonar technology, which became the driver for the development of modern fish finding instrumentation and other advanced scientific instrumentation (see Fernandes et al. 2002; Simmonds and MacLennan 2005). Similarly, when the demand for monitoring of demersal fish emerged, the obvious first approach was to develop methodology around the catch per unit effort (CPUE) of the fleet as a measure of fish density (see, e.g. Beverton and Holt 1957). However, the evident difficulties in interpreting CPUE information, due to the continuous development of trawl gears and the increased fishing power of vessels, prepared the ground for standardised trawl surveys (see Doubleday and Rivard 1981). Compensation models, e.g. using engine power or other technology-driven effort measures, have been applied (Maunder and Punt 2004; Bishop et al. 2004). A classical example is found in the shrimp fisheries where double and triple trawls have been introduced without possibilities for logging this in the fishermen's logbooks. In contrast, the standardised bottom trawl surveys provided time series of data, unaffected by technology creep. Such surveys support new understanding of the dynamics of exploited fish stocks, and a fundamentally better basis for their assessment and manage ment (NRC 1998). Thus, scientists have not been inclined to adopt technological improvements in survey gears, fearing that new or more efficient capture methods might distort long time series. Indeed, many survey trawls are inefficient at capturing recruits even when this is a major scientific objective (see, e.g. Engås and Godø 1989; Godø and Walsh 1992; Somerton et al. 2007). Also, survey time series can be corrupted because technology for monitoring trawl geometry and performance has not been included in the standard protocols (see, e.g. Godø and Engås 1989; http://www.nefsc.noaa.gov/survey_gear/). Thus, standardisation is a manifestation of what present knowledge suggests is going on below the sea surface, while this in many cases will be far from reality. The literature is replete with information about surprising observations uncovered by new underwater technology. There is often a conflict of interest between the assessment scientists, who want comparable time series to stabilise assessment models, and the survey scientists who want to collect more informative datasets for the future. Thus, two alternatives exist: we can either keep on using the same equipment and routines, or alternatively, adopt new gears and procedures that have well-documented performance in relation to the old standard. In the first case, the original time series is maintained, but new knowledge can be implemented through adjustment obtained in calibration studies (see, e.g. Somerton, et al. 2007). In the latter case, the whole time series is maintained through efficient correction of past results based on calibration studies, while new data are collected with new gears and procedures. This implies some quality reduction in the overall time series, but in most cases the new time series will offer much better monitoring capability for the future (see, e.g. Godø 1994).

21.2.3 Technology for Development of EAF: Practice and Gaps

The EAF sets new data demands for stock assessment and requires additional restrictions on exploitation to prevent the degradation of fish stocks, the environment and ecosystems. In the past 30 years there has been substantial research on fish stock conservation, including novel fishing gear construction to enhance species and size selection (see, e.g. Valdemarsen 2001 and references therein). Video and acoustic observation techniques have given new insights to fish behaviour in the catching phase leading to the construction of novel gears for more responsible fishing (see, e.g. Fernö and Olsen 1994). Also, the literature presents convincing evidence of the degree of change in the benthic communities (Frid et al. 2000; Jennings et al. 2001). Effects like the destruction of deep sea coral reefs (Fossa et al. 2002) and ghost nets fishing for years after being lost by fishermen (Matsuoka et al. 2005) have upset the public leading to increasing concern and drive towards more environmentally friendly fishing technologies.

Interestingly enough, the focus on technology-driven research to promote new observation and monitoring methods under the EAF has been limited. Some scientists occasionally bring this issue into discussion (see, e.g. Amaratunga and Lassen 1998; AIRFB, this volume) but it seems difficult to bridge the assessment and management demands with the potential offered by new technology. As commented above, much work has been done on how to utilise simple information obtained with current scientific-survey methodology to establish new ecosystem indicators (Cury and Christensen 2005; Jennings et al. 2001). An important instrument in implementing the EAF is the establishment of MPAs. However, little is done to incorporate adequate monitoring methodology, which is imperative for such schemes, although that could interfere with present assessment techniques (see Field et al. 2006). Due to the need for routine assessments and management advice, the involved scientists seem to prioritise short-term adjustments before major changes that in the long term might be more appropriate. An example is the noteworthy increased effort and interest in applying modern technology in general marine science, for example, the development of advanced observatory technology (see, e.g. http://www.neptunecanada.com/science/index.html; Favali and Beranzoli 2006) although this has received little attention in the discussion of EAF.

Thus, in relation to the EAF, there are three obvious gaps where technology has to play a central role:

- *Technology as an avenue to new information and knowledge about the ecosystem.* There is general agreement that the current state of knowledge concerning the ecosystem function and dynamics is a limiting factor for realising MPAs, and according to Tudela and Short (2005) the concept can be undermined due to lack of scientific substance. New technology is definitely needed to open avenues to deeper understanding of marine life (see, e.g. www.coml.org; Godø 1998).
- *Technology to validate assumptions and estimate parameters in models.* Advanced ecosystem models often suffer from non-validated assumptions that limit the information needed to estimate model parameters. New underwater observation systems could in many cases shed light on these problems and over time improve the foundation of existing models (see, e.g. Johansen et al. 2007) as well as preparing the ground for new models.
- *Technology-driven design of advanced coordinated observation-modelling systems.* New technology could enable scientists to collect data in time and space far beyond what we see today. Integrated observation systems may in the future provide advanced models with the information needed both to assess the state of the ecosystem and set the level of responsible harvest. This would be consistent with current developments in oceanographic modelling.

In Section 21.3 I will try to demonstrate some of the opportunities offered by new technology for the solution of scientific questions as well as better monitoring related to EAF requirements.

21.3   Technologies with Potential

21.3.1   Observation Methods for Marine Life

No technology is perfect, but there is a suite of techniques available for studying marine resources and their physical and biological environment. Among these there is no universal tool that provides all the information needed for EAF. It is of utmost importance to be aware of the strengths and weaknesses of the various technologies, to avoid misuse of effort, for example, that bottom trawls are restricted to fish distributed close to the seabed while acoustic methods fail in the near-bottom deadzone (Simmonds and MacLennan 2005; Aglen 1996). Further, when new technology is adopted, cost-efficiency aspects should be an integral part of the evaluation, particularly with respect to expensive long-term monitoring programmes. The complexity of marine ecosystems is well recognised. To avoid spending time and effort on the unsolvable, we should always keep in mind that there are known, unknown and the unknowable features of the ecosystem (see Levin 2003). Here I will focus on presently underutilised technologies and how they may play a role in the next generation of observation and modelling systems (see, e.g. Fig. 21.2).

An applied technology consists normally of one or several sensors mounted on a platform or sensor carrier. It is essential to choose the sensors and platform according to the task to be solved. For example, a noisy vessel with a hull-mounted echo sounder may fail to detect pelagic resources in shallow water due to their unavailability in the surface blind zone (Aglen 1994; Fig. 21.3) and avoidance reactions to the approaching vessel (see Fig. 21.3; Ona et al. 2007a).

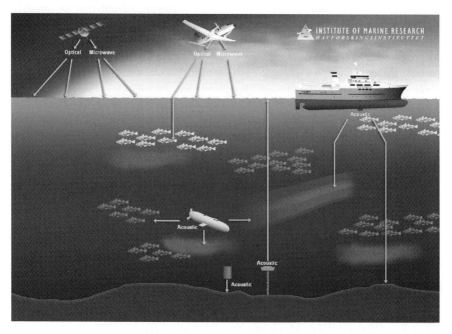

Fig. 21.2 Remote sensors (optical, microwaves, acoustics) with their most common carrier and tasks in the water column like densities of fish and environment conditions (blue clouds). The yellow lines indicate the sampling ranges and thus the limitations (from Godø and Tenningen 2009).

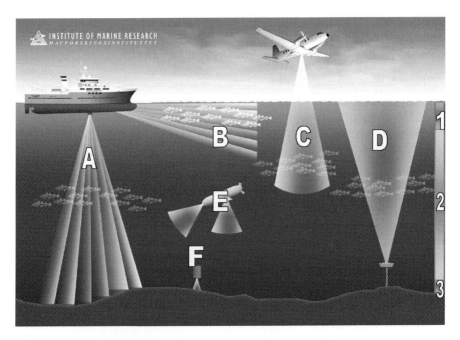

Fig. 21.3 Observation zones for different sensors. 1) Upper echo sounder blind zone observed by lidar (C) and bottom mounted transducers (D) as well as vessel sonar (B). 2) Mid water zone observed by all platforms. 3) Bottom "dead zone" partly inaccessible for multi beam vessel based echo sounders (A) (multibeam as well as split and sigle beam systems). An AUV (E) can inspect closer any volume of water and report about the acoustic properties of the target species. Autonomous buoys with high frequency echo sounder and cameras on mission at constant distance to bottom (F), inform about the details in the vessel dead zone (from Godø and Tenningen 2009).

In the following text, I will give particular attention to acoustics, optics and biological sampling gears as sensor systems. Other remote sensing techniques will be briefly covered. Further, the sensor holders (platforms) will be given special attention as a key part of sensor systems in general.

21.3.2   Acoustics

Acoustic techniques include both passive and active sensors. Acoustic applications in ecosystem monitoring are covered in more detail in Chapters 17 and 18. Here I concentrate on applications alone or in conjunction with other technologies closely related to EAF. Active sensors transmit well-defined, often calibrated, signals and they record and interpret the received echoes (see Simmonds and MacLennan 2005). Passive sensors listen to sounds in the sea over a broad spectrum of frequencies, and try to distinguish biological sources from ambient noise. Both approaches are presently used but underutilised. Compared with light, sound transmits over a far longer range under water; it has various functions and is in many cases more important for marine species compared to terrestrial life. Numerous marine species, included many fishes, detect and produce sound and use it for communication (Popper and Fay 1993). Some marine mammals have developed sophisticated sound sources, used not only for communication, but also for the detection, identification and navigation/capture of prey (Johnson et al. 2004). There is every reason to study the production and use of sound by marine life. These animals have developed high-performance acoustic tools through long evolution, and an obvious goal for our own efforts should be to become as good in this field as are the whales.

21.3.2.1 Active Acoustics

Whales hunt in three stages: *detection, identification* and *feeding*. When large-toothed whales generate low frequency clicks (<20 kHz) they can search large volumes of water for prey concentrations (see, e.g. Madsen et al. 2002; Makris et al. 2006). At shorter distances, higher frequencies are used while at very close range whales may change to even higher frequencies and a broader band signal (Johnson et al. 2004). This provides more detailed information about the prey (*identification* phase) and probably helps the whales to choose appropriate targets and optimise the feeding effort. Fishermen and marine scientists collecting their data follow essentially the same stages as the whales: *detection, identification* and *catching/ biological sampling*. In addition, there is the *quantification* phase; this is important for fishermen to plan their catching strategy to avoid gear damage and discards, and for scientists to assess the abundance of ecosystem components.

Detection: In modern pelagic fisheries, the fishermen use medium-frequency sonars (18–30 kHz) to detect areas with high biomass (see, e.g. Misund 1997). Detection ranges are normally 3–6 km depending on oceanographic conditions. Once a fish school is detected, many skippers have the possibility to change to higher-frequency sonar (80–200 kHz), which provides a more detailed picture of the school and records its size and behaviour. This is essential information for the success of the catching phase in pelagic fisheries. Sometimes the vessels pass over the school using echo sounders to further enhance the information needed for correct decision making. Echo sounders are also standard tools used by scientists during acoustic surveys, but while horizontal sonars are essential fishing aids, they are seldom used by scientists for ecosystem studies. Why this is so will be further discussed in the section on Quantification. Figure 21.3 illustrates the different observation zones of various instruments. Note particularly that the vessel-mounted echo sounder fails to detect organisms distributed in the surface blind zone and the near-bottom deadzone (see also Aglen 1994). The detection capabilities are limited by physical considerations (Ona and Mitson 1996). Although higher frequencies overcome the problems to some extent, more adequate solutions call for the use of alternative platforms (see later in this chapter). While fishery sonars seldom go below 18 kHz, limited by the fact that transducer size is inversely related to frequency, Makris et al. (2006) have demonstrated the detection of fish schools at distances of 20–40 km using a military sonar system operating at about 400 Hz (Fig. 21.4). These systems are presently cumbersome to operate and are only useful

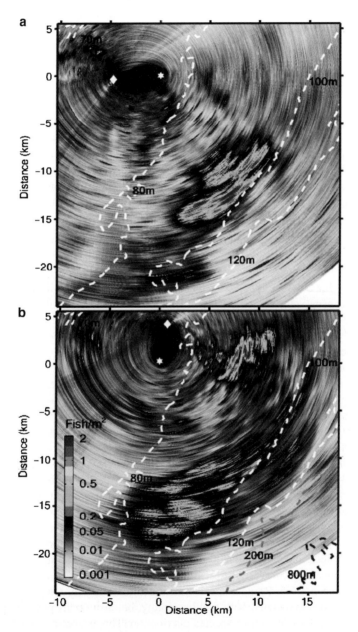

Fig. 21.4 Long range detection with low frequency acoustics as demonstrated by Makris et al 2006.

in shelf areas with waveguide properties. Nevertheless, they represent a tool that enables the monitoring of ecosystem distribution and dynamics at local population scales with a temporal resolution of seconds, and thus offer improved possibilities to study dynamics and interactions at the population level.

Horizontally beamed sonars can search large volumes quickly with a spatial resolution appropriate for detecting and mapping of highly aggregated and patchily distributed pelagic and semipelagic fishes. The backscattered signal is, however, highly dependent on the acoustic properties of the targets. First, in most fishes the main backscattering comes from the swim bladder. Thus, fish without a swim bladder have reduced "visibility" with acoustic systems. Further, the fish orientation strongly affects backscattering strength due to high directivity especially at higher frequencies (see, e.g. Johannesson and Mitson 1983; Simmonds and MacLennan 2005). This creates difficulties when using horizontally beamed high-frequency acoustics for abundance estimation, but can be informative when studying the behavioural dynamics of fish (Freon and Misund 1999). But sonars are essential tools for mapping pelagic resources that are highly aggregated and patchily distributed (Misund et al. 1996; Gerlotto et al. 1999) as well as facilitating efficient commercial fisheries on these species (Misund 1994).

Identification: A major problem in acoustic mapping and quantification is the identification of species and size of the observed fish. These biological characteristics, as well as their fat content, maturity stage, etc., determine the backscattering properties (Ona 2003; Horne 2003; Chu et al. 2003). A major issue in acoustic developments for ecosystem studies is remote identification. How can we use the biological variability and other target characteristics to distinguish species and sizes? This is a relatively new field where substantial development can be expected in the near future. Presently, identification is normally done by trawl sampling, or using other gears appropriate to the target species (see, e.g. Gunderson 1993). Biological sampling is often difficult for three main reasons. First, fishing gears are selective, and strong size and species selection will occur even for gears designed as representative sampling tools (see, e.g. Godø 1994). Second, the species/size composition is often highly variable, creating difficulties in extrapolating information from one catch to the population in larger areas. Third, the capture efficiency is also variable. Analysis indicates that for semidemersal species, there is a need to trawl at several stations (4–8) to characterise the population at a particular location (Godø 1998).

Presently, the most promising approach to remote species identification appears to be multi-frequency acoustics. Differences in the frequency dependence of backscattering strength are used to distinguish groups of species, sometimes to the level of individual species (Korneliussen and Ona 2000; Horne 2000; Kloser et al. 2002; Jech and Michaels 2006). The varying response to the available spectrum of frequencies provides sufficient data for discrimination between species as shown in Fig. 21.5. Remote species identification not only eliminates some of the uncertainties associated with trawl sampling, but also opens possibilities for a more continuous resolution of species distribution. This is crucial for understanding species interactions and may facilitate better knowledge of their distribution in relation to the physical and biological environment. This development is at an

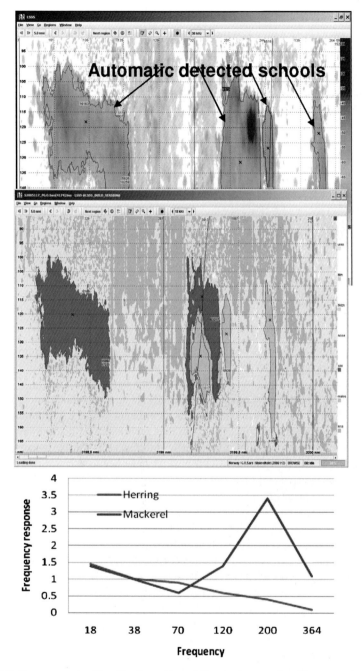

Fig. 21.5 Remote species identification with multifrequancy acoustics. Upper panel shows a normal echogram at 38 kHz. In the middle panel pixels of herring (red) and mackerel (yellow) has been identified based on their characteristic differences in frequency response relative to 38 kHz as shown in lower panel. Presented with courtesy from Rolf Korneliussen, Institute of Marine Research Bergen, Norway.

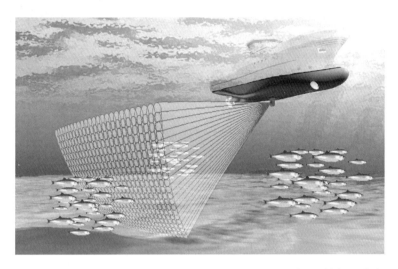

Fig. 21.6 The Simrad MS70 multibeam sonar is a 500 beam sonar that at high resolution may include a medium sized school of pelagic fish in one ping. Courtesy H.P. Knudsen, Institute of Marine Research, Bergen.

early stage and substantial improvements are expected in the near future. While researchers presently have only a limited number of discrete frequencies available with calibrated scientific sounders, we expect that extension of the technique to a continuous bandwidth, perhaps from below 20 kHz up to several hundred kilohertz will improve the quality of remote species identification (see, e.g. Simmonds et al. 1996). These studies also suggest possibilities for rough size discrimination using wideband techniques. Similarly, systematic differences in the frequency response of large and small fish with no swim bladder have been observed using discrete frequencies (Pedersen and Ona 2007).

For a given species the fish size is normally proportional to the swim bladder size. At certain frequencies the backscattering increases substantially due to resonance of the swim bladder. If a wideband signal that covers the resonance frequency of the targeted fish is transmitted, then it is theoretically possible to estimate the size of the fish by detecting the resonance in the received echo spectrum (McCartney and Stubbs 1971; Thompson and Love 1996; Nero et al. 2004). This method of remotely sizing fish is not yet a practical tool. The resonance of fish with swim bladders, for most sizes except juveniles, occurs at frequencies below those normally used by fisheries research vessels (38 or 18 kHz). Transducers transmitting at lower frequencies are large, difficult to handle and impractical to operate together with the other equipment used in ecosystem studies. Also, transmitting at frequencies of hundreds of hertz or a few kilohertz might affect the studied fish as most species can hear these or even higher frequencies (Popper et al. 2004). The technological challenge is to develop a sonar system that will enable practical operation and logging of resonance frequencies without disturbing the target.

The so-called parametric systems, which are smaller and operate with very low source levels at low frequencies (Dybedal 1993), have not yet been extensively tested for this purpose. If a parametric device could transmit an omnidirectional signal, it would be an extremely useful tool for the sizing of schooling fish, with important applications in both ecosystem studies and commercial fishing. Remote non-invasive sizing and species identification of fish might help to reduce discards and improve efficiency in commercial fishing, and thus should also be seen as an important management tool for responsible fishing.

The frequency response of backscattering over a wide frequency range gives detailed information about the ecosystem on fine temporal and spatial scales. As species, genera, orders and classes of biological targets have different acoustic characteristics, there are good reasons to believe that this information can be used to, for example, characterise trophic levels and associated variations in space and time. This concept remains a subject for further study. Of particular interest is the potential to establish efficient ecosystem indicators based on the backscattering spectra of the ecosystem components.

Quantification: When active acoustic techniques have detected and identified biota, biomass quantification is the next step. The standard approach is echo integration, where acoustic densities revealed by vertical echo sounding are converted to biomass, based on knowledge of the acoustic properties of the identified species (see, e.g. Simmonds and MacLennan 2005).

Biomass quantification has been the primary application of echo sounder technology in fisheries monitoring. To enable full water column coverage, there is now increasing interest in using multibeam horizontal sonars. The new Simrad ME70 sonar system is an important technological step in that direction, being the first sonar to have high-quality calibration capabilities (Ona et al. 2006, 2007b). Thus, new technology is in place, but still there is substantial work to be done to enable true density estimation of fish aggregations at any angle of insonification. A similar echo sounder system is also developed (ME70). This sounder improves assessment of fish close to bottom and enables studies of fish reaction under the vessel during survey situations.

The counting of single-fish echoes is an alternative method for quantification when the fish can be acoustically resolved as individual targets. It has mostly been applied in fresh water (Kubecka et al. 1992), but also occasionally in the sea (Aksland 2006). A more common technique in the marine environment is school counting using horizontally directed sonar beams (see, e.g. Gonzalez and Gerlotto 1998).

In an ecosystem perspective, the unique possibilities of behaviour quantification offered by acoustic systems are of utmost importance for improved understanding of both species interactions and fish behaviour in relation to the environment. Individual fish can be tracked as they swim and their behaviour revealed (Handegard 2007); similarly, schools can be followed to demonstrate their migration and behaviour patterns in relation to environmental features (Brehmer et al. 2006; Hjellvik et al. 2004). Thus, acoustic methods provide data on the behavioural dynamics of single fish and schools that may support modelling of behaviour and species interactions at the population level.

Acoustic cameras, for example, the Didson dual frequency system, have proved to be useful tools for detecting, identifying and quantifying fish at close range. They have been used in ecological studies under extreme conditions (Mueller et al. 2006) as well as research on fish behaviour, and avoiding the need for any artificial light disturbance (Rose et al. 2005).

Multibeam echo sounders are also used for bottom mapping. In the EAF, habitat mapping is important both to protect vulnerable habitats (Fossa et al. 2002) and for monitoring effects of fishing on the natural environment. Acoustic methodology in this area is under rapid development, and in addition to the usual bottom depth indication, it could provide detailed information on the properties of the seabed and potential risks to animals living on and within the bottom substrate.

21.3.2.2 Passive Acoustics

Passive acoustic instruments enable us to hear and distinguish biologically produced sound from other sources like vessels, wind and waves. The application of passive acoustics in an EAF context is characterised in the same way as active acoustics, namely for detection, identification and quantification of the sound producers. The most common applications are the detection, identification and counting of marine mammals and fish (Horne 2000). Also, given appropriate knowledge, analysis of sound production gives insights to biological processes like mating, rivalry and feeding behaviour. While detection is instrument-dependent (sensitivity, frequency spectrum, etc.) identification is knowledge-based, i.e. it depends on recognition of known sounds from earlier recordings. Thus, identification involves a learning process based on sound recordings and visual observation of the sound producer. When an adequate database is available, long-term monitoring with sensitive listening instruments may uncover information about the occurrence and activity patterns of marine life such as snapping shrimp (Watanabe et al. 2002), breeding fish (Finstad and Nordeide 2004) and hunting whales (see Fig. 21.7; Madsen et al. 2002). An obvious limitation of passive acoustics is that detection depends on sounds being produced, and this may or may not occur according to the season, species, size and sex of the animal. Passive acoustics may offer even more potential when combined with other observation methods like active acoustics and optical techniques.

21.3.3 Optics

Optical methods in the EAF context include remote sensing of sea-surface properties by space satellites, lidar measurements made from aeroplanes (possible down to about 50 m depth) and underwater photographic and video techniques that have a range of only a few meters from the camera (see Godø and Tenningen [2009] for more details). In particular, remote sensing by satellite gives frequently updated pictures of the sea surface over a large area, with, e.g. a pseudo-colour representation

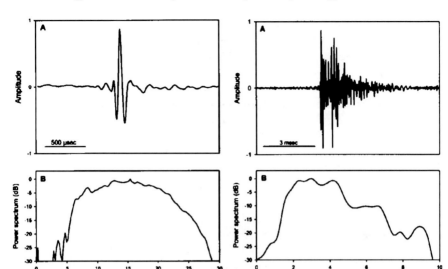

Fig. 21.7 Sperm whale vocalisation. Power spectrum and amplitude at creak click (left) and slow click (right) (Madsen et al 2002).

of temperature variations. This information is presently used in ocean circulation models, and can be an important tool, for example, in understanding fish distribution and migration. Zagaglia et al. (2004) provide a good example of the use of such data in their study of yellowfin tuna.

Lidar is an acronym for "light detection and ranging". An airborne lidar uses laser pulses to detect fish and plankton in much the same way as an echo sounder does with acoustic pulses (see Chapter 19). Laser light in the mid-visible portion of the spectrum (532 nm wavelength) is capable of penetrating the sea surface and propagating through the upper layer of the ocean down to 25–50 m depending on visibility conditions. Reflections of the laser light are detected, measured and used to generate lidargrams similar to the echograms used in fisheries acoustics (Fig. 21.8; Churnside et al. 1997). Also, as with acoustics, the lidar recordings may be converted to densities when the backscattering properties of the targets are known (Tenningen et al. 2006). Lidar is an ideal tool for mapping fish and other marine organisms in the near-surface layer. These organisms are often inaccessible to vessel acoustics due to the surface blind zone or they might be disturbed during measurements and react to avoid the vessel (Aglen 1994). Further, the rapid coverage offered by airborne lidar is an advantage when monitoring fast-migrating pelagic species. Lidar development for marine monitoring is in its infancy. Substantial advances along the lines seen in acoustics are expected, including expanding the beam width through scanning and increasing the bandwidth for identification purposes.

Conventional optical techniques like still and video cameras have wide-ranging applications in marine studies. These systems have become smaller and more

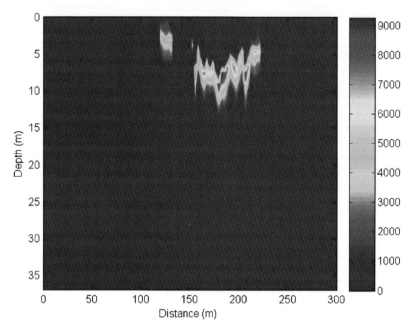

Fig. 21.8 Lidargram of mackerel school at surface as recorded in the Norwegian Sea (Adopted from Godø and Tenningen 2009).

robust, and can be deployed many kilometres from the human observer using ROVs (Lorance and Trenkel 2006; see also Fig. 21.9) or landers (Priede and Bagley 2000); they offer an intuitive understanding of the underwater reality. Their limitations are just as obvious. The use of light affects the behaviour of illuminated animals, which may be attracted or repelled, and the short-range penetration of light in water restricts the observation volume. Video and photographic tools are today and will continue in the future to be indispensable in marine research. These traditional techniques are particularly important for the "teaching" of newer systems like lidar and acoustics with respect to species and size identification. Expanding the overall capability of light cameras with acoustic methods, e.g. by including an acoustic camera, is a good example of how sensors can be efficiently combined (see Mueller et al. 2006).

21.3.4   Catching and Biological Sampling

Trawls and other fishing gears have been major tools for collecting information about marine resources. These gears are all very selective and the catch will represent a size-dependent fraction of the local population (see, e.g. Engås 1994; Somerton et al. 2007). Standard techniques have been developed to maintain the comparability of catch results within time series (see, e.g. Doubleday and Rivard 1981).

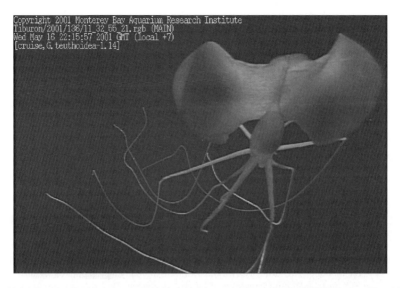

Fig. 21.9 Picture of a deep water squid from a video recording taken by ROV Tiburon (Vecchione et al. 2001).

Towed gears sample a small fraction of the water volume, and operational conditions like rough ground restrict the area that can be covered. Thus, the efficiency of these gears depends on the vertical and horizontal distribution patterns of the targeted organisms, as well as their behavioural responses to the gear (Godø 1994). Nevertheless, many trawl surveys have proved fully adequate for tracking trends in population development. Under the EAF, the small observation volumes, the high variability and low spatial resolution of fishing samples limit their possibilities to provide basic information about ecosystem dynamics; nevertheless, in the future it will still be essential to catch biological samples for ground truthing other technological approaches.

In contrast, responsible fishing calls for selective gears that remove target species within a specified size range, without damaging other species or sizes of fish or their habitat, e.g. the bottom substrate. Thus, development of more selective and environmentally friendly gears is important for an EAF. Instrumented trawls that hardly touch the seabed and mechanically select fish of given species and size are possible. The Icelandic underwater tagging device demonstrates the possibilities offered by instrumented trawls (Sigurdsson et al. 2006). They handle and tag fish at deep water and plan further development of the technology for the purpose of size and species selectivity. Another example is the novel technique used for trawling Antarctic krill (Fig. 21.10). In the latter example, living krill is brought onboard the vessel with far better quality than conventional fishing techniques can achieve and unwanted by-catch of fish and marine mammals are avoided. These examples demonstrate the availability of new solutions that select target species and sizes, as well as improved onboard processing, setting new standards for the catch selectivity and quality.

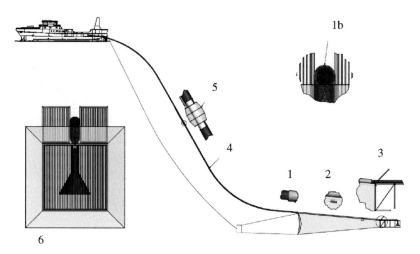

Fig. 21.10 The continuous trawling system developed by Aker Sea Food for catching krill (courtesy Aker Seafood, AS, Oslo, Norway). 1) Stabilizing/buoyancy regulated by air (1b), 2) Monitoring, 3)Adjustable grating for release of unwanted bycatch, 4)Conveyor hose, 5) Regulator, 6) Inside the trawl, seen backwards against cage and grading grate. The system fundamentally improve quality of catch and prevent unwanted bycatch.

21.3.5   Platforms and Carriers for Optimal Sensor Positioning

Marine resource monitoring is presently done with a limited range of platforms and sensors. The catch rates of commercial and research trawlers, and acoustic back-scattering measurements are the dominant sources of information. Some additional platform–sensor combinations, like photographic and lidar surveys done from aeroplanes, and photographic surveys from ROVs, are occasionally used. Compared to the suite of platforms and sensors used in general marine biology, fishery resource monitoring is simplistic. The main reason, as discussed in Section 21.2.2 is the need for a strictly standardised sampling regime to reduce uncertainty. Nevertheless, recent technological developments have produced new sensors and platforms that in combination create possibilities that have not yet been seriously evaluated by assessment and management scientists. Further, the extended demands of the EAF require data that traditional approaches cannot provide. In that context, alternative platforms need to be considered.

Modern technology has, in recent years, overcome two important factors that limited applications in the EAF. Firstly, acoustic and optical instrumentation and other sensor systems have become small and robust, so they can easily be implemented in small autonomous platforms like ROVs and landers (Fernandes et al. 2000; Patel et al. 2004; Godø et al. 2005; Fig. 21.11). Their robustness also enables operation in difficult environments like fishing vessels (Karp 2007). Thus, scientific instrumentation no longer prevents the collection of data at fundamentally different temporal and spatial scales compared to traditional methodology. The various

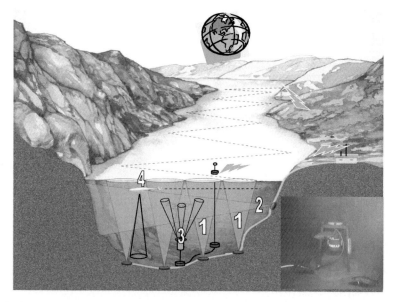

Fig. 21.11 Acoustic observatory installed at the entrance of the Ofotfjord in northern Norway. Two vertical split beam echo sounders (1) (of four originally planned, one of the bottom mounted systems shown in lower right corner), observed densities from bottom to surface. A horizontal sonar (2) watched a section crossing the fjord and was used to evaluate how representative the sounder volumes are for what cross section of the fjord. The Doppler current profiler (3) monitored flux of water masses and migration of fish schools through the section. Data were transferred through cables to the cabin on shore and further by radio link to the nearest Internet connection. The AUV (4) was occasionally used as a "watch dog" to inspect the section with echo sounder. The Observatory was established in 2003 and is still running (Godø et al. 2005).

platforms all have strengths and weaknesses (Table 21.1). A key question is whether we are able to benefit from the excellence of the individual platforms and simultaneously combine the data they produce in a constructive way? Are we able to design a cost-efficient operational monitoring system much better than the present regime with respect to resource assessment, improving the information flow on key ecosystem indicators as well as monitoring anthropogenic pressures? This will be dealt with in the following sections.

21.4 Demands and Technology Solutions for the EAF

21.4.1 Demands Set by EAF

There is a clear difference in perspective when looking at demands and solutions with and without the technological potential for progress. Ecosystem indicators are normally sought among data collected with traditional technology. Taking a technological perspective, we may focus more on the gaps and problems that need

Table 21.1 Various types of platforms that are presently underutilised in ecosystem and fisheries monitoring. Their potential application for specific tasks in relation to the EAF is ranked from 1 to 4, 4 being the most important. Numbered references are given below

Task platforms	Temporal resolution	Spatial resolution	Undisturbed observation	Habitat mapping	References
Silent R/V	1	2	4	3	1, 2
ROV	1	3	1	4	3, 4
AUV	1	3	4	3	5–7
Fishing vessels	4	4	1	2	8, 9
Cabled platforms	4	1	4	1	10–12
Aeroplanes	3	4	4	1	13, 14
Other vessels, e.g. coast guard, liners	4	1	1	1	15, 16

1. Mitson (1995)
2. Ona et al. (2007a)
3. Lorance and Trenkel (2006)
4. Adams et al. (1995)
5. Fernandes et al. (2000)
6. Patel et al. (2004)
7. Fernandes et al. (2002)
8. Karp (2007)
9. Mackinson and Van Der Kooij (2006)
10. Godø et al. (2005)
11. Horne (2005)
12. Farmer et al. (1999)
13. Churnside and Okumura (2001)
14. Tenningen et al. (2006)
15. Knutsen et al. (2005)
16. Holley and Hydes (2002)

to be overcome, and consider what technological approaches are available for the task. We need to consider three specific questions.

1. How to gain adequate ecosystem information?

The most common observation and monitoring systems have very limited spatial and temporal resolution with respect to monitoring the dynamics of marine life. This is due to the limited vessel resources available to carry out, e.g. trawl or acoustic surveys. Also, the sensors and platforms traditionally used cannot disentangle mixed time and space effects during surveys. New technology offers possibilities for distributing sensors in time and space with shorter temporal scales that may help to fill this gap (Fig. 21.11 and 21.12). Traditional methods for mapping bottom habitats are even more problematic. There is a strong demand for detailed mapping of bottom habitats to a level where we can monitor the disturbances caused by human activity such as bottom trawling. Acoustic systems, multibeam as well as single beam, enable high-resolution coverage over limited amount of time. This methodology allows us to characterise sediment types and perhaps even the biota living in the sediments. Advanced technology is more prominently applied in studies of the physical properties of the oceans. The autonomous moorings, drifters and gliders along with observations from traditional vessel platforms used in that field could fulfil many of the demands set by the EAF.

2. How can we detect and evaluate human impacts relevant to the EAF?

Within EAF, management actions are guided by the precautionary principle. Thus, fishing or industrial activities might be stopped or restricted due to uncertainty about their environmental impact. Consequently, there is a strong need for adequate monitoring to inform decisions that will satisfy the precautionary principle while avoiding unnecessary actions that are disadvantageous in economic and social terms. The only realistic cost-efficient solutions are found in new technologies as described above, and in particular, remote sensing techniques based on acoustics and optics.

3. Is there need for a technology-driven strategic shift?

It is possible to gradually evolve new technology and methods for ecosystem research and monitoring. However, at some stage I think that a step change is necessary to provide sufficient temporal and spatial resolution in the analysis and evaluation of ecosystem integrity and development. This implies a fundamental change in demands for data collection and monitoring strategies, as well as the sensor and platform technologies involved in the monitoring. Modelling has to be done with the available data, and when knowledge is limited, this must be compensated by making assumptions, which can seriously affect the performance achieved. New technologies open possibilities for testing specific assumptions and collecting data for the estimation of model parameters. This has so far received little attention in fisheries research. I expect advances in this area to become an integral part of the development of advanced models when state-of-the-art technology is properly applied (see, e.g. in Johansen et al. 2007). Due to the complexity of advanced spatio-temporal models, it is unrealistic to expect good results without a substantial effort to calibrate the models against real observations, and to replace the most doubtful assumptions with empirical observations. Thus, new models to support the EAF must be developed with full regard to the technologies available for collecting the basic input data. Such organic communication between models and observation systems is a prerequisite for improving models from merely interesting to fully operational tools for the management of ecosystems.

The following text presents an example of how an operational system can be established based on the participation of stakeholders, users and scientific organisations in a joint long-term effort. This example is presently under consideration by a scientific and technology consortium and represents a possible way forward based on the technologies partly described in this chapter.

21.4.2 Synthesis of Information (Modelling): The Barents Sea 2015 Scenario

The Barents Sea (Fig. 21.12), its environment and commercially exploited fish stocks have been studied for more than 100 years; some of the archived time series go back to the 1800s (see, e.g. Sakshaug et al. 1994; Godø 2003). The region is probably one

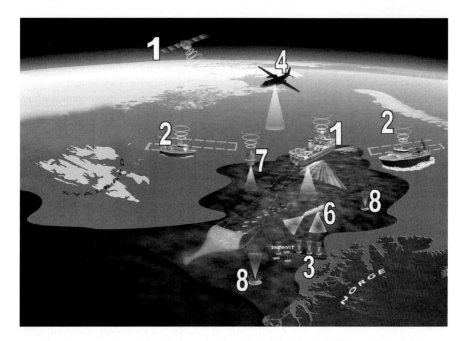

Fig. 21.12 An Open Monitoring System (OMS) taking advantage of data collected from a variety of sensors and platforms form all stakeholders. 1) Research vessels, 2) Fishing vessel equipped with scientific instruments, 3) Petroleum industry infrastructure scientifically instrumented, 4) Aeroplanes collected surface and subsurface information with lidar and radar, 5) Satellite data on surface properties, 6) AUVs with acoustic, optic and various environmental sensors, 7) Drifting instrumented buoys, 8) bottom mounted autonomous systems (Godø and Tenningen 2009).

of the best studied large marine ecosystems. The fish stocks are heavily exploited although some ecosystem factors (e.g. cod–capelin interactions) are incorporated in the assessment and management advice (ICES 2007). Presently, oil and gas exploration and offshore field development are in progress. This represents additional pressures on the environment and may have long-term effects on the ecosystem integrity and the productivity of renewable resources. Thus, there is political pressure and interest from all the stakeholders who see the need for improved understanding of the ecosystem as well as an adequate monitoring programme that includes marine resources, ecosystem integrity and pollution problems. Finally, the Barents Sea borders the Arctic Ocean and thus is susceptible to the effects of climate change. The idea behind the following proposal is that the Barents Sea case can serve as a demonstration of advanced ecosystem monitoring because:

1. Large economic interests are involved in the exploitation of both petroleum and renewable resources.
2. It is a sensitive and vulnerable area that is liable to suffer from both climate change and anthropogenic disturbances.

3. Through the petroleum industry we have access to the best subsea technology and expertise.
4. There is political will and unified interest among all stakeholders to establish a rigorous scientific basis for the responsible utilisation of the Barents Sea and its resources in the future.

Figure 21.12 illustrates the idea of the new Open Monitoring System (OMS). The technological solutions have been discussed earlier in this chapter. All stakeholders are expected to participate in the monitoring, contributing their own competence and resources. The idea is that, e.g. oil companies will allow sensors to be attached to their platforms or other installations, and their presence at a location of interest provides a sampling opportunity for other purposes. All information will be available online as soon as sensor data have been retrieved and quality checks completed. The participants will benefit from having access to the information collected through the OMS via the Internet, excepting any restrictions due to national security interests, military considerations or industrial property rights.

Governmental research institutions (Institute of Marine Research and Norwegian Polar Institute):
Present responsibility: These institutions are presently responsible for most of the monitoring of the physical and biological environment as well as advice on the exploitation of renewable resources.

Infrastructure and platforms: They have several research vessels equipped with scientific instrumentation operating in the area throughout the year. They are developing portable stationary systems for long-term deployment and monitoring.

Tasks in OMS: Routine quasi-synoptic surveying of the whole area to update the status of the environment and its exploited resources. Responsible for maintaining up to date instrumentation and methodology, including associated R&D programmes.
Examples of gain from OMS: Extension of present activities into ecosystem monitoring. Improved separation of climatic and anthropogenic effects. Better platforms and quality of advice on ecosystem management at all levels.

Defence (Including Norwegian Defence Research Establishment)
Present responsibility: Military security and enforcement of fishery regulations through the coastguard.

Infrastructure and platforms: Several coastguard vessels, helicopters and aeroplanes operate in the area year around. The vessels can be equipped for scientific acoustic surveying and the collection of biological samples from fishing vessels after inspections at sea; potentially, they could also use multibeam echo sounders for bottom/habitat mapping. The aeroplanes will be equipped with lidar and Synthetic Aperture Radar for sea-surface and sub-surface observations (see, e.g. Godø and Tenningen 2009).

Tasks in OMS: Report data taken during routine operations, in particular, extended reporting from the fisheries inspections, and participate in integrated campaigns as circumstances permit. Security exception will be situation dependent.

Examples of gain from OMS: Better physical monitoring of the marine environment. Improved methods for analysing underwater sound spectra, e.g. for accurate classification of signals as biological or military in origin.

Fishing industry
Present responsibility: Standard catch reporting
Infrastructure and platforms: Modern fishing vessels are advanced platforms with a multitude of possibilities similar to research vessels. The vessels can be fitted with advanced acoustic instrumentation, and the trawl can be equipped with various acoustic and hydrographic sensors, with automatic reporting of data once the gear is hauled. There will be a segregated data collection regime where some vessels will have advanced equipment and more substantial reporting responsibilities than others. See Karp (2007) for more details of the type of instrumentation that can be deployed.

Tasks in OMS: Keep instrumentation working and serviced, including calibration procedures. Collect biological samples according to agreed protocols. Check that the automatic reporting system via satellite communication is operational.

Examples of gain from OMS: Continuous flow of biological and physical information that can be used in strategic planning of fishing operations. Better understanding of scientific data and the background to management advice. Strengthen their image as a responsible stakeholder in the EAF.

Petroleum industry
Present responsibility: Mainly limited to environmental and habitat monitoring associated with offshore field installations.

Infrastructure and platforms: Large fixed and floating offshore platforms linked to seabed installations by sophisticated data transmission systems. Operational vessels, ROVs and AUVs.
Tasks in OMS: Provide facilities within their infrastructure for OMS instrumentation. Support development of OMS with technological expertise. Give access to available windows in AUV/ROV operations for research.

Examples of gain from OMS: Improve security in field developments. Limit effects of small and large accidents during sea operations. Avoid confusion between industrial and climatic effects on ecosystems. Strengthen their image as a responsible stakeholder in the EAF.

National and international academic institutions
Present responsibility: Limited to specific contracts or agreements made at governmental level.

Infrastructure and platforms: Education facilities and scientific laboratories.

Tasks: Support through advice and sharing scientific knowledge during the design of the system. Develop new concepts and models based on the information flow from OMS. Carry out laboratory and field experiments to support the establishment and improvements of the OMS. Secure appropriate competence to operate OMS in the future through education of personnel at all levels.

Examples of gain from OMS: Access to quality data at all levels relevant to the ecosystem. Coordinated activity to achieve ground-breaking science.

21.5 The Known, the Unknown and the Unknowable

The Census of Marine Life (www.CoML.org), which is a technology-driven initiative to expand knowledge of marine life, has emphasised the importance of the above header statement. Much can be done with new technology, but it is easy to waste time and effort on the "unknowable". Particularly, with regard to quantitative knowledge it is important to realise that sometimes the complexity of the natural world goes beyond what can realistically be observed and monitored (Levin 2003). Present monitoring systems are in many cases inadequate for responding to the tough demands set by the EAF. The suggested use of new tools like ecosystem indicators and MPAs is a move that will improve the basis for good decision making. But are these innovations sufficient to cope with the challenging demand? In this chapter I have tried to demonstrate that technology offers a suite of new options that can fill gaps in knowledge as well as providing adequate monitoring systems. There is much potential in the integration of models and observation systems to optimise sampling, parameterise models and remove unnecessary assumptions. Further development needs international coordinated scientific efforts to learn efficient ways to utilise new technology and establish adequate models. Any renewal of marine science and monitoring approaches calls for joint efforts among the stakeholders; success will depend on realising the possibilities for establishing appropriate consortia that have the necessary scientific, technological and management capacity. An overarching need, however, is that such efforts are done with appropriate humility towards the complexity of marine ecosystems, to avoid wasting time and effort on the "unknowable".

Acknowledgements The writing of this chapter has been supported by the project EcoFish partially financed by the Norwegian Research Council. David MacLennan is thanked for careful reading and comments to the manuscript.

References

Adams PB, Butler JL, Baxter CH, Laidig TE, Dahlin KA, Wakefield WW (1995) Population estimates of pacific coast groundfishes from video transects and swept-area trawls. Fish Bull 93:446–455

Aglen A (1994) Sources of error in acoustic estimation of fish abundance. In: Fernö A, Olsen S (eds) Marine Fish Behaviour in Capture and Abundance Estimation. Fishing News Books, Blackwell Science, Oxford, pp 107–133

Aglen A (1996) Impact of fish distribution and species composition on the relationship between acoustic and swept-area estimates of fish density. ICES J Mar Sci 53:501–506

Aksland M (2006) Applying an alternative method of echo-integration. ICES J Mar Sci 63: 1438–1452

Amaratunga T, Lassen H (1998) What future for capture fisheries – a shift in paradigm: visioning sustainable harvest from the Northwest Atlantic in the twenty-first century. J Northw Atl Fish Sci 23:1–275

Anderson ED (2002) 100 Years of science under ICES. ICES Mar Sci Symp 215

Ben-Tuvia A, Dickson W (1968) Proceedings of the Conference on Fish behaviour in relation to Fishing Techniques and Tactics. FAO Fish Rep 62

Beverton RJH, Holt SJ (1957) On the Dynamics of Exploited Fish Populations. Her Majesty's Stationary Office London, 577 pp

Bishop J, Venables WN, Wang YG (2004) Analysing commercial catch and effort data from a penaeid trawl fishery – a comparison of linear models, mixed models, and generalised estimating equations approaches. Fish Res 70:179–193

Bombeo-Tuburan I, Coniza EB, Rodriguez EM, Agbayani RF (2001) Culture and economics of wild grouper (*Epinephelus coioides*) using three feed types in ponds. Aquaculture 201:229–240

Brehmer P, Do Chi T, Mouillot D (2006) Amphidromous fish school migration revealed by combining fixed sonar monitoring (horizontal beaming) with fishing data. J Exp Mar Biol Ecol 334:139–150

Browman HI, Stergiou KI (2004) Marine protected areas as a central element of ecosystem-based management: defining their location, size and number. Mar Ecol-Prog Ser 274:271–272

Chu DZ, Wiebe PH, Copley NJ, Lawson GL, Puvanendran V (2003) Material properties of North Atlantic cod eggs and early-stage larvae and their influence on acoustic scattering. ICES J Mar Sci 60:508–515.

Churnside JH, Okumura K S (2001) A comparison of airborne LIDAR and echo sounder performance in fisheries. J Mar Acoust Soc Jpn 28(3):49–61

Churnside JH, Wilson JJ, Tatarskii VV (1997) Lidar profiles of fish schools. Appl Opt 36:6011–6020

Cury P, Christensen V (2005) Quantitative ecosystem indicators for fisheries management – introduction. ICES J Mar Sci 62:307–310

Doubleday WG, Rivard D (1981) Bottom trawl surveys. Can Sp Publ Fish Aquat Sci 58:1–273

Dybedal JI (1993) TOPAS: parametric end-fire array used in offshore applications. In: Hobaek H (ed) Advances in Nonlinear Acoustics. World Scientific, Singapore, pp 264–275

Engås A (1994) The effects of trawl performance and fish behaviour on the catching efficiency of demersal sampling trawls. In: Fernö A, Olsen S (eds) Marine Fish Behaviour in Capture and Abundance Estimation. Fishing News Books, Blackwell Science, Oxford, pp 45–68

Engås A, Godø OR (1989) Escape of fish under the fishing line of a Norwegian sampling trawl and its influence on survey results. Journal du Conseil International pour l'Exploration de la Mer 45:269–276

FAO (2001) Declaration of the Reykjavik Conference on Responsible Fisheries. Reykjavik Iceland 2001. C 2001/INF/25

Farmer DM, Trevorrow MV, Pedersen B (1999) Intermediate range fish detection with a 12-kHz sidescan sonar. J Acoust Soc Am 106(5):2481–2490

Favali P, Beranzoli L (2006) Seafloor observatory science: a review. Ann Geophys 49:515–567

Fernandes PG, Brierley AS, Simmonds EJ, Millard NW, McPhail SD, Armstrong F, Stevenson P, Squires M (2000) Fish do not avoid survey vessels. Nature 407:152

Fernandes PG, Gerlotto F, Holliday DV, Nakken O, Simmonds EJ (2002) Acoustic applications in fisheries science: the ICES contribution. ICES Mar Sci Symp 215:483–492

Fernö A, Olsen S (1994) Marine fish behaviour in capture and abundance estimation. Fishing News Books, Oxford, 222 pp

Field JC, Punt AE, Methot RD, Thomson CJ (2006) Does MPA mean 'major problem for assessments'? Considering the consequences of place-based management systems. Fish Fisheries 7:284–302

Finstad JL, Nordeide JT (2004) Acoustic repertoire of spawning cod, *Gadus morhua*. Environ Biol Fish 70:427–433

Fossa J, Mortensen PB, Furevik DM (2002) The deep-water coral *Lophelia pertusa* in Norwegian waters: distribution and fishery impacts. Hydrobiology 471:1–12

Fréon P, Misund OA (1999) Dynamics of pelagic fish distribution and behaviour: effects on fisheries and stock assessment. Fishing New Books, Oxford

Frid CLJ, Harwood KG, Hall SJ, Hall JA (2000). Long-term changes in the benthic communities on North Sea fishing grounds. ICES J Mar Sci 57:1303–1309

Gerlotto F, Soria M, Fréon P (1999) From two dimensions to three: the use of multibeam sonar for a new approach in fisheries acoustics. Can J Fish Aquat Sci 56:6–12

Godø OR (1994) Factors affecting the reliability of groundfish abundance estimates from bottom trawl surveys. In: Fernö A, Olsen S (eds) Marine Fish Behaviour in Capture and Abundance Estimation. Oxford Fishing News Books, Blackwell Science, Oxford, pp 166–199

Godø OR (1998) What can technology offer the future fisheries scientist – possibilities for obtaining better estimates of stock abundance by direct observations. J Northw Atl Fish Sci 23:105–131

Godø OR (2003) Fluctuation in stock properties of North-east Arctic cod related to long-term environmental changes. Fish Fisheries 4:121–137

Godø OR, Engås A (1989) Swept area variation with depth and its influence on abundance indices of groundfish from trawl surveys. J Northw Atl Fish Sci 9:133–139

Godø OR, Tenningen E (2009) Remote sensing. In: Megrey B, Moksnes E (eds) Computers in Fisheries Research. Springer, London (in press)

Godø OR, Walsh SJ (1992) Escapement of fish during bottom trawl sampling – implications for resource assessment. Fish Res 13:281–292

Godø OR, Patel R, Torkelsen T, Vagle S (2005) Observatory technology in fish resources monitoring. Proceedings of the International Conference "Underwater Acoustic Measurements: Technologies & Results". 28 June–1 July 2005. Heraklion, Crete, Greece.

Gonzalez L, Gerlotto F (1998) Observation of fish migration between the sea and a Mediterranean Lagoon (Etang De L'or, France) using multibeam sonar and split beam echo sounder. Fish Res 35:15–22

Gunderson DR (1993) Surveys of fisheries resources. Wiley, New York, 248 pp

Handegard NO (2007) Observing individual fish behavior in fish aggregations: tracking in dense fish aggregations using a split-beam echosounder. J Acoust Soc Am 122:177–187

Hjellvik V, Godø OR, Tjøstheim D (2004) Diurnal variation in acoustic densities: why do we see less in the dark? Can J Fish Aquat Sci 61:2237–2254

Holley SE, Hydes DJ (2002) 'Ferry-Boxes' and data stations for improved monitoring and resolution of eutrophication-related processes: application in Southampton Water UK, a temperate latitude hypernutrified Estuary. Hydrobiology 475:99–110

Horne JK (2000) Acoustic approaches to remote species identification: a review. Fish Ocean 9:356–371

Horne JK (2003) The influence of ontogeny, physiology, and behaviour on the target strength of walleye Pollock (*Theragra chalcogramma*). ICES J Mar Sci 60:1063–1074

Horne JK (2005) Fisheries and marine mammal opportunities in ocean observatories. In: Proceedings of Underwater Acoustic Measurements: Technologies & Results. Heraklion, Crete, 8 pp

ICES (2007) Report of the Arctic Fisheries Working Group (SFWG) 18–27 April 2007, Vigo, Spain. ICES CM 2007/ACFM 16:1–651

Jech JM, Michaels WL (2006) A multifrequency method to classify and evaluate fisheries acoustics data. Can J Fish Aquat Sci 63:2225–2235

Jennings S (2005) Indicators to support an ecosystem approach to fisheries. Fish Fisheries 6:212–232

Jennings S, Dinmore TA, Duplisea DE, Warr KJ, Lancaster JE (2001) Trawling disturbance can modify benthic production processes. J Anim Ecol 70:459–475

Johannesson KA, Mitson RB (1983) Fisheries acoustics – a practical manual for aquatic biomass estimation. FAO Fish Tech Pap 240:1–249

Johansen GO, Skogen MD, Godø OR, Torkelsen T (2007) Predicting recruitment of 0-group gadoids in the Barents Sea critical interaction between models and observations. ICES CM/B07:1–17

Johnson M, Madsen PT, Zimmer WMX, De Soto NA, Tyack PL (2004) Beaked whales echolocate on prey. Proc Roy Soc Lond Ser B Biol Sci 271:383–386. Notes: Suppl. 6

Karp W (2007) Collection of acoustic data from fishing vessels. ICES Coop Res Rep 287:1–90

Kloser RJ, Ryan T, Sakov P, Williams A, Koslow JA (2002) Species identification in deep water using multiple acoustic frequencies. Can J Fish Aquat Sci 59:1065–1077

Knutsen O, Svendsen H, Osterhus S, Rossby T, Hansen B (2005) Direct measurements of the mean flow and eddy kinetic energy structure of the upper ocean circulation in the NE Atlantic. Geophys Res Lett 32(14):L14604

Korneliussen RJ, Ona E (2000) Some applications of multiple frequency echo sounder data. ICES J Mar Sci 59:291–313

Kubecka J, Duncan A, Butterworth AJ (1992) Echo counting or echo integration for fish biomass assessment in shallow waters. In: Weydert M (ed) Proceedings of the European Conference on Underwater Acoustics. Elsevier, London, pp 129–132

Levin SA (2003) Complex adaptive systems: exploring the known, the unknown and the unknowable. Bull Am Math Soc 40:3–19

Livingston PA (2005) Pices' role in integrating marine ecosystem research in the North Pacific. Mar Ecol-Prog Ser 300:257–259

Lorance P, Trenkel VM (2006) Variability in natural behaviour, and observed reactions to an ROV, by mid-slope fish species. J Exp Mar Biol Ecol 332:106–119

Mackinson S, Van Der Kooij J (2006) Perceptions of fish distribution, abundance and behaviour: observations revealed by alternative survey strategies made by scientific and fishing vessels. Fish Res 81:306–315

Madsen PT, Wahlberg M, Mohl B (2002) Male sperm whale (*Physeter macrocephalus*) acoustics in a high-latitude habitat: implications for echolocation and communication. Behav Ecol Sociobiol 53:31–41

Makris NC, Ratilal P, Symonds DT, Jagannathan S, Lee S, Nero RW (2006) Fish population and behavior revealed by instantaneous continental shelf-scale imaging. Science 311:660–663

Matsuoka T, Nakashima T, Nagasawa N (2005) A review of ghost fishing: scientific approaches to evaluation and solutions. Fish Sci 71:691–702

Maunder MN, Punt A (2004). Standardizing catch and effort data: a review of recent approaches. Fish Res 70:141–159

McCartney BS, Stubbs AR (1971) Measurements of acoustic target strengths of fish in dorsal aspect, including swimbladder resonance. J Sound Vib 15:397

Misund OA (1994) Swimming behaviour of fish schools in connection with capture by purse seine and pelagic trawl. In Fernö A and Olsen S (eds) Marine Fish Behaviour in Capture and Abundance Estimation. Fishing News Books, Blackwell Science, Oxford, pp 84–106

Misund OA (1997) Underwater acoustics in marine fisheries and fisheries research. Rev Fish Biol Fish 7:1–34

Misund OA, Skjoldal HR (2005) Implementing the ecosystem approach: experiences from the North Sea, ICES and the Institute of Marine Research, Norway. Mar Ecol-Prog Ser 300:260–265

Misund OA, Aglen A, Hamre J, Ona E, Røttingen I, Skagen DW, Valdemarsen JW (1996) Improved mapping of schooling fish near the surface: comparison of abundance estimates obtained by sonar and echo integration. ICES J Mar Sci 53:383–388

Mitson R (1995) Underwater noise of research vessels: review and recommendations. ICES Coop Res Rep 209:1–61

Mous J, Sadovy Y, Halim A and Pet JS (2006) Capture for culture: artificial shelters for grouper collection in SE Asia. Fish Fisheries 7:58–72

Mueller RP, Brown RS, Hop H, Moulton L (2006) Video and acoustic camera techniques for studying fish under ice: a review and comparison. Rev Fish Biol Fish 16:213–226

Murray J, Hjort J (1912) The depths of the ocean: a general account of the modern science of oceanography based largely on the scientific researches of the Norwegian steamer Michael Sars in the North Atlantic. Macmillan, London

Nero RW, Thompson CH, Jech JM (2004) In situ acoustic estimates of the swimbladder volume of Atlantic herring (*Clupea harengus*). ICES J Mar Sci 61:323–337

NRC (1998) Improving Fish Stock Assessments. National Academic Press, Washington, DC, 177 pp

Ona E (2003). An expanded target-strength relationship for herring. ICES J Mar Sci 60:493–499

Ona E, Mitson RB (1996) Acoustic sampling and signal processing near the seabed: the deadzone revisited. ICES J Mar Sci 53:677–690

Ona E, Dalen J, Knudsen H, Patel R, Andersen LN, Berg S (2006) First data from sea trials with the new MS70 multibeam sonar. J Acoust Soc Am 120:3017–3018

Ona E, Godø OR, Handegard NO, Hjellvik V, Patel R, Pedersen G (2007a) Silent research vessels are not quiet. J Acoust Soc Am 121:EL145-EL150

Ona E, Andersen LN, Knudsen HP, Berg S (2007b) Calibrating multibeam, wideband sonar with reference targets. Proceedings of OCEANS 2007, Aberdeen, Scotland: 1-5 [doi:10.1109/OCEANSE.2007.4302473]

Patel R, Handegard NO, Godø OR (2004) Behaviour of herring (*Clupea harengus* L.) towards an approaching autonomous underwater vehicle. ICES J Mar Sci 61:1044–1049

Pedersen R, Ona E (2007) Size-dependent frequency response in sandeel schools. ICES J Mar Sci (in preparation)

Popper AN, Fay RR (1993) Sound detection and processing by fish – critical-review and major research questions. Brain Behav Evol 41:14–38

Popper AN, Plachta DTT, Mann DA, Higgs D (2004) Response of clupeid fish to ultrasound: a review. ICES J Mar Sci 61:1057–1061

Priede IG, Bagley PM (2000) *In situ* studies on deep-sea demersal fishes using autonomous unmanned lander platforms. Oceanogr Mar Biol Ann Rev 38:357–392

Rose CR, Stoner AW, Matteson K (2005) Use of high-frequency imaging sonar to observe fish behaviour near baited fishing gears. Fish Res 76:291–304

Sakshaug E, Bjørge A, Gulliksen B, Loeng H, Mehlum F (1994) *Økosystem Barentshavet*. Universitetsforlaget, Oslo, 303 pp

Sigurdsson T, Thorsteinsson V, Gústafsson L (2006) *In situ* tagging of deep-sea redfish: application of an underwater, fish-tagging system. ICES J Mar Sci 63:523–531

Simmonds EJ, Armstrong F, Copland PJ (1996) Species identification using wideband backscatter with neural network and discriminant analysis. ICES J Mar Sci 53:189–195

Simmonds J, MacLennan DN (2005) Fisheries Acoustics. Blackwell Science, Oxford, 437 pp

Sissenwine M, Murawski S (2004) Moving beyond 'intelligent tinkering': advancing an ecosystem approach to fisheries. Mar Ecol-Prog Ser 274:291–295

Somerton DA, Munro PT, Weinberg KL (2007) Whole-gear efficiency of a benthic survey trawl for flatfish. Fish Bull 105:278–291

Stergiou KI, Browman HI (2005) Bridging the gap between aquatic and terrestrial ecology – introduction. Mar Ecol-Prog Ser 304:271–272

Tenningen E, Churnside JH, Slotte A, Wilson JJ (2006) Lidar target-strength measurements on northeast atlantic mackerel (*Scomber scombrus*). ICES J Mar Sci 63:677–682

Thompson CH, Love RH (1996) Determination of fish size distributions and areal densities using broadband low frequency measurements. ICES Int Symp Fish Plank Acoust 53:197–202

Toresen R, Østvedt O (2000) Variation in abundance of Norwegian spring – spawning herring (*Clupea harengus*, Clupeidae) throughout the 20th century and the influence of climatic fluctuations. Fish Fisheries 1:231–256

Tudela S, Short K (2005) Paradigm shifts, gaps, inertia, and political agendas in ecosystem-based fisheries management. Mar Ecol-Prog Ser 300:282–286

UN Declaration of the *World Summit on Sustainable Development* (WSSD) (or the *Earth Summit 2002*). Johannesburg, South Africa, 26 August 4–September 2002

Valdemarsen JW (2001) Technological trends in capture fisheries. Ocean Coast Manage 44:635–651

Walsh SJ, Engas A, Ferro R, Fonteyne R, Marlen B (2002) To catch or conserve more fish: the evolution of fishing technology in fisheries science. ICES Mar Sci Sympos 215

Watanabe M, Sekine M, Hamada E, Ukita M, Imai T (2002) Monitoring of shallow sea environment by using snapping shrimps. Water Sci Technol 46:419–424

Zagaglia CR, Lorenzzetti JA, Stech JL (2004) Remote sensing data and longline catches of yellowfin tuna (*Thunnus albacares*) in the equatorial Atlantic. Remote Sens Environ 93:267–281

Chapter 22
Accounting for Spatial Population Structure in Stock Assessment: Past, Present, and Future*

Steven X. Cadrin and David H. Secor

Abstract Stock identification has been an important prerequisite for stock assessment throughout its history. The earliest evaluations of recruitment variability recognized that understanding the spatial scale of a fishery resource is essential for studying population dynamics. A paradigm of stock structure was based on closed migration circuits and geographic variation of phenotypic traits and formed a premise for fishery modeling conventions in the mid-1900s. As genetic techniques developed in the late 1900s, the "stock concept" was refined to include a degree of reproductive isolation. Realization that there was no single method that addressed the various assumptions of stock assessment and needs of fishery management prompted a more holistic view of population structure that called for multiple sources of demographic and genetic data. Recent applications of advanced techniques challenge the traditional view of populations as geographically distinct units with homogeneous vital rates and isolation from adjacent resources. More complex concepts such as metapopulations and "contingent theory" may be more applicable to many fishery resources with sympatric population structure. These more complex patterns of population structure have been incorporated into some advanced stock assessment techniques and metapopulation models that account for movement among areas and sympatric heterogeneity. Wider application of spatially explicit models in future stock assessments will require clear identification of stock components, evaluating movement rates and determining the degree

S.X. Cadrin
NOAA/UMass Cooperative Marine Education and Research Program,
School for Marine Science & Technology, 838 South Rodney French Boulevard,
New Bedford, MA 02744-1221, USA
e-mail: Steven.Cadrin@noaa.gov

D.H. Secor
University of Maryland Center for Environmental Science,
Chesapeake Biological Laboratory, 1 William St., Solomons, MD 20688, USA
e-mail: Secor@CBL.UMCES.edu

*Manuscript submitted for Proceedings of the AIFRB 50th Anniversary Symposium "Future of Fishery Science in North America", February 13–15, 2007, Seattle WA

of reproductive isolation. Because spatial structure affects how populations respond to fisheries, incorporation of heterogeneous patterns and movement in stock assessment models should improve advice for fishery management.

Keywords Mixed-stock fisheries · migration · spatial heterogeneity · stock assessment · stock identification

> "The theory of exploitation of homogeneous stocks of a fish species has had considerable attention in recent years"

> W.E. Ricker (1958)

22.1 Introduction

Stock assessment of fishery resources is an exercise in simplification of complex population processes and patterns of variation (Hilborn and Walters 1992; Quinn and Deriso 1999). Even the most elaborate population models are gross abstractions that hopefully retain the general properties of the population while avoiding many subtle complications. Stock assessment modeling can be viewed as a progression of increasing demographic complexity, with successive extensions incorporating information on size, age, gender, and maturity. Despite some compelling demonstrations of the importance of spatial structure for population dynamics (e.g., Ricker 1958; Sinclair 1988; MacCall 1990; NRC 1994), spatial aspects of demographic structure have been relatively ignored.

The history of stock identification of fishery resources is marked with technological milestones that represent advances in methodologies, providing new perspectives on what defines a "stock" and revised concepts of stock structure (Cadrin et al. 2005). However, depicting the historical development of population structure concepts as an overly simplistic beginning and a gradual modification to more and more complex views would be a revisionist history. A more accurate review would recognize that recent advocacy for more complex views of population structure revive concepts from some of the earliest research on the subject (Secor 2005). Similarly, some of the pioneering work on stock assessment modeling explicitly considered spatial structure. For example, spatial variation of parameter values and movement of fish are two "extensions of the simple theory of fishing" offered by Beverton and Holt (1957).

We review the historical development, state of the art, and future considerations of population structure in the context of stock assessment modeling. This summary is neither a comprehensive history of stock identification research nor a complete review of spatial models for other applications like general ecology or conservation biology. Our objectives are to describe the mutual relationship between the disciplines of stock assessment and stock identification by presenting their parallel histories and to discuss the research needed to advance spatial aspects of population modeling.

22.2 The Past: Coevolution of Stock Assessment and Stock Identification

A convenient and common starting point for a retrospective review of population structure is the late nineteenth century, when the Platonic paradigm of typology was being replaced with a more dynamic view of populations and greater appreciation of variability among populations (Lebedev 1969; Sinclair 1988; Sinclair and Smith 2002). The typological concept emphasized homogeneity, so much so that populations were represented by a single "type specimen," and variability within populations was considered developmental noise that obscured perceptions of the underlying type (Mayr 1982). In the context of population genetics, homogeneity is associated with panmixia (i.e., random mating of individuals within a population). Although such essentialist views of natural populations have been largely abandoned, vestiges of typology persist in the form of parametric statistics (Gould 1981), which describe populations using simple means and test for differences among populations by comparing differences in group means in the context of variability within groups (e.g., Fisher 1932). Similar vestiges of typology also pervade stock assessment modeling despite the advancement of stochastic methods.

Population structure was a central theme in the overfishing debate at the turn of the twentieth century (Smith 1994). Homogeneity and panmixia are often inferred as a corollary of the nineteenth-century perceptions that some abundant fishery resources were "inexhaustible" (Huxley 1882) and not vulnerable to local depletions. The "migration theory" explained local fluctuation in fishery yields as the result of large-scale movements of a single, expansive population of each species. However, the migration theory and the implicit panmictic view of fish populations were not consistent with the distinct patterns of geographic variation in morphology among local groups of Atlantic herring observed by Heinke (1898; reviewed by Sinclair, 1988). Early explorations of phenotypic variation eventually refuted the single-population migration theory, and supported the recognition of population structure and local recruitment patterns for explaining local fluctuations in fishery resources (Smith 1988, 1994). Sinclair and Smith (2002) describe the notion of fish stocks as a component of the paradigm shift from "migration thinking" to "population thinking."

Just as the overfishing debate formed a view of population structure of fishery resources in the northeast Atlantic, a coincident scientific debate over the "parent stream theory" of Pacific salmon was a significant development for understanding stock structure and had many common elements (Secor 2005): a premier biologist described migration patterns as random dispersion (Jordan 1905), but a field-oriented fishery scientist observed local phenotypic variation and natal homing (Gilbert 1914) that challenged this Pacific version of the migration theory. Unlike the coastal fishery resources in the eastern Atlantic, that could effectively be described with allopatric population structure (i.e., spatially distinct components), Pacific salmon populations exhibited sympatry (i.e., spatially overlapping components) of oceanic stages and philopatry (i.e., natal homing). Resolving the parent stream theory required nearly a century of research on oceanic migrations (Royce et al. 1968), and many aspects remain unresolved. In the context of stock structure and stock assessment, the overfishing issue and parent stream theory demonstrated that determining the appropriate geographic scale is necessary for understanding population dynamics (Secor 2005).

22.2.1 Phenotypic Stocks

Continued research on morphological variation and local movements set the stage for definition of phenotypic stocks and the development of conventions in fish population dynamics to model groups of fish with similar vital rates. When Russell (1931) addressed the issue of overfishing by expressing sustainable yield as the sum of recruitment and individual growth minus mortality (Fig. 22.1), "he effectively defined the field of fisheries biology as one concerned primarily with defining unit stocks and fisheries and their sizes" (Pauly 1986). The characters used to define local "races" in early studies were particularly suited for the operational definition of stock needed to apply Russell's equation. Hjort (1914) defined groups of Atlantic herring, Atlantic cod, and haddock, and Gilbert (1914) classified sockeye salmon to spawning location using growth patterns recorded on fish scales. Groups of fish with different circuli patterns on their scales had different individual growth rates. Furthermore, scale annuli could be used to determine age, and were used to study geographic patterns of recruitment and to estimate mortality rates. Therefore, each of Russell's components of production (growth, recruitment, and mortality) was indicated by fish scales, and scale patterns were effectively used to identify stocks that could be accurately modeled using Russell's harvest equation.

In the early 1900s, fish populations were considered to have homogeneous genetic constitutions, and variations in vital rates or morphology among stocks were regarded as a result of varying environments (Ricker 1972). This phenotypic stock definition was the context in which Thompson and Bell (1934) developed yield-per-recruit, Graham (1935) derived maximum sustainable yield, and Ricker (1954) and Beverton and Holt (1957) quantified stock–recruit relationships and many other

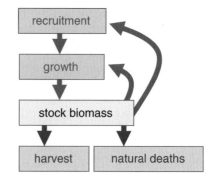

Fig. 22.1 Schematic depiction of the components of population growth, renewal, and fishery production as described by Russell (1931) (Modified from Pauly 1986)

foundations of stock assessment. These advances in assessment modeling were accompanied by the development of conventions for defining phenotypic stocks, as described in Marr's (1957) comprehensive review of stock identification methodology. However, such conventions were most appropriately applied to allopatric groups that could be effectively delineated in space and time.

The paradigm of allopatric structure was reasonable for some marine populations that inhabit distinct estuaries, continental shelves, offshore banks or open ocean basins. However, anadromous populations presented a challenge to this view of population structure. Ricker (1958) and others recognized the challenge of monitoring and managing fisheries that harvest mixed stocks, illustrating how less-productive components are vulnerable to depletion by mixed-stock fisheries. Despite the clear reproductive isolation of anadromous populations, the sympatric nature of their oceanic stages continues to be a challenge for considering stock composition in stock assessment models.

22.2.2 The Migration Triangle

Definition of phenotypic stocks for population modeling helped to determine the appropriate spatial scale for studying population processes, allowing fishery scientists to explain fluctuations in fishery yields primarily as the result of variable recruitment. That consensus prompted another thrust in fisheries research toward explaining the causes of recruitment variability. Research on recruitment dynamics included advances in fisheries oceanography, climate effects, reproductive processes, and most importantly for the subject of population structure, early life-history stages (Sinclair 1988). One question raised in this research agenda was "why do fish spawn where they do?" Information on spawning migrations and distributions of early life stages was depicted by Harden Jones' (1968) migration triangle (Fig. 22.2a), which illustrates "denatant" drift of planktonic stages to nursery habitat, ontogenetic "recruitment" to adult habitat, and seasonal migrations to spawning grounds. Although this frequently used form of the triangle represents a single migration circuit, Harden Jones also presented a more flexible schematic to allow a diversity of patterns, including the overlapping circuits of spring-spawning, autumn-spawning, and winter-spawning herring in the North Sea (Fig. 22.2b).

The migration triangle has been adopted widely in the fisheries literature, in most instances with greater emphasis on life-cycle closure (i.e., the maintenance of a population from one generation to the next) than Harden Jones initially intended. For example, concepts related to the recruitment problem formulated by Cushing (the match–mismatch concept; 1982) and Sinclair (the member-vagrant hypothesis; 1988) were entirely dependent upon life-cycle closure, yet Harden Jones recognized that straying was a necessary form of "biological insurance" against environmental change (Secor 2002). The simple version of the triangle (Fig. 22.2a) has become somewhat iconic in that researchers assume that a single migration circuit represents all populations. For example, tagging studies proliferated in the late

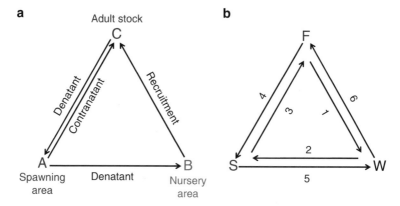

Fig. 22.2 Two forms of the migration triangle depicted by Harden Jones (1968): (**a**) a simple migration circuit involving passive drift of pelagic phases from the spawning area to the nursery area with the prevailing current (denatant), recruitment to the adult stock and seasonal migration to and from the spawning area; and (**b**) a flexible schematic for more complex movement patterns among spawning (S), feeding (F), and wintering (W) habitats, describing sympatric groups with different movement patterns (e.g., spring, autumn, and winter spawning herring in the North Sea)

1900s (Jakobsson 1970; Thorsteinsson 2002), many aimed at describing a single migration circuit for each population. Ironically, the migration triangle became an icon for allopatric stock structure, despite its original intent to illustrate a diversity of patterns, including sympatric structure.

22.2.3   The Stock Concept

Formal definition of the term "stock" is commonly attributed to the "Stock Concept Symposium" (STOCS), which was organized to address the concern that hatchery-based stocking of fish in the Great Lakes was altering genetic diversity and fitness of wild populations (Fetterolf 1981) and included some of the first direct observations of fish genotypes. However, a decade earlier in Seattle, a workshop on "The Stock Concept in Pacific Salmon" identified the need to understand the genetic basis of variability among salmon stocks (Larkin 1972). Despite an absence of genotypic observations, Ricker (1972) summarized information on philopatry, patterns of variation, and breeding experiments to support a compelling argument that phenotypic differences are heritable.

The development of electrophoretic methods to study allozyme frequencies allowed researchers to test genetic differences among stocks and shifted the emphasis of stock identification from phenotypic to genotypic methods. Booke's (1981) definitions of "genetic stock" and "phenotypic stock" gave clear preference to genetics, allowing that phenotypic stock identification could be used when genotypic stock characterization was not possible. Ihssen et al.'s (1981) methodological review included a wide range of traditional and newly developed techniques, but awarded electrophoresis the "primary position among methods used for stock identification." Redefinition of stocks and advocacy for genetic methods prompted a proliferation of electrophoretic studies of fish stocks (e.g., Hedgecock 1984; Utter 1991). However, unlike the distinct patterns of genetic variation among anadromous salmon stocks, allozymes indicated little intraspecific variation of most marine species and supported the previous concept of genetically homogeneous populations (Wirgin and Waldman 2005).

Perhaps as a result of the different electrophoretic results between anadromous and marine species, stock assessment and management of salmon explicitly recognized sympatric population structure, but population modeling of other fishery resources generally continued an allopatric, phenotypic approach (e.g., Brown et al. 1987). The last few decades of the 1900s have been described as the "golden age" of fisheries stock assessment (Quinn 2003). With the foundations for stock assessment modeling laid in the mid-twentieth century, applications of basic models proliferated, and estimation techniques advanced through developments in computers and statistics. Formal definition of the stock concept represented a milestone in considering population structure for fishery management, but impact of the stock concept was limited to populations that demonstrated genetic divergence (e.g., Pacific salmon). For many other fishery resources with apparently less genetic variation (or a lack of information on genetics), the earlier view of phenotypic stocks and allopatric structure persisted. Despite advances in modeling and development of new stock identification techniques, there was little progression in stock concepts in the northeast Atlantic and elsewhere during the last half of the twentieth century (Stephenson 2002). Unlike the earlier coevolution of stock identification and stock assessment, population modeling has been slow to adapt to the recognition of complicated population structures.

22.2.4 An Interdisciplinary Approach

Since the application of electrophoresis to stock identification in the early 1980s, a series of technological advances further changed the approach to identifying stocks (and associated definition of "stock"). A series of molecular genetic markers, including mitochondrial DNA, minisatellites, microsatellites, random amplified polymorphic DNA, amplified fragment length polymorphisms and single nucleotide polymorphisms were developed, each having potentially greater sensitivity to genetic variation (Wirgin and Waldman 2005). Similar to the advocacy for allozymes as the "primary" method to detect reproductive isolation in the 1980s, each newly developed technique was claimed to be the next best thing for determining stock structure. As new data on genetic variation became available, there was a desire to reconcile new results with previous information from traditional techniques like tagging or morphology (Waples 1998). When results from multiple stock identification approaches were compared, interdisciplinary perspectives emerged (e.g., Waldman et al. 1997; Waldman 1999).

Begg and Waldman (1999) reviewed the various techniques used to identify stocks and advocated a method that integrates results from different approaches. Information from genetic, phenotypic, and environmental approaches can be complementary, because the definition of a stock includes all three components (Dizon et al. 1992; Coyle 1998). Using information from multiple methods also increases the likelihood that stocks are correctly identified (Hohn 1997). In addition to the advances in molecular genetics, other technological advances such as image analysis, microchemistry, electronic tags, and geostatistics have improved the ability to detect population structure. The state of the art in stock identification is to apply multiple approaches to stock identification, ideally on the same samples, and compare all results to achieve an interdisciplinary perspective (e.g., Abaunza et al. 2004; Hatfield et al. 2005). A recent development in stock identification that involves interdisciplinary synthesis of genetic and phenotypic approaches is the definition of management units on the basis of demographic independence (Palsbol et al. 2006; Waples and Gaggiotti 2006).

Recent application of advanced technologies and interdisciplinary approaches reveal complex patterns of spatial structure of many marine populations, similar to the complex patterns of anadromous species detected much earlier (Conover et al. 2006). These heterogeneous, sympatric patterns present challenges to stock assessment modeling. New conventions in sampling fish populations, fishery monitoring, population modeling, and fishery management are needed to account for new perspectives on population structure.

22.3   The Present: Rethinking "Population Thinking"

Despite the successive advances in theory and practice regarding population structure, problems with stock definition are common. For example, Stephenson (2002) reported a mismatch between population structure and management units for approximately one third of the stock units in the northeast Atlantic, and many of these problems involved subunits within stocks. Apparently, sympatric structure is a problem for the current "stock concept."

22.3.1   Alternative Concepts of Population Structure

Ironically, the "population thinking" that replaced typological perspectives of biological species still has inaccurate vestiges of typology in the common assumption of homogeneity within allopatric stocks. Although the migration triangle can be used to represent a single migration circuit, Harden Jones (1968) doubted that a single migration pattern would persist over many generations, and speculated that a "level of straying and multiplicity of spawning grounds" provide insurance against unfavorable conditions. Harden Jones's discussion on multiple behavioral groups was prescient of the complex patterns currently revealed by advanced technologies (Secor 2005).

In an attempt to reconcile the spatial structure of Atlantic herring spawning groups and management of a coast-wide fishery, Stephenson et al. (2001) advocated a schematic depiction of spatial patterns for each life-history stage that allows for a range of reproductive patterns, from philopatry to mixing (Fig. 22.3). The approach was generalized by Smedbol and Stephenson (2001) to be applied to any species with spatial structure. For example, Hare (2005) illustrated the spatial patterns of weakfish reported by Thorrold et al. (2001) using the schematic approach to visualize population structure. This simple visualization offers an important conceptual break from the single circuit of ontogenetic movements depicted by the migration triangle (Fig. 22.2a), because it allows for heterogeneous patterns and groups within a population and a fishery.

The high rate of straying among local groups and the general lack of genetic variation observed in most marine species from electrophoresis prompted challenges to the widespread application of the stock concept (Smith and Jamieson 1986; Smith et al. 1990). Although these authors recognized spatial heterogeneity, a

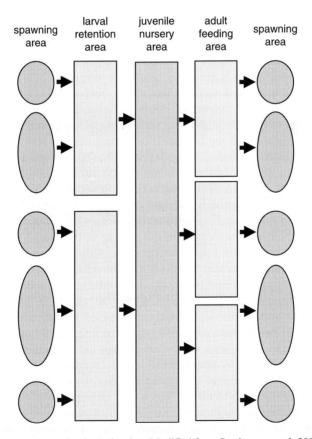

Fig. 22.3 Spatial patterns of Atlantic herring (Modified from Stephenson et al. 2001)

mechanism for maintaining spatial structure was not presented (McQuinn 1997). Several fishery scientists have since applied metapopulation concepts developed for terrestrial populations (Levins 1968) to explain persistence of local spawning groups and extensive mixing of marine populations (McQuinn 1997; Policansky and Magnuson 1998; Smedbol and Wroblewski 2002; Smedbol et al. 2002). However, the local extinctions and recolonizations implicit in the theoretical metapopulations described by Levins (1968) may not apply to many marine populations, and applying metapopulation models may not be appropriate (Smedbol et al. 2002). Still, some suggestions of local extinction and recolonization have been found (e.g., Georges Bank herring, Overholtz and Friedland 2002), and the emphasis on extinction and recolonization may not be necessary (Kritzer and Sale 2004). Perhaps the most important distinction is that the metapopulation concept recognizes spatial distribution of behavioral groups, whereas the discrete stock concept recognizes spatial patterns of genetic variation (McQuinn 1997).

A similar concept of population structure is contingent theory (Secor 1999, 2002, 2005). Hjort (1914) developed the term to describe behavioral groups of Norwegian herring with distinct migration circuits that mix during certain seasons and life-history stages. Contingent theory allows for sympatric structure within populations in which contingents have potentially different life histories (e.g., movement patterns, habitats, and productivity). Advances in methodologies that can determine an individual's environmental history (e.g., otolith microchemistry, parasite assemblages, and electronic tagging) allow researchers to discriminate and track contingents. Clark (1968) revived Hjort's contingent theory to describe groups of striped bass with distinguishable patterns of seasonal movement. Secor (1999) extended the view of population structure depicted by Stephenson, and showed how the simple migration triangle represents one contingent in a sympatric complex of contingents (Fig. 22.4). Contingent theory has been applied to a variety of fishery resources, and much like the term "stock," it may have slightly different definitions for each application. A recent workshop on the role of behavioral groups on population dynamics defined contingents as "groups of fish with different capabilities and life-cycle patterns" (ICES 2007). For our purposes, a contingent is a cohesive group of individuals within a population that share a common migrational pattern. Although these groups are not necessarily distinct genetically, the divergent movement patterns may serve as a mechanism of reproductive isolation, and contingent structure may lead to genetic structure.

An elusive aspect of developing a robust concept of population structure is understanding the mechanisms of natal homing. The intent of the migration triangle was to explain why fish spawn where they do, but the mechanism of philopatry, its primary premise, remains entirely unexplained. Evaluating the relative amounts of fidelity to spawning ground or straying to other spawning grounds is critical to modeling population dynamics of connected subpopulations (e.g., Porch and Turner 1998; Fromentin and Powers 2005).

Despite the long history of stock identification and the many methodological advancements, the definition of management units remains a practical decision,

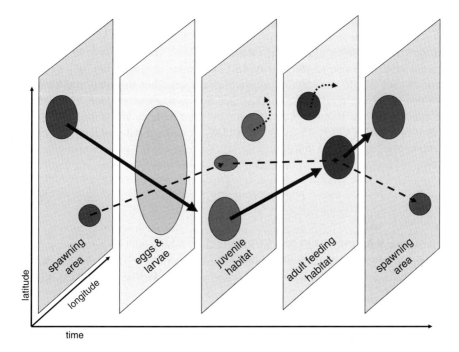

Fig. 22.4 Geographic distributions of successive life-history stages, with one "migration triangle" depicted as solid arrows and, a second contingent depicted as dashed arrows and "vagrants" as dotted arrows (Modified from Secor 1999)

because it depends on the management objective (Cadrin et al. 2007). Managing the recovery of species at risk of extinction requires a different operational definition of a stock than the definition needed for managing sustainable yield. Therefore, conservation biology tends to focus more on long-term reproductive isolation and selective adaptations than stock assessment, which focuses more on demographic independence in the short term.

22.3.2 Advances in Spatial Population Modeling

The continuing problems of exploiting sympatric populations, local depletions and recent issues of conserving essential fish habitats and designation of marine-protected areas require new information from stock assessments, and impose new challenges for population modeling (e.g., Field et al. 2006; Tuckey et al. 2007). These new challenges require the incorporation of spatial patterns in sampling and stock assessment modeling. Considering spatial structure in stock assessments takes three general forms: spatial heterogeneity, movement, and reproductive isolation.

22.3.2.1   Spatial Heterogeneity

In their text on stock assessment methods, Hilborn and Walters (1992) describe spatial models as one of the four typical extensions to simple biomass dynamics models. For sessile or sedentary invertebrates, spatial heterogeneity can be incorporated relatively easily. As demonstrated for Tasmanian abalone, conventional relationships, like Baranov's catch equation (that equates abundance, N, at time t as a function of catch, C, fishing mortality, F, and natural mortality, M) can be decomposed into spatial units (i):

$$N_{t,i} = \frac{C_{t,i}\left(F_{t,i} + M_i\right)}{\left[1 - e^{-\left(F_{t,i} + M_i\right)}\right]F_{t,i}} \tag{22.1}$$

All components of biological or fishery production (fishing mortality, natural mortality, individual growth, and recruitment; Fig. 22.1) can be similarly partitioned into discrete spatial units. For example, survival of a cohort over time can be spatially explicit:

$$N_{t+1,i} = N_{t,i}e^{-Z_{t,i}} \tag{22.2}$$

where total mortality (Z) is allowed to vary in time and space. Assuming homogeneity in vital rates, when spatial patterns exist, results in biased estimates from conventional dynamic pool models (e.g., yield-per-recruit, biomass-per-recruit, and age-based production models). In a spatial analysis of Atlantic sea scallop, Hart (2001) recommended calculating per-recruit expectations in sub-stock areas and weighting expectations by local recruitment.

A milestone in spatial modeling was the consideration of heterogeneous environments by MacCall (1990), who modified a conventional biomass dynamics model to include realized population growth rate for each habitat. Density-dependent habitat use was depicted as a basin, in which relatively large populations expand their geographic range, inhabiting less-productive "fringe" habitats (Fig. 22.5). MacCall's simulations of the basin model suggest that the optimal harvesting strategy is to fish the less-productive, fringe habitats. However, one critical assumption of the model is conformance to an ideal free distribution, in which distribution of individuals equalizes the net rate of growth rate of the entire population. Shepherd and Litvak (2004) warn that many populations may not conform to the ideal free distribution. Furthermore, the basin model implicitly assumes no barriers or restrictions to movement or mating.

22.3.2.2   Movement

Modeling spatial patterns of mobile populations requires reliable observations and estimates of movement rates. The cohort model described above (Equation 22.2) can be modified to include movement:

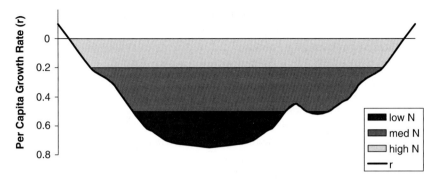

Fig. 22.5 MacCall's (1990) basin model of density-dependent habitat use, depicting three levels of population abundance (N), where r is intrinsic rate of increase

$$N_{t+1,i} = \left(\sum_j N_{t,j} \alpha_{ji} \right) e^{-Z_{t,j}} \tag{22.3}$$

where α_{ji} is the proportional movement of fish from area j to i (or the probability of movement [Hilborn 1990]). All possible movements among k areas can be considered a matrix, in which diagonal elements are proportional residence, off-diagonal elements are movements, and rows sum to 1:

$$A = \begin{bmatrix} \alpha_{1,1} & \alpha_{1,2} & \cdots & \alpha_{1,k} \\ \alpha_{2,1} & \alpha_{2,2} & \cdots & \alpha_{2,k} \\ \cdots & \cdots & \cdots & \cdots \\ \alpha_{k,1} & \alpha_{k,1} & \cdots & \alpha_{k,k} \end{bmatrix} \tag{22.4}$$

Among the first quantitative approaches to evaluating movement rates of fishery resources was that of Beverton and Holt (1957), who extended their framework for stock assessment models by considering fish movement as diffusion (D) in dimensions x (e.g., latitude) and y (e.g., latitude):

$$\frac{dN}{dt} = \frac{d}{dx}\left(D\frac{dN}{dx} \right) + \frac{d}{dy}\left(D\frac{dN}{dy} \right) \tag{22.5}$$

This continuous model was used to derive movement across discrete boundaries, termed the "box-transfer model." This pioneering work on modeling movement was extended by including directional movement (u and v, in the same dimensions) and total mortality (Z; Sibert et al. 1999):

$$\frac{dN}{dt} = \frac{d}{dx}\left(D\frac{dN}{dx} \right) + \frac{d}{dy}\left(D\frac{dN}{dy} \right) - \frac{d}{dx}(uN) - \frac{d}{dy}(vN) - ZN \tag{22.6}$$

Movement parameters D, u, and v can be estimated from tagging data, if area and time are discretized, and the number of recaptures in each period and area ($r_{t,i}$) can be predicted by rearranging Equation 22.1, replacing total abundance with number of marked fish (n) and including a reporting rate (β):

$$r_{t,i} = n_{t,i}\left[1 - e^{-(F_{t,i}+M_i)}\right]\frac{F_{t,i}}{\left(F_{t,i}+M_i\right)}\beta_{t,i} \tag{22.7}$$

Movement rates can be estimated using a series of continuous and discrete models (Hilborn 1990; Schwarz 2005).

As the number of areas and the possibility of local extinction within an area increase, these movement models approach the structure and behavior of meta-population models. Levins (1968) developed a simple metapopulation model, and Smedbol and Wroblewski (2002) modified the model to include the natural rate of extinction and the added effect of fishing. However, their applications to Atlantic cod were limited by the inability to distinguish "commercial extinction" from true extinction, recolonization from rebuilding of a remnant subpopulation, and large subpopulations from small subpopulations. Quantitative models of metapopulations have diversified to the extent that Smedbol et al. (2002) advocate a clear definition of terms and assumptions when using the term "metapopulation." Kritzer and Sale (2004) describe more complex metapopulation models that are more relevant to fishery resources (e.g., allow changes in population size). Complex metapopulation models can also include demographic structure (e.g., age, maturity) and age-dependent movement rates among subpopulations. Application of an age-based metapopulation model to yellowtail flounder showed that population dynamics of some stocks were highly sensitive to rates of planktonic drift and directed movements at older ages (Fig. 22.6; Hart and Cadrin 2004).

22.3.2.3 Reproductive Isolation

MacCall's (1990) basin model, which implicitly assumes that mating occurs among individuals from all areas, suggests that the optimal harvest strategy is to fish the less-productive, fringe habitats. In contrast, Ricker's (1958) modeling of a mixed fishery of reproductively isolated components supports the opposite conclusion (i.e., conserve the least productive components), illustrating the importance of determining the degree of reproductive isolation for spatial modeling. Unfortunately, all of the spatially explicit models described above are essentially allopatric, and consideration of sympatric groups requires information on stock composition of mixed harvests and samples.

Incorporation of movement patterns in the stock assessment of Atlantic bluefin tuna demonstrates the importance of reproductive isolation for population dynamics (NRC 1994). Porch (2003) reviews the development of age-based methods for

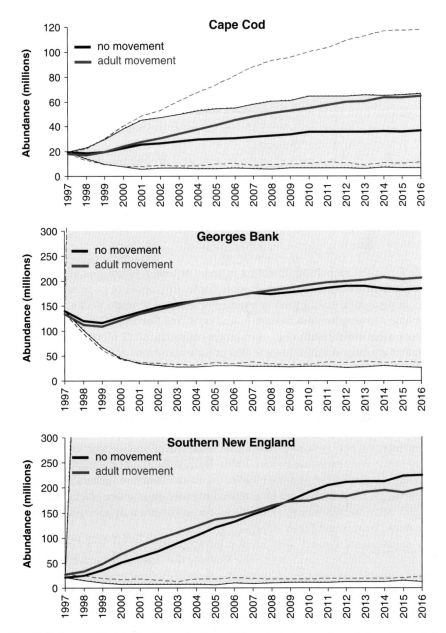

Fig. 22.6 Projected stock abundance of yellowtail flounder in three subpopulations (Cape Cod, Georges Bank, and southern New England), assuming a fishing mortality rate of $F_{0.1}$ for all subpopulations with no movement (black line and shaded confidence region) and adult movement (red line and dashed confidence region) illustrating the sensitivity of including movement for the smaller Cape Cod subpopulation (projections from Hart and Cadrin 2004)

two intermixing stocks using Beverton and Holt's (1957) box-transfer model (Equation 22.5). Simulation analyses showed that stock assessment models that included movement among eastern and western management areas performed better than separate assessments of each area (Porch et al. 1998). Subsequent model developments included two assumptions of reproductive isolation:

- *Diffusion*: Fish from one area move to another and spawn there (i.e., reproductive mixing). The process defined by the cohort model with movement (Equation 22.3) is used to model "diffusion."
- *Overlap*: Fish from one area move to another, but return to their natal area to spawn (i.e., complete philopatry, sympatry). Overlap involves a different process for tracking cohort abundance:

$$N_{t+1,i} = \left(\sum_s N_{t,s} P_{s,i,t} \right) e^{-Z_{t,i}} \tag{22.8}$$

where $P_{s,i,t}$ is the proportion of stock s in area i at time t. Estimation of P requires stock composition analysis in which identifiable differences (e.g., genetic, phenotypic, or environmental signals) in baseline samples of separate stocks are used to determine composition of a mixture (e.g., Prager and Shertzer 2005).

Stock assessment results (e.g., recruitment, spawning stock biomass, and fishing mortality) are quite sensitive to these alternative assumptions of overlap or diffusion. Estimates of Atlantic bluefin tuna abundance were relatively insensitive to movement rates when assuming philopatry, but were sensitive to movement estimates when assuming diffusive movement (Porch and Turner 1999), illustrating the importance of philopatry in modeling spatial dynamics. Indeed, several reviews advocate more elaborate population structure for Atlantic bluefin tuna, such as partial reproductive isolation, contingent structure or metapopulation dynamics (Secor 2001; Fromentin and Powers 2005). Despite the relatively simple two-stock modeling developed for Atlantic bluefin, the model structure (generalized here in Equations 22.3 and 22.10) can be extended to more areas, more stocks, and more complex relationships among stocks (e.g., partial isolation or age-specific movement rates [Powers and Porch 2004]).

A synthesis of current methods for incorporating spatial structure in stock assessment modeling remains somewhat Platonic (i.e., assuming random variability around a population average). Although population dynamics can be disaggregated to account for spatial heterogeneity in vital rates (e.g., Equations 22.1 and 22.2), the processes within subareas (i) are assumed to be homogeneous. Similarly, movement patterns can be incorporated into models (e.g., Equations 22.3–22.7), but movement is expressed as a net rate, assuming homogeneity within areas (α_{ij}), ignoring different behavioral groups within areas that have divergent movement patterns (e.g., contingents). Perhaps the promising development of genetic stock identification, otolith microchemistry, and electronic tagging will support the development of stock assessment models that account for the spatial diversity apparent in many fishery systems.

22.4 The Future: Prospects for Spatial Monitoring and Modeling

Challenges in modeling and managing fishery resources with complex spatial patterns will become increasingly important as migration behaviors are discovered that do not conform to patterns of allopatry in coastal populations (e.g., Block et al. 2005; Kraus and Secor 2005). The recurrent problem posed by sympatric distributions of populations and contingents indicates that improvements in determining the stock composition of mixed samples are needed. The sensitivity of population dynamics to movement and philopatry suggests that research should continue for estimating individual variation in migration and for a better understanding of natal homing.

Stock composition analysis has improved over the last few decades from valuable developments in genetic markers (e.g., Waples et al. 1990) and statistical mixing models (e.g., Prager and Shertzer 2005; Pella and Masuda 2006). The progress made in the recovery of Pacific salmonid stocks (Banks 2005) demonstrates that appropriately designed sampling of mixed-stock fisheries can support accurate estimates of stock composition and provide information for reliable stock assessment and real-time management decisions.

The effective stock composition analysis demonstrated for Pacific salmonids should be applied to more coastal and marine species that support mixed-stock fisheries. As genetic markers or other natural tags of natality (e.g., otolith growth patterns; Campana and Thorrold 2001) become more affordable to analyze, tissue collection should be incorporated into regular fishery and research sampling programs that currently measure biological attributes (e.g., size, gender, maturity, and age). If such programmatic sampling of genetics and other natural tags becomes common, fisheries agencies will need to invest in analytical facilities to process large volumes of tissue samples in support of regular stock composition analyses.

As more elaborate population models are developed to include spatial structure, the dependence on accurate movement rates will also increase. For example, an area-disaggregated, Bayesian state–space model was developed to estimate population parameters for Atlantic sardine, but estimates of movement were too uncertain to provide reliable estimates (Stratoudakis 2006). Similarly, Kritzer and Sale (2004) show how metapopulation models rely on accurate rates of exchange among subpopulations. Most importantly, movement rate is a critical criterion to determine genetically discrete populations (Waples and Gagiotti 2006).

The prospects for advancement in the estimation of movement rates are promising, because of recent developments in tagging technology and statistical estimation procedures. As reviewed by Schwarz (2005), estimation of mortality from tagging models improved substantially in the last decade because of advances in statistical procedures, analytical designs, and the development of conventions for best practices. The likelihood-based mortality estimators that were initially developed to estimate mortality have been extended to include movement, but further improvements are likely as they are applied to different systems. In addition, electronic tagging can be used to estimate fishing and natural mortality rates (Hightower et al. 2001).

The rapid development of electronic tagging technologies expands the capacity to understand fine-scale movement patterns. Such individual-based observations can be used to estimate movement rates and the mode of movement (e.g., overlap or diffusion). Similar to the trend in genetic techniques, electronic tags are continually becoming more affordable, as well as smaller, more reliable and longer-lasting (Thorsteinsson 2002). Continued advances in the analysis of natural tags (e.g., scale microstructure and otolith microchemistry) also promise to improve the understanding of ontogenetic movement patterns and identification of past spawning and nursery habitats (e.g., birth certificates).

Our focus has been on consideration of spatial structure in stock assessments, but advances in spatial modeling have wide applicability in other aspects of fishery science, such as ecological theory (e.g., population stability, biocomplexity, and climate effects); more effective fishery management plans (e.g., rebuilding and mixed-stock sustainable yield); design, implementation, and evaluation of marine-protected areas, defining the appropriate spatial scale of ecosystem processes and management; designation and protection of essential fish habitat; improved assessment of environmental and economic impacts of management alternatives; and optimization of spatial harvest strategies.

Advances in global positioning, vessel monitoring systems, and geostatistics offer powerful tools for monitoring spatially explicit information from fisheries and research samples. Population concepts have evolved to the point of accepting complex spatial patterns that can accommodate problems related to how mixed-stock fisheries and climate affect recruitment and resource sustainability. The extension of conventional stock assessment methods to incorporate spatial patterns is relatively straightforward and available (e.g., the most recent version of stock-synthesis assessment software allows for spatial heterogeneity and movement; Methot 2005). It appears that the missing link in applying spatially explicit population models is the lack of information on movement rates and patterns, reproductive isolation, and stock composition. Therefore, advancement in modeling spatial population structure for stock assessment requires more extensive sampling of stock composition (genetic sampling of mixed stocks, analysis of environmental signals for contingents), and tagging studies designed to estimate movement rates and patterns of movement with respect to natal homing.

Acknowledgments We appreciate the contributions of the past, present, and future that helped us in this work. We greatly benefited from the historical reviews of Mike Sinclair and Tim Smith. We thank our many current collaborators for shaping our thoughts on spatial structure and contemporaries who are developing new methods and technologies. We also thank our students (the future stewards of these ideas) for stimulating our work. Fred Serchuk, Robin Waples, and Dick Beamish reviewed the draft manuscript and contributed many constructive thoughts. Finally, we thank the American Institute of Fishery Research Biologists and the organizers of the symposium for inviting this review.

References

Abaunza P, Murta AS, Mattiucci R, et al. (2004) Stock discrimination of horse mackerel (*Trachurus trachurus*) in the northeast Atlantic and Mediterranean Sea: integrating the results from different stock identification approaches. ICES CM 2004/EE:19

Banks MA (2005) Stock identification for conservation of threatened or endangered species. In: Cadrin SX, Friedland KD, Waldman JR (eds) Stock identification methods. Elsevier Academic Press, Burlington, MA

Begg GA, Waldman JR (1999) A holistic approach to fish stock identification. Fish Res 43:37–46

Beverton RJH, Holt SJ (1957) On the dynamics of exploited fish populations. UK Ministry of Agriculture and Fisheries, Fish Invest Ser 2:19

Block BA, Teo LLH, Walli A, et al. (2005) Electronic tagging and population structure of Atlantic bluefin tuna. Nature 434:1121–1127

Booke HE (1981) The conundrum of the stock concept – are nature and nurture definable in fishery science? Can J Fish Aquat Sci 38:1479–1480

Brown BE, Darcy GH, Overholtz W (1987) Stock assessment /stock identification: an interactive process. In: Kumpf HE, Vaught RN, Grimes CB, et al. (eds) Proceedings of the stock identification workshop. NOAA Technical Memorandum NMFS-SEFC-199

Cadrin SX, Friedland KD, Waldman J (2005) Stock identification methods: applications in fishery science. Elsevier Academic Press, San Diego, CA

Cadrin, SX, Rothschild BJ, Wirgin I (2007) To Lump, or to Split? … maybe the wrong question for stock identification of fishery resources. ICES CM 2007/L:01.

Campana SE, Thorrold SR (2001) Otoliths, increments, and elements: keys to a comprehensive understanding of fish populations? Can J Fish Aquat Sci 58(1):30–38

Clark J (1968) Seasonal movements of striped bass contingents of Long Island Sound and the New York Bight. Trans Am Fish Soc 97:320–343

Conover DO, Clarke LM, Munch SB, et al. (2006) Spatial and temporal scales of adaptive divergence in marine fishes and the implications for conservation. J Fish Biol 69(Suppl C):21–47

Coyle T (1998) Stock identification and fisheries management: the importance of using several methods in a stock identification study. In: Hancock DA (ed) Taking stock: defining and managing shared resources. Australian Society for Fishery Biology, Sydney, pp 173–182

Cushing DH (1982) Fisheries biology: a study in population dynamics, 2nd edn. University of Wisconsin Press, Madison, WI

Dizon AE, Lockyer C, Perrin WF, et al. (1992) Rethinking the stock concept – a phylogeographic approach. Conserv Biol 6(1):24–36

Fetterolf CM Jr (1981) Foreword to the Stock Concept Symposium. Can J Fish Aquat Sci 38:iv–v

Field JC, Punt AE, Methot RD, et al. (2006) Does MPA mean "major problem for assessments"? Considering the consequences of place-based management systems. Fish Fisheries 7(4):284–302

Fisher RA (1932) Statistical methods for research workers. Oliver and Boyd, London

Fromentin JM, Powers JE (2005) Atlantic bluefin tuna: population dynamics, ecology, fisheries and management. Fish Fisheries 6(4):281–306

Gilbert CH (1914) Contributions to the life history of the sockeye salmon. Report to British Columbia Fisheries Department, Vancouver, BC

Gould SJ (1981) Mismeasure of man. W.W. Norton, New York

Graham M (1935) Modern theory of exploiting a fishery and application to North Sea trawling. J Cons Int Explor Mer 10:264–274

Harden Jones FR (1968) Fish migration. St. Martin's, New York

Hare JA (2005) The use of early life stages in stock identification studies. In: Cadrin SX, Friedland KD, Waldman JR (eds) Stock identification methods. Elsevier Academic Press, San Diego, CA

Hart D, Cadrin SX (2004) Yellowtail flounder (*Limanda ferruginea*) off the northeastern United States, implications of movement among stocks. In: Akçakaya HR et al. (eds) Species conservation and management: case studies. Oxford University Press, New York

Hart DR (2001) Individual-based yield-per-recruit analysis, with an application to the Atlantic sea scallop, *Placopecten magellanicus*. Can J Fish Aquat Sci 58:2351–2358

Hatfield EMC, Zuur AF, Boyd J, et al. (2005) WESTHER: a multidisciplinary approach to the identification of herring (*Clupea harengus* L.) stock components west of the British Isles using biological tags and genetic markers. ICES CM 2005/K:01

Hedgecock D (ed) (1984) Identifying fish subpopulations. California Sea Grant College Program Publication, Rep T-CSGCP-013

Heinke, F (1898) Naturgeschichte des herings. Teil 1. Deutscher Seefischerei Verein. Band III

Hilborn R (1990) Determining fish movement patterns from tag recoveries using maximum likelihood estimators. Can J Fish Aquat Sci 47:635–643

Hilborn R, Walters CJ (1992) Quantitative fisheries stock assessment: choice, dynamics, and uncertainty. Chapman & Hall, Boca Raton, FL

Hightower JE, Jackson JR, Pollock KH (2001) Use of telemetry methods to estimate natural and fishing mortality of striped bass in Lake Gaston, North Carolina. Trans Am Fish Soc 130:557–567

Hjort J (1914) Fluctuations in the great fisheries of northern Europe. Rapp p-v Réun Cons Int Explor Mer 20:1–228

Hohn AA (1997) Design for a multiple-method approach to determine stock structure of bottlenose dolphins in the mid-Atlantic. NOAA Tech Mem NMFS-SEFSC 401

Huxley THH (1882) Inaugural address. Fisheries exhibition, London (from Blinderman C, Joyce D (1998) The Huxley file (available at: http://aleph0.clarku.edu/huxley/SM5/fish.html)

ICES (2007) Report of the Workshop on Testing the Entrainment Hypothesis (WKTEST), 4–7 June 2007, Nantes, France. ICES CM 2007/LRC:10

Ihssen PE, Booke HE, Casselman JM, et al. (1981) Stock identification: materials and methods. Can J Fish Aquat Sci 38:1838–1855

Jakobsson J (1970) On fish tags and tagging. Oceanogr Mar Biol Ann Rev 8:457–499

Jordan DS (1905) A guide to the study of fishes. Henry Holt and Co. New York

Kraus RT, Secor DH (2005) Application of the nursery-role hypothesis to an estuarine fish. Mar Ecol Prog Ser 291:301–305

Kritzer JP, Sale PF (2004) Metapopulation ecology in the sea: from Levins' model to marine ecology and fisheries science. Fish Fisheries 5:131–140

Larkin PA (1972) The stock concept and management of pacific salmon. In: Simon RC, Larkin PA (eds) The stock concept in Pacific salmon. H.R. MacMillan Lectures in Fisheries, University of British Columbia, Vancouver, BC

Lebedev NV (1969) Elementary populations of fish (translated from Russian: Elementarnye populyatsii ryb, Izdatel'sto "Pishchevaya Promyshlennost" Moskva 1967). Israel Program for Scientific Translation, Jerusalem

Levins R (1968) Evolution in changing environments: some theoretical explorations. Princeton University Press, Princeton, NJ

MacCall AD (1990) Dynamic geography of marine fish populations. University of Washington Press, Seattle, WA

Marr JC (1957) Contributions to the study of subpopulations of fishes. US Dept Int Fish Wildl Serv Sp Sci Rep Fish 208

Mayr E (1982) The growth of biological thought. Harvard University Press, Cambridge

McQuinn IH (1997) Metapopulations and the Atlantic herring. Rev Fish Biol Fish 7:297–329

Methot RD (2005) Technical description of the stock synthesis II assessment program (available at: http://nft.nefsc.noaa.gov/)

National Research Council (NRC) (1994) An assessment of Atlantic bluefin tuna. National Academy Press, Washington, DC

Overholtz W, Friedland K (2002) Recovery of the Gulf of Maine-Georges Bank Atlantic herring (*Clupea harengus*) complex: perspectives based on bottom trawl survey data. Fish Bull 100:593–608

Palsbol PJ, Berube M, Allendorf FW (2006) Identification of management units using population genetic data. Trends Ecol Evol 22:11–16

Pauly D (1986) Concepts that work: some advances in tropical fisheries research. In: Dizon JL, Hosillos LV (eds) The first Asian fisheries forum. Asian Fisheries Society, Manila, Philippines

Pella J, Masuda M (2006) The Gibbs and split–merge sampler for population mixture analysis from genetic data with incomplete baselines Can J Fish Aquat Sci 63:576–596

Policansky D, Magnuson JJ (1998) Genetics, metapopulations, and ecosystem management of fisheries. Ecol Appl 8:S119–S123

Porch CE (2003) VPA 2-Box version 3.01 user's guide. Sustainable Fisheries Division Contribution SFD-01/02–151

Porch CE, Turner SC (1999) Virtual population analyses of Atlantic bluefin tuna with alternative models of trans-Atlantic migration. ICCAT Coll Vol Sci Pap 49:291–305

Porch C, Kleiber P, Turner S, et al. (1998) The efficacy of VPA models in the presence of complicated movement patterns. ICCAT Coll Vol Sci Pap 50(2):591–622

Powers PE, Porch CE (2004) Approaches to developing management procedures which incorporate mixing. ICCAT Coll Vol Sci Pap 56(3):1144–1152

Prager MH, Shertzer KW (2005) An introduction to statistical algorithms useful in stock composition analysis. In: Cadrin SX, Friedland KD, Waldman JR (eds) Stock identification methods. Elsevier Academic Press, San Diego, CA

Quinn TJ (2003) Ruminations on the development and future of population dynamics models in fisheries. Nat Res Model 16:341–392

Quinn TJ II, Deriso RB (1999) Quantitative fish dynamics. Oxford University Press, New York

Ricker WE (1954) Stock and recruitment. J Fish Res Bd Canada 11:559–623

Ricker WE (1958) Maximum sustainable yields from fluctuating environments and mixed stocks. J Fish Res Bd Canada 15:991–1006

Ricker WE (1972) Hereditary and environmental factors affecting certain Salmonid populations. In: Simon RC, Larkin PA (eds) The stock concept in Pacific salmon. H.R. MacMillan Lectures in Fisheries, University of British Columbia, Vancouver, BC

Royce WF, Smith LS, Hartt AC (1968) Models of oceanic migrations of Pacific salmon and comments on guidance mechanisms. US Fish Bull 66:441–462

Russell ES (1931) Some theoretical considerations on the "overfishing" problem. Cons Perm Int Explor Mer 6:3–20

Schwarz CJ (2005) Estimation of movement from tagging data. In: Cadrin SX, Friedland KD, Waldman JR (eds) Stock identification methods. Elsevier Academic Press, San Diego, CA

Secor DH (1999) Specifying divergent migrations in the concept of stock: the contingent hypothesis. Fish Res 43:13–34

Secor DH (2001) Is Atlantic bluefin tuna a metapopulation? ICCAT SCRS/2001/056

Secor DH (2002) Historical roots of the migration triangle. ICES J Mar Sci 215:329–335

Secor DH (2005) Fish migration and the unit stock: three formative debates. In: Cadrin SX, Friedland KD, Waldman JR (eds) Stock identification methods. Applications in fishery science. Elsevier Academic Press, San Diego, CA

Shepherd TD, Litvak MK (2004) Density-dependent habitat selection and the ideal free distribution in marine fish spatial dynamics: considerations and cautions. Fish Fisheries 5:141–152

Sibert JR, Hampton J, Fournier DA, et al. (1999) An advection–diffusion-reaction model for the estimation of fish movement parameters from tagging data, with application to skipjack tuna (*Katsuwonus pelamis*). Can J. Fish Aquat Sci 56:925–938

Sinclair M (1988) Marine populations: an essay on population regulation and speciation. Washington Sea Grant Program, Seattle, WA

Sinclair MM, Smith TD (2002) The notion that fish species form stocks. ICES Mar Sci Symp 215:297–304

Smedbol RK, Stephenson RL (2001) The importance of managing within-species diversity in cod and herring fisheries of the north-western Atlantic. J Fish Biol 59(Suppl A):109–128

Smedbol RK, Wroblewski JS (2002) Metapopulation theory and northern cod population structure: interdependency of subpopulations in recovery of a groundfish population. Fish Res 55:161–174

Smedbol RK, McPherson A, Hansen MM, et al. (2002) Myths and moderation in marine "metapopulations." Fish Fisheries 3:20–35

Smith PJ, Jamieson A (1986) Stock discreteness in herrings: a conceptual revolution. Fish Res 4:223–234

Smith PJ, Jamieson A, Birley AJ (1990) Electrophoretic studies and the stock concept in marine teleosts. J Cons Int Explor Mer 47:231–245

Smith TD (1988) Stock assessment methods: the first fifty years. In Gulland JA (ed) Fish population dynamics, 2nd edn. Wiley, New York

Smith TD (1994) Scaling fisheries, the science of measuring the effects of fishing, 1855–1955. Cambridge University Press, Cambridge

Stephenson RL (2002) Stock structure management and structure: an ongoing challenge for ICES. ICES Mar Sci Symp 215:305–314

Stephenson RL, Clark KJ, Power MJ, et al. (2001) Herring stock structure, stock discreteness, and biodiversity. In: Funk F, Blackburn J, Hay D, et al. (eds) Herring, expectations for a new millennium. Lowell Wakefield Fisheries Symposium Series No. 18. Alaska Sea Grant College Program, Fairbanks

Stratoudakis G (2006) SARDYN: Sardine dynamics and stock structure in the north-eastern Atlantic. Final report Q5RS – 2002 – 000818 (available at: http://ipimar-iniap.ipimar.pt/sardyn)

Thompson WF, Bell FH (1934) Biological statistics of the Pacific halibut fishery. (2) Effect of changes in intensity upon total yield and yield per unit of gear. Int Fish Comm Rep 8

Thorrold SR, Latkoczy C, Swart PK, et al. (2001) Natal homing in a marine fish metapopulation. Science 291:297–299

Thorsteinsson V (2002) Tagging methods for stock assessment and research in fisheries. Report of concerted action FAIR CT.96.1394 (CATAG). Reykjavik. Mar Res Inst Tech Rep 79

Tuckey T, Yochum N, Hoenig J, et al. (2007) Evaluating localized vs. large-scale management: the example of tautog in Virginia. Fisheries 32:21–28

Utter FM (1991) Biochemical genetics and fishery management: an historical perspective. J Fish Biol 39:1–20

Waldman JR (1999) The importance of comparative studies in stock analysis. Fish Res 43(1–3):237–246

Waldman JR, Richards RA, Schill WB, et al. (1997) An empirical comparison of stock identification techniques applied to striped bass. Trans Am Fish Soc 126:369–385

Waples RS (1998) Separating the wheat from the chaff: patterns of genetic differentiation in high gene flow species. J Heredity 89:438–450

Waples RS, Gaggiotti O (2006) What is a population? An empirical evaluation of some genetic methods for identifying the number of gene pools and their degree of connectivity. Mol Ecol 15:1419–1439

Waples RS, Winans GA, Utter FM, et al. (1990) Genetic approaches to the management of Pacific salmon. Fisheries 15(5):19–25

Wirgin I, Waldman JR (2005) Use of nuclear DNA in stock identification: single-copy and repetitive sequence markers. In: Cadrin SX, Friedland KD, Waldman JR (eds) Stock identification methods. Elsevier Academic Press, San Diego, CA

Chapter 23
Genetic and Evolutionary Considerations in Fishery Management: Research Needs for the Future

Robin S. Waples and Kerry A. Naish

Abstract Genetic methods have become indispensable for sound fishery management and will become even more so in the twenty-first century. Selectively neutral genetic markers are widely used in stock identification, mixed-stock fishery analysis, monitoring levels of genetic diversity within populations and levels of connectivity among populations, and for a range of other applications. We expect that future research will continue to provide incremental improvements in the number and type of genetic markers available, as well as in the methods for data analysis and the necessary computational resources. Topics that will merit special consideration include: (1) developing a better understanding of the various flavors of demographic independence and how genetic markers can provide relevant insights; (2) more powerful ways to deal with the low signal-to-noise ratio of population differentiation found in many marine species (the signal of genetic differences is small compared to various sources of random noise); (3) better integration of genetic information, biological information, and information about physical features of the habitat to provide a fuller picture of dynamic marine ecosystems; (4) whether the tiny effective size to census size ratios reported for some marine species are accurate, and if so what this means for conservation of large marine populations. In contrast to the situation with neutral genetic markers, evolutionary changes involving traits related to fitness have only recently attracted much attention in fishery management. In general, any changes to marine ecosystems alter the selective regimes that component species experience and hence can be expected to produce an evolutionary response. Three general topics are particularly important in this regard: harvest, artificial propagation, and climate change. One key challenge is to disentangle the effects of genetics versus environment in determining observed patterns of phenotypic change. Quantitative genetic and molecular genetic approaches can accomplish this, and our capabilities for examining functional parts of the genome are rapidly expanding. However, these methods are logistically challenging and resource-intensive, so in the near future will be feasible

R.S. Waples (✉)
Northwest Fisheries Science Center, 2725 Montlake Blvd. East, Seattle, WA 98112, USA
e-mail: robin.waples@noaa.gov

K.A. Naish
School of Aquatic and Fishery Sciences, University of Washington, Seattle, WA 98195, USA

R.J. Beamish and B.J. Rothschild (eds.), *The Future of Fisheries Science in North America*,   427
Fish & Fisheries Series, © Springer Science + Business Media B.V. 2009

for only a fraction of the species of interest to fishery management. Therefore, in the short term, inferences about evolutionary changes in marine species will have to draw on information for model species and other better-studied aquatic organisms.

Keywords Climate change · effective population size · genetic markers · harvest · hatcheries · mixed-stock fisheries · stock identification

23.1   Introduction

The historical approach to fishery management was characterized by independent assessments for each species, but two major developments in the last half of the twentieth century substantially modified this paradigm. First, the gradual recognition that most species are composed of multiple independent populations or stocks, and subsequent refinement of the stock concept (Simon and Larkin 1972; Ihssen et al. 1981; Waldman 2005), led to a greater appreciation of the importance of genetic considerations in conservation and management (e.g., Ryman and Utter 1987). Fishery managers are now routinely aware of, if not necessarily expert in, the importance of conserving fitness and genetic diversity within and among populations. Second, and more recently, the concept of "ecosystem management" has been increasingly adopted in fishery and conservation applications (Mooney 1998; Lubchenco et al. 2003). As a result, stock assessments often include evaluations of effects of management actions on marine food webs, interspecific interactions (competition, predation), and community structure.

  Although integration of ecosystem considerations represents an advance over historical, one-dimensional approach to stock assessments, virtually all evaluations conducted under this framework have focused exclusively on ecological factors. In contrast, evolutionary considerations of anthropogenic changes to aquatic ecosystems have received little attention. This is a serious omission, because any changes to aquatic ecosystems will also change selective regimes experienced by component species, and we can expect evolution (genetic change over time) to occur as a result. Furthermore, evolutionary changes are an important concern for fishery management. Not only can these changes affect productivity and sustainability of living natural resources; they also are generally irreversible on human time frames, if at all. Irreversible changes are a particular concern under another major advancement in natural resource management over the past decade – increasing application of the precautionary principle (articulated in the 1992 Rio Declaration, Anonymous 1992; Restropo 1999). The essence of the precautionary principle is to guide management toward actions that will not cause irreversible harm. Because biological systems are complex, it is often difficult to predict with certainty the consequences of any proposed action. Application of the precautionary principle involves shifting the burden of proof from asking, "Is there convincing evidence that the proposed action will cause irreparable harm?" to "Is there convincing evidence that the proposed action will *not* cause irreparable harm?" Answering the last question in a comprehensive way requires consideration of evolutionary as well as ecological consequences of proposed management actions.

In this chapter, we consider research that is needed to fully integrate genetic and evolutionary considerations into twenty-first century fishery management (see Waples et al. 2008 for a discussion of organizational, conceptual, and technical barriers that have hampered full use of genetic data in fishery management). For some well-established, genetically based programs, such as stock identification and mixed-stock fishery analysis, these future research needs are primarily incremental improvements to existing technologies and analyses. Evaluating the consequences of evolutionary changes in human-altered ecosystems will be more challenging, as this topic has been relatively neglected. Here, we focus on three major types of anthropogenic changes that can be expected to elicit a profound evolutionary response in natural populations: selective harvest, artificial propagation, and climate change.

Genetically based methods have a wide range of additional applications to fishery management, even though the problems they address are not necessarily evolutionary in nature. Although a full treatment of this topic is beyond the scope of what we can cover in this chapter, we expect these applications to be increasingly important in the future. Examples include forensics and enforcement (Baker et al. 2007), monitoring the status of individuals (von Schalburg et al. 2005; Purcell et al. 2006), monitoring the health of marine ecosystems (Niemi et al. 2004; Suttle 2005), and parentage analysis and the elucidation of mating systems (Emery et al. 2001; Avise et al. 2002).

23.2 "Traditional" Applications Using Molecular Genetic Markers

In this section we consider what might be called "traditional" applications of genetic markers to fishery management, which date back at least to the 1950s. The primary management issues addressed by these methods have been stock identification and population structure, analysis of mixed-stock harvests, and assessments of levels of genetic variation within populations. In the future, all of these approaches will benefit from incremental improvements in the number and quality of molecular markers, as well as continued refinements in methods of data analysis. Specific considerations are outlined below.

23.2.1 Stock Identification and Population Structure

Because stock assessments typically assume that the unit under consideration is demographically homogeneous, stock identification is a necessary precursor to scientifically based fishery management (see Cadrin et al. 2005 for a review). The revelation that protein electrophoresis could uncover abundant genetic variation in natural populations led to an explosion of studies on natural populations of fish (reviewed by Ward et al. 1994). More recently, mitochondrial DNA (mtDNA) and nuclear DNA (primarily microsatellites and single nucleotide polymorphisms [SNPs]) methods have seen widespread application (Carvalho and Hauser 1998; DeYoung and Honeycutt 2005; Ward 2006).

Genetic methods have provided a great deal of insight into stock structure in marine species, but they have some inherent limitations for drawing inferences about stock structure. This is because levels of migration or gene flow that can be most effectively studied using genetic techniques are low compared to the levels that are typically of interest in stock identification (Carvalho and Hauser 1994; Waples 1998; Waples and Gaggiotti 2006). For example, stock assessments generally assume demographic independence of the units under consideration, which means that population dynamics are driven more by local births and deaths than by immigration. The transition from demographic independence to dependence has not been rigorously studied, but available information suggests it occurs in the range of $m = 0.1$ (Hastings 1993; McElhany et al. 2000) – that is, if the rate of exchange of individuals among populations (m) is greater than about 10%, the populations tend to have correlated demographic trajectories. This poses a challenge for use of genetic markers, since 10% migration represents a very high rate of gene flow in evolutionary terms. As a consequence, genetic markers typically have low statistical power to distinguish between migration rates (say 5% vs 20%) that could have very different management implications.

In the future, more work is needed in several areas. First, it is important to clarify the management question(s) more precisely than simply calling for "stock identification." Just as there is no single, universally accepted definition of "species" or "subspecies,'" there is no single biological definition of "stock." Rather, the appropriate concept of "stock" or "population" should be related to the management goals one is trying to achieve (Waples and Gaggiotti 2006). This point can be illustrated by considering three federal laws in the United States that mandate conservation of subspecific units of marine species: the Marine Mammal Protection Act (MMPA), the Endangered Species Act (ESA), and the Magnuson-Stevens Fisheries Conservation and Management Act (MSA). The MMPA defines a biological stock as a "group of animals in common spatial arrangement that interbreeds when mature." Although this definition is rather vague, the stipulation of the MMPA to avoid localized depletions has led to identification of stocks on a fine geographic scale in many instances. Under the SFA, the term "stock of fish" means a "species, subspecies, geographical grouping, or other category of fish capable of management as a unit." A key driving force in implementation of the SFA is the requirement to rebuild overfished populations to sustainable levels within 10 years. Under this management framework, therefore, demographic linkages with another stock would have to be very strong to experience any significant rescue effect over this short time period. The ESA allows listing of subspecies and "distinct population segments" (DPSs) of vertebrates, as well as full species. In Pacific salmon (*Oncorhynchus* spp.) status determinations under the ESA, most DPSs have included relatively large geographic areas and a substantial number (~20–30) of component stocks. To provide a framework for ESA recovery planning for salmon, McElhany et al. (2000) developed the concept of viable salmonid populations (VSP) and defined a demographically independent population (within a DPS) as one for which immigration from other populations does not appreciably affect extinction risk over a 100-year time period. Thus, all these statutes require consideration of population linkages, but each entails a different concept of demographic independence and a different standard for assessing how strong the isolation must be to warrant recognition of separate management units.

Second, it is a curious fact that what would appear to be a central question in population ecology (how much migration among populations is required before they behave as a coupled system?) has received relatively little attention. The Hastings (1993) study, which is the only published paper that directly addresses this topic, provides only a rough guide for certain definitions of dependence/independence. More rigorous evaluations are needed of the levels of migration that lead to various flavors of demographic dependence/independence.

Third, the future will see incremental increases in the power of genetic markers to provide useful information for high-gene flow species. With arbitrarily large numbers of genetic markers, in theory it is possible to resolve arbitrarily small genetic differences among populations. However, the increases in power will be asymptotic, so at some point the marginal benefits of adding additional markers will decline sharply. Furthermore, along with greatly increased power comes greatly increased sensitivity of the results to data artifacts (Waples 1998). This topic has only recently been treated seriously in the scientific literature (Bonin et al. 2004; Pompanon et al. 2005). In the future, it will be increasingly important to rigorously assess data quality, quantify rates of genotyping errors, refine methods to keep error rates as low as possible, and evaluate performance of genetic methods for data that (inevitably) include some low level of errors.

Fourth, when the signal-to-noise ratio is low, as it often is in studying genetic structure of marine species, it is also important to pay particular attention to experimental design, because small departures from random sampling can be mistaken for a biological signal (Waples 1998). In addition, dynamic and chaotic processes in the ocean that govern dispersal and survival of larvae often produce patterns of genetic variation that are not temporally stable and are difficult to interpret in terms of population structure (Johnson and Wernham 1999; Pujolar et al. 2006; Selkoe et al. 2006). These challenges mean that adequate spatial and temporal replication is generally required to develop rigorous information about population structure in marine species.

Fifth, recent advances in physical oceanography and remote sensing open up exciting new possibilities to more fully integrate inferences about connectivity based on genetic markers with life-history traits and the dynamic nature of the marine environment (e.g., Galindo et al. 2006; Selkoe et al. 2008).

23.2.2 Mixed-Stock Harvests and Individual Assignments

Mixed-stock harvests represent one of the most pervasive and challenging problems facing fishery managers. Naturally occurring genetic marks have some inherent advantages for the analysis of mixed-stock fisheries compared to more traditional methods: they do not require large efforts to manually mark individuals every generation; genetic markers are temporally stable compared to some other stock identification methods (such as scale pattern analysis) that need to be surveyed in baseline populations every year (Cadrin et al. 2005); they can provide information for natural populations as well as cultured ones. A maximum-likelihood method for genetic stock identification (GSI) was developed in the late 1970s that estimates the stock composition of a mixed harvest, based on genetic samples from populations that potentially contribute to the mixture and from the mixture itself. Subsequently, the basic approach has been modified a number of times (reviewed by Pella and Milner 1987; Manel et al. 2005) and applied broadly to a wide range of species and geographic areas (reviewed by Shaklee et al. 1999). Challenges for the future include the following:

- Increases in the number of genetic markers will improve overall power but will make the assumption that the markers are independent increasingly untenable. Studies to elucidate the physical linkages among markers will be important to maximize their usefulness. If properly accounted for, physical linkages can actually improve power of genetic markers to resolve mixtures of gene pools (Falush et al. 2003; Verardi et al. 2006).
- Current methods for assessing the power of GSI to resolve mixtures of specific natural populations are biased towards an overly optimistic assessment of power (Anderson et al. 2008). Improvements are needed in this area, not only in assessing uncertainty in stock origin of sampled fish, but also in incorporating uncertainty in sampling from the fishery into an overall assessment of precision and accuracy.
- Increasingly, conservation concerns focus on stocks that are minor contributors to fisheries but can be the triggers for harvest management decisions. For example, in the analysis of a large fishery, it can be crucial to determine whether a particular stock is totally absent, or contributes a small fraction (say <1%) of the catch. This is a very challenging statistical problem that requires advances in two areas: (1) power to distinguish stock-of-origin of fish from genetically similar stocks, and (2) power to draw reliable inferences about minor contributors from relatively small samples from a large fishery.
- Although genetic markers are relatively stable over time, gene frequencies change stochastically in all finite populations, and evaluations of the effects of these changes on GSI estimates have been rather limited (Waples 1990). This issue will be increasingly important as increased power allows resolution of ever-more-closely related populations.
- The standard GSI model is referred to as "constrained maximum likelihood" because it assumes that all stocks in the mixture are represented in the baseline. An unconstrained model, which allowed for the possibility that unsampled populations might contribute to the mixture, was proposed by Smouse et al. (1990), but only recently have statistical refinements, development of more variable genetic markers, and vastly increased computational power combined to make this feasible to implement for large-scale management problems. The recent increased interest in genetic clustering methods (e.g., Pritchard et al. 2000; Pella and Masuda 2006) holds promise for significant advances in this area in the future.
- A management problem related to GSI involves identification of stock-of-origin for individuals, using genetic "assignment tests." This can be important, for example, in

forensic applications (where it might be necessary to determine whether a particular individual was taken from a protected population), or to screen individuals in a broodstock program to ensure genetic purity (Olsen et al. 2000). These applications involve consideration of trade-offs between Type I and Type II error rates, and this topic needs further development (Paetkau et al. 2004; Manel et al. 2005).

23.2.3 Levels of Genetic Variation

Genetic variation within populations provides the raw material for evolution and the resilience for populations to respond to future environmental (and anthropogenic) challenges. Studies of marine and anadromous populations have provided a great deal of information about natural levels of genetic variability, primarily at gene loci thought to be selectively neutral. Two related topics will be important to resolve in the future, both involving the effective size (N_e) of natural populations. N_e, which determines the rate of evolutionary change, is generally smaller than the census size (N).

"Bottlenecks" in large populations: an overlooked risk? – Conventional wisdom holds that most marine species are largely immune to genetic effects of bottlenecks, because after a population "crash" a widespread marine species still might have very large numbers (>10^6) of individuals. It is true that dramatic declines that still left a large N_e would have essentially no effect on the rate of allele frequency change (which is already negligible in large populations), and it is also true that such declines would not necessarily have an appreciable affect on the most commonly used measure of genetic variability, average heterozygosity (H). Population genetics theory tells us that, at equilibrium, the expected value of H for selectively neutral alleles is given by

$$H = 1 - 1/(1 + 4N_e\mu), \tag{1}$$

where μ is the per-generation mutation rate to neutral alleles (Kimura and Crow 1964). Table 23.1 shows that for μ in the range 10^{-3}–10^{-5} (as might apply, e.g., to

Table 23.1 At equilibrium, the amount of selectively neutral genetic diversity in a closed population is a function of effective population size and mutation rate (μ). The number of alleles is much more sensitive than heterozygosity to "crashes" in large populations (Based on Kimura and Crow 1964)

	Effective population size		
(A) Heterozygosity			
	10^5	10^7	10^9
$\mu = 0.001$	0.998	~1.0	~1.0
$\mu = 0.00001$	0.800	0.998	~1.0
(B) Effective number of alleles			
	10^5	10^7	10^9
$\mu = 0.001$	400	40,000	4,000,000
$\mu = 0.00001$	5	400	40,000

microsatellites), H is nearly 1 for very large populations and still very high if N_e drops to 10^5. In contrast, the effective number of alleles that can be maintained in a population ($n = 1/(1 - H)$; Kimura and Crow 1964) is nearly a linear function of N_e for this range of N_e and μ values. Thus, whereas a population that declines from $N_e = 10^9$ to 10^5 (four orders of magnitude) experiences very little loss of heterozygosity, the same population can be expected to lose the vast majority of alleles, most of which occur at very low frequency (Ryman et al. 1995). So what? What happens if a population loses a large fraction of its standing crop of rare alleles, as long as it maintains adequate levels of heterozygosity?

One of the challenges in the future will be to answer this question. Although immediate consequences for the population might not be severe, longer-term prospects are less certain. Rare alleles lost during a population crash could take a very long time to regenerate by mutation. Certain genes, such as those associated with the major histocompatibility complex (MHC), are extremely variable in vertebrates. These loci play crucial roles in immune response, and the high levels of variability appear to be maintained by natural selection (Bernatchez and Landry 2003; Garrigan and Hedrick 2003). If the ratio N_e/N is very low in some marine species (see Section 23.3), these "bottleneck" effects on genetic diversity might be apparent at much higher population sizes than has previously been appreciated.

The N_e/N ratio in marine species: can we believe the tiny estimates? – At least a half-dozen published studies of N_e in marine species have estimated tiny ratios of effective size to census size ($N_e/N \sim 10^{-3}$–10^{-5}; Turner et al. 2002; Hauser et al. 2002; Hauser and Carvalho 2008). These estimates contrast sharply with theoretical considerations that suggest the N_e/N ratio should only rarely be less than 0.5 (Nunney 1993) and empirical data for a wide range of species, for which most estimates fall in the range of 0.1–0.5 (Frankham 1995). Hedgecock (1994) proposed a possible mechanism (sweepstakes recruitment) to explain low N_e/N ratios in marine species. According to this hypothesis, chance events in the chaotic and patchy marine environment cause most families to produce no offspring at all that survive to reproduce; instead, the next generation is derived from the relatively few sweepstakes winners, whose offspring happen to arrive in the right places at the right times to promote feeding, predator avoidance, dispersal, and other vital life-cycle events. If this hypothesis is generally true, and as a result N_e/N ratios in many marine species are orders of magnitude lower than typically found for other organisms, it means that serious genetic concerns might apply to populations that would otherwise appear to be immune. For example, Hauser et al. (2002) estimated that for the New Zealand snapper (*Pagrus auratus*) in Tasman Bay, harmonic mean N for the period 1950–1998 was several million fish but $\hat{N}_e$ was only 176 – over four orders of magnitude smaller. A population with $N_e > 10^6$ experiences essentially no genetic drift, while a population with $N_e = 176$ is in the range where random genetic processes are a real conservation concern. It is therefore important to consider the question, "Are the tiny N_e/N ratios estimated for some marine species reliable?"

All of the tiny N_e/N estimates are based on genetic estimates of N_e and demographic estimates of N. Although estimates of N are often highly uncertain, it seems unlikely that errors in estimating population abundance could be large enough to explain this discrepancy. Genetically based estimates of N_e typically also have a high stochastic variance, and the distribution is very skewed toward high values because of the inverse relationship between N_e and the genetic quantities being measured. As a result, the upper bounds of confidence limits (CIs) to $\hat{N}_e$ are often very high. For example, for many of the point estimates of N_e reported by Hauser et al. (2002), the upper bounds of the CI included infinity, even when the point estimate was in the range of 10^2-10^3. Issues that need more attention in the future include the following:

- Effects of small departures from random sampling
- Effects of genotyping errors
- Small biases in N_e estimators that might become large under certain conditions
- Effects of migration and selection
- Optimal ways to combine information from multiple methods for estimating N_e

23.3   Evolutionary Changes in Aquatic Populations

The applications of genetic markers discussed above all take advantage of natural evolutionary processes that have occurred among and within populations and which produce a signal that can help inform fishery management. Here we consider evolutionary changes to natural populations that are influenced by anthropogenic factors (see Smith and Bernatchez 2008 for extensive treatment of this general topic). Whereas the approaches discussed above generally utilize neutral markers, the changes considered here involve traits directly related to fitness. Most fitness traits are coded by a number of genes – known as quantitative trait loci (QTL) – that interact with each other and with the environment to produce a range of phenotypes. In this section, we first discuss the potential for (and some general limitations of) the study of quantitative traits in aquatic populations, and we conclude by focusing in more detail on three specific topics: selective fisheries, artificial propagation, and climate change.

23.3.1   Establishing the Genetic Basis of Trait Variation

A difficult challenge is to discriminate evolutionary (genetic) changes from phenotypic plasticity, which occurs when the same genotype exhibits different phenotypes in diverse environments (Naish and Hard 2008). The range of phenotypes a particular population exhibits under different environmental conditions is known as its norm of reaction (Pigliucci 2005). A population can respond plastically (without genetic change) to different environmental conditions by moving along its norm of reaction, or it can respond by evolving a different relationship between genotype and environment (i.e., by changing its norm of reaction). Figure 23.1 illustrates these points schematically, using hypothetical mean trait values for four populations across five different environments. Population 4 has a flat norm of reaction (no variation in trait value among environments) and hence no phenotypic plasticity. Populations 1 and 2

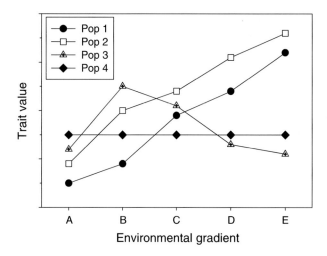

Fig. 23.1 Norms of reaction for four hypothetical populations across an environmental gradient. Datapoints are mean population trait values in each environment. See text for discussion

have roughly parallel norms of reaction, indicating considerable phenotypic plasticity. Population 3 shows a complex reaction norm, with a peak trait value in the middle of the environmental gradient. The different reaction norms imply genetically based adaptations to local environmental conditions in different populations. Populations 1–3 have considerable ability to respond to environmental change by modifying their phenotypes, but Population 4 does not. One possible response to environmental change of a population like Population 4 is to evolve a different norm of reaction to provide more opportunity for phenotypic plasticity.

Two general approaches are used to discriminate genetic and environmental effects on the phenotype. The first, particularly useful for detecting evidence of local adaptation, relies on reciprocal transplants or common garden rearing (Kawecki and Ebert 2004). Because environmental variation is controlled, differentiation between individuals or populations is assumed to be genetic. The second approach partitions the observed phenotypic variance of a fitness trait, V_p, into the genetic (V_g) and environmental (V_e) variances ($V_p = V_g + V_e$) by studying the co-inheritance of traits between relatives. If one can estimate the heritability (h^2, the fraction of total trait variation that is due to additive genetic variance) and the strength of selection S for a trait, the population response R can be predicted from the breeder's equation $R = h^2 S$ (Falconer and Mackay 1996). However, both approaches have limited applications to natural populations. Transplant experiments are logistically challenging, even if a permit could be obtained, and common garden experiments typically require at least two generations of breeding to reduce maternal effects. Estimating h^2 in fitness-related traits requires careful tracking of relatives in a common environment (Falconer and Mackay 1996), which is difficult in nature. As a consequence, estimates of heritability in fishes are dominated by production-related traits relevant to aquaculture (Gjedrem 2000). Improvements in methods of pedigree analysis using molecular markers have made it possible to study the inheritance of traits between relatives, which in turn has led to estimates of h^2 in some wild populations (Keller et al. 2001; Garant et al. 2003). Use of these methods to track rate, direction, and mode of selection in wild populations is promising, but will necessarily involve research over many generations.

It is becoming increasingly apparent that quantitative traits cannot be evaluated in isolation. For example, body size and growth rate typically affect other traits such as age at maturity, number of offspring, and survivorship. Evolution of correlated traits can be studied by evaluating changes in the genetic variance – covariance matrix (the **G** matrix; McGuigan 2006). Future research needs in this area are the generation of empirical measures of additive genetic variances and covariances of correlated traits in wild populations, which can then be used to parameterize models investigating the likely consequences of anthropogenic influences on fish populations.

23.3.2   Molecular Approaches to the Study of Adaptive Traits

The many logistical challenges associated with traditional quantitative genetic methods have fostered interest in molecular approaches aimed at studying functionally important genetic variation, which is the raw material on which selection acts. If the genes underlying a quantitative trait can be identified, then a more mechanistic approach to predicting evolutionary responses can be developed. Efforts towards identifying this variation is most advanced in organisms whose genomes have been sequenced (e.g., humans (Akey et al. 2004), the fruit fly, *Drosophila* (Schlotterer et al. 2006), and the plant *Arabidopsis* (Mitchell-Olds and Schmitt 2006)). Most marine species will probably not be the target of sequencing efforts for many years, and even if sequences were generated, the task of identifying functionally important variation requires large-scale studies at the population level, across a significant number of individuals. Therefore, identifying this variation in non-model organisms (including most fish species) will require adapting the most promising results from model organisms.

Approaches for detecting adaptive genetic variation in non-model organisms can be broadly classified into "top-down" methods (which start with the phenotype and work downward to elucidate the underlying genetic factors) and "bottom-up" methods (which start with DNA polymorphisms and work upwards to the phenotype) (Fig. 23.2; see also Vasemagi and Primmer 2005).

Detection of quantitative trait loci (QTL) – An example of a top-down approach that has been rapidly expanding in fishes (Danzmann and Gharbi 2001) is the use of genome mapping to detect QTL underlying life-history traits. Genome mapping of QTL involves tracking the co-inheritance of markers and genes in pedigrees segregating for the fitness traits of interest. If trait genes and markers are inherited together, the latter can be used as proxies for the former. QTL mapping is dependent on populations that segregate for the trait of interest. Such populations are readily available in model organisms but are rare in marine species – even in aquaculture

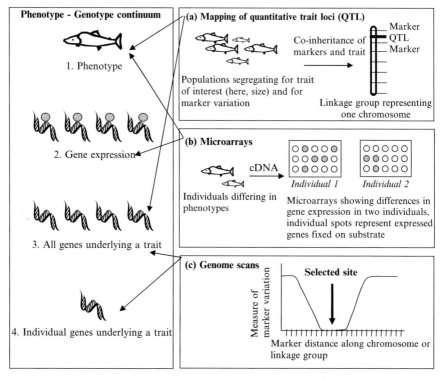

Fig. 23.2 Schematic illustrating methods for detecting functional genetic variation in non-model organisms whose genome sequences are unknown. Methods are initially aimed either at the top of the phenotype continuum (**a** and **b**), or at the bottom (**c**). Approaches correspond with those described in the text

populations that have been cultured for several generations. In the future, attention should be focused on creating and maintaining such experimental lines, or on performing informative crosses between populations that differ in life-history traits. Such efforts will be considerably enhanced by development of genome maps with sufficient marker coverage. One disadvantage of QTL studies is that markers found to be linked to QTL are only useful if recombination has not occurred between these two regions, so this approach is most useful for recently diverged populations (Lynch and Walsh 1998).

Microarrays and mapping of expression variation – Microarrays, which consist of many expressed or short DNA sequences spotted on chips or membranes (Stearns and Magwene 2003), represent a second type of top-down approach. Messenger RNA extracted from samples can be hybridized to these arrays to identify genes that are either up- or down-regulated under different conditions. Most microarray studies have focused on physiological or disease status rather than evolutionary considerations; however, Roberge et al. (2006) showed that a subset of genes in two aquaculture strains of Atlantic salmon of different origin have similar patterns of directional divergence from their progenitor wild strains, suggesting a generalized genetic response to the effects of domestication.

In a related approach that integrates QTL mapping and expression analyses (eQTL mapping; Schadt et al. 2005), individuals are assayed with microarrays and expression levels are treated as phenotypes and mapped to chromosome regions. In this way, gene regulation can be attributed to polymorphism in the gene itself (if segregation in expression levels map to the same position as the gene) or to other genomic regions controlling the transcription of a given gene (if the eQTL maps to another position), and interactions between genes in a pathway underlying a complex trait can be teased apart.

Microarray studies require careful experimental design. Differences in gene expression can arise from environmental as well as genetic variation, and extensive replication is necessary to accommodate environmental and life-history differences between test subjects. In addition, such experiments do not discriminate between partially and fully transcribed genes. At present microarray experiments are expensive to implement on a large scale, so technological improvements would be needed to allow implementation in large experiments. Such improvements would also require advances in complex analytical methods for the extensive datasets that would be generated.

Population genomics – Population genomics is a bottom-up approach that involves sampling the genome for polymorphisms that might help elucidate the dominant evolutionary processes affecting populations. If neutral markers are physically linked to genes under selection, the neutral markers will also be affected by selection (through a process known as hitchhiking), and their behavior will deviate from models of neutrality (Luikart et al. 2003; Storz 2005). The neutral markers thus can be used as proxies for selected regions to address a wide array of questions. The basic population genomics approach involves comparing divergence between two or more populations at a large number of marker loci, in a process known as a "genome scan." Genes whose behavior does not conform to a neutral model are subjected to additional tests to determine the type of selection involved. Inferences can be drawn about the population's demographic history (drift and gene flow) and weighed against the relative role of fitness and adaptation in determining population divergence.

Population genomic approaches are rapidly evolving (Goetz and MacKenzie 2008), and the statistical power to detect the existence and mode of selection is still unclear. A handful of theoretical studies have shown that the likelihood of detecting genomic regions under selection will depend on mutation rate of the marker, the recombination rate between that marker and the selected region, the time since divergence between the populations, the strength of selection, the sizes of the populations since divergence, and whether selection has acted upon new, or standing, variation within a population (Beaumont 2005; De Kovel 2006). Therefore, a strong research need exists for analytical approaches that will provide a realistic understanding of what can and cannot realistically be accomplished using population genomics. One additional disadvantage is that the exact link between the phenotype and the chromosome region under selection will not be known without further study. However, population genomics can help in targeting regions of the genomes that have played an important role in evolution for further characterization by sequencing.

23.3.3 Selective Harvest

Harvest management typically focuses on the numbers of fish captured or the harvest rate. However, any mortality imposed upon a population also has the potential to elicit an evolutionary response (genetic change) in the population if the mortality is selective (non-random) with respect to a genetically controlled trait. A wide range of traits might be subject to this type of selection, including size and age, migration timing, and behavior.

Size and age – Ricker's (1981, 1995) early work on this topic involved Pacific salmon, but more recently the potential evolutionary consequences of size-selective harvest for marine fish have attracted a good deal of attention. In a widely publicized example, Olsen et al. (2004, 2005) showed that in the decades prior to the collapse of cod (*Gadus morhua*) populations off Newfoundland and Labrador, a continual shift occurred toward maturation at an earlier age and smaller size. In an attempt to control for various confounding environmental factors, these cod studies focused on reaction norms for maturation and concluded that there was evidence for rapid fishery-induced evolution of maturation patterns. Because fecundity and gamete quality in cod (and other marine species) are positively correlated with size (Trippel 1998), the observed changes in life history substantially reduce productivity (and hence resilience) of the affected populations. Furthermore, evolution of early maturation at small size in response to selective take of larger fish can have long-lasting consequences for a population. Even if selective harvest is terminated, it might take a long period of time (perhaps much longer than the period of change driven by selective harvest) for the population to re-evolve traits consistent with high productivity. During this period, the population must repay the 'Darwinian debt' created by rapid evolution toward a new fitness peak in an altered environment (Walsh et al. 2006).

The approach used to identify fisheries-induced evolutionary change in cod ("probabilistic maturation reaction norms" [PMRNs]) recognizes the important effect of evolution on correlated traits (Grift et al. 2003). The approach is retrospective in nature and assumes that one correlated trait (age at maturity) is under genetic control, while the other (length at maturity) is environmentally influenced (Grift et al. 2003; Olsen et al. 2004). This assumption is controversial. Many quantitative genetic studies have shown that both traits are influenced by genetic variation and the environment, and some have shown that growth rate rather than size at age determines maturation (Morita and Fukuwaka 2006). Observed changes in the maturation schedule in cod were rapid following a fishery moratorium – faster than might be expected if evolution were occurring – suggesting that the changes in maturation reaction norms might have responded to shifting environmental conditions, rather than to genetic change (Law 2007). However, proponents of PMRNs have argued that the approach is valid as long as growth has an environmental component (Dieckmann and Heino 2007). It has also been pointed out that the two-dimensional PMRN approach used in the cod example does not separate phenotypic plasticity from changes in maturation schedule, although a multidimensional approach might (Kraak 2007). The maturation reaction norm is only one of many traits that might respond to harvest selection (Law 2007), so there is a strong need to experimentally verify the PMRN approach across a wide range of harvested species and traits.

Migration timing – Diadromous fishes are legendary for their long migrations, but many marine species also make considerable geographic movements between feeding and breeding grounds, and these migration routes often provide key harvest opportunities. Harvest of migrating fish stocks has a strong potential to elicit an evolutionary response; unless the harvest rate (chance that an individual is captured) is constant over the entire migration, certain portions of the population will experience higher mortality than others. Furthermore, temporally selective harvest could affect other traits correlated with migration timing. For example, if harvest rates are higher in the first part of the migration, and if older, larger fish tend to migrate early, then this type of fishery would have the same consequences for age and size as one whose gear actively selected for large fish.

An empirical example from Pacific salmon illustrates how management strategies can have evolutionary consequences for migration timing. The Bristol Bay, Alaska, fishery for sockeye salmon has remained productive for over a century, in part because of strong safeguards to ensure that an adequate number of spawners reach the spawning grounds. One way to accomplish this is to keep harvest rates low until the escapement goals are reached, in which case the fishery actively selects against late-migrating fish. Quinn et al. (2007) found empirical evidence that over a 35-year period under this type of harvest regime, sockeye from the Egigik district have shifted toward an earlier migration timing – the type of response predicted by an evolutionary model, although environmental effects cannot be ruled out.

Behavior (catchability) – Harvests will also be selective if the probability that an individual is caught (its catchability) depends on its behavior (cautious vs curious; more vs less aggressive; preference for deep vs shallow habitat; strong vs weak diurnal vertical migrations; etc.). To the degree that such differences affect catchability and have an underlying genetic basis, they can be expected to produce an evolutionary response in the population. This topic has had little treatment in the literature, but an unpublished study over four generations in largemouth bass (*Micropterus salmoides*) showed a strong evolutionary response to selection for vulnerability to angling (Philipp et al. in press).

Research needs for the future – If (as will almost inevitably be the case) harvest rates are not uniform across all types of individuals in a population, and if the traits subject to selective harvest have at least in part a genetic basis, the key questions are not *whether* evolutionary change will occur, but rather the following: How large will these evolutionary changes be? What effect will these changes have on fishery yield and population viability? If selective harvest is terminated, can evolutionary changes be reversed, and if so, how long will it take? The following types of research can provide insight into these questions.

- Characterize the genetic basis of traits under selection (especially size, growth, age structure, and fertility), as well as the genetic correlations among traits.
- Measure the degree of selectivity of harvests, and monitor changes in life-history traits in exploited populations.
- Determine how to distribute fishing mortality over fish of different sizes (or other traits) to minimize selectivity.
- Evaluate how shaping fishery impacts (controlling time and location of harvests; gear types) affects harvest selectivity.
- Consider scaling harvest rate and selectivity to ocean/habitat productivity.

Integration of the **G** matrix into harvest models can contribute to an understanding of the evolutionary consequences of fishing (Law 2000) and the effects of measures taken to reduce these changes (Hard 2004). This approach can be parameterized using empirical measures of heritabilities, variances, and covariances derived for individual species. Further work in this area should seek to integrate these approaches into standard harvest and population viability models.

23.3.4   Artificial Propagation

Evolutionary consequences of fish culture on natural fish populations, especially for salmonids, have been extensively treated in the literature, and this topic will only briefly be summarized here. Interested readers can find more detailed treatments of these topics in Busack and Currens (1995), Waples and Drake (2004), Naylor et al. (2005), Ward (2006), Naish et al. (2008), and Fraser (2008). It is important to distinguish two general kinds of artificial propagation: (1) aquaculture, which involves raising individuals to market size in what are intended to be closed systems; and (2) sea ranching, which involves intentional release of individuals into the wild. Similarly, two general types of sea-ranching programs can be identified: those designed for harvest augmentation (which, like aquaculture programs, might produce societal benefits, but also entail risks to natural populations); and those designed for supplementation of natural populations (which therefore involve intentional integration of hatchery and natural production). Aquaculture and harvest augmentation programs provide little or no benefit to wild populations, but (in at least some circumstances) the risks can be contained within an acceptable level. Supplementation programs have the potential to (at least temporarily) improve the status of wild populations, but also can dramatically change the selective regimes they experience. Careful articulation and consideration of program goals is therefore necessary in evaluating the overall risks and benefits of artificial propagation.

The principle evolutionary changes that natural populations can experience directly or indirectly as a result of artificial propagation are loss of genetic diversity within and among populations and loss of fitness. A number of strategies can be used to help reduce genetic risks of artificial propagation (see references cited above), but certain facts must be realized at the outset and should be factored into management plans. First, evolutionary changes cannot be avoided in cultured populations. The whole *raison d'etre* of an artificial propagation program is to provide a benign environment to promote high survival. Dramatic changes in the mortality profile of a population cannot occur without attendant evolutionary changes; at best one can hope to minimize these changes. Second, it is not possible to simultaneously minimize all risks associated with artificial propagation; many of the risks are inversely correlated, such that reducing one inevitably increases another (Waples and Drake 2004). To take just one example, in supplementation or sea-ranching programs, releasing juveniles early helps to minimize the scope for selective changes due to culture and provides more opportunity for selection to act in the wild. However, early-release juveniles typically have lower survival (which might affect program goals), and releasing large numbers of early life-stage juveniles into the wild increases opportunities for competitive interactions with wild fish and also affects the selective regimes experienced by the wild population.

Although much of the literature on effects of cultured fish on natural populations involves salmonids, the number of marine species that are actively propagated is in the hundreds and rapidly increasing (Leber et al. 2004). Nearly half of all the fish consumed around the world are now farmed (FAO 2006), and expectations for future increases in marine protein production rely heavily on aquaculture. Recently, Woods Hole Oceanographic Institution convened a Marine Aquaculture Task Force to recommend standards and practices for US marine aquaculture to protect the health of marine ecosystems. If the recommendation of the Panel (Anonymous 2007) to shift culture toward local, native populations is followed, it would considerably alleviate concerns regarding spread of disease and invasive species. However, this change could also greatly increase opportunities for genetic interactions with wild populations, with uncertain consequences. Some research topics that will be important in the future include the following:

- What are the relative evolutionary consequences for natural populations interacting with cultured populations of local versus foreign origin? Genetic interactions with cultured, non-native populations almost certainly will be deleterious, but their frequency might be greatly reduced by physical, behavioral, or management isolating mechanisms. Genetic interactions with cultured populations of local origin should be less harmful *per interaction*, but the frequency of interactions might be greatly increased. The net effect on the wild population will be a function of both the frequency and severity of the interactions when they do occur, and this topic has not been treated rigorously in the literature.
- What are the genetic mechanisms involved in domestication? (see Araki et al. 2008). To what extent is domestication reversible, and on what time frames? What culture practices can effectively minimize effects of domestication?
- What is the reproductive success in the wild of individuals reared in culture? What are the long-term effects of supplementation on the ability of populations to sustain themselves in the wild?
- What is the relative importance of inbreeding depression and outbreeding depression, and how does this vary among species and among populations within species?

A variety of methods can provide insight into these questions. Genomic and quantitative genetic studies will be feasible for some species. Ready availability of numerous, highly polymorphic molecular markers allows reliable parentage analysis of progeny, which provides high power to studies of reproductive success and selection differentials in the wild. Genomic approaches can identify genes that might respond to selection, inbreeding depression, and outbreeding depression, and changes in their population frequencies can be monitored. However, it should be recognized that many of these questions can only be answered with long-term studies, since many fitness effects will only become apparent over many generations.

23.3.5   Climate Change

Evidence is rapidly accumulating to demonstrate that global warming has caused genetic changes in animal species as diverse as birds, squirrels, and mosquitoes (Bradshaw and Holzapfel 2006). To date, the documented changes have primarily involved timing of seasonal events (e.g., reproduction or migration), although it is expected that further research will uncover morphological, behavioral, and physiological adaptations, such as increased thermal tolerance. The effects of climate on fish population dynamics has long been recognized (Cushing 1983; Tolimieri and Levin 2005), but only recently has attention focused on the likely future consequences of global warming (Overland and Wang 2007). Aquatic species have evolved mechanisms to allow persistence in variable environments, but long-term, directional changes (such as are predicted under many climate change scenarios) will represent a new type of challenge. Here are the some current and future research needs that would increase our understanding of this key topic.

- Document nature, extent, and rate of environmental changes expected to be associated with future climate change. This would involve translating general predictions from the most recent Intergovernmental Panel on Climate Change report (IPCC 2007) into more focused predictions for marine ecosystems.
- Identify key species and their vulnerable life stages that are most likely to be affected by global warming. A key element here will be integration of biological information for the species with physical information about their changing habitats.
- Evaluate capacity of the species to respond to environmental changes through phenotypic plasticity. Initially, such experiments can be conducted in the laboratory by exposing individuals from genetically uniform populations to different temperature environments. Ideally, these experiments would then be extended to the wild by transplanting related individuals into different environments and studying the range of their phenotypic responses, such as levels of expression at relevant genes, changes in correlated fitness traits, and their reproductive success.
- Evaluate the capability of the species to adapt to new environmental conditions. Can the species evolve fast enough to keep pace with the rate of environmental change? In the absence of experimental data for the species in question, insights can be obtained from literature surveys that show rates of evolutionary change typically

found in natural populations (Hendry and Kinnison 1999), from long-term studies of evolution of quantitative traits in controlled environments, or from models based on empirical measures of genetic variation underlying fitness traits.

- Look for mismatches between anticipated magnitude and rate of change in the physical environment and the capacity for a plastic and/or an evolutionary response. These mismatches identify the species and life stages that are most likely to be strongly affected by global warming.

23.4   Discussion and Conclusions

Genetic methods have become indispensable to twenty-first-century fishery management, and their relative importance is likely to increase in the coming decades, particularly as genetic approaches become better integrated with more traditional ones. In this chapter we have considered only a fraction of the types of questions that can and will be addressed by genetic approaches, and we have chosen to focus on two areas: (1) questions that utilize neutral genetic markers and provide insights into population genetic processes that are relevant to conservation and management, and (2) questions that rely on methods that study traits under selection, which are more directly related to population fitness.

One general topic that should get more serious consideration in the future is the relationship between statistical significance and biological significance (Waples 1998; Hedrick 1999; Palsbøll et al. 2007). Although the traditional hypothesis-testing framework has many advantages, it is safe to say that no aquatic population ever suffered any consequences directly as the result of a P-value associated with a statistical test. Rather, biological consequences depend on the magnitude and direction of an effect (the effect size). That is, it is not enough to determine that a (nonzero) effect can be detected; it is also important to determine how large the effect is, and what the biological consequences are for the species of interest. The importance of this point for identification of stocks and conservation units has been emphasized in the literature (Waples 1998; Palsbøll et al. 2007), but it applies more generally to the field of fishery management.

This is an exciting time for use of genetic methods in applied conservation, as a confluence of several factors has conspired to provide unprecedented opportunities. First, technical advances in the laboratory have uncovered an essentially unlimited number of highly variable genetic markers that can be utilized to study natural populations. These methods can extract DNA nonlethally from increasingly small amounts of biological material, which allows routine, non-invasive monitoring as well as retrospective analyses using historic samples (Schwartz et al. 2007). Second, numerous powerful analytical methods have been developed in the past decade that provide new opportunities to test hypotheses about contemporary evolutionary and ecological processes in populations (Pearse and Crandall 2004; Manel et al. 2005). Finally, in accordance with Moore's law, computational power continues to increase rapidly, and these increases have made feasible implementation of many of the new likelihood-based methods that are computationally demanding. We expect that incremental improvements in each of these areas will continue into the future, and these improvements will continue to refine and improve the applications of neutral genetic markers to fishery management.

Collectively, these and other advances have brought us to the point where it is feasible to contemplate the rigorous study of genetic variation at fitness-related traits in natural populations. These studies, however, are logistically challenging and resource-intensive, so in the near future they will be feasible only for a small fraction of marine species. It would be useful, therefore, if researchers working on marine species would coordinate selection of species on which to conduct quantitative genetic or population genomic studies. If this were done, it should be possible to arrange coverage across taxa in such a way to maximize the inferences that can be drawn from the studied species to the large majority that will not be studied directly. Both types of studies (whether they focus on neutral markers or genes under selection) could benefit from better integration with other types of information for the species in question (life history, physiology, behavior) and the environments they inhabit (current patterns, physical features of the habitat).

It is easy to see that whereas the ecological consequences of anthropogenic changes on aquatic populations have received a great deal of attention, the evolutionary consequences of these changes have been relatively neglected. It is much more difficult to determine exactly what these evolutionary consequences will be, and how important they will be to conservation and management. This is now an active area of research for terrestrial as well as marine species (e.g., a new journal, *Evolutionary Applications*, began publication in 2008), and we expect important new insights will emerge over the next decade.

Acknowledgments We thank Jeff Hard, Lorenz Hauser, Jeff Hutchings, Rick Methot, and Dave Philipp for useful discussions and suggestions.

References

Akey JM, Eberle MA, Rieder MJ, Carlson CS, Shriver MD, Nickerson DA, Kruglyak L (2004) Population history and natural selection shape patterns of genetic variation in 132 genes. PLoS Biol 2: e286

Anderson EC, Waples RS, Kalinowski S (2008) An improved method for predicting the accuracy of genetic stock identification. Can J Fish Aquat Sci 65:1475–1486

Anonymous (1992) Rio Declaration on Environment and Development. Environ Conserv 19:366–368

Anonymous (2007) Sustainable Marine Aquaculture: Fulfilling the Promise; Managing the Risks. Report of the Marine Aquaculture Task Force, January 2007. Pew Charitable Trust. Available from: http://pewtrusts.com/pdf/Sustainable_Marine_Aquaculture_final_1_07.pdf

Araki H, Berejikian BA, Ford MJ, Blouin M (2008) Fitness of hatchery-reared salmonids in the wild. Evol Appl 1:342–355

Avise JC, Jones AG, Walker D, Dewoody JA (2002) Genetic mating systems and reproductive natural histories of fishes: lessons for ecology and evolution. Annu Rev Genet 36:19–45

Baker CS, Cooke JG, Lavery S, Dalebout ML, Ma Y-U, Funahashi N, Carraher C, Brownell RL (2007) Estimating the number of whales entering trade using DNA profiling and capture-recapture analysis of market products. Mol Ecol 16:2617–2626

Beaumont MA (2005) Adaptation and speciation: what can F-st tell us? Trends Ecol Evol 20:435–440

Bernatchez L, Landry C (2003) MHC studies in nonmodel vertebrates: what have we learned about natural selection in 15 years? J Evol Biol 16:363–377

Bonin A, Bellemain E, Eidesen PB, Pompanon F, Brochmann C, Taberlet P (2004) How to track and assess genotyping errors in population genetics studies. Mol Ecol 13:3261–3273

Bradshaw WE, Holzapfel CM (2006) Climate change – evolutionary response to rapid climate change. Science 312:1477–1478

Busack CA, Currens KP (1995) Genetic risks and hazards in hatchery operations: fundamental concepts and issues. Am Fish Soc Symp 15:71–80

Cadrin SX, Friedland KD, Waldman JR (eds) (2005) Stock identification methods: applications to fishery science. Elsevier Academic Press, Amsterdam

Carvalho GR, Hauser L (1994) Molecular genetics and the stock concept in fisheries. Rev Fish Biol Fish 4:326–350

Carvalho GR, Hauser L (1998) Advances in the molecular analysis of fish population structure. Ital J Zool 65:21–33

Cushing DH (1983) Climate and Fisheries. Academic, London

Danzmann RG, Gharbi K (2001) Gene mapping in fishes: a means to an end. Genetica 111:3–23

De Kovel CGF (2006) The power of allele frequency comparisons to detect the footprint of selection in natural and experimental situations. Genet Select Evol 38:3–23

DeYoung RW, Honeycutt RL (2005) The molecular toolbox: Genetic techniques in wildlife ecology and management. J Wildl Manage 69:1362–1384

Dieckmann U, Heino M (2007) Probabilistic maturation reaction norms; their history, strengths, and limitations. Mar Ecol Prog Ser 335:249–251

Emery AM, Wilson IJ, Craig S, Boyle PR, Noble LR (2001) Assignment of paternity groups without access to parental genotypes: multiple mating and developmental plasticity in squid. Mol Ecol 10:1265–1278

Falconer DS, Mackay TFC (1996) Introduction to Quantitative Genetics. Prentice Hall, Harlow, England, pp 464

Falush D, Stephens M, Pritchard JK (2003) Inference of population structure using multilocus genotype data: Linked loci and correlated allele frequencies. Genetics 164:1567–1587

FAO (2006) State of world aquaculture. FAO Fisheries Technical Paper 500. Food and Agriculture Organization of the United Nations. Rome

Frankham R (1995) Effective population size/adult population size ratios in wildlife: a review. Genet Res 66:95–107

Fraser D (2008) How well can captive breeding programs conserve biodiversity? A review of salmonids. Evol Appl 1:535–586

Galindo HM, Olson DB, Palumbi SR (2006) Seascape genetics: a coupled oceanographic-genetic model predicts population structure of Caribbean corals. Curr Biol 16:1622–1626

Garant D, Dodson JJ, Bernatchez L (2003) Differential reproductive success and heritability of alternative reproductive tactics in wild Atlantic salmon (*Salmo salar* L.). Evolution 57:1133–1141

Garrigan D, Hedrick PW (2003) Perspective: Detecting adaptive molecular polymorphism: lessons from the MHC. Evolution 57:1707–1722

Gjedrem T (2000) Genetic improvement of cold-water fish species. Aquat Res 31:25–33

Goetz WF, MacKenzie S (2008) Functional genomics with microarrays in fish biology and fisheries. Fish and Fisheries 9:378–395

Grift RE, Rijnsdorp AD, Barot S, Heino M, Dieckmann U (2003) Fisheries-induced trends in reaction norms for maturation in North Sea plaice. Mar Ecol Prog Ser 257:247–257

Hard, JJ (2004) Evolution of chinook salmon life history under size-selective harvest. In: Hendry A, Stearns S (eds) Evolution illuminated: salmon and their relatives. Oxford University Press, New York, pp. 315–337

Hastings A (1993) Complex interactions between dispersal and dynamics: lessons from coupled logistic equations. Ecology 74:1362–1372

Hauser L, Adcock GJ, Smith PJ, Bernal Ramirez JH, Carvalho GR (2002) Loss of microsatellite diversity and low effective population size in an overexploited population of New Zealand snapper (*Pagrus auratus*). Proc Nat Acad Sci USA 99:11742–11747

Hauser L, Carvalho GR (2008) Paradigm shifts in marine fisheries genetics: ugly hypotheses slain by beautiful facts. Fish and Fisheries 9:333–362

Hedgecock D (1994) Does variance in reproductive success limit effective population sizes of marine organisms? In: Beaumont AR (ed) Genetics and evolution of aquatic organisms. Chapman & Hall, London, pp. 122–134

Hedrick PW (1999) Perspective: highly variable loci and their interpretation in evolution and conservation. Evolution 53:313–318

Hendry AP Kinnison MT (1999) The pace of modern life: measuring rates of contemporary microevolution. Evolution 53:637–1653

Hoarau G, Boon E, Jongma1 DN, Ferber S, Palsson J, Van der Veer HW, Rijnsdorp AD, Stam WT, Olsen JL 2005 Low effective population size and evidence for inbreeding in an overexploited flatfish, plaice (*Pleuronectes platessa* L.) Proc R Soc B 272:497–503

Ihssen PE, Bodre HE, Casselman JM, McGlade JM, Payne NR, Utter F (1981) Stock identification: materials and methods. Can J Fish Aquat Sci 38:1838–1855

IPCC (Intergovernmental Panel on Climate Change) (2007) Climate change 2007: 4th Assessment Report, http://www.ipcc.ch/

Johnson MS, Wernham J (1999) Temporal variation of recruits as a basis of ephemeral genetic heterogeneity in the western rock lobster *Panulirus cygnus*. Mar Biol 135:133–139

Kawecki TJ, Ebert D (2004) Conceptual issues in local adaptation. Ecol Lett 7:1225–1241

Keller LF, Grant PR, Grant BR, Petren K (2001) Heritability of morphological traits in Darwin's Finches: misidentified paternity and maternal effects. Heredity 87:325–336

Kimura M, Crow JF (1964) The number of alleles that can be maintained in a finite population. Genetics 49:725–738

Kraak SBM (2007) Does the probabilistic maturation reaction norm approach distentangle phenotypic plasticity from genetic change? Mar Ecol Prog Ser 335:295–300

Law R (2000) Fishing, selection, and phenotypic evolution. ICES J Mar Sci 57:659–668

Law R (2007) Fisheries-induced evolution; present status and future directions. Mar Ecol Prog Ser 335:271–277

Leber KM, Kitada S, Blankenship HL, Svåsand T (eds) (2004) Stock enhancement and sea ranching. Blackwell, Oxford, pp. 562

Lubchenco J, Palumbi SR, Gaines SD, Andelman S (2003) Plugging a hole in the ocean: the emerging science of marine reserves. Ecol Appl 13: S3–S7

Luikart G, England PR, Tallmon D, Jordan S, Taberlet P (2003) The power and promise of population genomics: From genotyping to genome typing. Nat Rev Genet 4:981–994

Lynch M, Walsh B (1998) Genetics and Analysis of Quantitative Traits. Sinauer, Sunderland, MA, 980pp

Manel S, Gaggiotti O, Waples RS (2005) Assignment methods: matching biological questions with appropriate techniques. Trends Ecol Evol 20:136–142

McElhany P, Ruckelshaus MH, Ford MJ, Wainwright TC, Bjorkstedt EP (2000) Viable salmon populations and the recovery of evolutionarily significant units. Northwest Fisheries Science Center, Seattle, 156pp. NOAA Technical Memorandum, NMFS-NWFSC-58 http://www.nwfsc.noaa.gov/publications/

McGuigan K (2006) Studying phenotypic evolution using multivariate quantitative genetics. Mol Ecol 15:883–896

Mitchell-Olds T, Schmitt J (2006) Genetic mechanisms and evolutionary significance of natural variation in *Arabidopsis*. Nature 441:947–952

Mooney HA (1998) Ecosystem management for sustainable marine fisheries. Ecol Appl 8: S1–S1

Morita K, Fukuwaka MA (2006) Does size matter most? The effect of growth history on probabilistic reaction norm for salmon maturation. Evol 60:1516–1521

Naish KA, Hard JJ (2008) Bridging the gap between the genotype and the phenotype: linking genetic variation, selection and adaptation in fishes. Fish and Fisheries 9:396–422

Naish KA, Taylor J, Levin P, Quinn TP, Winton JR, Huppert D, Hilborn R (2008) An evaluation of the effects of conservation and fishery enhancement hatcheries on wild populations of salmon. Adv Mar Biol 53:61–194

Naylor RL, Hindar K, Fleming IA, Goldberg R, Mangel M, Williams S, Volpe J, Whoriskey F, Eagle J, Kelso D (2005) Fugitive salmon: assessing risks of escaped farmed fish from aquaculture. Bioscience 55:427–437

Niemi G, Wardrop D, Brooks R, Anderson S, Brady V, Paerl H, Rakocinski C, Brouwer M, Levinson B, McDonald M (2004) Rationale for a new generation of indicators for coastal waters. Environ Health Perspect 112:979–986

Nunney L (1993) The influence of mating system and overlapping generations on effective population size. Evolution 47:1329–1341

Olsen JB, Bentzen P, Banks MA, Shaklee JB, Young S (2000) Microsatellites reveal population identity of individual pink salmon to allow supportive breeding of a population at risk of extinction. Trans Am Fish Soc 129:232–242

Olsen EM, Heino M, Lilly GR, Morgan MJ, Brattey J, Ernande B, Dieckmann U (2004) Maturation trends suggestive of rapid evolution preceded collapse of northern cod. Nature 428:932–935

Olsen EM, Lilly GR, Heino M, Morgan M J, Brattey J, Dieckmann U (2005) Assessing changes in age and size at maturation in collapsing populations of Atlantic cod (*Gadus morhua*). Can J Fish Aquat Sci 62:811–823

Overland JE, Wang M (2007) Future climate of the North Pacific Ocean. EOS 88:178–180

Palsbøll PJ, Bérubé M, Allendorf FW (2007) Identification of management units using population genetic data. Trends Ecol Evol 22:11–16

Paetkau D, Slade R, Burden M, Estoup A (2004) Genetic assignment methods for the direct, real-time estimation of migration rate: a simulation-based exploration of accuracy and power. Mol Ecol 13:55–65

Pearse DE, Crandall KA (2004) Beyond Fst: population genetic analysis for conservation. Cons Genet 5:585–602

Pella J, Milner GB (1987) Use of genetic marks in stock composition analysis. In: Ryman N, and Utter F (eds) Population genetics & fishery management. University of Washington Press, Seattle, WA, pp. 247–276

Pella J, Masuda M (2006) The Gibbs and split-merge sampler for population mixture analysis from genetic data with incomplete baselines. Can J Fish Aquat Sci 63:576–596

Philipp DP, Cooke SJ, Clausen JE, Koppelman JB, Suski CD, Burkett DP (in press) Selection for vulnerability to angling in largemouth bass. Trans Am Fish Soc (was cited in the text as a personal communication)

Pigliucci M (2005) Evolution of phenotypic plasticity: where are we going now? Trends Ecol Evol 20:481

Pompanon F, Bonin A, Bellemain E, Taberlet P (2005) Genotyping errors: causes, consequences and solutions. Nat Rev Genet 6:847–859

Pritchard JK, Stephens M, Donnelly P (2000) Inference of population structure using multilocus genotype data. Genetics 155:945–959

Pujolar JM, Maes GE, Volckaert FAM (2006) Genetic patchiness among recruits in the European eel *Anguilla anguilla*. Mar Ecol Prog Ser 307:209–217

Purcell MK, Nichols KM, Winton JR, Kurath G, Thorgaard GH, Wheeler P, Hansen JD, Herwig RP, Park LK (2006) Comprehensive gene expression profiling following DNA vaccination of rainbow trout against infectious hematopoietic necrosis virus. Mol Immunol 43:2089–2106

Quinn TP, Hodgson S, Flynn L, Hilborn R, Rogers DE (2007) Directional selection by fisheries and the timing of sockeye salmon (*Oncorhynchus nerka*) migrations. Ecol Appl 17:731–739

Restropo VR (ed) (1999) Providing scientific advice to implement the precautionary approach under the Magnuson-Stevens Fishery Conservation and Management Act. Proceedings of the fifth national NMFS stock assessment workshop. US Department of Commerce, NOAA Tech. Memo. NMFS F/SPO-40, 161p

Ricker WE (1981) Changes in the average size and average age of Pacific salmon. Can J Fish Aquat Sci 38:1636–1656

Ricker WE (1995) Trends in the average size of Pacific salmon in Canadian catches. In: Beamish RJ (ed.) Climate change and northern fish populations, vol 121. Canadian Special Publication in Fisheries and Aquatic Sciences 121, Ottawa, pp. 593–602

Roberge C, Einum S, Guderley H, Bernatchez L (2006) Rapid parallel evolutionary changes of gene transcription profiles in farmed Atlantic salmon. Mol Ecol 15:9–20

Ryman N, Utter F (eds) (1987) Population genetics & fishery management. University of Washington Press, Seattle, WA

Ryman N, Utter F, Laikre L (1995) Protection of intraspecific biodiversity of exploited fishes. Rev Fish Biol Fish 5:417–446

Schadt EE, Lamb J, Yang X, Zhu J, Edwards S, GuhaThakurta D, Sieberts SK, Monks S, Reitman M, Zhang C, Lum PY, Leonardson A, Thieringer R, Metzger JM, Yang L, Castle J, Zhu H, Kash SF, Drake TA, Sachs A, Lusis AJ (2005) An integrative genomics approach to infer causal associations between gene expression and disease. Nat Genet 37:710–717

Schlotterer C, Neumeier H, Sousa C, Nolte V (2006) Highly structured Asian *Drosophila* melanogaster populations: A new tool for hitchhiking mapping? Genetics 172:287–292

Schwartz MK, Luikart G, Waples RS (2007). Genetic monitoring: a promising tool for conservation and management. Trends Ecol Evol 22:25–33

Selkoe KA, Gaines SD, Caselle JE, Warner RR (2006) Current shifts and kin aggregation explain genetic patchiness in fish recruits. Ecology 87:3082–3094

Selkoe KA, Henzler CM, Gaines SD (2008) Seascape genetics and the spatial ecology of marine populations. Fish and Fisheries 9:363–377

Shaklee JB, Beacham TD, Seeb L, White BA (1999) Managing fisheries using genetic data: case studies from four species of Pacific Salmon. Fish Res 43:45–78

Simon RC, Larkin PA (eds) (1972) The stock concept in Pacific salmon. HR MacMillan Lectures in Fisheries, University of British Columbia, Vancouver

Smith TB, Bernatchez L (2008) Evolutionary change in human-altered environments. Mol Ecol 17:1–499

Smouse PE, Waples RS, Tworek JA (1990) A mixed fishery model for use with incomplete source population data. Can J Fish Aquat Sci 47:620–634

Stearns SC, Magwene P (2003) The naturalist in a world of genomics. Am Nat 161:171–180

Storz JF (2005) Using genome scans of DNA polymorphism to infer adaptive population divergence. Mol Ecol 14:671–688

Suttle CA (2005) Viruses in the sea. Nature 437:356–361

Tolimieri N, Levin PS (2005) The roles of fishing and climate in the population dynamics of bocaccio rockfish. Ecol Appl 15:458–468

Trippel EA (1998). Egg size and viability and seasonal offspring production of young Atlantic cod. Trans Am Fish Soc 127:339–359

Turner TF, Wares JP Gold JR (2002) Genetic effective size is three order of magnitude smaller than adults census size in an abundant, estuarine-dependant marine fish (*Scianops ocellatus*). Genetics 162:1329–1339

Vasemägi A, Primmer CR (2005) Challenges for identifying functionally important genetic variation: the promise of combining complementary research strategies. Mol Ecol 14:3623–3642

Verardi A, Lucchini V, Randi E (2006) Detecting introgressive hybridization between free-ranging domestic dogs and wild wolves (*Canis lupus*) by admixture linkage disequilibrium analysis. Mol Ecol 15:2845–2855

von Schalburg KR, Rise ML, Brown GD, Davidson WS, Koop BF (2005) A comprehensive survey of the genes involved in maturation and development of the rainbow trout ovary. Biol Reprod 72:687–699

Waldman JR (2005) Definition of stock: an evolving concept. In: Cadrin SX, Friedland KD, Waldman JR (eds) Stock identification methods: applications to fishery science. Elsevier Academic Press, Amsterdam, pp. 7–16

Walsh MR. Munch SB, Chiba S, Conover DO (2006) Maladaptive changes in multiple traits caused by fishing: impediments to population recovery Ecol Lett 9:142–148

Waples R (1998) Separating the wheat from the chaff: patterns of genetic differentiation in high gene flow species. J Hered 89:438–450

Waples RS (1990) Conservation genetics of Pacific salmon. II. Effective population-size and the rate of loss of genetic-variability. J Hered 81:267–276

Waples RS, Drake J (2004) Risk-benefit considerations for marine stock enhancement: a Pacific salmon perspective. In: Leber KM, Kitada S, Blankenship HL, Svåsand T (eds) Stock enhancement and sea ranching: developments, pitfalls and opportunities, 2nd edn. Blackwell, Oxford, pp. 260–306

Waples RS, Gaggiotti O (2006) What is a population? An empirical evaluation of some genetic methods for identifying the number of gene pools and their degree of connectivity. Mol Ecol 15:1419–1439

Waples RS, Punt AE, Cope J (2008) Integrating genetic data into fisheries management: how can we do it better? Fish and Fisheries 9:423–449

Ward RD, Woodward M, Skibinski DOF (1994) A comparison of genetic diversity levels in marine, freshwater, and anadromous fishes. J Fish Biol 44:213–232

Ward RD (2006) The importance of identifying spatial population structure in restocking and stock enhancement programmes. Fish Res 80:9–18

Chapter 24
Trends in Fishery Genetics

Marc Kochzius

Abstract Fishery genetics has a history of half a century, applying molecular biological and genetic techniques to answer fisheries-related questions in taxonomy and ecology of fishes and invertebrates. This review aims to provide an overview of the developments in fishery genetics of the last decade, focussing on DNA-based species and stock identification. Microsatellites became the 'gold standard' in genetic stock identification, but accumulating sequence information for commercially important species opens the door for genome-wide SNP (Single Nucleotide Polymorphism) analysis, which will support or even displace microsatellites in the future. Recent advancements in DNA analytics, such as DNA microarrays and pyrosequencing, are highlighted and their possible applications in fishery genetics are discussed. Emphasis is also given to DNA barcoding, a recently advocated concept using a fragment of the mitochondrial cytochrome oxidase I (COI) gene as a standard marker for the identification of animals. DNA barcoding becomes more and more accepted in the scientific community and the international initiative Fish-BOL (Fish Barcoding of Life) aims to barcode all fish species. These novel technologies and concepts will enable a tremendous progress in fishery genetics.

24.1 Introduction and Historical Overview

Fishery genetics bridges the gap between fisheries research and molecular biology by applying molecular methods to fisheries-related questions in taxonomy and ecology of fishes and invertebrates. In a broader sense, fisheries genetics can also deal with the evolution of species relevant in fisheries, as well as breeding in aquaculture.

The application of molecular genetic approaches in fisheries research started in the 1950s, investigating blood group variants in tunas, salmonids and cod to analyse population structure (review by Ligny 1969). However, these methods have not been broadly adapted in fisheries research, because genetic variation could be detected

M. Kochzius
Centre for Applied Gene Sensor Technology (CAG), University of Bremen, Germany

R.J. Beamish and B.J. Rothschild (eds.), *The Future of Fisheries Science in North America*, 453
Fish & Fisheries Series, © Springer Science + Business Media B.V. 2009

much easier by determining protein polymorphism using starch gel electrophoresis (Ward and Grewe 1994). Studies on the variation of protein polymorphism in fish started in the early 1960s by investigating hemoglobin variants in cod (*Gadus morhua*) and whiting (*Merlangius merlangus*) (Sick 1961). The focus was then changed to enzymatic proteins (so-called allozymes). Because they were easy to handle, results were reproducible, and the method was reasonably inexpensive (Ferguson 1994; Ward and Grewe 1994). Allozymes are differing in electrophoretic mobility due to allelic differences at a single gene. In the 1970s, allozyme analysis by protein electrophoresis was widely applied, providing new knowledge about populations, such as size, migration and isolation (Utter 1994). This is reflected in the number of fish- and fishery-related publications, which started to grow in the late 1970s (Fig. 24.1). The prime use of allozyme variation was the application as a molecular marker for stock identification, in the sense of identifying reproductively isolated populations for fisheries management, as well as species identification. However, such genetic markers have to be neutral, because polymorphism between populations caused by selection will not provide information if stocks will be reproductively isolated. There has been a long debate whether allozyme variation is subject to selection and it has been shown that at least some protein coding loci are not neutral (Ferguson 1994). Additionally, allozyme analysis is only an indirect measure for genetic variation in deoxyribonucleic acid (DNA), and therefore has a lower resolution (= variation) than direct assessment of DNA (Ward and Grewe 1994). Another disadvantage of protein electrophoresis is the handling of samples, which have to be

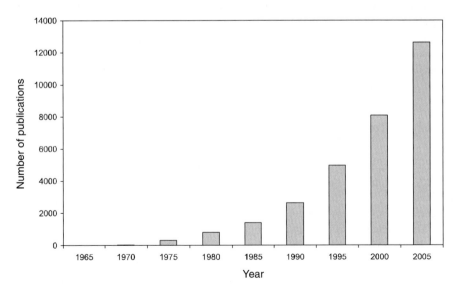

Fig. 24.1 Cumulative number of publications related to fishery genetics in the literature data base 'Aquatic sciences and fisheries abstracts (ASFA)' with the search string: (Title = fish* or Keywords = fish* or Abstract = fish*) and (Title = genetic* or Keywords = genetic* or Abstract = genetic*)

analysed fresh immediately after sampling or have to be kept frozen until analysis. This is not problematic on modern research or commercial fishing vessels, but is a logistic problem in land-based surveys in remote areas with no infrastructure, e.g., in many developing countries. Problematic is also the large amount of tissue needed, which can usually only be obtained by sacrificing the specimens. This is especially problematic when the species under study is endangered or protected (Park and Moran 1994). In contrast, only minute amounts of tissue are needed for DNA analysis and such tissue samples can be stored at ambient temperature in 96% ethanol, but for long-term preservation storage of such samples at 4°C is recommended (Zhang and Hewitt 1998). Therefore, direct analysis of DNA is advantageous.

Even though it was revealed already in the 1940s that DNA is the substance of inheritance (Avery et al. 1944) and the molecular structure was known since the 1950s (Watson and Crick 1953), first attempts to analyse DNA directly in population genetics have only been made in the 1970s by studying the Restriction Fragment Length Polymorphism (RFLP) of mitochondrial DNA (e.g., Brown and Vinograd 1974; Upholt and Dawid 1977). At the same time, other studies important for the application of mtDNA RFLPs in population genetics were the discovery of the predominantly maternal inheritance of mtDNA in higher animals (Dawid and Blackler 1972) and their high rate of evolution (Brown et al. 1979), as well as the development of a statistical method to calculate the divergence between mtDNA haplotypes from RFLP banding pattern (Upholt 1977). The basis for direct analysis of DNA sequences was the development of the 'dideoxy' method by Sanger et al. (1977), but routine analysis of DNA sequences was finally accelerated by the invention of the polymerase chain reaction (PCR) by Mullis and colleagues (Saiki et al. 1985, 1988). Since the 1990s, the application of PCR and DNA sequencing became a routine method in fisheries genetics and the number of publications in that area, which started with only four in 1965, was rapidly increasing to >12,000 in 2005 (Fig. 24.1). The development of these methods lead to a growing interest in DNA sequencing for the study of the evolution and population genetics of fishes, which is reflected in a growing number of DNA sequences derived from bony fishes. Starting with seven sequences from bony fishes in 1993, the number of sequences in international sequence databases reached more than 5.5 million by the end of 2007 (Fig. 24.2). Most of data entries (64%) are nuclear EST (expressed sequence tag) sequences, derived from gene expression studies with model fish species such as zebra fish (*Danio rerio*), but also from species important in fisheries and aquaculture; e.g., Atlantic cod (*Gadus morhua*), turbot (*Psetta maxima*), European eel (*Anguilla anguilla*), and Atlantic salmon (*Salmo salar*). GSS (genome survey sequences) entries represent 18% and are mainly obtained from nuclear genome sequencing projects on zebra fish (*Danio rerio*) and Japanese pufferfish (*Takifugu rubripes*). CoreNucleotide sequences also have a proportion of 18% and comprise mitochondrial as well as nuclear DNA sequences, many of them obtained for studies on phylogenetics, species identification, and population genetics. These sequences sum up to one million and are the ones most users are interested in.

This review aims to provide an overview of recent developments and trend in genetics and their utilisation and potential application in fisheries research. Even though

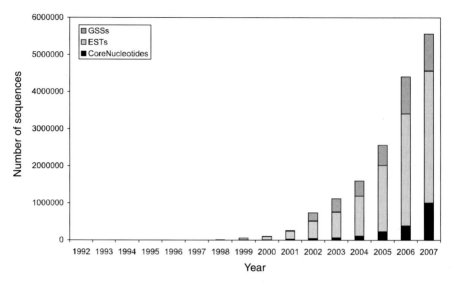

Fig. 24.2 Number of DNA sequences for bony fishes (Teleostei) in international DNA databases (derived from GenBank, December 2007). GSSs: Genome survey sequences; ESTs: Expressed sequence tags; CoreNucleotides: all nucleotide sequences that are not GSSs or ESTs

fishery genetics deal with a variety of freshwater as well as marine fishes and invertebrates, this review will be restricted to species and stock identification of marine fishes. Since the direct analysis of DNA sequences is the 'gold standard' in the era of genomics, protein analysis will not be considered in this review. Excellent overviews on the state of the art in fishery genetics, including protein analysis, at the end of the last century are provided in a special issue of *Reviews in Fish Biology and Fisheries* (e.g., Carvalho and Hauser 1994; Park and Moran 1994; Ward and Grewe 1994) and by Ward (2000). This review will cover established DNA-based methods and examples of their application, as well as recent advancements, namely DNA barcoding, DNA microarrays, and pyrosequencing, which have the potential to revolutionise species and stock identification in fisheries research.

24.2 DNA Marker

In fisheries genetics, mitochondrial and nuclear markers are utilised. In order to understand advantages and disadvantages of the different marker systems, a brief overview is given.

Mitochondrial DNA (mtDNA) sequences are the most used genetic markers in phylogenetics, phylogeography, as well as population genetics, and have replaced allozymes. In the late 1990s more than 80% of phylogeographic studies used mtDNA as a genetic marker (Avise 1998). Mitochondrial markers are widely used in studies on fishes and a large number of 'universal' primers are available to amplify different fragments of the mitochondrial genome of many fish species (Meyer 1993, 1994). However, no marker is completely universal, and therefore, multiplex-PCRs or 'cocktails' with several 'universal' primers are needed to ensure that a certain fragment can be amplified from most fish species (Ivanova et al. 2007; Sevilla et al. 2007).

Mitochondria are the power plants of the cell and have a central role in cell metabolism. A vertebrate mitochondrial genome is a double-stranded, circular molecule of about 16,000 bp length that contains 13 protein, 2 rRNA and 22 tRNA genes. Compared to the nuclear genome it is extremely compact, about 93% of mtDNA are genes, whereas non-coding DNA has only a proportion of about 3% in the nuclear genome (Fig. 24.3). In higher animals, mitochondrial DNA is maternally inherited, does not appear to undergo recombination, and evolves about ten times faster than the nuclear genome. It is also believed that mtDNA is a neutral marker, not undergoing selection. These features made mtDNA extremely important in molecular systematics and population genetics.

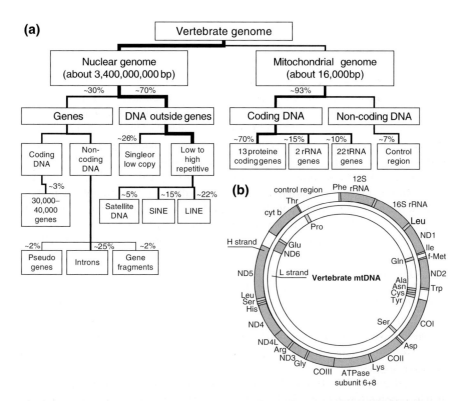

Fig. 24.3 (a) Vertebrate genome (Adapted from Haeseler and Liebers 2003); (b) Vertebrate mitochondrial genome. Protein and rRNA (dark grey) and tRNA (light grey) genes are located on the outer H (heavy) strand and the inner L (light) strand (Adapted from Page and Holmes 1998)

However, there are some pitfalls, because not all of the features mentioned above can be generalised. In some animals, recombination of mtDNA is known, e.g. in the mussel *Mytilus galloprovincialis* (Ladoukakis and Zouros 2001), and also biparental inheritance was observed (Ballard and Whitlock 2004).

Nuclear mitochondrial pseudogenes (Numts) can be another problem. Numts are copies of mitochondrial genes or fragments of them that have been transferred to the nuclear genome. These Numts evolve much differently than mtDNA and if accidentally utilised as mtDNA in data analysis, the results are flawed (Bensasson et al. 2001). The general notion is that fishes rarely have or do not have Numts at all. However, based on recent complete nuclear genomes of three fish species, the presence of Numts is under debate. An analysis of the complete nuclear genomes of *Fugu rubripes*, *Tetraodon nigroviridis*, and *Danio rerio* showed the presence of several Numts (Antunes and Ramos 2005). However, subsequent analysis of a new release of the complete nuclear genomes of *Fugu rubripes* did not confirm this result (Venkatesh et al. 2006). The authors concluded that methodological problems in shotgun sequencing and sequence assembly lead to the artificial incorporation of mitochondrial fragments in the consensus sequence. Other recent studies, however, indicate the presence of Numts in nuclear fish genomes (Teletchea et al. 2006; Knudsen et al. 2007).

Closely related species can hybridise and if these hybrids backcross, the maternally inherited mitochondrial genome can be passed from one species into the other by introgression (Ballard and Whitlock 2004). These species cannot be differentiated by mtDNA analysis, which is a potential problem in species identification.

The general assumption that mtDNA is a neutral marker is also questioned and it appears that mitochondria are often under selection (Ballard and Whitlock 2004; Bazin et al. 2006). A comparative study on genetic diversity in populations obtained from allozymes, nuclear and mitochondrial sequences did not find a correlation between mitochondrial genetic diversity and population size. In contrast, allozyme and nuclear sequence data showed such a correlation, which is congruent to the assumption in population genetics theory that genetic diversity is proportional to the effective population size. The authors concluded that mtDNA is not neutral and is subject to adaptive evolution (Bazin et al. 2006).

The most applied nuclear markers in population genetics are microsatellites, which are short tandem arrays composed of di-, tri- or tetranucleotide repeats with a length of tens to hundreds of base pairs (Tautz 1989). They are dispersed throughout the nuclear genome, are highly abundant and have a proportion of about 5% together with minisatellites (Fig. 24.3). Supposed to be non-coding and neutral, microsatellites are highly variable in length and are therefore an ideal tool to study intraspecific variation (Wright and Bentzen 1994; O'Connell and Wright 1997). When the flanking regions are known, locus-specific primers can be design and amplification of several microsatellite loci is possible in a multiplex-PCR (O'Connell and Wright 1997). However, microsatellites also have some drawbacks, such as problems in scoring, null alleles, and in some microsatellite loci selection. Problems in scoring can arise when so-called stutter bands are observed that are experimental artefacts caused by slipped-strand mispairing during PCR. It can be difficult to differentiate the true band from the stutter band. Another problem is null alleles, which are not amplified during PCR due to mutations at the primer biding site. They may contribute significantly to the observed patterns in variation of microsatellites (O'Connell and Wright 1997). An excellent review on microsatellites in fish, including technical details, is provided by O'Connell and Wright (1997).

24.3   Species Identification

24.3.1   Identification of Eggs and Larvae

The identification of species is the prerequisite for any meaningful biological research. The identification of adults, especially of commercially important fish species, is usually not problematic, even though marine cryptic species are frequently revealed by genetic studies (Knowlton 2000). However, the morphological identification of eggs and larval stages of fishes and invertebrates can be extremely problematic and in many cases even impossible. Classical microscopy methods require a high degree of taxonomic expertise, which is currently falling short, and are very time consuming. Therefore, the identification of species is a major bottleneck, hampering the necessary monitoring of marine biodiversity. For instance, about one third of the specimens in 138 studies on invertebrate diversity in European seas were not identified to species level (Schander and Willassen 2005). Even more problematic are the species-rich tropical seas. Even though the adult fish fauna of Indo-Pacific coral reefs is very well investigated (e.g. Randall 1983; Smith and Heemstra 1991; Randall 1995; Randall et al. 1997; Allen 2000), the larval stages of these species are poorly known. The authors of the three standard works for the identification of Indo-Pacific coral reef fish larvae state that the aim of the books is the identification of families, not genera or species (Leis and Rennis 1983; Leis and Trnski 1989; Leis and Carson-Ewart 2000). Also eggs of many fish species are impossible to be distinguished by morphological character (Moser et al. 1984). However, the correct identification of fish eggs and larvae to species level is a prerequisite for proper fish stock assessment based on ichthyoplankton surveys.

A solution for this problem is the application of DNA-based identification methods. They are powerful tools with an unprecedented accuracy due to their inherently highest possible resolution, which can reach even the level of single base changes in a whole genome. Minute amounts of template from eggs or larvae can be amplified by PCR and sequenced or analysed by several other methods.

24.3.1.1   DNA Sequencing

One of the first studies using sequencing for the identification of fish larvae utilised a fragment of the mitochondrial cyt *b* gene (Hare et al. 1994). However, this study showed drastically possible pitfalls of this approach if PCR is not conducted carefully. The universal primers used did not only bind to the DNA of the fish larvae, but also to human DNA, which lead to contaminations in this study. The PCR with templates from one adult specimen as well as all larvae investigated were contaminated and human cyt *b* sequences were obtained, leading to wrong conclusions about larvae identity as well as phylogenetic relationships (Hare et al. 1996). However, after processing the samples again, sequences from the species under study could be obtained, and larvae were identified by a phylogenetic analysis (Hare et al. 1998). Due to the large amount of sequence data available in international data bases, it is now possible to check if obtained sequences are the result of a contamination.

Other studies also used the approach of sequencing mtDNA for the identification of eel (16S rRNA; Aoyama et al. 1999), rockfish (cyt *b*; Rocha-Olivares et al. 2000) and coral reef fish larvae (control region; Pegg et al. 2006), as well as eggs of alfonsino (16S rRNA; Akimoto et al. 2002). A study on the identification of tuna and billfish larvae based on cyt *b* sequencing implemented an automated high-throughput system with a liquid-handling robot, reducing manual pipetting to a minimum. This system is able to sequence more than 800 specimens per week, reaching an identification success rate of 89% (Richardson et al. 2006).

24.3.1.2   Length Polymorphisms of PCR Products

Another approach is the amplification of a molecular marker with species-specific differences in length that can be detected by electrophoresis. Such species-specific differences in length occur, for example, in the 16S ribosomal RNA (16S rRNA) gene of the mitochondrial genome. This gene has a well-characterised secondary structure (Meyer 1993; Ortí et al. 1996) and especially the highly variable loop regions exhibit many insertions and deletions, so-called indels, which cause species-specific differences in length. Additionally, the mitochondrial 16S rRNA gene exhibits a very low intraspecific, but sufficient interspecific variation to discriminate different species. This was, for example, shown in a study on lionfishes (Kochzius et al. 2003), which revealed that individuals of the same species exhibit identical 16S rRNA haplotypes even though they were sampled at sites thousands of kilometres apart, but clear differences could be detected between closely related lionfish species. The detection of length differences with an automated sequencer was successfully applied for the identification of the very similar eggs of European hake (*Merluccius merluccius*), megrim (*Lepidorhombus whiffiagonis*), and fourspotted megrim (*Lepidorhombus boscii*) (Perez et al. 2005). This was even possible with formaldehyde-fixed eggs, which is usually very problematic due to degradation of DNA and cross-linking of DNA with denatured proteins.

24.3.1.3   Species-Specific Primers

The above-mentioned method utilises the natural variation in length of ribosomal genes by amplifying a certain fragment with a single primer pair. However, differences in length of PCR products that are detectable in a gel electrophoresis can also be obtained by using species-specific primer pairs in a multiplex-PCR. Species-specific primers for a cyt *b* fragment allowed the identification of eggs and larvae of blue marlin (*Makaira nigricans*), dolphinfish (*Coryphaena equiselis*), shortbill spearfish (*Tetrapturus angustirostris*), swordfish (*Xiphias gladius*) and wahoo (*Acanthocybium solandri*). This methodology is rather simple and does not require sophisticated laboratory equipment and can be therefore set up on a research vessel for shipboard identification of samples within 3 h, allowing researchers to adopt sampling protocols for more efficient study of egg and larval distribution (Hyde et al. 2005). In other studies, multiplex-PCRs amplifying fragments of mtDNA allowed the differentiation of two garfish (control region; *Hyporhamphus* spp; Noell et al. 2001) and four rockfish species (cyt *b*; *Sebastomus* spp; Rocha-Olivares 1998).

24.3.1.4   PCR-SSCP

An even simpler method is the detection of single strand confirmation polymorphism (SSCP) by a polyacrylamide gel electrophoresis (PAGE), which does not even require prior knowledge of the sequence. The detection of SSCP is based on differences in the sequence of the PCR-amplified molecular marker which results in a different secondary structure and mobility in electrophoresis (Sunnucks et al. 2000). Species can be again discriminated by different banding pattern. The feasibility of PCR-SSCP for the identification of fish eggs based on a fragment of the 16S rRNA gene was shown for formaldehyde-fixed eggs of European hake (*Merluccius merluccius*), megrim (*Lepidorhombus whiffiagonis*), Atlantic mackerel (*Scomber scombrus*) and longspine snipefish (*Macrorhamphosus scolopax*) (García-Vásquez et al. 2006).

24.3.1.5   PCR-RFLP

Differences in banding pattern after electrophoresis are also utilised by the RFLP (*restriction fragment length polymorphism*), even though the principle behind it is different. Here, a certain genetic marker is amplified by PCR and the product is digested with restriction enzymes. Due to sequence variation, the restriction enzymes will cut the PCR product at different positions, resulting in species-specific fragments of different lengths that can be visualised by gel electrophoresis. This approach was for example used to identify larvae of five gobiid fishes (Lindstrom 1999) and eggs of three European horse mackerel (*Trachurus* spp) species (Karaiskou et al. 2003) based on restriction digests of PCR-amplified cyt *b* fragments. Using the nuclear locus BM32-2 in a PCR-RFLP, larval billfishes of four species could be identified. A single eye of a 3 mm billfish larvae yielded sufficient DNA for analysis (McDowell and Graves 2001; Luthy et al. 2005).

24.3.1.6   Molecular Probes

An even more difficult task is the identification of digested larval remains in the gut of predatory fish. Such studies can be important to reveal if the recovery of a fish stock is negatively influenced by predation on larval fish. Even though fishing pressure on north-western Atlantic cod (*Gadus morhua*) was reduced after the collapse of the fisheries, stocks did not recover, which could be due to predation on their larvae. However, identification of larval remains is extremely difficult and many times impossible. In addition, larvae of cod can not easily be distinguished from other gadid fish species based on morphological characters. Therefore, an assay based on PCR to amplify a fragment of the 16S rRNA gene and a dot-blot hybridisation procedure was developed to identify larval remains of cod. In a first step, DNA was extracted from homogenised stomach content. Then, a PCR with gadid-specific primers for a fragment of the 16S rRNA gene was conducted and the PCR product was fixed to a nylon membrane. Afterwards, a species-specific probe was hybridised to the immobilised PCR product. Hybridisation of the biotin-labelled probe to the sample was visualised by chemiluminescence. Identification of cod larvae from stomach content was possible with this methodology (Rosel and Kocher 2002). However, the described methodology only provides a qualitative answer and can not quantify the amount of detected larvae. Additionally, this method only unfolds its advantages completely if probes for the detection of several species are utilised. Much more powerful is the species-specific detection of PCR products by immobilised probes on a DNA microarray, which will be discussed later.

24.3.1.7   Real-Time PCR

The latest development in identifying fish eggs and larvae is the use of the TaqMan™ technology, which is a PCR monitored in real time by detecting the signal of a fluorophore (reporter dye) that is covalently bound at the 5'-end of a species-specific oligonucleotide probe. Signal emission of the reporter dye (e.g. FAM, 6-carboxyfluorescein) is suppressed by a so-called quencher dye (TAMRA, 6-carboxy-tetramethylrhodamine) at the 3'-end of the oligonucleotide probe. Since Taq polymerase has a 5' nuclease activity, it cleaves the non-extendible hybridisation probe during the extension phase of PCR. The reporter dye is released from the quencher dye and its fluorescence signal can be detected, which is a direct measure of the amplification rate (Heid et al. 1996). Using a probe for Japanese eel (*Anguilla japonica*) in a TaqMan™ assay based on the 16S rRNA gene, identification of eggs and leptocephali larvae on board of a research vessel was possible within 3–4 h (Watanabe et al. 2004). Since fluorophores with different emission spectra are available, a multiplex reaction with different probes is possible. This approach was utilised in a survey on the abundance and distribution of cod eggs (*Gadus morhua*) that are difficult to distinguish from whiting (*Merlangius merlangus*) and haddock (*Melanogrammus aeglefinus*). These 'cod-like' eggs are usually believed to be cod eggs in ichthyoplankton surveys conducted to estimate the spawning stock biomass by the annual egg production method (AEPM). Probes for these three species, carrying different reporter dyes, were used in a multiplex PCR. The study revealed that only 34% of 'cod-like' eggs were actually cod, and that 58% were whiting and 8% were haddock, inflating estimates of the cod stock biomass (Fox et al. 2005).

24.3.2   Species Identification in Food Control

Most seafood products are processed and diagnostic features are removed, making species identification based on morphology impossible. Therefore, DNA-based identification methods are the only possibility to identify species. Species identification in food control is important to prevent commercial fraud, because substitution of lower value species for high price species frequently occurs (Sweijd et al. 2000). For instance, the European Union (EU) has strict regulations for seafood labelling, which must include, e.g. the species name (EU Council Regulation No 104/2000; EU Commission Regulation No 2065/2001). However, about 420 species of fish are on the German market alone, making a reliable identification urgently necessary to protect the customer.

24.3.2.1   Isoelectric Focusing of Proteins

Even though this review focuses on DNA-based identification methods, isoelectric focusing (IEF) will be briefly mentioned, because it is a well-established technique for the identification of fish species and regularly used in food control (Rehbein 1990). IEF separates proteins in an electrophoresis along a pH-gradient. Depending on their electric charge, the proteins will move to the anode or cathode through the gel. Due to the pH-gradient the proteins will change their electrical charge until they reach the isoelectric point. At the isoelectric point the protein no longer has a net electrical charge and stops moving through the gel. Therefore, species-specific banding pattern of muscle proteins can be produced with this method. In several countries catalogues for commercial species with IEF banding patterns, as well as photographs and a description of the fish species are available, e.g. France (Durand et al. 1985), Belgium (Bossier and Cooreman 2000), Australia (Yearsley et al. 2001; Yearsley et al. 2003), the United States (Tenge et al. 1993) and Germany (Rehbein and Kündiger 2005). The German data base also contains PCR-RFLP and PCR-SSCP pattern, as well as DNA sequences. However, the banding pattern produced by IEF can be influenced by the freshness of the fillet or fish, type of muscle (light or dark), and conditions of frozen storage (Rehbein 1990), hampering the identification by comparing them with reference banding pattern. Heat-sterilised canned fish, for example can not be identified by IEF, because proteins are severely denatured. Another problem is that protein profiles are not able to differentiate closely related species, e.g. in tuna, sardine and salmon (Mackie et al. 1999). Additionally, IEF requires a certain amount of material, which might not be available in all cases.

Such problems are not encountered in DNA analysis, which requires only minute amounts of DNA that can be amplified by PCR. Therefore, nowadays DNA-based identification is the method of choice, providing a good alternative to protein electrophoresis (Mafra et al. 2007)

24.3.2.2 DNA Sequencing

A study on commercial fraud on the American fish market revealed by sequencing of a 953 bp cyt *b* fragment that three quarters of fish sold as red snapper (*Lutjanus campechanus*) were mislabelled and belonged to other species (Marko et al. 2004). In another study, flatfishes were differentiated successfully based on sequences of a 464 bp cyt *b* fragment (Sotelo et al. 2001). However, both studies showed also the general limitation of such an approach: exact identification is only possible if corresponding reference sequences are available. In order to differentiate four species of anchovies (*Engraulis* spp), a cyt *b* fragment of 540 bp length was sequenced. Analysis of sequences obtained from canned and frozen anchovies identified the species correctly in all commercial samples (Santaclara et al. 2006). A study aiming to identify canned tuna showed that even a short cyt *b* sequences of 126 bp is sufficient to differentiate six tuna species (Quinteiro et al. 1998).

24.3.2.3 Length Polymorphisms of PCR Products

In order to detect the substitution of Greenland halibut (*Reinhardtius hippoglossoides*) for sole (*Solea solea*) fillets, the length polymorphism of the nuclear 5S rRNA gene was utilised. The two species could be clearly distinguished by the size of the PCR fragments in a gel electrophoresis (Céspedes et al. 1999). Length polymorphism in the same gene was also used to identify the three horse mackerel species (*Trachurus* spp) occurring in European seas (Karaiskou et al. 2003).

24.3.2.4 Species-Specific Primers

An identification assay based on species specific primers for a mitochondrial control-region fragment was developed for four Mediterranean grey mullet species (Mugilidae) in order to identify the origin of bottarga (salted and semi-dried ovary product). High-quality bottarga is produced from *Mugil cephalus* in Sardinia, but might be substituted by lower quality products from other species and regions. The developed assay differentiated the four grey mullet species and was even able to confirm the Sardinian origin of bottarga from *Mugil cephalus* (Murgia et al. 2002). Another study on six grouper (*Epinephelus aeneus*, *E. caninus*, *E. costae*, *E. marginatus*, *Mycteroperca fusca* and *M. rubra*) and two substitute species (nile perch, *Lates niloticus* and wreck fish, *Polyprion americanus*) showed that group specific primers targeting the 16S rRNA gene in a multiplex PCR produced fragments of different length. Grouper species were represented by a 300 bp fragment and its substitute species nile perch and wreck fish by 230 bp and 140 bp, respectively. These fragments could be detected in a gel electrophoresis. By this method, the groupers could be differentiated from the two substitute species, and no cross-reaction with was observed with DNA samples from 41 marketed fish species. However, identification of the grouper species was not possible (Trotta et al. 2005).

A special case in the application of specific primers is the detection of genetically modified coho salmon (*Oncorhynchus kisutch*), which contains an 'all-salmon' gene-construct. The genetic alteration was detected by the presence of a PCR product that was amplified with specific primers annealing within the gene-construct (Rehbein et al. 2002).

24.3.2.5 PCR-SSCP

High-priced tuna species are subject to commercial fraud, especially in canned products (Mackie et al. 1999). Due to the processing, DNA of canned fish is degraded and therefore, only short fragments can be amplified. PCR-SSCP analysis of a 123 bp fragment from the cyt *b* gene allowed the correct identification of eight tuna species in 90% of the cases, even in mixed samples (Rehbein et al. 1999).

PCR-SSCP was also successfully used to differentiate ten salmon species of the genera *Salmo*, *Oncorhynchus* and *Salvelinus* in raw and cold-smoked salmon, as well as salmon caviar. The identification was based on PCR amplified fragments of 300–460 bp length of the mitochondrial cyt *b* and nuclear parvalbumine and growth hormone genes (Rehbein 2005).

24.3.2.6 PCR-RFLP

The most widely applied PCR-based method for the identification of fish species in food control is PCR-RFLP. Several genetic markers are utilised, such as nuclear 5S rRNA (Aranishi 2005) or mitochondrial 16S rRNA (Chakraborty et al. 2007), but species identification is mainly based on fragments of the mitochondrial cyt *b* gene (Mafra et al. 2007). All following examples have utilised this gene.

A fragment of 464 bp length was amplified from a variety of smoked, pickled and heat-treated salmon products. RFLP pattern of the PCR products in a framework of ten species allowed the identification of all commercial samples (Hold et al. 2001). In order to differentiate the three European horse mackerel species, a fragment of about 370 bp length was digested with restriction enzymes, detecting mislabelling of blue jack mackerel (*Trachurus picturatus*) as Mediterranean horse mackerels (*T. mediterraneus*) on the Greek and Italian market (Karaiskou et al. 2003). Based on an amplified 126 bp fragment, six canned tuna species could be identified by RFLP pattern (Quinteiro et al. 1998). A study on flatfishes developed an assay for the identification of 24 species. PCR-RFLP was based on a fragment of 464 bp length, allowing the identification of 11 commercial samples of frozen fish to species and of three to family level (Sotelo et al. 2001). Investigations on six different sardines and four allied species based on short fragments of 142 and 147 bp allowed the unambiguous differentiation of *Sardina pilchardus* from all other studied species in canned and raw products (Jérôme et al. 2003). Analysis of the closely related anchovies (*Engraulis* spp) by PCR-RFLP based on a fragment of 284 bp length allowed only the discrimination of two species (*E. anchoita* and *E. ringens*) and a species pair (*E. japonicus/Engraulis encrasicolus*), showing the shortcomings of the method. The latter two species could only be identified by DNA sequencing (Santaclara et al. 2006).

In all examples mentioned above, restriction fragments have been separated by conventional gel electrophoresis. In order to enhance the resolution and reproducibility, fragment size analysis was carried out by a lab-on-a-chip capillary electrophoresis. Using this system, the identification of ten fish species based on a 464 bp fragment was possible, also in mixtures of two species (Dooley et al. 2005a, b).

24.3.2.7 Real-Time PCR

As described already above in the paragraph on species specific primers, the aim of a study on six groupers was their discrimination from two substitute species. In order to avoid gel electrophoresis, a conventional real-time PCR assay based on a fragment of the mitochondrial 16S rRNA gene was developed. The different-sized fragments (groupers: 300 bp; nile perch 230 bp; wreckfish: 140 bp) could be detected at the end of the multiplex PCR according to different melting temperatures (Trotta et al. 2005). A dye (e.g. SYBR Green I) is added to the PCR reaction, which only emits a continuously monitored fluorescence signal when intercalated to double-stranded DNA. By gradually heating up the double-stranded PCR products, the two strands will dissociate at a specific melting temperature. The dye will be released from the DNA and the reduced fluorescence signal can be measured. Such a real-time PCR assay could be used for an automated high-throughput system.

24.3.3 Species Identification in Fisheries and Trade

Identification of species is also very important to enforce fishery regulations and international agreements (Sweijd et al. 2000). On the one hand, for many species fishery regulations are implemented that restrict fishing activity in order to reduce the fishing pressure on exploited stocks. On the other hand, for some species the level of exploitation is not even known due to problems in morphological identification, preventing the development of management strategies.

An example is the fishery on sharks, because the catch is usually processed on board of the fishing vessels and morphological features important for identification such as head, fins and tails are removed in order to save space for storage

(Pank et al. 2001; Greig 2005). On the contrary, another problem is the identification of shark fins that receive high prices on the East Asian markets. After removal of the fins, the animals are discarded; making an identification based on morphological characters impossible (Hoelzel 2001; Pank et al. 2001). In order to solve this identification problem for the monitoring of the US Atlantic shark fishery, a 1,400 bp fragment spanning the 3′-end of the 12S rRNA gene, the complete valine tRNA gene, and 5′-end of the 16S rRNA gene was sequenced for 35 shark species, based on archived voucher specimens. This mitochondrial marker showed sufficient variation to discriminate the species under study (Greig et al. 2005) and the data could be used to develop easy-to-handle identification assays. In another study, a multiplex PCR method using the nuclear ribosomal ITS2 region was developed to distinguish two Atlantic shark species (*Carcharhinus plumbeus* and *C. obscurus*) based on size differences of the PCR products (Pank et al. 2001). In order to identify the origin of processed shark products such as dried fins, fin soup or cartilage pills in international trade, the above mentioned approaches using DNA fragments of more than 1,000 bp are not applicable. DNA in these processed products is degraded and therefore such long fragments can not be amplified. In these cases, only shorter fragments of less than 200 bp can be amplified by PCR and used for identification. This approach was chosen in a study by Hoelzel (2001) who sequenced small fragments of 188 bp from the mitochondrial cyt *b* gene to identify the origin of fin soup (hammerhead shark, *Sphyrna lewini*) and cartilage pills (basking shark, *Cetorhinus maximus*).

Other examples for species identification in fisheries assessment and control are North Atlantic sandeels as well as Patagonian and Antarctic toothfish. Sandeels are targeted by an industrial fishery in the North Sea and catches mainly consist of lesser sandeel (*Ammodytes marinus*). However, the occurrence of the other two sandeel species *A. tobianus* and *Gymnammodytes semisquamatus* is not generally quantified, which would be important in assessing the impact of the fishery on these species. In order to address this problem, a PCR-RFLP assay based on a fragment of the mitochondrial 16S rRNA/ND1 gene was developed that could discriminate all North Atlantic sandeel species (Mitchell et al. 1998).

In the Southern Ocean a large fishery with growing catches of toothfish has developed in recent years. Most of the catches are believed to consist of the Patagonian toothfish (*Dissostichus eleginoides*), but since the fishery is extending further South to Antarctic waters, there is also an unregulated fishery on the Antarctic toothfish (*D. mawsoni*), probably exceeding the total allowable catch (TAC) set by the Commission for the Conservation of Antarctic Marine Living Resources. Additionally, toothfish are mislabelled as hake or bass on the market. Since head, gut and tails are removed already on board of the fishing vessels before freezing, and the fish is often processed to filets on land, identification by morphological characters is impossible. Therefore, three molecular identification methods have been developed in order to monitor the fishery on the Antarctic toohfish. On the one hand, two mtDNA-based methods have been utilised: PCR-RFLP of a 16S rRNA fragment and length polymorphism in the control region, both discriminating the two species. On the other hand, isoelectric focusing (IEF) of muscle proteins also enabled the identification of the two species (Smith et al. 2001).

24.4 Stock Identification

A basic concept in fisheries management is the 'sustainable yield', which is a harvestable surplus that can be exploited by fisheries without jeopardising the stock. However, the main problem with this concept is the identification of a stock and the question how to utilise information on stock structure in fisheries management (Carvalho and Hauser 1994). In this review the term 'stock' always refers to the 'biological' or more precisely 'genetic stock' concept, which is defined as a reproductively isolated unit that is genetically distinct.

Several methods for stock identification have been used, such as parasite distribution, morphometrics and meristics, allozymes and DNA analysis. This review will focus on state-of-the-art PCR-based DNA analytical methods such as sequencing and microsatellite analysis. A comprehensive overview on allozyme and RFLP analysis is, for example, provided by Carvalho and Hauser (1994) and Park and Moran (1994).

For many centuries cod (*Gadus morhua*) was an important protein resource for Europe's population, armies and, naval as well as merchant fleets (Kurlansky 1997), but due to steadily increasing fishing pressure the stocks in the North Atlantic declined since the 1970s (Marteinsdottir et al. 2005) and the north-western Atlantic stocks even collapsed in the late 1980s and, early 1990s (Pauly et al. 2002). Atlantic cod is one of the best studied marine fish of commercial importance and genetic studies on stock structure have been conducted for more than 40 years on both sides of the Atlantic Ocean (Marteinsdottir et al. 2005). Therefore, examples of genetic investigations on cod are chosen to highlight the development, power and pitfalls of genetic tools to study genetic stock structure. Emphasis is also given to novel tools for genetic data analysis that go beyond genetic stock identification.

First studies on cod genetic stock structure began in the 1960s by investigating haemoglobin variants (Sick 1961, 1965a, b), but the discovery of selection on this locus made it unreliable for such studies (Mork and Sundnes 1985). A large-scale study throughout the species range utilising allozymes showed a very low amount of genetic differentiation and the most divergent population was from the Baltic Sea (Mork et al. 1985). Allozyme studies in the north-western Atlantic also did not show a significant genetic structure (Pogson et al. 1995).

However, does a very low amount of genetic differentiation reflect a high level of gene flow or limited resolution of the genetic marker? This question could only be answered by studying genetic variation directly on DNA level. The most basic technique to study genetic variation directly on DNA level is RFLP-analysis of mtDNA. A study utilising this technique did not detect a higher level of genetic differentiation and supported former results on the differentiation of Arctic and coastal cod in the north-eastern Atlantic obtained by haemoglobin and allozyme analyses (Dahle 1991). Analysis of partial cyt *b* sequences could detect significant genetic structure between populations in the north-western (Newfoundland) and north-eastern Atlantic (Greenland, Iceland, Faroe Islands, Norway, Baltic and White Sea). Within the north-eastern Atlantic, only the population from the Baltic Sea was significantly different (Árnason 2004), which was also detected by an earlier study using allozymes (Mork et al. 1985). Samples from the Baltic Sea as well as Faroe Islands showed no significant genetic structure within these regions using cyt *b* sequences (Árnason et al. 1998; Sigurgíslason and Árnason 2003). No significant genetic structure was also found by utilising cyt *b* sequences of cod from several sites in the north-western Atlantic (Carr and Crutcher 1998). Since mtDNA did not provide a sufficient resolution to reveal predicted population structure in cod, nuclear markers, such as nuclear RFLP loci and microsatellites, were applied.

The first utilised nuclear DNA markers were DNA fingerprints (Dahle 1994). They were studied by Southern blot analysis to screen for 17 nuclear restriction fragment length polymorphism (RFLP) loci scored by 11 anonymous cDNA clones (Pogson et al. 1995). In contrast to previous allozyme studies, significant genetic differences between all sites in the northern Atlantic (Nova Scotia, Newfoundland, Iceland, North Sea, Balsfjord and Barents Sea) were detected. A regional study in the north-eastern Atlantic (Nova Scotia and Newfoundland) using the same methodology also revealed significant but weak genetic structure (Pogson et al. 2001). The mean F_{ST}-value, which can reach values from 0 (no structure) to 1 (complete separation of populations), of 10 loci was only 0.014, but the mean value for significant differences was 0.068. Both studies revealed isolation-by-distance (IBD), indicating limited dispersal that contrasts previous results based on allozymes and mtDNA analysis. Another large scale analysis also based on nuclear DNA RFLP variation detected a significant genetic difference between the Barents Sea and all other sites in the north-eastern Atlantic (Celtic Sea, Loftstaahraun, North Sea, Trondheimsfjorden; F_{ST}-values ranging from 0.48 to 0.67) and north western Atlantic (Scotian Shelf; F_{ST}-values ranging from 0 to 0.60). There was also a separation between north-western and north-eastern Atlantic, but pairwise comparison of samples from the Scotian Shelf and Celtic Sea did not show a significant differentiation. No significant differentiation was revealed between samples from the Celtic Sea, Loftstaahraun, North Sea, and Trondheimsfjorden (Jónsdóttir et al. 2003). A small-scale study in southern Iceland indicated sub-structuring of cod populations, which supported earlier tagging experiments (Imsland et al. 2004).

One of these nuclear RFLP loci, pantophysin I (*Pan* I) (originally called GM798 or *Syp* I; Pogson et al. 1995), has two main alleles (*Pan* I^A and *Pan* I^B) that show different frequencies in cod from the Barents Sea and Norwegian coast. The *Pan* I^A allele is predominant among Norwegian coastal cod (allele frequency up to 0.91), whereas the *Pan* I^B dominates in cod from the Barents Sea (allele frequency up to 0.89). F-statistics between groups showed significant values between populations from the Barents Sea and Norwegian coast (0.36) and between populations within the two groups (0.04), indicating a strong genetic stock structure (Pogson and Fevolden 2003). In order to allow rapid and cost-effective genotyping of these two alleles, a PCR with allele specific, fluorescent labelled primers was developed to detect length differences of PCR products from the two alleles by an automated sequencer. This assay can be included in a multiplex PCR for microsatellite analysis in order to reduce costs (Stenvik et al. 2006a). Concordance of the *Pan* I locus with microsatellites supports that genetic structuring observed in the *Pan* I locus is due to restricted gene flow, even though this locus might be to some extent under selection (Skarstein et al. 2007).

The most used nuclear markers in population genetics of fish are microsatellites. These markers were first developed for cod in the early 1990s (Brooker et al. 1994) and since then a large number of studies using microsatellites for genetic stock analysis of cod were published. Even though many loci are known and utilised, newly developed microsatellites are recently published (Stenvik et al. 2006b; Westgaard et al. 2007), based on data mining of published EST (expressed sequence tags) sequences in Genbank. This approach is very elegant, because time consuming and expensive laboratory work for the development of microsatellites is avoided.

Microsatellite studies on the genetic stock structure of cod on different spatial scales in the north-western Atlantic revealed genetic heterogeneity even on a small geographic scale. Analysis of a larval cod aggregation on the Western Bank of the Scotian Shelf revealed a high genetic heterogeneity, but single cohorts on the basis of age-at-length were genetically homogenous. These results indicate that the different cohorts of larvae originated from different spawning events and that oceanographic processes, such as eddies, retain eggs and larvae within well defined geographic areas on the Scotian Shelf. Comparison with adult cod showed that the genetic structure of the cod larvae was most similar to adults from Western Bank. This study gave interesting insights into reproduction and larval dispersal of cod, suggesting differential reproductive success among spawning groups and local retention of eggs and larvae by eddies (Taggart et al. 1998). A spatio-temporal study showed that a coastal population of cod was significantly different from an offshore population and that the genetic structure of the inshore population was temporally stable over the study period of 4 years (Taggart et al. 1998). A similar pattern was found by nuclear RFLP analysis in the Norwegian Barents Sea (Pogson and Fevolden 2003). A study investigating cod on Georges and Browns Bank as well as Bay of Fundy revealed a small but significant genetic differentiation, which could be attributed to existing gyres and separation of the banks by the deep Fundian Channel (Taggart et al. 1998). Summarising the above-mentioned results, Taggart et al. (1998) concluded that genetic stock structure can be revealed on scales of 60–100 nautical miles. However, a genetic differentiation between Georges and Browns Bank was not detected in another study using microsatellites (Lage et al. 2004). This difference can have two reasons: On the one hand, this difference might reflect a temporal variation; on the other hand, it might be due to methodological differences, because different microsatellite loci have been used in the two studies. Additionally, Lage et al. (2004) applied SNP analysis in the *Pan* I locus. Sampling was also extended further south, showing that the population from Nantucket Shoals was significantly different from Georges and Browns Bank (Lage et al. 2004). A detailed study off Newfoundland and Labrador tested the hypothesis of discrete 'bay stocks' of cod, which could not be confirmed. However, the significant genetic differentiation of coastal and off-shore populations, as well as between different banks was supported (Beacham et al. 2002).

In contrast to tagging experiments, previous genetic studies on cod suggested high dispersal and limited structuring in European Seas. Therefore, genetic stock structure was re-analysed using microsatellites and revealed formerly undetected genetic differentiation. All European populations were significantly different from populations on the New Scotian shelf and Barents Sea, and within the North Sea, for example, four genetically distinct stocks could be detected (Hutchinson et al. 2001). A large-scale analysis, including samples from the species' whole distribution range revealed a significant difference (F_{ST} = 0.03) between samples from the Scotian Shelf and Baltic Sea to all other sample sites (West and East Greenland, West and East Iceland, Faeroes Ridges, Barents Sea, and Celtic Sea), which indicates three major population groupings: western Atlantic, mid and east North Atlantic, and Baltic Sea (O'Leary et al. 2007). Analysis of samples collected in an area stretching from Spitzbergen to the North Sea using microsatellites and the *Pan* I locus was concordant for both markers in detecting genetic differentiation of north-east Arctic cod (NEAC), Norwegian costal cod (NCC) and North Sea cod (NSC). Both markers also revealed sub-structuring in NCC, but only microsatellites detected genetic differentiation in NEAC and NSC (Skarstein et al. 2007). A significant genetic structure was also revealed between populations from southern Iceland compared to populations from eastern Iceland and the Faroe Islands by using microsatellites and SNP analysis in the *Pan* I locus. The lack of genetic differentiation between cod from eastern Iceland and the Faroe Islands indicated larval exchange between these two regions (Pampoulie et al. 2008). Analysis of stock structure on a scale of 300 km along the Norwegian coast in the Skagerrak revealed a low (F_{ST} = 0.0023) but significant genetic structure (Knutsen et al. 2003). Another study in the same area even detected significant structure on a much smaller scale of only 30 km (Jorde et al. 2007). However, the question remains if these genetically distinct populations, e.g. coastal and offshore populations, are temporally stable. Microsatellite studies on juvenile cod collected in coastal waters of the Skagerrak showed that they are predominantly of North Sea origin in a year with high inflow of North Sea water, whereas in another year they were of local origin. These results indicate that current-mediated larval transport from offshore populations influences the composition of coastal cod populations (Knutsen et al. 2004).

The power of microsatellites in genetic stock structure analysis was demonstrated by assigning cod specimens from the Baltic Sea, North Sea and north-eastern Arctic Ocean to their origin with an accuracy of 97–100% (Nielsen et al. 2001). This application of microsatellite analysis can be a valuable tool in fisheries control to reveal poaching and can also aid control of correct labelling of seafood products.

Another innovative approach in the application of microsatellites and SNP analysis for fisheries genetics is the study of temporal genetic stock structure over several decades by the analysis of DNA extracted from archived otoliths using ancient-DNA techniques. A study covering three decades (1964–1994) revealed a temporally stable genetic stock structure of cod off Newfoundland and Labrador (Ruzzante et al. 2001) using microsatellites. Analysis of the genetic diversity of a population based on microsatellites in the North Sea between 1954 and 1998 showed reduced values between 1954 and 1970, followed by a recovery until 1998 (Hutchinson et al. 2003). Another study applying microsatellites compared the genetic composition of a cod population in the North Sea in 1965 and 2002, as well as a population in the Baltic Sea in 1928 and 1997. No significant change in allele frequencies was found for the North Sea population, but a small but significant change in the population from the Baltic Sea. However, there was no evidence for a genetic bottleneck in both populations (Poulsen et al. 2006). This study was expanded with samples from the Faroe Bank (1978 and 1992) and Faroe Plateau (1969 and 2002) and additionally to microsatellites, the *Pan* I locus was used. Both markers showed temporal stability in allele frequencies for all studied populations (Nielsen et al. 2007).

Estimation of population size is also an important task in fisheries research and can be supported by fishery genetics. Based on genetic data and genetic models the effective population (N_e) size can be estimated, which is the number of individuals contributing to the gene pool by reproduction. This number can be relatively small compared to the total population, which seems to be the case in cod. A study on two populations in the North Sea and Baltic Sea estimated an effective population size of more than 500, but most probably ranging in the thousands, which is concordant to the reproduction biology of the species. The authors conclude that this effective population size is sufficient to maintain the evolutionary potential of the species and that this number is not likely to be of general concern (Poulsen et al. 2006). In contrast, N_e for another population in the North Sea was estimated to be as low as 69 individuals from 1954–1960 and 121 individuals from 1960–1970 (Hutchinson et al. 2003).

However, there are also some pitfalls in microsatellite analysis. The general notion that microsatellites are neutral markers was questioned by several recent studies, showing that the loci Gmo132, Gmo37, Gmo34 and Gmo8 in cod are not neutral (Nielsen et al. 2006; Skarstein et al. 2007; Westgaard and Fevolden 2007). These findings strongly question some of the results summarised above, especially cases where genetic differentiation was only or mainly found in these loci. This is, for example, the case in the above-mentioned 300 km-scale and 30 km-scale studies at the Norwegian coast, where most of the significant genetic differentiation was attributed to Gmo132 and Gmo37 (Knutsen et al. 2003) or Gmo132 and Gmo34 (Jorde et al. 2007). These examples show that a large number of microsatellite loci should be utilised to detect loci under selection and to ensure that they do not flaw the analysis. Caution is also recommended in cases where genetic differentiation is only detected by one or two loci (Nielsen et al. 2006).

However, the advances in molecular genetics described above completely changed the view on dispersal capability and gene flow in cod from large-scale panmixing to strong genetic stock structure on even very small scales. A comprehensive overview on utilised markers and genetic stock structure of cod is provided in O'Leary et al. (2007). Analysis of more SNPs (Wirgin et al. 2007) and automation (Stenvik et al. 2006a) will enhance genetic studies on cod and enables regular large scale screening of genetic stock structure. Findings and experiences based on cod genetics could be also transferred to other fish species.

As shown above, analysis of genetic stock structure provides information on the level of differentiation between populations. Based on these data, the amount and direction of migration between populations can be estimated by applying genetic models and statistical methods. The computer programme MIGRATE (Beerli 2008) is a maximum likelihood estimator based on the coalescent theory. It uses a Markov chain Monte Carlo approach to investigate possible genealogies with migration events (Beerli and Felsenstein 2001). Migration and effective population size can be estimated from sequence, SNP, microsatellite and electrophoresis data. In order to study migration patterns of marine ornamental fish from the Red Sea, this method was applied to mitochondrial control region sequence data of lionfish (*Pterois miles*) and fourline wrasse (*Larabicus quadrilineatus*) in the Red Sea. The study on gene flow between populations of the coral reef dwelling lionfish *P. miles* indicated panmixia between the Gulf of Aqaba and northern Red Sea, but analysis of migration patterns showed an almost unidirectional migration originating from the Red Sea proper (Kochzius and Blohm 2005). The genetic population structure of the fourline wrasse *L. quadrilineatus* indicted limited larval dispersal distance of only about 5 km in the Red Sea. Analysis of molecular variance (AMOVA) detected the highest significant genetic variation between northern and central/southern populations ($\Phi_{ct} = 0.012$; $p < 0.001$), and migration analysis revealed several folds higher northward than southward migration, which could be linked to oceanography and spawning season. In order to enable a sustainable ornamental fishery on the fourline wrasse, the results of this study suggest managing populations in the northern and southern Red Sea separately as two different stocks. The rather low larval dispersal distance needs to be considered in the design of marine protected areas to enable connectivity and self seeding (Froukh and Kochzius 2007).

24.5   Emerging Approaches and Technologies

24.5.1   DNA Barcoding

Molecular genetic methods have been widely applied for species identification and phylogenetics of animals, but due to the application of different molecular markers it is impossible to implement a unifying identification system. As also shown above, another problem is the huge variety of utilised methods, ranging from banding pattern in gel electrophoresis to DNA sequences. The different zoological disciplines developed their own traditions, utilising different genetic markers. Ichthyologists focused mainly on the mitochondrial cyt *b* gene, but also used 16S and 12S rRNA genes frequently (Meyer 1993, 1994), whereas research on invertebrates was mainly based on COI (Folmer et al. 1994). This was due to the pioneering works of Kocher et al. (1989) and Folmer et al. (1994), being the first to publish universal primers for a plethora of animal taxa of vertebrates and invertebrates, respectively.

The first 'standard gene' for identification and phylogenetics was implemented more than 10 years ago in microbiology, utilising the sequences of the small subunit rRNA gene and setting up the data base ARB (Ludwig et al. 2004). The idea of using a 'standard gene' for a global bioidentification system of animals was later proposed by Hebert et al. (2003a, b), showing that the 'Folmer fragment' (Folmer et al. 1994) of the mitochondrial COI gene can serve as a 'DNA barcode' to identify closely related species and higher taxa in all animal phyla, except cnidarians. The approach of DNA taxonomy, which gives DNA sequences a central role in defining species (Blaxter 2003; Tautz et al. 2002, 2003) provoked a controversial debate and was especially criticised by classical taxonomists (Lipscomb et al. 2003; Mallet and Willmot 2003; Seberg et al. 2003). Currently, taxonomy is in a crisis, facing the lack of prestige and resources (Godfrey 2002) and therefore classical taxonomists feared that funding will only be provided for DNA barcoding (Ebach and Holdrege 2005). However, DNA barcoding rather offers a unique opportunity of bringing together classical taxonomy, genetics and ecology in joint projects – and funding (Gregory 2005). DNA barcoding of metazoans is not reasonable without the expert knowledge in classical taxonomy based on morphology, because a DNA sequence that is not related to a precisely described voucher specimen is not of much value (Schindel and Miller 2005).

Exceptions are very diverse taxa that are not studied well yet taxonomically. In such cases 'molecular operational taxonomic units' (MOTU), also called 'phylospecies' or 'genospecies', can help to retrieve an overview on the genetic diversity, which can be translated in certain limits to taxon diversity (Blaxter 2004; Blaxter et al. 2004, 2005; Floyd et al. 2002). In the marine realm, this approach was for example successfully applied for estimating the diversity of nematods (Blaxter 2004) and stomatopods (Barber and Boyce 2006). However, in order to describe our planet's biodiversity properly in an integrative taxonomy approach, morphological, ecological and genetic data have to be combined (Dayrat 2005). Striking examples of this approach are the discoveries of two new whale species (Dalebout et al. 2002; Wada et al. 2003), a new dolphin species (Beasley et al. 2005) and a new giant clam species (Richter et al. 2008).

DNA barcoding utilises a short DNA sequence from an agreed-upon standard position in the genome for the identification of species. Such DNA barcode sequences usually have a length of 500–700 bp and can be obtained quickly and cheaply. Using a high-throughput sequencing system, the costs for a DNA barcode sequence are less than 1 Euro. Following the suggestion of Hebert et al. (2003a), the 5′-end of the mitochondrial COI gene, which is usually 648 bp long, is the standard barcode region for higher animals. Studies on various groups, such as birds (Hebert et al. 2004a; Kerr et al. 2007), moths and butterflies (Hebert et al. 2003a, 2004b; Hajibabaei et al. 2006a), bats (Clare et al. 2007) and fishes (Ward et al. 2005; Spies et al. 2006) have shown the feasibility of this approach. However, the suitability can not be generalised, because the 5′-end of the mitochondrial COI gene seems to be not well suited for amphibians (Vences et al. 2005), cnidarians (Hebert et al. 2003b) and sponges (Erpenbeck et al. 2005).

DNA barcode sequences are stored in the Barcode of Life Data System (BOLD, www.barcodinglife.org, Hajibabaei et al. 2005; Ratnasingham and Hebert 2007), but also in Genbank (www.ncbi.nlm.nih.gov) where such sequences receive the keyword 'barcode'. Formal barcode sequences have to be related to the following data: (1) species name (although this can be interim), (2) voucher data (catalogue number and institution storing), (3) collection record (collector, collection date and location with GPS coordinates), (4) identifier of the specimen, (5) COI sequence of at least 500 bp, (6) PCR primers used to generate the amplicon, and (7) trace files (Ratnasingham and Hebert 2007). In January 2008, BOLD contained 335,000 barcode sequences of 47,000 animal, fungi, protist and plant species. However, of these only about 128,000 barcodes of 12,000 species are validated. In Genbank, 11,000 sequences with the keyword 'barcode' were recorded.

Specimens can be identified with the BOLD Identification System (IDS) by submitting a sequence of the 5′end from the COI gene online. This query sequence is aligned to the global alignment of all reference sequences in BOLD. A positive identification is given, if the query sequence matches a reference sequence with less that 1% difference. If such a match could not be found, the query sequence will be assigned to a genus if the difference to a reference sequence is less than 3%. In cases where a query sequence can also not be related to a certain genus, a list with the 100 closest reference sequences is provided, giving information to which higher taxon level the query sequence belongs. Until BOLD is filled with barcodes of all taxa, such cases will appear frequently. However, with increasing taxon coverage, the number of precise identifications will increase. Additionally, a phylogenetic neighbour joining (NJ) tree with the 100 most similar sequences is provided (Ratnasingham and Hebert 2007).

In processed products and forensic samples DNA degradation can prevent the amplification of DNA barcodes that usually have a length of 650 bp. However, 'minimalist barcodes' of about 100 bp length can be obtained from samples with degraded DNA and these short sequences still contain enough information for species assignment (Hajibabaei et al. 2006b).

One of the large barcoding projects is Fish-BOL (Fish barcode of life initiative, www.fishbol.org), aiming to barcode all of the 30,000 known fish species. This initiative was launched in June 2005 on a workshop at the University of Guelph (Canada) with the aim to barcode all marine fish species until 2010. The programme and all presentations can be downloaded at www.fishbol.org/meeting_june05.php. In January 2008, 23,000 barcodes of 4,300 fish species have been generated and about 3,000 are public at BOLD. In Genbank, 1,400 COI sequences of bony fishes with the keyword 'barcode' are available.

However, other projects are also collecting sequence information of fishes and setting up sequence data bases, such as MitoFish (http://mitofish.ori.u-tokyo. ac.jp), FishTrace (www.fishtrace.org), and Fish & Chips (www.fish-and-chips. uni-bremen.de). The aim of MitoFish is the compilation of mitochondrial DNA sequences for evolutionary research (e.g. Saitoh et al. 2006; Lavoué et al. 2007). In January 2008, 82,000 mtDNA sequences from 9,800 fish species, including complete mitochondrial genomes of 395 species, were available in the MitoFish data base. The FishTrace data base provides 2,700 complete mitochondrial cyt *b* and nuclear rhodopsin sequences of 220 commercial fish species from Europe, all of them linked to voucher specimens. In the framework of 'Fish & Chips' (Kochzius et al. 2007a, b), more than 1,400 fragments of the mitochondrial 16S rRNA, cyt *b*, and COI genes from about 80 species have been sequenced as the basis for the development of DNA microarrays for the identification of commercial fishes from European seas (Kochzius et al. 2008). Additionally, the Fish & Chips data base compiled 4,750 16S rRNA, cyt *b*, and COI sequences of European marine fishes obtained from Genbank. Fish & Chips, FishTrace and MitoFish allow the comparison of query sequences to the reference sequences by using the algorithm BLAST (Altschul et al. 1990), which guides species identification. Currently, the Fish & Chips data base is not public, but it will be freely available as soon as the development of the microarrays is finished.

24.5.2 DNA Microarrays

A DNA microarray is a systematic arrangement of oligonucleotide probes that are immobilised on a solid surface. These probes are complementary to DNA target sequences to be detected (Pirrung 2002). There are many technologies available, differing in fabrication and signal detection. The surface material can be glass, silicon, plastic or metal (Pirrung 2002; Dufva 2005), different surface chemistries for the immobilisation of the probes are available (Benters et al. 2002; Pirrung 2002), and the application of the probes can be done by spotting or in situ synthesis (Pirrung 2002; Dufva 2005). A microarray can contain thousands and even upto many hundred thousands of spots with different oligonucleotide probes, enabling a high redundancy (Dufva 2005). Also different detection systems are available, using electrochemistry (Metfies et al. 2005) or fluorescent dyes (Relógio et al. 2002).

However, currently glass microscope slides with a chemically modified surface on which oligonucleotide probes are spotted are very common. Usually, the DNA target is labelled during PCR amplification with a fluorophore and its hybridisation to the probe on the microarray can be detected with a fluorescence scanner (Fig. 24.4).

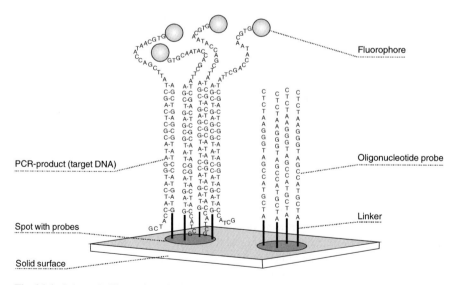

Fig. 24.4 Schematic illustration of a DNA microarray

The application of DNA microarrays for gene expression analysis has already reached the routine level of high-throughput systems (Blohm and Guiseppi-Elie 2001; Hoheisel 2006), but they have been only recently used for the identification of organisms, such as microbes (Wang et al. 2002; Call et al. 2003; Korimbocus et al. 2005; Loy and Bodrossy 2005), plants (Rønning et al. 2005), and animals (Pfunder et al. 2004). In the marine realm, DNA microarrays are used for the identification of bacteria (Peplies et al. 2003; Peplies et al. 2004), phytoplankton (Metfies and Medlin 2004; Metfies et al. 2005; Godhe et al. 2007), invertebrates (Chitipothu et al. 2007), and fishes (Kochzius et al. 2008). Microarrays are also used for gene expression analysis (Williams et al. 2003; Lidie et al. 2005; Wang et al. 2006; Cohen et al. 2007; Jenny et al. 2007) and genotyping of marine organisms in population genetics (Moriya et al. 2004, 2007).

In a fisheries research context, DNA microarrays can be applied for the identification of ichthyoplankton, processed fish in fishery and food control, as well as genotyping for stock identification. There is also a great potential for the application in gene expression analysis for breeding in aquaculture (e.g. Panserat et al. 2008), but this is not in the scope of this review. As shown above in many examples, DNA-based identification is a very important task for fishery genetics. In many cases, especially when analysing environmental samples, a very high number of target species needs to be detected and discriminated against a even much higher number of other species. DNA microarrays are believed to have the potential of identifying hundreds of species in parallel, making them a promising tool. However, the different microarray platforms are still error prone and quantification is difficult (Shi et al. 2006). Even receiving a clear qualitative result, i.e. presence or absence of a certain species, can be sometimes difficult. One methodical limitation is the design of species-specific probes that do not always exhibit the hybridisation properties they were selected for in silico. Therefore, the probes must be empirically tested in hybridisation experiments to ensure that they do not give false-negative or false-positive signals.

An important point in probe design is the choice of the molecular marker. On the one hand, intraspecific variation should be as low as possible to ensure that the designed probes match all individuals of a species and not only a certain populations. Therefore, it is important to obtain sequence information from a wide geographic range. On the other hand, interspecific variation has to be large enough to differentiate closely related species. Since all probes on a microarray will be exposed to the same experimental conditions, their features, such as length (usually 20–30 bp), melting temperature (T_m), and GC content (usually about 50%), have to be more or less identical. Additionally, secondary structures and dimer formations have to be avoided. A comparison of probes for the identification of fishes based on sequences of the mitochondrial 16S rRNA, cyt b, and COI genes have shown a very different performance at the same experimental conditions (Kochzius et al. 2007a). Inter- as well as intra-marker signal intensities are very variable, making quantification currently impossible (Kochzius et al. 2007a, 2008). However, a quantification of different species in a mixed sample would be desirable, because this would enable for example the identification and quantification of eggs in plankton samples. Experiments with known mixtures of amplified target DNA and multiplex-PCRs amplifying the three markers from a mixture of species showed that the parallel identification of fish species is potentially feasible (Hauschild 2008).

DNA-microarrays can also be used for screening genotypes in fish stock identification. In a study on chum salmon (*Oncorhynchus keta*), a microarray was developed to detect 30 known mtDNA control region haplotypes (Moriya et al. 2004). This microarray was applied to screen 2,200 chum salmon to investigate the stock composition in the Bering Sea and North Pacific Ocean (Moriya et al. 2007). The study showed that a rapid and accurate identification of haplotypes with DNA microarrays is possible on board of a fisheries research vessel. Based on this data, the proportion of Japanese, Russian and North American stocks and their geographic distribution was estimated (Moriya et al. 2007). Another application for such a technology is the localisation of the geographic origin of the catch in fisheries control. If a fishery on a species is closed in a certain region, but not in another, fisheries authorities could use such a tool to control the origin of the catch. For instance, the European Union (EU) has strict regulations for seafood labelling, which must include for example geographic origin (EU Council Regulation No 104/2000; EU Commission Regulation No 2065/2001). In cases of a known strong genetic population structure, where certain genotypes can be assigned to certain geographic areas, such an approach is realistic. The recently launched EU-funded research project 'FishPopTrace' (https://fishpoptrace.jrc.ec.europa.eu) aims to obtain such data and to develop methods to trace the geographic origin of commercially important species.

24.5.3 New Sequencing Technologies

Modern sequencing methods were introduced about 30 years ago, with the development of the dideoxy method of Sanger et al. (1977) and the chemical method of Maxam and Gilbert (1977). Since then several other sequencing methods have been introduced, but the refined Sanger sequencing method (Smith et al. 1986; Prober et al. 1987) still remains as the 'gold standard' used in genome sequencing projects and other applications (Marziali and Akeson 2001; Shendure et al. 2004; Metzker 2005; Hudson 2007; Hutchison 2007). However, this 'gold standard' is now challenged by new sequencing technologies, such as 454 sequencing (pyrosequencing; www.454.com), Solexa/Illumina 1G SBS technology (sequencing by synthesis; www.illumina.com), and Agencourt/ABI SOLiD technology (sequencing by oligonucleotide ligation and detection; www.appliedbiosystems.com) (Hudson 2007).

A common principle of these three technologies is the random fragmentation of genomic DNA that is immobilised on a solid support. This could be either microscopic beads (454 sequencing and SOLiD) or a flow cell (SBS technology). Afterwards the immobilised fragments are amplified by PCR in an emulsion phase (emPCR) that prevents cross-contamination in order to receive clonal PCR products. The single beads are afterwards either placed in microscopic wells (454 sequencing) or are immobilised on a planar surface (SOLiD) and the sequencing chemistry is applied, which differs between platforms. The 454 platform is based on pyrosequencing and detects chemiluminescence signals, Solexa/Illumina applies reversible-terminator sequencing by synthesis, measuring different fluorescence signals for each base, and SOLiD ligates sequence specific labelled oligonucleotides, also using four different fluorophores (Hudson 2007). The main disadvantage of these technologies is the short read length of about 100 bp for the 454 Life Sciences Genome Sequencer GS20, 35 bp for the Solexa SBS technology and only 25 bp for SOLiD. This makes an assembly of hundreds of thousands of short fragments necessary, which is a difficult task that needs new bioinformatic tools to be developed (Jeck et al. 2007; Warren et al. 2007). The new 454 Genome Sequencer FLX has an enhanced read length of 200 bp and if this can be extended to >500 bp, it might succeed Sanger sequencing (Hudson 2007).

Since 454 Life Sciences' Genome Sequencer GS20 was launched in 2005, 100 scientific papers using the technology were published until November 2007 (www.454.com), ranging from de novo genome sequencing of viruses (Thomas et al. 2007) and bacteria (Goldberg et al. 2006) to metagenomics in marine environmental research (Huber et al. 2007) and ancient Neandertal DNA (Green et al. 2006). In comparison, only a few studies using either the Solexa/Illumina 1G SBS or Agencourt/AB SOLiD technology are published and therefore this review will focus on 454 pyrosequencing. Comprehensive general overviews on DNA sequencing technologies are provided by Marziali and Akeson (2001), Shendure et al. (2004), Metzker (2005), Hudson (2007), and Hutchison (2007).

Pyrosequencing is a sequencing-by-synthesis method that detects the incorporation of nucleotides by the enzymatic luminometric inorganic phyrophosphate detection assay (ELIDA) that emits light (Hymann 1988; Nyrén et al. 1993). Since the nucleotides are added subsequently, the light signal can be related proportional to each type of nucleotide (A, T, G or C) and the sequence can be assembled. This method was enhanced in recent years (Ronaghi et al. 1996, 1998) and finally developed into a highly parallel 454 Genome Sequencer, capable of sequencing about 25 million bases in only 4 h, which is about 100 times faster than the conventional Sanger sequencing using capillary-based electrophoresis systems (Margulies et al. 2005). Reagents for one run of the 454 Genome Sequencer cost about US\$ 5,000, reducing the cost per base ten times compared to Sanger sequencing (Hudson 2007). Whole genome shotgun sequencing is possible without cloning of DNA fragments into bacterial cells, because fragmented genomic DNA is ligated to linker sequences and single fragments are captured on the surface of a 28-μm bead. Single beads are isolated in an emulsion droplet and amplification is performed. It is also possible to capture PCR-products on those beads. The beads are arrayed in the 1.6 million 75-pl wells of a fibre-optic slide and sequencing is carried out with DNA polymerase by primed synthesis. As described above, incorporation of nucleotides produces a light signal, which is detected by a CCD imager that is coupled with the fibre-optic array. As mentioned earlier, read length reaches about

100 bp, but recent advancements enabled read length of up to 200 bp. In comparison to Sanger sequencing, which reaches a read length of about 700 bp, this is rather short, but due to the extreme high number of parallel sequencing reactions, this disadvantage can be compensated to some degree, because complete sequences can be assembled from an extremely high number of redundant and overlapping fragments (Margulies et al. 2005), demanding considerable computational power. Other sources of error in pyrosequencing are homopolymeric stretches of the DNA, e.g. seven 'C's. The incorporation of seven nucleotides will give a single light signal that will only differ in its intensity compared to a single nucleotide. Up to three identical nucleotides can be detected, but measurements of longer homopolymeric stretches become increasingly inaccurate (Hudson 2007).

Even though 454 sequencing is capable to sequence whole prokaryotic genomes de novo, the technology can still not sequence whole eukaryotic genomes de novo, which are by far more complex. Nevertheless, the 454 technology is well-suited for re-sequencing, because if a 'master sequence' is already known, the small re-sequenced fragments can be aligned to that. Re-sequencing can be utilised for model species, of which whole genomes are available, or their close relatives. Such approaches can give valuable insights into genome-wide intraspecific variation. However, the cost of several hundreds of thousands of dollars still limits the application of whole genome re-sequencing. A much cheaper alternative is the sequencing of expressed sequence tags (ESTs), which focuses on messenger RNA that encodes proteins. Re-sequencing of these specific loci opens an avenue to genome wide SNP screening and genotyping in non-model species (Hudson 2007), which can be used as high resolution markers for population genetics.

The high capacity of parallel sequencing of the 454 technology could also be utilised for simultaneous sequencing of multiple homologous PCR products in stock identification (population genetics) and species identification (DNA barcoding). However, these PCR products can not exceed the current maximum read length of 200 bp, which usually do not contain enough variability for studies in population genetics. An exception is the highly variable mitochondrial control region, showing high haplotype diversity in a fragment of less than 200 bp in lionfish (Kochzius and Blohm 2005). As mentioned above, 'minimalist barcodes' of about 100 bp length are sufficient to identify species (Hajibabaei et al. 2006b) and 454 sequencing is currently tested for barcoding and identification of mixed samples (Hajibabaei 2007). Such an approach could be applied to complete plankton samples in order to identify ichthyoplankton by extracting the complete DNA and performing a multiplex-PCR with universal primer cocktails for fishes (Ivanova et al. 2007; Sevilla et al. 2007). Species identification is then conducted by comparing the sequences to corresponding sequence data bases, such as BOLD or FishTrace. The number of retrieved sequences per species could be used as an indication of its relative abundance or biomass. Such an approach is frequently used in environmental metagenomic studies on microbial communities (Huber et al. 2007). In order to exploit the highly parallel pyrosequencing even further, multiple environmental samples or pooled PCR products can be analysed in a single run. Assignment of the sequences to the different environmental samples or single PCR products from the pool can be done by specific tags that are part of the PCR primers. These tags are combinations of 2–4 nucleotides that will be sequenced together with the primer and PCR product. The feasibility of this approach was shown for short 16S rDNA sequences that can identify 13 mammal species (Binladen et al. 2007) and multiple environmental samples (Huber et al. 2007).

Even though 454 pyrosequencing is currently too expensive for routine application in fisheries research, it has the potential to revolutionise fisheries genetics in the future, if the technology becomes more cost-effective and cheaper.

24.6   Conclusions

DNA analytics have made tremendous progress in the last decade, facilitating sequencing of the human and other organisms' genome (Hutchison 2007). DNA microarray technology developed to a flourishing field (Blohm and Guiseppi-Elie 2001; Hoheisel 2006) and application of this technology was extended from model organisms to non-model species in gene expression analysis and species identification. PCR technology was advanced to real time PCR (Heid et al. 1996), including the development of the TaqMan™ assay. Just recently, novel DNA sequencing technologies such as 454 pyrosequencing emerged, producing huge amounts of sequence data in a single run by highly parallel sequencing (Margulies et al. 2005; Hudson 2007). Even though 'traditional' PCR-based techniques (e.g. PCR-RFLP, PCR-SSCP) are still used in species identification, the above mentioned novel technologies are more and more utilised (Fox et al. 2005; Kochzius et al. 2008). The concept of DNA barcoding is getting established in the scientific community and the tremendous effort of barcoding all fishes is currently undertaken by an international consortium in the framework of the Fish-BOL initiative. In the field of genetic stock identification, microsatellites have been established as the 'gold standard', but due to enhancements of sequencing technologies and a steadily increasing amount of sequence data, discovery and analysis of SNPs will become more and more important. Today, numerous novel tools and technologies are available to be used in fishery genetics and their application will enable a tremendous progress in this field.

References

Akimoto S, Kinoshita S, Sezaki K, Mitani I, Watabe S (2002) Identification of alfonsino and related fish species belonging to the genus *Beryx* with mitochondrial 16S rRNA gene and its application on their pelagic eggs. Fisheries Science 68: 1242–1249

Allen G (2000) Marine fishes of South-East Asia. Periplus, Hong Kong, 292 pp

Altschul SF, Gish W, Miller W, Myers EW, Lipman DJ (1990) Basic local alignment search tool. Journal of Molecular Biology 215: 403–410

Antunes A, Ramos MJ (2005) Discovery of a large number of previously unrecognized mitochondrial pseudogenes in fish genomes. Genomics 86: 708–717

Aoyama J, Mochioka N, Otake T, Ishikawa S, Kawakami Y, Castle PHJ, Nishida M, Tsukamoto K (1999) Distribution and dispersal of anguillid leptocephali in the western Pacific revealed by molecular analysis. Marine Ecology Progress Series 188: 193–200

Aranishi F (2005) Rapid PCR-RFLP method for discrimination of imported and domestic mackerel. Marine Biotechnology 7: 571–575

Árnason E (2004) Mitochondrial cytochrome *b* DNA variation in the high-fecundity Atlantic cod: trans-Atlantic clines and shallow gene genealogy. Genetics 166: 1871–1885

Árnason E, Petersen PH, Pálsson S (1998) Mitochondrial cytochrome *b* DNA sequence variation of Atlantic cod, *Gadus morhua*, from the Baltic and the White Seas. Hereditas 129: 37–43

Avery OT, MacLeod CM, McCarty, M (1944) Studies of the chemical nature of the substance inducing transformation of pneumococcal types. Induction of transformation by a desoxyribonucleic acid fraction isolated from *Pneumococcus* Type III. Journal of Experimental Medicine 79: 137–158

Avise JC (1998) The history and purview of phylogeography: a personal reflection. Molecular Ecology 7: 371–379

Ballard JWO, Whitlock MC (2004) The incomplete natural history of mitochondria. Molecular Ecology 13: 729–744

Barber P, Boyce SL (2006) Estimating diversity of Indo-Pacific coral reef stomatopods through DNA barcoding of stomatopod larvae. Proceedings of the Royal Society of London B 273: 2053–2061

Bazin R, Glémin S, Galtier N (2006) Population size does not influence mitochondrial genetic diversity. Science 312: 570–572

Beacham TD, Brattey J, Miller KM, Le KD, Withler RE (2002) Multiple stock structure of Atlantic cod (*Gadus morhua*) off Newfoundland and Labrador determined from genetic variation. ICES Journal of Marine Science 59: 650–665

Beasley I, Robertson KM, Arnold P (2005) Description of a new dolphin, the Australian snubfin dolphin *Orcaella heinsohni* sp. n. (Cetacea, Delphinidae). Marine Mammal Science 21: 365–400

Beerli P (2008) MIGRATE-N: estimation of population sizes and gene flow using the coalescent. Distributed over the Internet, http://popgen.scs.fsu.edu/Migrate-n.html

Beerli P, Felsenstein J (2001) Maximum likelihood estimation of a migration matrix and effective population size in n subpopulations by using a coalescent approach. Proceedings of the National Academy of Sciences of the United States of America 98: 4563–4568

Bensasson D, Zhang D-X, Hartl DL, Hewitt GM (2001) Mitochondrial pseudogenes: evolution's misplaced witness. Trends in Ecology and Evolution 16: 314–321

Benters R, Niemeyer CM, Drutschmann D, Blohm D, Wöhrle D (2002) DNA microarrays with PAMAM dendritic linker systems. Nucleic Acids Research 30: e10

Binladen J, Gilbert MTP, Bollback JP, Panitz F, Bendixen C (2007) The use of coded PCR primers enables high-throughput sequencing of multiple homolog amplification products by 454 parallel sequencing. PLoS ONE 2: e197

Blaxter M (2003) Counting angels with DNA. Nature 421: 122–124

Blaxter M (2004) The promise of a DNA taxonomy Philosophical Transactions of the Royal Society B 359: 669–679

Blaxter M, Elsworth B, Daub J (2004) DNA taxonomy of a neglected animal phylum: an unexpected diversity of tardigrades. Proceedings of the Royal Society of London B 271(Suppl 4): S189–S192

Blaxter M, Mann J, Chapman T, Thomas F, Whitton C, Floyd R, Abebe E (2005) Defining operational taxonomic units using DNA barcode data. Philosophical Transactions of the Royal Society B 360: 1935–1943

Blohm D, Guiseppi-Elie A (2001) New developments in microarray technology. Current Opinion in Biotechnology 12: 41–47

Bossier P, Cooreman K (2000) A databank able to be used for identifying and authenticating commercial flatfish (Pleuronectiformes) products at the species level using isoelectric focusing of native muscle proteins. International Journal of Food Science and Technology 35: 563–568

Brooker AL, Cook D, Bentzen P, Wright JM, Doyle RW (1994) Organization of microsatellites differs between mammals and cold-water teleost fishes. Canadian Journal of Fisheries and Aquatic Sciences 51: 1959–1966

Brown WM, Vinograd J (1974) Restriction endonuclease cleavage maps of animal mitochondrial DNAs. Proceedings of the National Academy of Sciences of the United States of America 71: 4617–4621

Brown WM, George M Jr, Wilson AC (1979) Rapid evolution of animal mitochondrial DNA. Proceedings of the National Academy of Sciences of the United States of America 76: 1967–1971

Call DR, Borucki MK, Loge FJ (2003) Detection of bacterial pathogens in environmental samples using DNA microarrays. Journal of Microbiological Methods 53: 235–243

Carr SM, Crutcher DC (1998). Population genetic structure in Atlantic cod (*Gadus morhua*) from the North Atlantic and Barents Sea: contrasting or concordant patterns in mtDNA sequence and microsatellite data? In: Herbing IH von, Kornfield I, Tupper M, Wilson J (eds) The Implications of localized fishery stocks. Northeast Regional Agricultural Engineering Service, Ithaca, NY, pp 91–103

Carvalho GR, Hauser L (1994) Molecular genetics and the stock concept in fisheries. Reviews in Fish Biology and Fisheries 4: 326–350

Céspedes A, García T, Carrera E, Gonzáles I, Fernández A, Hernández PE, Mart'n R (1999) Identification of Sole (*Solea solea*) and Greenland Halibut (*Reinhardtius hippoglossoides*) by PCR amplification of the 5S rDNA gene. Journal of Agriculture and Food Chemistry 47: 1046–1050

Chakraborty A, Aranishi F, Iwatsuki Y (2007) Polymerase chain reaction–restriction fragment length polymorphism analysis for species identification of hairtail fish fillets from supermarkets in Japan. Fisheries Science 73: 197–2001

Chitipothu S, Cariani A, Venugopal MN, Landi M, Mählck MT, Jaeger J, Kochzius M, Arvanitidis C, Weber H, Nölte M, Tinti F, Magoulas A, Blohm D (2007) Towards microarray-based DNA-barcoding of marine invertebrates. Second international barcode of life conference, 18–20 September 2007, Taipei, Taiwan, conference abstracts, p. 82

Clare EL, Burton KL, Engstrom MD, Eger JL, Hebert PDN (2007) DNA barcoding of Neotropical bats: species identification and discovery within Guyana. Molecular Ecology Notes 7: 184–190

Cohen R, Chalifa-Caspi V, Williams TD, Auslander M, George SG, Chipman JK, Tom M (2007) Estimating the efficiency of fish cross-species cDNA microarray hybridization. Marine Biotechnology 9: 491–499

Dahle G (1991) Cod, *Gadus morhua* L., populations identified by mitochondrial DNA. Journal of Fish Biology 38: 295–303

Dahle G (1994) Minisatellite DNA fingerprints of Atlantic cod (*Gadus morhua*). Journal of Fish Biology 44: 1089–1092

Dalebout ML, Mead JG, Baker CS, Baker AN, Helden AL van (2002) A new species of beaked whale *Mesoplodon perrini* sp. n. (Cetacea: Ziphiidae) discovered through phylogenetic analyses of mitochondrial DNA sequences. Marine Mammal Science 18: 577–608

Dawid IB, Blackler AW (1972) Maternal and cytoplasmic inheritance of mitochondrial DNA in *Xenopus*. Developmental Biology 29: 113–253

Dayrat B (2005) Towards integrative taxonomy. Biological Journal of the Linnean Society 85: 407–415

Dooley JJ, Sage HD, Brown HM, Garrett SD (2005a) Improved fish species identification by use of lab-on-a-chip technology. Food Control 16: 601–607

Dooley JJ, Sage HD, Clarke M-AL, Brown HM, Garrett SD (2005b) Fish species identification using PCR-RFLP analysis and lab-on-a-chip capillary electrophoresis: application to detect white fish species in food products and an interlaboratory study. Journal of Agriculture and Food Chemistry 53: 3348–3357

Dufva M (2005) Fabrication of high quality microarrays. Biomolecular Engineering 22: 173–184

Durand P, Landrein A, Quero J-C (1985) Catalogue electrophoretic des poissons commerciaux. IFREMER, Centre de Nantes, Nantes

Ebach MC, Holdrege C (2005) DNA barcoding is no substitute for taxonomy. Nature 434: 697

Erpenbeck D, Hooper JNA, Wörheide G (2005) CO1 phylogenies in diploblasts and the 'Barcoding of Life' - are we sequencing a suboptimal partition? Molecular Ecology Notes 6: 550–553

Ferguson A (1994) Molecular genetics in fisheries: current and future perspectives. Reviews in Fish Biology and Fisheries 4: 379–383

Floyd R, Abebe E, Papert A, Blaxter M (2002) Molecular barcodes for soil nematode identification. Molecular Ecology 11: 839–850

Folmer O, Black M, Hoeh W, Lutz R, Vrijenhoek R (1994) DNA primers for amplification of mitochondrial cytochrome c oxidase subunit I from diverse metazoan invertebrates. Molecular Marine Biology and Biotechnology 3: 294–299

Fox CJ, Taylor MI, Pereyra R, Villasana MI, Rico C (2005) TaqMan DNA technology confirms likely overestimation of cod (*Gadus morhua* L.) egg abundance in the Irish Sea: implications for the assessment of the cod stock and mapping of spawning areas using egg-based methods. Molecular Ecology 14: 879–884

Froukh T, Kochzius M (2007) Genetic population structure of the endemic fourline wrasse (*Larabicus quadrilineatus*) suggests limited larval dispersal distances in the Red Sea. Molecular Ecology 16: 1359–1367

García-Vázquesz E, Alvarez P, Lopes P, Karaiskou N, Pérez J, Tei A, Martínez JL, Gomes L, Triantaphyllidis C (2006) PCR-SSCP of the 16S rRNA gene, a simple methodology for species identification of fish eggs and larvae. Scientia Marina 70S2: 13–21

Godfrey HC (2002) Challenges for taxonomy. Nature 417: 17–19

Godhe A, Cusack C, Pedersen J, Andersen P, Anderson DM, Bresnan E, Cembella A, Dahl E, Diercks S, Elbrächter M, Edler L, Galluzzi L, Gescher G, Gladstone M, Karlson B, Kulis D, LeGresley M, Lindahl O, Marin R, McDermott G, Medlin LK, Naustvoll L-J, Penna A, Kerstin Töbe (2007) Intercalibration of classical and molecular techniques for identification of *Alexandrium fundyense* (Dinophyceae) and estimation of cell densities. Harmful Algae 6: 56–72

Goldberg SM, Johnson J, Busam D, Feldblyum T, Ferriera S, Friedman R, Halpern A, Khouri H, Kravitz SA, Lauro FM, Li K, Rogers YH, Strausberg R, Sutton G, Tallon L, Thomas T, Venter E, Frazier M, Venter JC (2006) A Sanger/pyrosequencing hybrid approach for the generation of high-quality draft assemblies of marine microbial genomes. Proceedings of the National Academy of Sciences of the United States of America 103: 11240–11245

Green RE, Krause J, Ptak SE, Briggs AW, Ronan MT, Simons JF, Du L, Egholm M, Rothberg JM, Paunovic M, Pääbo S (2006) Analysis of one million base pairs of Neanderthal DNA. Nature 444: 330–336

Gregory TR (2005) DNA barcoding does not compete with taxonomy. Nature 434: 1067

Greig TW, Moore MK, Woodley CM, Quattro JM (2005) Mitochondrial gene sequences useful for species identification of western North Atlantic Ocean sharks. Fishery Bulletin 103: 516–523

Haeseler A von, Liebers D (2003) Molekulare Evolution. Fischer Taschenbuch Verlag, Frankfurt am Main, Germany

Hajibabaei M (2007) Exploring archival and environmental samples through minimalist barcodes. Second international barcode of life conference, 18–20 September 2007, Taipei, Taiwan, conference abstracts, p 55

Hajibabaei M, deWaard JR, Ivanova NV et al. (2005) Critical factors for the high volume assembly of DNA barcodes. Philosophical Transactions of the Royal Society of London Series B 360: 1959–1967

Hajibabaei M, Janzen DH, Burns JM, Hallwachs W, Hebert PDN (2006a) DNA barcodes distinguish species of tropical Lepidoptera. Proceedings of the National Academy of Sciences of the United States of America 103: 968–971

Hajibabaei M, Smith MA, Janzen DH, Rodriguez JJ, Whitfield JB, Hebert PDN (2006b) A minimalist barcode can identify a specimen whose DNA is degraded. Molecular Ecology Notes 6: 959–964

Hare JA, Cowen RK, Zehr JP, Juanes F, Day KA (1994) Biological and oceanographic insights from larval labrid (Pisces: Labridae) identification using mtDNA sequences. Marine Biology 118: 17–24

Hare JA, Cowen RK, Zehr JP, Juanes F, Day KA (1996) Erratum. Marine Biology 126: 346

Hare JA, Cowen RK, Zehr JP, Juanes F, Day KA (1998) A correction to: biological and oceanographic insights from larval labrid (Pisces: Labridae) identification using mtDNA sequences. Marine Biology 130: 589–592

Hauschild J (2008) Hybridisierungsexperimente für die Entwicklung eines Mikroarrays zur Artidentifizierung europäischer Meeresfische. Diploma Thesis, University of Bremen, Germany

Hebert PDN, Cywinska A, Ball SL, deWaard JR (2003a) Biological identifications through DNA barcodes. Proceedings of the Royal Society of London B: 270, 313–321

Hebert PDN, Ratnasingham S, deWaard JR (2003b) Barcoding animal life: cytochrome c oxidase subunit 1 divergences among closely related species. Proceedings of the Royal Society of London B 270(Suppl 1): S96–99

Hebert PDN, Stoeckle MY, Zemlak TS, Francis CM (2004a) Identification of bird through DNA barcodes. PLoS Biology 2: e312

Hebert PDN, Penton EH, Burns JM, Janzen DH, Hallwachs W (2004b) Ten species in one: DNA barcoding reveals cryptic species in the neotropical skipper butterfly *Astrates fulgerator*. Proceedings of the National Academy of Sciences of the United States of America 101: 14812–14817

Heid CA, Stevens J, Livak KJ, Williams PM (1996) Real time quantitative PCR. Genome Research 6: 986–994

Hoheisel JD (2006) Microarray technology: beyond transcript profiling and genotype analysis. Nature Reviews Genetics 7: 200–210

Hoelzel AR (2001) Shark fishing in fin soup. Conservation Genetics 2: 69–72

Hold GL, Russell VJ, Pryde SE, Rehbein H, Quinteiro J, Rey-Mendez M, Sotelo GC, Pérez-Martin RI, Santos AT, Rosa C (2001) Validation of a PCR-RFLP based method for the identification of salmon species in food products. European Food Research and Technology 212: 385–389

Huber JA, Mark Welch D, Morrison HG, Huse SM, Neal PR, Butterfield DA, Sogin ML (2007) Microbial population structures in the deep marine biosphere. Science 318: 97–100

Hudson ME (2007) Sequencing breakthroughs for genomic ecology and evolutionary biology. Molecular Ecology Notes 8: 3–17

Hutchinson WF, Carvalho GR, Rogers SI (2001) Marked genetic structuring in localised spawning populations of cod *Gadus morhua* in the North Sea and adjoining waters, as revealed by microsatellites. Marine Ecology Progress Series 223: 251–260

Hutchinson WF, Oosterhout C van, Rogers SI, Carvalho GR (2003) Temporal analysis of archived samples indicates marked genetic changes in declining North Sea cod (*Gadus morhua*). Proceedings of the Royal Society of London B 270: 2125–2132

Hutchison CA III (2007) DNA sequencing: bench to bedside and beyond. Nucleic Acids Research 35: 6227–6237

Hyde JR, Lynn E, Humpfreys R Jr, Musyl M, West AP, Vetter R (2005) Shipboard identification of fish eggs and larvae by multiplex PCR, and description of fertilized eggs of blue marlin, shortbill spearfish, and wahoo. Marine Ecology Progress Series 286: 269–277

Hyman ED (1988) A new method of sequencing DNA. Analytical Biochemistry 174: 423–436

Imsland AK, Jónsdóttir ÓDB, Daníelsdóttir AK (2004) Nuclear DNA RFLP variation among Atlantic cod in south and south-east Icelandic waters. Fisheries Research 67: 227–233

Ivanova NV, Zemlak TS, Hanner RH, Hebert PDN (2007) Universal primer cocktails for fish DNA barcoding. Molecular Ecology Notes 7: 544–548

Jeck WR, Reinhardt JA, Baltrus DA, Hickenbotham MT, Magrini V (Magrini, Vincent), Mardis ER, Dangl JL, Jones CD (2007) Extending assembly of short DNA sequences to handle error. Bioinformatics 23: 2942–2944

Jenny MJ, Chapman RW, Mancia A, Chen YA, McKillen DJ, Trent H, Lang P, Escoubas JM, Bachere E, Boulo V, Liu ZJ, Gross PS, Cunningham C, Cupit PM, Tanguy A, Guo X, Moraga D,

Boutet I, Huvet A, De Guise S, Almeida JS, Warr GW (2007) A cDNA microarray for *Crassostrea virginica* and *C. gigas*. Marine Biotechnology 9: 577–591

Jérôme M, Lemaire C, Bautista JM, Fleurence J, Etienne M (2003) Molecular phylogeny and species identification of sardines. Journal of Agriculture and Food Chemistry 51: 43–50

Jónsdóttir ÓDB, Imsland AK, Atladóttir ÓY, Daníelsdóttir AK (2003) Nuclear DNA RFLP variation of Atlantic cod in the North Atlantic Ocean. Fisheries Research 63: 429–436

Jorde PE, Knutsen H, Espeland SH, Stenseth NC (2007) Spatial scale of genetic structuring in coastal cod *Gadus morhua* and geographic extent of local populations. Marine Ecology Progress Series 343: 229–237

Karaiskou N, Triantafyllidis A, Triantaphyllidis C (2003) Discrimination of three *Trachurus* species using both mitochondrial- and nuclear-based DNA approaches. Journal of Agriculture and Food Chemistry 51: 4935–4940

Kerr KCR, Stoeckle MY, Dove CJ, Weigt LA, Francis CM, Hebert PDN (2007) Comprehensive DNA barcode coverage of North American birds. Molecular Ecology Notes 7: 535–543

Knowlton N (2000) Molecular genetic analyses of species boundaries in the sea. Hydrobiologia 420: 73–90

Knudsen SW, Moller PR, Gravlund P (2007) Phylogeny of the snailfishes (Teleostei: Liparidae) based on molecular and morphological data. Molecular Phylogenetics and Evolution 44: 649–666

Knutsen H, Jorde PE, André C, Stenseth NC (2003) Fine-scaled geographical population structuring in a highly mobile marine species: the Atlantic cod. Molecular Ecology 12: 385–394

Knutsen H, André C, Jorde PE, Thuróczy E, Stenseth NC (2004) Transport of North Sea cod larvae into Skagerrak coastal populations. Proceedings of the Royal Society of London B 271: 1337–1344

Kocher TD, Thomas WK, Meyer A, Edwards SV, Pääbo S, Villablanca FX, Wilson AC (1989) Dynamics of mitochondrial DNA evolution in animals: amplification and sequencing with conserved primers. Proceedings of the National Academy of Sciences of the United States of America 86: 6196–6200

Kochzius M, Blohm D (2005) Genetic population structure of the lionfish *Pterois miles* (Scorpaenidae, Pteroinae) in the Gulf of Aqaba and northern Red Sea. Gene 347: 295–301

Kochzius M, Söller R, Khalaf MA, Blohm D (2003) Molecular phylogeny of the lionfish genera *Dendrochirus* and *Pterois* (Scorpaenidae, Pteroinae) based on mitochondrial DNA sequences. Molecular Phylogenetics and Evolution 28: 396–403

Kochzius M, Antoniou A, Botla S, Campo Falgueras D, Garcia Vazquez E, Hauschild J, Hervet C, Hjörleifsdottir S, Hreggvidsson GO, Kappel K, Landi M, Magoulas A, Marteinsson V, Nölte M, Planes S, Seidel C, Silkenbeumer N, Tinti F, Turan C, Weber H, Blohm D (2007a) Fish and Chips: microarray-based DNA-barcoding of European marine fishes. Second international barcode of life conference, 18–20 September 2007, Taipei, Taiwan, conference abstracts, p 24

Kochzius M, Kappel K, Döbitz L, Silkenbeumer N, Nölte M, Weber H, Hjörleifsdottir S, Marteinsson V, Hreggvidsson G, Planes S, Tinti F, Magoulas A, Garcia Vazquez E, Turan C, Medlin L, Metfies K, Gescher C, Cariani A, Landi M, Hervet C, Campo Falgueras D, Antoniou A, Bertasi F, Chitipothu S, Blohm D (2007b) The "Fish & Chips" project: microarrays as a tool for the identification of marine organisms in biodiversity and ecosystem research. Oceans'07 IEEE Aberdeen, 18–21 June 2007, Aberdeen, Conference Proceedings, ISBN: 1-4244-0635-8, pp 4

Kochzius M, Nölte N, Weber H, Silkenbeumer N, Hjörleifsdottir S, Hreggvidsson GO, Marteinsson V, Kappel K, Planes S, Tinti F, Magoulas A, Garcia Vazquez E, Turan C, Hervet C, Campo Falgueras D, Antoniou A, Landi M, Blohm D (2008) DNA microarrays for identifying fishes. Marine Biotechnology 10: 207–217

Korimbocus J, Scaramozzino N, Lacroix B, Crance JM, Garin D, Vernet G (2005) DNA probe array for the simultaneous identification of herpesviruses, enteroviruses, and flaviviruses. Journal of Clinical Microbiology 43: 3779–3787

Kurlansky M (1997) Cod. A biography of the fish that changed the world. Walker & Company, New York

Ladoukakis ED, Zouros E (2001) Direct evidence for homologous recombination in mussel (*Mytilus galloprovincialis*) mitochondrial DNA. Molecular Biology and Evolution 18: 1168–1175

Lage C, Kuhn K, Kornfield I (2004) Genetic differentiation among Atlantic cod (*Gadus morhua*) from Browns Bank, Georges Bank, and Nantucket Shoals. Fishery Bulletin 102: 289–297

Lavoué S, Miya M, Saitoh K, Ishiguro NB, Nishida M (2007) Phylogenetic relationships among anchovies, sardines, herrings and their relatives (Clupeiformes), inferred from whole mitogenome sequences. Molecular Phylogenetics and Evolution 43: 1096–1105

Leis JM, Carson-Ewart BM (2000) The larvae of Indo-Pacific coastal fishes: an identification guide to marine fish larvae. Fauna Malesiana Handbooks 2, Brill, Leiden, 850 pp

Leis JM, Rennis DS (1983) The larval stages of Indo-Pacific coral reef fishes. New South Wales University Press, Kensington; University of Hawaii Press, Honolulu, 269 pp

Leis JM, Trnski T (1989) The larvae of Indo-Pacific shorefishes. University of Hawaii Press, Honolulu, 371 pp

Lidie KB, Ryan JC, Barbier M, Van Dolah FM (2005) Gene expression in Florida red tide dinoflagellate *Karenia brevis*: analysis of an expressed sequence tag library and development of DNA microarray. Marine Biotechnology 7: 481–493

Ligny W de (1969) Serological and biochemical studies inn fish populations. Oceanography and Marine Biology - An Annual Review 7: 533–542

Lindstrom DP (1999) Molecular species identification of newly hatched Hawaiian amphidromous gobioid larvae. Marine Biotechnology 1: 167–174

Lipscomb D, Platnick N, Wheeler Q (2003) The intellectual content of taxonomy: a comment on DNA taxonomy. Trend in Ecology and Evolution 18: 65–66

Loy A, Bodrossy L (2005) Highly parallel microbial diagnostics using oligonucleotide microarrays. Clinica Chimica Acta 363: 106–119

Ludwig W, Strunk O, Westram R, Richter L, Meier H, Yadhukumar, Buchner A, Lai T, Steppi S, Jobb G, Förster W, Brettske I, Gerber S, Ginhart AW, Gross O, Grumann S, Hermann S, Jost R, König A, Liss T, Lüßmann R, May M, Nonhoff B, Reichel B, Strehlow R, Stamatakis A, Stuckmann N, Vilbig A, Lenke M, Ludwig T, Bode A, Schleifer K-H (2004) ARB: a software environment for sequence data. Nucleic Acids Research 32: 1363–1371

Luthy SA, Cowen RK, Serafy JE, McDowell JR (2005) Towards identification of larval sailfish (*Istiophorus platypterus*), white marlin (*Tetrapturus albidus*), and blue marlin (*Makaira nigricans*) in the western North Atlantic Ocean. Fishery Bulletin 103: 588–600

Mackie IM, Pryde SE, Gonzales-Sotelo C, Medina I, Peréz-Martín R, Quinteiro J, Rey-Mendez M, Rehbein H (1999) Challenges in the identification of species of canned fish. Trends in Food Science and Technology 10: 9–14

Mafra I, Ferreira IMPLVO, Oliveira MBPP (2007) Food authentication by PCR-based methods. European Food Research and Technology 227: 1438–2377

Mallet J, Willmott K (2003) Taxonomy: renaissance or Tower of Babel? Trend in Ecology and Evolution 18: 57–59

Margulies M, Egholm M, Altman WE, Attiya S, Bader JS, Bemben LA, Berka J, Braverman MS, Chen YJ, Chen ZT, Dewell SB, Du L, Fierro JM, Gomes XV, Godwin BC, He W, Helgesen S, Ho CH, Irzyk GP, Jando SC, Alenquer MLI, Jarvie TP, Jirage KB, Kim JB, Knight JR, Lanza JR, Leamon JH, Lefkowitz SM, Lei M, Li J, Lohman KL, Lu H, Makhijani VB, McDade KE, McKenna MP, Myers EW, Nickerson E, Nobile JR, Plant R, Puc BP, Ronan MT, Roth GT, Sarkis GJ, Simons JF, Simpson JW, Srinivasan M, Tartaro KR, Tomasz A, Vogt KA, Volkmer GA, Wang SH, Wang Y, Weiner MP, Yu PG, Begley RF, Rothberg JM (2005) Genome sequencing in microfabricated high-density picolitre reactors. Nature 437: 376–380

Marko PB, Lee SC, Rice AM, Gramling JM, Fitzhenry TM, McAlister JS, Harpert GR, Moran AL (2004) Misslabelling of a depleted reef fish. Nature 430: 309–310

Marteinsdottir G, Ruzzante D, Nielsen EE (2005) History of the North Atlantic cod stocks. ICES CM 2005/AA

Marziali A, Akeson M (2001) New DNA sequencing methods. Annual Review of Biomedical Engineering 3: 195–223

Maxam AM, Gilbert W (1977) A new method for sequencing DNA. Proceedings of the National Academy of Sciences of the United States of America 74: 560–564

McDowell JR, Graves JE (2001) Nuclear and mitochondrial DNA markers for specific identification of istiophorid and xiphiid billfishes. Fishery Bulletin 100: 537–544

Metfies K, Medlin L (2004) DNA microchips for phytoplankton: the fluorescent wave of the future. Nova Hedwigia 79: 321–327

Metfies K, Huljic S, Lange M, Medlin LK (2005). Electrochemical detection of the toxic dinoflagellate *Alexandrium ostenfeldii* with a DNA-biosensor. Biosensors and Bioelectronics 20: 1349–1357

Metzker ML (2005) Emerging technologies in DNA sequencing. Genome Research 15: 1767–1776

Meyer A (1993) Evolution of mitochondrial DNA in fishes. In: Hochachka PW, TP Mommsen (eds) Biochemistry and molecular biology of fishes, volume 2. Amsterdam, Elsevier, pp 1–38

Meyer A (1994) DNA technology and phylogeny of fish. In: Beaumont AR (ed) Genetics and evolution of aquatic organisms. London, Chapman & Hall, pp 219–249

Mitchell A, McCarthy E, Verspoor E (1998) Discrimination of the North Atlantic lesser sandeels *Ammodytes marinus*, *A. tobianus*, *A. dubius* and *Gymnammodytes semisquamatus* by mitochondrial DNA restriction fragment patterns. Fisheries Research 36: 61–65

Mork J, Sundnes G (1985) Haemoglobin polymorphism in Atlantic cod (*Gadus morhua*): Allele frequency variation between yearclasses in a Norwegian fjord stock. Helgoländer Meeresuntersuchungen 39: 55–62

Mork J, Ryman N, Ståhl G, Utter F, Sundnes G (1985) Genetic variation in Atlantic cod (*Gadus morhua*) throughout its range. Canadian Journal of Fisheries and Aquatic Sciences 42: 1580–1587

Moriya S, Urawa S, Suzuki O, Urano A, Abe S (2004) DNA microarray for rapid detection of mitochondrial DNA haplotypes of chum salmon. Marine Biotechnology 6: 430–434

Moriya S, Sato S, Azumaya T, Suzuki O, Urawa S, Urano A, Abe S (2007) Genetic stock identification of chum salmon in the Bering Sea and North Pacific Ocean using mitochondrial DNA microarray. Marine Biotechnology 9: 179–191

Moser HG, Richards WJ, Cohen DM, Fahay MP, Kendall, AW Jr, Richardson SL (1984) Ontogeny and systematics of fishes. Special Publication Number 1; American Society of Ichthyologists and Herpetologists; Allen Press, Lawrence, 759 pp

Murgia R, Tola G, Archer SN, Vallerga S, Hirano J (2002) Genetic identification of grey mullet species (Mugilidae) by analysis of mitochondrial DNA sequence: application to identify the origin of processed ovary products (Bottarga). Marine Biotechnology 4: 119–126

Nielsen EE, Hansen MM, Schmidt C, Meldrup D, Grønkjær P (2001) Population origin of Atlantic cod. Nature 413: 272

Nielsen EE, Hansen MM, Meldrup D (2006) Evidence of microsatellite hitch-hiking selection in Atlantic cod (*Gadus morhua* L.): implications for inferring population structure in nonmodel organisms. Molecular Ecology 15: 3219–3229

Nielsen EE; MacKenzie BR, Magnussen E, Meldrup D (2007) Historical analysis of Pan I in Atlantic cod (*Gadus morhua*): temporal stability of allele frequencies in the southeastern part of the species distribution. Canadian Journal of Fisheries and Aquatic Sciences 64: 1448–1455

Noell CJ, Donnellan S, Foster R, Haigh L (2001) Molecular discrimination of garfish *Hyporhamphus* (Beloniformes) larvae in southern Australian waters. Marine Biotechnology 3: 509–514

Nyrén P, Pettersson B, Uhlén M (1993) Solid phase DNA minisequencing by an enzymatic luminometric inorganic pyrophosphate detection assay. Analytical Biochemistry 208: 171–175

O'Connell M, Wright JM (1997) Microsatellite DNA in fishes. Reviews in Fish Biology and Fisheries 7: 331–363

O'Leary DB, Coughlan J, Dillane E, McCarthy TV, Cross TF (2007) Microsatellite variation in cod *Gadus morhua* throughout its geographic range. Journal of Fish Biology 70: 310–335

Ortí G, Petry P, Porto JIR, Jégu M, Meyer A (1996) Patterns of nucleotide change in mitochondrial ribosomal RNA genes and the phylogeny of Piranhas. Journal of Molecular Evolution 42: 169–182

Page RDM, Holmes EC (1998) Molecular evolution: a phylogenetic approach. Blackwell Science, Oxford, UK

Pampoulie C, Steingrund P, Stefánsson MÖ, Daníelsdóttir AK (2008) Genetic divergence among East Icelandic and Faroese populations of Atlantic cod provides evidence for historical imprints at neutral and non-neutral markers. ICES Journal of Marine Science 65: 65–71

Pank M, Stanhope M, Natanson L, Kohler N, Shivji M (2001) Rapid and simultaneous identification of body parts from the morphologically similar sharks *Carcharhinus obscurus* and *Carcharhinus plumbeus* (Carcharhinidae) using multiplex PCR. Marine Biotechnology 3: 231–240

Panserat S, Ducasse-Cabanot S, Plagnes-Juan E, Srivastava PP, Kolditz C, Piumi F, Esquerré D, Kaushik S (2008) Dietary fat level modifies the expression of hepatic genes in juvenile rainbow trout (*Oncorhynchus mykiss*) as revealed by microarray analysis. Aquaculture 275: 235–241

Park LK, Moran P (1994) Developments in molecular genetic techniques in fisheries. Reviews in Fish Biology and Fisheries 4: 272–299

Pauly D, Christensen V, Guénette S, Pitcher TJ, Sumaila U, Walters CJ, Watson R, Zeller D (2002) Towards sustainability in world fisheries. Nature 418: 689–695

Pegg GG, Sinclair B, Briskey L, Aspden WJ (2006) MtDNA barcode identification of fish larvae in the southern Great Barrier Reef, Australia. Scientia Marina 70S2: 7–12

Peplies J, Glöckner FO, Amann R (2003) Optimization strategies for DNA microarray-based detection of bacteria with 16S rRNA-targeting oligonucleotide probes. Applied and Environmental Microbiology 69: 1397–1407

Peplies J, Lau SC, Pernthaler J, Amann R, Glöckner FO (2004) Application and validation of DNA microarrays for the 16S rRNA-based analysis of marine bacterioplankton. Environmental Microbiology 6: 638–645

Perez J, Álvarez P, Martinez JL, Garcia-Vazquez E (2005) Genetic identification of hake and megrim eggs in formaldehyde-fixed plankton samples. ICES Journal of Marine Science 62: 908–914

Pfunder M, Holzgang O, Frey JE (2004) Development of microarray-based diagnostics of voles and shrews for use in biodiversity monitoring studies, and evaluation of mitochondrial cytochrome oxidase I vs. cytochrome *b* as genetic markers. Molecular Ecology 13: 1277–1286

Pogson GH, Fevolden S-E (2003) Natural selection and the genetic differentiation of coastal and Arctic populations of the Atlantic cod in northern Norway: a test involving nucleotide sequence variation at the pantophysin (*Pan* I) locus. Molecular Ecology 12: 63–74

Pogson GH, Mesa KA, Boutilier RG (1995) Genetic population structure and gene flow in the atlantic cod *Gadus morhua*: a comparison of allozyme and nuclear RFLP loci. Genetics 139: 375–385

Pogson GH, Taggart CT, Mesa KA, Boutilier RG (2001) Isolation by distance in the Atlantic cod, *Gadus morhua*, at large and small geographic scales. Evolution 55: 131–146

Poulsen NA, Nielsen EE, Schierup MH, Loeschcke V, Grønkjær P (2006) Long-term stability and effective population size in North Sea and Baltic Sea cod (*Gadus morhua*). Molecular Ecology 15: 321–331

Prober JM, Trainor GL, Dam RJ, Hobbs FW, Robertson CW, Zagursky RJ, Cocuzza AJ, Jensen MA, Baumeister K (1987) A system for rapid DNA sequencing with fluorescent chain-terminating dideoxynucleotides. Science 238: 336–341

Pirrung MC (2002) How to make a DNA chip. Angewandte Chemie International Edition, 41: 1276–1289

Quinteiro J, Sotelo CG, Rehbein H, Pryde SE, Medina I, Pérez-Martín RI, Rey-Méndez M, Mackie| IM (1998) Use of mtDNA direct polymerase chain reaction (PCR) sequencing and PCR-restriction fragment length polymorphism methodologies in species identification of canned tuna. Journal of Agriculture and Food Chemistry 46: 1662–1669

Randall JE (1983) Red Sea reef fishes. Immel Publishing, London, 192 pp

Randall JE (1995) Coastal fishes of Oman. University of Hawaii Press, Honolulu, 432 pp

Randall JE, Allen GR, Steene, RC (1997) Fishes of the Great Barrier Reef and Coral Sea. Crawford House Publishing, Bathurst, 564 pp

Ratnasingham S, Hebert PDN (2007) BOLD: the barcode of life data system (www.barcodinglife. org). Molecular Ecology Notes 7: 355–364

Rehbein H (1990) Electrophoretic techniques for species identification of fishery products. Zeitschrift für Lebensmittel-Untersuchung und -Forschung 191: 1–10

Rehbein H (2005) Identification of the fish species of raw or cold-smoked salmon and salmon caviar by single-strand conformation polymorphism (SSCP) analysis. European Food Research and Technology 220: 625–632

Rehbein H, Kündiger R (2005) FischDB – die Fischdatenbank. Bundesforschungsanstalt für Ernährung und Lebensmittel, Forschungsbereich Fischqualität, Germany; www.fischdb.de

Rehbein H, Mackie IM, Pryde S, Gonzales-Sotelo C, Medinac I, Perez-Martin R, Quinteiro J, Rey-Mendez M (1999) Fish species identification in canned tuna by PCR-SSCP: validation by a collaborative study and investigation of intra-species variability of the DNA-patterns. Food Chemistry 64: 263–268

Rehbein H, Devlin RH, Rüggeberg H (2002) Detection of a genetic alteration and species identification of coho salmon (*Oncorhynchus kisutch*): a collaborative study. European Food Research and Technology 214: 352–355

Relógio A, Schwager C, Richter A, Ansorge W, Valcárcel (2002) Optimization of oligonucleotide-based DNA microarrays. Nucleic Acids Research 30: e51

Richardson DE, Vanwye JD, Exum AM, Cowen RK, Crawford DL (2006) High-throughput species identification: from DNA isolation to bioinformatics. Molecular Ecology Notes 7: 199–207

Richter C, Roa-Quiaoit HA, Jantzen C, Zibdeh M, Kochzius M (2008) Collapse of a new living species of giant clam in the Red Sea. Current Biology 18: 1349–1354

Rocha-Olivares A (1998) Multiplex haplotype-specific PCR: a new approach for species identification of the early life stages of rockfishes of the species-rich genus *Sebastes* Cuvier. Journal of Experimental Marine Biology and Ecology 231: 279–290

Rocha-Olivares A, Moser GH, Stannard J (2000) Molecular identification and description of pelagic young of the rockfishes *Sebastes constellatus* and *Sebastes ensifer*. Fishery Bulletin 98: 353–363

Ronaghi M, Karamohamed S, Pettersson B, Uhlén M, Nyrén P (1996) Real-time DNA sequencing using detection of pyrophosphate release. Analytical Biochemistry 242: 84–89

Ronaghi M, Uhlén M, Nyrén P (1998) A sequencing method based on real-time pyrophosphate. Science 281: 363–365

Rønning SB, Rudi K, Berdal KG, Holst-Jensen A (2005) Differentiation of important and closely related cereal plant species (Poaceae) in food by hybridization to an oligonucleotide array. Journal of Agricultural and Food Chemistry 53: 8874–8880

Rosel PE, Kocher TD (2002) DNA-based identification of larval cod in stomach contents of predatory fishes. Journal of Experimental Marine Biology and Ecology 267: 75–88

Ruzzante DE, Taggart CT, Doyle RW, Cook D (2001) Stability in the historical pattern of genetic structure of Newfoundland cod (*Gadus morhua*) despite the catastrophic decline in population size from 1964 to 1994. Conservation Genetics 2: 257–269

Saiki RK, Scharf S, Faloona F, Mullis KB, Horn GT, Erlich HA, Arnheim N (1985) Enzymatic amplification of β-globin genomic sequences and restriction site analysis for diagnosis of sickle cell anemia. Science 230: 1350–1354

Saiki, RK, Gelfland DH, Stoffel S, Scharf SJ, Higuchi R, Horn GT, Mullis KB, Erlich HA (1988) Primer-directed enzymatic amplification of DNA with a thermostable DNA polymerase. Science 239: 487–491

Saitoh K, Sado T, Mayden RL, Hanzawa N, Nakamura K, Nishida M, Miya M (2006) Mitogenomic evolution and interrelationships of the Cypriniformes (Actinopterygii: Ostariophysi): the first evidence towards resolution of higher-level relationships of the world's largest freshwater-fish clade based on 59 whole mitogenome sequences. Journal of Molecular Evolution 63: 826–841

Sanger F, Nicklen S, Coulson AR (1977) DNA sequencing with chain-terminating inhibitors. Proceedings of the National Academy of Sciences of the United States of America 74: 5463–5467

Santaclara FJ, Cabado AG, Vieites JM (2006) Development of a method for genetic identification of four species of anchovies: *E. encrasicolus*, *E. anchoita*, *E. ringens* and *E. japonicus*. European Food Research and Technology 223: 609–614

Schander C, Willassen E (2005) What can biological barcoding do for marine biology? Marine Biology Research 1: 79–83

Schindel DE, Miller SE (2005) DNA barcoding a useful tool for taxonomists. Nature 435: 17

Seberg O, Humphries CJ, Knapp S, Stevenson DW, Petersen G, Scharff N, Andersen NM (2003) Shortcuts in systematics? A commentary on DNA-based taxonomy. Trend in Ecology and Evolution 18: 63–65

Sevilla RG, Diez A, Norén M, Mouchel O, Jérôme, Verrez-Bagnis V, Pelt H van, Favre-Krey L, Krey G, The Fishtrace Consortium, Bautista JM (2007) Primers and polymerase chain reaction conditions for DNA barcoding teleost fish based on the mitochondrial cytochrome *b* and nuclear rhodopsin genes. Molecular Ecology 7: 730–734

Shendure J, Mitra RD, Varma C, Church GM (2004) Advanced sequencing technologies: methods and goals. Nature Reviews Genetics 5: 335–344

Shi L, Reid LH, Jones WD, Shippy R, Warrington JA, Baker SC, Collins PJ, de Longueville F, Kawasaki ES, Lee KY, Luo Y, Sun YA, Willey JC, Setterquist RA, Fischer GM, Tong W, Dragan YP, Dix DJ, Frueh FW, Goodsaid FM, Herman D, Jensen RV, Johnson CD, Lobenhofer EK, Puri RK, Scherf U, Thierry-Mieg J, Wang J, Wilson M, Wolber PK, Zhang L, Amur S, Bao W, Barbacioru CC, Bergstrom Lucas A, Bertholet V, Boysen C, Bromley B, Brown D, Brunner A, Canales R, Megan Cao XM, Cebula TA, Chen JJ, Cheng J, Chu T-M, Chudin E, Corson J, Corton JC, Croner LJ, Davies C, Davison TS, Delenstarr G, Deng X, Dorris D, Eklund AC, Fan X-H, Fang H, Fulmer-Smentek S, Fuscoe JC, Gallagher K, Ge W, Guo L, Guo X, Hager J, Haje PK, Han J, Han T, Harbottle HC, Harris SC, Hatchwell E, Hauser CA, Hester S, Hong H, Hurban P, Jackson SA, Ji H, Knight CR, Kuo WP, LeClerc JE, Levy S, Li Q-Z, Liu C, Liu Y, Lombardi MJ, Ma Y, Magnuson SR, Maqsodi B, McDaniel T, Mei N, Myklebost O, Ning B, Novoradovskaya N, Orr MS, Osborn TW, Papallo A, Patterson TA, Perkins RG, Peters EH, Peterson R, Philips KL, Pine PS, Pusztai L, Qian F, Ren H, Rosen M, Rosenzweig BA, Samaha RR, Schena M, Schroth GP, Shchegrova S, Smith DD, Staedtler F, Su Z, Sun H, Szallasi Z, Tezak Z, Thierry-Mieg D, Thompson KL, Tikhonova I, Turpaz Y, Vallanat B, Van C, Walker SJ, Wang SJ, Wang Y, Wolfinger R, Wong A, Wu J, Xiao C, Xie Q, Xu J, Yang W, Zhang L, Zhong S, Zong Y, Slikker Jr W (2006) The microarray quality control (MAQC) project shows inter- and intraplatform reproducibility of gene expression measurements. Nature Biotechnology 24: 1151–1161

Sick K (1961) Haemoglobin polymorphism in fishes. Nature 192: 894–896

Sick K (1965a) Haemoglobin polymorphism of cod in the Baltic and Danish Belt Sea. Hereditas 54: 19–48

Sick K (1965b) Haemoglobin polymorphism of cod in the North Sea and the North Atlantic Ocean. Hereditas 54: 49–69

Sigurgíslason H, Árnason E (2003) Extent of mitochondrial DNA sequence variation in Atlantic cod from the Faroe Islands: a resolution of gene genealogy. Heredity 91: 557–564

Skarstein TH, Westgaard J-I, Fevolden S-E (2007) Comparing microsatellite variation in northeast Atlantic cod (*Gadus morhua* L.) to genetic structuring as revealed by the pantophysin (*Pan* I) locus. Journal of Fish Biology 70 (Suppl C): 271–290

Smith LM, Sanders JZ, Kaiser RJ, Hughes P, Dodd C, Connell CR, Heiner C, Kent SB, Hood LE (1986) Fluorescence detection in automated DNA sequence analysis. Nature 321: 674–679

Smith MM, Heemstra PC (1991) Smith's Sea fishes. Southern Book Publishers, Johannesburg, 1048 pp

Smith PJ, Gaffney PM, Purves M (2001) Genetic markers for identification of Patagonian toothfish and Antarctic toothfish. Journal of Fish Biology 58: 1190–1194

Sotelo CG, Calo-Mata P, Chapela MJ, Pérez-Martín RI, Rehbein H, Hold GL, Russell VJ, Pryde S, Quinteiro J, Izquierdo M, Rey-Méndez M, Rosa C, Santos AT (2001) Identification of flatfish (Pleuronectiforme) species using DNA-based techniques. Journal of Agriculture and Food Chemistry 49: 4562–4569

Spies IB, Gaichas S, Stevenson DE, Orr JW, Canino MF (2006) DNA-based identification of Alaska skates (*Amblyraja, Bathyraja* and *Raja*: Rajidae) using cytochromec oxidase subunit I (COI) variation. Journal of Fish Biology 69(Suppl B): 283–292

Stenvik J, Wesmajervi MS, Damsgård B, Delghandi M (2006a) Genotyping of pantophysin I (*Pan* I) of Atlantic cod (*Gadus morhua* L.) by allele-specific PCR. Molecular Ecology Notes 6: 272–275

Stenvik J, Wesmajervi MS, Fjalestad KT, Damsgård B, Delghandi M (2006b) Development of 25 gene-associated microsatellite markers of Atlantic cod (*Gadus morhua* L.). Molecular Ecology Notes 6: 1105–1107

Sunnucks P, Wilson ACC, Beheregaray LB, Zenger K, French J, Taylor AC (2000) SSCP is not so difficult: the application and utility of single-stranded conformation polymorphism in evolutionary biology and molecular ecology. Molecular Ecology 9: 1699–1710

Sweijd NA, Bowie RCK, Evans BS, Lopata AL (2000) Molecular genetics and the management and conservation of marine organisms. Hydrobiologia 420: 153–164

Taggart CT, Ruzzante DE, Cook D (1998) Localised stocks of cod (*Gadus morhua* L.) in the northwest Atlantic: the genetic evidence and otherwise. In: Herbing IH von, Kornfield I, Tupper M, Wilson J (eds) The Implications of localized fishery stocks. Northeast Regional Agricultural Engineering Service, Ithaca, NY, pp 65–90

Tautz D (1989) Hypervariability of simple sequences as a general source for polymorphic DNA markers. Nucleic Acids Research 17: 6463–6471

Tautz D, Arctander P, Minelli A; Thomas RH; Vogler AP (2002) DNA points the way ahead in taxonomy. Nature 418: 479

Tautz D, Arctander P, Minelli A, Thomas RH, Vogler AP (2003) A plea for DNA taxonomy. Trend in Ecology and Evolution 18:70–74

Teletchea F, Laudet V, Hanni C (2006) Phylogeny of the Gadidae (sensu Svetovidov, 1948) based on their morphology and two mitochondrial genes. Molecular Phylogenetics and Evolution 38: 189–199

Tenge B, Dang N-L, Fry F, Savary W, Rogers P, Barnett J, Hill W, Rippey S, Wiskerchen J, Wekell M (1993) The regulatory fish encyclopaedia: an Internet-based compilation of photographic, textural and laboratory aid in species identification of selected fish species. U.S. Food & Drug Administration, U.S.A, www.cfsan.fda.gov/~frf/rfe0.html

Thomas JA, Hardies SC, Rolando M, Hayes SJ, Lieman K, Carroll CA, Weintraub ST, Serwer P (2007) Complete genomic sequence and mass spectrometric analysis of highly diverse, atypical *Bacillus thuringiensis* phage. Virology 368: 405–421

Trotta M, Schönhuth S, Pepe T, Cortesi ML, Puyet A, Bautista JM (2005) Multiplex PCR method for use in real-time PCR for identification of fish fillets from grouper (*Epinephelus* and *Mycteroperca* species) and common substitute species. Journal of Agriculture and Food Chemistry 53: 2039–2045

Upholt WB (1977) Estimation of DNA sequence divergence from comparison of restriction endonuclease digests. Nucleic Acids Research 4: 1257–1265

Upholt WB, Dawid IB (1977) Mapping of mitochondrial DNA of individual sheep and goats: rapid evolution in the D loop region. Cell 11: 571–583

Utter FM (1994) Perspectives of molecular genetics and fisheries into the 21st century. Reviews in Fish Biology and Fisheries 4: 374–378

Vences M, Thomas M, Meijden A van der, Chiari Y, Vieites DR (2005) Comparative performance of the 16S rRNA gene in DNA barcoding of amphibians. Frontiers in Zoology 2: 5, doi:10.1186/1742-9994-2-5

Venkatesh B, Dandona N, Brenner S (2006) Fugu genome does not contain mitochondrial pseudogenes. Genomics 87: 307–310

Wada S, Oishi M, Yamada TK (2003) A newly discovered species of living baleen whale. Nature 426: 278–281

Wang D, Coscoy L, Zylberberg M, Avila PC, Boushey HA, Ganem D, DeRisi JL (2002) Microarray-based detection and genotyping of viral pathogens. Proceedings of the National Academy of Sciences of the United States of America 99: 15687–15692

Wang B, Li F, Dong B, Zhang X, Zhang C, Xiang J (2006) Discovery of the genes in response to white spot syndrome virus (WSSV) infection in *Fenneropenaeus chinensis* through cDNA microarray. Marine Biotechnology 8: 491–500

Ward RD (2000) Genetics in fisheries management. Hydrobiologia 420: 191–201

Ward RD, Grewe PM (1994) Appraisal of molecular genetic techniques in fisheries. Reviews in Fish Biology and Fisheries 4: 300–325

Ward RD, Zemlak TS, Innes BH, Last PR, Hebert PDN (2005) DNA barcoding Australia's fish species. Philosophical Transactions of the Royal Society B 360: 1847–1857

Warren RL, Sutton GG, Jones SJM, Holt RA (2007) Assembling millions of short DNA sequences using SSAKE. Bioinformatics 23: 500–501

Watanabe S, Minegishi Y, Yoshinaga T, Aoyama J, Tsukamoto K (2004) A quick method for species identification of Japanese eel (*Anguilla japonica*) using real-time PCR: an onboard application for use during sampling surveys. Marine Biotechnology 6: 566–574

Watson JD, Crick FHC (1953) A structure for deoxyribose nucleic acid. Nature 171: 737–738

Westgaard J-I, Fevolden S-E (2007) Atlantic cod (*Gadus morhua* L.) in inner and outer coastal zones of northern Norway display divergent genetic signature at non-neutral loci. Fisheries Research 85: 306–315

Westgaard J-I, Tafese T, Wesmajervi MS, Nilsen F, Fjalestad KT, Damsgård B, Delghandi M (2007) Development of ten new EST-derived microsatellites in Atlantic cod (*Gadus morhua* L.) Conservation Genetics 8: 1503–1506

Williams TD, Gensberg K, Minchin SD, Chipman JK (2003) A DNA expression array to detect toxic stress response in European flounder (Platichthys flesus). Aquatic Toxicology 65: 141–157

Wirgin I, Kovach AI, Maceda L, Roy NK, Waldman J, Berlinsky DL (2007) Stock identification of Atlantic cod in US waters using microsatellite and single nucleotide polymorphism DNA analyses. Transactions of the American Fisheries Society 136: 375–391

Wright JM, Bentzen P (1994) Microsatellites: genetic markers for the future. Reviews in Fish Biology and Fisheries 4: 384–388

Yearsley GK, Last PR, Ward RD (2001) Australian seafood handbook (domestic species) - an identification guide to domestic species. FRDC/CSIRO Marine Research, Australia, 469 pp

Yearsley GK, Last PR, Ward RD (2003) Australian seafood handbook (imported species) - an identification guide to imported species. FRDC/CSIRO Marine Research, Australia, 224 pp

Zhang D-X and Hewitt GM (1998) Field collection: Animals. In: Karp A, Isaac PG, Ingram DS (eds) Molecular tools for the screening of biodiversity. Chapman & Hall, London, pp 46–48

Chapter 25
The Uncertain Future of Assessment Uncertainty

Robert Mohn

Abstract Uncertainty in assessments has historically taken a backseat to the point estimates. Managers and industry seem to want "the number" without all the qualifiers and caveats. Recently measurement error and process error have been more commonly reported, but both of these are predicated by models. However, the uncertainty in model choice is much more difficult to quantify. Inter-model comparisons have been done in the past (e.g., Several ICES Methods Working Groups of the 1980s or National Research Council 1998) with the goal of picking the best model, mine *versus* yours. The determination of model uncertainty requires a different paradigm, the simultaneous contemplation of mine *and* yours. Consequently, quantifying model uncertainty requires some replication of work. A fundamental problem in quantifying model uncertainty is determining the universe of plausible models. This approach means that multiple and divergent runs are routinely required. Using multiple models on a single stock is resonant to Rosen's definition of a complex system (Rosen 1985, p 322): "Namely, we are going to define a system to be complex to the extent that we can observe it in non-equivalent ways." Our challenge is to navigate the way through the complexity of resource assessment and develop, utilize, review, and communicate the resultant uncertainty from the consideration of multiple models.

25.1 Introduction

The provision of advice for resource management addresses three principal questions: (1) What is the status of the resource? (2) Where is it going? (3) How sure are we? This manuscript focuses on the third question. To accomplish this, three aspects of uncertainty will be explored: measurement, process, and model uncertainty. These concepts have been defined in a number of ways by various

R. Mohn
Population Ecology Division, Bedford Institute of Oceanography,
Department of Fisheries and Oceans, P.O. Box 1006,
Dartmouth, Nova Scotia, Canada, B2Y 4A2

R.J. Beamish and B.J. Rothschild (eds.), *The Future of Fisheries Science in North America,*   495
Fish & Fisheries Series, © Springer Science + Business Media B.V. 2009

authors, so definitions will be developed here to avoid confusion. I will take a Newtonian-based point of view to develop a framework for these three sorts of uncertainty. A dynamic system may be described by the following equation:

$$x = f(x, a) \qquad (25.1)$$

where x denotes the state variables of interest, a the parameters, and f is some function describing x's dynamics. An example from classical mechanics would be a mass suspended from a spring. The state variable is the position of the mass, the parameter is the stiffness of the spring, and the model is that the force is proportional to the stretching of the spring (Hooke's Law). The measurement error is in determining the position of the mass, the process uncertainty is in ascertaining the stiffness of the spring, and the model uncertainty is whether or not the force is indeed proportional to the distension of the spring. To translate to a fish stock, the biomass, yield effort, and F may be thought of as the state variables of interest and their uncertainty can be thought of as a measurement uncertainty. The parameters are those associated with life history and the fishery, for example natural mortality, growth rates, and selectivity. Many models have been proposed to describe fisheries, and this chapter only investigates single-species models, examples of which can be found in Ricker (1975) and Quinn and Deriso (1999) among others.

To illustrate the proposed multi-model approach, three assessment models will be applied to the Eastern Scotian Shelf cod stock (Mohn et al. 1998, Fanning et al. 2003). The state variables of interest will be constrained to spawning stock biomass (SSB) and the average fishing mortality over the ages of 5–9, (F_{5-9}). These fisheries models use catch and survey data to reconstruct the population. This fishery has been closed since 1993, so there is little catch information in recent years. The stock also appears to have undergone a change in natural mortality which confounds the analysis (Trzcinski et al. 2006). For our purposes, I have chosen simple models and assumed that the natural mortality (M) is constant for all ages and years. Two of the models are age-structured and the third is an age aggregated stage-structured model. For the purposes of illustration, the only process error examined is natural mortality, and each model profiles this parameter. These models were not chosen because they perform particularly well or even would be considered for the provision of advice on this stock. Their purpose is only to show how simultaneous nonequivalent models may be used to gain insights into model uncertainty and how that may be compared to other types of uncertainty. The discussion will comment on model results and their associated uncertainty but also more broadly on the implications of a multiple model analysis.

25.2 Methods

The three models presented are standard assessment models: a virtual population model (VPA), details of which are in Mohn et al. (1998); a non-age-structured (NAS) model, which is described in Collie and Sissenwine (1983); and a version of a stock synthesis (SS) model (Methot, 1990). In all cases the data are from the Eastern Scotian Shelf cod stock (NAFO 4VsW). Data range from 1970 to 2003 for the age-structured models and from 1970 to 2005 for the NAS model. The longer data series was chosen for the NAS model as it should improve the 2003 estimates. Abundance indices are from annual trawl surveys performed in July. This stock was closed to directed fishing in 1993 although some low levels of removals via by-catch are still reported annually.

All the models were run in the ADMB (Fournier 1994) environment. In the VPA, measurement uncertainty was estimated from MCMC generated posteriors of the state variables of interest, spawning stock biomass (SSB), and fishing mortality averaged over ages 5 through 9 (F_{5-9}) in 2003, which is the most recent year shared by the models. The Markov chains were not stable for either the NAS or SS models. More work might have produced stable versions, but for expediency measurement error was quantified using different methods. Although this is not an ideal treatment of measurement error, it can still be used to obtain the overall conclusions about uncertainty. For the NAS model measurement error was estimated from bootstrapping the residuals of the abundance index, whereas for the SS model the estimates from the Hessian approximation of the multivariate normal distribution were used. The only process parameter investigated was natural mortality and it was assumed to be constant over all years and ages. The process uncertainty was displayed as scaled symbols in the SSB-F space reflecting the likelihood for each trail M. The largest symbol denotes M giving the best fit. Decreasing circles are shown down to two likelihood points and those more than two points away are shown as a small square.

The overlay of these models in the same state-space is used to assess model uncertainty and is shown on the same SSB-F plot. A quantification of this error was done in terms of the distance in the SSB dimension. For each model, the mean of the best estimates was calculated and the distance from this mean to each estimate was chosen as the model uncertainty for each model.

The first model was a standard VPA which reconstructs cohorts from the oldest to the youngest age, and the catch was assumed to be known without error (Mohn et al. 1998). The data range from 1970 to 2003 and over ages 1 to 15. The oldest members of each cohort need to be either estimated or derived by model constraints from estimated values. Those parameters directly estimated will be called explicit while those that are inferred from model specification will be termed implicit. The eight explicit parameters are the terminal numbers at age for ages 3 through 10. The model is tuned to survey abundance at age data for ages 3–6 over the 34-year period. The other Fs in the terminal year were constrained with a prespecified selectivity. Similarly, the Fs on the oldest ages were constrained to be a function of a prespecified selectivity pattern. In all models a range of Ms (constant over age and time) was specified to provide a profile to illustrate process error. The measurement error is depicted by integrating the posterior distribution from the mode outward and then finding the contour, which encloses 50% of the distribution. The 50% contour was chosen as it approximates a single-standard deviation in each dimension.

Model	# Explicit parameters	# Observations	Observations/parameter
VPA	8	136	17.0
SS	83	1020	12.3
NAS	72	106	1.47

The second model is a simplified version of the stock synthesis model presented in Trzcinski et al. (2006). This is a forward-projecting model which estimates the initial standing stock and annual recruitment. It also estimates selectivity parameters for the catch and survey. While it has slightly lower ratio of observations per explicit parameter, in fact many of the observations are weakly informative, especially at older ages. The MCMC did not stabilize so the measurement error is taken from the Hessian approximation.

The NAS model used aggregated catch and indices of recruited and recruiting biomass and is based on the age-aggregated model in Collie and Sissenwine (1983). It is essentially a flow balance model in which recruits flow in annually and the outflows are natural mortality and fishing. The catch data are the aggregated numbers from ages 4 to 15. It is tuned to two abundance series: one (the fully recruited series) is the sum of the survey from 4 to 15; the other is the pre-recruitment series which is the survey at age 3. Age 4 was chosen as it approximates the age of the onset of maturity and is fully recruited to the survey gear. It has about an order of magnitude of fewer observations per parameter than the age-structured models under consideration. These resultant population numbers are converted to biomass by multiplying by the mean weight in the survey for ages 4–15. F is found from catch and adult biomass. As such, the SSB average F on the biomass is not exactly comparable with the results from the age-structured models. Because there are few older cod in the period under consideration both the F and B are quite similar to the F_{5-9} and SSB from the age-structured models. Using age 4 for recruitment in the NAS model adds a bit more biomass than SSB from the age-disaggregated models that has maturity set at age 5. On the other hand, the selectivity for the age-disaggregated was at a lower age than 4, so using 4 for the data preparation is deemed to be a reasonable compromise. As the modeled selectivity in the VPA and SS are domed, the F from this model will track lower. Of course, in an assessment or a more rigorous investigation, care would be needed to assure comparability among models. The MCMC did not stabilize for this model; consequently, measurement error was estimated by bootstrapping. A program resampled the residuals from abundance data fits, a method often called conditioned, non-parametric bootstrapping. The distributions were treated as above in that they were integrated from the mode out and the 50% contour determined. Similar to the VPA and SS model, M trajectories in SSB-F space were determined by profiling M over a range but in this case the range was from 0.1 to 0.35 in steps of 0.05. The range was reduced from the other models as the biomass went to unrealistic levels at about $M = 0.35$.

25.3   Results

Figure 25.1 shows the trajectory of the SSB-F for 2003 for a range of Ms from 0.1 to 0.7. The trajectory is seen to be a two-valued function with the upper limb related to lower Ms. To keep the plot less cluttered, measurement uncertainties are only given for Ms of 0.2, 0.5, and 0.7. Each ellipse encloses 50% of the posterior and show a slight negative correlation. The symbol size shows that $M = 0.5$ has the highest likelihood and that 0.4 is within two likelihood units of it. It is interesting to note that a natural mortality of 0.5 is also the point at which the curve doubles back to the right as M increases.

Figure 25.2 shows the same results for the SS model. Again the trajectory is two-valued and the reversal point is near the optimum fit. Unlike the VPA results, the lower limb is now associated with lower values of M. The measurement error is shown as a single-standard deviation on either side of the point estimate and is only given for an M of 0.4. Unlike the case above, that the measurement error is larger relative to the process error (the extent of the M trajectory) than in the VPA results. The process error remains within two likelihood points of $M = 0.4$ when profiling from 0.3 to 0.5. Although the errors are plotted as being orthogonal, they are again slightly negatively correlated.

Figure 25.3 shows the measurement and process error for the NAS model and it follows a hyperbolic shape frequently seen in SSB-F space. The M profile was truncated at 0.35 because the biomass became unrealistically large

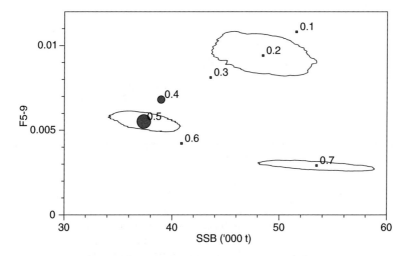

Fig. 25.1 Process and measurement errors from the VPA model. The contours at Ms of 0.2, 0.5, and 0.7 enclose 50% of the posterior distribution to represent measurement error. The largest circle is the most likely level of M and points within two likelihood units are shown as proportionally smaller circles. Points farther than two likelihood points away are shown as dots

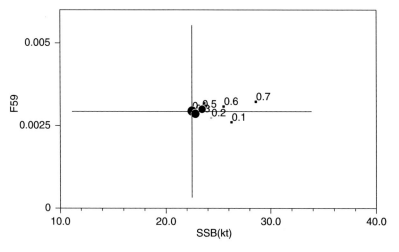

Fig. 25.2 Process and measurement errors from the SS model. The measurement error is shown as single-standard deviations from the Hessian approximation. The largest circle is the most likely level of *M* and points within two likelihood units are shown as proportionally smaller circles. Points farther than two likelihood points away are shown as dots

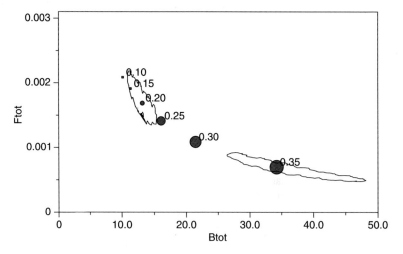

Fig. 25.3 Process and measurement errors from the VPA model. The contours at *M*s of 0.2 and 0.35 enclose 50% of bootstrap distribution to represent measurement error. The largest circle is the most likely level of *M* and points within two likelihood units are shown as proportionally smaller circles. Points farther than two likelihood points away are shown as dots

for higher *M*s. As the MCMC did not stabilize in this model, the measurement error was estimated from bootstrapping and displayed as 50% contours for *M*s of 0.2 and 0.35. These contours follow the same general path as the points in the *M* profile.

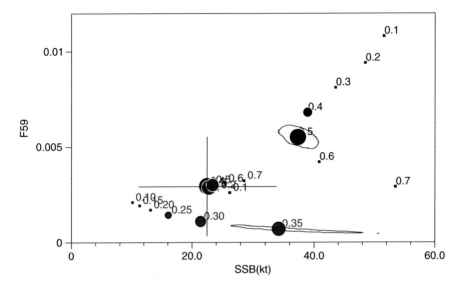

Summary of uncertainties (SSB kt)				
	VPA	SS	NAS	Average
Measurement	6.7	22.7	24.3	17.9
Process	3.5	1.9	22.9	9.4
Model	6.0	8.9	2.9	5.9

Fig. 25.4 Examples of measurement, process, and model uncertainty for three models of ESS cod. Black symbols are for the VPA, blue for SS, and red for the NAS model. The profiles are from Figs. 25.1–25.3 and a measurement uncertainty is given for the best fit on each profile

Figure 25.4 places the previous three figures onto a single plot and retains the measurement error for the Ms with greatest likelihood. Now we can compare measurement, process, and model uncertainties. The model uncertainty is inferred from the spread among model trajectories. Note that the model profiles are separated and at no point in the state–space graph do they overlap.

The following table summarizes the relative size of the measurement, process, and model errors in the SSB dimension. The process error is the SSB distance between the best M and the point that is farthest away but within two likelihood units. The model error is quantified as the distance from the best M in each profile and the mean of them. The mean may not have any relevance except for the sake of a reference for the model uncertainty. This is because it may fall in a region that none of the models would inhabit. The summary could have been done in terms of F distances or even a normalized Euclidean distance, but because of the large discrepancy in F's between the NAS model and the other two it was not included in this summary table.

In this example, the average process uncertainty, was about half of the measuremet error, and model uncertainty was about a third of measurement error. The NAS model was worse than the average in terms of both measurement and process error. Model error cannot be ascribed to any model as it is a system attribute, but the NAS had the smallest contribution to the model uncertainty.

25.4 Discussion

Although this manuscript is a plea for giving more attention to the quantification and communication of uncertainty in general, it has focused mainly on the inclusion of model uncertainty. More attention should be given to the estimation and communication of all sources of uncertainty. The multiple model method was chosen to illustrate and compare the types of uncertainty. This is a fairly simple proposed method, but it does require more work to execute and more work to review. This general approach provides better depictions of the sources of uncertainty and a more consistent basis for comparison among model systems. Although many of the clients of resource assessments do not appear to want more of the uncertainties to be communicated, presumably because it is ignored in, or complicates, the decision process, it is our responsibility and in the public's interest to do so anyhow. The third question "How sure are we?" cannot be ignored. The future of quantifying uncertainty is that it will require more resources to do it well, but it is increasingly needed to avoid what appears to be increasingly common undesired effects of fishing on the productivity of a fishery.

The proposed approach is in contrast to model averaging and to some degree to formal or even informal model selection. The proposal is that authors and the review process consider diverse models and then choose the best one as a "base case." In addition, they need to retain and communicate the divergence among predictions. Some degree of subjectivity is left in this approach, but can be minimized by a careful description of the criteria by which the base model is chosen

The 4VsW cod example used to illustrate the estimation of measurement, process, and model uncertainties presented some unexpected results. An unexpected result was the two-valued (folded) trajectories of the M-profiles for the VPA and SS models in SSB-F space. Even more unexpected was that one trajectory was inverted compared to the other. Process uncertainty could have been estimated as the uncertainty in M directly in AD Model builder, but the trajectories give some insight into the model behavior as well as quantifying the uncertainty. When M was estimated in the VPA model it was found to be 0.46 with a CV of about 6%, but this is an incomplete description of this process uncertainty. When the SSB-F contour was compared between variable M and constant M models, it had no influence on the SSB dimension, but stretched the F dimension by about 20%. The M trajectory and contour information complement the more traditional measures of uncertainty such as the CV.

We have used three different methods to estimate the measurement uncertainty, partially because the MCMC failed to stabilize in two of the models. In the spirit of using multiple models for the same calculation, all three were applied to the VPA model. The Hessian and MCMC estimates of the uncertainty in M were quite similar, both being about 7%, while the bootstrapping estimate was considerably larger at 18%. If this is a general bias of bootstrapping some correction will have to be made when it is compared to other methods. Also, the measurement errors in this case study for the NAS based on bootstrapping may also be biased upwards.

It is difficult to know how to span the space of plausible models. What is its extent; is it continuous? The more divergent the models are, the more likely they are to explore wider regions of the space of plausible conceptual models. Simply changing the steepness of the stock recruit relationship or changing a selectivity from an asymptotic to a domed form is unlikely to yield much information, except the explicit sensitivity to such a simple change. As the multiple model approach is applied to more stocks, it may result in the definition of a small set of trial models or model types for a first investigation.

In addition to yielding insights into uncertainty, the use of multiple models helps fight a reluctance among assessment teams to adopt new models. Often groups will develop assessment tools, develop a protective mentality, and hold on to them for too long. Then they suddenly switch to a new model in the face of overwhelming evidence – sort of a punctuated equilibrium. If more models were routinely used on individual stocks, the adoption and rejection of models would be met with less resistance. Furthermore, the task of reviewing the work of other institutions would be easier because of increased familiarity with a number of methods and the higher probability of find methods with which one has experience.

I realize the resource implications of multiple models for individual assessments – more work to do and more results to communicate. Furthermore, often our clients often do not want to see all uncertainties, it just makes things more difficult to assimilate. They feel it is our job to pick the best model (or base case) as part of the assessment preparation. While this is true, it is also our obligation to include the uncertainty.

There is a paradox with the quantification of uncertainty. The more uncertainty we quantify and hence the more that is known about the resource, the larger the clouds grow around each point and process. This is because some of the uncertainty from the unquantifiable category which cannot be depicted has been added to the quantifiable. Measurement and process errors can only be reduced by more and/or better data. Including more sources of error in the analysis will tend to increase the apparent uncertainty, but it more completely answers the questions about stock status.

The examples in this study are for fished single-species models. There is increasing interest in ecosystem modeling. The multiple model approach is still applicable but will probably be much more difficult, but will likely be more important as well. Consider an example from Trzinscki et al. (2006) of a fished two-species model (cod and seal). Figure 25.5 shows the CVs of the seal population, its energetics, and finally the consumption of cod. The interaction term has a CV of about 150% in recent years. In light of this magnitude, a complete description of the various sources of uncertainty, including the model uncertainty, is going to be important. The component cod, seal, and interaction models are all candidates for alternative models. Finding dimensions for representing the contributions of the various sources of uncertainty analogous the SSB-F space used in this study will be a challenge, let alone exploring that space. These issues will be exacerbated in multispecies or ecosystem models.

Resource management is a complex system in the sense of Rosen and needs to be described in nonequivalent ways. This view embraces a quantification of the model uncertainty in terms similar to measurement and process uncertainty and some other benefits as well. Yogi Berra is attributed with "If you come to a fork in

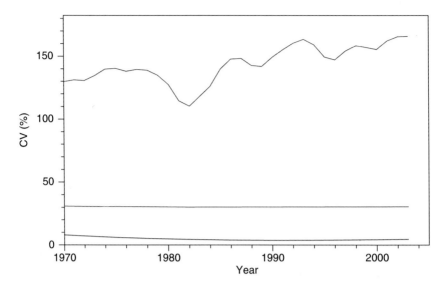

Fig. 25.5 Annual estimates of the coefficient of variation for the gray seal–cod interaction taken from Trzcinski et al. (2006.) The black line is the uncertainty in the estimation of the seal population. The blue line is the combined uncertainty of the seal population and seal energetics. The red line is seal population, energetics, and cod consumption

the road, take it." The multi-model approach may be described as "If you come to a fork in the road, take them both." Or better yet, take them all.

References

Collie, J.S. and M.P. Sissenwine. 1983. Estimation of population size from relative abundance data measured with error. Can. J. Aquat. Fish. Sci. 401871–401879.

Fanning, L.P., R.K. Mohn and W.J. MacEachern. 2003. Assessment of 4VsW cod to 2002. DFO-CSAS Res. Doc. 2003/027. 41 pp.

Fournier, D. 1994. An introduction to AD Model Builder for use in nonlinear modeling and statistics. Otter research. 51 pp.

Methot, R.D. 1990. Synthesis model: an adaptable framework for analysis of diverse stock assessment data. Int. North Pac. Comm. Bull. 50:259–277.

Mohn, R.K., L.P. Fanning and W.J. Maceachern. 1998. Assessment of 4VsW cod in 1997 incorporating additional sources of mortality. Can. Stock Asses. Sec. Res. Doc. 98/78. pp. 49.

National Research Council. 1998. Improving Fish Stock Assessments. National Academy Press. Washington DC. pp. 177.

Quinn, T.J. II and R.B. Deriso. 1999. Quantitative Fish Dynamics. Oxford Univ. Press. New York.

Ricker, W.E. 1975.Computation and Interpretation of Biological Statistics of Fish Populations. Fish. Res. Bd. Can. Bull. No. 191. 382 pp.

Rosen, R. 1985. Anticipatory Systems: Philosophical, Mathematical and Methodological Foundations. Pergamon Press. Oxford. pp. 436.

Trzcinski, M.K., R. Mohn and W.D. Bowen. 2006. Continued decline of an Atlantic cod population: how important is grey seal predation. Ecol. Appl. 16:2276–2292.

Chapter 26
Model Selection Uncertainty and Bayesian Model Averaging in Fisheries Recruitment Modeling

Yan Jiao, Kevin Reid, and Eric Smith

Abstract Stock recruitment (SR) modeling is the central part of fisheries population dynamics analysis. Models and modeling techniques on SR relationships have been evolving for decades, and have moved from traditional SR models, such as the Ricker and Beverton-Holt models to measurement error models, and Kalman filter time series models. Though SR models are evolving, people still typically select a specific model and then proceed as if the selected model had generated the data. This approach ignores the uncertainty in the model selection, leading to overconfident inferences and decisions with higher risk than expected. Bayesian model averaging (BMA) provides a coherent mechanism for accounting for this model uncertainty. In this study, Lake Erie walleye (Sander vitreus) fishery was used as an example. Six mathematical models were developed, which included a Ricker model, a hierarchical Ricker model, a residual auto-regressive model, a Kalman filter random walk model, a Kalman filter autoregressive model, and a Ricker measurement error model. The posterior distributions of estimated productivity and recruitment from these models were weighted based on the Deviance Information Criterion (DIC) to provide our predictive posterior distribution of population productivity and recruitment over time. To test the efficiency of the Bayesian averaging approach and the uncertainty from model selection, a further simulation study was done based on the example fishery. Our results showed that model selection uncertainty is high and BMA explained the data reasonably well. We suggest that BMA is more appropriate in simulating SR models. The framework developed here can be used for other species population SR analysis. We also suggest that the model selection uncertainty be considered and the

Y. Jiao(✉)
Department of Fisheries and Wildlife Sciences, Virginia Polytechnic
Institute & State University, Blacksburg, VA, 24061-0321, USA
e-mail: yjiao@vt.edu

K. Reid
Ontario Commercial Fisheries Association, Box 2129, 45 James Street,
Blenheim, ON, N0P 1A0, Canada

E. Smith
Department of Statistics, Virginia Polytechnic Institute & State University,
Blacksburg, VA 24061, USA

R.J. Beamish and B.J. Rothschild (eds.), *The Future of Fisheries Science in North America*,   505
Fish & Fisheries Series, © Springer Science + Business Media B.V. 2009

BMA be applied to other stock assessment models and even in the fisheries management decision making in the future.

Keywords Bayesian model averaging · Deviance Information Criterion · recruitment · time series · walleye

26.1 Introduction

Stock recruitment (SR) modeling is the central part of fisheries population dynamics analysis. However, it is also the bottleneck of many fisheries population dynamics and stock assessment. Concerns about SR modeling have never decreased. Models and modeling techniques have been evolving for decades, moving from traditional SR models, such as Ricker model and Beverton-Holt (BH) model, generalized Ricker model with environmental variables, to measurement error models, to autoregressive error models, and then extended to other Kalman-filter time series models (Walters and Ludwig 1981; Quinn and Deriso 1999; Peterman et al. 2003). Though SR models are evolving, the typical analysis is to select a single model from some class of models and then analysis proceeds as if the selected model had generated the data. This approach ignores the uncertainty in the model selection, leading to overconfident inferences and decisions that are more risky than one thinks (Draper 1995). Bayesian model averaging (BMA) provides a coherent mechanism for accounting for this model uncertainty. Framework development and application of BMA in SR modeling is necessary and will help SR modeling and fishery stock assessment and management.

In this study, a walleye (*Sander vitreus*) fishery from Lake Erie is used as an example. Its dynamic has been observed to be high and may be heavily influenced by environmental changes (Walleye Task Group 2004). In some years, the productivity can be very high while in some other years it can be very low in spite of the spawner stock size. A commonly used stock recruitment model obviously cannot satisfy the mission in modeling the dynamic variations of productivity and/or the environmental noise.

Traditional SR models and model selection for analyzing the SR relationship and the productivity of a stock can be inadequate because of the measurement errors in the SR data and/or the noisy signals transferred to the SR data. Besides a commonly used SR regression model, we included five other mathematical models developed based on available time series data for the example fishery, which includes a measurement error model, an autoregressive residual model, a random walk model, a Kalman filter autoregressive productivity model, and a hierarchical Ricker model (see the method section). Measurement error can be very important in analyzing the SR relationship because of the possible measurement error in the spawner stock size (Walters and Ludwig 1981). Recent research found that noise in the nature may not be white, but are colored in many cases (Caswell and Cohen 1995; Halley 1996; Vasseur and Yodzis 2004). The classification of noise by spectral density is given "color" terminology, with different types named after different colors. It can also be classified as how the noise signal is temporally autocorrelated. Spectral density is constant for white noise and the noise signal is independent; while spectral density for colored noise is changing with changing frequency and the noise signal is autocorrelated (Halley 1996; Petchey 2000). These colored noises can be more dangerous to species when population size is low (Halley and Kunin 1999; Morales 1999; Schwager et al. 2006). We developed these four models to simulate the possible colored noises in the example fishery. The autoregressive residual and population productivity models were used to simulate the colored noise for the population itself and for population productivity (Morales 1999; Schwager et al. 2006). The random walk model is a special case of the autoregressive population productivity model and is reasonable to investigate because it has fewer parameters and can simulate the nonwhite noise at the same time (Peterman et al. 2003). The hierarchical Ricker model was used to simulate the hierarchy of possible productivities, which has been discussed related to regime shifts, changes of productivity regimes, etc. (Beamish et al. 1999; Glantz 1992).

We used a Bayesian approach to analyze the time series models for population productivity and recruitment. WinBUGS was used for this purpose (Spiegelhalter et al. 2004). A Bayesian approach has been very useful in dealing with time series model, which uses observations to update prior models of process noise, measurement noise, and state variables. Deviance Information Criterion (DIC) was used to compare different models. It is difficult to say what constitutes an important difference in DIC (Spiegelhalter et al. 2004). After the Bayesian analysis of each model, a Bayesian model averaging was used to balance model goodness of fit and model selection uncertainty. The posterior distributions of estimated productivity and recruitment from these models were weighted based on their DIC to provide our predictive posterior distribution of population productivity and recruitment over time.

To test the efficiency of the Bayesian averaging approach in modeling SR data and the uncertainty from model selection, a further simulation study was done based on the example walleye fishery. Our goal in this study is to assess the model selection uncertainty, and to find an appropriate method for understanding the productivity and the SR relationship of different fisheries. The framework developed here can be used for other species productivity and SR analysis.

Although six models were used in this study, they have included most of the types of the recruitment models. Models can be exchanged or modified to other similar models, e.g., BH model can replace Ricker model, hierarchical BH model can replace the hierarchical Ricker model, etc. The BMA framework can also extend from six to seven or, however, many models are of interest. The new recruitment modeling approach developed in this study improved our understanding of recruitment dynamics as well as raised the quality of stock assessment and management of the fisheries.

26.2   Materials and Methods

26.2.1   Example Fishery Used in This Study

The recruitment of walleye is the 2-year-old walleye in Lake Erie. Walleye older than age 3 are regarded as the spawning stock. The walleye recruitment (2-year-old fish) and spawning stock (3 + fish) are estimated using a statistical catch-at-age method (Fig. 26.1, Walleye Task Group 2004).

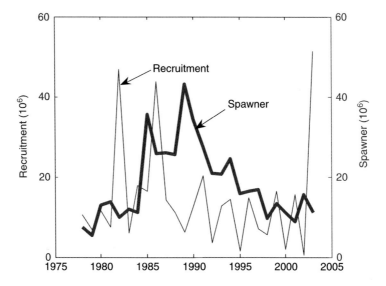

Fig. 26.1 Stock and recruitment of walleye fisheries over time

26.2.2 Models Used

According to the stock recruitment pattern (Fig. 26.2), Ricker model (RM) was selected to fit the walleye stock recruitment data. The Ricker model can be written as:

$$R_t = \alpha S_{t-2} e^{-\beta S_{t-2}^{\varepsilon_1}} \text{ or}$$
$$Ln(R_t) = Ln(\alpha) + Ln(S_{t-2}) - \beta S_{t-2} + \varepsilon_1 \tag{26.1}$$

where R_t is the recruitment numbers and S_t is the spawner abundance at year t, error ε_1 is independent and normally distributed with mean 0 and variance $\sigma_{\varepsilon_1}^2$. t ranges from 1978 to 2003.

The second model that we used was

$$Ln(R_t) = Ln(\alpha) + Ln(S'_{t-2}) - \beta S'_{t-2} + \varepsilon_2$$
$$Ln(S) = Ln(S') + \varepsilon_3 \tag{26.2}$$

In this model, measurement error in spawner abundance S is considered; S' is the true spawner abundance, and S is the measurement of S' with error ε_3. Errors ε_2 and ε_3 are independent and normally distributed with mean 0 and variance $\sigma_{\varepsilon_2}^2$ and $\sigma_{\varepsilon_3}^2$. We called this model the measurement error model (MEM).

The third model that we used was

$$Ln(R_t) = Ln(\alpha) + Ln(S_{t-2}) - \beta S_{t-2} + u_t$$
$$u_{t+1} = \phi u_t + \varepsilon_4 \tag{26.3}$$

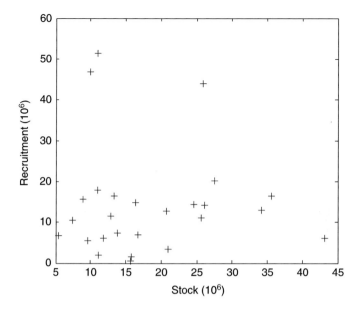

Fig. 26.2 Spawner stock and the corresponding recruitment

In this model, the residual error u_t is modeled as a first-order autoregressive process. ϕ is the autocorrelation coefficient, and the error ε_4 is independent and normally distributed with mean 0 and variance $\sigma_{\varepsilon_4}^2$. We called this model the residual autoregressive model (RAM).

The forth model that we used was

$$Ln(R_t) = Ln(\alpha_t) + Ln(S_{t-2}) - \beta S_{t-2} + \varepsilon_5$$
$$Ln(\alpha_{t+1}) = Ln(\alpha_t) + \varepsilon_6 \qquad (26.4)$$

In this model, productivity α_t is modeled as a random walk process; and errors ε_5 and ε_6 are independent and normally distributed with mean 0 and variances $\sigma_{\varepsilon_5}^2$ and $\sigma_{\varepsilon_6}^2$. We called this model the random walk model (RWM).

The fifth model that we used was

$$Ln(R_t) = Ln(\alpha_t) + Ln(S_{t-2}) - \beta S_{t-2} + \varepsilon_7$$
$$Ln(\alpha_{t+1}) = Ln(\bar{\alpha}) + \varphi[Ln(\alpha_t) - Ln(\bar{\alpha}) + \varepsilon_8 \qquad (26.5)$$

where productivity α_t is modeled as a first-order autoregressive process, and ϕ is the autocorrelation coefficient. We called this model the Kalman filter autoregressive model (KFAM).

The sixth model that we used was

$$Ln(R_t) = Ln(\alpha_t) + Ln(S_{t-2}) - \beta S_{t-2} + \varepsilon_9$$
$$\alpha_t \in N(a,b) \qquad (26.6)$$
$$a \in U(c,d)$$

Table 26.1 Notations used in this paper

Symbols	Meaning
t	Year when recruitment data is measured
R_t	The observed recruitment in year t
S_t	The observed spawner stock size in year t
α_t	Productivity parameter in the SR models
β	Density-dependent parameter in the SR models
$\sigma_{\varepsilon_1}^2$ to $\sigma_{\varepsilon_9}^2$	Variance of residual errors in different models
ϕ	Autocorrelation coefficient in the residual autoregressive model
φ	Autocorrelation coefficient in the K-F autoregressive model
a	Mean of α in the hierarchical Bayesian model
b	Variance of α in the hierarchical Bayesian model
c	Lower bound of the uniform distribution of a
d	Upper bound of the uniform distribution of a

where error ε_9 is independent and normally distributed with mean 0 and variance $\sigma_{\varepsilon_9}^2$. α_t is modeled to follow a hierarchical distribution, i.e., α_t follows a normal distribution $N(a, b)$ with mean a and variance b; a, the mean of α follows a uniform distribution between c and d. The $N(a, b)$ distribution is truncated to make sure that α has positive values. This is a Bayesian hierarchical model, so we called it the hierarchical Ricker model (HRM).

The above models were used to predict the recruitment and to determine the productivity. A summary of the notation used in this chapter can also be found in Table 26.1.

26.2.3 Bayesian Method and Priors

A Bayesian method was used to estimate the parameters in these models. WinBUGS software was used. WinBUGS is numerically intensive software package that implements general Bayesian models using "Metropolis-Hasting within Gibbs sampling" (Gilks 1996; Spiegelhalter et al. 2004). Bayesian implementation of these models requires specification of prior distributions on all unobserved quantities. In general, noninformative priors (here, wide uniform distribution) were used for variances $\sigma_{\varepsilon_1}^2$, $\sigma_{\varepsilon_2}^2$, and so on.

A uniform distribution was used for the prior of $Ln(\alpha)$, i.e., $U(Ln(\alpha_{min})$, $Ln(\alpha_{max}))$, where α_{min} was determined as the $Min(R_t / S_{t-2})$ and α_{max} was determined as the $Max(R_t / S_{t-2})$.

Equation (26.1) can be written as $\beta = [Ln(\alpha) + Ln(S_{t-2}) - Ln(R_t)]/S_{t-2}$. A uniform distribution was used for the prior of β, i.e., $U(0.0001, \beta_{max})$, where β_{max} was determined as the $Max([Ln(\alpha) + Ln(S_{t-2}) - Ln(R_t)]/S_{t-2}) = Max([Ln(\alpha_{max}) + Ln(S_{t-2}) - Ln(R_t)]/S_{t-2})$.

Priors for ϕ and φ were assumed to be between -1 and 1 with uniform distributions. A lognormal distribution was used for the prior of the spawner stock size with measurement error, i.e., $Ln(S) \sim N(Ln(\bar{S}_t), var(Ln(S_t)))$, where $var(Ln(S_t))$ is the variance of the $Ln(S_t)$.

Table 26.2 Priors used for the Bayesian time series models and their posterior median, and standard deviation (Std) and the Deviance Information Criteria (DIC)

Model	Parameters	Prior	Median	Std	DIC
Ricker model	$Ln(\alpha)$	$U(-3.32, 1.55)$	0.29	0.45	198.45
	β	$U(0.0001, 0.31)$	0.04	0.02	
	$\sigma_{\varepsilon_1}^2$	$U(0.0001, 10)$	1.24	0.43	
Measurement error model	$Ln(\alpha)$	$U(-3.32, 1.55)$	0.22	0.46	327.70
	β	$U(0.0001, 0.31)$	0.04	0.02	
	$Ln(S_j)$	$N(2.78, 0.27)$	2.81	0.21	
	$\sigma_{\varepsilon_2}^2$	$U(0.0001, 10)$	1.93	1.71	
	$\sigma_{\varepsilon_3}^2$	$U(0.0001, 10)$	1.20	0.42	
Residual auto-regressive model	$Ln(\alpha)$	$U(-3.32, 1.55)$	0.10	0.35	197.35
	β	$U(0.0001, 0.31)$	0.04	0.02	
	ϕ	$U(-1, 1)$	-0.42	0.24	
	$\sigma_{\varepsilon_4}^2$	$U(0.0001, 10)$	1.17	0.41	
Random walk model	$Ln(\alpha_t)$	$U(-3.32, 1.55)$	-0.1–0.51^*	0.47–0.59	200.38
	β	$U(0.0001, 0.31)$	0.04	0.02	
	$\sigma_{\varepsilon_5}^2$	$U(0.0001, 10)$	1.16	0.43	
	$\sigma_{\varepsilon_6}^2$	$U(0.0001, 10)$	0.06	0.11	
Kalman filter auto-regressive model	$Ln(\bar{\alpha})$	$U(-3.32, 1.55)$	0.24	0.24	178.35
	β	$U(0.0001, 0.31)$	0.04	0.01	
	φ	$U(-1, 1)$	-0.51	0.43	
	$\sigma_{\varepsilon_7}^2$	$U(0.0001, 10)$	0.54	0.46	
	$\sigma_{\varepsilon_8}^2$	$U(0.0001, 10)$	0.43	0.37	
Hierarchical Ricker model	$Ln(\alpha_t)$	$N(a, b)$	-0.64–0.62^a	0.53–0.77	192.99
	β	$U(0.0001, 0.31)$	0.03	0.02	
	$\sigma_{\varepsilon_9}^2$	$U(0.0001, 10)$	0.90	0.48	
	a	$U(-3.32, 1.55)$	0.10	3.68	
	b	$U(0.0001, 10)$	0.31	0.36	

[a] See Fig. 26.4 for the mean values

Besides the above priors, noninformative uniform distributions were used for other variance parameters. Although a common choice of inverse-gamma distribution for variance parameters we used uniform distributions as these have better properties with multilevel models (Gelman 2005). A summary of the priors used in the models above can be found in Table 26.2.

26.2.4 Convergence Diagnostics

A critical issue in using Markov Chain Monte Carlo (MCMC) methods is how to determine when random draws have converged to the posterior distribution. Here, three methods were considered: monitoring the trace, diagnosing the autocorrelation plot, and Gelman and Rubin statistics (Spiegelhalter et al. 2004). In this study, three chains were used. After several sets of analysis, for each chain, the first 20,000 iterations with a thinning interval of 5 were discarded, and another 50,000 iterations were used in the Bayesian analysis.

26.2.5 Model Selection and Bayesian Model Averaging

The DIC is used as a model selection criterion in this study.

$$DIC = 2\bar{D} - \hat{D} \, or \, \bar{D} + p_D$$
$$D(y,\theta) = -2\log \text{Likelihood}(y|\theta) \qquad (26.7)$$
$$p_D = \bar{D} - \hat{D}$$

D is the deviance, a measurement of prediction goodness for our models. p_D is called the effective number of parameters in a Bayesian model. The DIC is a hierarchical modeling generalization of the Akaike information criterion (AIC) and Bayesian information criterion (BIC), also known as the Schwarz criterion. It is particularly useful in Bayesian model selection problems, where the posterior distributions of the models have been obtained by MCMC simulation. Like AIC and BIC it is an asymptotic approximation as the sample size becomes large. It is only valid when the posterior distribution is approximately multivariate normal (Spiegelhalter et al. 2002, 2004).

After the above models were solved using Bayesian approaches. DIC values were used to compare different models and to weight different model outcomes when averaging the posterior distribution of the averaged model result (Hoeting et al. 1999; Spiegelhalter et al. 2004). Both MATLAB and WinBUGS languages were used for this purpose. Equation (26.8) showed how the weight of each model is decided, where k is the kth model.

$$\Delta_{DIC_k} = DIC_k - \min(DIC)$$
$$weight_k = e^{-2\Delta_{DIC_k}} \bigg/ \sum_i e^{-2\Delta_{DIC_k}} \qquad (26.8)$$

26.3 Simulation Study

A simulation study was designed to test the performance of the proposed Bayesian model averaging approach and evaluate model selection uncertainty. The following simulation algorithm was used: (1) estimate population productivity parameter $Ln(\alpha)$ and the other parameters from these models using the example species; treat these estimates as true parameters; (2) generate data with uncertainties using a Monte Carlo simulation approach with the uncertainty levels the same as the "true" uncertainties (variances here) estimated from the example population; (3) analyze the generated data set by using different models (the six above, always pick the best model based on model selection criteria and a model averaging approach); (4) evaluate the uncertainty arising from model selection and the performance of Bayesian model averaging (see Table 26.3 for the simulation results).

Procedures (2)–(4) described above were repeated for 500 times to yield 500 sets of estimated population productivity and recruitment over time from each of the models. The sum of square errors (SSE) for log-transformed population productivity in the ith simulation $Ln(\hat{\alpha}_i)$ can be calculated as

$$\text{SSE}(Ln(\hat{\alpha}_i)) = \sum_{t=1}^{k} [Ln(\hat{\alpha}_{t,i}) - Ln(\alpha_t')]^2 \qquad (26.9)$$

where i indicates the ith simulation run and k is the numbers of years. The $\hat{\alpha}_{t,i}$ is the estimated α in year t and in the ith simulation; α_t' is the "true" α in year t. The relative estimation error (REE) for estimated log-transformed population productivity in the ith simulation, $\text{REE}(Ln(\hat{\alpha}_i))$, was used here, which was calculated as

$$\text{REE}(Ln(\hat{\alpha}_i)) = \sum_{t=1}^{k} \{[Ln(\hat{\alpha}_{t,i}) - Ln(\alpha_t')] / Ln(\alpha_t')\}^2 \qquad (26.10)$$

The REE calculated in Equation (26.10) measures the overall estimation errors including both estimation biases and variations in estimates among the 500 simulation runs. A boxplot was used to summarize the REE_s derived in the 500 simulation runs. An estimation procedure with small REE suggests that it performs well and tends to have smaller estimation error in estimating population productivity and recruitment. The same method was used to calculate $\text{REE}(Ln(\hat{R}_i))$, relative estimation error for estimated log-transformed recruitment in the ith simulation.

Model selection uncertainty was evaluated through the probability of choosing the true model. For example, when the Ricker model was used as the true model, in each of these 500 runs, the simulation algorithm would pick up the best model based on the DIC values (smallest DIC means the best model); the best model would be recorded in each of the simulation runs. After the 500 runs, the probability of each model chosen as the best model was counted. For example, if the Ricker model was chosen as the best model in 100 of 500 runs, then the probability is 20%.

After the simulation study with standard deviation $\sigma_{\varepsilon_{1:9}}$ used to define the "true" uncertainty levels, a further simulation study with uncertainty levels of 30% of the true uncertainty level were developed, i.e., the 30% of the $\sigma_{\varepsilon_{1:9}}$. The purpose of this study was to investigate whether the model selection uncertainty was controlled by the uncertainty levels.

26.4 Results

The posterior means of the fitted recruitment were different when different models were used (Fig. 26.3). The KFAM, the HRM, and the RWM tended to describe the data better than the other models. The posterior mean and medians

Table 26.3 Relative estimation error (REE) of population productivity estimates and the probability of being the best model among 500 simulation runs when the "true" error values were used. CI is the posterior credible interval

True model used to generate data with uncertainty	models used to fit the simulated data	Estimates						
		REE			DIC			P^*
		Median	5% CI	95% CI	Median	5% CI	95% CI	
Ricker model (RM)	RM	**3.58**	**0.29**	**8.54**	**200.33**	**176.63**	**214.17**	**0.024**
	MEM	3.71	0.38	8.66	325.08	301.98	339.08	0.000
	RAM	3.53	0.28	8.83	202.94	177.97	216.61	0.008
	RWM	5.30	2.20	10.83	201.93	177.59	215.96	0.006
	KFAM	6.19	4.73	9.37	194.80	164.24	210.67	0.636
	HRM	5.75	3.11	10.92	196.53	165.72	211.99	0.326
Measurement error model (MEM)	RM	5.58	0.53	12.62	201.93	179.80	214.96	0.068
	MEM	**5.20**	**0.57**	**11.54**	**333.98**	**306.40**	**352.30**	**0.000**
	RAM	5.23	0.59	12.63	203.79	180.45	217.44	0.012
	RWM	6.30	2.89	13.28	203.59	179.84	218.15	0.000
	KFAM	8.30	6.51	12.22	196.41	166.22	212.43	0.638
	HRM	6.49	4.24	13.63	198.02	166.01	213.58	0.282
Residual autoregressive model (RAM)	RM	7.05	0.76	19.04	211.84	193.35	221.37	0.000
	MEM	6.79	0.71	18.81	336.24	316.71	345.85	0.000
	RAM	**6.55**	**0.62**	**14.89**	**197.99**	**182.91**	**207.35**	**0.292**
	RWM	6.62	2.53	14.30	215.59	197.09	225.04	0.000
	KFAM	29.73	17.95	43.22	193.05	161.90	207.96	0.698
	HRM	12.52	7.99	20.49	210.02	187.35	220.81	0.010

							P^*
Random walk model (RWM)	RM	11.60	2.81	95.23	213.09	139.61	254.60	0.100
	MEM	11.55	2.81	95.41	338.01	264.26	376.72	0.000
	RAM	10.67	2.72	83.92	214.62	139.33	252.96	0.056
	RWM	**7.10**	**2.34**	**39.66**	**212.39**	**131.73**	**254.03**	**0.068**
	KFAM	10.01	2.28	58.82	207.87	126.68	250.59	0.612
	HRM	8.91	2.62	56.00	211.19	129.51	255.67	0.164
K-F autoregressive model (KFAM)	RM	12.46	5.09	70.69	201.02	178.88	213.09	0.008
	MEM	11.79	5.09	69.76	325.63	302.34	337.75	0.000
	RAM	12.56	5.07	68.56	202.01	180.91	214.11	0.004
	RWM	10.16	5.15	47.60	203.69	180.56	216.17	0.000
	KFAM	**13.45**	**5.26**	**57.97**	**193.27**	**165.73**	**208.13**	**0.740**
	HRM	10.26	4.66	45.35	196.27	166.85	210.91	0.248
Hierarchical Ricker model (HRM)	RM	24.52	8.51	108.88	251.45	227.92	262.94	0.522
	MEM	23.87	8.20	105.28	375.70	350.27	387.97	0.000
	RAM	21.78	7.80	98.66	253.05	227.69	264.36	0.066
	RWM	13.94	5.38	59.08	254.47	229.20	266.02	0.004
	KFAM	15.24	5.63	70.86	251.60	222.43	263.41	0.390
	HRM	**13.05**	**4.99**	**57.86**	**252.30**	**227.04**	**264.22**	**0.018**

P^*: probability of being the best model

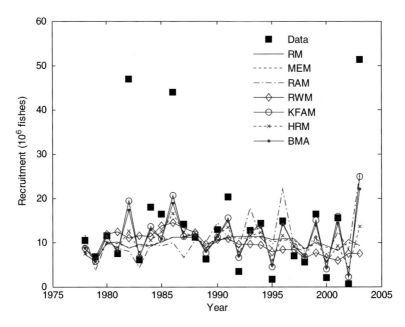

Fig. 26.3 Recruitment data and the estimated recruitment using different models and the Bayesian model averaging results

of the population productivity and the other parameters were different when different models were used (Fig. 26.4 and Table 26.2). The estimated productivity over time from the RWM model showed shift in two regimes. Before mid-1980s, productivity was high but decreased entering into the 1990s. The estimated productivity over time from the KFAM and HRM showed shift in two regimes. Before mid-1980s, productivity was high but decreased entering into the 1990s. This approach also showed that the variation in the productivity is increasing (Fig. 26.4). Among the six models, the KFAM resulted in the lowest DIC value, and the resulting posterior means of the $Ln(\alpha)$ values over time were very different. The HRM and RAM, and the Ricker model resulted in DIC values relatively lower than models other than KFAM (Table 26.2).

From the simulation study based on the example walleye fishery data, we can see that the model selection uncertainty is high because in only 2.4% of the cases the true model was found when the true model was the Ricker model; and it is 0%, 29.2%, 6.8%, 74%, and 1.8% when the true models were the Ricker model, MEM, RAM, RWM, KFAM, and HRM separately (Table 26.3). In general, KFAM tended to be selected as the best model no matter what the true models were. MEM had no chance to be selected as the best model in this study. The Ricker model and RWM have very limited chance to be selected as the best model in the simulation study. The same pattern of model selection uncertainty could be observed from the simulation study when the variance of the errors used in the

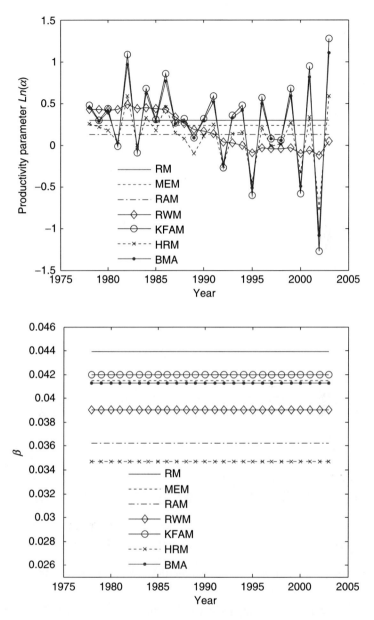

Fig. 26.4 Productivity ($Ln(\alpha)$ here) estimation and parameter β estimation using different models and the Bayesian model averaging results

data simulation is 30% of the "true" variance (Table not shown). KFAM has an extremely high chance to be selected as the best model even when the magnitudes of the noises are low.

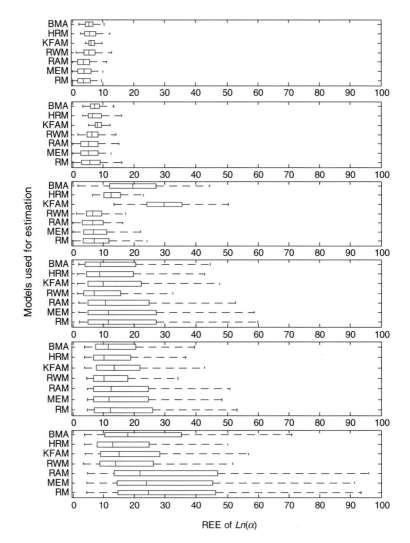

Fig. 26.5 Boxplot of the relative estimation error (REE) of population productivity $Ln(\alpha)$ estimation in the simulation study when the "true" error values were used. From top to the bottom, the true models used in the simulation are RM, MEM, ARM, RWM, KFAM, and HRM

The REE$_s$ of $Ln(\alpha)$ were higher when the true models were KFAM, RWM, and HRM (Table 26.3, Fig. 26.5). The magnitude differences of REE$_s$ were caused by the errors used in the simulation study. Though the error values were from the same example fishery, the colored-noise models tended to generate higher noises in magnitude over time (Halley and Kunin 1999). $Ln(\alpha)$ tended to be highly influenced by the colored noise or the multilevel productivity. This is an important implication in fisheries because the "noise" might not be white and there have been multilevel productivity observed in many fisheries under different environmental regimes (Beamish et al. 1999).

The simulation study also showed that the "true" model tended to give estimates with lower or lowest REE of $Ln(\alpha)$ and $Ln(R)$ (Table 26.3 and Figs. 26.5 and 26.7). MEM tended not to work well, having higher REE values regardless of the true model except when the Ricker model was used as the true model in the simulation study. The REE estimates when the BMA approach was used were low and were very close to the REE estimates when the KFAM was used (Figs. 26.5–26.7). This is because of the low DIC values for KFAM, which also resulted in an extremely high probability of it to be selected as the best model.

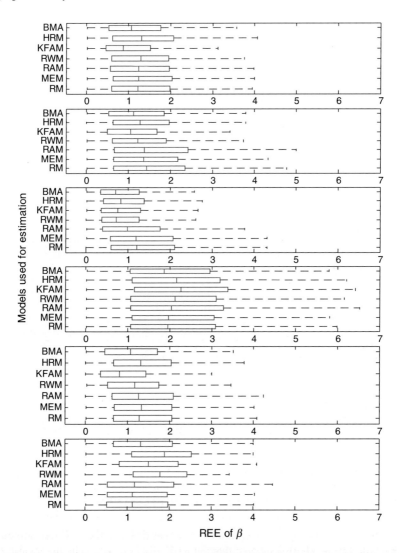

Fig. 26.6 Boxplot of the relative estimation error (REE) of β estimation in the simulation study when the "true" error values were used. From top to the bottom, the true models used in the simulation are RM, MEM, ARM, RWM, KFAM, and HRM

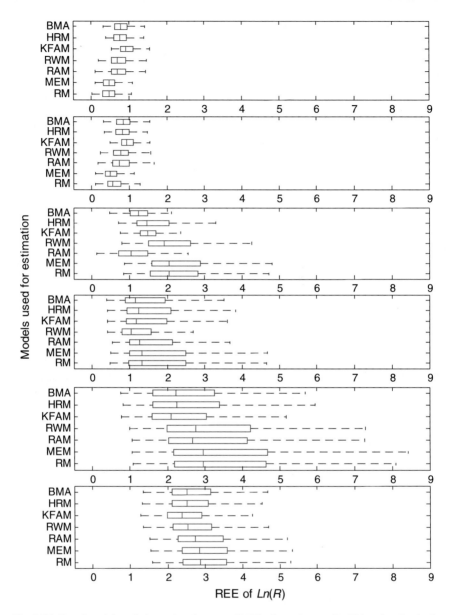

Fig. 26.7 Boxplot of the relative estimation error (REE) of recruitment ($Ln(R)$) estimation in the simulation study when the "true" error values were used. From top to the bottom, the true models used in the simulation are RM, MEM, ARM, RWM, KFAM, and HRM

In this specific example fishery, it looks like that BMA framework was dominated by KFAM. The result also showed that the REE_s did not have to be the lowest even when the model averaging approach was used. However, we need to realize that the simulation study was based on comparing the estimates with the "true" values, such as the "true" $Ln(\alpha)$ and the "true" recruitment, but the real fisheries fact in the last 100 years indicated that noise may be as important as the default "true" models. REE_s for $Ln(R)$ was obviously lower when simulated R was used instead of "true" R and when KFAM was used for estimation. We also need to realize that the corresponding risks of extinction are different when different models are used (Halley and Kunin 1999; Morales 1999; Schwager et al. 2006).

Parameters $Ln(\alpha)$ and β are correlated, which explains some of the REE_s patterns. For example, when the true models were MEM in generating data, the REE_s of $Ln(\alpha)$ were low when RM, MEM, and RAM were used for estimation, and the REE_s of $Ln(\alpha)$ were high when RWM, KFAM, and HRM were used for estimation. However, the REE_s of β were high when RM, MEM, and RAM were used for estimation, and the REE_s of β were low when RWM, KFAM, and HRM were used for estimation. When the true models were KFAM and HRM, the REE_s of $Ln(\alpha)$ were high when RM, MEM, and RAM were used for estimation, and the REE_s of $Ln(\alpha)$ were low when RWM, KFAM, and HRM were used for estimation. However, the REE_s of β were low when RM, MEM, and RAM were used for estimation, and the REE_s of β were high when RWM, KFAM, and HRM were used for estimation.

The analysis indicated that estimated $Ln(\alpha)$ and other parameter values and their creditable intervals (CI) were robust to priors for variances of $\sigma_{\varepsilon 1}^2$ to $\sigma_{\varepsilon 7}^2$ in the Bayesian time series models, e.g., when uniform distribution or inverse-gamma distribution was used. Little difference was observed, so we did not show the results here because of their robustness to the noninformative priors. However, the informative priors of $Ln(\alpha)$, β, and others do influence the results, and we think that it is necessary to use these informative priors, such as the prior of $Ln(\alpha)$, β, and the spawner stock size in the measurement error model. Information is based on the 26 years of observations and helped to constraint the parameter estimates to a degree.

26.5 Discussion

In this example fishery, the top model was KFAM, according to the model selection criterion DIC. In another population dynamics analysis on an endangered freshwater mussel we found that the top models were the hierarchical Bayesian model, the random walk model, and the exponential growth model through real data analysis and a simulation study (Jiao et al. 2008). However, the simulation study we have discussed showed that for this specific example fishery the KFAM tended to be selected as the best model regardless of the default true model, after adding noise to these models. A further study on other SR data set is suggested to look at the goodness of fit of the KFAM and the other models, which will help us to understand whether this is caused by the data pattern of this walleye fishery or a general phenomenon for all the fisheries. This simulation study also suggests that DIC may not be efficient enough as a model selection criterion in model selection and model averaging. The performance of DIC as a model selection criterion is an ongoing topic in statistics. It is beyond the scope of this study. Considering the fact that no other criterion is currently in use other than DIC, we used this criterion.

In this study, we chose the class of model for BMA based on a good model supported by the data, here the Ricker model, and then averaged over an expanded class of models centered around the good model, such as the residual autoregressive Ricker model, RWM based on the structure of the Ricker model. How to choose the class of models for BMA is discussed by many scientists (Draper 1995; Madigan and Raftery 1994; Hoeting et al. 1999), which incorporate the approach we used here and over the entire class of models or over models with different error structures. We would support all these approaches if they can explain the model from the biological data, i.e., any models supported by the data. The BMA approach provides a very flexible framework to incorporate different models and it balances the model selection uncertainty and the goodness of fit of different models incorporated in the BMA framework. It is important to incorporate good models in the BMA framework and at the same time weak models do not need to be worried about because they will be less weighted if they are less supported by the data.

Traditional SR models such as the Ricker model and the BH model are widely used because of their simplicity while models such as RAM, RWM, KFAM, and HRM are rarely used in current fisheries because of the difficulty in estimating parameters. MCMC provides a very convenient way to solve these models and Bayesian approaches are the only computational, possible, or convenient method to solve above models (De Valpine and Hasting 2002). Models such as RAM, RWM, KFAM, and HRM are highly recommended to be incorporated into such a model averaging framework because they are much better in capturing the noise or stochastic component than the traditional SR models. The RWM and KFAM were said to be better at describing the trend of the population productivity changes over time (Peterman et al. 2000, 2003).

Random noise caused by environmental variation can be more important than the "true" population dynamic pattern. This is especially true and crucial for populations with small stock size (Halley and Kunin 1999; Morales 1999; Petchey 2000; Schwager et al. 2006). A BMA approach is suggested to incorporate these models describing various possible noise components. A simulation study is recommended in studying SR modeling and risk analyses, which would help to understand the model selection uncertainty. The approach should also help lower the risk of misunderstanding the population status, and help to improve the problems involved in our modeling approach and better manage our natural resources.

High model selection uncertainty implies that estimating population parameters and evaluating the population status based on one model or one best model is dangerous. In theory, BMA provides better average predictive performance than any single model that could be selected, and this theory has been supported by many examples (Hoeting et al. 1999). The application of BMA in fisheries is still very limited. Based on the fact that the model selection uncertainty is extremely high in fisheries and the model averaging approach works well, a model averaging approach is recommended in dealing with these SR relationship and the other models in fisheries, such as the catchability model used in age-aggregated models and the age- or size-structured models, the productivity function used for the surplus production models, and the error structure used in both the age-aggregated models and the age- or size-structured models (Polacheck et al. 1993; Harley et al. 2001; De Valpine and Hasting 2002; Jiao and Chen 2004; Jiao et al. 2006). A lot of discussions are currently going on among scientists as to which model should be used. The answer may be different for different fisheries. However, it is very difficult to assess because of the uncertainties in the data and in the model selection. The BMA approach balances the needs for best models and model selection uncertainty, takes into account these important sources of uncertainty and provides better inferences about parameters. It eventually lowers the risk of mis-managing our fisheries through better stock assessment.

Acknowledgement This research was supported by Virginia Polytechnic Institute & State University's new faculty start-up funding to Y. Jiao, the USDA Cooperative State Research, Education and Extension Service, Hatch project #0210510 to Dr. Y. Jiao, and grant for Decision Analysis and Adaptive Management of Lake Erie Percid Fisheries awarded to Y. Jiao by the Ontario Commercial Fisheries Association.

References

Beamish RJ, Noakes DJ, McFarlane GA, Klyashtorin L, Ivanov VV, Kurashov V (1999) The regime concept and natural trends in the production of Pacific salmon. Can J Fish Aquat Sci 56:516–526

Caswell H, Cohen JE (1995) Red, white and blue: environmental variance spectra and coexistence in metapopulations. J Theor Biol 176:301–316

De Valpine P, Hasting A (2002) Fitting population models incorporating process noise and observation error. Ecol Monogr. 72(1):57–76

Draper D (1995) Assessment and propagation of model uncertainty (with discussion). J Roy Statist Soc Ser B 57:45–97

Gelman A (2005) Prior distributions for variance parameters in hierarchical models. Bayesian Anal 2:1–19

Gilks WR (1996) Full conditional distributions. In: Gilks WR, Richardson S, Spiegelhalter DJ (eds) Markov Chain Monte Carlo in practice. Chapman & Hall, London, pp 75–78

Glantz MH (1992) Climate variability, climate change and fisheries. Cambridge University Press, Cambridge

Halley JM (1996) Ecology, evolution and 1/f-noise. Trends Ecol Evol 11:33–37

Halley JM, Kunin WE (1999) Extinction risk and the 1/f family of noise models. Theor Pop Biol 56:215–230

Harley SJ, Myers RA, Dunn A (2001) Is catch-per-unit-effort proportional to abundance? Can J Fish Aquat Sci 58:1760–1772

Hoeting JA, Madigan D, Raftery AE, Volinsky CT (1999) Bayesian model averaging: a tutorial. Statist Sci 14:382–417

Jiao Y, Chen Y (2004) An application of generalized linear model in production model and sequential population analysis. Fish Res 70:367–376

Jiao Y, Reid K, Nudds T (2006) Variation in the catchability of yellow perch (*Perca flavescens*) in the fisheries of Lake Erie using a Bayesian error-in-variable approach. ICES J Mar Sci 63:1695–1704

Jiao Y, Neves R, Jones J (2008) Models and model selection uncertainty in estimating growth rates of endangered freshwater mussel populations. Can J Fish Aquat Sci 65:0000–0000 (in press)

Madigan D, Raftery AE (1994) Model selection and accounting for model uncertainty in graphical models using Occam's window. J Am Statist Assoc 89:1535–1546

Morales JM (1999) Viability in a pink environment: why "white noise" models can be dangerous. Ecol Lett 2:228–232

Petchey OL (2000) Environmental colour affects aspects of single-species population dynamics. Proc Roy Soc Lond B 267:747–754

Peterman RM, Pyper BJ, Grout JA (2000) Comparison of parameter estimation methods for detecting climate-induced changes in productivity of Pacific salmon (*Onchorhynchus* spp.). Can J Fish Aquat Sci 57:181–191

Peterman RM, Pyper BJ, MacGregor BW (2003) Use of the Kalman filter to reconstruct historical trends in productivity of Bristol Bay sockeye salmon (*Oncorhynchus nerka*). Can J Fish Aquat Sci 60(7):809–824

Polacheck T, Hilborn R, Punt AE (1993) Fitting surplus production models: comparing methods and measuring uncertainty. Can J Fish Aquat Sci 50:2597–2607

Quinn II TJ, Deriso RB (1999) Quantitative fish dynamics. Oxford University Press, Oxford, 542 pp

Schwager M, Johst K, Jeltsch F (2006) Does red noise increase or decrease extinction risk? Single extreme events versus Series of unfavorable conditions. Am Nat 167:879–888

Spiegelhalter DJ, Best NG, Carlin BP, van der Linde A (2002) Bayesian measures of model complexity and fit (with discussion). J Roy Statist Soc B 64:583–640

Spiegelhalter DJ, Thomas A, Best N, Lunn D (2004) WinBUGS user manual (version 1.4.1). MRC Biostatistics Unit, Cambridge, UK

Vasseur DA, Yodzis P (2004) The color of environmental noise. Ecol 85:1146–1152

Walleye Task Group (WTG) (2004) Report of the walleye task group to the standing technical committee, Lake Erie Committee of the Great Lakes Fishery Commission. 26 pp

Walters CJ, Ludwig D (1981) Effects of measurement errors on the assessment of stock–recruit relationship. Can J Fish Aquat Sci 38:704–710

Chapter 27
Feedback Control in Pacific Salmon Fisheries

Carrie A. Holt and William K. de la Mare

Abstract Feedback control, as defined in the field of control systems engineering, is the property of a system that permits the output (i.e., performance) to be compared with the input (i.e., objectives) when choosing an appropriate control action. Although similar principles are used in adaptive control, adaptive management, and management procedures, there are cases where it is not obvious how to implement feedback control, thereby limiting its use in practice. There are several advantages to incorporating feedback control in fisheries. It forces managers to be explicit about objectives and directly links management actions to those objectives. Furthermore, management performance is less sensitive to perturbations and long-term persistent changes because management actions are adjusted in response to those changes in order to achieve objectives. We apply feedback control to one case example, a sockeye salmon (*Oncorhynchus nerka*) fishery, by using simulation modeling to identify target harvest rules that achieve management objectives, accounting for uncertainties in the outcomes from implementing those harvest rules. This type of feedback control system can be applied to other fisheries where outcomes of management regulations do not match targets in order to improve managers' abilities to achieve their objectives.

Keywords feedback control · Pacific salmon · simulation modeling · outcome uncertainty

27.1 Introduction

Control systems theory was developed from the field of systems engineering to design rules that achieve specified objectives. Control systems link objectives with information gathered on the state of the system, so as to hold an attribute or attributes

C.A. Holt(✉)
Fisheries and Oceans Canada, Pacific Biological Station, Nanaimo,
British Columbia, V9T6N7, Canada
e-mail: Carrie.Holt@dfo-mpo.gc.ca

W.K. de la Mare
Commonwealth Scientific and Industrial Research Organisation (CSIRO),
Cleveland, Queensland, 4163, Australia

of the system at desired levels (de la Mare 2005). The term feedback is commonly used to refer to cause-effect relations, but it has a more specific meaning in control systems theory. It is the property of a system that permits the outputs, or performance, to be compared with the inputs or objectives when identifying control actions (Distefano et al. 1967). When applied to fisheries, the actions taken by managers are adapted according to the current state of the fishery, the management objectives, and a decision algorithm in a feedback control loop (Walters and Buckingham 1975), typically aiming to stabilize annual catches and population abundances at desired levels (Utne 2006; Loehle 2006).

The concept of feedback control has been applied in the fisheries literature in the forms of adaptive control (Walters and Hilborn 1976), adaptive management (Walters 1986), and management procedures (de la Mare 1996), but simple ways to implement feedback control are not always obvious, limiting their use in actual practice. In the 1970s and 1980s, a computer modeling technique, stochastic dynamic programing, was first applied to adaptive control problems in fisheries to identify management actions that maximized harvest accounting for future learning about uncertain states in the fisheries system (e.g., parameters of stock-assessment models). Those studies typically recommended constant annual catches or exploitation rate, depending upon management objectives and environmental drivers, among other factors (e.g., Walters 1986; Walters and Parma 1996). Dynamic programming is currently not widely used because simpler, simulation modeling approaches are now available. In addition, management efforts have moved, to an extent, away from identifying optimal regulations towards effective ways to implement those regulations (Walters and Martell 2004).

Similar to the field of adaptive control, adaptive management is based on the "dual effect of control," which combines the direct value of harvest with the indirect value of information (Walters 1986; Alexander et al. 2006). By manipulating population abundances through changes in fishing effort or observing natural perturbations in abundances, stock-assessment scientists may be able to observe greater contrast in population status, potentially resulting in improved estimates of parameters used in stock assessments and therefore long-term management performance. Deliberate manipulations (i.e., management experiments) have been applied to several fisheries (e.g., sockeye salmon fisheries in British Columbia [Walters et al. 1993], fisheries on the Great Barrier Reef [Campbell et al. 2001], and fisheries on the Northwest Shelf of Australia, [Sainsbury et al. 1997]). However, limits to implementing this approach include high costs of monitoring programs during experimentation and difficulty in estimating the effects of experimental perturbations separately from environmentally driven changes or other exogenous factors (Walters and Martell 2004).

Management procedures are the third area where feedback control has been applied to fisheries. This approach to management uses rules, more recently derived by simulation, which specify how data are collected, analyzed, and used, and how regulations are determined from those data (de la Mare 1996). Feedbacks may be incorporated into management procedures if they are updated periodically (e.g., every 3–5 years, as suggested by Johnston and Butterworth [2005]). However, the difference between what actually happens in the fishery (i.e., the outcome from implementing fishing regulations) and management objectives is often not fully accounted for when setting management targets (McAllister et al. 1999; Holt and Peterman 2006).

Feedback control systems have five attributes that may improve fisheries management. First, they encourage managers to identify measurable objectives that are used to evaluate management performance (Utne 2006). Second, management actions are explicitly tied to those objectives and are adjusted according to how well objectives are achieved. Third, when objectives and control actions are explicitly stated, decision making is more transparent, enhancing credibility of the decision-making process among stakeholders (as shown for various South African fisheries, Geromont et al. 1999). Fourth, feedback control can increase accuracy in achieving objectives because management actions are adjusted to reduce deviations between metrics of performance (outputs) and targets (inputs) (DiStefano 1967). Fifth, control systems that incorporate feedback are less sensitive to perturbations or persistent changes, if management adjustments are implemented in time to attain control over output metrics (DiStefano 1967). By responding to the current state of the system (i.e., allowing learning), feedback control systems can adapt to short-term disturbances and long-term shifts in system components.

27.2 Sockeye Salmon Fisheries on the Fraser River, British Columbia

Here, we describe a simulation-based application of feedback control to a case example of sockeye salmon fisheries on the Fraser River, British Columbia. Although this example has some of the attributes of feedback control listed above, some are missing or not completely incorporated into the fishery. Specifically, the differences between management targets and what actually happens in the fishery are often ignored when identifying those targets, potentially resulting in reduced accuracy in achieving objectives (attribute 4). Furthermore, although decision algorithms used by managers to set fishing regulations are based in part on escapement goals, other objectives (e.g., maximizing yield, maintaining annual yields above a threshold level, or minimizing interannual variability in yields) may exist, but are not explicitly stated or accounted for when setting management targets (attributes 1 and 2). First, we describe the current management practices for that fishery, and then propose an alternative management arrangement that incorporates feedback control. We conclude with challenges to applying this approach in this and other fisheries.

Managers of sockeye salmon on the Fraser River currently choose annual target harvest rates based on a combination of forecasts of recruitment (mature adults that return to the coast toward natal spawning areas), estimated productivity (potential rate of change) of sockeye populations, prespecified guidelines regarding harvest rate plus natural mortality rate, and escapement targets (Fraser River Sockeye Spawning Initiative 2005). Although prespecified harvest rules have not been explicitly stated for this fishery, historical target harvest rates can be described retrospectively in the form of a target harvest rule (as in Cass et al. 2003; Holt and Peterman 2006). In other words, target harvest rates have been chosen in a manner analogous to the harvest rule shown in Fig. 27.1 (solid line). Once target harvest rates are selected (Fig. 27.2, Boxes 2 and 3), they are implemented in the fishery

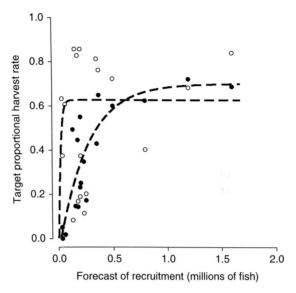

Fig. 27.1 Data and best-fit curves for the target harvest rule (i.e., the relation between forecasts of recruitment and target harvest rates, solid circles and solid lines) and the realized harvest function (i.e., what actually happens in the fishery, the relation between forecasts of recruitment and realized harvest rates, hollow circle and dashed line) for the Early Stuart stock aggregate of sockeye salmon (Fraser River, British Columbia) and return years 1986 through 2003 (Adapted from Fig. 2a, Holt and Peterman 2006)

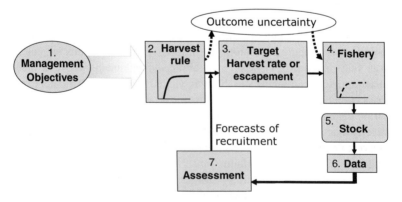

Fig. 27.2 Schematic representation of the sockeye salmon fishery on the Fraser River, British Columbia. Each component of the system is depicted by a box, numbered 1–7. Two limitations to applying feedback control are shown: differences between management targets and realized outcomes, i.e., "outcome uncertainties" are not accounted for when setting targets (relation between Boxes 2 and 4), and management objectives are only incompletely accounted for when identifying those targets (relation between Boxes 1 and 2)

(Fig. 27.2, Box 4), which in turn affects salmon population dynamics (Fig. 27.2, Box 5). Data on catch and escapement are then collected (Fig. 27.2, Box 6) and applied to stock assessments (Fig. 27.2, Box 7). One component of those stock assessments is forecasts of recruitment, which are used to set target harvest rates (Fig. 27.2, Boxes 2 and 3).

The sockeye salmon fishery on the Fraser River incorporates feedback in the broad sense because information on the state of the system is used to generate forecasts and set target harvest rates. However, this does not represent a feedback control system because it does not account for the coupling between harvest rules and multiple management objectives for Fraser River sockeye salmon. In particular, the extent to which harvest rules can achieve objectives other than escapement goals has not been quantified.

An additional limitation is that the target harvest rates or escapement goals often differ from actual or realized outcomes in the fishery (outcome uncertainties) (Holt and Peterman 2006) due to, for example, uncertainty in forecasts of recruitment and variability in compliance of harvesters to regulations, but these differences have no effect on management in subsequent years. Despite large uncertainties in the ability of managers to implement regulations exactly as planned, management targets for this fishery are usually set assuming that they can be achieved exactly.

When quantifying bias in outcome uncertainty for the sockeye salmon fishery on the Fraser River, Holt and Peterman (2006) compared the target harvest rule, i.e., the relation between forecasts of recruitment and historical target harvest rates (from 1987 through 2003), with another relation, the realized harvest function. That function described realized harvest rates along a gradient in forecasts of recruitment over the same time period. For instance, for one stock aggregate, the Early Stuart run, at low forecasts of recruitment, the realized harvest function (dashed line in Fig. 27.1) was higher than the target harvest rule (solid line in Fig. 27.1), which may lead to conservation concerns from overharvesting at low stock abundance. Alternatively, at high forecasts of recruitment, the realized harvest function was higher than the target rule, potentially leading to lost fishing opportunities. Deviations between these two functions represent an average difference over 18 years, but the magnitude and nature of outcome uncertainties may have changed over that period. In fact, deviations between target and realized harvest rates increased over the last 5 years of that time series for most Fraser River stock aggregates (Holt and Peterman 2006), suggesting that the magnitude of outcome uncertainty can systematically change over time. In addition to bias, large observed stochasticity around the realized harvest function suggested that in the short term for that stock aggregate, outcome uncertainty will be dominated by stochastic variation, but in the long term those biases may become important.

In order to fully incorporate feedback control into this fishery, we propose an alternative management arrangement (Fig. 27.3). This approach is similar to current practices in that forecasts of recruitment are used to determine target harvest rates, but information from assessments (e.g., historical harvest rates, escapement, and catch) are also used to parameterize a simulation model of the fisheries system that contains biological and management components (described in detail in Holt and Peterman 2007). That simulation model evaluates the performances of various trial

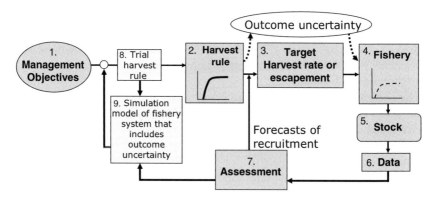

Fig. 27.3 Schematic representation of an alternative management arrangement for the sockeye salmon fishery on the Fraser River, British Columbia. This arrangement includes feedback control by adjusting target harvest rules to account for multiple objectives, and accounting for uncertainty in the ability to implement target harvest rules exactly as planned. The hollow circle is where management objectives (Box 1) are compared with the performance of trial harvest rules (Box 8)

harvest rules for their ability to achieve multiple management objectives. Through an optimization procedure, the harvest rules that best achieve management objectives can then be identified and applied to the fishery. This management arrangement addresses the previously mentioned limitations to applying feedback control by coupling those harvest rules to multiple objectives, and by accounting for outcome uncertainties in the simulation model used to identify harvest rules. Furthermore, the harvest rule that best achieves objectives can be updated periodically as knowledge of the parameters describing population dynamics and outcome uncertainties improves (i.e., when the simulation model is reparameterized with newly collected data). Although this type of simulation model has been used in a wide variety of other fisheries to evaluate performance of management procedures (e.g., South African sardine [De Oliveira et al. 1998], South African West Coast rock lobster [Johnston and Butterworth 2005], whales in the Antarctic [de la Mare 1996]), outcome uncertainties, i.e., differences between actual harvest levels and targets, are usually not fully considered in those models (McAllister et al. 1999; Kell et al. 2005).

27.2.1 Application of Feedback Control to the Summer Run Stock

Harvest rules were identified for one stock aggregate of sockeye salmon on the Fraser River, the Summer run (Holt and Peterman 2007). Those harvest rules were chosen to achieve multiple management objectives, accounting for outcome uncertainties. The management objectives considered here were to maximize yield under two constraints: less than a 10% chance that mean annual catches

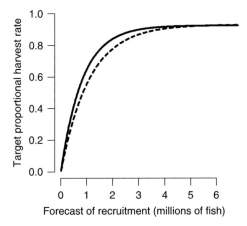

Fig. 27.4 Target harvest rules that ignore differences between management targets and outcomes (solid line), and account for those differences (dashed line), for the Summer run stock aggregate of sockeye salmon on the Fraser River, British Columbia. These target harvest rules are generated for a simulation model of the fisheries system that includes biological and management components

fall below 10% of maximum sustainable yield, and less than a 10% chance that spawner abundances fall below 10% of spawner abundances at maximum recruitment. These objectives were adapted from current management initiatives on the Fraser River (Cass et al. 2003). Although they represent only one possible trade-off among harvesting and conservation goals, others could also be evaluated using the same methods. Trial harvest rules with the same structural form as Fig. 27.1 (solid line), but different parameter values were considered (e.g., different values for the maximum harvest rate at high recruitment, recruitment below which target harvest rates are zero, and parameter representing the shape of the curve). The harvest rule shown in Fig. 27.4 (solid line) best achieved those objectives when outcome uncertainties were accounted for in the simulation model. Alternatively, when outcome uncertainties were assumed to be negligible, the harvest rule that best achieved objectives was more precautionary; the target harvest rates were lower for any given forecast of recruitment (Fig. 27.4, dashed line).

The revised management arrangement has the attributes of a feedback control system described above. Multiple objectives are explicitly stated and directly tied to management actions. Decisions about the choice of target harvest rates are transparent because they are based on a harvest rule. Accuracy in achieving objectives increases when differences between targets and outcomes are accounted for in the simulation that identifies target harvest rules. Finally, by periodically updating the simulation model with new information on parameter values and structural form, the recommended management actions can respond to changes that occur in the system due to, for example, climate variability and changes in harvesting patterns, among other factors.

27.3   Challenges

There are several challenges to implementing feedback control in fisheries systems to close the feedback loop using simulation models, two of which are: setting objectives and communicating management advice. Developing objectives and identifying trade-offs among conflicting objectives are difficult and time-consuming when stakeholder input is required (e.g., through a series of workshops) (Punt 2006). Identifying trade-offs may be especially challenging if objectives extend beyond single-species to multispecies and ecosystem-based goals (Sainsbury et al. 2000; Picktich et al. 2004). For example, in one management procedure for a multispecies fishery in South Africa, total allowable catches for sardine were determined in part based on allowable catches in the sardine fishery and the expected bycatch of sardines in the anchovy fishery. That expected bycatch was calculated from total allowable catches of anchovies in the anchovy fishery (Cochrane et al. 1998; De Oliveira 1998; Geromont et al. 1999). Therefore, trade-offs occurred between maximizing catches of sardines and anchovies, which required careful consideration of risks in both fisheries.

In addition, some decision makers may prefer flexibility when setting objectives rather than being constrained by recommendations generated from prespecified objectives (Punt 2006). For example, prior to establishing the Australian Fisheries Management Authority in 1991, some managers in fisheries under federal jurisdiction resisted the introduction of explicit objectives because it allowed management practices to be audited more easily than previously possible (Smith et al. 1999). Furthermore, resistance to explicit objectives may arise due to possible constraints placed on managers limiting their ability to respond to new and emerging issues (e.g., due to changes in stakeholder preferences and/or composition) (Smith et al. 1999).

To address this challenge, managers can adopt objectives that emerge from analyses of historical data (e.g., the target harvest rule identified here the Summer run stock aggregate of sockeye salmon on the Fraser River, British Columbia). Those objectives can be updated periodically through an iterative process of negotiations among stakeholders and implementation. During negotiations, simulation modeling exercises, where stakeholders simulate the outcomes from various actions can help elicit objectives by giving stakeholders a better understanding of possible outcomes, and allowing them to identify attributes of those outcomes that they value (Cooke 1999). This tool may reduce the number of iterations of negotiations necessary before reaching consensus. For example, stakeholders in fisheries on the eastern stock of gemfish in Australia were better able to quantify objectives, which initially consisted of only vague notions of recovery of depleted stocks, when a simulation modeling approach was adopted (Punt and Smith 1999).

One challenge when using simulation models to evaluate management procedures is difficulty in communicating technical details to a variety of stakeholders with diverse backgrounds, about alternative management procedures, modeling approaches for evaluating them, and their relative performances (Punt et al. 2001, 2006; Peterman 2004). For example, during negotiations for the fisheries on South African sardine and anchovy, difficulties in communicating the stochastic nature of models used to evaluate management procedures and metrics describing risks were key obstacles to agreement (Cochrane et al. 1998).

One approach to addressing this challenge that has been suggested by several previous authors (e.g., Walters 1986; Butterworth et al. 1997) involves incorporating stakeholders in the process of developing simulation models and evaluating management procedures. Stakeholder participation in those processes permits in-depth engagement when generating recommendations for management, and therefore increases acceptance of consequent management actions. For example, in a synthesis of management procedures for five marine resources in South Africa, Geromont et al. (1999) found that cooperation between scientists and industry during model development was critical for achieving agreement on a final management procedure, and ensuring effective implementation. In addition, the involvement of an industry-funded scientist in the assessment group for fisheries on the eastern stock of gemfish in Australia strengthened industry support of the simulation modeling approach; support which persisted despite recommendations for severe restrictions to the fishery (Smith et al. 1999). Furthermore, general workshops on simulation modeling approaches to evaluate management procedures can be useful for communicating across hierarchical levels of fisheries management that are required in decision making. For example, for gemfish fisheries in Australia, workshops helped bridge a communication gap between stock-assessment advisory groups, management advisory groups, and decision makers (Smith et al. 1999).

Although involving key stakeholders in the development of simulation models for evaluating management procedures has been successful for fisheries dominated by a few large industries (e.g., the fishery on South African anchovy), the feasibility of participation by a large number of stakeholders in negotiations (e.g., in fisheries consisting of many small operators) is not clear (Butterworth et al. 1997). In such cases, a gradual approach to communicating the methodology, alternative management procedures, and performance of those procedures may be especially critical. This could occur through a series of workshops each addressing only one component of the methodology and involving small groups of stakeholders. However, stakeholders will need to be convinced of the potential benefits of a new approach to management and opportunity costs of the status quo before they will invest time and resources into such workshops.

In summary, incorporating feedback control into fishery systems can improve management by identifying management actions that explicitly achieve stated objectives. Management actions are adaptive to change (e.g., those related to climate variability and variability in harvester's behavior) when feedback control systems are used because those actions are adjusted according to their ability to achieve management objectives. Although similar concepts exist in the fisheries literature on adaptive control, adaptive management, and management procedures, simple ways to implement feedback between management targets and outcomes in the fishery are not always obvious. Here we present one approach to closing the "feedback loop" that accounts for differences between target harvest rules and realized outcomes. Although management objectives and procedures differ among fisheries, this approach to feedback control is broadly applicable. For example, Kell et al. (1999) found that management procedures for North Sea plaice that incorporated feedbacks in the ability to achieve targets performed better relative to those without feedbacks when discarding was present in the fishery. However, that study did not include stochastic variation in achieving management targets as we suggest in our example.

Acknowledgments We thank Randall Peterman and Sean Cox for their valuable contributions to this work. Ian Guthrie from the Pacific Salmon Commission, and Al Cass and Jeff Grout from Fisheries and Oceans Canada supplied data on the Fraser River fishery. Funding for this study was provided by grants from the Natural Sciences and Engineering Council of Canada, NSERC, the Canada Foundation for Innovation, the British Columbia Knowledge Development Fund, and Fisheries and Oceans Canada, all awarded to R.M. Peterman, plus an NSERC Postgraduate Scholarship to C.A. Holt.

References

Alexander, C. A. D., Peters, C. N., Marmorek, D. R., and Higgins, P. 2006. A decision analysis of flow management experiments for Columbia River mountain whitefish (*Prosopium williamsoni*) management. Can. J. Fish. Aquat. Sci. **63**: 1142–1156.

Butterworth, D. S., Cochrane, K. L., and de Oliveira, J. A. A. 1997. Management procedures: a better way to manage fisheries? the South African experience. *In* Global Trends: Fisheries Management. *Edited by* E. K. Pikitch, D. D. Huppert, and M. P. Sissenwine. America Fisheries Society, Bethesda, MD, pp. 83–90.

Campbell, R. A., Mapstone, B. D., and Smith, A. D. M. 2001. Evaluating large-scale experimental designs for management of coral trout on the Great Barrier Reef. Ecol. Appl. **11**: 1763–1777.

Cass, A., Folkes, M., and Pestal, G. 2003. Methods for assessing harvest rules for Fraser River sockeye salmon. Pacific Scientific Advice Review Committee Working Paper S2003–14. Fisheries and Oceans Canada, Nanaimo, British Columbia.

Cochrane, K. L., Butterworth, D. S., de Oliveira, J. A. A., and Roel, B. A. 1998. Management procedures in a fishery based on highly variable stocks and with conflicting objectives: experiences in the South African pelagic fishery. Rev. Fish Biol. Fish. **8**: 117–214.

Cooke, J. G. 1999. Improvement of fishery-management advice through simulation testing of harvest algorithms. ICES J. Mar. Sci. **56**: 797–810.

de la Mare, W. K. 1996. Some recent developments in the management of marine living resources. *In* Frontiers of Population Ecology. *Edited by* R. B. Floyd, A. W. Sheppard, and P. J. De Barro. CSIRO Publishing, Melbourne. pp. 599–616.

de la Mare, W. K. 2005. Marine ecosystem-based management as a hierarchical control system. Marine Policy **29**: 57–68.

De Oliveira, J. A. A., Butterworth, D. S., Roel, B. A., Cochrane, K. L., and Brown, J. P. 1998. The application of a management procedure to regulate the directed and bycatch fishery of South African sardine *Sardinops sagax*. *In* Benguela Dynamics: Impact of Variability on Shelf-Sea Environments and their Living Resources. *Edited by* S. C. Pillar, C. L. Moloney, A. I. L. Payne, and F. A. Shillington. Sea Fisheries Research Institute, Cape Town, South Africa, pp. 449–469.

DiStefano, J. J. I., Stubberud, A. R., and Williams, I. J. 1967. Schaum's Theory and Problems of Feedback and Control Systems. McGraw-Holl Book Company, Toronto.

Fraser River Sockeye Spawning Initiative. 2005. Report of the Technical Working Group on the Fraser River Sockeye Spawning Initiative, Vancouver, British Columbia.

Geromont, H. F., De Oliveira, J. A. A., Johnston, S. J., and Cunningham, C. L. 1999. Development and application of management procedures for fisheries in South Africa. ICES J. Mar. Sci. **56**: 952–966.

Holt, C. A. and R. M. Peterman. 2006. Missing the target: uncertainties in achieving management goals in fisheries on the Fraser River, British Columbia, sockeye salmon (*Oncorhynchus nerka*). Can. J. Fish. Aquat. Sci. **63**: 2722–2733.

Holt, C. A. and R. M. Peterman. 2007. Uncertainties in population dynamics and outcomes of regulations in sockeye salmon fisheries: implications for management. Can. J. Fish. Aquat. Sci. **65**: 1459–1474.

Johnston, S. J. and Butterworth, D. S. 2005. Evolution of operational management procedures for the South African west coast rock lobster (*Jasus Ialandii*) fishery. N. Z. J. Mar. Freshwater Res. **39**: 687–702.

Kell, L. T., O'Brien, C. M., Smith, M. T., Stokes, T. K., and Rackham, B. C. 1999. An evaluation of management procedures for implementing a precautionary approach in the ICES context of North Sea plaice (*Pleuronectes platessa* L.). ICES J. Mar. Sci. **56**: 834–845.

Kell, L. T., Pastoors, A. A., Scott, R. D., Smith, M. T., Van Beek, F. A., O'Brien, C. M., and Pilling, G. M. 2005. Evaluation of multiple management objectives for Northeast Atlantic flattish stocks: sustainability vs. stability of yield. ICES J. Mar. Sci. **62**: 1104–1117.

Loehle, C. 2006. Control theory and the management of ecosystems. J. Appl. Ecol. **43**: 957–966.

McAllister, M. K., Starr, P. J., Restrepo, V. R., and Kirkwood, G. P. 1999. Formulating quantitative methods to evaluate fishery-management systems: what fishery processes should be modelled and what trade-offs should be made? ICES J. Mar. Sci. **56**: 900–916.

Peterman, R. M. 2004. Possible solutions to some challenges facing fisheries scientists and managers. ICES J. Mar. Sci. **61**: 1331–1343.

Punt, A. 2006. The FAO precautionary approach after almost 10 years: have we progressed towards implementing simulation-tested feedback-control management systems for fisheries management? Nat. Resour. Model. **19**: 441–464.

Punt, A. E. and Smith, A. D. M. 1999. Harvest strategy evaluation for the eastern stock of gemfish (*Rexea solandri*). ICES J. Mar. Sci. **56**: 860–875.

Punt, A. E., Smith, A. D. M., and Cui, G. 2001. Review of progress in the introduction of management strategy evaluation (MSE) approaches in Australia's South East Fishery. Mar. Freshwater Res. **52**: 719–726.

Sainsbury, K. J., Campbell, R. A., Lindholm, R., and Whitelaw, A. W. 1997. Experimental management of an Autralian mulitispecies fishery: examining the possibility of trawl-induced habitat modification. *In* Global Trends: Fisheries Management. *Edited by* E. L. Pikitch, D. D. Huppert, and M. P. Sissenwine. American Fisheries Society Symposium, Bethesda, MD.

Sainsbury, K. J., Punt, A. E., and Smith, A. D. M. 2000. Design and operational management strategies for achieving fishery ecosystem objectives. ICES J. Mar. Sci. **57**: 731–741.

Smith, A. D. M., Sainsbury, K. J., and Stevens, R. A. 1999. Implementing effective fisheries-management systems – management strategy evaluation and the Australian partnership approach. ICES J. Mar. Sci. **56**: 967–979.

Utne, I. B. 2006. Systems engineering principles in fisheries management. Mar. Policy **30**: 624–634.

Walters, C. J. 1986. Adaptive Management of Renewable Resources. MacMillan, New York.

Walters, C.J. and Buckingham, S. 1975. A control system for intraseason salmon management. *In* Proceedings of a Workshop on Salmon Management, February 24–28, 1975. International Institute for Applied Systems Analysis, Laxenburg, Austria. CP-75-2, pp. 105–137.

Walters, C. J. and Hilborn, R. 1976. Adaptive control of fishing systems. J Fish. Res. Bd. Can. **33**: 145–159.

Walters, C. J. and Martell, S. J. D. 2004. Fisheries Ecology and Management. Princeton University Press, Princeton, NJ.

Walters, C. J. and Parma, A. M. 1996. Fixed exploitation rate strategies for coping with effects of climate change. Can. J. Fish. Aquat. Sci. **53**: 148–158.

Walters, C. J., Goruk, R. D., and Radford, D. 1993. Rivers Inlet sockeye salmon: an experiment in adaptive management. N. Am. J. Fish. Man. **13**: 253–262.

Chapter 28
The Future of Fisheries Science: Merging Stock Assessment with Risk Assessment, for Better Fisheries Management

Daniel Goodman

Abstract The future of fisheries science, as a science, can be expected to follow the largely self-correcting and improving trajectory of any science with an empirical basis. As long as we continue our monitoring, surveys, and experiments, the science will progress, giving us somewhat increased predictive power with time, and also giving us better-documented predictive power in the form of realistic confidence intervals for the predictions. The open question that is explored in this chapter is how fisheries *management* in the future might capitalize on the capabilities of the fisheries science of the future. Our premise is that predictive power, though it increases, will still remain limited in practice, so that much of the potential for improved management will rest with the decision-process making use of the quantification of uncertainty along with making use of the predictions themselves. The key to effective use of rigorous but partially uncertain scientific predictions in decision making is the adoption of properly crafted decision rules. The proper crafting (optimization) of a decision rule requires a utility function – a metric which honestly represents the desirability or undesirability of the possible outcomes. The content of a utility function expresses social values, which is not the domain of the science, but the correct use of the utility function definitely is within the domain of science. Furthermore, a commitment to operate within this kind of science-triggered decision framework would be a policy decision in its own right, and that decision has not yet been made. The scientific community will have an important educational role to play in the discussions which might lead to such a sea change. In this chapter, we review the theory of a science-triggered decision framework and illustrate a number of the issues with a concrete case history.

D. Goodman
Department of Ecology
Bozeman, MT 59717-0346
e-mail: goodman@rapid.msu.montana.edu

R.J. Beamish and B.J. Rothschild (eds.), *The Future of Fisheries Science in North America*, 537
Fish & Fisheries Series, © Springer Science + Business Media B.V. 2009

28.1 Preamble

The future of fisheries science, as best we can foresee, holds promise of many things to be optimistic about, and some things whose prospects are more in doubt.

There is an old maxim of economics that bad currency will drive good currency out of circulation when both must be accepted as legal tender. As an element of actual behavior, this was noted in antiquity. As a predictable dynamic this is generally known as Gresham's Law, having been so named by a nineteenth-century English economist in honor of a sixteenth-century financial agent and advisor to the Tudor monarchy, who warned of the phenomenon in his advice to Elizabeth I. The principle had been clearly formulated by Copernicus, a few decades earlier than Gresham, while working on monetary reform in his capacity as a provincial administrator within the kingdom of Poland.

Fortunately, the dynamic of bad coinage driving out good does not have a parallel in the progress of science. Quite the opposite; the dynamics of science are self-correcting. Because of the empirical, observational frame of reference of science, and the demand for reproducibility, good science will drive out bad science.

Our empirical base in fisheries science is expanding dramatically. New instrumentation, new remote sensors, and the accumulating long data series will provide new opportunities for exciting increases in knowledge. And new computers and statistical methods will provide new and refined ways to analyze the wealth of new data. So, there are no reasons to have misgivings about the scientific core of the fisheries science of the future. We can expect that we will see more environmental change, obtain more data, analyze those data, get a little bit better at predicting, and get a lot better at knowing how uncertain our predictions are.

The application of fisheries science is fisheries management, so we might also ask about the future of fisheries management. Here policy enters the picture. If good science drives out bad science, can we count on good policy automatically to drive out bad policy? Probably not. Adoption of good policy requires an act of will in the face of many countervailing forces.

The successful integration of stock assessment and risk assessment in fisheries management will definitely require a new policy context. Stock assessment estimates the size and productivity of a population. We know of course that the estimates are uncertain. But we should know enough about what we are doing in the technicalities of a stock assessment to estimate the uncertainty itself, if the stock assessment is carried out by processing data according to a defined methodology, with an underlying model, and some statistical basis, to get the estimates. Often enough, then, by fully utilizing the statistical basis of the assessment, the method can deliver, along with the assessment estimate, some measure of the uncertainty. If the statistical approach is Bayesian, the measure of uncertainty is conveyed in probabilities, which can form the technical basis for a risk assessment.

Risk assessment considers a spectrum of possible developments, some of which are threatening, and calculates their probabilities. This is now a common undertaking for evaluating the status of populations in connection with the

Endangered Species Act and the Marine Mammal Protection Act. This should become common practice for implementing a "precautionary approach" under the Magnuson-Stevens Act.

Conceptually, merging stock assessment with risk assessment is easy: report the results of the stock assessment as a spectrum of possibilities, and assign each a probability based on the statistical appraisal of uncertainty, which should have been built into the stock assessment. We already have the tools.

Then what? How does this scientific report get used for management? How does it lead to *better* decisions? How indeed.

28.2   Example: Decision Analysis for Conservation of the Cook Inlet Beluga Whale Population

28.2.1   Introduction

The beluga whale (*Delphinapterus leucas*) is a coastal, shallow water species, with habitat requirements which lead to disjunct spatial distribution of populations. The Cook Inlet (Alaska) population is a demographically, geographically, and genetically distinct stock, which in the late 1990s underwent a documented marked decline, and range contraction, primarily due to overharvest (O'Corry-Crowe and Lowry 1997; O'Corry-Crowe et al. 1997; Frost and Lowry 1990; Harrison and Hall 1978; Laidre et al. 2000; Mahoney and Shelden 2000; Rugh et al. 2000; Speckman and Piatt 2000). The population then became a subject of conservation concern, and an object of regulatory rule making and assessments.

The earliest estimate of population size for an aerial survey of beluga over the entire Cook Inlet was conducted by Don Calkins of the Alaska Department of Fish and Game in 1979 (and reported for a US National Oceanographic and Atmospheric Administration Environmental Assessment process in 1989). The estimate was 1,293 individuals. By then, the population already had a long, but unquantified, history of exploitation. Systematic annual surveys of the population, by the National Marine Fisheries Service (NMFS), began in 1994, by which time the population estimate had dropped to about half the earlier estimate. Estimation of harvest also began in 1994.

It is generally believed that the average population growth rate for a healthy cetacean population that is far below its carrying capacity would be around 4% (Wade 1998). This value is actually used as a matter of policy as the default assumed growth rate when calculating a tolerable incidental mortality for whales in some regulatory settings. If we accept at face value the NMFS annual population estimates and harvest mortality estimates for the Cook Inlet beluga population, and multiply the point estimate of population size by this default 4% figure for a back-of-the-envelope calculation of maximum nominally sustainable harvest, we may obtain an annual comparison of actual and nominally sustainable harvest mortality, starting in 1994. The results of this rough calculation for 1994–1998 are shown in Fig. 28.1. The 1995–1998 harvests appear to be excessive by a wide margin.

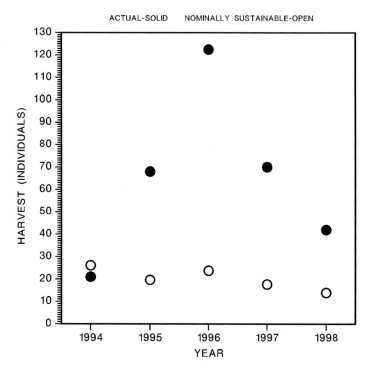

Fig. 28.1 Comparison of estimates of the actual harvest mortality (landed plus struck-and-lost), by year, shown as the solid dots, and rough estimates of the nominal sustainable harvest, assuming "normal" dynamics, shown as the open circles.

28.2.2 Chronology of Decisions

In 1999 direct federal legislative action imposed a 1-year moratorium on subsistence hunting of Cook Inlet beluga except as negotiated under agreement, which was not then in place, between the NMFS and a recognized Alaska Native organization involved in the harvest. The population was petitioned for Endangered Species listing in 1999 (it had been on NMFS candidate list since 1991), but in 2000 NMFS decided against listing. The population was determined to be *depleted* under the Marine Mammal Protection Act in 2000. NMFS began discussing a proposed harvest plan with the Cook Inlet Marine Mammal Council, a native organization, starting in 2000. In 2000, under authority of the National Environmental Policy Act, NMFS issued a draft Environmental Impact Statement on federal actions affecting Cook Inlet beluga, and proposed a continuing, but low, subsistence harvest as the *preferred alternative*. Later in 2000, direct federal legislation extended the provisional moratorium on subsistence hunting of Cook Inlet beluga, until such time as NMFS through formal rule making should approve a harvest regulation plan in consultation with an Alaska Native group.

At the end of 2000 an administrative law hearing – with representation of, among others, NMFS, the US Marine Mammal Commission, several native groups, two environmental NGOs, and an oil and gas organization – resulted in an agreement, through formal stipulations, on a "framework" for developing a detailed subsistence harvest regulation procedure, and an agreement on low harvests for the next 4 years, with the court to retain jurisdiction for the interim until adoption of a longer term "science-based" plan. The interim 4-year harvest agreement had only one data-driven contingency, and that would respond to any observed "unusual mortality." The discussions between parties continued, punctuated in particular by two public meetings outside court in 2003 to consider progress in developing a science-based plan, a rule making hearing in court in 2004, and a series of briefings demanded by the administrative law judge in 2005. A harvest quota was adopted under formal rule making each year.

In 2006, the court, based on briefs received from the respective parties, decided on a plan which would continue a low constant harvest through 2009, with a complicated data-driven decision rule to take effect in 2010 and thereafter. That data-driven decision rule would respond to information bearing on the population size and trend.

The reported harvests beginning in 1999 have averaged less than one per year. With near cessation of the harvest, the population, nevertheless, has failed to recover as expected, intensifying the conservation concern. The continuing annual census estimates, in fact, are trending downward. NMFS in 2005 issued a draft Conservation Plan for Cook Inlet beluga. The population was again petitioned for Endangered Species Act listing, and in 2006 NMFS initiated a new status review on that question. The Cook Inlet beluga was listed as *critically endangered* by the International Union for the Conservation of Nature (IUCN) in 2006 (Lowry et al. 2006), under their formal scientific peer review system, but this determination has no regulatory force in the United States.

In April of 2007, NMFS announced a zero subsistence harvest quota for the 2007 season, departing from the earlier low harvest agreement negotiated for the period till 2010, and three days later NMFS announced a recommendation for ESA listing of the Cook Inlet beluga. The agency is obliged to make a determination on listing within one year after a petition is concluded to warrant consideration. The 2007 population point estimate was slightly higher than that for 2006, but not by enough to alter the statistical conclusion of a downward trend. The slight short term run up from 2005 to 2007 looked rather like the short term runs for 1998–2000 and from 2002–2004, both of which were followed by declines ratcheting the series of point estimates downward overall. All these year to year changes in the point estimates were well within the noise level of the individual estimates. In April of 2008, NMFS invoked a clause of the ESA allowing a six month extension of the one year decision period "for purposes of soliciting additional data" if "there is substantial disagreement regarding the sufficiency or accuracy of the available data." The 2008 population point estimate was identical to that for 2007. On Oct. 17, 2008, NMFS announced the decision to list the Cook Inlet Beluga Whale population as endangered under ESA.

So far, the only overt regulatory action, consequent upon the Marine Mammal Protection Act determination that the population is *depleted*, has been the regulation of subsistence harvest. By Alaska standards, there is considerable human activity in and around Cook Inlet: the city of Anchorage and its port lie at the head of the inlet, there is ongoing oil and gas exploration in the inlet, and there are plans for a bridge across the inlet within the current range of the beluga population. Endangered Species Act listing will sharpen the focus on habitat-related issues.

The Cook Inlet beluga population now numbers around 300 individuals, based on the recent census estimates. Except for the federal legislatively imposed harvest moratorium in 1999, the 2007 quota is the first that has been set to zero on grounds that the population is too small or showing too little growth. In 1 year, though, under the terms of the court supervised stipulations, zero harvest was triggered by an observed "unusual mortality" for that year.

28.2.3   Perspective

The drift toward increasing protection for the Cook Inlet beluga seems reasonable, on the face of it, but the pace of the decisions may appear a little slow. One might wonder at the possible false starts and the amount of rework in the various assessments and the formal decision junctures. The procedural overhead certainly has been high. Why doesn't the process work more smoothly and efficiently? And, couldn't science help?

Much of the legislation authorizing the kinds of regulation under consideration here mandates the use of "best science," and the history recounted here transpired during an era of avowed governmental commitment to "science-based" environmental regulation. For the case of the Cook Inlet beluga decisions, we will consider here what facts the science was able to bring to the forum, and what policy was ready to use those facts. To the extent that the policy wasn't ready, we will consider how science could help the evolution of policy.

Bear in mind that during the period covered by this history Cook Inlet beluga was not the only marine mammal conservation problem in the world, and, though serious, it arguably was not the most pressing. Nor was it the most complicated or most contentious. This case has not yet involved commercial fisheries or the Magnuson-Stevens Act, for example.

28.3   The Science

28.3.1   The Data

Aerial surveys, conducted annually by NMFS since 1994, have been used to generate statistically defined annual population estimates (Hobbs et al. 2000). These census estimates by themselves provided the basis for concluding that a steep population decline was underway through the 1990s, and that by 2000 the population was down to around 500 individuals.

NMFS has compiled reports of harvest landings, and estimates of "struck-and-lost mortality" associated with the native harvest since 1994 (Mahoney and Shelden

2000). These data were sufficient to establish that the harvest in the 1990s constituted an overharvest, which could not be sustained by a whale population of this size with typical whale population dynamics.

NMFS also kept records of all reported strandings, and verified those that involved mortalities. The stranding information depends on reports of opportunity, and is not the result of a deliberate survey with known statistical properties.

28.3.2   The Analysis

The set of population estimates, in conjunction with estimates of harvest mortality, present an opportunity for statistical modeling to refine the inferences about the population size and trend, and about the underlying population growth rate for the period starting in 1994.

The intensive formal regulatory decision process has been going on since 2000. (I joined as an expert for the US Marine Mammal Commission at that time, and have been involved in more or less annual assessments since then.) At the onset of the decision process, the data trajectory was short – 1994 through 2000 – but each passing year has added another pair of data points on census and harvest, and therefore each year has allowed an assessment that is statistically more secure than the assessment preceding. In retrospect, now, we can document a detailed narrative of: what did we know, when did we know it, and, as the evidence in the case unfolded, what was (or should have been) predictable.

Figure 28.2 shows the relation to the data of a Bayesian analysis of a simple population model with constant (but unknown) intrinsic population growth rate, and superimposed harvest, using data for 1994–2006. The NMFS survey population estimate for each year is shown as a diamond, and the vertical thin bars give the reported 95% confidence interval for each NMFS survey. The harvest data are represented in terms of the estimates of harvest landed plus struck-and-lost shown as thick vertical bars from the x-axis.

The Bayesian model analysis is represented in Fig. 28.2 in terms of the posterior mode and the posterior 95% interval for the true population size for each year, shown as the one heavy and the two thin trajectories connecting those quantities for each year. The data, the model, and all the statistical analyses reported below, are documented in detail in the Appendix.

The Bayesian posterior mode for true population size tracks the survey point estimates in Fig. 28.2, as expected. The assumption of a constant underlying growth rate does not give rise to any apparent systematic lack of fit. The trajectory of the posterior modes clearly responds to the high harvests in the early years. The uncertainty of the model inferences in the early years is high: in those years the uncertainty in the NMFS survey estimates was high (because of lower survey effort in those years).

Overall, the model 95% posterior intervals for the true population size are considerably narrower than the 95% confidence intervals from the individual

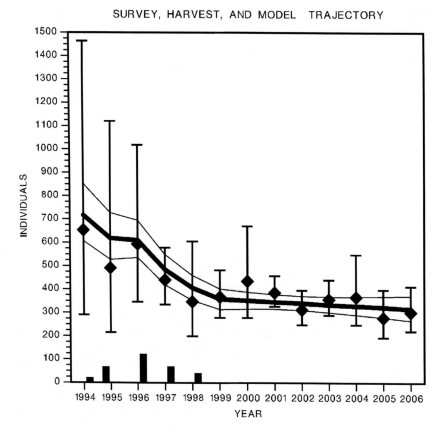

Fig. 28.2 Comparison of the census estimates and the fitted population model. The solid diamonds are the NMFS population survey point estimates, and the thin vertical lines are the 95% confidence limits on those estimates. The heavy trajectory connects the posterior modes for population size from Bayesian fitting of a population model; and the thin trajectories on either side of that show the 95% interval from the model. The thick vertical bars extending up from the x-axis are the NMFS estimates of harvest mortality

annual surveys. This is because the model is jointly supported by all the data, whereas the survey estimates themselves are not mutually reinforcing.

I presented the first such analysis to the Cook Inlet beluga decision process at the August 2003 meeting of technical experts working on the science-based harvest management plan; using data through 2002 (the 2003 estimate was not yet available). The disconcerting result, at that time, was the position of the posterior marginal on the inference for the underlying (intrinsic) population growth rate R. This distribution is shown in Fig. 28.3 as the broad, bell-shaped distribution (shaded).

The presumed "normal" range for growth rate of a whale population that is not experiencing the restrictions of carrying capacity is between 2% and 6% (Wade 1998).

INFERENCE FROM 1994-2002 DATA, FILLED; 1994-2006 DATA, OPEN

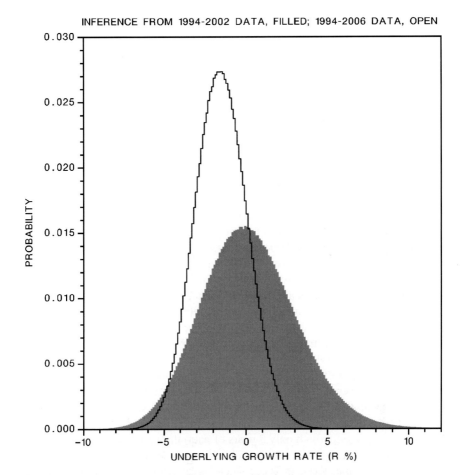

Fig. 28.3 Posterior distribution of underlying growth rate R, as percent, from the joint inference on underlying growth rate and true population size. The shaded distribution is based on the 1994–2002 data; the open distribution is based on the 1994–2006 data

The historic carrying capacity for Cook Inlet beluga is of course unknown, but we do know that the present population estimate is about one quarter of the estimate from 25 years ago. This naturally raises the question whether the population growth rate exhibited by the Cook Inlet beluga is abnormally low, compared to what would be expected for a healthy generic cetacean population.

The analysis based on the data through 2002 showed a posterior mode ("best estimate") for the underlying (harvest corrected) growth rate at around zero growth, rather than in the 2–6% "normal" range, and it showed a 74% probability that the growth rate is below 2%. This should raise serious suspicion that something, in addition to the harvest, is not right with this population. It also should cast doubt on the legitimacy of using an assumed intrinsic growth rate in the 2–6% range for modeling the effects of proposed harvest regimes on the future dynamics of the population – which is what the modelers (myself included) had been doing up till that point in the proceedings.

Other parties initially were reluctant to use this new, data-derived estimate of the population's underlying growth rate, citing the "uncertainty" of the estimate. One year later, in the administrative law hearing of 2004, the estimates of growth rate, obtained in this way, were, with the encouragement of the judge, adopted in the proceedings, replacing the assumption of 2–6% intrinsic growth.

The inference on underlying growth rate did show considerable uncertainty, as shown in the spread of the distribution (Fig. 28.3). But, of course, uncertainty is in the nature of statistical inference from limited data. The point is to quantify that uncertainty, and take it into account by allowing appropriate margins of safety in the decision process. The range of values for the growth rate encompassed by the uncertainty in the inference did not extend beyond the plausible: growth rates as high as 7% and 11% have been documented for two particular recovering whale populations (Bannister 1994; Branch et al. 2004), and rates of decline as steep as 8% are not out of the question if something is wrong in the circumstances of the population.

The narrower distribution (open outline) in Fig. 28.3 is the current inference on underlying growth rate using the data from 1994 through 2006. This showed a posterior mode at −1.5%, and a 98% probability that the underlying growth rate is less than 2%. The added data, 2003–2006, reduced the uncertainty (narrower distribution), and concentrated the probability in the lower half of the distribution inferred from just the 1994–2002 data.

One specific quantification of uncertainty in the inference on the underlying growth rate is the standard deviation of the posterior distribution of that growth rate. The sequence of open circles in Fig. 28.4 shows annual snapshots of how, as the data accrued, the uncertainty declined, as measured by posterior standard deviation in $R\%$, starting in 1998 (with only 5 years of data) through 2006. Interestingly, the reduction in uncertainty was itself predictable, and could have been anticipated based just on the data through 2002, when there was concern that the inference might be "too uncertain." The band of thin lines from 2003 to 2006 in Fig. 28.4 shows the 95% interval for the predicted posterior standard deviation based on the inference from the data through 2002; the band of thick lines shows the interior 50% interval. (Technically, these predictions are the result of a preposterior analysis, described in the Appendix.) The agreement between predicted and realized was obviously quite close.

Another view of the progression of the inference on the growth rate is given in Fig. 28.5, where a summary of the inference is plotted against the last year of data used, and the summary is in terms of the mode, the interior 50% interval and the 95% interval for the underlying growth rate. This makes clear the progression of shrinkage of the posterior distribution of R. The inferences earlier than 2001 span an unreasonable posterior range. Technically, this is owing to an interaction of the paucity of data in that period and the simple prior distribution used. With so little data up until 2001, a more sophisticated and complicated analysis should be done, as explained in the Appendix. From 2002 on, the amount of data is sufficient that the simple prior is acceptable.

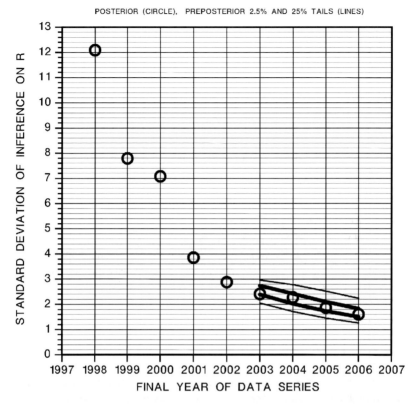

Fig. 28.4 Standard deviation of the posterior distribution of the underlying growth rate $R\%$, plotted against the final year used in the data sequence beginning in 1994. The circles are the standard deviations from the inference on the actual data. The heavy lines show the inner 50% interval of the predicted (preposterior) standard deviation based on the data from 1994 to 2002; and the thin lines show the 95% interval

Finally, Fig. 28.6 restates the progression of the inference on the growth rate by plotting, against the last year of data used, the probability that the growth rate is less than the reference values that label the contour lines.

Notwithstanding the uncertainty in the early years, the series of analyses tell a consistent story that there is considerable posterior probability of the growth rate being abnormally low. In the early years, this might have been discounted as due essentially to the very broad distribution reflecting the high uncertainty. By 2002, though, the resolution is good enough that the evidence of abnormally low growth rate cannot be dismissed, and by 2006 the resolution is good enough to rule out, as extremely improbable, the possibility that the growth rate is within the normal range. For the most part, the changes in the inference over time are simply a refinement (shrinking of the distribution) as a consequence of more data that are reasonably consistent with the previous data.

Indeed, there are now enough data to consider the question whether the growth rate might be changing, as reflected in evidence of change during the course of

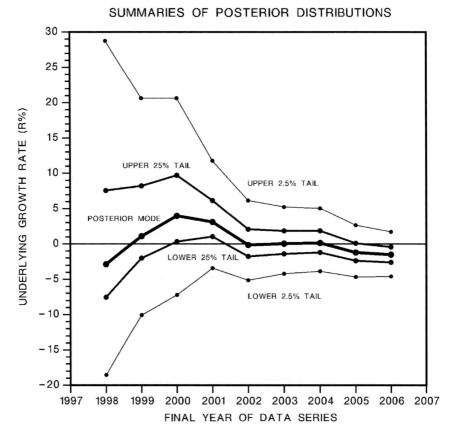

Fig. 28.5 Mode (thickest line) and 50% (medium lines) and 95% (thin lines) intervals of the posterior distribution of the underlying growth rate $R\%$, plotted against the final year used in the data sequence beginning in 1994

the observed period. This also gets at the concern that there may be uncertainties in the earliest part of the data series that are not fully taken into account in the analysis: the population abundance surveys are statistically characterized and their uncertainty is formally taken into account in the Bayesian analysis, but the struck-and-lost component of the harvest, which was large in the early years, was not reliably known and has not been statistically characterized. Figure 28.7 shows a summary of the inference on underlying growth rate plotted against the first year of data used, starting with 1999, where all analyses use data through 2006. Excluding the 1994–1998 data removes all the years with large harvests and large struck-and-lost mortality. As in Fig. 28.5, the summary is in terms of the mode, the interior 50% interval and the 95% interval for the underlying growth rate.

In Fig. 28.7 the sample size of years included decreases from left to right, and the uncertainty as revealed in the spread increases accordingly. Nevertheless it is

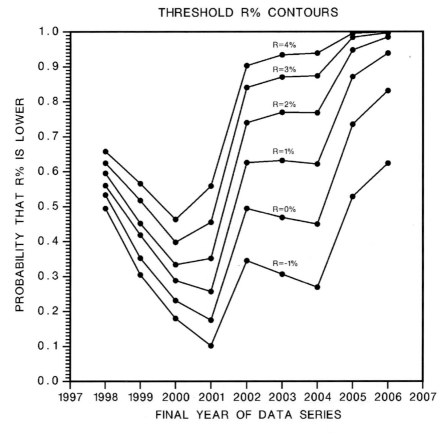

Fig. 28.6 Cumulative posterior probability that the underlying growth rate $R\%$ is less than the labeled contours, plotted against the final year used in the data sequence beginning in 1994

clear that the inference based on the years after 1998 indicates a *lower* underlying growth rate than the full data series (compare Figs. 28.5 and 28.7), and as the analyses are confined to the more recent data within the 1999–2006 interval the inference stays centered at this low mode.

The joint distribution of inferred true population size and underlying growth rate from the various analyses may be recast by displaying it in terms of the projected future population size, assuming that the current underlying growth rate persists unchanged, and assuming no further harvest. The projections for 100 years past the last census estimate used, with these assumptions, are shown in Fig. 28.8. By the time of the assessment using the 2006 census, the projected probability, under these assumptions, of fewer than ten individuals remaining in 100 years is 13.7%; the probability of fewer than 2 is 2.1%. This kind of projection constitutes one component of a "population viability analysis" (PVA), but it falls short of a full appraisal of the risk of extinction, which would be considerably higher than the 2.1% resulting here. The differences are explained in the Appendix.

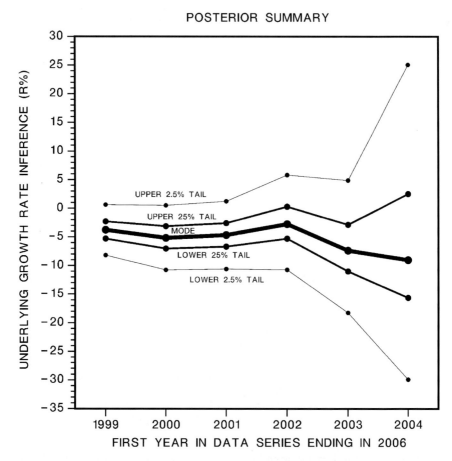

Fig. 28.7 Mode (thickest line) and 50% (medium lines) and 95% (thin lines) intervals of the posterior distribution of the underlying growth rate R%, plotted against the first year used in the data sequence ending in 2006

Because the statistical properties of the stranding reporting program are unknown, and the search effort is unquantified and probably not constant, the stranding data do not lend themselves to modeling. Inspection of those data does raise some pointed questions. Figure 28.9 shows the reported stranding mortalities by year, expressed as a percent of the population size (NMFS survey estimate for that year). The appearance of an upward trend catches the attention. Also, the recent tendency toward values around 3% annually is sobering, in that it could make up a fair portion of the estimated deficit in the population's underlying growth rate. If nothing else, this should raise the stakes for implementing a statistically designed carcass survey, or determining (calibrating?) the statistical properties of the existing program, and investigating the causes of the observed strandings.

CONTOURS FOR POPULATION SIZE 100 YEARS LATER

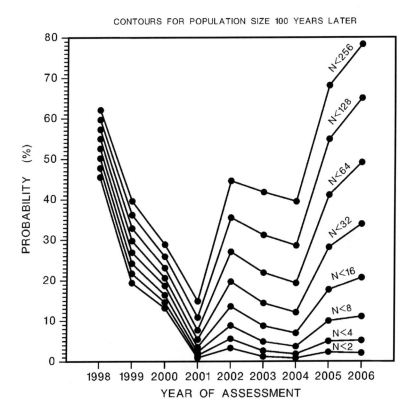

Fig. 28.8 Cumulative posterior probability that the population size 100 years later will be less than the labeled contours, plotted against the final year used in the data sequence, assuming constant exponential growth according to the inference on underlying growth rate $R\%$ and assuming no harvests after the year of the assessment

28.3.3   The Trajectory of Analyses in Relation to the Chronology of Decisions

Analyses of the sorts described in this section have been part of the continuing mix of information considered in the regulatory discussions. But, with the exception of the decision in 2000 to declare the Cook Inlet beluga *depleted*, the decision in 2007 to recommend Endangered Species Act listing, and the 2008 decision to list, it is not easy to pinpoint how these kinds of inferences have directly determined regulatory decisions made so far. The regulatory decisions made so far have only regulated native subsistence harvest, even though these analyses point to the conclusion that some other unknown factor must also be depressing the population growth rate. The kinds of analyses described here *were* used to quantify the expected performance of the subsistence harvest decision rule that was slated to be used starting in 2010.

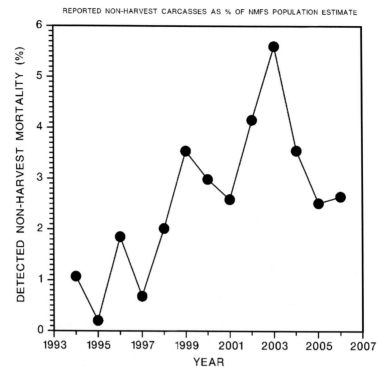

Fig. 28.9 Reported (NMFS) stranding mortalities shown as percent of the concurrent estimate (NMFS) of population size, by year

The fact that analyses of the sort described here did bear on the NMFS 2007 recommendation for Endangered Species Act listing, and the 2008 listing decision itself, allows us to revisit the impression that the decision process may have been "slow." Theses analyses have been done more or less annually starting in 2003 (when the 2002 survey estimate became available), and the graphs presented in this chapter show how the results of the analyses have changed through time as the data accumulated. The degree of certainty that the population merits Endangered Species Act listing in part is reflected in the certainty that the underlying population growth rate is abnormally low and in the certainty that current trends will lead to extinction. These certainties have increased over time from 2003, so perhaps it is a good thing that the slow decision has led to a more certain decision. On the other hand, during this time the actual condition of the population has deteriorated: as the numbers declined, the risk of extinction has increased. Investment in new research and monitoring to determine why the population is not recovering still has not been initiated. From this standpoint, the slow decision is not a good thing. So coming to a conclusion whether there is a problem with the pace of the decision process requires some basis for judging the balance between certainties of the evidence for the decision versus degree of risk to the resource. This balance will determine how much certainty should be, or should have been, enough for the decision to be made.

28.4   Policy Guidance

In order to make consistent, effective use of scientific population data and statistical assessments, regulatory decision rules must be stated in terms of three clearly specified elements: (1) a quantitative target for the desired state of the population, (2) a quantitative (probabilistic) confidence threshold for whether that target is being attained based on the available data, and (3) specification of actions that will be taken if this condition is not met (Goodman 2005). That is, a data-driven decision rule should have the form: the estimatable population characteristic C should be determined to be above target level T with probability P (at least); and if this cannot be demonstrated, action A is triggered.

The decisions under consideration for Cook Inlet beluga have occurred in a context where these three elements for a decision rule are not fully established in legislation or regulation or compelling court precedent. The National Environmental Policy Act specifies none of the three. Under the Marine Mammal Protection Act (MMPA), the target status defining *depleted* is quantitatively defined, but the confidence threshold is not. Alaska native subsistence harvest is exempted from many of the otherwise automatic prohibitions on *intentional take* and quantitative limitations on *incidental take* from *depleted* populations under the MMPA. NMFS has the authority under the MMPA to restrict native subsistence harvest when "necessary" for conservation, but no quantitative criteria have been promulgated.

Incidental take from *depleted* populations by commercial fisheries is governed by a quantitative target and a quantitative confidence threshold as defined by the "potential biological removal" (PBR) section of the MMPA. In particular, the PBR was developed to ensure with 95% probability that incidental take not delay the time to recovery by more than 10%, and to ensure with 95% probability that recovery be achieved within 100 years (Barlow et al. 1995), where *recovery* is defined quantitatively as a target state under the criteria for *depleted*. The PBR standards were discussed as potential models in developing a science-based subsistence harvest decision rule for the Cook Inlet beluga, but ultimately these were not adopted as binding standards for subsistence harvest of Cook Inlet beluga.

The MMPA is primarily directed at *take* and is not explicit about protecting habitat. The Endangered Species Act (ESA) definitely is concerned with habitat, where necessary. The ESA criteria for its most important operational decisions (*listing*, *jeopardy*, and *critical habitat*) are not defined quantitatively in legislation or regulation. Most individual ESA *recovery plans* do define case-specific quantitative target states for up-, down-, and de-listing; only a very few define an accompanying confidence threshold. One NMFS report has recommended an ESA listing standard, with quantitative target state and confidence threshold, for large whales: 1% probability of extinction within 100 years (Angliss et al. 2002). But this recommendation has not been formally adopted as regulation.

The absence of applicable preexisting quantitative standards goes a long way toward explaining the seemingly slow and repetitive decision processes at work in this example. From the standpoint of quantitative standards or criteria, the system has been operating in a policy vacuum. This has left the decision makers the unenviable burden of forming their own interpretations, at each step, and of trying at the same time to be reasonable in weighing the competing claims of affected or interested parties.

The social costs are obvious. The absence of a standard will force the decision process to repeatedly cover the same ground, open the door to inconsistencies, encourage lobbying and litigation, and make inefficient use of the efforts of the scientists who are brought into the process but are not told what is the crucial thing that they are supposed to measure, estimate, or optimize.

28.5   Scientific Basis for Quantitative Standards

Science cannot make quantitative standards out of whole cloth, because environmental standards ultimately are policy, which is an expression of political values. Science can calculate the probabilities of consequences of chosen courses of action, in light of the available information. The key, then, should be to obtain a clear enough statement of values for scoring consequences, so that science can be employed to select the best course of action, where "best" is judged unambiguously by the value score of the consequences.

The ideal form of a decision rule is a standard expressed in terms of a quantitative description of a desired state, a quantitative threshold for certainty of attainment, and specification of the action that will be taken if the threshold is not met. For people without specific technical training in fields such as decision theory or game theory, the most mysterious element of the ideal decision rule is the confidence threshold.

Intelligent people are pretty good at knowing what they want; and all it takes is a belief in cause and effect to provide a foundation for evaluating the probable efficacy of actions. But correct intuitions about probabilities are not a common trait, even among the well educated. How, then, can science help the people responsible for policy statements to correctly formulate a confidence threshold for a decision rule?

Oddly, examining case studies and hypothetical scenarios, or reviewing the details of a given case under consideration, are *not* good ways to elicit the information for establishing the appropriate confidence threshold. Understandably, participants to a decision are prone to have a preference as to how it will turn out, and this will color their opinions as to appropriate case-specific margins of safety. Similarly, the elicited responses to information about expected frequencies of decision outcomes are unreliable, most especially since, it so happens, that these frequencies are affected by many factors besides the decision rule itself, including the quantity and quality of available data and the frequencies of the actual states of nature.

Fortunately, and rather remarkably, the crucial information for determining the correct confidence threshold is not dependent on the case-specific details of the available data or the expected frequencies of outcomes. Therefore, it is possible to elicit this information in a context more removed from wishful thinking and the pressures of interested parties. The crucial information is simply a quantification of the relative desirability (or undesirability), per occurrence, of each of the possible outcomes.

28.5.1   *The Utility Function*

In the terminology of decision theory (Berger 1985), the measure of desirability is called a *utility function*. Alternatively it may be called a cost function or a loss function. (*Cost* is negative *utility*.) This measure is commonly applied to economic decision making, though the measure need not be monetary. All that is necessary about the units is that they be consistent across the spectrum of outcomes being evaluated, and that we know whether the units are a measure of net benefit (which we will want to maximize) or net cost (which we will want to minimize).

Consider a generic binary decision: (1) that action A is needed, or (2) that action A is not needed. Assume that the critical target state T, and its relation to the correct decision in the absence of uncertainty, is known: if the actual state S is known to be below T, then action is needed. The question is: when there is uncertainty about the actual state, what *probability* that the actual state is below T should justify the decision for action?

Imagine that there are four relevant possible outcomes to the binary decision:
(A) Decided that action was not needed, when it actually was not needed.
   (Called generically a "true negative.")
(B) Decided that action was not needed, when it actually was needed.
   (Called generically a "false negative.")
(C) Decided that action was needed, when it actually was needed.
   (Called generically a "true positive.")
(D) Decided that action was needed, when it actually was not needed.
   (Called generically a "false positive.")

Let the relative, per occurrence, *costs* of these respective outcomes be designated C_a, C_b, C_c, and C_d. The costs reflect net costs of implementation, yield, and social benefits, political costs of errors, and practical costs of emergency action or damage control which may be required in the event of an error.

Figure 28.10 illustrates one example set of these costs as the y-axis positions, labeled as the dots a, b, c, and d, while the x-axis positions are 0% or 100% probability that action really is needed as the case may be for respective "true" and "false" decisions. The line labeled "decide action is needed" connects the two dots for that decision, while the line labeled "decide action is not needed" connects the two dots for the other decision. The two endpoints of each line represent the costs of the decisions under circumstances of no uncertainty.

Intermediate positions on each line represent the average (technically, the "expected") costs of the labeled decision where the probability that action truly is

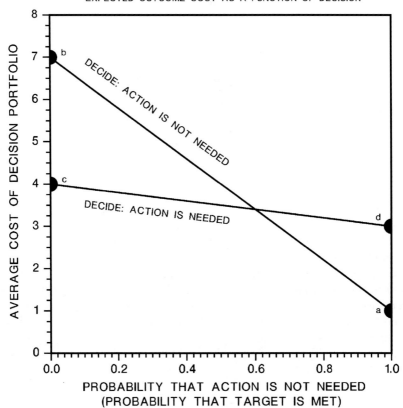

Fig. 28.10 Decision analysis for a binary decision with a cost structure given by the y-positions of the labeled points at the 0.0 and 1.0 positions on the x-axis: a is "true negative," b is "false negative," c is "true positive," and d is "false positive." The lines labeled "decide action is needed" and "decide action is not needed" show on the y-axis the expected cost of taking that decision as a function of the certainty that action is not needed, shown on the x-axis

not needed is given by the x-position. The costs change linearly with the probability of the true state, because for any given decision (line), the average cost is a mixture of the two costs corresponding to the endpoints, where the mixing proportion is the probability of the true state (i.e., the mixture is a linear combination).

For the realistic configuration where

$$C_b > C_c$$

and

$$C_d > C_a$$

the lines will cross within the domain of certainty between 0% and 100%. To the left of the intersection, the expected cost of deciding that action is not needed exceeds the expected cost of deciding that action is needed. To the right of the intersection, the reverse is true. Therefore, the x-axis projection of the intersection point (60% in the example in Fig. 28.10) is the critical certainty level. The algebraic solution for the critical probability P is

$$P = \frac{C_b - C_c}{C_b - C_c + C_d - C_a}$$

If the probability that action is not needed (as calculated from the case-specific data) does not meet or exceed this threshold, P, the decision (in order to minimize expected cost) should be to take that action. *So the cost structure alone is sufficient to determine the correct confidence threshold for the decision rule.*

Note that the confidence threshold does not, by itself, determine frequencies of decisions, frequencies of decision outcomes, or the frequencies of outcomes conditional on decisions or conditional on the state of nature. All these depend on frequencies of the states of nature and/or on quantity and quality of available case-specific data. *Therefore, the optimal (cost minimizing) decision rule cannot be deduced from statements of the desired frequencies of decisions or decision outcomes.*

The above comments extend to decisions with higher than binary dimensionality. The general formulation is that the optimal decision minimizes the integral of probability of outcome times net cost of outcome, where the cost is fully specified just in the utility function (not context-dependent), and the probability of outcome follows directly from the inference from data about the state of nature in the case being decided, and the integral is over all states of nature. The search for the minimum may become more complicated than just closed form algebraic solution, or simple geometry, but we have marvelous computer tools for finding the minimum by numerical means. For N possible actions and M relevant states of nature, the table of outcomes becomes N by M, each cell with its own value for cost, with row designating the action chosen and column designating the state of nature. The optimal decision is to choose the row for which the dot product of the cost values in that row and the inferred vector of probabilities of the states of nature, in this case, is minimum. The key point, for the purpose of this discussion, is that even with the more complicated set of possible choices and possible outcomes, the optimization of the decision still requires a utility function.

28.5.2   Performance Testing

"Performance testing" and "management strategy evaluation" are analyses of predicted performance of a decision rule in the context of specified scenarios. Based on the preceding section, such methods should *not* be used for selecting a confidence threshold for a decision rule, since the optimal confidence threshold for a decision rule should follow directly from the utility function. What these performance testing and management strategy evaluation explorations can offer, properly done, is a cost-benefit analysis of the possible investment in more or better data. In some literature this is called a "value of information" analysis. Here I will treat this kind of analysis in its Bayesian setting as preposterior analysis.

Rather than evaluate an arbitrarily selected scenario of the state of nature, preposterior analysis averages the evaluation over a distribution of scenarios whose probabilities are given by the actual inference from the existing data. The mathematical definition is given in Raiffa and Schlaifer (1961). Computationally, preposterior analysis consists of sampling the joint posterior distribution for the primary parameters, simulating future data conditional on the sampled parameter values, carrying out a "future" Bayesian inference on the parameters with the simulated data augmenting the existing data, and cumulating the "preposterior distribution" of the result of interest from the inference. This gives the probability distribution of the outcome of future inference based on the design of the future data collection, and taking into account the information derived from the data that are already in hand.

The preposterior inference has access to no more real information about the primary parameters than what resides in the existing data, so the preposterior distribution of the primary parameters will simply recapitulate the original inference with the existing data. What the preposterior inference does provide that is new is the option for conditional decomposition of the distribution of the future inference. The decomposition can be conditioned, for example, on the true value of the unknown parameters, thus showing the distribution of how decisions based on the future inferences will relate to truth.

Since the computational process of the preposterior analysis tracks the probability of the true values of the parameters as well as the probability of the inference, this allows for the calculation of the probabilities of the various possible outcomes of a decision rule in terms of the distribution of the relationship between decisions (based on the future inference and the decision rule) and the true state of nature. Here we analyze performance of an illustrative, but unofficial and entirely hypothetical, decision rule concerning whether the Cook Inlet beluga population growth rate is abnormally low compared to expectations for a normal healthy cetacean population.

Consider the decision rule that when there is less than some critical threshold probability that the underlying growth rate is above 1% the population will be classified as showing "abnormally low growth rate." If the probability that the growth rate is above 1% exceeds this threshold, the population will be classified as not having abnormally low growth rate. The premise of this made up example is that the determination of abnormally low growth rate would lead to some regulatory action or management intervention.

As in the discussion of the binary decision rule in the preceding section, there are four possible outcomes of applying the decision rule. These are:

1. "True negative" for a decision "negative" when R is genuinely above 1%
2. "False negative" for a decision "negative" when R is actually below 1%
3. "True positive" for a decision "positive" when R is genuinely below 1%
4. "False positive" for a decision "positive" when R is genuinely above 1%

Imagine that the cost structure for this set of outcomes is

Cost of true negative = 1
Cost of false negative = 30
Cost of true positive = 3.4
Cost of false positive = 2.4

With this cost structure, the critical threshold for certainty is 95%. So the optimal decision rule is: if there is not at least 95% certainty that the growth rate is above 1%, the population should be classified as showing abnormally low growth.

When there is perfect information about the growth rate, the decision will only be true positive or true negative, as the case may be. With uncertainty about the growth rate, some fraction of the decisions will be false positive or false negative.

The analysis that follows is based on the Cook Inlet beluga data through 2004. At that time the inference on the growth rate of the population showed a probability of 38% that the growth rate is at least 1%. Therefore, with the decision rule under consideration, the population would then have been classified as a "positive" and there would have been a 62% probability that this was a true positive and a 38% probability that this was a false positive. With increasing future data, the uncertainty about the growth rate will diminish, improving the accuracy of the determination of whether the growth rate is above or below 1%. As the accuracy of the determinations improves, application of the decision rule will convert some fraction of the current false positives to true negatives.

The projected change in performance of the decision rule as a function of the expected accrual of future data is graphed in Fig. 28.11. This prediction of performance

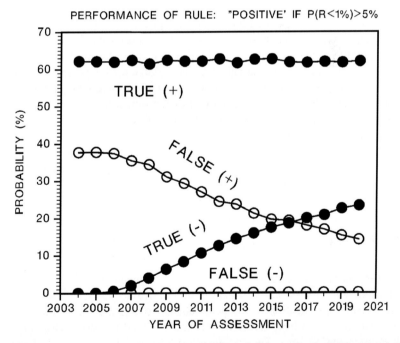

Fig. 28.11 Projected performance of the hypothetical decision rule: decide positive (action needed) if there is not at least 95% certainty that the underlying growth rate is greater than 1%. The projection is based on the data from 1994 to 2004, assuming that the coefficients of variation in the survey estimates after 2004 are a random sample of the values realized from 1998 to 2004

is intriguing. At first sight, it is something of a marvel that this kind of prediction can be done at all. But it is important to understand that this prediction of performance does not provide rational grounds for revising the decision rule.

There may be good reasons to dislike the prospects of deciding "positive," and there are surely good reasons for disliking false positives. But all this should have already been captured in the utility function which was used to calculate the confidence threshold for the decision rule in the first place (which could have been done before the data were obtained and before the decision rule was tentatively applied). The decision based on the actual data is simply the playing out of the decision rule with the available data; and the calculated (and predicted) probabilities of the respective outcomes are merely the automatic consequences of the decision rule and those same data. The fact of the decision itself, and the fact of the resulting calculated frequencies of outcome, does not provide any additional information that bears on the merits of the decision rule or the confidence threshold. Initially, the intuition may rebel at this fact, but that is part of why decision theory warrants the status of a distinct field of technical knowledge, to which thick textbooks are dedicated.

28.5.3   Value of Information

The analysis of performance as a function of the quantity and quality of future data does provide a basis for planning the investment in future data. One can calculate, for example, the probability that a specified increment in future data will change the current decision. Or one can calculate the reduction in expected costs of the future decision that will be realized when a specified increment of future data revises the probability distribution of the decision outcomes. The cost reduction in expected outcome can be compared to the cost of obtaining the specified increment of future data in an explicit cost-benefit analysis. Note that none of these value-of-information applications provide a substitute for the utility function. In fact they require that the decision rule and utility function already be known in order for the analysis to be carried out.

28.6   Epilogue

Unless there is a retreat from investment in fishery independent surveys and fishery monitoring, stock assessment will gain in descriptive accuracy, predictive power, and statistical sophistication in the years ahead. Expressing the results of stock assessments in probabilistic terms will naturally extend the stock assessments to "risk assessments" of sorts. The qualification, "of sorts," has to do with the metric for scoring risk.

From a decision theoretic standpoint, "the risk" is the expected undesirability of the outcome, where undesirability is measured in terms of net cost, in the broad and not necessarily monetary sense. The expected cost is the integral, over possible outcomes, of the product of the probability of the outcome and the per occurrence cost of the outcome.

In the absence of a utility function for scoring the per occurrence costs, there is nothing to integrate. The "risk assessment" then will be just the list of possible outcomes and their probabilities. This list may represent the result of high quality science, but it doesn't constitute a very definite guide to management decisions, and it doesn't offer scope for the science to optimize the decision.

With a utility function, the risk assessment delivers a single number, which quantifies the risk. If there are several alternative courses of action under consideration, each course of action leads to a different set of probabilities of outcomes, so each course of action gives rise to a different risk assessment and a different quantified risk. Assuming the utility function correctly represents value, the course of action associated with the lowest risk then is the best. In this context, the risk assessment provides an obvious guide to management decisions, and makes fullest use of the available science to optimize the decision.

The policy setting of fisheries management is likely to become more challenging in the foreseeable future, since there are growing demands for more considerations (notably ecosystem and protected species and cultural) and more constituencies to be taken into account, somehow. The operational goals for these new mandates tend not to be very precisely specified, and there often is little guidance on priorities for deciding among competing goals. Such a setting diminishes the role of science, which is a pity, since some of the issues, ecosystem in particular (Harwood 2007), involve many preeminently scientific questions that are very interesting and very important.

Effective engagement of science in the decision process depends on decision rules that respond directly to statistical estimates of specified quantities. The canonical form for such a decision rule would be: the estimatable quantitative system characteristic C should be determined to be above target level T with probability P (at least); and if this cannot be demonstrated, action A is triggered. Decision rules of this form are few and far between in present-day fisheries management.

Because Gresham's Law in reverse is not known to operate on the evolution of policy, we cannot expect good decision rules to arise spontaneously. The adoption of good decision rules will probably require some active promotion by the science community, since other constituencies may not find such rules naturally congenial or readily understandable. While it is perfectly legitimate for scientists to press for a particular *form* of decision rule (most especially on technical grounds), there is a delicate line which should be honored in pressing for the *content* of the decision rules, so as not to blur the distinction between science and policy.

As regards the choice of C, the critical system attribute, and T the target state, science can add clarity to the discussion by describing objective consequences.

The threshold confidence, P, poses a more difficult communication challenge. Individuals not trained specifically in the technicalities of these matters tend to have misguided intuitions about probabilities, and their instinctive request, for a menu of concrete scenarios to choose among, will, if complied with, generally lead to the wrong choice for the wrong reason.

The optimal (cost minimizing) confidence threshold for a decision rule is determined solely by the cost structure of the possible outcomes. If the cost structure is known, the optimal confidence threshold is obtained simply by calculation. The frequencies of outcomes, on the other hand, which are what the uninitiated ask about, are influenced by the quantity and quality of case-specific data, and the frequencies of states of nature, as well. Preferences about these frequencies, while real, are beside the point (except for weighing the merits of investment in more or better data), and do not provide a basis for optimization, or a basis for consistent decisions when confronting a range of circumstances presenting different amounts of data and different frequencies of true state.

Forcing a focus on the utility function, namely the per occurrence net cost of each possible outcome, will require some highly disciplined, patient, and persistent communication between scientific experts and the authorities responsible for policy. If this communication is successful in eliciting a utility function, it will expand the scope for useful, and legitimate, involvement of future fisheries science in the decision process, and lead to more consistent, more efficient, potentially more rapid, and hopefully less contentious, decision making.

References

Angliss RP, Lodge KL (2004) Alaska Marine Mammal Stock Assessments, 2003. U.S. Department of Commerce NOAA Tech Memo NMFS-AFSC-144, 230 p

Angliss RP, Silber GD, Merrick R (2002) Report of a workshop on developing recovery criteria for large whale species. NOAA Tech Memo NMFS-OPR-21

Bannister JL (1994) Continued increase in humpback whales off Western Australia. Rep Int Whal Comm 44:309–310

Barlow J, Swartz SL, Eagle TC, Wade PR (1995) U.S. Marine Mammal Stock Assessments: Guidelines for Preparation, Background, and a Summary of 1995 Assessments. NOAA Tech Memo NMFS-OPR-956, September 1995, 76 pp

Berger JO (1985) Statistical Decision Theory and Bayesian Analysis. Springer Verlag, New York

Branch TA, Matsuoka K, Miyashita T (2004) Evidence for increases in Antarctic blue whales based on Bayesian modeling. Mar Mam Sci 20:726–754

Frost KJ, Lowry LF (1990) Distribution, abundance, and movements of beluga whales, *Delphinapterus leucas*, in coastal waters of western Alaska. Can Bull Fish Aquat Sci 224: 39–57

Goodman D (2002) Uncertainty, risk and decision: the PVA example. In: Berkson JM, Kline LL, Orth DJ (eds) Incorporating Uncertainty into Fisheries Models. Am Fish Soc Symp 25:171–196

Goodman D (2004) Taking the prior seriously: Bayesian analysis without subjective probability. In: Taper M, Lele S (eds) The Nature of Scientific Evidence. University of Chicago Press, pp 379–409

Goodman D (2005) Adapting regulatory protection of marine mammals to cope with future change. In: Reynolds JE, Perrin WF, Reeves R, Montgomery S, Ragen TJ (eds) Marine Mammal Research: Conservation Beyond Crisis. Johns Hopkins University Press, Baltimore, MD, pp 165–178

Harrison CS, Hall JD (1978) Alaskan distribution of the beluga whale, *Delphinapterus leucas*. Can Field-Natural 92:235–241

Harwood J (2007) Is there a role for ecologists in an ecosystem approach to the management of marine resources? Aquat Conserv: Mar Freshwater Ecosyst 17:1–4

Hobbs RC, Rugh DJ, DeMaster DP (2000) Abundance of beluga, *Delphinapterus leucas*, in Cook Inlet, Alaska, 1994–2000. Mar Fish Rev 62(3):37–45

Laidre KL, Shelden KEW, Rugh DJ, Mahoney BA (2000) Beluga, *Delphinapterus leucas*, distribution and survey effort in the Gulf of Alaska. Mar Fish Rev 62(3):27–36

Lowry L, O'Corry-Crowe G, Goodman D (2006) *Delphinapterus leucus* (Cook Inlet Population). In: IUCN, 2006 IUCN Red List of Threatened Species.

Mahoney BA, Shelden KEW (2000) Harvest history of belugas, *Delphinapterus leucas*, in Cook Inlet, Alaska. Mar Fish Rev 62(3):124–133

O'Corry-Crowe GM, Lowry LF (1997) Genetic ecology and management concerns for the beluga whale (*Delphinapterus leucas*). In: Dizon AE, Chivers SJ, Perrin WF (eds) Molecular Genetics of Marine Mammals. Soc Mar Mammalogy, Spec Pub No. 3, pp. 249–274

O'Corry-Crowe GM, Suydam RS, Rosenberg A, Frost KJ, Dizon AE (1997) Phylogeography, population structure and dispersal patterns of the beluga whale *Delphinapterus leucas* in the western Nearctic revealed by mitochondrial DNA. Mol Ecol 6:955–970

Raiffa H, Schlaifer R (1961) Applied Statistical Decision Theory. Division of Research, Harvard Business School, Boston (reprinted by Wiley)

Rugh DJ, Shelden KEW, Mahoney BA (2000) Distribution of belugas, *Delphinapterus leucas*, in Cook Inlet, Alaska, during June/July 1993–1998. Mar Fish Rev 62(3):6–21

Speckman SG, Piatt JF (2000) Historic and current use of lower Cook Inlet, Alaska, belugas, *Delphinapterus leucas*. Mar Fish Rev 62(3):22–26

Taylor BL, DeMaster DP (1993) Implications of non-linear density dependence. Mar Mam Sci 9: 360–371

Wade PR (1998) Calculating limits to the allowable human-caused mortality of cetaceans and pinnipeds. Mar Mam Sci 14(1):1–37

Appendix: Details of the Cook Inlet Beluga Modeling Analysis

Data

The data used were those supplied by NMFS in the course of reporting to the US Marine Mammal Commission and the administrative law review of the regulation of subsistence harvest for this population. The population estimates, and their reported error coefficients of variation, were from NMFS surveys conducted as per Hobbs et al. (2000); the harvest estimates were obtained as per Mahoney and Shelden (2000) except that the value used for struck-and-lost in 1996 was the mid-point of the reported range of the estimate. The non-harvest carcass reports are recorded by NMFS, but are not the result of a designed survey. These data are given in Table 28.1.

Table 28.1 Data for the Cook Inlet Beluga population

Year	Population estimate	CV of population estimate	Landed harvest report	Struck and lost report	Non-harvest carcass report
1994	653	0.430	19	2	7
1995	491	0.440	42	26	1
1996	594	0.280	49	73.5	11
1997	440	0.140	35	35	3
1998	347	0.290	21	21	7
1999	967	0.140	0	0	13
2000	435	0.230	0	0	13
2001	386	0.087	1	0	10
2002	313	0.120	1	0	13
2003	357	0.0107	1	0	20
2004	366	0.200	0	0	13
2005	278	0.180	2	0	7
2006	302	0.160	0	0	8

Population Model

The population dynamics for most years were modeled simply as

$$N_{t+1} = (N_t - H_t)\left(1 + \frac{R}{100}\right),$$

where N is the population size at the time of that year's survey, H is that year's harvest including mortality ascribed to struck-and-lost, and R is the presumed constant underlying fractional rate of increase in units of percent. The non-harvest carcass reports are not used in this model, so the mortality that they represent is absorbed into the non-harvest mortality component of R. The value of R does not include the effect of harvest, so it represents the underlying (harvest corrected) growth rate.

This model applies where the harvest takes place in a short season, just after the annual survey, which was the case for all years except 1995 in the data record.

In 1995, the harvest took place just before the survey, so

$$N_{1995} = (N_{1994} - H_{1994})\left(1 + \frac{R}{100}\right) - H_{1995},$$

and

$$N_{1996} = N_{1995}\left(1 + \frac{R}{100}\right).$$

This basic population dynamics model is not identical to the one used throughout most of the analyses used in the administrative law proceedings and regulatory assessments. That model, in part for consistency with earlier, and other, modeling used in the regulatory process (Angliss and Lodge 2004), represented the dynamics as density dependent, using the Taylor and DeMaster (1993) equation. Because the time series of census data could not be used to estimate the density dependence parameters (carrying capacity and the shape parameter) these were fixed at "stipulated" values in the administrative law proceeding. Because the population sizes are far below the stipulated carrying capacity, the more complicated density-dependent model made almost no difference.

Likelihood Model

The likelihood model accepts at face value the reported coefficients of variation from the NMFS survey, and calculates σ, the error standard deviation of the population estimates, as the CV for that year times n, the point estimate of population size for that year.

$$\sigma_t = n_t CV_t$$

Simulations conducted by Rod Hobbs, and reported during the course of the 2003 administrative law review of the regulation of subsistence harvest for this population, indicated that the population estimate error is normal, or possibly t, distributed. The likelihood model adopted here treats the population estimate as normally distributed about the true population size, with a known standard deviation for that year calculated from the reported CV of the population estimate. So, the likelihood contribution for year t is

$$L = \left(\frac{1}{\sqrt{2\pi}\sigma_t} \right) e^{-\frac{1}{2}\left(\frac{N_t - n_t}{\sigma_t} \right)^2}$$

Accordingly, treating error in population estimates in the respective years as independent, the joint likelihood of the data series of m years is the product of m such likelihood terms. In the absence of the simulations indicating Gaussian error, it would have been conventional to assume log normal census error, and the corresponding likelihood model would be log normal.

Bayesian Inference

Because the series of true population sizes is assumed to follow the deterministic population model, the entire trajectory of true population sizes can be calculated from knowledge of R and the true population size in any one year. Therefore, the statistical inference was carried out as a joint inference on two unknown parameters, N_{1994} and R. This was conducted as a Bayesian inference, using the 1994–2006 data series, the joint likelihood model, the population model, and independent uniform priors on N_{1994} and R, broad enough so that the breadth of the priors did not constrain the posterior distributions. The data series contained enough information by 2001 that this casual approach to defining a prior was innocuous. For shorter segments of the data series an empirical prior would have been necessary to get good inferences (Goodman 2004). If decision rules using Bayesian inference are intended for use in very data poor situations, correct choice of an informative prior will be very important.

The posterior distribution of true population size in any year after 1994 can be calculated as a derived parameter from the joint posterior distribution on N_{1994} and R. The preposterior analyses were carried out by calculating the reporting quantity as a derived parameter, obtained from the distribution of future data conditional on N and R, and computing the reporting quantity as a result of future inference on the future data. The Fortran programs used for carrying out the inferences are documented under the "Bayesian Algorithm Project" available at www.esg.montana.edu.

Population Viability Analysis

Population viability analysis (PVA), in a Bayesian setting, attempts a probabilistic appraisal of the risk of extinction, most generally presented as a posterior distribution of the time to extinction or the time to decline below some other specified population size threshold. The distribution reflects uncertainties in the values of dynamic parameters as well as the playing out of random processes in the dynamics (Goodman 2002). Further, the analysis attempts to take into account biological (especially genetic) and environmental mechanisms which make the dynamics progressively less favorable as the population becomes smaller. The probabilistic projection shown in Fig. 28.7 captures only the parameter uncertainty of the population trend component, and does not incorporate other pertinent mechanisms affecting the population dynamics or their uncertainties. To this extent, then, a thorough PVA for the Cook Inlet beluga would give higher probabilities of extinction than are shown in Fig. 28.7 of the main text.

Chapter 29
The Future of Fisheries Science on Canada's West Coast Is Keeping up with the Changes

Richard J. Beamish and Brian E. Riddell

Abstract A look back at the issues in fisheries management on Canada's Pacific coast identifies a history of surprises. "Expect the unexpected," was the advice of W.E. Ricker. Surprises will always occur, but fisheries science needs to be in a position to minimize their economic impacts and explain the causes to the people who manage and care about fish and fisheries. For example, we now recognize the critical role of climate and the ocean in the regulation of recruitment. However, we do not understand the mechanisms that link climate to the life-history strategies of key commercial species. We know that marine ecosystems off British Columbia are warming and we know that marine ecosystems can change rapidly. However, without a better understanding of the processes that link climate to fish abundance, fisheries science will be restricted in the advice it provides to managers and patrons. We also know that human populations will continue to grow and increase the demand for seafood. Expansion of marine aquaculture and ocean ranching is the only way to meet this demand. There are excellent aquaculture-related opportunities in the coastal communities of British Columbia, but the impacts on natural resources are not clear and poorly researched. Wild fish and shellfish, properly harvested and properly managed, will likely continue to command premium prices. These conditions and others will affect the kind of fisheries science we do on the Pacific coast of Canada. Effective fisheries science organizations in the future will be the ones whose leaders adapt the fastest to new knowledge and new issues by forming teams of researchers that combine experience, curiosity, and new thinking. An independent fisheries science research advisory board will help make the best use of all available fisheries science in the province. With this approach, surprises may become learning experiences.

Keywords British Columbia · pacific coast · fisheries management

R.J. Beamish and B.E. Riddell
Fisheries and Oceans Canada, Pacific Biological Station, 3190 Hammond Bay Road, Nanaimo, BC V9T 6N7 Canada
e-mail: Richard.Beamish@dfo-mpo.gc.ca; Brain.E.Riddell@dfo.mpo.gc.ca

R.J. Beamish and B.J. Rothschild (eds.), *The Future of Fisheries Science in North America*, 567
Fish & Fisheries Series, © Springer Science + Business Media B.V. 2009

29.1 Introduction

In the early 1900s much of the early science related to marine fisheries on the west coast of Canada was conducted by scientists at the Pacific Biological Station in Nanaimo, British Columbia. By the 1960s, the reputation of the science and scientists was highly regarded around the world. However, despite the strong scientific support, the history of Canada's Pacific coast fisheries has been punctuated by major surprises. In this chapter, we describe a few of the key surprises to show that there is an unexpected side to fisheries science that will become more problematic as climate change compounds an already variable environment, and future environmental conditions interact with the cumulative effects of fishing. We speculate about the future issues in fisheries science on Canada's Pacific coast as a way of encouraging colleagues, managers, fishermen, and the general public to prepare for an uncertain future and the complexity associated with sustaining both fisheries and natural resources.

The hindsight and foresight relating to the complexities of fisheries science and management may not be clear, but it is good enough to show that there are two distinct approaches for the future. One approach would be to stay the course, hoping that future impacts of climate on fisheries and ecosystems can be managed with little change in how we do our science. According to this scenario, whatever science is available (sometimes called "the best available science" or "the weight of evidence") should be adequate to make timely biological and social adjustments when making management decisions. The second approach, which we think evolves logically from the lessons of the past, is to learn from the past but recognizes that future environments and social pressures will be different from those in the past. We suggest that it is necessary to rethink what science is needed, how we manage and conduct our science, and that a more adaptive approach is needed to navigate fisheries science through the uncharted waters of the future.

29.2 Examples of Surprises

The major fisheries on Canada's Pacific coast have been targeted on spiny dogfish (*Squalus acanthias*), Pacific herring (*Clupea pallasi*), Pacific sardine (*Sardinops sagax*), the aggregate of Pacific salmon species (*Oncorhynchus* spp.), Pacific halibut (*Hippoglossus stenolepis*), and recently Pacific hake (*Merluccius productus*) (Beamish et al. 2008a). Most fisheries research has focussed on Pacific salmon because this group of six major species are highly valued as commercial and recreational species. They provide food, and social and ceremonial values to First Nations, and are iconic as general indicators of ecosystem health. It is difficult to assign budgets to species, but a good guess is that about 70% of the fisheries research dollars assigned to Pacific coast fisheries have been spent on Pacific salmon.

29.2.1   Pacific Salmon

In 1954, W.E. (Bill) Ricker published his famous paper on stock and recruitment of Pacific salmon (Ricker 1954). Yet despite the improved understanding of the dynamics of Pacific salmon that resulted from Ricker's insights, Canadian catches of Pacific salmon did not increase in the 1960s and 1970s (Fig. 29.1a). This was sufficiently puzzling that in 1973 Bill published another famous paper in which he wrote, "a puzzling problem of Pacific salmon ecology is why the runs of major river systems, when brought under the best available management, rather consistently fail to produce levels close to what has generally been expected of them on the basis of their past history" (Ricker 1973). The concern was that Pacific salmon abundances and the resulting catches were not rebuilding as expected from the stock and recruitment models. We now know that the ocean rearing areas were in a period of reduced capacity to produce Pacific salmon (Beamish and Bouillon 1993; Beamish et al. 2004, 2008b). It was not until after the 1977 regime shift (Beamish and Bouillon 1993, 1995; Francis and Hare 1994; Mantua et al. 1997; Zhang et al. 1997; Minobe 2000) that the productivity and catches of Pacific salmon increased off the west coast of Canada (Fig. 29.1a). In fact, the highest Canadian catch in history occurred in 1985 which surprised everyone.

At about the same time as Ricker published his paper in 1973, other scientists studying Pacific salmon suggested that the failure of Pacific salmon to rebuild to historic abundances was a consequence of insufficient juveniles being produced in freshwater. One well-known researcher wrote that there seemed to be lots of head-room for expansion of salmon populations before the rearing areas of the ocean became a limiting factor. This consensus that juvenile Pacific salmon abundance was limiting total abundance led to the establishment of the Salmon Enhancement Program (SEP), a massive hatchery and artificial salmon rearing program (Fisheries and Environment Canada 1978) that still exists. However, after 1985, Pacific salmon catches declined steadily and now are approximately one half the historic average. As Canadian catches declined, the total catch by all other countries increased (Fig. 29.1b). Canadian commercial catches declined from average levels in the late 1960s and early 1970s of about 19% of total catches of all countries before SEP (1977) to about 3% of the total catches from the mid-1990s to 2005 (Fig. 29.1c). Included in this decline was the total collapse of a major recreational fishery for coho salmon in the Strait of Georgia (Beamish et al. 1999).

It is now recognized that climate profoundly affects the ocean carrying capacity for Pacific salmon (Beamish and Bouillon 1993; Francis and Hare 1994; Mantua et al. 1997; Beamish et al. 2000; Finney et al. 2000, 2002; Ruggerone and Goertz 2004; Briscoe et al. 2005). The climate effects are shown as trends (Trenberth 1990; Trenberth and Hurrell 1994; Mantua et al. 1997; Thompson and Wallace 1998; Minobe 2000; Yasunaka and Hanawa 2002) and are not random as was originally assumed. It is also recognized that adding artificially reared salmon to the ocean is tricky (Beamish et al. 2008a) and may even result in the replacement of wild stocks (Hilborn and Eggers 2000). The target of SEP to double the Pacific salmon catch has not been met and

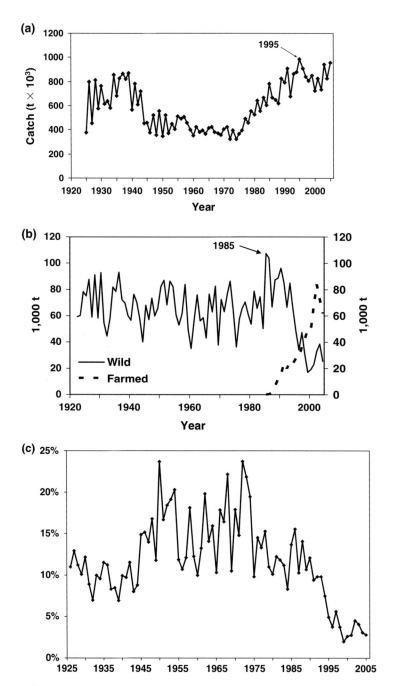

Fig. 29.1 (**a**) Total Canadian catch of Pacific salmon from 1925 to 2005. The annual production of farmed salmon is shown separately. (**b**) Total production of farmed salmon produced and wild salmon caught in British Columbia from 1924 to 2005. The largest total catches occurred in 1995, 2003, and 2005 with 985,100, 942,400, and 959,000 t, respectively (for 1925–1992 data email Chrys.Neville@dfo-mpo.gc.ca; for 1993–2005 data, see http://www.npafc.org/new/pub_statistics. html). (**c**) Percentage of Canadian salmon in total Pacific salmon catch from 1925 to 2005

the program has not really adapted to the recognition that the original assumptions of unused and/or stable ocean capacity are invalid. Perhaps the biggest surprise is that everyone remains puzzled by the current low Pacific salmon catches in Canada when the total Pacific salmon catches by all other countries are at historic high levels. After decades of research, monitoring, and analysis, our ability to explain the determinants of Pacific salmon production in British Columbia remains surprisingly limited.

29.2.2   Aquaculture

As the wild and hatchery Pacific salmon catches were declining, production from salmon farming was increasing (Fig. 29.1b). There was no vision of the potential of Atlantic salmon farming in the 1960s and 1970s and the concept of culturing fish off the coast of British Columbia was unheard of. A well-known researcher advised in the late 1960s that aquaculture might equal the efficiency of rearing chickens, but it could scarcely surpass it. Today, the food conversion ratio (dry weight of food to wet weight of fish) for salmon farming is 1.2:1, compared to 2–3:1 for chickens (Brown et al. 2001). Atlantic salmon (*Salmo salar*) production in British Columbia is now two and one half times larger than wild Pacific salmon catches and the comparative value (Fig. 29.2) is about four times greater (BC Seafood Industry Year in Review 2006, available at www.env.gov.bc.ca/omfd/reports/YIR-2006.pdf). The industry in 2002 provided about 4,700 full-time jobs in many areas of the province needing employment (BCSFA 2003).

Aquaculture generally and salmon farming in particular are controversial for a variety of reasons, but the main reason is that the rapid development of the industry was a surprise to most people. As the industry developed, management agencies did not establish monitoring programs and advisory bodies. Impacts of salmon farming on wild Pacific salmon were not researched until environmental groups alarmed the public (Morton et al. 2004). In the absence of an established research program,

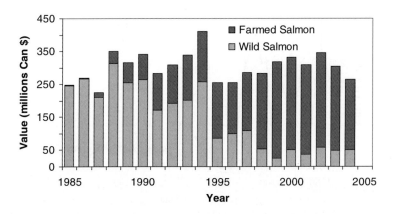

Fig. 29.2 Value of wild and farmed salmon in British Columbia from 1985 to 2004

ideological debates flourished. Hindsight shows us that it was a mistake not to take aquaculture and salmon farming seriously when Atlantic salmon farming was approved in British Columbia in the early 1980s. It was unfortunate that the salmon farming industry did not recognize that culturing an exotic species of salmon in British Columbia would be so controversial. It was equally unfortunate that fisheries scientists, who have a responsibility to provide management advice, did not have research programs that evaluated the impacts of salmon farming or other forms of aquaculture.

29.2.3   Pacific Herring

The Pacific herring fishery has always been a major fishery on Canada's Pacific coast. After the collapse of the sardine fishery in the late 1940s (MacCall 1979; Beamish et al. 2008b; McFarlane and Beamish 1999), Pacific herring were actively fished for reduction to fish meal and fish oil (Fig. 29.3). The annual increases in catch did not appear to affect recruitment resulting in several well-known scientists reporting that it was not possible to overfish herring. In less than 10 years after this statement was made, the Pacific herring fishery was closed (Fig. 29.3). The collapse of the Pacific herring fishery is now recognized as resulting from too high a fishing mortality at a time of continual poor recruitment into the fishery (Beamish et al. 2001). The Pacific herring fishery resumed in the early 1970s (Fig. 29.3) as a roe fishery that has been sustained to the present. The recent fishery is considered to be well managed with a series of regulations that prevent overfishing. However, the mechanisms that caused the herring recruitment failure are still not clearly understood. In fact, after about 70 years of research, the factors affecting year-class strength of Pacific herring off Canada's Pacific coast are still very poorly understood.

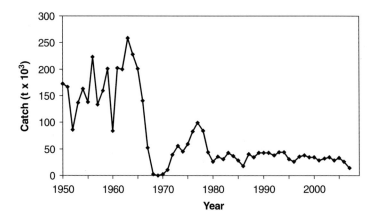

Fig. 29.3 Canadian catch of Pacific herring from 1950 to 2007

29.2.4   *Groundfish*

In the 1960s, groundfish were considered a relatively minor fishery. The rapid increase in catches beginning in the mid-1980s (Fig. 29.4) was not expected. Thus, it was a surprise that the value of the groundfish fishery today is about $300 million, exceeding the combined value of Pacific salmon and Pacific herring fisheries (BC Seafood Industry Year in Review 2006, available at www.env.gov.bc.ca/omfd/reports/YIR-2006.pdf). We know of no study or even an opinion that recognized this possibility.

There are 59 species in the current groundfish fishery and it is noteworthy that about 30 species have maximum ages exceeding 30 years (Beamish et al. 2006). It was not until the 1980s that it was recognized that many species were substantially older than previously thought (Beamish et al. 1983). Leaman and Beamish (1984) proposed that the length of life of a species is related to the length of time over the evolutionary history of a species that climate-related conditions were unfavorable for reproduction. Thus, the importance of climate and climate trends in both the life-history strategy and the stock and recruitment relationships for groundfish species have been recognized only recently. However, it is still unknown whether older fish in a population are needed for a species to adapt to climate changes. Fishing down the age composition or "longevity overfishing" is still a developing concept (Beamish et al. 2006).

There were a number of surprises associated with the success of the groundfish fishery. One example worth noting relates to spiny dogfish, as they could play a key role in the future. One of the first major fisheries off Canada's Pacific coast was for spiny dogfish with landings from the Strait of Georgia from 1870 to 1950 that exceeded the landings of Pacific salmon (Beamish et al. 2008c). The fishery collapsed in 1950 when synthetic vitamin A became available (Ketchen 1986).

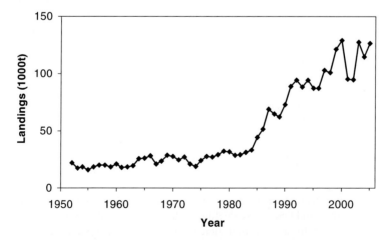

Fig. 29.4 Groundfish catches off the coast of Canada from 1951 to 2005

From 1950 to 1972, there were a number of attempts by both government and industry to eradicate spiny dogfish, largely because people did not like them (Beamish et al. 2008c). There were stories about the harm spiny dogfish did to more charismatic species, although there was no scientific proof. The facts are that spiny dogfish are very slow growing and thus consume only between 1.5 and 2 times their body weight a year (Brett and Blackburn 1978), thus they are not voracious predators. Spiny dogfish mature at an average age of 30 years for females and produce 3–11 "pups" about every 2 years (Beamish et al. 2008c; McFarlane and Beamish 1986; Ketchen 1975). Maximum ages are about 100 years (Beamish et al. 2006). Thus, some spiny dogfish that survived the "liver fisheries" in the 1930s and 1940s could still be alive today. The scientific issue today is that after more than 100 years of exploiting this slow-growing, long-lived fish, we know little about the animal. It is surprising that one of the major species in our west coast fisheries has survived massive fishing pressures and 20 years of eradication attempts, yet there is no understanding of why they are currently so abundant, what regulates their abundance and what their role is in the marine ecosystem.

29.3 Future Issues in Fisheries Science on Canada's Pacific Coast

Species evolve life-history strategies to survive in a naturally changing environment, but how do these strategies adapt when humans intervene and/or their environment changes? Unexpected events in fisheries science are to be expected as Bill Ricker advised. We think that fisheries science should be prepared to expect even more surprises as the impacts of global warming and climate change increasingly affect the poorly understood life-history strategies.

In Section 29.4 we present our vision of the future issues in fisheries science on Canada's Pacific coast. We begin with the key species and a brief assessment of their potential responses to changes in climate. We then select ten issues that we suggest will affect fisheries research on Canada's Pacific coast. We conclude by suggesting how to make more efficient use of the scientific effort that is available.

29.4 The Species

29.4.1 Pacific Salmon

Pacific salmon will remain the most important group of species even though their commercial catch and value have declined. The values of wild commercially caught and sport caught salmon may well increase with the development of new markets and specialty products, but we expect the catch to remain much reduced from historic values, a result of biological production and management/allocation decisions.

Beamish and Noakes (2004) and Beamish et al. (2008b) examined the impact of climate on the past, the present, and the future of the key Pacific salmon species in British Columbia. Beamish and Noakes (2004) speculated that there would be a general increase in Pacific salmon production in the subarctic Pacific, but production in British Columbia would decline. They also suggested that Pacific salmon will move into the Canadian Arctic in increasing numbers. Pacific salmon stocks south of about 55°N will probably be most adversely affected by a warmer climate as there appears to be an oscillation of productivity about this latitude (Hare et al. 1999). Sockeye salmon (*Oncorhynchus nerka*) in the Fraser River will be severely affected as the species is virtually at its southern limit (Rand et al. 2006). Pink (*O. gorbuscha*) and chum (*O. keta*) salmon will likely be less affected because of their reduced dependence on freshwater ecosystems and their use of mainstem channels. However, the marine ecosystem will also have an important influence on these two species (Beamish et al. 2008a). Coho (*O. kisutch*) and chinook (*O. tshawytscha*) salmon are not at their southern limits in British Columbia, but they enter the ocean later than the other salmon species and could be impacted by a trend toward earlier plankton production. In general, over the next 30 years, there will be much greater variability in Pacific salmon production in British Columbia that, in contrast to the past, will be mainly related to climate impacts in the ocean and in freshwater rather than the effects of fishing. The impacts of economic development and water use will likely exacerbate the effects of climate, again increasing uncertainty and competition among resource industries.

29.4.2   Pacific Herring

Beamish et al. (2008b) concluded that Pacific herring would likely continue to fluctuate in abundance over the next 30 years. Earlier plankton production should favor improved juvenile survival, but an increase in predators offshore, such as Pacific hake, would reduce recruitment. Pacific herring fisheries are conservatively managed (Schweigert 2001), so it is quite unlikely that overfishing will be a direct factor in Pacific herring production. The warming trend in the Strait of Georgia could become a problem (Fig. 29.5). Research is needed to determine how an additional degree Celsius will affect spawning behavior and prey availability for first-feeding larvae. Pacific herring off the west coast of Vancouver Island may be in low abundance because of predation from Pacific hake, which may increase in abundance as the ocean warms off Vancouver Island.

29.4.3   Pacific Halibut

Pacific halibut off British Columbia are virtually all recruited from waters off Alaska. Thus, it is the conditions in the Gulf of Alaska and perhaps the Bering Sea that most affect the abundance of Pacific halibut off British Columbia. Furthermore,

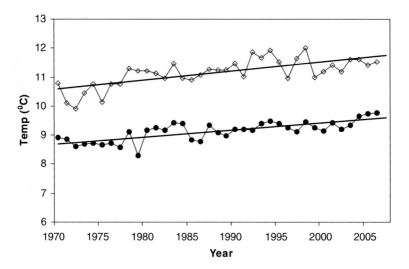

Fig. 29.5 Annual bottom (solid dot) and surface (open diamond) temperatures from 1970 to 2007 at the Nanoose Station in the Strait of Georgia

changing ocean conditions may reduce the southward migration of Pacific halibut into the waters off Vancouver Island. If there is a strengthening of the Aleutian Low as identified by Mote et al. (1999), then Pacific halibut production may be either similar to that of the past 30 years or even increase. By contrast, if there is a weakening of the Aleutian Low (Overland and Wang 2007), Pacific halibut production would be expected to decline. Pacific halibut are managed conservatively through the International Pacific Halibut Commission and it is unlikely that overfishing in the directed fishery would occur. However, the mortality of juvenile Pacific halibut as bycatch off Alaska remains high (Salveson et al. 1992). The inability to reduce this bycatch along with climate-induced changes in behavior may require a rethinking of Pacific halibut management strategies in Canada.

29.4.4 Sablefish

Sablefish (*Anoplopoma fimbria*) currently are the second most valuable of the west coast groundfish fisheries ($30.5 million in 2006), behind Pacific halibut ($53.9 million in 2006), with Pacific hake ranking third ($26.9 million in 2006; BC Seafood Industry Year in Review 2006, available at www.env.gov.bc.ca/omfd/reports/YIR-2006.pdf). Sablefish are a long-lived (a maximum of 113 years, McFarlane and Beamish 1983, 1995), deepwater species (Beamish et al. 2006) and thus have survived periods of unfavorable ocean conditions over evolutionary time. If care is taken in their management and older age classes are protected (Beamish et al. 2006), then it is possible that sablefish will continue to sustain a fishery as they have over the past 30 years. However, relatively little is known about the linkages between climate and recruitment. It would be wise to acquire this information as quickly as possible in case there is a "sablefish surprise" around the corner.

29.4.5   Pacific Hake

Pacific hake currently support the largest fishery, by weight, in British Columbia (about 100,000 t [BC Seafood Industry Year in Review 2006, available at www.env.gov.bc.ca/omfd/reports/YIR-2006.pdf]). Most Pacific hake are caught off the west coast of Vancouver Island. There is a relatively large and separate population of smaller Pacific hake in the Strait of Georgia that receives very little fishing pressure. Over the past 50 years the Pacific hake population in the Strait of Georgia has increased and is now the dominant biomass there. It is likely that they will continue as the dominant species over the next 30 years. This is important because they are also the major food source for harbor seals (*Phoca vitulina*; P. Olesiuk, personal communication 2007 Nanaimo, BC) that can be significant predators on other species including Pacific salmon. An emphasis on ecosystem-based management in the Strait of Georgia will require a much improved understanding of the relationship among the species that have direct trophic relationships with Pacific hake.

The impacts of climate change on the offshore population of Pacific hake will affect the production and the number that migrate into waters off British Columbia. It is possible that a warmer climate will result in larger spawning populations off the coast of California and more migration into the Canadian zone. If this occurs, there will be greater predation by Pacific hake on species such as Pacific herring and salmon as well as larger fisheries for Pacific hake.

29.4.6   Pacific Ocean Perch and Other Rockfish

It is remarkable how little we know about climate impacts on Pacific ocean perch and the large group of other rockfish species in the groundfish fishery. These species are generally long-lived and thus are able to survive long periods of ocean conditions unfavorable for reproduction. However, they are relatively easy to overexploit because of their schooling behavior. It is probable that a management strategy may be to close large areas to fishing. Such a strategy, would at a minimum, require more monitoring of the rockfish fishery. Preferably, more research into how climate affects recruitment would be needed. When wild fish command premium prices in the future, rockfish could become one of the most valuable groups of species in British Columbia. Thus, it is in the long-term interest to manage these species conservatively now. Their future will be determined mainly by the amount of fishing and our ability to manage the harvest at least over the next 30 years.

29.4.7   Pacific Cod

Pacific cod (*Gadus macrocephalus*) used to be one of the key groundfish species. We know that warmer bottom temperatures are detrimental to the early development of juvenile Pacific cod (Alderdice and Forrester 1971). It is possible that the current bottom temperatures in the Strait of Georgia (Fig. 29.5) are already too warm for their continued survival and it is likely that Pacific cod will all but disappear in the Strait of Georgia and off the west coast of Vancouver Island over the next 30 years.

The Species at Risk Act (SARA) may be imposed as a way to provide protection for these disappearing Pacific cod stocks. If this were the case, other fisheries would be affected. Clearly, there is a need for research to determine if the current and expected declines are a natural response to climate change, in which case it is unlikely that stocks could be rebuilt unless climate change impacts are reversible in the long term, which is also unlikely.

29.4.8   Spiny Dogfish

Spiny dogfish currently are worth more per kilogram than pink salmon. Who would have thought that a hated pest would become a sought-after species? There is no indication that spiny dogfish populations are overexploited; but they need to be carefully managed, because SARA can be used to protect stocks that are in low abundance. If this happened, other fisheries for other species could also be impacted. One difficulty is that very little is known about how to manage spiny dogfish. We do not understand the mechanisms that affect recruitment, control abundance, or even where and when reproduction occurs. If the demand for spiny dogfish continues, we have a lot of catching up to do to understand its population ecology.

29.4.9   Shellfish

Geoduck clams (*Panope abrupta*), Dungeness crabs (*Cancer magister*), and spot prawns (*Pandalus platyceros*) are the major species in Pacific shellfish fisheries. There is very little information about the factors affecting their production. Larval crabs are a key prey species of many species including Pacific salmon, particularly in the Strait of Georgia. A focus on ecosystem-based management will require a much improved understanding of the dynamics of shellfish, in general, and crab and shrimp species, in particular. The landed value of shellfish (wild and cultured) in British Columbia was $127 million in 2006 which represented 16% of the total landed value of all seafood. Prawns, geoducks, and crabs were the most valuable species, with landed values in 2006 of $38.7, $33.0, and $23.1 million, respectively. Harvests have remained relatively stable for prawns and geoducks from 2004 to

2006 (BC Seafood Industry Year in Review 2006, available at www.env.gov.bc.ca/ omfd/reports/YIR-2006.pdf). Annual catches of Dungeness crabs have been relatively stable in the Strait of Georgia, but variable in other areas. The reasons for the stability and variability are not known. Dungeness crabs, spot prawns, and geoducks are at about their center of distribution in British Columbia; thus, any initial increases in temperature would not be expected to have a major impact. It is the change in currents and the availability of bottom habitat that could have the greatest impact through the larval stages. In general, there is not enough known about the factors that regulate recruitment of these species to determine how they will respond to climate changes over the next 30 years. It is possible that there may be greater variability in recruitment and continued uncertainty about the causes of such variability.

29.5   Future Issues in Fisheries Research

We selected ten issues that we think will drive fisheries science on Canada's Pacific coast over the next few decades and possibly longer. Issues in ocean and climate sciences also need to be considered and integrated into fisheries science, but these considerations are outside the scope of this chapter.

29.5.1   Issue 1: Climate Change

Climate change is clearly the key factor affecting the future of fisheries science and resource trends along Canada's Pacific coast. In British Columbia, for example, the Strait of Georgia has warmed almost 1°C over the past 40 years (Fig. 29.5). Currently, the impacts of this warming are not understood much beyond speculation.

Only over the last 20 years has the importance of climate in the dynamics of fish and fisheries been fully appreciated. In the future, climate will rule. It probably is fair to conclude that we cannot confidently identify how any one species in the Pacific coast fishery will respond to climate change. Informed speculation is useful, but impact assessments need to become more quantitative if we are to move from alarming the public and ourselves to identifying management actions. An extensive effort is needed to understand how climate will impact fish and fisheries. Initially, retrospective analyses using statistical models will help to identify possible responses of species and stocks to change, but mechanistic models are needed to predict climate change impacts. Retrospective studies are helpful in providing insights into past relationships, but it is unclear how the records from the past will represent the future.

Pacific salmon may be the first species to be impacted because of their relatively short life span. There is some understanding of climate-related impacts on salmon in freshwater but, even though Pacific salmon spend most of their life in the ocean, our understanding of the linkages between climate and productivity in the marine environments is at best rudimentary. The importance of understanding this linkage has never been more evident than when one examines the salmon returns to British Columbia in 2006 and 2007. Pacific salmon that entered the sea during 2005 suffered extremely poor survival as evidenced by: (a) very poor returns of pink salmon to British Columbia and southeast Alaska in 2006; (b) poor spawning returns of coho salmon to southern British Columbia in 2006; (c) extremely poor returns of age-4 sockeye salmon from the Fraser River north to the Skeena River in 2007; and (d) poor returns of age-4 chinook salmon to the Fraser River and southern British Columbia in 2007. However, while British Columbia had the poorest ever returns of Pacific salmon, catches of Pacific salmon in Japan, Russia, and most of Alaska are, in aggregate, at record high levels (Fig. 29.1b). Understanding the spatial and temporal impacts of climate change on the population ecology of Pacific salmon is a challenge to the scientific community. The overwhelming problem is the inadequate logistical capability to conduct coastal marine ecosystem research, let alone participate in any international, open-ocean studies.

29.5.2 Issue 2: Wild Salmon Policy

A policy for the conservation of wild Pacific salmon was adopted in 2005 (Fisheries and Oceans Canada 2005). The policy was developed as a management framework to conserve Pacific salmon in an uncertain future while continuing to provide a range of benefits to Canadians. Under the Wild Salmon Policy (WSP), wild salmon will be maintained by identifying and managing Conservation Units that reflect both their geographic and genetic diversity, and defining two benchmarks to assess the status of each Conservation Unit. The upper benchmark or reference point defines a unit that is healthy and capable of sustaining harvest; the lower benchmark defines a Conservation Unit that may be at risk and should be managed with conservation in mind. The lower benchmark is precautionary through a significant "buffer" that will differentiate the benchmark from conditions under which a Conservation Unit is at risk of extinction. The policy also begins to implement ecosystem-based management for salmon and requires the development of habitat and ecosystem (freshwater and marine) assessment frameworks in the Yukon and British Columbia. While the WSP required nearly a decade of drafts and public consultations, its completion is timely and its effective implementation is likely to dominate the management of Pacific salmon for the next decade.

We have identified climate change as the key issue in the immediate future; however, no one can predict either the rate or magnitude of change. The WSP includes the elements needed for effective conservation, but requires a commitment to full implementation, including: assessment and monitoring programs (Strategies 1–3), development of an effective regional advisory structure (Strategy 4) for communities that participate in resource decisions, and an independent review process (Strategies 5 and 6). Implementation is a logistical challenge by itself. Climate change is likely to exacerbate this challenge by increasing debate over resource uses (e.g., forestry, water allocation, fishing), access to salmon, and how risk-averse society chooses to be. The first public review is scheduled for 2010 and the outcome of that review will determine the midterm agenda for the WSP.

The hatchery/wild Pacific salmon interaction debate (Meffe 1992; Hilborn and Eggers 2000; Levin et al. 2001; Myers et al. 2004; Zaporozhets and Zaporozhets 2004) is one issue that will continue, particularly if social science decisions are made that support recreational fisheries using artificially produced Pacific salmon. The definition of "wild" in the WSP was problematic, but an aggregation of Pacific salmon that spawns naturally and has offspring that spawn naturally is considered wild. In other words, enhancement could be used to increase the spawning abundance of Pacific salmon, but the effectiveness of enhancement to the WSP will be assessed based on its net benefit (including interactions) to natural production. Implementation of the policy requires a significant commitment to better monitoring and support for science.

29.5.3   Issue 3: Pacific Salmon Hatcheries

Our Canadian Pacific salmon hatcheries will continue to produce salmon for the next 30 years, and will require resources to repair an aging infrastructure. Most biologists now recognize that adding more juvenile Pacific salmon to the ocean will neither double the total catch nor even guarantee sustainable catches. The use of major hatcheries and spawning channels is a complicated issue with both successes and failures, but with extraordinary public support (PFRCC 2005). Even recently, with increasing evidence of concerns for chinook and coho hatchery production, the idea of using hatchery fish to sustain recreational fisheries through mark-selective fisheries is increasingly popular. A recent study of hatchery and wild coho salmon in the Strait of Georgia showed that climate is linked to marine survival through the amount of early marine growth (Beamish et al. 2008a). It was proposed that the impacts were greater on hatchery coho salmon because of their later release dates. There were other differences in the population ecology of hatchery and wild coho salmon, indicating that the two types are not identical in the ocean. As climate change negatively impacts the production of natural chinook and coho salmon in the southern part of British Columbia, we foresee that hatchery fish will be used experimentally to support recreational fisheries in fishing zones that depend on hatchery fish, particularly in the Strait of Georgia. The difficulty, however, continues to be that the impact of hatchery fish on wild fish is not even close to being understood. Future research may show that there is competition between wild and hatchery salmon that affects early marine growth and ultimately marine survival and that the effect of interactions varies with environmental conditions. When marine conditions are good for the production of salmon, there may be no concern for such interactions. However, when conditions are poor such interactions could be substantial, with lasting impacts on wild stocks.

After 30 years of the Salmonid Enhancement Program in Canada very little has been learned about hatchery/wild salmon interactions. The debate between enhanced and natural salmon has been largely fueled by beliefs *versus* knowledge, a result of a lack of research investment. This situation will need to change in the immediate future if the determinants of marine survival are to be understood within an increasingly uncertain climate, particularly within the WSP. Possibly, the social science decision to enhance the expectation of catching coho and chinook salmon in the recreational fishery will result in additional resources for research. In the absence of a strategic plan, however, efforts to support such recreational fisheries with marked hatchery fish constitutes nothing more than costly trial and error. Studies are also urgently needed to determine the causes of the early marine mortality in salmon generally. It is remarkable that most salmon that enter the ocean die, yet we know very little about what kills them.

29.5.4   Issue 4: Certification

Certification is a relatively new initiative in world fisheries but its short history suggests some success involving consumer purchasing power to generate change and promote environmentally responsible stewardship of the world's fisheries. Certainly, the most widely known certification process is that of the Marine Stewardship Council (MSC) (http://eng.msc.org/) established in the late 1990s. In a bid to reverse the continued decline in the world's fisheries, the MSC developed environmental standards for sustainable and well-managed fisheries and a product label (eco-labeling) to identify environmentally responsible fisheries and management practices. Consumers concerned about overfishing and its environmental and social consequences can choose seafood products, which have been independently assessed against the MSC standard and identified if the standards are met. The MSC's mission statement succinctly summarizes their concept: "To safeguard the world's seafood supply by promoting the best environmental choice" (web site above). Their web site now reports that as of September 2007 there are 857 MSC-labeled seafood products sold in 34 countries worldwide. Over 7% of the world's edible wild-capture fisheries are now engaged in the program, either as certified fisheries (22 fisheries) or in full assessment (30 other fisheries) against the MSC standard for a sustainable fishery. In British Columbia, four sockeye salmon fisheries received conditional certifications in 2007.

The potential power of consumers and marketing has been further developed by specification of seafood products from "sustainable" fisheries as identified by environmental organizations. Examples are: SeaChoice (http://seachoice.org/) and Seafood Watch (http://www.mbayaq.org/cr/seafoodwatch.asp).

While we believe that the educational value of these initiatives will continue to be important, we suggest that a better understanding of the state of the world's fishery resources combined with the response of management agencies will actually reduce the market value of certification and eco-labeling. These initiatives have been important in bringing greater exposure of overfishing and ecosystem impacts of fishing to the public, but there are clear indications that the risks of overfishing and related ecosystem impacts are being recognized and acted upon. Market values may very well adjust to availability, quality, and preference for products but we think that fisheries will be adjusted before market pressures require it. It is certainly possible, however, that the profitability of some fisheries will depend on market responses and how well those fisheries succeed in becoming more sustainable. We suggest this for three reasons. First, the state of marine resources, the risks associated with overfishing, and the need for change have been prominent in recent legislation in Canada and the United States, particularly in the US re-authorization of the Magnuson-Stevens Fishery Conservation and Management Act of 2006 (http://www.nmfs.noaa.gov/sfa/magact/), and the Oceans Act and the Species at Risk Act in Canada. It is particularly notable that the Magnuson-Stevens Act sets a deadline for ending overfishing in the United States by 2011. Second, the emergence of ecosystem-based management as a new paradigm in resource management has significantly broadened debate about the impact of overfishing and the impacts of fishing on habitats. While the literature on ecosystem-based management is expanding rapidly, papers describing the concept include the National Research Council (1999), Pikitch et al. (2004), and Sinclair et al. (2002). In general, we expect the need to consider ecosystem-based impacts will reduce fishing rates that were previously based on single-species assessments and improve sustainability. However, there is a major concern regarding the latter – the likelihood of recovery of overfished marine resources. Third, reviews of the recovery of overfished marine fish populations indicate that recovery may be protracted and highly uncertain (Hutchings and Reynolds 2004; Hutchings 2000). Consequently, the impacts of overfishing may be more prolonged than previously assumed. We suggest that this will become a significant issue providing support for more precautionary approaches toward fishing marine resources.

Assuming that resource managers and policy makers are similarly motivated to improve the state of our marine resources, there is an adequate scientific basis for acting to improve the sustainability of fisheries, long before the market pressures discussed above stimulate the industry to change. The importance of certification to Canadian west coast fisheries may very well rise and fall over the next 30 years.

29.5.5   Issue 5: Species at Risk Act

In Canada the Species at Risk Act (SARA, Government of Canada 2003) is now law. Under this legislation, any citizen can request a review of the status of a wildlife species (a species, subspecies, variety, or a geographically or genetically distinct population; paragraph 2 of the Act). If through a prescribed peer-review process, the species is recommended for listing (i.e., protection) and if the Federal government accepts the recommendation, then the responsible management agency must by law protect the species and its habitat, and establish procedures to restore its abundance. For British Columbia fishes and fisheries, given our history of development and fishing, SARA will increasingly pose limitations on future development, land and water uses, and fishing. This may be particularly true for Pacific salmon, because within the WSP over 400 Conservation Units will likely be defined. For other important species such as sablefish or Pacific halibut, there are few definitions of stocks or conservation unit subgroups. As these stock units are identified there will be major impacts on traditional fishing practices as SARA is used to provide protection.

Under a climate change scenario, however, SARA may become increasingly restrictive without any guarantee of successful protection. We expect that the number of wildlife species requiring protection will increase as climate impacts affect their survival. As it may not be possible to reverse the impacts of climate change, government officials will have to produce a recovery plan for any listed species, but with less and less assurance that the species can be secure, even with full protection of its habitat. Thus, if a species or stock declines in abundance because of climate change, the decline should not be any more reversible than climate change itself. SARA may need to be revisited to account for the impacts of climate change.

SARA is directed at single wildlife species and its associated habitat. Such a focus is inconsistent with the development of ecosystem-based management strategies and directives. Whether through development effects, climate change, or both, ecosystems are dynamic and will change continually (e.g., mountain pine beetle impacts, aquatic invasive species, water budgets). As ecosystems change, we can also anticipate human populations and cultures to adjust. The singular focus of a SARA listing will undoubtedly come into conflict with climate and ecosystem changes and human responses. Even if the rationale for SARA continues to be accepted, the variability associated with climate-related impacts will most likely result in mind-numbing debates about causes and effects. If the listings continue under the legal requirement to restore species, then there will need to be many more biologists working to understand the population ecology of a diversity of stocks and species. It appears that fisheries science is just beginning to feel the effects of SARA. In the medium term, however, we expect that SARA will be rethought and revised. An ecosystem approach that establishes marine protected areas could be integrated into SARA. For example, instead of establishing small areas for protection, larger areas such as the entire Strait of Georgia could be a marine protected area, but with fishing allowed in specific areas. A species needing protection would simply not be made available and would respond naturally to the changing ecosystem.

29.5.6   Issue 6: Aquaculture and Ocean Ranching

In September 2006, the Food and Agriculture Organization (FAO) reported that "While in 1980 just 9 per cent of the fish consumed by human beings came from aquaculture, today 43 per cent does, ..." (FAO Technical Report No. 500, 2006; available at http://www.fao.org/docrep/009/a0874e/a0874e00.htm).

Where will anybody who eats seafood caught on Canada's Pacific coast get their seafood in the future? We propose that the public acceptance of cultured seafood will facilitate increased investment and provide employment in aquaculture in coastal areas where jobs are needed. The FAO reports that there will be a world increase in seafood consumption of about 25% over the next 30 years (FAO 2007). Fish farming is the world's fastest growing food production sector, exceeding the annual rate of livestock production by a factor of 3 (FAO 2007). Approximately half of all fish consumed by humans is now raised on farms. Within 25 years, the world population is expected to consume about 83 million tons of farmed fish, up from 46 million tons in 2004 (FAO 2007). A limitation to the growth of aquaculture and the aquafeeds industry is the virtually fixed supply of fish oil and fish meal. We speculate that technology will be developed to genetically engineer plants to produce the proteins needed in the various diet formulations. This technological advance will reduce the cost of aquaculture resulting in a supply of inexpensive seafood that is certified as safe to eat and safe for the environment. With an affordable and plentiful supply of safe seafood, management agencies will be able to reduce fishing rates and thus rebuild overfished stocks. In British Columbia, for example, wild seafood could become a premium product, changing the nature of fisheries for many species, but for rockfish in particular. It is also possible that the added value for wild species may result in more fisheries research being supported by the private sector. Salmon farms will continue to be the major supplier of salmon to the world as it becomes a more affordable and safe food source. Salmon farming will also achieve ecological sustainability through reduced ecological impacts. However, like all farming, there will be new challenges related to climate impacts, diseases, and perhaps invasive species. All challenges will require research support and if the Canadian Department of Fisheries and Oceans is not seen to be providing the appropriate research, then aquaculture will be moved to another agency. Certification of products from salmon farming will become more important than eco-labeling of wild seafood products.

We think that private ocean ranching will be authorized. It is probable that the initial pilot projects will be approved in areas with limited natural Pacific salmon production, perhaps associated with First Nations' groups. Russia, Japan, and the United States already have major hatchery programs that release billions of young salmon into the common feeding areas for Pacific salmon on the high seas. These countries recognize the opportunities of ocean ranching. Canada will eventually follow their lead. There are opportunities for Canada to culture species and stocks such as "summer" chum salmon. The science in support of these pilot projects may be conducted outside of government, but government experts will be needed to evaluate proposals and review reports. There is an obvious linkage between hatcheries and the WSP that needs to be agreed upon and managed.

29.5.7   Issue 7: Ecosystem-Based Management Will Lead to Regional Management

Assessments of the abundance of commercially important species will continue to be made at the single-species level, but there will be a switch from single-species management to ecosystem-based management. Ecosystem-based management will need to include an evaluation of the role of a species within its ecosystem. This is a perspective that will require communication among management agencies, fishing interests, and those who want to see natural resources protected. It is likely that some areas will be closed to all fishing, except perhaps recreational fishing, for a few species. As previously indicated, we think that within the medium term, the Strait of Georgia probably will become a marine protected area, perhaps with fishing limited to First Nations and some recreational users. The strategy would be to close all fishing and open specific fisheries for species that are known to have surplus production within the ecosystem. A major new research activity will be the development of reliable models linking climate to the production of plankton through to fish production. The extension of fishery models to many non-fishery-related interests (e.g., industry, community development) will significantly broaden the range of groups involved in resource assessments, fishing allocations, and decision making. Stock assessment even at single-species levels will become more open and transparent via the use of web sites, standardized assessment models, and common data systems. It is likely that there will be virtual stock assessment agencies that use this standard software to produce internationally accepted stock assessments.

It would be the height of presumption to imply that humans can manage an entire ecosystem, but we can certainly alter the components of it. Ecosystem-based management is considered a new management paradigm and it explicitly involves human activities, communities, and values (Christensen et al. 1996). With continued human population growth and a changing climate we can anticipate increased conflict among resource users and between these users and the natural systems. However, if ecosystem-based management successfully involves more people in resource assessments and decisions, then more people may understand the complexities of the interaction of climate and recruitment, and fewer may immediately blame government officials when stocks decline. Improved public awareness will help promote what Robert Feynman (Feynman 1998) called, "honesty in science." According to Feynman, honesty in science is telling intelligent people what they need to know to make intelligent decisions, rather than giving them information that encourages them to support a particular message. An informed public will also recognize the inefficiencies of government agencies that maintain separate and even competing departments with overlapping responsibilities. A movement towards more regional management structures and processes (e.g., Strategy 4 of the Wild Salmon Policy; Fisheries and Oceans Canada 2005) will need to ensure that there are broader overall policies to resolve the trade-offs between resources and between users. The improved awareness of what we actually know and do not know probably would have the consequence of reducing the amount of fishing to account for uncertainty. However, as previously mentioned, we anticipate a substantial added value to the price of wild fish that are properly handled and processed, with a net result of less spent on fuel, more sustainable exploitation rates, and an increase in earnings.

29.5.8   Issue 8: Improved International Cooperative Research to Make Use of the Best Science

All species in the commercial fisheries off Canada's Pacific coast occur in the territorial waters of other Pacific Rim countries. Pacific salmon also share a common pasture in the waters beyond 200 miles. The science conducted by each country is shared among all scientists through the peer-reviewed literature, conferences, and workshops. We think that it is time to coordinate this science to ensure that information needed to manage exploited species in a changing climate becomes available to all countries, faster and cheaper. Existing organizations such as the North Pacific Anadromous Fish Commission (NPAFC 2001) and the North Pacific Science Organization (PICES) offer opportunities to share information and, in some cases, to plan research (also see Armstrong et al. [1998] for Atlantic salmon). In the future, these two organizations as well as others such as the International Pacific Halibut Commission need to integrate their activities to help understand how key species are responding to a changing climate.

In the future, we suggest that more Pacific salmon research be directed to the common feeding areas of Pacific salmon in cooperation with scientists from Russia, Japan, and the United States. Tremendous advances in genetic stock identification, new archival and acoustic tags, and effective capture methods provide the technologies needed to understand how climate affects the survival, distribution, and productivity of Pacific salmon stocks in the ocean, and issues of competition among salmon from the member countries. NPAFC has established a team approach through BASIS (Bering-Aleutian Salmon International Survey; NPAFC 2001) that can be further developed to improve greatly the ability to forecast returns reliably. The initial steps to extend the studies into the North Pacific will begin in 2008. Ultimately, NPAFC will have to establish principles of applying the best available science (as recently outlined by Nugent and Profeta [2006] for the United States) in the resolution of debates among these salmon-producing countries.

Canada and the United States communicate through the Pacific Salmon Commission (PSC, www.psc.org) to manage some stocks of mutual interest. The Commission supports a number of research projects, although there is no overall research plan. As climate impacts become more worrisome, it will become apparent that it is in each country's interest to have a science plan and a proposal granting process that recognizes the contributions from integrated teams. Further, the information that is collected from teams needs to be used according to a new code of ethics. As well a new reward system needs to be put in place for scientists in the teams.

29.5.9   Issue 9: The "Watson Effect"

Bill Ricker noted that everything appears simple once it is discovered. He called this the "Watson effect" after Mr. Sherlock Holmes' trusted assistant. Major discoveries often follow from major technological advances such as new kinds of microscopes or the use of coded-wire tags. In recent years, there have been a number of major advances in fisheries science. Satellites became operational, computers arrived and increased their power according to Moore's Law that computing power doubles every 18–24 months (see http://en.wikipedia.org/wiki/Moore's_Law), the Internet came into being, and genetic stock identification became DNA-based. In addition to these remarkable new technologies and techniques, we speculate that there will be new discoveries in the future that will change our thinking about the processes that regulate the organization of marine ecosystems.

Climate impacts occur on different scales, but the decadal scale is the most frequently observed. Decadal-scale shifts in climate are called regime shifts, which can be defined as climate-forced persistent changes in the marine ecosystems. The most dramatic regime shift in recent years occurred in 1977 (Trenberth 1990; Mantua et al. 1997; Zhang et al. 1997; Thompson and Wallace 1998; Beamish et al. 2000, 2004; Minobe 2000; Benson and Trites 2002; Yasunaka and Hanawa 2002). In British Columbia, this regime shift affected the trends in abundance of a number of fish species (Beamish and Bouillon 1993, 1995). Despite the major physical, biological, and economic impacts of this shift, there is still no understanding of the processes that caused the sudden shift in atmospheric circulation and wind intensity over the subarctic Pacific. The discovery of this mechanism will alert managers and scientists to expect changes in ecosystems years earlier than in the past. This information is important for a number of reasons but, for example, Beamish et al. (1999) suggested that periods of decreasing length of day (LOD) or a speeding up of the solid earth was associated with increased Pacific salmon production and a more intense Aleutian Low resulting in stormier winters.

One possible way to detect a regime shift is associated with energy transfer among the four shells (atmosphere, hydrosphere, solid earth, and core) of the planet. It is now possible to measure planetary processes accurately. As energy in a body rotating in a frictionless environment is conserved, and because the four shells of the planet rotate at different speeds, the energy lost from one shell must be transferred to one of the other three shells (Eubanks 1993). The index of energy transfer is the length of day (LOD). The LOD is the difference between the astronomically determined duration of the day and the standard LOD, which was established as exactly 86,400 s on 1 January 1958. Changes in the LOD are expressed as the difference between the measured LOD relative to the standard LOD. It is generally believed that the energy associated with decadal-scale changes reflect core-mantle energy transfers (Eubanks 1993). Seasonal changes in the LOD (Fig. 29.6) are closely linked to the atmosphere, but the shifts in the trends of the seasonal changes may indicate when a decadal-scale shift occurred. If the next regime shift, which may be in 2008, is associated with a shift in the trend in the LOD, it may be possible to use the pattern of energy transfers to forecast regime shifts. Once the discovery is made it will then be necessary to understand if global warming affects the mechanism. One thing is clear; the discovery of the mechanism causing regime shifts will eventually be made and it will then become another example of the Watson effect.

29.5.10 Issue 10: A New Approach to Fisheries Science

We suggest that it is time to rethink how we do fisheries science. Today and in the past, fisheries science was carried out mainly in universities, government agencies, and some private companies. University science was more curiosity-based and

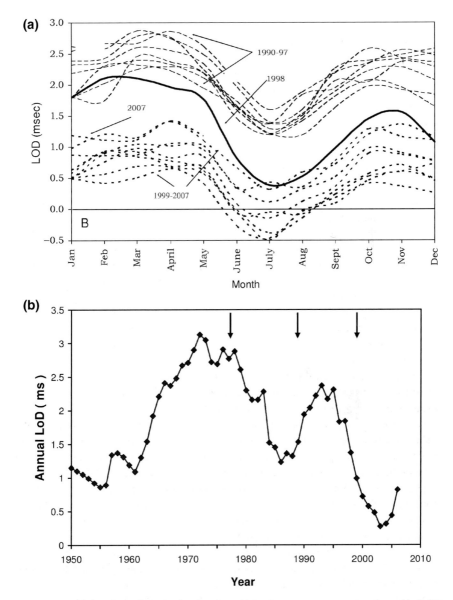

Fig. 29.6 (**a**) Seasonal length of day (LOD). The dashed lines show the seasonal trend in LOD with a slowing down of the solid earth in the northern hemisphere winter and a speeding up in the summer. The regime from 1990 to 1997 shifts in 1998 (thick solid line) about May and a new regime begins 1998–2007. (**b**) The annual trend in LOD. *Arrows* identify the 1977, 1989, and 1998 regime shifts. It is possible that regime shifts occur shortly after a change in the trend of the annual LOD

represented a number of independent areas of research. Government science was directed at solving management issues, maintaining long-term databases and providing advice on resource utilization. Both types of science contributed about equally to new knowledge. Some integration occurred, but multidisciplinary teams that operated for a number of years were hard to find.

We think that the science organizations that move faster and smarter in the future will provide the best advice. This means that more large-scale, multidisciplinary research needs to be carried out. It is unlikely that this will happen on its own. Universities and university researchers pride themselves on their independence; and federal and provincial organizations are at the mercy of government budgets. It is rare to find examples of fisheries research in British Columbia that has been integrated into a team. Consequently, it has been difficult to create teams of scientists, as the individual investigators do what is necessary to survive. Organizations do this too and are commonly thought of as "silo" organizations.

We suggest a new model. The new model does not touch existing structure; rather it adds an independent fisheries research advisory board that reports to the general public annually through the Federal and provincial ministers, perhaps not unlike the old Fisheries Research Board of Canada, but on a provincial scale. Initially a Board of Management was formed in 1898 with two major tasks: to prove its value to the Canadian government as an instrument of research in aid of Canadian fisheries, and to prove to the scientific community that it could operate a valuable laboratory for biological and fisheries research. The Fisheries Research Board of Canada replaced the Biological Board by an Act of Parliament in 1937 essentially to manage marine and freshwater research programs in Canada with a focus on fisheries-related research. The Board lasted until 1972 and the intriguing history was finally documented by Kenneth Johnstone (1977) in *The Aquatic Explorers*. Our fisheries science advisory board is not a management board. Its principal function is to identify the short-term and long-term research and monitoring requirements that will produce the advice needed by the managers of the marine ecosystems and associated fisheries. The board would review what is accomplished each year and identify what needs to be done in the future. The recommended research would apply to all fishery researchers in the province, including university researchers. The intent is to think strategically, but to recognize that the science needed to respond to climate change must be flexible and responsive. The fisheries science advisory board would be small, chaired by a prominent business leader, with three senior Fisheries & Oceans Canada scientists and three from universities (Fig. 29.7). The Head of federal government fisheries science and a Dean of Science should be on the board. Add a retired judge or someone who is experienced at understanding what is really going on. Members would rotate on 3-year terms, have no funds other than direct expenses, and be supported by government and universities. The board would ensure that the maximum use of all available fisheries research is available to work on the critical issues of the future. The board needs to have some teeth for government to listen which would come from the high profile of its members. The board would report annually to both the federal and provincial ministers responsible for fisheries and aquatic ecosystems in a manner originally envisaged for the Pacific Fisheries

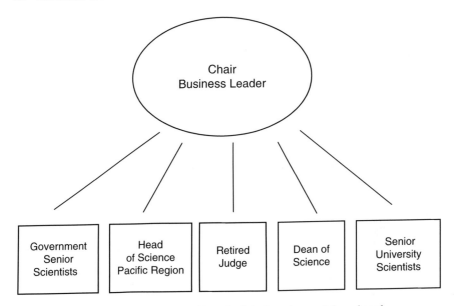

Fig. 29.7 Possible structure and composition of a fisheries science advisory board

Resource Conservation Council (2005; see http://www.fish.bc.ca/about_the_pfrcc). There are obvious difficulties with such a board, but we think that given the imminent impacts of climate change on the west coast fisheries and marine ecosystems, it is time to rethink how we do fisheries science.

29.6   Summary

We identified ten issues that we think will affect fisheries science on Canada's Pacific coast over the next 30 years. We could identify more but we think that the future of fisheries science in British Columbia will focus on the issues included here. All issues will be related to the impacts of climate on fisheries. We speculate that there will be greater variability in the populations of many species and stocks and that there will be continued uncertainty about the causes of such variability. A combination of modeling, monitoring, and a stable research program will be needed to manage resources on an ecosystem basis and to minimize the risk of overfishing. Improved climate models may eventually provide more accurate forecasts of regional climate changes, but the linkages with the population ecology of a species will remain elusive until the reasons for a particular life-history strategy are better understood. For some fisheries it may be necessary to change fishing methods altogether as the need to adapt to the impacts of climate change may challenge traditional approaches. It is unlikely that progress will be made quickly without changing how we do fisheries science. We think it is possible to have more of a business model for all fisheries science. We think that a new model for fisheries research will also provide the level funding that will be essential to ensure that good stewardship decisions are made.

Acknowledgments Lana Fitzpatrick, Chrys Neville, and Rusty Sweeting helped with the preparation of this manuscript.

References

Alderdice DF, Forrester CR (1971) Effects of salinity and temperature on embryonic development of the Petralesole (*Eopsetta jordani*). J Fish Res Bd Can 28:727–744

Armstrong, JD, Grant JWA, Forsgren HL, Fausch KD, DeGraaf RM, Fleming IA, Prowse TD, Schlosser IJ (1998) The application of science to the management of Atlantic salmon (*Salmo salar*): integration across scales. Can J Fish Aquat Sci 55 (Suppl 1):303–311

BCSFA (British Columbia Salmon Farmers Association) (2003) Catch. Vol 1(2), 8 p

Beamish RJ, Bouillon DR (1993) Pacific salmon production trends in relation to climate. Can J Fish Aquat Sci 50:1002–1016

Beamish RJ, Bouillon DR (1995) Marine fish production trends off the Pacific coast of Canada and the United States. In: Beamish RJ (ed) Climate change and northern fish populations. Can Spec Pub Fish Aquat Sci 121:585–591

Beamish RJ, Noakes DJ (2004) Global warming, aquaculture, and commercial fisheries. In: Leber KM, Kitada S, Blankenship HL, Svasand T (eds) Stock enhancement and sea ranching: developments, pitfalls and opportunities. Blackwell, Oxford, pp 25–47

Beamish RJ, McFarlane GA, Chilton DE (1983) Use of oxytetracycline and other methods to validate a method of age determination for sablefish. In Proceedings of the International Sablefish Symposium, 29–31 March 1983, Anchorage. Alaska Sea Grant Report 83-3, pp 95–116

Beamish RJ, McFarlane GA, Thomson RE (1999) Recent declines in the recreational catch of coho salmon (*Oncorhynchus kisutch*) in the Strait of Georgia are related to climate. Can J Fish Aquat Sci 56:506–515

Beamish RJ, Noakes DJ, McFarlane GA, Pinnix W, Sweeting R, King J (2000) Trends in coho marine survival in relation to the regime concept. Fish Ocean 9:114–119

Beamish RJ, McFarlane GA, Schweigert J (2001) Is the production of coho salmon in the Strait of Georgia linked to the production of Pacific herring? In: Funk F, Blackburn F, Hay D, Paul AJ, Stephenson R, Toresen R, Witherell D (eds) Herring: expectations for a new millennium. Alaska Sea Grant College Program, AK-SG-01-04, Fairbanks, pp 37–50

Beamish RJ, Schnute JT, Cass AJ, Neville CM, Sweeting RM (2004) The influence of climate on the stock and recruitment of pink and sockeye salmon from the Fraser River, British Columbia, Canada. Trans Am Fish Soc 133:1396–1412

Beamish RJ, McFarlane GA, Benson A (2006) Longevity overfishing. Prog Ocean 68:289–302

Beamish RJ, Sweeting RM, Lange KL, Neville CM (2008a) Changing trends in the percentages of hatchery and wild coho salmon in the Strait of Georgia and the implications for management. Trans Am Fish Soc 137:503–520

Beamish RJ, King JR, McFarlane GA (2008b) Canada. In: Beamish RJ, Yatsu A (eds) Impacts of climate and climate change on the key species in the fisheries in the North Pacific. A report of the North Pacific Marine Science Organization (PICES) Working Group 16 (in press)

Beamish RJ, McFarlane GA, Sweeting RM, Neville C (2008c) The sad history of dogfish management. In: Galucci V, McFarlane G, Bargman G (eds) Biology and management of spiny dogfish. Am Fish Soc Pub (accepted Sept 2007)

Benson AJ, Trites AW (2002) Ecological effects of regime shifts in the Bering Sea and eastern North Pacific. Fish Fisheries 3:95–113

Brett JR, Blackburn JM (1978) Metabolic and energy expenditure of the spiny dogfish, *Squalus acanthias*. J Fish Res Bd Can 35:816–821

Briscoe RJ, Adkison MD, Wertheimer A, Taylor SG (2005) Biophysical factors associated with the marine survival of Auke Creek, Alaska, coho salmon. Trans Am Fish Soc 134:817–828

Brown L, Hindmarsh R, McGregor R (2001) Dynamic agriculture book three, 2nd edn. McGraw-Hill, Sydney.

Christensen NL, Bartuska AM, Brown JH, Carpenter S, D'Antonio C, Francis R, Franklin JF, MacMahon JA, Noss RF, Parsons DJ, Peterson CH, Turner MG, Woodmansee RG (1996) The report of the Ecological Society of America committee on the scientific basis for ecosystem management. Ecol Appl 6(3):665–691

Eubanks TM (1993) Variations in the orientation of the Earth. In: Smith DE, Turcotte DL (eds) Contributions of space geodesy to geodynamics: earth dynamics. Geophysical Series 24, American Geophysical Union, Washington, DC, pp 1–54

FAO (Food and Agriculture Organization) (2007) The state of the world fisheries and aquaculture 2006. Food and Agriculture Organization of the United Nations, Rome, 180p (available at ftp://ftp.fao.org/docrep/fao/009/a0699e/a0699e.pdf/)

Feynman RP (1998) The meaning of it all. Helix Books, Addison-Wesley, Reading, MA, 133p

Finney BP, Gregory-Eaves I, Douglas MSV, Smol JP (2000) Impacts of climate change and fishing on Pacific salmon abundance over the past 300 years. Science 290:795–799

Finney BP, Gregory-Eaves I, Douglas MSV, Smol JP (2002) Fisheries productivity in the northeastern Pacific Ocean over the past 2000 years. Nature 416:729–733

Fisheries and Environment Canada (1978) The salmonid enhancement program. A public discussion paper. Information Branch, Fisheries and Marine Service, Vancouver, 36p

Fisheries and Oceans Canada (2005) Canada's policy for conservation of wild Pacific salmon. Fisheries and Oceans Canada, Vancouver, 49p

Francis RC, Hare SR (1994) Decadal-scale regime shifts in the large marine ecosystem of the northeast Pacific: a case for historical science. Fish Ocean 3:249–291

Government of Canada (2003) Species at Risk Act. Assented to 12 December 2002. Canada Gazette 25(3) Chap 29. Queens Printer for Canada, 104 p. (available at www.sararegistry.ca/the_act/)

Hare SR, Mantua NJ, Francis RC (1999) Inverse production regimes: Alaskan and west coast salmon. Fish 24(1):6–14

Hilborn R, Eggers D (2000) A review of the hatchery programs for pink salmon in Prince William Sound and Kodiak Island, Alaska. Trans Am Fish Soc 129:333–350

Hutchings JA (2000) Collapse and recovery of marine fishes. Nature 406:882–885

Hutchings JA, Reynolds JD (2004) Marine fish population collapses: consequences for recovery and extinction risk. BioScience 54:297–309.

Johnstone K (1977) The Aquatic Explorers: A history of the Fisheries Research Board of Canada, University of Toronto Press, Toronto, 342 p

Ketchen KS (1975) Age and growth of dogfish, *Squalus acanthias*, in British Columbia waters. J Fish Res Bd Can 32:43–59

Ketchen KS (1986) The spiny dogfish (*Squalus acanthias*) in the northeast Pacific and a history of its utilization. Can Spec Pub Fish Aquat Sci 88:78

Leaman BM, Beamish RJ (1984) Ecological and management implications of longevity in some northeast Pacific groundfishes. INPFC Bull 42:85–97

Levin PS, Zabel RW, Williams JG (2001) The road to extinction is paved with good intentions: negative association of fish hatcheries with threatened salmon. Proc Roy Soc B Biol Sci 268(1472):1153–1158

MacCall AD (1979) Population estimates for the waning years of the Pacific sardine fishery. Cal Coop Fish Invest Rep 20:72–82

Mantua NJ, Hare SR, Zhang Y, Wallace JM, Francis RC (1997) A Pacific interdecadal climate oscillation with impacts on salmon production. Bull Am Meteorol Soc 78:1069–1079

McFarlane GA, Beamish RJ (1983) Biology of adult sablefish (*Anoplopoma fimbria*) in waters off western Canada. In: Proceedings of the International Sablefish Symposium, 29–31 March 1983, Anchorage. Alaska Sea Grant Report 83-8, pp 59–80

McFarlane GA, Beamish RJ (1986) A tag suitable for assessing long-term movements of spiny dogfish and preliminary results from use of this tag. N Am J Fish Manage 6:69–76

McFarlane GA, Beamish RJ (1995) Validation of the otolith cross-section method of age determination for sablefish (*Anoplopoma fimbria*) using oxytetracycline. In: Secor DH, Dean JM, Campana SE (eds) Recent developments in fish otolith research. The Belle W. Baruch Library in Marine Science, Vol 19, pp 319–329. University of South Carolina Press, Columbia

McFarlane GA, Beamish RJ (1999) Sardines return to British Columbia waters. In: Freeland HJ, Peterson WT, Tyler A (eds) Proceedings of the 1998 Science Board Symposium on the impacts of the 1997/98 El Niño event on the North Pacific Ocean and its marginal seas. North Pacific Marine Science Organization. PICES Scientific Report No. 10, pp 77–82

Meffe GK (1992) Techno-arrogance and halfway technologies: salmon hatcheries on the Pacific Coast of North America. Conserv Biol 6:350–354

Minobe S (2000) Spatio-temporal structure of the pentadecadal variability over the North Pacific. Prog Ocean 47:381–408

Morton A, Routledge R, Peet C, Ladwig A (2004) Sea lice (*Lepeophtheirus salmonis*) infection rates on juvenile pink (*Oncorhynchus gorbuscha*) and chum (*Oncorhynchus keta*) salmon in the nearshore marine environment of British Columbia, Canada. Can J Fish Aquat Sci 61:147–157

Mote P, Canning D, Fluharty D, Francis R, Franklin J, Hamlet A, Hershman M, Holmberg M, Gray-Ideker R, Keeton WS, Lettenmaier D, Leung E, Mantua N, Miles E, Noble B, Parandvash H, Peterson DW, Snover A, Willard S (1999) Climate variability and change, Pacific Northwest. NOAA Office of Global Programs, and JISAO/SMA Climate Impacts Group, Seattle, 110 p

Myers RA, Levin SA, Lande R, James FC, Murdoch WW, Paine RT (2004) Hatcheries and endangered salmon. Science 303:1980

NRC National Research Council (1999) Sustaining marine fisheries. National Academy Press, Washington, DC

NPAFC North Pacific Anadromous Fish Commission (2001) Plan for Bering-Aleutian Salmon International Survey (BASIS) 2002–2006. NPAFC Doc 579, Rev 2, 27p

Nugent I, Profeta T (2006) Pathway to ocean ecosystem-based management: design principles for regional ocean governance in the United States. Nicholas Institute for Environmental Policy Solutions Duke University (available at http://www.env.duke.edu/institute/oceansm.pdf)

Overland JE, Wang M (2007) Future climate of the North Pacific Ocean. Eos, Trans Am Geophys Union 88(16):178

PFRCC Pacific Fisheries Resource Conservation Council (2005) Perspectives on salmon enhancement and hatcheries: what the council heard. Vancouver BC, 12 p (available at http://www.fish. bc.ca/files/HatchHeard_2005_0_Complete.pdf)

Pikitch EK, Santora C, Babcock EA, Bakun A, Bonfil R, Conover DO, Dayton P, Doukakis P, Fluharty D, Heneman B, Houde ED, Link J, Livingston PA, Mangel M, McAllister MK, Pope J, Sainsbury KJ (2004) Ecosystem-based fishery management. Science 305: 346–347

Rand PS, Hinch SG, Morrison J, Foreman MGG, MacNutt MJ, MacDonald JS, Healey MC, Farrell AP, Higgs DA (2006) Effects of river discharge, temperature, and future climates on energetics and mortality of adult migrating Fraser River sockeye salmon. Trans Am Fish Soc 135:655–667

Ricker WE (1954) Stock and recruitment. J Fish Res Bd Can 11:559–623

Ricker WE (1973) Two mechanisms that make it impossible to maintain peak-period yields from stocks of Pacific salmon and other fishes. J Fish Res Bd Can 30:1275–1286

Ruggerone GT, Goetz FA (2004) Survival of Puget Sound chinook salmon (*Onchorhynchus tshawytscha*) in response to climate-induced competition with pink salmon (*Oncorhynchus gorbuscha*). Can J Fish Aquat Sci 61:1756–1770

Salveson S, Leaman BM, Low L-L, Rice JC (1992) Report of the Halibut Bycatch Work Group. International Pacific Halibut Commission Technical Report 25, 29 p

Schweigert JF (2001) Stock assessment for British Columbia herring in 2001 and forecasts of the potential catch in 2002. Can Stock Assess Res Doc 2001/140

Sinclair M, Arnason R, Csirke J, Karnicki Z, Sigurjonsson J, Rune Skjoldal H, Valdimarsson G (2002) Responsible fisheries in the marine ecosystem. Fish Res 58:255–265

Thompson DWJ, Wallace JM (1998) The Arctic Oscillation signature in the winter time geopotential height and temperature fields. Geophys Res Lett 25:1297–1300

Trenberth KE (1990) Recent observed interdecadal climate changes in the Northern Hemisphere. Bull Am Meteorol Soc 71:988–993

Trenberth KE, Hurrell JW (1994) Decadal ocean-atmosphere variations in the Pacific. Clim Dyn 9:303–319

Yasunaka S, Hanawa K (2002) Regime shifts found in the northern hemisphere SST field. J Meteorol Soc Jpn 80:119–135

Zaporozhets OM, Zaporozhets GV (2004) Interaction between hatchery and wild Pacific salmon in the Far East of Russia: a review. Rev Fish Biol Fisher 14:305–319

Zhang Y, Wallace M, Battisti, DS (1997) ENSO-like interdecadal variability: 1900–93. J Clim 10:1004–1020

Chapter 30
Climate and Fisheries: The Past, The Future, and The Need for Coalescence

Anne Babcock Hollowed and Kevin M. Bailey

Abstract In this chapter we review the history of fisheries science with respect to climate impacts on fisheries and prognosticate the future of this type of research. Our review of the development of climate and fisheries research reveals that advances in our discipline emerge from the coalescence of four factors: shifts in fisheries economics and policy; developments in theoretical ecology; innovations in small-scale field and laboratory studies; and progress in large-scale fisheries statistics and modeling. Major advances have occurred when scientists interacted in multidisciplinary forums. We find that efforts to understand the impact of climate on the annual production and distribution of fish have produced a primary level of understanding of the processes underlying stock structure, production, and distribution of fish species. We find that ecosystem-based approaches to management have been advocated to a greater or lesser degree throughout the last century. In the future, we expect that advances in scientific understanding and improved computing power will allow scientists to explore the complex nature of environmental interactions occurring at different spatial and temporal scales. New field programs will develop to support the development of spatially explicit models of fish that include complex interactions within and between species, and fish behavior. Field sampling programs will benefit from continuing innovations in technology that improve collection of information on the abundance, distribution of fish, and the environment. New technologies will also be utilized in laboratory studies to rapidly assess the reproductive potential, food habits, and genetic history of fish under different environmental conditions. We expect that interdisciplinary training will continue to serve as a catalyst for new ideas in climate and fisheries. However, as researchers shift their focus from retrospective studies and now-casts to long-term implications of fishing and climate on the ecosystem we expect that training in oceanography, ecological theory, and environmental policy will be needed to provide a foundation for the development of models that depict the trade-offs of nature and human use in a realistic manner. Finally, we challenge fisheries scientists to track the accuracy of long- to medium-term forecasts of future states of nature and the potential impact of climate and fisheries on them.

A.B. Hollowed and K.M. Bailey
Alaska Fisheries Science Center, 7600 Sand Point Way NE, Seattle, WA 98115, USA

R.J. Beamish and B.J. Rothschild (eds.), *The Future of Fisheries Science in North America*, 597
Fish & Fisheries Series, © Springer Science + Business Media B.V. 2009

Keywords Climate · fisheries · oceanography · complexity · recruitment · history of fisheries

30.1   Introduction

The common definition of coalescence is the coming together of different units. In the varied disciplines of science, coalescence can take on many different meanings. In genetics, it can mean how lineages merge backwards in time, while in ecology, it has a more forward-looking description of how groups of organisms come together to forge a community. Here we embrace both meanings of coalescence. We look backwards to identify the factors that merged to mold the development of fisheries science as we know it today. We build on this knowledge to prognosticate the new advances in our field. We find that climate and fisheries interactions need the coming together, or merging, of climate scientists, oceanographers, and fisheries scientists in the broadest sense.

A new era of fisheries science emerged over a century ago with the formation of the International Council for the Exploration of the Seas (ICES) in 1902 at a time when European scientists realized the need to coordinate their research and management efforts across international boundaries (Rozwadowski 2002). Around the same time, Johan Hjort (1914) published his work asserting the concept that the dynamics of fish populations were caused by varying recruitment levels rather than large-scale geographic displacements, and pointed at natural fluctuations in populations caused by varying survival rates of young fishes as the probable cause. These products were the result of a "golden age" of fisheries science in Europe and North America, a remarkable proliferation of research and a coalescence of scientific technology and concepts, driven by the economic and political pressures of overfishing and market demand (Smith 1994). Now, near the beginning of a new century when the world faces new challenges of increasing demand for seafood resulting from increasing global population levels, high market value, overfishing, and uncertain climate conditions, more than ever we need to understand the complexity of processes governing population dynamics of marine fishes. Our discussion will trace the historical development of ideas linking climate and fisheries and how the legacy of the past might forge the future of fisheries science. In our long-range vision of the future we think scientists will be able to better forecast ecosystem responses to climate change and shifting demands for seafood because they more fully understand the processes controlling species interactions, and dynamics and the role of climate on these processes. Forecasting tools will allow scientists to inform managers on the impacts of their actions on society and the ecosystem, as well as test scientific concepts. In the short term, we encourage research to understand how animals respond to spatial and temporal changes in environmental conditions in order to construct a foundation for forecasting patterns of ecosystem change due to shifts in climate and the impacts of commercial fishing. Because of unknown future technological advances, it is imprudent to predict future developments much beyond the span of a career, so with that in mind, we discuss some areas where advances are feasible within the coming decades.

30.1.1   A Brief History of Climate and Fisheries Studies: Where We Have Been

There have been three major approaches for scientists studying climate and fisheries: (1) small-scale field observations and laboratory experimentation, (2) large-scale analysis and modeling of survey and commercial harvest data, and (3) development of theory. While these approaches are conducted at different scales and are often pursued independently, when they interface, significant advances occur. Several factors may serve as catalysts at the interface of scientific approaches (Fig. 30.1). The first catalyst is interdisciplinary coalescence and training (Wooster 1988). Individuals trained in more than one discipline communicate more effectively and often introduce new concepts and techniques to different branches of fisheries science. The second catalyst is the initiation of advances in science resulting from changes in marine policies, which directs public interest and the flow of resources for research. For example, considerable research has been generated by social pressure to stop overfishing and understand climate change. However, a change in economic and political pressure can also shift resources away from research, sometimes before answers are reached (Smith 1994). This occurred in the 1930–1940s when the declining economic market for fish and lack of interest by authorities ended the golden age of Norwegian fisheries research (Solhaug and Saetersdal 1972). The third catalyst is a shift in procedures, including methodology driven by technology (computing power, satellites, molecular biology) that opens new insights, and philosophy that results in new ways of viewing how science should be done. These catalysts accelerate paradigm shifts in research. Our brief review of the past

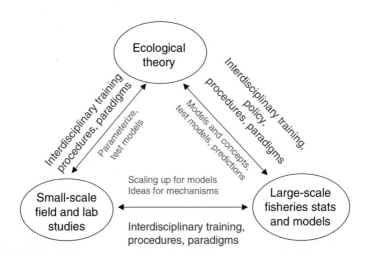

Fig. 30.1 Representation of three major approaches (circled corners) in fisheries science, with catalysts (blue letters) that stimulate interactions between approaches, and feedback mechanisms (red letters)

reveals that these catalysts have acted in concert to precipitate several new avenues for research.[1]

30.1.2 Climate and Fisheries: The Early Years, Setting the Stage

Early research on fisheries and climate variability focused on local and seasonal changes in availability to the fishery (Fig. 30.2). As early as 1832 a Swede named Nilsson reported that the collapse of the Bohlusan herring fishery was due to poor local conditions (Smith 1994). Later, A. Ljungman in 1880 attributed changes in the Bohsulan fishery to changes in weather and solar activity. Due to the importance of fisheries in the Norwegian economy, the modern era of fisheries research

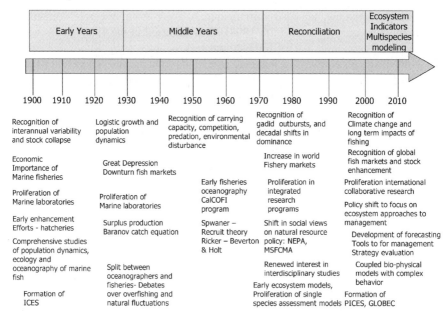

Fig. 30.2 Time line of major events influencing fisheries science (see text for more additional references)

---

[1]We recognize that our brief review of the history of fisheries science will inevitably omit some of the classic papers written in the twentieth century. For example, although we cite mainly temperate and subarctic studies, we recognize that many major advancements in fisheries oceanography were accomplished by biologists studying reef fishes (e.g., Sale 1991). Our purpose was not to be comprehensive in our review but to demonstrate the events that resulted in the major advances in fisheries science with respect to climate impacts.

began in Norway in 1864 when the Norwegian government asked G.O. Sars to examine why catches of cod around the Lofoten Islands fluctuated (Smith 1994). Soon afterwards, a number of factors came together in the late nineteenth century to initiate a proliferation of marine laboratories and fisheries research programs not only in Norway, but in Russia, Scotland, Canada, England, Germany, Denmark, the Netherlands, and the United States. One of these factors, the HMS *Challenger* expedition in 1872–1876, clearly had an impact on emphasizing a more global view of marine biology and oceanography and set a precedent for scientific surveys (Kesteven 1972).

In the United States the US Fish Commission was founded in 1871 to address industry conflicts, largely due to the efforts of Spencer Baird (Smith 1994). Partly because of public antagonism towards legislation that regulated harvests, augmentation programs were developed to offset fisheries harvests. Thus, in the 1880s popular programs to propagate cod were started in the United States, Norway, and Canada to enhance cod's declining abundance in the sea. Up to 2.5 billion yolk-sac cod larvae were released annually by American hatcheries and several hundred million were released by Norwegians (Solemdal et al. 1984).

Interest in studying the effectiveness of hatchery releases in enhancing natural populations partly motivated scientific approaches to studying larval survival in the ocean. But as well, large-scale field programs to study fish eggs and larvae in relation to ocean conditions emerged from an interest in the mechanisms behind natural fluctuations in fish populations (Kendall and Duker 1998). In particular, the young Norwegian scientist Johan Hjort along with other Scandinavians, notably C.G. Petersen, F. Nansen, G. Ekman, M. Knudsen, B. Helland-Hansen, and O. Petterson (Solemdal et al. 1984; Smith 1994) initiated studies on hydrographic and fisheries interactions. Around the same time, Baird conceived a research program involving comprehensive studies of population dynamics, ecology, and oceanography of marine fishes (Kendall and Duker 1998). Many of the concepts underlying modern fisheries research came to be developed during this time. These advances were made possible by parallel landmark developments, including the demonstration of racial strains, or local stocks, of herring by Heincke in 1875, and methods for quantitative sampling of plankton by Hensen in the 1880s. A relatively unheralded Danish scientist named C.G.J. Petersen developed techniques for tagging fish, introduced the concept of density dependence in fisheries, and refined methods of demonstrating yearly cohorts. The foundation of the plankton cycle in the ocean was worked out by Hensen and collaborators in Kiel in 1875–1920. Around 1909, Fisheries Oceanography studies started in Japan, and by 1919 Kitihara established that fish aggregate around frontal zones (Uda 1972). These developments contributed to a series of classic and enduring papers by Hjort (1914, 1926), which demonstrated that major fisheries fluctuations were due to irregularities in recruitment of year classes, with the cause of such irregularities occurring early in the life history, which was rooted in hydrographic and planktonic conditions. He also touched on contemporary issues including migrations and mixing of stocks, age and growth, larval drift, variations of quality such as lipid content, and recruitment variations.

The early legacy of ICES, international cooperation in research, information and data exchange, and coordinated transboundary management has been critical to successful fisheries management, and the focus of ICES has been followed in the Pacific by the North Pacific Marine Science Organization (PICES). However, ICES also created barriers to the integration of fisheries and oceanography. After the initial meeting in 1902, two committees were formed: a hydrographic committee and a biological program committee. The biological component was further split into a tagging subcommittee and an overfishing subcommittee. Hjort and his colleagues on the tagging committee became involved in studying fisheries fluctuations due to natural conditions. In our opinion the schism between climate variability and overfishing proponents officially started about then and has been a chasm ever since that has been difficult to bridge.

30.1.3 Climate and Fisheries: The Middle Years

Between the two World Wars and during the Great Depression the economic market for fish products took a downturn and European research on marine fisheries fluctuations reached a low point due to lack of social pressure to capture the interest of politicians (Fig. 30.2). Fisheries research took a new turn that lasted through the 1970s, which focused on developing of the theory of sustainable harvest levels and fisheries harvest models. During this period, scientific investigations by fisheries biologists, mathematical ecologists, and oceanographers developed in parallel rather than in concert.

By the middle of the twentieth century, ecologists had published papers on the importance of carrying capacity, competition for limited resources, the role of predation, and environmental disturbance (Lotka 1925; Volterra 1926; Nicholson 1933; Andrewartha and Birch 1954; MacArthur and Wilson 1967). The well-known concepts of logistic growth (Von Bertalanffy 1938) and allometry were utilized in early models of fish populations. Biologists recognized the importance of considering different sources of mortality when estimating catch (Baranov 1918). During this period, marine fish populations were modeled using surplus production models (Graham 1935; Schaefer 1954; Fox 1970) and Leslie matrix applications that allowed biologists to track the influence of fishing and natural mortality on the whole population. In fact, with the availability of readily available information on year-class strength from fisheries statistics, correlative approaches between fisheries and climate have proliferated from the 1930s to the present day. One major breakthrough was realized through the coalescence of ecological theory and fish population dynamics modeling. Ricker (1954) and Beverton and Holt (1957) adapted the theoretical concepts of carrying capacity, competition, predation, and environmental disturbance into well-known governing equations for the relationship between spawners and recruitment to commercial fisheries.

While most fisheries research was on overfishing to the exclusion of climate as a forcing function, there were bright spots in fisheries oceanography. Reasoning that most of the variability occurred at young life stages, interdisciplinary research teams studied the impact of environmental factors on survival of eggs and larvae

(Sette 1943; Walford 1938). Predation, competition, and the direct and indirect impact of environmental forcing on recruitment and spawning, as well as the impact of fishing on spawning stock biomass and early life history were factors considered in these investigations. An early example of a fisheries oceanography-based program is the California Cooperative Oceanic Fisheries Investigations (CalCOFI) program in the northeast Pacific Ocean. This program has its historical roots in the inspiration of D.S. Jordan and W.F. Thompson to study linkages of natural fluctuations in fisheries with overfishing, and after the collapse of the California sardine the program built on the strengths of scientists at the Scripps Institution of Oceanography, the California Department of Fish and Game, and the National Marine Fisheries Service.

Great debates concerning the role of overfishing versus natural fluctuations were repeated over different stocks in different areas of the world. Among the most famous were the Thompson-Burkenroad debates of the late 1940s (Thompson and Bell 1934; Thompson 1950; Burkenroad 1951).

30.1.4   Climate Variability and Fisheries: A Period of Reconciliation

Several key findings put the climate and fisheries topic back on the menu in the latter part of the twentieth century (Fig. 30.2). The Russell cycle described a 60-year cycle of plankton and fish abundance related to warming and cooling trends in the English Channel (Russell et al. 1971). The dramatic environmental effect of the 1958–1959 El Nino event in the Pacific and collapse of the world's largest fishery, the Peruvian achovetta, were impossible to ignore (Barber et al. 1985). The publication of a landmark paper by Andrew Soutar showed that dramatic large fluctuations occurred in fish populations prior to industrial fisheries (Soutar and Isaacs 1974). Finally, in an influential treatise, Cushing (1975) hypothesized that patterns of fish production were linked to multiple factors including that match–mismatch of the seasonal production cycle and readiness of larvae to feed, larval growth, density-dependence, and the temporal and spatial overlap of predators. By the 1980s, the pioneering concepts of Wooster (1961) and others gave acceptance to fisheries oceanography as a new interdisciplinary research field.

Systematic surveys and time trends in catches revealed that marked outbursts or abrupt collapses in fish stocks occurred throughout the world (Murphy 1961; Skud 1982; Cushing 1984). These findings and a better economic picture renewed interest in process-oriented research. Scientists endeavored to understand the mechanisms underlying these shifts and several hypotheses were resurrected from the past as a result. Among these, advances in physical oceanographic tools resulted in hypotheses regarding the role of interannual variation in wind on ocean currents and the influence of these factors on the transport of larvae to suitable nursery grounds (Nelson et al. 1977; Parrish et al. 1981). Sinclair's (1988) member vagrant hypothesis linked the concepts of larval drift, fidelity to spawning location, and subpopulation structure. New teams of scientists were formed to study recruitment processes (e.g., the MARMAP[2] program in the northwest Atlantic Ocean, Georges Bank GLOBEC[3], W.C. Leggett's research team at MacGill University, Canada's OPEN[4] and NOAA's[5] FOCI[6] programs). Interdisciplinary research teams were not limited to the field, a parallel effort occurred in the evolution of laboratory studies where chemists, physiologists, and behavioral ecologists conducted studies on environmental factors influencing predation and feeding. These studies fostered the development of biochemical indicators for evaluation of fish condition, diet, and predation. Innovative combinations of lab experiments applied to field observations have given insight to survival processes (Blaxter and Hempel 1963; Lasker 1981; Houde 1989). Discovery of increments deposited daily on larval otoliths also provided a major new tool for these studies (Campana and Neilson 1985). Likewise, interdisciplinary collaborations have resulted in the development and application of advanced laboratory techniques to assess stock structure including otolith chemistry and molecular genetics. Physiological studies have provided information on the environmental requirements of fishes, and behavioral studies have provided critical information on the complexity of responses of fishes to environmental conditions.

Theoretical breakthroughs in fisheries science also resulted from the application of ecological theory to the study of exploited aquatic or marine systems. Studies in theoretical ecology provided the foundation for cross-disciplinary research in the study of climate impacts on fish population dynamics. Concepts of trophic cascades (Hairston et al. 1960) and keystone species (Paine 1969) had also taken root in fisheries ecology (e.g., Frank et al. 2005; Kawasaki 1993). In the 1980s, Connell's (1985) work in supply side ecology led to new interest in recruitment, particularly in reef fish systems. In the 1990s, theoretical ecologists focused attention on the role of complexity and hierarchical organization and the role of shifting spatial dimensions in ecosystems (Odum 1992; Levin 1992), which have led to similar studies in marine systems (e.g., Bailey et al. 2005). Pioneering work in population genetics (Wright 1943) expanded into fisheries genetics (e.g., Doherty et al. 1995). Ecologists have explored the role of stock structure, or metapopulation structure, and species diversity as attributes of ecosystems that contribute to overall population stability (Hanski 1991) and likewise fisheries has followed this lead (e.g., Smedbol and Wroblewski 2002). Chesson (1984) introduced the concept of longevity and the storage effect where the reproductive potential of a population can be stored in a few strong year classes. The importance of this concept was recognized as a strategy for preserving reproductive potential in marine organisms when conditions conducive to recruitment success are rare (Leaman and Beamish 1984; Beamish et al. 2006).

In the 1960s and 1970s ecologists endeavored to describe the functional relationships governing species interactions. These concepts were adapted for use in early computer-generated simulation models. Holling (1965) introduced functions to represent interactions of predators to increasing prey density. These early computer

---

[2] Marine Resources Monitoring Assessment and Prediction.

[3] Global Ocean Ecosystem Dynamics.

[4] Open Production Enhancement Network.

[5] National Oceanic and Atmospheric Administration.

[6] Fisheries Oceanographic Coordinated Investigations.

simulations included functional relationships between predators and prey and the role of the environment in mediating these relationships (Anderson and Ursin 1977; Laevastu and Larkins 1981).

In the 1970s, Pope (1972) introduced a statistical modeling approach founded on Baranov's (1926) catch equations that tracked the impact of fishing on the individual cohorts within the population. The widespread application of cohort analysis to commercial fish population provided a technique to reconstruct time trends in recruitment, which resulted in numerous correlative studies linking environment and distribution.

In the 1990s, individual-based models (IBMs) provided a basis for tracking environmental forcing on survival during the early life history at fine spatial and temporal scales (Rose et al. 1993; Hermann et al. 1996). In some regions, scientists have coupled these two modeling approaches into a complex marine ecosystem model (e.g., the European Regional Seas Ecosystem Model; Baretta et al. 1995). As our knowledge of the ocean has expanded beyond seasonal effects on local availability, so has the horizon for research expanded beyond the local and seasonal scales of global fisheries dynamics. Fisheries scientists discovered decadal scale variations in fish populations resulting in renewed interest in the mechanisms linking ocean conditions and fish production (Hollowed and Wooster 1992; Beamish and Bullion 1993; Omori and Kawasaki 1995; Hollowed et al. 2001; Steele 2004; Cury and Shannon 2004). At the same time fisheries scientists are realizing that multiple factors including the environment and overfishing interact to cause fluctuations in fish stocks (Fogarty et al. 1991; Stenseth et al. 1999; Rothschild 2007).

Major breakthroughs in fisheries also occurred in response to shifts in societal views regarding the use of natural resources. In the 1970s, fisheries scientists began to recognize the interdependence of ecology, production, and management of our Nation's fisheries (McEvoy 1996; Pascoe 2006). This recognition led to a more formal articulation of the goals of society's use of natural resources (Sanchirico and Hanna 2004). In the United States and Canada, these shifts resulted in increased public demand for environment-friendly policies. The growing scientific evidence of human impacts on managed ecosystems, and subsequent public awareness of these impacts, prompted the US Government to pass the National Environmental Policy Act (NEPA), the Marine Mammal Protection Act, the Fisheries Conservation and Management Act, and the Clean Water Act in the 1970s. These acts established limits to human impacts on marine ecosystems. The new laws challenged fisheries biologists and population dynamics modelers to work together to define limits of impact for use in management of marine resources (Fluharty 2005).

With the passage of the Magnuson-Stevens Fishery Conservation and Management Act (MSFCMA) in 1996, scientists were faced with defining concepts like sustainability and overfishing (Restrepo 1999). As a result, with all their flaws and deficiencies the concepts central to depicting population growth used in the 1950s were resurrected to define targets such as maximum sustainable yield (MSY) and the biomass associated with MSY (B_{MSY}), and limits such as the fishing mortality associated with the overfishing level (F_{OFL}). Armed with these biological reference points management measures were designed and adopted to prevent overfishing and rebuild depleted stocks (Mace 2001). The emphasis on stock rebuilding and evaluation of the performance of management strategies relative to reference points lead to a renewed interest in simulation modeling. A simulation modeling approach was introduced to evaluate the performance of management strategies when assessed using uncertain observations and variable climate conditions (management strategy evaluations [MSE]; De la Mare, 1996).

At the end of the twentieth century, fisheries scientists and policy makers emerged from a decade of research designed to revise harvest polices to prevent overfishing and rebuild overfished stocks, to refocus their attention on developing tools to forecast the long-term implications of fishing on marine ecosystems (Ecosystem Principles Advisory Panel 1999; DeMaster et al. 2006; Arkema et al. 2006). This shift in focus called for the development of ecosystem approaches to management (EAM) and renewed effort to identify the biological and physical mechanisms underlying fish interactions, distribution and production, and the role of fishing on these processes. At around the same time, a growing awareness of the impact of human-derived changes in climate that are likely to impact physical properties of the ocean (IPCC 2007), ocean acidity, and sea ice extent (Overland and Wang 2007) has led to an elevation of the priority of climate–fisheries interactions. We see this shift in focus as the harbinger of future directions of fisheries science in the coming decades and the catalyst for coalescence of fisheries scientists focused on a common research problem.

30.2 The Next Generation of Fisheries Science

While there has been considerable progress over the last century there remain many opportunities for research in the coming decades. We identify four major research themes in climate–fisheries where we expect progress, including: (1) expansion of theory to include complex processes in recruitment, thereby enabling scientists to better forecast impacts of changing climate and fishing; (2) enhanced recognition of the spatial scale of key processes and the role of climate in adjusting these processes; (3) increased emphasis on behavioral and foraging responses of adult and juvenile fishes to changes in local environmental conditions; and (4) development of modeling tools to assess the performance of management strategies under changing environmental conditions. Research in these four areas will assist efforts to define the role of climate change and interaction with fishing on marine ecosystems.

30.2.1 Complexity in Recruitment

Some argue that the climate–recruitment interactions are too complex to effectively contribute to management of stocks, and correlations of recruitment with environmental factors always fall apart eventually. We perceive recruitment as a complex process (rather than a "problem" with a simple answer), and such a view opens understanding that is critical to effective management of stocks. For example, advocates of ecosystem approaches to management (EAM) hope to forecast performance of management strategies under future states of nature. To perform this analysis, scientists will need to develop realistic forecasts of fish production, distribution, and market demands and constraints. Predicting ecosystem responses to environmental disturbance and fishing requires the integration of complex processes influencing organisms over different time and space scales.

There is no simple factor that is going to dominate recruitment, and factors will shift in importance over space and time as biological players, environmental regimes, and history change. Interacting factors influence important parameters in recruitment, such as larval feeding (Porter et al. 2005). In the future, development of statistical models can help with understanding the interactions between conditions (e.g., Ciannelli et al. 2004). Enhanced understanding of processes, comparative knowledge of population interactions under different conditions, and methods of scaling up to metapopulations should be productive avenues of research. We envision that combining the probabilistic nature of the many lower scale-level interacting factors with the different types of constraining and boundary or higher scale-level factors will lead to better forecasting models. These can be linked to real-time information on current recruitment status (such as current juvenile abundance) and combined with multispecies models with contemporary estimates of spatial distribution and interactions to better define the arc of a year class while continually updating and refining predictions of its recruitment level (Fig. 30.3).

There are many questions to confront in the future. How much do we need to know to forecast and how far ahead? In the early 1900s, scientists were already thinking about forecasts, but believed definitive predictions were premature (Hjort 1914, p. 227). Sette was making formal predictions of the Atlantic mackerel fishery based on incoming year classes as early as 1928 (Smith 1994). But are short-term forecasts and correlations good enough? What is the appropriate scale and how much do we need to know about fine-scale processes, such as larval behavior? With the growing concerns regarding the long-term implications of climate change on marine ecosystems, we expect that the required time frame for forecasting will be extended to decades. We expect that as fisheries scientists shift their focus from prevention of overfishing to assessing the performance of management strategies relative to benchmarks of ecosystem status under different climate scenarios (Fig. 30.2). This shift in focus will increase the need for a mechanistic understanding of processes underlying recruitment to enable scientists to forecast fish reproductive success under different states of nature.

As in other scientific disciplines, we expect that there will be continued debate regarding whether the merits of holistic (e.g., correlative) or reductionist (e.g., individual based) approaches are best. A criticism of a holistic approach is that there is little confidence without understanding mechanisms. A criticism of the reductionist approach is that while it leads to understanding mechanisms, it also can lead down a narrow alley, sometimes without a good perspective of how this path fits into the bigger roadmap. On the other hand, having detailed mechanistic

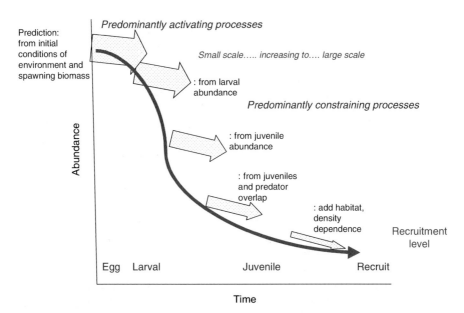

Fig. 30.3 Representation of a scheme for forecasting recruitment, continually refining the forecast with new information on the year class as it develops. The direction of the patterned block arrows reflects the accuracy relative to the true recruitment trajectory (red line) and the width of the block represents precision of the forecast at each stage. We envision a first prediction based on spawning biomass, environmental conditions, and possibly regime state. The forecast gets refined with information on larval and juvenile abundances, abundance and overlap of predators, density-dependence, and habitat availability

knowledge presents far greater opportunities for engineering solutions, in our case a forecast model. We expect that the greatest discoveries will be made through the development of techniques that scale local responses to population-scale events. For example, physiologists have gained great understanding of metabolic pathways, but it takes a more holistic approach to understand what goes wrong in a cancer cell and how to engineer a cure, the objective of integrated systems biology models. In parallel, we have learned much about mechanisms and processes like density-dependence, dispersal, larval feeding, and predation, but an integrated approach is needed to understand how these processes interact to shape a year class. As noted by Levin and Pacala (1997), fisheries scientists should also explore the possibility that there are ecological principles that might govern reproductive potential of marine fish. For example, landscape limits on abundance and recruitment by the amount of available habitat (Rijnsdorp et al. 1992; Bailey et al. 2005), and emergent scaling and power laws (Marquet et al. 2005; Taylor 1961) could be utilized in spatially explicit simulation models.

30.2.2 *Spatial Ecology*

In the next decade, fisheries biologists are likely to make progress understanding the mechanisms underlying the role of oceanography in governing the boundaries of suitable habitat, and the role of varying habitat volumes on competition between species for limited resources, and the spatial overlap of predators and prey. Fisheries biologists have recognized the role of stock density on habitat use (MacCall 1990). Recent studies demonstrate that shifts in ocean conditions alter the distribution and volume of pelagic ocean habitats and the spatial distribution of suitable habitat and partitioning between competing species (Rooper et al. 2006; Agostini et al. 2006; Hollowed et al. 2007), as well as how changes in the seascape influence predator–prey interactions (Ciannelli and Bailey 2005). Next-generation fisheries forecasts must address the role of climate on the quality and quantity of suitable habitat and its influence on the distribution and abundance of predators and prey. To accomplish this goal, fisheries scientists will need to study landscape effects such as corridors, connectivity, and patch structure and how spatial variation affects ecological processes. In these studies there are combined influences of a fixed landscape, such as bathymetry, coastline morphology and geology, and labile components such as currents and fronts, to form the seascape.

Enhanced near-synoptic sampling of the three-dimensional properties of ocean habitats on a seasonal basis is required to adequately monitor the role of environmental disturbance on the quantity and use of ocean habitats by marine fish and shellfish. We anticipate that progress in data collection will come from technological innovations coupled with data collection partnerships between commercial and recreational fishers, universities, and state and federal agencies responsible for stock assessments. The US National Integrated Ocean Observing System and its regional representatives provide the foundation for storing and distributing information obtained through the partnership (http://www.ocean.us/what_is_ioos). What is currently lacking is the mechanism to coordinate, fund, and standardize the collection of information needed to utilize ships of opportunity as platforms for ocean monitoring. Processing and analyzing the wealth of data also requires considerable resources and development.

Along with environmental monitoring, parallel effort is needed to monitor seasonal patterns of habitat use and the association between fish and their habitats. Meeting this challenge will require studies focused on behavioral ecology. Enhanced research on fish movement, and factors influencing foraging responses will be productive. Technological advances in acoustic tags, archival tags, and tag deployment are likely to continue and we expect that future biologists will be able to record fish movements in many regions of the northern hemisphere (Sheridan et al. 2007).

Another aspect of biocomplexity and spatial ecology is stock structure (Hilborn et al. 2003). Genetic stock structure implies little movement between geographically separated populations, whereas ecological metapopulation structure, where there is potential of movement to the degree of affecting demographic rates, is an adaptation that contributes to overall population stability (Hanski 1991). Climate effects on the interaction of landscape, fish movement, and population structure are important topics of future research. Population structure adds great complexity to the management of fisheries and sometimes managers often do not want to hear that there is fine-scale structure in their populations, but when multiple stocks are managed as one, vulnerable populations are in danger of overexploitation (Fu and Faning 2004). Local adaptation of different populations to environmental conditions and movement between populations are key issues in the interaction of stock structure and population dynamics.

30.2.3   *Fisheries Interactions and Local Ecology*

As fisheries scientists strive to develop sufficient understanding of the ecosystem to model the environmental and economic trade-offs associated with fishing, there will be an increased demand for improved understanding of the factors governing species co-existence in a variable environment. There will be a need for new types of field experiments designed to assess the foraging response of fish to changing habitat conditions and to predict how these changes will influence competition and predation. One approach is to establish a network of focused field locations for detailed behavioral studies. Scientists can scale up findings from these local regions to inform whole ecosystem models. These focused sites should build on the Before After Control Impact (BACI) framework to utilize the comparative approach for understanding fish behavioral responses to different types of environmental disturbance (Smith et al. 1993). Wilson et al. (2003) provide an example of this type of study where scientists attempted to utilize the comparative approach to assess the response of fish to fishing.

Our current understanding of food–web interactions, life-history strategies, and trophic effects of fisheries is based primarily on large-scale analyses and models with relatively little consideration of the explicit effects of spatial variability (e.g., Aydin 2004; Hollowed et al. 2000; Christensen and Walters 2004). While the development of multispecies and whole ecosystem models represents advancements in fisheries modeling, these models rely on extremely simple interaction terms between predator and prey. While these parameterizations may be adequate for evaluating energy pathways in food webs, they fail to address the complex issues of behavioral responses of predator and prey (schooling), competition and resource limitation, and environmental disturbance (Walters and Kitchell 2001; Bakun 2001). Patterns of interaction between species (e.g., predator–prey), and the strength of these interactions are mediated by behavior, abundance of alternative predators and prey, and environmental disturbance (Chesson 2000; Rice 2001). New research programs are needed to address these deficiencies.

Advances in our understanding of foraging behavior will also require careful review of the spatial and temporal scales of the interaction. Species interactions involve complex processes between climate, predators, and prey that occur on short time scales and small space scales. Fisheries scientists are beginning to recognize these complex processes. For example, Fauchald et al. (2006) found evidence that schooling fish migrations were influenced by the age structure, and the density of the school relative to available prey. Next-generation studies will focus on the role of climate in determining these interrelated factors. There may be critical foraging interactions that happen at local scales, particularly for central place foragers, or at foraging hotspots (Croll et al. 1998). Behavioral responses to predator and prey densities may reveal new functional interactions between species. Walters and Kitchell (2001) hypothesize that juvenile trophic interactions and behavior at local scales can cause depensatory recruitment dynamics in target species. They further suggest that we study juvenile survival rates and recruitment performance, abundance trends of potential competitors in juvenile rearing areas, and diet compositions of juveniles and competitors. In addition, it is important to assess the spatial and temporal dynamics of juvenile foraging behavior (i.e., determine the dimensions of the "foraging arena"). These types of localized process-oriented studies of fish behavior are examples of the research envisioned by the initiative described here.

30.2.4 Modeling Tools to Integrate

In the future, there will be increased demand for models that accurately assess the economic and ecological trade-offs of natural resource use. How we manage our fisheries and how we view the impact of environmental changes like climate effects on fisheries is bound by our concept of how ecosystems are organized. Is it hopelessly complex and chaotic, or ordered and hierarchical? One approach to understand how ecosystems work is a "top-down" modeling approach, which in the vernacular of systems biology is hypothesis-driven, thereby making a model of how we think the system works and comparing it to the data. Another approach, again in the terminology of systems biology is a "bottom-up" approach, which combines as much information that we know about the system as possible and then try to reconstruct it in a grand ecosystems model. Probably in the long run, both approaches are needed in a push–pull dynamical interaction to craft the most predictive description.

Four-model modifications could be incorporated into existing models in the near future. First, current individual-based models often lack the validation that is applied in stock assessments. Analysts need to revise coupled biophysical models to enable them to tune parameters governing fish responses to their environment to field observations in a manner similar to stock assessment procedures. To achieve this, long-term commitments to egg, larval, and juvenile surveys will be required. We expect that technological developments will enable scientists to collect many observations through remote sensing making this added requirement for data collection affordable. Adapting existing models to track observed and predicted outcomes will enable scientists to track uncertainty in biophysical interactions and to adapt interagency research efforts to target the most important processes governing recruitment.

Second, quantitative fisheries biologists could accelerate research on functional responses by empirical studies. This would be accomplished through a coordinated effort involving field and modeling exercises. Humston et al. (2004) provide a framework for conducting this type of analysis where different patterns of dispersion are generated by applying different foraging behaviors. Simulated larval distributions are statistically compared to observed data to select the foraging behavior that provides the best fit to the observations.

Third, if ocean habitat volumes could be measured, then an index of volume of habitat could be incorporated in spawner recruit relationships to account for interannually varying probability of intraspecific competition for resources (Iles and Beverton 2000).

In the long term, we see a future marked by innovations in spatial models that track the ontogeny of fish and shellfish in response to seasonal shifts in environmental conditions in three dimensions. Given the creativity of modelers and the technological advances occurring in computational power, new types of models will be developed and tested. Next-generation models should be able to forecast the impact of climate variability and fishing on fish and shellfish. Coupled biophysical models are currently used to track larval dispersal pathways, lower trophic-level production, and habitat characteristics (Ito et al. 2004, Baretta et al. 1995). We expect that the next generation of individual-based models will likely incorporate Bayesian treatment of parameter selection to identify what model formulation best fits the data.

Finally, we expect that lower trophic level and larval dispersal and survival models will be coupled to spatially explicit population dynamics models to track the full life cycle of marine fish and shellfish. In the short term, we expect that modelers will debate issues of scale and complexity in an attempt to resolve the trade-offs between simplifying the system to capture the main controlling processes influencing population dynamics (e.g., spawner–recruit relationships modified with ecosystem indices) and an attempt to capture biological realism by tracking individual behavior through the full life cycle. First-generation versions of these full life cycle, coupled biophysical, predator–prey models include ERSEM (Baretta et al. 1995) and NEMEROFISH (Megrey et al. 2007). Likewise, stock assessment biologists have been simulating future population dynamics for decades and several examples of techniques for incorporating environmental forcing in spawner–recruit relationships exist. We expect that the most useful model configuration will draw from both simulation approaches. Simulations of older life states can track complex behavioral responses of fish and shellfish to regional ocean conditions, fish density, predator abundance, and prey availability. Next-generation models should allow the user to assess the performance of different management strategies under different climate scenarios. This will require interactions between ecosystem modelers, climatologists, oceanographers, and stock assessment scientists. This interface will extend existing management strategy evaluation (MSE) techniques to include multispecies interactions to evaluate implications of harvest strategies under a variable climate.

Friedman (2005) predicts that as the world becomes more connected through the Internet the generation of new ideas will accelerate. We expect that the Internet will serve as a catalyst for rapid development of next-generation models in fisheries science as well. Some modelers are already providing software for common use by the fisheries community (Schunte et al. 2007). We expect that virtual laboratories will emerge where fisheries modelers from around the world will collaborate to develop techniques for incorporating biological complexity into ecosystem simulations. In the short term, international marine science organizations such as International GLOBEC, ICES, and PICES will continue to serve as catalysts for the exchange of information between scientists.

30.3 Conclusions

Our review of the evolution of scientific thought regarding climate and fisheries revealed that breakthroughs in thinking were often achieved through interdisciplinary research. Through the leadership of a few key individuals the barriers that once partitioned fisheries biologists and oceanographers have disappeared and fisheries oceanography is now an accepted field of study. While some barriers have fallen, there continues to be a need to merge the disciplines of fisheries management and fisheries science, and fisheries science and ecology. Mangel and Levin (2005) encourage the inclusion of ecology in the teaching of fisheries science. Quinn and Collie (2005) extend this recommendation to include some training in fisheries management to allow future scientists to recognize the difficulties facing managers as they attempt to balance the competing goals of resource conservation and use. In the context of climate and fisheries questions, training in oceanography, ecology, and management will be needed to develop scenarios that forecast changes in ocean conditions, the demand for fish, the constraints to the resource, and the responses of fish to these factors.

As we look to the future, we expect that major breakthroughs in recruitment studies will come from a better understanding of the concept of scale, and the properties that exist at the frontiers between scales (breaks). Such enhanced understanding includes the potential hierarchical ordering of scale. The interaction between individual processes and the organisms of interest and how those interactions lead to regulation and control of populations are key questions in population biology. Studies of cybernetics are needed to examine the processes of control and regulatory feedback, and whether the interactions of parts of the system result in self-organization. For that matter, are marine ecosystems self-organizing at all?

We envision a future where mathematical ecologists and modelers will develop forecasting tools to assess hypotheses regarding climate impacts on fishing. In the latter half of the twentieth century, fisheries biologists had sufficient data to recognize decadal patterns in fish production. These findings, and their potential impact on commercially exploited fish populations, resulted in the formation of interdisciplinary research teams to understand the processes underlying fish responses to climate shifts. These two events represent only part of the scientific method where a general hypothesis has been proposed based on retrospective studies and the hypothesis has been refined through field and laboratory research. The scientific method requires that we begin to test the hypothesis, and future fisheries scientists are likely to utilize a combination of field observations and modeling to make these tests. We see the shift to hypothesis testing as an evolution where models are used to predict fish responses to climate and these predictions are evaluated against observations. We anticipate that the collection of biological and physical observations will be enhanced through technological developments that allow underway sampling from ships of opportunity and remote sensing.

Our review also reveals the need for renewed interest in fish behavior. Throughout the twentieth century, fisheries scientists were primarily focused on estimating the abundance of fish stocks. This was a natural first step needed to ensure conservation of the resource. However, knowledge of fish behavior will be needed to address questions of ecosystem impacts of climate and fishing. We expect that this area of research will grow considerably in the next few decades. Major progress will likely come through the development of a few well-monitored sites where real-time or seasonal patterns of fish behavior will be monitored. These sites will serve as at-sea laboratories where conditions can be manipulated to assess the responses of fish to different environmental conditions.

We hope that this chapter, and volume in general, will be viewed as our prediction of future trends in the field of fisheries science. As the disciplines coalesce, we look forward to watching the outcome of our collective experiment in climate and fisheries.

Acknowledgments We thank Phil Mundy and an anonymous reviewer for their comments. We also thank Warren Wooster for his mentoring, numerous discussions and insights. Finally, we acknowledge inspiration derived from reading Smith (1994), which should be required reading for all fisheries scientists.

References

Agostini, V.N., Francis, R.C., Hollowed, A.B., Pierce, S., Wilson, C., and Hendrix, A.N. (2006) The relationship between hake (*Merluccius productus*) distribution and poleward sub-surface flow in the California Current System. Canadian Journal of Fisheries and Aquatic Sciences **63**, 2648–2659.

Anderson, K.P. and Ursin, E. (1977). A multispecies extension to the Beverton and Holt theory, with accounts of phosphorous circulation and primary production. Meddeleser af Danmarks Fiskeri- og Havundersoegelser. – og Havunders. N.S. **7**, 319–435.

Andrewartha, H.G. and Birch, L.C. (1954) The Distribution and Abundance of Animals. University of Chicago Press, Chicago, IL. 782p.

Arkema, K.K., Abramson, S.C., and Dewsbury, B.M. (2006) Marine ecosystem-based management: from characterization to implementation. Frontiers in Ecology and the Environment **4**(10), 525–532.

Aydin, K. (2004) Age structure or functional response? Reconciling the energetics of surplus production between single-species models and Ecosim. African Journal of Marine Science **26**, 289–301.

Bakun, A. (2001) School-mix feedback: a different way to think about low frequency variability in large mobile fish populations. Progress in Oceanography **49**(1–4), 485–512.

Bailey, K.M., Ciannelli, L., Bond, N., Belgrano, A., and Stenseth, N.C. (2005) Recruitment of walleye pollock in a complex physical and biological ecosystem. Progress in Oceanography **67**, 24–42.

Baranov, F.I. (1918). On the question of the biological basis of fisheries. Nauchin Issled. Ikhtiologicheskii Inst. Izv **1**, 81–128 (in Russian, Translation by W.E. Ricker, 1945).

Baranov, F.I. (1926). On the question of the dynamics of the fishing industry (in Russian, Translation by W.E. Ricker, 1945).

Barber, R.T., Chavez, F.P., and Kogelschatz, J.E. (1985) Biological effects of El Nino, pp. 399–438. In: Vegas, M. (ed.) Seminario Regional Ciencas Tecnologica Agresion Ambiental: El Fenomeno "El Nino." Contec Press, Lima, Peru.

Baretta, J.W., Ebenhoh, W., and Ruardij, P. (1995) The European Regional Seas Ecosystem Model, a complex marine ecosystem model. Netherlands Journal of Sea Research **33**, 233–246.

Beamish, R.J. and Bouillon, D.R. (1993) Pacific salmon production trends in relation to climate. Canadian Journal of Fisheries and Aquatic Sciences, **50**, 1002–1016.

Beamish, R.J., McFarlane, G.A., and Benson, A. (2006) Longevity overfishing. Progress in Oceanography **68**, 289–302.

Beverton, R.J.H. and Holt, S.J., (1957) On the dynamics of exploited fish populations. United Kingdom Ministry of Agriculture, Food, and Fisheries Investigations Series 2, 19: 533pp.

Blaxter, J.H. and Hempel, G. (1963) The influence of egg size on herring larvae (Clupea harengus L.). Journal du Conseil International pour Exploration de la Mer, **28**, 211–240.

Burkenroad, M.D. (1951). Some principles of marine fishery biology. Journal du Conseil, **18**, 300–310.

Campana, S.E. and Neilson, J.D. (1985) Microstructure of fish otoliths. Canadian Journal of Fisheries and Aquatic Science **42**, 1014–1032.

Ciannelli L. and Bailey, K.M. (2005) Landscape (seascape) dynamics and underlying species interactions: the cod-capelin system in the southeastern Bering Sea. Marine Ecolology Progressive Series **291**, 227–236.

Ciannelli, L., Chan, K.S., Bailey, K.M., and Stenseth, N.C. (2004) Non-additive effects of environmental variables on the survival of a large marine fish population. Ecology **85**, 3418–3427.

Chesson, P., 2000. General theory of competitive coexistence in spatially varying environments. Theoretical Population Biology **58**, 211–237.

Chesson, P.L. 1984. The storage effect in stochastic population models. Lecture Notes on Biomathematics, **54**, 76–89.

Christensen, V. and Walters, C.J. (2004) Ecopath with Ecosim: methods, capabilities, and limitations. Ecological Modeling **172**, 109–139.

Connell, J.H., (1985) The consequences of variation in initial settlement vs. post-settlement mortality in rocky inter-tidal communities. Journal Exploration Marine Biololgy and Ecology **93**, 11–45.

Croll, D.A., Tershy, B.R., Hewitt, R.P., et al. (1998) An integrated approach to the foraging ecology of marine birds and mammals. Deep Sea Research II **45**, 1353–1371.

Cury, P. and Shannon, L. (2004) Regime shifts in upwelling ecosystems: observed changes and possible mechanisms in the northern and southern Benguela. Progress in Oceanography **60**, 223–243.

Cushing, D.H. (1975) Marine Ecology and Fisheries. Cambridge University Press, Cambridge, 278 pp.

Cushing, D.H. (1984) The gadoid outburst in the North Sea. Journal Conseil International pour l'Exploration de la Mer **41**, 159–166.

De la Mare, W.K. (1996). Some recent developments in the management of marine living resources, pp. 599–616. In: Floyd, R.B., Shepherd, A.W., De Barro, P.J. (eds) Frontiers of Population Ecology. CSIRO Publishing, Melbourne, Australia.

DeMaster, D., Fogarty, M., Manson, D., Matlock, G., and Hollowed, A. (2006) Management of living resources in an ecosystem context, pp. 15–28. In: Murawski, S.A. and Matlock, G.C. (eds) Ecosystem Science Capabilities Required to Support NOAA's Mission in the Year 2020. US Department of Commerce National Oceanic and Atmospheric Administration Technical Memorandum NMFS-F/SPO-**74**, 97p.

Doherty, P.J., Planes, S., and Mather, P. (1995) Gene flow and larval duration in seven species of fish from the Great Barrier Reef. Ecology **76**, 2373–2391.

Ecosystem Principles Advisory Panel (1999). Ecosystem-based Fishery Management: A report to congress by the Ecosystem Principles Advisory Panel. US Department of Commerce, National Oceanic and Atmospheric Administration, National Marine Fisheries Service, Silver Spring, MD, 54 pp.

Fauchald, P., Mauritzen, M., and Gjosaeter, H. (2006) Density-dependent migratory waves in the marine pelagic ecosystem. Ecology **87**(11), 2915–2924.

Fluharty, D. (2005) Evolving ecosystem approaches to management of fisheries in the USA. Marine Ecology Progressive Series **300**, 248–253.

Fogarty, M.J., Sissenwine, M.P., and Cohen, E.B. (1991) Recruitment variability and the dynamics of exploited marine populations. Trends Ecology and Evolution **6**, 241–246.

Fox, W.W. (1970) An exponential yield model for optimizing fish populations. Transactions of the American Fisheries Society **99**, 80–88.

Frank, K.T., Petrie, B., Choi, J.S., and Leggett, W.C. (2005) Trophic cascades in a formerly cod-dominated ecosystem. Science **308**, 1621–1623.

Friedman, T. (2005) The World Is Flat: A Brief History of the Twenty-First Century. Farrar Status and Giroux, New York

Fu, C. and Fanning, L.P. (2004) Spatial considerations in the management of Atlantic cod off Nova Scotia, Canada. North American Journal of Fisheries Management **24**, 775–784.

Graham, M. (1935) Modern theory of exploiting a fishery and applications to North Sea trawling. Journal Conseil International pour l'Exploration Mer **10**, 264–274.

Hairston, N.G., Smith, F.E., and Slobodkin, L.B. (1960) Community structure, population control and competition. American Naturalist **94**, 421–425.

Hanski, I. (1991) Single-species metapopulation dynamics: concepts, models and observations, pp. 17–38. In: Gilpin, M. and Hanski, I. (eds) Metapopulation Dynamics. Academic, San Diego, CA.

Hermann, A.J., Hinckley, S., Megrey, B.A., and Stabeno, P. (1996) Interannual variability of the early life history of walleye pollock near Shelikof Strait as inferred from a spatially explicit, individual-based model. Fisheries Oceanography **5**(Suppl. 1), 39–57.

Hilborn, R., Quinn, T.P., Schindler, D.E., and Rogers, D.E. (2003) Biocomplexity and fisheries sustainability. Proceedings of the National Academy of Sciences of the United States of America **100**, 6564–6568.

Hjort, J. (1914) Fluctuations in the great fisheries of northern Europe, viewed in the light of biological research. Rapp. P.-V. Reun. Conseil Permanent International pour Exploration **20**, 1–228.

Hjort, J. (1926) Fluctuations in the year classes of important food fishes. Journal du Conseil **1**, 5–38.

Holling (1965) The functional response of predators to prey density and its role in mimicry and population regulation. Memoirs of the Entomology Society of Canada 1–85.

Hollowed, A., Ianelli, J.N., and Livingston, P.A. (2000) Including predation mortality in stock assessments: a case study for Gulf of Alaska walleye pollock. ICES Journal Marine Science **57**, 279–293.

Hollowed, A.B. and Wooster, W.S. (1992) Variability of winter ocean conditions and strong year classes of northeast Pacific groundfish. ICES Marine Science Symposia, ICES Journal of Marine Science **195**, 433–444.

Hollowed, A.B., Hare, S.R., and Wooster, W.S. (2001) Pacific basin climate variability and patterns of Northeast Pacific marine fish production. *Progress in Oceanography* **49**, 257–282.

Hollowed, A.B. Wilson, C.D., Stabeno, P., and Salo, S. (2007) Effect of ocean conditions on the cross-shelf distribution of walleye pollock (*Theragra chalcogramma*) and capelin (*Mallotus villosus*). Fisheries Oceanography **16**(2), 142–154.

Houde, E.D. (1989) Subtleties and episodes in the early life history of fishes. Journal of Fish Biology, **35**(Supplement A), 29–38.

Humston, R., Olson, D.B., and Ault, J.S. (2004) Behavioral assumptions in models of fish movement and their influence on population dynamics. Transactions of the American Fisheries Society **133**, 1304–1328.

Iles, T.C. and Beverton, R.J.H. (2000) The concentration hypothesis: the statistical evidence. ICES Journal of Marine Science **57**, 216–227.

Intergovernmental Panel on Climate Change (IPCC). 2007. Working Group 1 Report, http://www.ipcc.ch/

Ito, S., Kishi, M.J., Yutaka, K., et al. (2004) Initial design for a fish bioenergetics model of Pacific saury coupled to a lower trophic ecosystem model. Fisheries Oceanography 13(Suppl. 1), 111–124.

Kawasaki, T. (1993) Recovery and collapse of the far eastern sardine. Fisheries Oceanography 2, 244–253.

Kendall, A.W. and Duker, G.J. (1998) The development of recruitment fisheries oceanography in the United States. Fisheries Oceanography 7, 69–88.

Kesteven, G.L. (1972) Science and sea fisheries. Proceedings of the Royal Society of Edinburgh Series B. 32, 325–332.

Laevastu, T. and Larkins, H.A. (1981) Marine Fisheries Ecosystem: Its Quantitative Evaluation and Management. Fishing News Books Ltd., Surrey, England, 159 pp.

Lasker, R. (1981) Factors contributing to variable recruitment of the northern anchovy (*Engraulis mordax*) in the California current: contrasting years, 1975–1978. Rapp. P. –v. Reun. Conseil Intternational pour l'Exploration de la Mer 178, 375–388.

Leaman, B.M. and Beamish, R.J. (1984). Ecological and management implications of longevity in some northeast Pacific groundfishes. International North Pacific Fisheries Commission Bulletin 42, 85–97.

Levin, S.A. (1992) The problem of pattern and scale in ecology. Ecology 73, 1943–1967.

Levin, S.A. and Pacala, S.W. (1997) Theories of simplification and scaling of spatially distributed processes, pp. 271–295. In: Tilman, D. and Kareiva, P. (eds) Spatial Ecology. Princeton University Press, Princeton, NJ.

Lotka, A. (1925) Elements of Physical Biology. Williams and Wilkins, Baltimore, MD (Reprinted 1956 by Dover, New York, as Elements of Mathematical Biology).

MacArthur, R.H. and Wilson, E.O. (1967) The Theory of Island Biogeography. Princeton University Press, Princeton, NJ.

MacCall, A.D. (1990) Dynamic Geography of Marine Fish Populations. University of Washington Press, Seattle, WA, 153 pp.

Mace, P.M. (2001) A new role for MSY in single-species and ecosystem approaches to fisheries stock assessment and management. Fish and Fisheries 2, 2–32.

Mangel, M. and Levin, P.S. (2005) Regime, phase and paradigm shifts: making community ecology the basic science for fisheries. Philosophical Transactions of the Royal Society of London Series B, Biological Sciences 360, 95–105.

Marquet, P.A., Quinones, R.A. Abades, S, et al. (2005) Scaling and power-laws in ecological systems. Journal of Experimental Biology 208, 1749–1769.

McEvoy, A.F. (1996) Historical interdependence between ecology, production and management in California fisheries, pp. 45–53. In: Bottom, D., Reeves, G. and Brookes, M. (eds) Sustainability Issues for Resource Managers. USDA Forest Service Technical Report PNW-GTR-370.

Megrey, B.A., Rose, K.A., Klumb, R.A., Hay, D.E., Werner, F.E., Eslinger, D.L., and Smith, S.L. (2007) A bioenergetics-based population dynamics model of Pacific herring (*Clupea harengus pallasi*) coupled to a lower trophic level nutrient–phytoplankton–zooplankton model: description, calibration, and sensitivity analysis. Ecological Modeling 202, 144–164.

Murphy, G.I. (1961) Oceanography and variations in the Pacific sardine population. California Cooperative Fisheries Investigations Report 8, 55–64.

Nelson, W.R., Ingham, M.C., and Schaaf, W.E. (1977) Larval transport and year-class strength of Atlantic menhaden, *Brevoortia tyrannus*. Fisheries Bulletin US 75, 23–41.

Nicholson, A.J. (1933) The balance of animal populations. Journal of Animal Ecology 2, 132–178.

Odum, E.P. (1992) Great ideas in ecology for the 1990s. BioScience 42, 542–545.

Omori, M. and Kawasaki, T. (1995) Scrutinizing the cycles of worldwide fluctuations in the sardine and herring populations by means of singular spectrum analysis. Bulletin of the Japanese Society of Fisheries Oceanography 59, 361–370.

Overland, J.E. and Wang, M. (2007). Future regional Arctic Sea ice declines. Geophysical Research Letters, 34, L17705, doi: 10.1029/2007GL030808.

Pascoe, S. (2006) Economics, fisheries and the marine environment. ICES Journal of Marine Science **63**, 1–3.

Paine, R.T. (1969) A note on trophic complexity and community stability. American Naturalist **103**, 91–93.

Parrish, R.H., Nelson, C.S., and Bakun, A. (1981) Transport mechanisms and reproductive success of fishes in the California Current. Biological Oceanography **1**, 175–203.

Pope, J.G. (1972) An investigation of the accuracy of virtual population analysis using cohort analysis. International Commission Northwest Atlantic Fisheries Research Bulletin **9**, 65–74.

Porter, S.M., Ciannelli, L., Hillgruber, N., Bailey, K.M., Chan, K.S., Canino, M.F., and Haldorson, L.J. (2005) Environmental factors influencing larval walleye pollock *Theragra chalcogramma* feeding in Alaskan waters. Marine Ecology Progressive Series **302**, 207–217.

Quinn, T.J. and Collie, J.S. (2005) Sustainability in single-species population models. Philosophical Transactions of the Royal Society Series B **360**, 147–162.

Restrepo, V.R. (1999) Proceedings of the Fifth National NMFS Stock Assessment Workshop. Providing Scientific Advice to Implement the Precautionary Approach Under the Manguson-Stevens Fishery Conservation and Management Act. NOAA Technical Memorandum NMFS-F/SPO-40, 160 pp.

Rice, J. (2001) Implications of variability on many time scales for scientific advice on sustainable management of living marine resources. Progress in Oceanography **49**(1–4), 189–210.

Ricker, W.E. (1954) Stock and recruitment. Journal of the Fisheries Research Board of Canada **11**, 559–623.

Rijnsdorp, A.D., van Beek, F.Z., Flatman, S., et al. (1992) Recruitment of sole stocks, *Solea solea* (L.), in the northeast Atlantic. Netherlands Journal of Sea Research **29**, 173–192.

Rooper, C.N., Gunderson, D.R., and Armstrong, D.A. (2006) Evidence for resource partitioning and competition in nursery estuaries by juvenile flatfish in Oregon and Washington. Fisheries Bulletin US **104**, 616–622.

Rose, K.A., Christensen, S.W., and DeAngelis, D.L. (1993) Individual-based modeling of populations with high mortality: a new method based on following a fixed number of model individuals. Ecological Modeling **68**, 273–292.

Rothschild, B.J. (2007) Coherence of Atlantic cod stock dynamics in the Northwest Atlantic Ocean. Transactions of the American Fisheries Society **136**, 858–874.

Rozwadowski, H.M. (2002) The Sea Knows No Boundaries. University of Washington Press, Seattle, WA.

Russell, F.S., Southward, A.J., Boalch, G.T., and Butler, E.I. (1971) Changes in biological conditions in the English Channel off Plymouth during the last half century. Nature **234**, 468–470.

Sale, P. (1991) The Ecology of Fishes on Coral Reefs. Academic, New York.

Sanchirico, J.N. and Hanna. S. (2004) Navigating US Fishery Management into the 21st century. Marine Resource Economics **19**, 395–406.

Schaefer, M.B. (1954) Some aspects of the dynamics of populations important to management of the commercial marine fisheries. Bulletin of the Inter-American Tropical Tuna Commission **1**(2), 27–56.

Schunte, J.T., Maunder, M.N., and Ianelli, J.N. (2007) Designing tools to evaluate fishery management strategies: can the scientific community deliver? ICES Journal of Marine Science **64**, 1077–1084.

Sheridan, S., Ferguson, J.W., and Downing S.L. (eds) (2007) Report of the National Marine Fisheries Service Workshop on Advancing the State of Electronic Tag Technology and Use in Stock Assessments August 23–25, 2005. NOAA Technical Memorandum, 93p.

Sette, O.E. (1943) Biology of the Atlantic mackerel (*Scomber scombrus*) of North America. Part 1: early life history, including growth, drift, and mortality of the egg and larval populations. US Fisheries Wildlife Service, Fisheries Bulletin **50**, 149–237.

Sinclair, M. (1988). The member vagrant hypothesis, pp. 67–77. In: Marine Populations An Essay on Population Regulation and Speciation. Washington Sea Grant Program, University of Washington Press, Seattle, WA.

Skud, B.E. (1982) Dominance in fishes: their relation between environment and abundance. Science **216**, 144–149.

Smedbol, R.K. and Wroblewski, J.S. (2002) Metapopulation theory and northern cod population structure: interdependency of subpopulations in a recovery of a groundfish population. Fisheries Research **55**, 161–174.

Smith, T.D. (1994) Scaling Fisheries. Cambridge University Press, Cambridge.

Smith, E.P., Orvos, D.R., and Cairns Jr., J. (1993) Impact assessment using before-after-control-impact (BACI) model: concerns and comments. Canadian Journal of Fisheries and Aquatic Science **50**, 627–637.

Solemdal, P., Dahl, E., Danielssen, D.S., and Moksness, E. (1984) Historic review, pp. 17–45. In: Dahl, E., Danielssen, D.S., Moksness, E and Solemdal, P (eds) The Propagation of Cod Gadus morhua L. Flodevigen rapportser 1, 1984. Institute of Marine Research, Flodevigen, Norway.

Solhaug, T. and Saetersdal, G. (1972) The development of fishery research in Norway in the nineteenth and twentieth centuries in the light of the history of the fisheries. Proceedings of the Royal Society of Edinburgh series B **32**, 399–412.

Soutar, A. and Isaacs, J.D. (1974) Abundance of pelagic fish during the 19th and 20th centuries as recorded in anaerobic sediments of the Californias. Fisheries Bulletin, US **72**, 257–273.

Stenseth, N.Chr., Bjornstand, O.N., Falck, W., et al. (1999) Dynamics of coastal cod populations: intra- and intercohort density dependence and stochastic processes. Proceedings of the Royal Society of London Series B **266**, 1645–1654.

Steele, J.H. (2004) Regime shifts in the ocean: reconciling observations and theory. Progress in Oceanography **60**, 135–141.

Taylor, L.R. (1961) Aggregation, variance and the mean. Nature **189**, 732–735.

Thompson, W.F. (1950) The Effect of Fishing on Stocks of Halibut in the Pacific. University of Washington Press, Seattle, WA.

Thompson, W.F. and Bell, F.H. (1934) Biological statistics of the pacific halibut fishery. 2. Effect of changes in intensity upon total yield and yield per unity of gear. Report International Fisheries (Pacific Halibut) Commission **8**, 49 pp.

Uda, M. (1972) Historical development of fisheries oceanography in Japan. Proceedings of the Royal Society of Edinburgh Series B **32**, 391–398.

Volterra, V. (1926) Variations and fluctuations of the number of individuals of animal species living together, pp. 409–448. In: Chapman, R.N. (ed.) Animal Ecology, McGraw-Hill, New York.

Von Bertalanffy, L., 1938. A quantitative theory of organic growth (inquiries on growth laws II). Human Biology **10**(2): 181–213.

Walford, L.A. (1938). Effect of currents on distribution and survival of the eggs and larvae of the haddock (*Melanogrammus aeglefinus*) on Georges Bank. Bulletin of the US Bureau of Fisheries **49**, 1–73.

Walters, C. and Kitchell, J. (2001) Cultivation/depensation effects on juvenile survival and recruitment: implications for the theory of fishing. Canadian Journal of Fisheries and Aquatic Sciences **58**, 39–50.

Wilson, C.D., Hollowed, A.B., Shima, M., Walline, P., and Stienessen, S. (2003) Interactions between commercial fishing and walleye pollock. Alaska Fishery Research Bulletin **10**, 61–77.

Wooster, W.S. (1961) Fisheries oceanography. California Cooperative Oceanic Fisheries Investigations Report **8**, 73–74.

Wooster, W.S. (1988) Immiscible investigators: oceanographers, meteorologists, and fishery scientists. Fisheries **13**, 18–21.

Wright, S. (1943) Isolation by distance. Genetics **28**, 114–138.

Chapter 31
Future Research Requirements for Understanding the Effects of Climate Variability on Fisheries for Their Management[*]

Franklin B. Schwing, William T. Peterson, Ned Cyr, and Kenric E. Osgood[3]

Abstract Climate variability is a key factor controlling the distribution and abundance of marine organisms and ecosystems structure. Climate science must be linked to ecosystem science and living marine resource management if we are to understand, quantify, and forecast the impacts of climate variability and future climate change on marine populations and ecosystem components. To effectively understand and incorporate into management the effects of climate variability on fisheries, we will need to expand greatly our capabilities in a number of research activities. Ecological *observations* must be maintained and enhanced to detect and increase our understanding of the impacts of climate variability and climate change on marine ecosystems. Climate-forced biophysical *models* must be developed and verified to increase understanding of ecosystem responses to climate and provide predictions to assist management. Ecological *indicators* that document ecosystem change and the impacts of climate variability on marine ecosystems must be developed and made operational. Regular *assessments* of ecosystem status must be prepared, distributed, and interpreted. Finally, climate *information* must be integrated into fisheries management plans and decisions. We recommend a series of initial priority activities, which include regional "proof of concept" demonstration projects for linking climate information with resource management; developing ecological indicators that document the state of, and impacts of climate variability

F.B. Schwing
NOAA Fisheries Service, Southwest Fisheries Science Center,
Environmental Research Division, Pacific Grove, CA, USA

W.T. Peterson
NOAA Fisheries Service, Northwest Fisheries Science Center,
Hatfield Marine Science Center, Newport, OR, USA

N. Cyr and K.E. Osgood
NOAA Fisheries Service, Office of Science & Technology, Silver Spring, MD, USA

*Submitted to "Future of Fisheries Science", Springer Fish & Fisheries Series,
AIFRB 50th Anniversary workshop

R.J. Beamish and B.J. Rothschild (eds.), *The Future of Fisheries Science in North America*, 621
Fish & Fisheries Series, © Springer Science + Business Media B.V. 2009

on, marine populations and their ecosystems; and creating web-based and dynamic regional integrated ecosystem assessments (IEAs).

Keywords Climate change · fisheries management · ecosystem-based management · ecological indicators · ecological assessments · physical-biological models

31.1 Introduction

Fisheries research and the management it supports are the subjects of shifting paradigms. A number of factors – the accumulation of environmental and fishery data via regular monitoring and repeated surveys, new technologies for making observations, and advances in computing capabilities – have improved our collective understanding of environmental processes and ecosystem functions, and facilitated the development of complex coupled biophysical models. These have all contributed to new insights about climate–fisheries interactions; moreover, this new work has mandated a new mindset.

Single-species stock management has proven to be insufficient for many populations, and we argue here that it needs to be replaced with a "big picture" ecosystem approach to management based on four new paradigms.

- Rather than short-term species assessments, multiyear assessments will allow long-range rebuilding and management plans.
- Where ocean variability was once thought to be small and thus unimportant in population dynamics, it is now acknowledged that climate variability affects ocean conditions and subsequently ecosystem structure and productivity.
- We are shifting from a "correlative" approach, to one that mechanistically relates environmental drivers to biological responses.
- Finally the magnitude of the ocean signals of climate variability is quantifiable, our future climate change is becoming more certain, and new focus is evolving toward understanding and mitigating the impacts of future climate change on ecosystem structure, production, and function.

The risk of not accepting these new paradigms is failure to recognize changes in the productivity of a system with respect to climate variability, which, without compensatory management decisions, could drive overfished stocks to a much longer recovery period or even to extinction. For example, recognizing that maximum sustainable yield (MSY) changes according to climate forcing and environmental state requires adjusting future quotas accordingly. However, it is incumbent that research addresses these new paradigms, and that science-based information be developed to aid the management decision process. This chapter outlines the components of a research program to relate climate variability to ecosystem-based management (EBM), and to develop the insight, products, and information for effective management from an ecosystem perspective. We argue that the new paradigms of fishery management require a new approach to the research that supports it.

Future research must address a sequence of critical questions:

- What are the key environmental drivers that shape ecosystem structure and productivity?
- How do climate events and future climate scenarios influence these drivers?
- How do marine ecosystems respond to climate variability?
- What are the ecological consequences of concern to our nation's economy and culture?

This new approach to research will provide a number of benefits. First, this research will supply scientific support for living coastal and marine resource management (assessment activities, e.g., stock assessments, ecosystem assessments). It will implement ecosystem approaches to management that consider multiple influences and outcomes. The work will synthesize knowledge about marine ecosystems in relation to human activities, recognize gaps, and improve understanding. Researchers will be able to (and will need to) communicate to managers, policy makers, and the public, the current state of marine ecosystems, the pressures they face, and the potential impacts and risks of management options.

In the long term, we will be in a position to produce regional *indicators* that show how climate variability affects ecological structure, productivity, and health, *assessments* of the status of living marine resources and ecosystems, and a *prognosis* of the effects of future climate change on these resources.

31.2 Rationale for Incorporating Climate Variability into Fisheries Management

Climate variability is a key factor controlling the distribution and abundance of marine organisms and ecosystems structure. The physical drivers related to climate can also impact the growth rates and reproductive success of marine species at all trophic levels. Climate shifts clearly perturb fisheries resulting in socioeconomic impacts. Therefore, for effective and proactive long-term management planning, climate variability must be considered. Should we fail to do so, we risk implementing management strategies that are inconsistent with evolving environmental conditions and thereby risk over- or underexploitation of harvested resources and similar mismanagement of non-harvested species.

There are many examples where climate change has been observed to have major impacts on living marine resource populations, and where insufficient appreciation of or accounting for climate forcing has resulted in drastic changes to ecosystems and impacts on human communities. Climate change ranges from interannual variations to long-term climate trends and its impacts may be expressed anywhere from local to global scales. Figure 31.1 gives an example of how shifts in ocean conditions, as defined by sea-surface temperature (SST), are reflected in nutrient levels, primary productivity, and upward through the food chain to commercially important fish stocks.

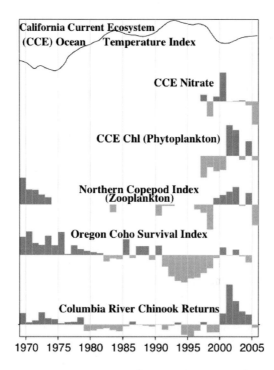

Fig. 31.1 Time series of physical, chemical, and biological variables in the California Current ecosystem, showing decadal variability associated with regime shifts

Climate forces directly influence regional temperature, wind, precipitation, snowpack, and streamflow patterns, which may impact the habitat suitability for species directly or indirectly. Many marine species respond directly to changes in temperature or salinity due to their physiological limits. A number of fish species have been observed to shift their distributions to deeper water (Weinberg 2005) or poleward in response to warming waters (Murawski 1993; Parker and Dixon 1998; Perry et al. 2005).

Physical climate change also affects ocean circulation and stratification, ecosystem productivity and structure, and subsequent habitat suitability. For example, interannual variability in the timing and strength of seasonal upwelling affects west coast zooplankton and fish populations. The "spring transition" to upwelling favorable conditions was delayed by as much as 2 months in portions of the California Current ecosystem during 2005 (Schwing et al. 2006) and 2006. This delay had severe consequences for many ecosystem components, including changes in biomass and species composition of zooplankton (Mackas et al. 2006), distribution, abundance, and recruitment of many pelagic nekton species (Brodeur et al. 2006), changed California sea lion foraging strategies (Wiese et al. 2006), and complete breeding failure of a dominant planktivorous marine bird (Sydeman et al. 2006). Similarly, El Niño events cause large ecosystem shifts on multiyear timescales, impacting ocean productivity and population distributions along the US west coast (Pearcy and Schoener 1987). Overall productivity is reduced during these events and individual species show marked responses. For example, market squid undergo severe population declines during El Niño events, California sea lions have greatly decreased pup production, rockfish show large reductions in reproductive success, and the abundance and distribution of Pacific hake are impacted.

On decadal timescales, changes in climate forcing have large impacts on the productivity of marine ecosystems. Rapid shifts in climate patterns, referred to as regime shifts, are reflected in ecosystems as sharp and sudden shifts in the dominant species and productivity levels. Decadal shifts between low and high production regimes in the California Current ecosystem are a reflection of changes in the large-scale ocean circulation and are indicated by climate indices such as the Pacific Decadal Oscillation (PDO) (Mantua et al. 1997), which is a measure of the north Pacific surface temperature variability. Table 31.1 shows how physical and biological changes in the California Current have been linked to positive and negative phases of the PDO, allowing hypotheses to be developed about the mechanisms of decadal ecosystem change.

The California Current and Gulf of Alaska ecosystems have historically shown a coupled, but opposite response; when the California Current is in the low productive phase, the Gulf of Alaska is highly productive and vice versa. The collapse of the California sardine fishery in the 1940s, which had large economic and social impacts, was partially due to a shift to cooler ocean conditions and the resulting changes in the ecosystem structure that the fishery management did not recognize at that time. Changes in species abundances reflect these decadal changes in ocean climate; e.g., zooplankton composition (Peterson and Schwing 2003; Hooff and Peterson 2006), Pacific salmon survival (Francis and Hare 1994; Peterson and Schwing 2003), sardine versus anchovy dominance in the California Current (Chavez et al. 2003), and the change from shrimp to groundfish dominance in the Gulf of Alaska in the late 1970s (Anderson and Piatt 1999; Benson and Trites 2002).

In the North Atlantic, decadal changes in large-scale atmospheric forcing, as measured by the North Atlantic Oscillation (NAO), have large impacts on marine populations (Drinkwater et al. 2003). The impact on populations depends upon how

Table 31.1 A comparison of general physical and biological changes in the California Current ecosystem associated with the phases of the Pacific Decadal Oscillation (PDO)

Positive PDO	Negative PDO
Upwelling weak; surface water warm	Upwelling strong; surface water cold
Zooplankton species dominants are small warm water species	Zooplankton species dominants are large cold water species from the subarctic Pacific
Euphausiids collapse	Euphausiids increase in numbers
Salmon survival declines	Salmon survival increases 5–10×
Small pelagic fish stocks collapse	Small pelagics boom cycle
Whiting migrate farther north into Canadian waters	Whiting migrate only to 45° N
Warm water predators enter coastal waters	Coastal waters predator-free

the NAO shapes the local conditions and their relation to the preferences of the populations. For example, cod recruitment in the western and eastern Atlantic is out of phase due to the opposite effects of the NAO on temperature in the two regions. Climate change may have contributed to the collapse of the northern cod stock off southern Labrador and northeastern Newfoundland in the 1990s.

31.3   Priority Research Activities

Our understanding of how the earth's climate system and marine ecosystems function and how they are linked has grown dramatically in recent years largely as a result of research carried out during the IGBP/GLOBEC programs. However, a number of intellectual and logistic gaps remain. We must link climate science to ecosystem science and living marine resource management if we are to understand, quantify, and forecast impacts of climate variability on marine populations and ecosystem components. Future research will use ecosystem observations, models, and indicators from climate and ecosystem elements to assess the impacts of climate on marine ecosystems and living marine resources and provide scientific information for management.

To effectively understand and incorporate into management the effects of climate variability on fisheries, we will need to expand greatly our capabilities in a number of research activities, detailed below.

1. Ecological *observations* at several trophic levels must be maintained and enhanced to detect and increase our understanding of the impacts of climate variability and climate change on marine ecosystems, at local and regional scales.
2. Climate-forced biophysical *models* must be developed and verified to aid understanding of ecosystem responses to climate and provide predictions to assist in the management of living marine resources.
3. Ecological *indicators* that document ecosystem change and the impacts of climate variability on marine ecosystems must be developed and made operational.
4. Regular *assessments* of ecosystem status must be prepared, distributed, and interpreted for management and decision making.
5. Climate *information* must be integrated into fisheries management plans and decisions. This will require increased communication between scientists and managers.

31.3.1   *Observations to Monitor Climate and Ecological Impacts*

Fisheries scientists cannot limit their observations to only the distribution and abundance of single species or guild of fishes. Rather, coastal and marine ecosystems must be continuously monitored through a network of in situ and remote observing systems, to track the changing state of environmental drivers and ecosystem components. The goal of this activity is to maintain and expand the suite of ecological observations to detect the impacts of climate variability and global change on entire marine ecosystems, not just on fish, at local and regional scales, on interannual, seasonal, and decadal timescales. The end result will be maps and time series of key ecosystem components, from nutrients to whales, and the physical variables that control their dynamics.

Existing survey and observing programs need to be maintained, and in many instances expanded spatially, temporally, and with additional measurements, but this must be done in an efficient and economical manner. Ocean sampling is an expensive and complex operation that must cover huge, dynamic regions. Thus, it is essential to coordinate survey programs and leverage on existing resources and observing activities, including those available from the fishing industry. Federal and state resource survey programs such as CalCOFI, and research programs such as GLOBEC are building blocks for these activities. Integrated and multinational efforts are needed in some systems to observe different sub-ecosystem types, and to monitor climate change and its ecological impacts across a population's domain (Fig. 31.2). Gap analyses of survey programs, using models and statistical methods, will ensure sampling is economical yet adequate in time and space. For example, many of the fishery surveys illustrated in Fig. 31.2 are conducted only once per year, thus although spatial coverage is impressive, the temporal resolution provides a single snapshot that is sufficient only to resolve changes in distribution or abundance at annual timescales.

A focus on selected "core" variables using standard methods is important. One approach is to monitor climate-sensitive "sentinel species." The features of these species should include: sensitivity to climate-driven environmental changes; a measurable response to climate change; relatively well-understood diagnostic features associated with change; a long observational history for supporting retrospective trend analyses; and representative of other trophically equivalent species. New technologies and sampling methods, including satellites, bio-loggers, acoustics, Autonomous underwater vehicles (AUVs), genomics, and gliders, expand the monitoring toolbox. Finally, Data Management and Communications (DMAC), which allows data to be accessed, subset, transported, and analyzed by scientists and managers, must be an essential component of all observing systems. Effective DMAC ensures common standards and interoperability for data sets, and provides a dividend on monitoring investments.

An important facet of this activity is field-based process research focused on gaining a better understanding of the *mechanisms and linkages* between climate forcing and ecosystem response at seasonal and interannual scales, including El Niño, La Niña, and regime shifts. If we do not understand the mechanisms by which environmental processes drive population variability, we will not be able to understand and project the consequences of future climate change with any reasonable success.

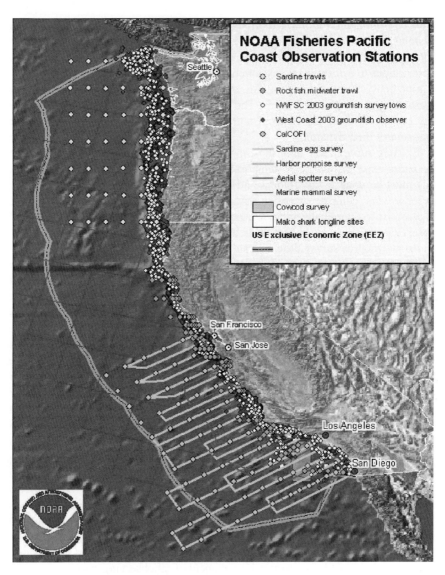

Fig. 31.2 Map of various NOAA NMFS west coast fish survey programs. Most of these surveys are done only once per year (Adapted from NMFS Office of Science and Technology)

31.3.2   Modeling to Guide the Process and Deliver Products and Forecasts

The primary goal of this facet of modeling is to develop and improve the suite of climate-forced biological models and coupled physical–biological models that *forecast* ecosystem production at seasonal and interannual timescales, as an aid in resource management at the regional scales (Fig. 31.3). The end result will be a set of fully integrated model systems that provide information for effective monitoring and management of ecosystems and their resources in light of climate variability, and accurate forecasts of significant ecological events and trends in the context of future climate change.

Models are a pivotal component to connecting monitoring with ecosystem assessments and forecasts. Regional-scale coupled physical–biological models need to be developed and tested that help managers incorporate climate variability into ecological forecasts, assessments, and conduct "if … then" scenario analyses, and decision support. Creating and disseminating model output products that are user-relevant to management is necessary.

While many modeling efforts are underway both at academic and Federal research facilities, linking them into a fully coupled, holistic model system is not straightforward. Environmental variability and climate projections must be incorporated into population assessments. Climate models must be linked and physical and biological models must be coupled. Ecosystem models need fish population components. One of the problems posed in linking disparate model types and efforts is scale mismatch (Fig. 31.4). The principal ecosystem process scales are days–weeks and meters–kilometers, while physical climate model scales of emphasis are intraseasonal and longer and 100 km – global. In addition to an inherent mismatch in the important scales for each, the respective models function on different scales. Before models can be coupled, the output scales of each must be made compatible. In anticipation of the holistic modeling system, physical climate models must be modified and operate to provide output on the scales needed to drive ecosystem models. For economically addressing the smaller scales where individuals and populations interact with each other and their environment, models should be nested in space and time.

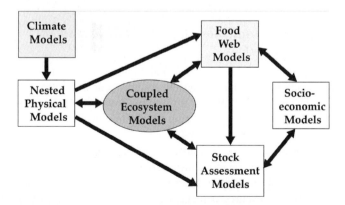

Fig. 31.3 Schematic of holistic modeling approach, connecting physical, biological, and social modeling efforts through a system of coupled ecosystem models

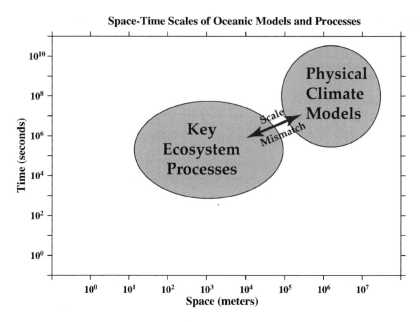

Fig. 31.4 Phase plot of time and space scales for dominant physical and biological ocean features and processes. Key ecosystem processes occur at much shorter time and space scales than emphasized in current physical climate models, leading to a potential scale mismatch

To advance from the old correlative approach to investigate cause-and-effect relationships, mechanistic-based diagnostic and prognostic models must be developed. Modeling should not be limited to physical, ecosystem, and resource models; rather socioeconomic modeling is a vital link to decision support and must be integrated into the holistic modeling effort. As with observational data, DMAC tools are necessary to provide adequate access to model output and products.

31.3.3 Ecological Indicators for Use in Management

A third activity is developing physical and biological indicators that track the changes in physical forcing and ecosystem response due to climate variability The end result will be a set of dynamic indicators that document and quantify the status and tendency of populations and their physical–biological environment, as an aid to improving fishery stock assessments.

Researchers need to produce a suite of *leading indicators* of ecological and oceanographic change based on modeling and observations, and making use of climate sentinel species to help determine the current and future status of the climate and ecological systems for resource management. These will be linked to *performance indicators* that will provide early warnings of major shifts in the productivity of key stocks. Such indicators are analogous to *leading economic indicators* and *stock indices* used extensively by the business community. Indicators are a means for detecting climate change in stock productivity and shifts in ecosystem structure, and for incorporating quantified environmental variability into assessment or management-related problems.

Indicators based simply on correlations are not sufficient, because the systems are complex and nonlinear; mechanistic-based indicators are required. While indicators are generally considered to be predominantly physical, developing lower trophic level indices (e.g., related to zooplankton) or fish-based indices (e.g., based on scales and otoliths to index variations in size at age) need to be emphasized. When related to physical and other ecological indicators, they can be used to evaluate the relationships between environmental pressures and population status. These biological indices are more directly connected to factors such as biodiversity, productivity, growth, and biomass, and thus are more direct and true indicators of potential operational use for resource management.

ICES has defined a set of desirable characteristics of ecosystem indicators. Effective indicators are said to be: easy to understand; responsive to manageable human activities; display responses linked in time to management action; based on easily and accurately measured quantities; responsive to one defined factor; measurable over a large area; and have existing data to provide historic dynamics to inform the selection of targets and thresholds.

31.3.4   Ecosystem Assessments

The goal of this activity is to produce and distribute regular assessments of regional marine ecosystem health and productivity, for scientific, managerial, and decision support needs for ecosystem management. Assessments document and quantify ecosystem status and tendency, and the impacts of past and future climate. They summarize and assess a number of aspects of marine ecosystems: the status of the ecosystem and specific components; the causes and consequences of the status; evaluation of past management actions; forecasts of future status with and without management action; and the costs and benefits of possible management actions.

Assessments include traditional assessments such as stock assessments, reports to fishery management councils, and biogeographic and environmental status reports, as well as new concepts such as integrated assessments (IAs). The US North Pacific Fishery Management Council (NPFMC) produces an annual Stock Assessment and Fishery Evaluation (SAFE) Document that includes an ecosystem assessment, updated status and trend indices, and ecosystem-based management indices and information for marine ecosystems in the North Pacific (NPFMC 2006). It provides updates of the trend and status of various ecological, oceanographic, and climate indicators, as well as ecosystem-level management indicators. It also delivers indices and information on marine ecosystem dynamics to stock assessment scientists and managers.

Canada's Department of Fisheries and Oceans (DFO) produces an annual "State of the Pacific Ocean" report, covering the state of Canadian Pacific marine ecosystems to "understand the natural variability of these ecosystems and how they respond to both natural and anthropogenic stresses" (DFO 2006). The report consists of a summary of conditions for the previous year, including key points for the period, brief descriptions of physical, ecological, and selected fish stock indices and fields, and some prognosis.

Integrated ecosystem assessments (IEAs) are now being promoted as a critical element of a strategy for an ecosystem approach to management. IEAs synthesize ecosystem components and functions – from the physical environment, living organisms, and human processes – in relation to specific ecosystem management objectives. They provide a comprehensive, "big picture" account of the baseline and current conditions, stressors, and drivers of an ecosystem, and evaluate and predict the risks and successes of various management actions (as well as no action), for a specific geographic area. IEAs are multifunctional. They are a *process* that identifies management priorities and objectives, makes quantified assessments, and evaluates management strategies; a *tool* that uses integrated analysis and modeling methods to integrate data and information; and a *product* that provides scientific support for policy and decision makers and an understanding of ecosystem dynamics to scientists.

The IEA concept is being taken one step further, evolving from a "put a staple through it" white-paper format, to a dynamic, web-based IEA system. This would include data sets, products, and indicators, including ecosystem- and community-specific assessments that are updated regularly. Researchers could access data products for models, indicators, and assessments, while managers and the public could find the current status of stocks and ecosystems with uncertainties quantified decision risk assessments. Visualization and analysis tools would be provided that allow scientists and managers to conduct more detailed data searches and analyses using their preferred software applications. In addition to having up-to-date information and assessments immediately accessible through the Internet, a dynamic IEA is flexible and can be customized for multiple stakeholders and management needs.

31.3.5 Integrating Climate Information into Fisheries Management

In the future, there must be a better and more coordinated effort of synthesizing and delivering ecosystem-relevant climate information and incorporating it into fisheries management plans and decisions. As with models, there is a critical but unfulfilled need for linking climate observation, models, and information into the fishery and ecosystem management process. As part of the evolution from climate change detection and attribution to mitigation and adaptation, the science of climate change needs to be integrated into areas where climate's impacts are important.

The climate and fishery science discipline groups are traditionally distinct, so efforts must be made to establish and strengthen dialog and cooperation. NOAA has established a Climate and Ecosystems Program that will increase the emphasis on understanding the consequences of climate variability and change on marine ecosystems, and forecast changes in fishery and other marine resources. This will be achieved by coupling climate and ecosystem observations with information from modeling and retrospective and process studies. The Canadian-Climate Impacts and Adaptation Research Network (C-CIARN) Fisheries has been created to "facilitate communication and research on the impacts of and adaptation to climate change on fish, fisheries and aquatic resources throughout Canada."

One potential activity is applying the results from global and regional climate projections, such as those used in the UN Intergovernmental Panel on Climate Change (IPCC) assessments, to fishery and ecosystem models, to assess the sensitivity and likely responses of marine populations to future climate scenarios.

31.4   Research and Products Loop

Figure 31.5 summarizes the components of this climate and ecosystem approach to fisheries research in a research and products loop, which illustrates the iterative research and information delivery process. Ecological observations and climate-forced biophysical models are coupled in the sense that models rely on observations for initializing and validating simulations, and both activities provide information ("data") over time and space for developing ecological indicators and regular fishery and ecosystem assessments. A DMAC component is essential for effective archiving, accessing, and application of observed and modeled data in these activities. Further synthesis and distillation may be necessary to provide science-based guidance for planning and decision information.

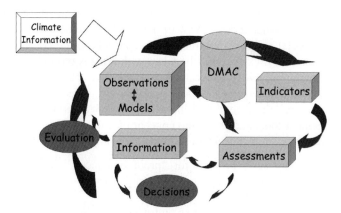

Fig. 31.5 Proposed "Research and Products Loop," the iterative process using observations and models to develop and produce indicators and assessments, providing information for science-based management decisions. Evaluation of information by its users contributes to improvements in future research and science advice

Evaluation and feedback of science-based products and information by their end users gives researchers critical input for modifying observing programs and models, and ultimately improving insight and information about climate and marine ecosystems. Because climate forcing is changing, the research and products supporting resource management must adapt to climate variability as our understanding about climate processes and impacts improves.

31.5   Near-Term Priorities

Our ultimate goal as researchers is to be able to predict the probable consequences of global climate change on ecological systems and their living resources, and to deliver to fisheries managers the knowledge and tools needed to incorporate climate variability into their decision making. Large investments have been made toward understanding the physical climate system and describing the mechanisms that have governed past climate variability and might control future change. However, relatively little work has been done to quantify the response of marine populations to climate forcing, understand the mechanisms linking climate drivers to ecological impacts, or the implications of future climate change for marine ecosystems. Future research must build a bridge between "physical forcing" and "ecosystem response" through observations, modeling, process studies, and analysis, leading to a better understanding of the critical factors that link climate and ecosystem variability.

Existing research, observations, and modeling activities must transition to an operational program to develop new physical and ecological indicators and ecosystem assessments so that managers and policy makers will be better enabled and equipped to evaluate the impacts of climate variability and change on fisheries management.

Moving to the ecosystem-based management (EBM) paradigm will be evolutionary (Field and Francis 2006). We will not be able to quickly change the way fisheries and coastal resources are managed, especially in the context of climate variability. Rather, small steps must be taken with a focus on regions with a high degree of climate variability that can impact sensitive marine ecosystems, and where an ecosystem management approach will be most effective. We recommend a series of initial priority activities in the near-term (<5 years) to develop science-based management tools and prove the concept of linking climate information to ecosystem management and decision support.

1. Regional demonstration projects for "proof of concept" for linking climate information with resource management (priority where climate signals and fishery responses are strongest)
2. Maintain core observation activities and data bases (include gap analysis and DMAC with standards)
3. Develop and validate climate-forced biophysical models to increase our understanding of climate impacts on ecosystems

4. Develop and test ecological indicators that document the state of, and impacts of climate variability on, marine populations and their ecosystems
5. Generate regional IEAs and make them web-based and dynamic

These initial activities can leverage a number of existing programs, but will need additional funding to succeed in a timely way. Nevertheless, they are essential if we are to understand the impacts of climate variability on fishery resources, recognize how these populations will respond to future climate change, and ultimately develop reliable long-term projections for managing resources from an ecosystem perspective with greater certainty. Great economic, social, and intrinsic benefits will result from investing in and applying this climate and ecosystem approach to research.

References

Anderson, P. J. and J. F. Piatt. 1999. Community reorganization in the Gulf of Alaska following ocean climate regime shift. *Mar. Ecol. Prog. Ser.* 189: 117–123.

Benson, A. J. and A. W. Trites.. Ecological effects of regime shifts in the Bering Sea and eastern North Pacific Ocean. *Fish Fisheries* 3: 95–113.

Brodeur, F. D., S. Ralston, R. L. Emmett, M. Trudel, T. D. Auth, and A. J. Phillips. 2006. Anomalous pelagic nekton abundance, distribution, and apparent recruitment in the northern California Current in 2004 and 2005. *Geophys. Res. Lett.* 33: L22S08, doi:10.1029/2006GL026614.

Chavez, F. P., J. Ryan, S. E. Lluch-Cota, and M. Ñiquen C. 2003. From anchovies to sardines and back: multidecadal change in the Pacific Ocean. *Science* 299: 217–221.

DFO, 2006. State of the Pacific Ocean 2005. DFO Science Ocean Status Report. 2006/001.

Drinkwater, K. F. 2002. A review of the role of climate variability in the decline of northern cod. *Am. Fish. Soc. Symp.* 32: 113–130.

Drinkwater, K. F., A. Belgrano, A. Borja, A. Conversi, M. Edwards, C. H. Greene, G. Ottersen, A. J. Pershing, and H. Walker. 2003. The response of marine ecosystems to climate variability associated with the North Atlantic Oscillation. In: The North Atlantic Oscillation: Climate Significance and Environmental Impacts, Amrican Geophysical Union, *Geophys. Mono.* 134: 211–234.

Field, J. and R. Francis. 2006. Considering ecosystem-based fisheries management in the California Current. *Mar. Policy* 30: 552–569.

Francis, R. C. and S. R. Hare. 1994. Decadal-scale regime shifts in the large marine ecosystems of the Northeast Pacific: a case for historical science. *Fish. Oceanogr.* 3: 279–291.

Hooff, R. C. and W. T. Peterson. 2006. Recent increases in copepod biodiversity as an indicator of changes in ocean and climate conditions in the northern California Current ecosystem. *Limnol. Oceanogr.* 51: 2042–2051

Mackas, D. L., W. T. Peterson, M. D. Ohman, and B. E. Lanviegos. 2006. Zooplankton anomalies in the California Current system before and during the warm ocean conditions of 2005. *Geophys. Res. Lett.* 33: L22S07, doi:10.1029/2006GL027930.

Mantua, N. J., S. R. Hare, Y. Zhang, J. M. Wallace, and R. C. Francis. 1997. A Pacific inter-decadal climate oscillation with impacts on salmon production. *Bull. Am. Meteorol. Soc.* 78: 1069–1079.

Murawski, S. A. 1993. Climate change and marine fish distributions: forecasting from historical analogy. *Trans. Am. Fish. Soc.* 122: 647–658.

North Pacific Fishery Management Council (NPFMC). 2006. Stock Assessment and Fishery Evaluation (SAFE) Document for the BSAI and GOA. Appendix C: Ecosystem Considerations for 2007 (J. Boldt, ed.). North Pacific Fishery Management Council, Anchorage, AK.

Parker, R. O., Jr. and R. L. Dixon. 1998. Changes in a North Carolina reef fish community after 15 years of intense fishing – global warming implications. *Trans. Am. Fish. Soc.* 127: 908–920.

Pearcy, W. G. and A. Schoener. 1987. Changes in the marine biota coincident with the El Niño in the northeastern subarctic Pacific. *J. Geophys. Res.* 92: 14417–14428.

Perry, A. L., P. J. Low, J. R. Ellis, and J. D. Reynolds. 2005. Climate change and distribution shifts in marine fishes. *Science* 308: 1912–1915.

Peterson, W. T. and F. B. Schwing. 2003. A new climate regime in northeast pacific ecosystems. *Geophys. Res. Lett.* 30(17): 1896, doi:10.1029/2003GL017528.

Schwing, F. B., N. A. Bond, S. J. Bograd, T. Mitchell, M. A. Alexander, and N. Mantua. 2006. Delayed coastal upwelling along the US West Coast in 2005: a historical perspective. *Geophys. Res. Lett.* 33: L22S01, doi:10.1029/2006GL026911.

Sydeman, W. J., R. W. Bradley, P. Warzybok, C. L. Abraham, J. Jahncke, K. D. Hyrenbach, V. Kousky, J. M. Hipfner, and M. D. Ohman. 2006. Planktivorous auklet (*Ptychoramphus aleuticus*) responses to the anomaly of 2005 in the California Current. *Geophys. Res. Lett.* 33: L22S09, doi:10.1029/2006GL026736.

Weinberg, J. R. 2005. Bathymetric shift in the distribution of Atlantic surfclams: response to warmer ocean temperature. *ICES J. Mar. Sci.* 62: 1444–1453.

Wiese, M. J., D. P. Costa, and R. M. Kudela. 2006. At-sea movement and diving behavior of male California sea lion (*Zalophus californianus*) during 2004 and 2005. *Geophys. Res. Lett.* 33: L22S10, doi:10.1029/2006GL027113.

Chapter 32
Opportunities in Social Science Research

Dale Squires

Abstract The opportunities for social science research change with developments in policy and social science, conservation biology, and ecological theory; population dynamics, quantitative methods, laws and current management or governance practices; industry operating procedures, social values, institutional change, and funding. This paper identifies opportunities for future social science research, and economics in particular, due to developments in economic theory and the shifting concerns of society. The opportunities lie in addressing the growing societal concerns over the environment, biodiversity, and sustainable resource use and bioeconomic modeling that begins to match advancements in population dynamics and ecology. The opportunities address multiple species, bioeconomic modeling that accounts for space, the heterogeneity of fishing industries and the need to address distributional issues and trade-offs. Future social science research will relate to impacts with different policies, incentives, and property and use rights, uncertainty, international management of transboundary stocks of fish and biodiversity conservation (whales, sea turtles, sea birds, dolphins, etc.), marine reserves, technical change, the shift in orientation from management of fisheries as a commercial fishery and a simple optimal harvest strategy to ecosystem management. Important ideas for the future include actual fisheries management of Pareto-improvements from a second-best situation rather than normative concerns that dominate most theoretical fisheries economics research.

32.1 Introduction

This chapter addresses opportunities for future social science research, and economics in particular, due to developments in economic theory, developments in other fields such as ecology and conservation biology, changes in fishing industries and social

D. Squires
NOAA Fisheries, Southwest Fisheries Science Center,
8604 La Jolla Shores Drive, La Jolla, California, 92037, USA
e-mail: Dale.Squires@noaa.gov

R.J. Beamish and B.J. Rothschild (eds.), *The Future of Fisheries Science in North America*, 637
Fish & Fisheries Series, © Springer Science + Business Media B.V. 2009

values, and changes in management measures. The perspective will be concerned first and foremost with what has resonated with the policy process rather than the different set of research objectives often addressed by academic economic research, which has its own, often independent and free-standing, set of objectives.

32.2 Economics

The economic analysis of fisheries as a field of serious study really began with the publication of what is still *the* seminal publication, that of Gordon (1954). (Warming (1911) initially introduced the basic ideas discussed by Gordon (1954), but because Warming wrote in Danish, his ideas languished for many years well past Gordon's writing.) Close on the heels of Gordon followed the two publications by Scott (1955a, b). Gordon was mostly predictive or positive, providing a model of rent dissipation under open access and Scott was mostly normative, addressing how society should optimally manage renewable resources. These publications defined what remains the central thrust of fisheries economics, that there is: (1) an economic or Pareto optimum, what is commonly known as the maximum economic yield (MEY), and an opportunity cost to not achieving this optimum, and (2) the fundamental reason for the overfishing and overcapacity found in fisheries is what is now understood to be the absence of fully formed property rights, such as open access.

The first theme, which includes economic harvesting strategies, has had minimal impact in practice, although considerable time and effort has been spent in developing this theme, suggesting an over-allocation of scarce economics resources. (Portions of the discussion on economics, especially bioeconomics, build upon an excellent review of renewable resource economics by Deacon et al. (1998) and of fisheries economics by Wilen (2000)). The second theme of property rights and establishing proper incentives to guide fisher behavior and align private incentives with socially desirable goals has been very influential, and this influence continues to grow. This second theme has been the central contribution of fisheries economics to fisheries management, and its concepts and ideas have widely diffused to other social sciences, fisheries science, conservation biology and ecology, industry, governments, and international organizations, and are even starting to make inroads into the thinking of conservationists.

32.3 Bioeconomics and Economically Optimum Harvest Strategies

Fisheries economics has traditionally focused on normative economically optimum harvest strategies through static and dynamic analysis of the harvest of a renewable fish stock. Fisheries economics approaches the stock of fish as a stock of natural capital, and applies capital theory to the natural and man-made stocks of capital to obtain the economically optimum exploitation rates or harvests and the corresponding economically optimum stocks of the natural and man-made capital. (Capital is any good, asset, capable of yielding a stream of economic returns to society through time, in contrast to a consumption good or service.) Key early and fundamental literature includes Crutchfield and Zellner (1962), Plourde (1970), Clark and Munro (1975), Clark et al. (1979), Smith (1968, 1969), Turvey (1964). Wilen (1985) and Brown (2000) provide reviews.

Smith (1968, 1969) was one of the first economists to discuss the dynamics underlying the overexploitation of an open-access resource (with three behavioral restrictions for the interactions of the resource stock, individual firms, and industry), the application of phase diagrams from the field of differential equations, and the possibility of extinction along the adjustment path due to overshooting even though the stock equilibrium is positive. Wilen (1976) applied the dynamic model of Smith (1968, 1969) to the Pacific fur seal, showing that the sealing industry followed a pattern similar to that predicted by Smith. Berck and Perloff (1984) considered how potential entrants to an open-access fishery form their expectations determines the fishery's adjustment path to a steady state but not the steady-state values themselves; the paper contrasts myopic and rational expectations. Bjørndal and Conrad (1987) applied the model of Smith (1968) to examine stock extinction under open access using a non-linear deterministic model for the North Sea Herring fishery. Smith (1968), Gould (1972), Clark (1973), and Berck (1979) considered conditions for extinction of an animal population.

Optimum utilization of the fishery resources implies managing the resources in such a manner as to ensure that they provide the maximum flow of economic benefits to society through time. In principle, these economic benefits range beyond simply economic rents in the commercial fishery (rents are revenues less the costs of economic inputs) to include the benefits to society from conservation, biodiversity, and other non-market uses. The theory of resource economics rapidly expanded with the publication of Pontryagin's book on optimal control theory in 1962 (Pontryagin et al. 1962). These techniques were brought to the task of describing optimal use paths for both renewable and nonrenewable resources (Wilen 2000). The notion of a discount rate or a time value to benefits and costs received at different points in time by society was introduced in the process and is now widely used by population biologists.

In short, fisheries economics, beginning with Scott (1955a), has principally focused on the normative "first-best Pareto optimum" that comes from economically optimal exploitation or harvest rates, i.e., on maximizing economic rents through harvest rates under alternative conditions. The principal message has been that there is an economic optimum (rent maximization) and an economic opportunity cost (the foregone benefits from not adopting the next best alternative) to not following these economic harvest strategies, i.e., a focus on MSY or some other biological optimum does not fully yield the fullest economic benefits to society that are possible, such as with the Maximum Economic Yield (MEY).

The first message has largely been ignored in practice by policy-makers, and the scientific basis of harvest strategies has remained firmly in the hands of population dynamics. Real world economic considerations have little impact when quotas are set (Homans and Wilen 1997). Moreover, Wilen (2000, p. 323) observes:

> My assessment is that the profession has probably been too preoccupied with abstract, conceptual, and normative analysis. While these types of contributions seem to be rewarded within the incentive systems of academia, they have not played important direct roles in the policy process. It is certain that we have reached negative returns to further demonstrations that open access dissipate rents compared with various versions of optimized fisheries.

Deacon et al. (1998) reiterate this conclusion, and further observe (p. 392), as noted by Deacon et al. (1998, p. 390):

> In hindsight, elaborating the basic conditions for optimal dynamic resource use absorbed an enormous amount of intellectual effort for a payoff whose practical importance has been relatively small. In fisheries, managers are virtually never concerned with getting biomass stocks close to dynamically optimal long run levels. Instead, fisheries managers raise questions like: how will the industry be affected by trip limits, mesh size changes, or limit entry? How will bycatch and discards be affected and is the biomass safe from stock collapse? Significantly, many of these 'management' questions are predictive rather than normative and closer in spirit to Gordon's focus. Ironically, they remain largely unanswered because economists chose to emphasize the optimization problem Scott posed instead.

Apparently, only in societies where the fishing sector provides an important contribution to gross domestic product and/or where property rights in fishing industries are more fully developed, such as with individual transferable quotas (ITQs), is much emphasis placed on economic rent. Otherwise, the policy emphasis tends to be placed on multiple objectives, including a biological optimum and social issues, such as employment and incomes, leading to optimum yield, and the distribution of costs and benefits in economically sub-optimum (second-best) allocations.

Bioeconomic models are the means by which economic harvest strategies have been analyzed. As a rule, single-species surplus production bioeconomic models based on Schaefer (1954) (that are sometimes more sophisticated in allowing for patchy resource environments, oceanographic dispersion of larvae, etc. as discussed below) form the workhorse bioeconomic model. Eggert (1998, p. 400) observes, "Analyzing the management of two or more competing species is more complex and, despite some progress, the single species approach still dominates the empirical work and simple stock-growth models are still practiced." Some progress in accounting for multiple species (Conrad and Adu-Asamoah 1986, Flaaten 1988, 1991; Clark 1990; Placenti et al. 1992; Herrera 2006) has been made in this area within the Schaefer and analytical framework, but much more is required outside of this framework as in Kjærsgaard and Frost (2008). Quirk and Smith (1970) examined ecologically interdependent fisheries, comparing the open-access equilibrium with the social optimum. Hannesson (1983b) extended these results to examine if there is a price at which it is economically sensible to switch from exploiting the prey to the predator.

Although surplus production bioeconomic models based on Schaefer (1954) have served as the workhorse, some attention has been given to models (especially empirical ones) incorporating demographic information, principally year classes, based on Beverton and Holt (1957). In the words of Eggert (1998, p. 402):

Dynamic optimization in the Beverton-Holt model quickly becomes complex and, including a stock–recruitment relationship, makes it almost incomprehensible from the dynamic viewpoint (Clark 1990). In empirical studies these problems are overcome by using some discrete instead of continuous variables, some strict assumptions are made, and optimization is solved by computer simulation (Hannesson 1993). The optimization problem is then to determine the efficient fishing mortality and mesh size, which depends on net growth rate and the real discount rate. In the simplest version, fishing mortality and cost per unit effort are assumed constant, but extensions are conveniently handled with a computer.

Steinshamn (1992) offers a comprehensive treatment of the Beverton-Holt model and Deacon (1989), Bjørndal and Brasãao (2006), and Kjærsgaard and Frost (2008) are excellent examples. Sumaila (1998b) uses a multicohort age-structured population model and a game theoretic framework in a predator–prey study. When year classes cannot be properly identified, another approach models growth according to the von Bertalanffy growth equation, such as Christensen and Vestergaard (1993) and Sparre and Vestergaard (1990).

Bioeconomic models have also always assumed time-invariant parameter values of the underlying growth functions except for an i.i.d. error term (Walters and Parma 1996; Castilho and Srinivasu 2005; Schlenker et al. 2007). Wilen (2004) and Schrank (2007) observe that fishing mortality is not likely to be constant, but is instead a function of economic and biological parameters. Schlenker et al. (2007) made innovative progress in allowing for cyclical growth parameters in both single and multispecies models. Neither optimal harvest rates nor optimal escapement remains constant as current bioeconomic models would predict. This approach shows that once the periodicity of the biological growth function is incorporated, many of the traditional policy prescriptions reverse. For example, periodic fluctuations in growth imply that it can be best to close a fishery during times when non-stationary biological growth parameters are improving most rapidly and the return from not fishing is highest (Schlenker et al. 2007).

A policy that derives the maximum sustainable harvest quota using the average growth rate will lead to overfishing and a crashing fish stock, as will an adaptive policy that utilizes a limited time-series of past data.

In sum, bioeconomic models have largely failed to keep pace with the very sophisticated and detailed population dynamics models that incorporate much more biological information, such as various forms of age-structured models and even more the modern synthetic models, time-varying biological parameters, and incorporation of uncertainty through Bayesian decision analysis (see Punt and Hilborn 1997). Nonetheless, for a countering view Hannesson (2007d, p. 699) recently observed:

> For stock assessment purposes, age-structured models are used for the Northeast Arctic cod. While more realistic, such models are also much more complex than aggregate biomass models. Furthermore, age-structured models introduce idiosyncratic elements of uncertainty, as parameters such as weight at age and natural mortality are not constant but variable and known only after the fact and with some uncertainty. The gains in validity from age structured models compared with aggregate biomass models will therefore be smaller than if their parameters were known with full certainty. This, and the fact that aggregate biomass models are computationally much simpler, is an argument for using them when they can be reconciled with reality.

Punt and Hilborn (1997) observe that the Bayesian approach to stock assessment determines the probabilities of alternative hypotheses using information for the stock in question and from inferences for other stocks/species. These probabilities are essential if the consequences of alternative management actions are to be evaluated through a decision analysis. Using the Bayesian approach to stock assessment and decision analysis, it becomes possible to admit the full range of uncertainty and use the collective historical experience of fisheries science when estimating the consequences of proposed management actions. In the words of Wilen (2000, p. 320): "While biologists developed new and richer depictions of more realistic population processes with simulation modeling, calibration, and statistical estimation, techniques, economists mostly continued to work with simpler models that could be analytically solved."

Little has substantively changed since these words of Wilen were written. There has been progress in addressing patchy resource abundance, dispersal, and oceanographic linkages, which has been applied to address bioeconomics of marine reserves and spatial regulation in fisheries, often in a surplus production framework, but not always (Sanchirico and Wilen 1999, 2001, 2005; Smith and Wilen 2003; Holland 2003; Holland et al. 2004; Janmaat 2005; Schnier and Anderson 2006; Herrera 2006; Smith 2006a; Kjærsgaard and Frost 2008). Feedback rules have been considered (Grafton et al. 2000b; Steinshamn 2002). One major conclusion that falls out from this literature is that economic incentives determine both participation and location choices, so that fishing effort is not spatially uniform and that optimistic conclusions about reserves ignore economic behavior. The extent the conclusions from this discussion are actually implemented, are believed, or form the basis of actual policies may be limited by the chasm in modeling techniques between the fields of fisheries economics and population biology.

Despite the very real progress that has been made in broadening bioeconomic models, until these models build off current biological best-practice and shift from an emphasis on normative analytical solutions (which necessarily restrict the complexity of the model) to prediction and stochastic dynamic simulations, the economics discussion in this area will have difficulty in informing actual policy decisions (as opposed to an internal debate among economists). Computer-based simulations and more realistic assumptions will be central to progress in bioeconomic modeling, as in other branches of economics (Beinhocker 2006). Recognizing that the steady-state equilibrium is not at all steady due to technical change and incorporating technical change into bioeconomic models will also extend the usefulness of these models given the importance of technical change in the fishery economy (Squires and Vestergaard 2004, 2007). The incorporation of technical change, however, means that the steady-state equilibrium does not exist and will require shifting from the aesthetically pleasing phase diagrams and approach paths to a non-existent steady-state equilibrium to continual disequilibrium (Squires and Vestergaard 2007). Bioeconomic models founded on surplus production are simply inconsistent with the stock assessment advice given by population dynamics biologists. It is also unclear how interested policy makers actually are in normative economically optimum harvest strategies.

Policy-makers and different constituents in practice are very interested in the predictive and distributional impacts of policies, given TACs, from economists (and other social scientists). An assessment of distributional impacts in turn requires firm (vessel)-level models that recognize the heterogeneity of catch (multiple outputs) and effort (variable inputs such as fuel consumption, ice, crew, and fixed inputs such as the capital stock of vessel and some gear), gear types and vessel size classes, regions, and other factors that contribute to the heterogeneity in fisheries. Some important economic work has been conducted in this area (Weitzman 1974b, Johnson and Libecap 1982; Karpoff 1987; Boyce 1992).

Capturing such concerns over distributional impacts in a bioeconomic framework requires dynamic disaggregated models rather than highly aggregated ones, such as those developed by Brazee and Holland (1996), Smith and Wilen (2003), Holland (2003), Bjørndal and Brasãao (2006), and Kjærsgaard and Frost (2008). The latter is illustrative of what can be hoped for from such bioeconomic models in that effort is disaggregated, the population dynamics is age-structured rather than surplus production, considers recruitment and selectivity, and spawning stock biomass, allows discarding, species and inputs are multiple, fleets may be multiple, different areas can be fished, allows dynamic numerical allocation, and can perform both optimizations and feedback simulations. In fisheries that are managed by ITQs or other forms of rights-based management, a top-down, centralized economic modeling approach such as the workhorse surplus production bioeconomic framework does not address the issues of concern to the policy process. Dynamic bioeconomic mathematical programming models such as Bjørndal and Brasãao (2006) and Kjærsgaard and Frost (2008), and discussed further below, are very promising in this regard, but their accuracy (as with any model) always remains questionable given the complexity and difficulty of the task.

As with virtually all empirical production modeling (such as bioeconomics, production functions and frontiers, and fishing capacity), the current state of technology is taken as given, and the results reflect regulations, policy-induced technology that developed under regulation, and a property rights regime. The results may differ sharply from the technology that occurs after rationalization and also the change of technology that occurs over the normal course of events. (Homans and Wilen (1997) and Wilen (2007) essentially make this point.) The entire composition and types of fleets that would occur in a bioeconomic optimum might well differ from the model that is reflected in the data and conceptual framework. Also shared with most production modeling is a failure to address what is endogenous and what is exogenous; for example, "fishing effort" as a concept is an endogenous intermediate product which itself is a function of exogenous market prices and state of technology, but instead is typically specified as immune to changes in markets, technology, resource, and environmental conditions. (In fact, the issue is far more complex. Fishing effort is a composite input formed under very rigorous conditions, that of input separability (Hannesson 1983a, Squires 1987b) or a non-separable two-stage production process (Kirkley 1990).

Bioeconomic models typically abstract from decentralized markets and over-look the technical, and allocative inefficiency that occurs among multiple outputs and among multiple inputs and simply evaluates a form of scale efficiency, i.e., it simply looks for a 'sweet spot' (Squires and Vestergaard 2004, 2007). The maintained behavioral hypotheses and policy objectives also differ from those found in actual fisheries. By overlooking technical inefficiency (identified with skipper skill by Kirkley et al. 1998), the potentially critical role of the skipper and the fishing firm's management function in general is overlooked. Moreover, the empirical bioeconomic optimum should perhaps best be called the regulated bioeconomic optimum because it implicitly accepts the regulatory structure that is in place at the time that the data were generated (a problem shared with all production modeling and any equilibrium, whether temporary and short-term or very long-term). Bioeconomics, along with most production modeling, largely overlooks the pervasive uncertainty found in fishing industries. One of the principle results is suggesting that solving fishery problems is as simple as removing fishing effort rather than addressing the importance of incentives and property rights through production processes reflecting fleet heterogeneity, institutions, governance, and distributional impacts.

The research challenge that is relevant to actual policy-making, as opposed to a normative and conceptual approach, will require a predictive orientation and stochastic dynamic simulation or a framework that explicitly builds upon the current state of population biology, disaggregation in production and industry (i.e., catch, effort, vessels, and geography), acknowledging the growing importance of the ecosystem and biodiversity, allowing for time-varying biological growth parameters, addressing stochasticity (perhaps by Bayesian decision analysis) and recognizing the multiple objectives of policy-makers and stakeholders, distributional impacts, incentives, governance, and pervasive uncertainty.

Finally, one of the most critical areas for bioeconomic research lies in extending these models to integrate in situ environmental benefits of natural resource stocks into optimizing models of natural resource (Perrings et al. 1992; Li and Löfgren 1998; Deacon et al. 1998). Since harvesting can impair the ecosystem, say through gear damage of the benthic habitat or bycatch, and other non-market services, dynamic analysis of intertemporal resource allocation needs to broaden to recognize that the flow of ecosystem and environmental services and biodiversity conservation are determined simultaneously with the flow and stock of the resource. As a consequence, the impact of environmental considerations on the optimal extraction decision can be far more complex than simply determining the MEY. Deacon et al. (1998, p. 387) observe:

> The fundamental insight is that the flow of ecosystem and environmental services is determined simultaneously with the flow and stock of the resource. As a consequence, the impact of environmental considerations can be far more complex than making a dichotomous choice between conservation and extraction. Moreover, accounting for the complex dynamics of ecosystem services is likely to amplify the importance of flow considerations. ... More generally, any environmental or ecosystem service provided by a natural resource stock can have important dynamic dimensions.

Li and Löfgren (1998) and Li et al. (2001) extend the basic bioeconnomic model to include non-market benefits from biodiversity conservation. An alternative approach, one that potentially serves as a rich source of research, is that by Finnoff and Tschirhart (2003).

32.4   Property Rights and Incentives

The second, and related, message from resource economics, one elegantly made by Warming (1911), Gordon (1954), and Scott (1955a, b), is that there is an enormous ecological and economic cost to society from open access, or more generally, from property rights that are not fully developed, and that property rights, markets, and other institutions coupled with public policies jointly create incentives (Grafton et al. 2006; Hilborn et al. 2005). The standard economic justification is that it facilitates the socially efficient exploitation of resources, enabling the owner (which may be an individual, group, or state) to exclude others from the resource and thereby internalize the externalities that would occur if access were free (de Meza and Gould 1992). The ability to exclude also provides incentives to invest in improving the quality of the resource and exploit at a socially optimum rate. Coase (1960) emphasized the importance of economic costs and discussed the distribution of property rights.

The importance and role of property rights is the key message that has soundly resonated with policy-makers, industry, environmental groups, ecologists, population biologists, conservation biologists, and others. Whole commercial fisheries economies in New Zealand, Iceland, and increasingly Australia, have been founded on rights-based management and represent the practical outcome of the concern with rights and incentives introduced by economics. In principle, economic incentives can be established through price controls (taxes and subsidies), quantity controls (quotas, rations), and property rights. The focus has largely rested on transferable shares of TACs or TAE, i.e., on transferable quantity controls upon which has been inferred a use right. Price controls have received little attention.

As a corollary, recognition is growing in fisheries, as well as other sectors of the economy and even globally, that traditional centralized "command-and-control" regulations, such as simple quantity controls on catch (trip limits) or fishing time (effort limits) can be counter-productive in achieving ecological and population objectives, much less economic ones. In the words of Wilen (2000, p. 309):

> One's view of the solution should follow from one's definition of the problem, of course, and from the start, economists viewed the policy problem differently from biologists, who defined the policy problem as one of excessive fishing mortality and hence one best addressed by reducing gear efficiency. Fisheries economists, in contrast, adopted Gordon's view, which was that excess fishing mortality was just one of several *symptoms* of the fundamental problem, a lack of property rights.

As a further corollary, recognition is growing that the problem is far more than the symptoms of overcapitalization, excessive fishing effort, or overcapacity (depending on the modeling and conceptual framework used).

The critical advance in introducing property and use rights in fisheries, through Individual Transferable Quotas (1TQs), is due to Francis Christy (1973). Christy, in effect, adapted ideas from the environmental economics literature on individual rights-based solutions as a mechanism for solving pollution problems by Crocker (1966) and Dales (1968). Wilen (2000, ft 22, p. 317) states, "Probably the strongest proponents of ITQs during the late 1970s were Anthony Scott and Peter Pearse. Scott's interests in property rights solutions went back to his earliest writings on resources policy and related to his long-standing interest in institutions and their influence on resource use. Pearse had similar conceptual interests in property rights solutions but, in addition, an astute appreciation for the politics of resource policy implementation, developed on several commissions he headed on Canadian forestry and water policy" (see Pearse 1980, 1981). Subsequent social science research has further developed and evaluated individual property and use rights in both theory and practice. After recognizing the importance of individual rights, social science research began to evaluate different forms of property and use rights, most notably forms of common rights (Ostrom 1990; Baland and Platteau 1996; Bromley 1991). Nonetheless, property rights solutions, particularly ITQs, are not a panacea and are not necessarily the appropriate regulatory instrument in all instances (Squires et al. 1995).

Further work remains in the area of property rights, both as new forms of rights emerge and for formal analytical evaluation. A prime example is fishing cooperatives or voluntary agreements as a form of use right that is increasingly important as a type of rights-based management (Townsend 2005; Pinto da Silva and Kitts 2006). The initial work has largely been descriptive, although the Coasian framework of transactions costs has been recognized (Townsend 2005; Edwards 2008). More critically, initial work is underway applying ideas from industrial organization, contract theory, economic theory of teams, and voluntary agreements in environmental economics (Segerson and Miceli 1998). The second area of property rights research with promise is that introducing spatial dimensions to capture externalities with a spatial component, such as fish and larval movement. These forms of territorial rights include TURFs (Territorial Use Rights in Fisheries, introduced by Francis Christy 1982) and a close cousin, ITQs with a spatial dimension. Interest is growing in the community management system of Japan (Yamamoto 1995) and inshore waters of Chile (Gonzalez 1996), which has a spatial dimension. An emerging issue requiring research is the conflict between the implicit spatial rights inherent in ITQs and explicit spatial management, as for example in New Zealand (Bess and Rallapudi 2007).

The critical but usually overlooked issue with spatial rights, the collective action problem of actually managing the common property, is perhaps the most important area for research here. After all, the US EEZs managed by fishery management councils or the North Sea managed by the European Commission are both a form of spatial management that captures many of the externalities with an enormous spatial dimension. As territorial rights expand, a larger role will have to be made for civil society and without a market or other form of decentralized allocation and decision-making mechanism, the form of institutions to actually manage the right remain an open question. If society has claims through existence value and public goods, then society will have to participate in the decision making, which will not be left solely to industry or spatial rights holders or environmentalists. Lifting our eyes from spatial externalities to the governance and broader public good concerns is critical. Emerging partnerships or hybrids such as with The Nature Conservancy and ownership of rights along the Central California Coast may point the way forward and is one of the most important events unfolding worldwide in fisheries biodiversity conservation. There is likely a trade-off between the geographic limit of spatial rights and the collective action problem, i.e., with the institutions that are necessary to manage rights that encompass more than direct use values associated with catching fish but which now include public good issues such as biodiversity, ecosystem health, and existence of species. The issue is also far more than capturing spatial externalities in the ecosystem, since biodiversity and existence value issues are growing concerns of the entire population and the problem is managing a fishery for the entire society rather than as a form of harvest strategy for optimum yield. Then there remains the very practical but difficult problems of how to allocate and then to actually organize and govern the collective owners of the spatial rights.

Although ITQs are largely viewed as improving economic efficiency, the role of ITQs in the conservation of resources and the ecosystems that support them and the ensuring equity in the use of resources remain topics of controversy and research (Sumaila 1998a, Munro and Sutinen 2007).

Another area of research follows up on research by Gary Libecap (2006a, b) and elsewhere (Libecap 1989; Johnson and Libecap 1982) and looks at the law and economics of property rights in greater detail, applying contract theory, and draws lessons from the use of resources in other sectors of the economy. Yet another issue is aboriginal rights and their interface with the rest of the economy; examples are the Makah Indian tribe, Eskimos, and the bowhead whale, the Inuit, Maori, and Torres Straits Islanders.

The two fundamental ideas behind fisheries management until quite recently have been harvest strategies and biological optimums from population dynamics and rights-based management from economics. A third fundamental concept, developed largely by sociologists and anthropologists is that of co-management (see the writings of Pinkerton, McCay, Jentoft, Pomeroy, and others) but is not discussed in this essay.

32.5   Environmental and Public Economics

A fourth, and more recent concept behind fisheries management, comes from ecology, that of ecosystems management and the importance of biodiversity. Simply put, recognition has grown to the point where it is part of conventional wisdom that commercial fisheries are embedded in marine ecosystems and that the entire food web and ecological linkages, both abiotic and biotic, need to be considered to maintain healthy and resilient ecosystems and following from this, ecological and economically sound commercial fisheries. Ecosystems management of fisheries includes the needs of predators and dependent species in their food web. Economics increasingly views the ecosystem as natural capital (Heal 2007).

A corollary to ecosystems management and ecology has recently emerged, that of conservation and conservation biology. This field has emphasized the ecological and intrinsic importance of biodiversity and extinction. As with all forms of capital, when these two components of ecosystems interact, they provide a flow of ecosystem services. The ecosystem services are a return on the natural capital, that is, these services are the return that comes from investing in rebuilding this natural capital. Heal (2007, p. 14) observes, "This newly emerging area of environmental economicsis concerned with the identification and analysis and valuation of these ESS (ecosystem services). What are they? How do they affect human societies? How do the actions of human societies affect them? In short, what are the *values* arising from ESS and why should humankind care about these values?" In addition, viewing the ecosystem as natural capital yield a flow of ecosystem services analytically allow research using techniques that are well-established elsewhere in economics, including welfare economics and capital theory, and provides a natural means of collaboration between economists and ecologists (National Research Council 2005; Heal 2007).

The key policy recommendation from this school of thought for fisheries management has been that of marine protected areas and spatial management. Economists have turned some attention to marine reserves and spatial management, observing that fisher behavior, fishing capacity, and discounting the future cannot be ignored, as discussed elsewhere. The real challenge here will be in designing the institutions required to develop and manage marine reserves, evaluating their costs and benefits, compliance, and enforcement, and in dealing with the overcapacity issues that arise when existing vessels are simply shoved out of an existing area to make room for reserves. The problem is compounded since reserves are often implemented, like much of fisheries management, when the fishery is already overfished with overcapacity.

Another economic approach to ecosystems management is illustrated by Hannesson et al. (2007b), who evaluate the ecological and economic factors entailed in the conservation and management of the California sardine. This sardine, which eats zooplankton in the California Current, serves as both a forage fish for pinnepeds, sea birds, baleen whales, albacore tunas, salmon, and thresher sharks, and as a direct take for a commercial fishery. Hannesson et al. (2007b) evaluate the economic and ecological trade-offs arising from sardines as both a predator with direct commercial takes and as a prey for other species within an ecosystem model (Field et al. 2006) and subject to low frequency, climate-driven changes in ocean conditions attributed to long-term (inter-decadal) variability in reproductive success and survival. This coupling of economics and ecosystems modeling to address the economic value of a species as a commercial catch and as a forage fish and other ecosystems-based questions should provide a very promising topic for future research.

Matching or complementing the emergence of ecosystems management and conservation biology are the fields of environmental and public economics. Key concerns of environmental economics, in addition to ecosystems as natural capital, are that of external costs and benefits and measures of total economic value of ecosystems as natural capital and their services. External costs are those costs that are not accounted for by producers and consumers in markets and are sometimes called market failure. External costs neatly match with ecological and conservation concerns that are not accounted for by existing markets. Environmental economists have spent considerable time and effort on policy tools that correct for market failure, and these approaches have considerable promise to contribute to fisheries management dealing with conservation and ecosystems, such the work of Segerson (2007) on policy instruments to tackle incidental takes of sea turtles. The concept of total economic value emerging out of environmental economics recognizes that non-market values are important and often sizeable and should be counted in any assessments of costs and benefits when developing management strategies. Thus, for example, markets for fish typically only account for direct use values associated with the production and consumption of fish, but the non-market values are important, including indirect use value from ecosystem services and recreational fisheries and existence value from biodiversity. Thus, non-market values capture concerns emerging from ecology and conservation biology and the concept of total economic value is spreading well beyond economics to become a useful concept and part of the vocabulary of others. To contribute to the design and implementation of real-world policies for ecosystem management, economists must also produce quantitative measures of ecosystems as stocks of natural capital and flows of ecosystem services with non-market economic values that allow decision-makers to trade-off extractive and non-extractive values (Wilen 2004; Smith 2006b). Work in this area includes Barbier and Strand (1998), Brock and Xepapadeas (2003), and Tilman et al. (2005). See also Natural Resource Council (2005).

ITQs, while going far in addressing the resource stock externality, remain incomplete in this regard because they address the flow or catch and not the stock itself (Scott 1988). Even further, ITQs do not begin to adequately address the remaining external costs associated with ecosystems and biodiversity. Spatial or territorial rights are advocated to address these issues, but as discussed elsewhere, spatial rights do not fully capture all of the relevant external costs and face important collective action and compliance and enforcement problems, although their potential remains tantalizing. The concepts from environmental and public economics are useful in analyzing the remaining economic analysis required.

Public economics is concerned with, among other things, public goods and bads. Public goods are those goods and services that are non-rivalrous (non-depletable by the consumption of one economic agent and thus available to others) and non-excludable (so that consumption of the public good is available to all who wish). Impure public goods, also known as mixed goods, are goods with characteristics of both public goods and private goods (excludable and rivalrous). Protected species are no longer common resources because they are now non-rivalrous rather than rivalrous as with common resources (which are rivalrous and non-excludable). Thus sea turtles under the Endangered Species Act or cetaceans under the Marine Mammal Protection Act have features of public goods, but because they are sometimes also exploited, even if sometimes inadvertently as incidental takes, are also rivalrous and hence form impure public goods. Similar considerations apply to the great whales. Ecosystems services are often considered public goods.

Public economics is concerned with the demand and the supply of public goods, including investment to generate this supply, the decision-making process and how these goods fit into a market economy, and the free-riding that arises when all those who enjoy the benefits of public goods do not provide their share of the costs of investment in, and supply of, public goods and services. External costs and market failure are sometimes seen as arising from public bads (Kolstad 1999). In this light, the question of healthy ecosystems and biodiversity can be viewed as the supply of public goods (or bads) and the issue arises of how to fully account for the investment in these public goods and the demand by humans for these public goods and services.

Mechanism design is a potential area of fruitful research in its application to ecological public goods and their services. When collective decisions must be made, individuals' actual preferences are not publicly observable. Individuals must nonetheless be relied upon to reveal this information (Groves and Ledyard 1977). Mechanism design is how this information is elicited and the extent to which the information revelation problem constrains the ways in which social decisions respond to individual preferences. Mechanism design is a well-established area in public economics, but has yet to be applied to analyze fisheries and especially public goods such as ecosystems and their services.

32.6 Industrial Organization and Information Theory

Fishing industries, from the perspective of general economics, are simply industries comprised of individual firms that are usually producing multiple products (often different species) from multiple inputs using joint production processes (Squires 1986; Kirkley and Strand 1988). Their unique feature, of course, is the exploitation of a renewable resource stock. But many other industries are unique in some manner, so fishing industries should not be immune from standard tools of economic analysis of firms and the industries in which they function. Capturing the heterogeneity of firms (vessels) and recognizing that a steady-state equilibrium and very long-term analysis is simply a normative concept that is conditional upon the states of technology and the environment further reinforces the importance of recognizing that fisheries are indeed industries, conditional upon levels of resource abundance and technology. Although the analysis may be limited to static rather than dynamic considerations, important insights into actual regulation and industry functioning will be gained.

Analyzing an industry comprised of individual multiproduct firms leads to the economics discipline of industrial organization and the recognition of the complexity and heterogeneity of fishing industries passed over by the focus on harvest strategies and aggregate inputs, outputs, and harvest technology. Industrial organization, along with the introduction of game theory as an analytical approach and a means of evaluating strategic interactions of firms and even nations, and the topic of contract theory, are one of the most promising and critical areas of economic research in addition to property rights. This approach also fits with the policy-making process that is seldom focused on some conceptual steady-state economic equilibrium, but is rather concerned with an ever-changing environment and market economy where information is limited and asymmetrically held by different parties, and uncertainty prevails, and industries are comprised of multiproduct firms with complex bundles of multiple inputs.

Early attempts in this area have largely been static and building off of the multiproduct analytical framework of Baumol et al. (1982) (Squires 1986, 1988; Kirkley and Strand 1988; Salvanes and Squires 1995; Weninger 1998; Lipton and Strand 1989, 1992) and limited to the application of static production economics to fishing vessels as multiproduct firms in a competitive industry. Such a static approach takes the resource stock as given and overlooks strategic behavior, changes in technology and the environment, and the spatial dimension. Disaggregation of fishing effort into individual inputs, such as capital, labor, and fuel, disaggregation of catch into individual species, and disaggregation of the aggregate production function into the individual production relations for individual firms precludes ready incorporation of biological growth processes into applications of industrial organization models and analysis of individual firms, where these applications have the analytical solutions that economists dearly love. As with future meaningful bioeconomic analysis that incorporates realistic and relevant biological relationships, meaningful, and relevant industrial organization models that incorporate biological growth functions and other biological information will necessarily have to move to simulation and away from analytical solutions.

More recent work in applying industrial organization and contract theory to fishing industries has recognized that that there is an asymmetric information issue, one of moral hazard, between the regulator (principal) and the vessels (agents). (Asymmetric information occurs when one party to a transaction has more or better information than the other party. Moral hazard refers to a problem of asymmetric information whereby the actions of one party to a transaction are unobservable. This information problem arises because the fishery manager does not have complete information about all variables relevant for regulation. Hence, the regulator cannot easily and at low cost monitor fisher behavior.)

Adverse selection problems, another form of asymmetric information, can also arise. (Adverse selection arises when an informed individual's trading decisions depend on that individual's privately held information in a manner that adversely affects uninformed market participants. In adverse selection models the ignorant party lacks information while negotiating an agreed understanding of or contract to the transaction, whereas in moral hazard theory the ignorant party lacks information about performance of the agreed-upon transaction or lacks the ability to retaliate for a breach of the agreement.) In a vessel-buyback market, for example, an individual is more likely to decide to sell his or her vessel when that owner knows that the vessel is not very good (Groves and Squires 2007). When adverse selection is present, uninformed traders, such as buyback agencies, may be more wary of any informed trader wishing to sell and the agency's willingness to pay for the vessel or permit offered may be lowered.

The area of asymmetric information can shed light on how fishing vessels respond to regulations and how to better design regulations in this regard. (In economics, the problem of motivating one party to act on behalf of another is known as 'the principal-agent problem.' The principal-agent problem arises when a principal compensates an agency for performing certain acts that are useful to the principal and costly to the agent, and where there are elements of the performance that are costly to observe. The solution to this information problem is to ensure the provision of appropriate incentives so agents act in the way principals wish.) Considerable work in this area has been accomplished in standard industrial organization economics, but fisheries economists have barely scratched the surface. Some of this work in fisheries has been fundamentally qualitative (Salvanes and Squires 1995; Squires et al. 2002; Kirkley et al. 2003; Ahmed et al. 2007), but increasingly, formal static and dynamic models are emerging centered around the work of Frank Jensen and Niels Vestergaard (Jensen and Vestergaard 2000, 2001, 2002a, b, c, 2007); see also Bergland and Pedersen (1997) and Herrera (2004). Homans and Wilen (1997) recognized the role of the regulator and the endogeneity of regulations, but overlooked the asymmetric information problem and contract theory in general, and extensions of their early insights to account for these would enhance their approach.

A contract is an agreement about behavior that is intended to be enforced, whether external enforcement or through self-enforcement. Contractual relationships occur between two or more economic agents, including fishers and regulators, if the parties, with some deliberation work together to set the terms of their relationship. Contract theory is an important part of the regulatory problem that has not received attention in fisheries economics (important exceptions include Cheung 1970 and Johnson and Libecap 1982), but is important because many of the formal and informal transactions and regulations in domestic and international fisheries can be examined from this perspective. Contract theory can be applied to analyze enforcement and regulatory compliance and the entire regulatory approach. Applications of contract theory should be one of the most promising areas of research. One application that comes to mind is to extend the work of Homans and Wilen (1997) in this direction.

The microeconomic theory of quotas, rations, and other quantity controls, including ITQs and ITEs (both of which are also forms of use rights) and limits on gear, fishing time, and vessel size, can be further applied to better understand their impact on fishing firms and their behavior. Moloney and Pearse (1979) and later Arnason (1990) and Boyce (1992) examined ITQs in a formal bioeconomic framework. The microeconomic theory of rationing and quotas for firms, which was initially developed in consumer theory (Neary and Roberts 1980) and international trade (Neary 1985), provides such as basis and was further developed for individual firms in an ex ante context by Fulginiti and Perrin (1993) and Squires (1994) and extended to production quotas by Squires and Kirkley (1991) and Segerson and Squires (1993) and to ITQs by Squires and Kirkley (1995, 1996) and Vestergaard (1999), with further developments by Vestergaard et al. (2005) and Hatcher (2005). Extending Neary (1985), Squires (2007) developed an ex post approach entailing virtual quantities, the dual to virtual prices (used in the ex ante approach), which can be applied to evaluate the effects of changes in existing quotas. The microeconomic theory of rationing and quotas has also been applied to address the substitution of unregulated inputs for regulated inputs in input-regulated fisheries (Wilen 1979; Squires 1986, 1994, 2007; Dupont 1991) and to the spillover effects between ITQ-regulated species and unregulated species (Asche et al. 2007). Boyce (2004), Heaps (2003), and Weitzman (2002) took a slightly different track. Potential research topics include fractional ITQs on either target species or bycatch, especially for the latter when bycatch, such as sea turtles, are rare events (see Haraden et al. 2004; Hannesson 2008, Bisack and Sutinen 2006). Little is known about ITQs or ITEs under uncertainty.

Formal modeling of ITEs has yet to be conducted. Important research questions include conditions under which ITEs are the preferable form of property right (e.g., some circumstances such as compliance and enforcement by Vessel Monitoring Systems as in the Western and Central Pacific), effects of continual productivity growth, substitution between the regulated components of the fictional fishing effort, such as days fished, and the unregulated inputs, and linkages between fishing effort and catch.

An important but under-researched research topic on quantity controls, or market- based instruments in general, is an assessment of actual markets for transferable quantity controls, including ITQs and vessel licenses. How competitive are such markets and do they convey the appropriate price signals? Batstone and Sharp (2003), investigating the relationship between fishing quota sale and lease prices and total allowable catch for the New Zealand red snapper fishery, found support for the relationship proposed by Arnason (1990), who observed that under the assumption of competitive markets, monitoring the effect of changing the TAC on quota prices could be used to determine the optimal TAC. Karpoff (1984a, b, 1985) and Huppert et al. (1996) examined the relationship between license prices and fishery rents in Alaska salmon fisheries. Newell et al. (2005, 2007) empirically addressed this issue for New Zealand ITQ markets – the most comprehensive dataset gathered to date for the largest system of its kind in the world, considered both permanent sales of quota and lease markets. Newell et al. (2005) investigated asset and lease markets separately and found that market activity appears sufficiently high to support a reasonably competitive market for most of the major quota species, that price dispersion decreased over time, evidence of economically rational behavior in each of the quota markets, and an increase in quota asset prices, consistent with increased profitability. Newell et al. (2007) found that quota asset prices were related to contemporaneous lease prices in the expected way, that stocks with higher growth rates of fish output prices tend to have higher quota asset prices, and that the New Zealand quota system as a whole has functioned reasonably well and the prices at which quota have sold appear to reflect expectations about future returns on specific fish stocks. Further research in this area in well-established markets is necessary to generalize these results and provide further evidence on the workings of ITQ programs.

Pigouvian taxes directly address the resource stock externality but have largely been passed over by economists in favor of rights-based management based on transferable quotas. Yet, recent and highly imaginative analyses suggest that the fisheries economics profession may still have more to learn about this subject (Rosenman 1986; Sanchirico and Wilen 2001; Weitzman 2002; Jensen and Vestergaard 2008). Jensen (2008) observes that a study of taxes versus ITQs under conditions of price uncertainty and asymmetric information about costs is a promising research area.

Wilen (2000) made an observation that still rings true, that little is known about the actual investment process and how investment responds to regulations, technical change, profits, property rights regimes, and the like. The literature on economic capacity utilization in fisheries, beginning with Squires (1987), Segerson and Squires (1990, 1993), Squires and Kirkley (1991, 1996), Weninger and Just (1996), and Vestergaard et al. (2005); Hannesson (1996), Jensen (1990), Bjørndal and Conrad (1987) examining different capital adjustment functions; and Weninger and McConnell (2000) using a Cournot-Nash framework starts in this direction, but the role of uncertainty, alternative expectations about possible future earnings, options, and many other factors remains insufficiently explored. Investment in an abstract, normative, and dynamic context in steady-state equilibrium was considered by Clark et al. (1979) and Boyce (1995). Lane (1988), using a panel of micro-level data on vessel upgrades gathered from accounting firms serving fishers, found that vessel investments were heterogeneous, discrete, and lumpy and not easily aggregated. Taking a different tack, Homans and Wilen (1997) assumed instantaneous rent dissipation or fast dynamics for the entry and exit of fishing capacity. Since entry and exit are often slow and include investment and/or disinvestment (either in vessels or even in gear and electronics such as embodied technical change), their model can be extended in this direction.

The concept of fishing capacity (FAO 1998; Kirkley and Squires 1999), an application of Johansen's (1968) plant capacity has loudly failed to resonate with many academic fishery economists (Andersen 2007; Wilen 2007), but has struck a resonant chord with policy makers and applied economists because of its ease of application and understanding, availability of data, use of TACs from population biologists, and emphasis on vessels (firms) rather than an aggregate industry approach, use of a multiple-input technology with capital and variable inputs or fishing time (effort), consistency with the approach to capacity in the general macro-economy and microeconomic theory, and most critically, the need for these types of quantitative measures that arises in the policy environment that can be used with TACs. In fact, measures of fishing capacity are one of the few quantitative products of economics actually desired by policy-makers and bureaucracies, precisely because it addresses the type of issues of concern for policy. The outcome is a moving target, which requires re-estimation on a regular basis, but policy-makers, stakeholders, and population biologists all understand that, like TACs, continual updating is required in an ever-changing, stochastic environment that is seldom, if ever, in a long-term, steady-state economic equilibrium. But without a scientifically rigorous approach to providing such an assessment, policy discussion and hard choices often grind to a halt. Policy makers and industry, if not the modelers, understand that economic and population models are simply models. As with all production models, including bioeconomic, the results are limited by conditioning upon the current state of technology, fleet configuration, regulatory conditions, existing data, and other factors discussed above. Current research includes evaluation with multiple objectives (Kjærsgaard 2007), undesirable outputs such as bycatch (Scott et al. 2008), which can include using directional distance functions to account for the undesirable outputs, and two-stage models in which optimum fleet size and structure is found (Kerstens et al. 2006; Scott et al. 2007).

32.7 Productivity Growth and Technical Change

Technical change has dramatically transformed virtually all industries, and fishing industries are no exception. Growth in productivity (or fishing power as it is called in the general fisheries literature), including technical change, is one of the most important areas for future research in fisheries economics, and may well be the single biggest contributor to the increases in fishing capacity and mortality that threaten many fishing industries. Productivity growth is comprised of many contributing components, including technical change, changes in technical efficiency, and changes in capacity utilization.

Beginning with the path-breaking paper by Hannesson (1983a) for a deterministic production frontier, an extension by Kirkley et al. (1995) for a stochastic production frontier, and Salvanes and Steen (1994) using the thick frontier approach, a veritable cottage industry has emerged that has analyzed (output-oriented) technical efficiency, and found a wide range of efficiencies. (Beginning with Kirkley et al. (1998), technical efficiency was also linked to skipper skill.) Squires (1987c), Segerson and Squires (1990, 1993), Weninger and Just (1996), Kirkley and Squires (1999), and Vestergaard et al. (2005) extended the economic theory of capacity and capacity utilization to fisheries. Earlier, a substantial literature arose in fisheries economics defining fishing capacity in terms of a maximum potential fishing effort that is then applied to the resource stock to produce a flow from the resource stock, i.e., the catch. This literature differs considerably from the microeconomic theory of capacity that is applied to all other industries (as expounded in Klein 1960; Morrison 1995).

Squires (1992) demonstrated that productivity growth must be separated from changes in the resource stock and extended standard analysis to renewable resource industries. Jin et al. (1992), Fox et al. (2003), and Hannesson (2007c) extended the measurement of productivity growth to profitability of a fishery and to overall fisheries in an economy using aggregate data and when new fisheries or products develop. Squires and Reid (2001), Felthoven and Paul (2004) and Squires et al. (2008) extended productivity growth to account for changes in the environment. Tara Scott is researching productivity growth when there are undesirable outputs and rare events, such as incidental takes and mortality of cetaceans, sea turtles, and pinnipeds in the California drift gillnet fishery. Ample scope exists for further methodological and empirical work on the measurement of productivity growth, including decompositions of productivity growth, inclusion of undesirable outputs, accounting for the state of the environment and resource stocks, different index number, and functional forms. Evidence is only now beginning to accumulate on actual rates of productivity growth in various fisheries.

Parallel to the economic analysis of productivity growth has been a steady series of studies on the growth in fishing power in the biological literature. Critically, population assessments involving the more disaggregated synthetic models often specify an assumed rate of growth in fishing power. (Comparably, macroeconomic models of climate change adopt a similar approach.)

The key problem for both economists and biologists remains distinguishing changes in productivity (and especially changes in technology) from changes in the resource stocks and the state of the environment (e.g., changes in temperature, thermoclines, etc.). No research has yet attempted to account for changes in ecosystem services as an outcome. Both economists and biologists largely rely on catch per unit effort-landings data for their source of information, and these data are confounded by all of these sources of variation. Fishery-independent data on biomass are important, and even stock assessments from fishery-dependent and other data are critical because of the exogenous information that is introduced (such as information on age structure, gender, length-weight and length-age, and recruitment) that helps to militate against the simultaneous bias and exogeneity statistical problems that otherwise emerge.

Remarkably little research has been conducted on the single biggest contributor to growth in productivity and fishing capacity, technical change. (Technical change can be classified as a product or process innovation, where product innovation or the creation of new products is far less important than process innovation in fish harvesting, which is concerned with new ways of producing existing products.) Remarkably, virtually all of natural resource economics has overlooked one of the most important driving forces in economic growth, technical change. In a positive framework, Squires and Grafton (2000) conducted the first formal econometric study of technical change in fishing industry. Kirkley et al. (2004) examined embodied technical change in fisheries. Jensen (2007) examined the impact of cell phones on artisanal fisheries in Kerala, India. Hannesson et al. (2007a) examined technical change in the Lofoten cod fishery of Norway and Gilbert et al. (2007) examined technical change in a Malaysian artisanal fishery. Squires (2007) examined technical change in a Malaysian purse seine fishery, finding that process innovations increased trip-level profits. Squires et al. (2008) measured the rate of exogenous technical progress and its diffusion in the Korean purse seine fleet for tunas in the Western and Central Pacific, but did not address more sophisticated rates of diffusion. Econometric studies can specify technical change as smooth and exponentially growing over time by using a time trend, but if there is panel data, then consideration can be given to the approach of Baltagi and Griffin (1988) that allows rates of technical change that are not smooth and exponential. Key empirical research questions include which type of innovations are adopted and why and by whom, their rates of diffusion, their impacts on input and output use and profits, their impact on catch per unit of effort, the catchability coefficient, overall resource abundance, site location, trip length, crew size, etc.

Research is only now beginning on technical change in a normative, bioeconomic framework. Murray (2005, 2006, 2007) examined the manner in which technical change can lead to stock collapse, which is a critical but vastly under-appreciated source of fishing mortality and capacity growth. Murray (2006) showed that technological change can lead to overestimation of natural growth in stock assessments. Squires and Vestergaard (2004, 2007) introduced exogenous technical change and exogenous and endogenous technical efficiency into the standard Gordon-Schaefer bioeconomic model. In the static model, they found that technical progress leaves maximum sustainable yield and the corresponding resource stock level unaffected but reduces the required effort to reach this point. Technical progress always reduces the static and dynamic open-access and Pareto optimum resource stock levels, at any level of fishing effort and resource stock, but only increases equilibrium sustainable yield for effort levels less than maximum sustainable yield; essentially, technical progress only expands sustainable yield when the marginal product of effort is positive. Technical progress increase rents up to the static open-access equilibrium level of effort. At the static Pareto optimum, technical progress reduces the effort required to reach the efficient scale of production. In a dynamic model, they developed a modified Golden Rule with technical change, in which there is a new term added beyond the marginal productivity of the resource and the marginal stock effect compared to the traditional rule, namely the marginal technical change and technical inefficiency effect. This term is positive, so that with technical change – all other things equal – the stock level is higher compared to the situation without technical change, beyond the marginal stock effect. However, over time the effect of technical progress will lead to lower stock levels, because the unit profit of harvest increases, meaning that the effect of these terms decline over time. They further found that in a dynamic context there is no longer a steady-state equilibrium. The effect of technical change is that the optimal level of the stock declines over time, because the unit profit increases due to technical progress. However, the short run effects of introducing technical progress is an increase in the stock size. In addition, technical progress after time sufficiently lowers costs to counterbalance the marginal stock effect in a stationary solution without technical progress. Technical change can lower the resource stock and there are not substitution possibilities between the resource stock, man-made capital, and technical progress in stock-flow production processes.

Most technology is actually embodied in the capital stock and in new investment in capital equipment particularly. However, little research in fisheries other than Kirkley et al. (2004) has been done specifying technical change as embodied rather than disembodied.

The relationship between exogenous technological change and extinction for pure compensation growth functions has been addressed in bioeconomic models by Murray (2007) through simulation and Squires and Vestergaard (2007) analytically, but has yet to be considered when there are Allee effects or more sophisticated forms of population dynamics, such as depensation or age-structured growth, or endogenous technological change.

The (Hicksian) biases in input usage and outputs (species) due to exogenous technical change have yet to be examined (and cannot be until fishing effort is disaggregated into individual inputs and total catch is disaggregated into individual outputs or species), although current work by Gillbert et al. (2007) is addressing this issue. Estimation of welfare gains (including all sources of total economic value) and rates of return from technical change, say in response to conservation requirements, has yet to be considered. Changes in property rights and regulatory regimes can be expected to generate input and output biases and alter the rates of substitution and product transformation, adoption, and diffusion.

Research on technical change has largely focused on target species or desirable outputs, but an important area of research is on reducing bycatch of undesirable outputs. Technical change research can also examine the impact of biotechnology and breeding through their impact on the intrinsic rate of growth and age structure of the population (McAusland 2005). The inter-relation between technical progress in the fisheries sector and the rest of the economy has yet to be examined, although recent but very limited progress has been made (Hannesson 2007c, Hannesson et al. 2007a).

Endogenous technical change is also important and has received no attention in the fisheries economics literature. What is unknown is the extent to which different property right regimes, regulations, market conditions, and other policies influence the development of new technology its rate of adoption and diffusion, and choice of technology to adopt. (Diffusion refers to the process by which a new technology gradually penetrates the relevant market.) Learning by doing (an alternative to modeling endogenous technical change as a function of research and development), such as the development of the back-down procedure to minimize dolphin mortality when harvesting tunas, is a very important part of the endogenous technological response to conservation issues, but has yet to be considered. What are the causal factors leading to learning by doing and how are costs lowered? There is no opportunity cost from learning by doing other than the cost of current production (since there is no crowding out of alternative research that might have been undertaken such as with research and development), which affects how this research will be conducted. A key challenge for research on learning by doing is disentangling the causal factors that lower costs (Pizer and Popp 2007). The statistical correlation between experience and lower costs is strong, but understanding the causes of cost reductions is necessary for policy decisions. Disentangling the various learning mechanisms is difficult and learning by doing can be confounded by research and development, which is often poorly measured even in research outside of fisheries.

Some new ideas – endogenous technical change – are developed through formal research and development (demand-pull responds to the market and technology-push responds to scientific advances), such as sonar to detect species composition below fishing aggregator devices in purse seine fisheries for tropical tunas. Studies could analyze the economic returns to private research and development, recognizing that knowledge spillovers result in a wedge between private and social rates of return, but the gap can be narrowed when such research is government financed. There is an opportunity cost to research and development from crowding out alternative research and this additional social cost needs to be considered. This wedge between private and social rates of return suggests that firms or even nations ignore potentially profitable technological developments since they are not able to capture a large share of the benefits to their research (which forms a public good). There is little or no evidence on returns to research and development, either private or government. Empirical evidence on private research and development is hampered by the lack of data that is not proprietary and on public research and development by the very nature of government projects, which are often more basic and long term.

A related question that arises is what is the optimal level of research and development for new technology to address ecosystem and biodiversity externalities, recognizing that due to the wedge between private and social returns that the existing level is likely to be suboptimal. The presence of two externalities complicates the matter and leads to second-best situations. This problem is aggravated for transboundary environmental issues, regardless of whether research is funded privately or by governments due to the transboundary externality and multiple governments, and the consequent creation of a second-best situation of two or more externalities (the transboundary one, the one arising from private and public good aspects of research, and in some instances the one arising from ill-structured property rights). Moreover, simply subsidizing new research and development may be insufficient and concern may need to be given to policies required for adoption of these technologies where resistance can sometimes be high. There is a growing body of empirical literature that links environmental policy to innovation in areas outside of the ocean (Pizer and Popp 2007). Most of these studies have focused on estimating the direction or magnitude of the relationship between policy and innovation (and use patent data). Since research on fisheries issues has not even begun to think about this topic, considerable opportunity may exist for future and important research.

Pizer and Popp (2007, p. 17) observe:

> In addition to correcting for underinvestment by private firms, many government research and development projects aim to improve commercialization of new technologies (or "transfer" from basic to applied research). Such projects typically combine basic and applied research and often are government/industry partnerships (National Science Board 2006). ... As such, this technology transfer can be seen as a step between the processes of invention and innovation.

This aspect of the development and diffusion of new technology has also yet to be researched in fisheries economics.

Endogenous technological change can also be classified as short- or long-term. For example, considerable technological change is in response to conservation and sustainability issues that are fairly immediate in nature, such as the development of turtle excluder devices or the replacement of circle hooks by J hooks in shallow set pelagic longline fisheries or the back-down procedure or Medina panel to reduce dolphin mortality in purse seine tuna fisheries. Other forms of technological change are longer term in nature and represent not process innovations to an existing production process as represented by the gear in use, but the introduction of an entirely new production process. New methods of fishing are one example, such as the development of trawl gear or fishing aggregator devices coupled with GPS, radio beacons, and sonar designed to determine the species composition below the fishing aggregator devices.

Diffusion of new technologies in fishing industries has barely been considered, with Squires et al. (2008) the only known study beginning to address the issue, although they only begin to scratch the surface. What are the factors that influence the rate of diffusion and the lag in general between invention and adoption? Many of the new technologies of importance to public policy are related to conservation and their diffusion is affected by public policy, such as various methods of reduce dolphin, sea bird, and sea turtle mortality, the use of pingers to reduce whale interactions on drift gill nets, or trawl mesh size and design research. Some technologies, such as floating aggregator devices, diffuse through fleets at different rates, with some skippers who are early adopters and other skippers adopting later. This diffusion even varies by national fleets with tropical tunas and purse seine vessels. Little is known about the rates of diffusion and its causal factors. For example, fishing aggregator devices appear to surpass competing technologies of finding tunas through schools in performance and cost, but are not immediately chosen, in part due to higher prices for the larger yellowfin tunas found in schools but also due to general fishing practices built up over time. Is this slow diffusion a result of rational choices responding to various incentives, market inefficiencies, or other factors?

The diffusion of new technologies takes time, which varies by the situation. Adoption of a new technology typically begins with a limited number of early adopters, followed by a period of more rapid adoption, in turn followed by a leveling off of the rate of adoption after most have adopted the technology. This process generates the well-known S-shaped diffusion curve: the rate of adoption rises slowly at first, speeds up, and then levels off as market saturation approaches (Pizer and Popp 2007). Pizer and Popp (2007, p. 19) observe:

> Early attempts to explain this process focused on the spread of information (e.g., epidemic models, such as Griliches [1957]) and differences among firms (e.g., probit models, such as David [1969]). In fisheries, Gilbert et al. (2007) examined technology adoption through a probit model. More recently, researchers combined these explanations while adding potential strategic decisions of firms. These papers find that firm-specific differences explain most variation in adoption rates, suggesting that gradual diffusion is a rational process in response to varying incentives faced by individual actors.

Pizer and Popp (2007) further note that environmental technologies can differ from many other technologies due to regulations, and that regulations dominate all firm-specific factors affecting the diffusion of such environmental technologies in the few empirical studies conducted in this field. Similar results, in which environmental regulations increase the probability of adopting environmentally friendly technologies, can be expected for endogenous technical change related to conservation, but not necessarily for commercial innovations such as fishing aggregator devices in the Western and Central Pacific Ocean, although to some degree regulations from dolphin conservation may have affected the deployment of such innovations in the Eastern Pacific Ocean. In contrast, innovations in response to expected permanent changes in market conditions, such as long-term increases in fuel prices or rising ex-vessel prices for some species from increasing scarcity, may be adopted more slowly, as it is cost savings, rather than a direct regulatory requirement, that matter. Innovations in fishing industries, whether in response to changes in conservation or market forces, can be expected to face the issue of diffusion across regions and nations when there are transboundary resources and transnational fishing fleets involved (see Keller 2004 for other industries).

Induced technical change has yet to be considered (see Ruttan 2000; Thirtle and Ruttan 1987). Constraints on public bads or undesirable outputs, such as bycatches of dolphins or turtles, create shadow prices. In output space, the ratio between the price of the private good or desirable output and the public bad or undesirable output alters and the fishing firm reduces its scale of production. Over the longer term, investment can change and even further, the change in product prices can induce technological change that shifts the production possibility frontier and is public bad-saving in its Hicksian bias.

Further research on technical change is one of the single most critical areas of research in fisheries economics because of the transformative power of technical change on the very nature of industries, the role of technical change as a key response to conservation needs, and the contribution of technical change to the growth in overcapacity, overfishing, and overfished resource stocks and depleted ecosystems. Much technological change, such as some forms of the electronics used on vessels to find fish, is exogenous to the fishery sector.

32.8   Mathematical Programming Models

Mathematical programming models, using linear, nonlinear, multi-objective, goal, and other approaches, may be among the most useful lines of research that provides policy makers what they tend to want. Specifically, such models are heterogeneous in vessels, regions, gears, and other defining factors, can incorporate age-structured population dynamics, recruitment, and selectivity, specify multiple outputs and multiple inputs, allow multiple objectives rather than a simply economically expedient objective function, and allow evaluating inter-temporal and intra-temporal policy trade-offs. As with all empirical production models, including bioeconomic and microeconomic production functions, the results are conditional upon the regulatory structure, data, fleet configuration, resource abundance, and technology, but these limitations cannot be overcome in empirical work and abstractions away from this simply lead to normative and conceptual models that are usually ignored by policy makers. Recent work in this area includes Enriquez-Andrade and Vaca-Rodriquez (2004), Mardle and Pascoe (1999, 2002), Kjærsgaard and Frost 2008). Some mathematical programming models are static (conditional upon the resource stock) and others are dynamic bioeconomic models. Kjærsgaard and Frost 2008) define the current research frontier and point the way for future applied research in multispecies, multiple inputs, and dynamic bioeconomic models with age-structured population dynamics, consideration of recruitment, selectivity, and spawning stock biomass. As they observe, the capital investment function requires further attention, allowing for trends in economic performance and stock levels, improved analyses regarding discards and selectivity, and incorporation of price and cost functions.

32.9 Technology Structure and Duality, Allocative and Technical Efficiency, Skipper Skill

Considerable intellectual effort has been allocated to this area of applied microeconomic research. This area of research focuses on individual vessels and a disaggregated production process that entails multiple inputs and outputs. This research area aims to better understand the harvesting process at the level of the individual vessel and is largely, although not entirely, empirical. Because this line of research deals with situations not normally assessed by standard microeconomics, it has also extended applied microeconomic theory in the area of rationing and quotas, capacity and capacity utilization, productivity growth, the multiproduct cost structure of firms from revenue and profit functions rather than directly from cost functions, and the individual firm's management. This line of research's principal message is that fishing industries are comprised of individual firms or vessels harvesting multiple outputs or species in a joint harvesting process and that, fisheries regulation that overlooks this central fact will be subject to under-performance.

Asche et al. (2005) and Jensen (2002) review dual approaches to modeling the harvesting technology. The dual approach is very suitable for providing knowledge of the disaggregated structure of production and costs based on a positive analysis and the theory of the firm (Asche et al. 2005). One of the questions of greatest interest addressed by the dual approach is the substitution between inputs in order to answer the question of how readily fishers can substitute between inputs in response to regulation. This is the question of the microeconomics of rationing and quotas discussed above, often loosely called "capital stuffing," which is addressed by the general model of input substitution under quantity controls by Squires (1994, 2007). Asche at al. (2005) and Jensen (2002) survey the numerous empirical studies of input substitution possibilities that use the dual approach. To date, although these studies have shed considerable light upon the structure of fishing technology, their impact upon policy has been negligible. Similar issues arise as with all production studies, as discussed elsewhere. The dual approach was used by Dupont (1990) to evaluate rent dissipation in a fishery. The dual approach has also been used to examine economic capacity utilization, with original extensions of the theory to profit maximization (Squires 1987a, Segerson and Squires 1993), revenue maximization (Segerson and Squires 1995), quotas and rations (Segerson and Squires 1993; Squires 1994), and multiple products (Segerson and Squires 1990). The dual approach has also been applied to examine the fishing vessel's multiproduct cost structure, with extensions of microeconomic theory to profit- and revenue-maximizing firms (Squires 1988; Squires and Kirkley 1991). The dual approach has also been applied to provide shadow prices for ITQs (Squires and Kirkley 1996; Dupont et al. 2005). The dual approach has also been applied to evaluate the multispecies issue, clarifying the issues of separability (conditions under which an aggregate input, fishing effort, and output, total catch) exist and joint and non-joint harvesting (Squires 1987; Kirkley and Strand 1988). An important but somewhat overlooked paper by Bjørndal (1987) develops an intertemporal profit function and examines the relationship between the optimal stock level and production technology. Extensions of this model represent an opportunity for further research. Additional research includes measuring potential resource rents under alternative regulatory regimes (Dupont 1991; Eggert and Tvererås 2007). Stephen Stohs is extending ration and quota theory from binding constraints with 100% probability to multiple binding constraints, each with an independent probability Poisson distribution because of count data, and collectively as a joint or combined probability with an associated distribution.

Considerable applied research has also been given to measures of (output-oriented) technical efficiency and skipper skill at the level of the individual vessel. Technical efficiency refers to the individual firm or vessel's level production given its bundle of inputs, such as vessel, gear and equipment, crew, fuel consumption, and states of technology, environment, and resource stocks, relative to the best-practice frontier established by the highest achieving firms or vessels. Hannesson (1983a) first measured technical efficiency through estimation of a deterministic production frontier, in which there is a one-sided deviation from an estimated frontier not accounting for stochastic disturbances; as with all deterministic frontiers, performance differences due to technical inefficiency and stochastic distur-bances cannot be distinguished. Salvanes and Steen (1994) measured technical efficiency through estimation of a thick frontier, in which rather than allowing for a one-sided disturbance term to account for inefficiency, the best-practice frontier is determined, after estimation of a production function, by grouping together the vessels with the smallest disturbances. Kirkley et al. (1995) first measured the frontier through a stochastic production frontier, thereby allowing for a one-sided disturbance term to account for technical inefficiency or deviations from the best-practice frontier and a two-sided i.i.d. disturbance term to account for stochastic variation above and below the best-practice frontier due to luck, measurement error, random variation in weather, excluded variables, and other factors. Squires and Vestergaard (2007) introduced output-oriented technical inefficiency into the Gordon-Schaefer bioeconomic model.

Kirkley et al. (1998) observed that technical inefficiency corresponds to the fishing captain's management of the vessel or skipper skill, where according to the good captain hypothesis some skippers display superior skill in finding and catching fish and thereby establish the best-practice frontier. They also related skipper skill or technical efficiency to various potential explanatory variables, including institutional and measures of the captain's human capital. Squires and Kirkley (1999) allowed for technical inefficiency through panel data methods, specifically fixed and random effects, with a standard production function rather than a production frontier using a one-sided disturbance term. Grafton et al. (2000a) first accounted for economic inefficiency, including both technical and cost inefficiency, in a study of the impact of ITQs in the British Columbia fishery for Pacific halibut. Kuperan et al. (2002) reviewed the anthropological fisheries literature on skipper skill and related it to technical efficiency in a Malaysian trawl fishery. Subsequently, numerous empirical studies have appeared in this area for fisheries around the world, demonstrating various ranges of technical efficiency or skipper skill and often, although not always, sometimes finding that measures of the captain's human capital can in part explain variations in skipper skill from vessel to vessel and sometimes not finding any relationship at all. Considerable work has been conducted in this area by Sean Pascoe and colleagues and recent econometric advances by Flores-Lagunas et al. (2007), Holloway and Tomberlin (2006), and others.

Potential research topics include more sophisticated studies of economic efficiency, based on profit, revenue, or cost efficiency, further assessment of the factors determining efficiency differences among vessels and skippers, extending the sophistication of econometric approaches, and in general accumulating the empirically based knowledge of efficiency differences and skipper skill. Promising areas of current research include the application of directional distance functions to allow for undesirable outputs such as bycatch and the impact upon firm or vessel efficiency and skipper skill; for example, are there some skippers that are better skilled at avoiding bycatch while maintaining high levels of efficiency in their target catches?

32.10   Consumption, Demand, and Price Analysis

Consumers are invariably left out of the picture, since policy actions seldom have a measurable immediate impact upon them, although they do on producers and the environment. The ready availability of close substitutes and imports can make the impacts of policy actions on consumer welfare negligible. Industry and environmental groups largely influence regulatory institutions with minimal representation by consumers at best. An analysis of consumers and consumer benefits requires vastly more attention than is given to the analysis of demand, which is an area of fertile research largely untouched. Issues that arise here are how to obtain measures of consumer welfare (especially compensating and equivalent variation) at different levels of the retail chain ranging from ex-vessel to retail. Such measures are useful in benefit–cost analysis, which in fishing industries largely ignores consumer welfare and concentrates on producer welfare. Responsiveness of price to ex-vessel landings can also be evaluated from such studies. Some attention has been given to this area, but more is required (Barten and Bettendorf 1989; Asche 1997; Salvanes and DeVoretz 1997; Holt and Bishop 2002; Fousekis and Revell 2004; Park et al. 2004; Asche et al. 2001; Wessells et al. 1999). One of the key research issues is whether at the ex-vessel level price is a function of quantity, yielding an inverse demand function, or quantity is a function of price, yielding a direct demand function, or whether it varies by species, yielding a mixed demand function. Additional research includes proper specification of functional forms, index numbers and aggregation of species into composites to reduce the number of species in the demand functions to manageable numbers, and separability issues to distinguish among types of fish and even between fish and other substitutes in consumption.

Measures of economic welfare for consumers (consumer surplus, compensating, and equivalent variation) and producers (producer surplus, rent) are a core component to cost–benefit analyses. Welfare analyses are often required in the United States and elsewhere, but considerable scope remains for research in this area. Important research is emerging (Park et al. 2004; Bockstael and McConnell 2006).

Horizontal and vertical price linkages between markets are important areas of economic research. Beginning with the first study by Squires (1986), a plethora of research has examined the nature of horizontal and vertical price linkages using time series econometric techniques, including Granger causality and co-integration; much of this research has been centered on the work of Frank Asche. The policy implications of such research are not always clear, although they help consumer demand specification by clarifying the extent of the market. Some analyses have been extended to policy research, evaluating the impact of spatial price linkages on a revenue-sharing scheme in the US tuna fleet and clarifying that all Regional Fishery Management Organization areas for highly migratory species are linked by price, so that policy actions in one area affecting the volume of landings can reverberate throughout the world (Jeon et al. 2008).

The impact of ecolabeling on demand and for conservation is an infant field of research (Gudmundsson andWessells 2000; Teisl et al. 2002). More research is justified in this area, including the relative costs and benefits for ecolabeling and certification of fisheries; the costs may outweigh the benefits in some smaller fisheries, especially artisanal ones.

32.11   Climate Change and Variation

Climate change and fisheries is emerging as an issue of growing importance. Climate change includes short-term, such as ENSO events, decadal, or medium-term, such as the Pacific Decadal Oscillation, and long-term, such as global warming. Biologists have established the importance of climate and the environment in general on abundance and catches, as for example in Pacific salmon and the PDO (Beamish and Bouillon 1993). Economists are beginning to address the impact of climate change on such issues as the change in harvesting strategies, fleet size, dynamics, and investment, risks of extinction, jurisdictional issues as fish stocks shift from one EEZ to another, variability in stocks and harvests, and other related issues (Costello et al. 1998; Dalton 2001; Hannesson et al. 2006; Sun et al. 2006; Herrick et al. 2007; Hannesson 2006, 2007a, b), and in 2007, special issues in *Marine Policy* and *Natural Resource Modeling*. Acidification of the ocean due to global warming and its economic impact upon fishing industries, biodiversity, and the ecosystem in general represents a potential research topic of growing importance. Nonetheless, there has been very little econometric work shedding light on the costs and benefits associated with different climates or on the costs of adjusting to different climate regimes. The damage function for climate change remains largely unknown. The role of technical change in adapting to climate change remains unexplored.

32.12 Economics of Transboundary Resources

Transboundary marine resources include global common resources, belonging to humanity as a whole, such as fish and sea turtles, and global public goods or bads, such as the protected great whales. Transboundary fish, such as highly migratory species including tunas, billfish, and oceanic sharks, and straddling stocks, such as coastal pelagics in Eastern Boundary Currents, are important in the international context. Multiple externalities arise in this context, including the transnational, "traditional" resource stock, and those related to public goods and common resources of biodiversity conservation and sustainable ecosystems.

Considerable modeling and empirical research, usually in a surplus production framework but sometimes in an age-structured framework, has been conducted in this area to consider the conditions under which a self-enforcing Pareto (economic) optimum can emerge and the opportunity cost of not reaching this optimum, i.e., in comparing the non-cooperative and cooperative equilibriums. The payoffs to players typically depend on the size of a state variable: the relevant resource stock. Brown (2000, pp. 897–898) observes that fishing games create special circumstances that must be considered:

> The existence of stock externalities casts the problem into the context of "dynamic games" in general, and not the special case of repeated games in particular. Fishing is not an infinitely repeated game because payoffs are state variable dependent. P. Dutta (1995) has demonstrated that the intuition developed from infinitely repeated games does not necessarily carry over to the more general category of dynamic games.

Research in this area began with important papers by Munro (1979) and Levhari and Mirman (1980). Munro (1979) was concerned with bargaining solutions in cooperative games. Munro combined the standard bioeconomic model of a fishery with cooperative game theory to show that if cooperative management is unconstrained, so that allowances are made for time-variant harvest shares and transfer payments, then optimal joint harvest requires the player with a lower discount rate to buy out the player with a higher discount rate entirely at the beginning of the program and manage the resource as a sole owner. Levhari and Mirman (1980) applied bioeconomic models to a two-state cooperative game and evaluated the Cournot-Nash equilibrium in which the policy function is linear in the population for non-cooperative, cooperative, and Stackelberg games and the role of side payments in reaching a cooperative agreement in steady-state equilibrium and in which each country has the same discount rate. Brown (2000, p. 897) observes:

Since open access is just the non-cooperative game with infinitely many players, the cooperative steady state solution with two players at the other end of the spectrum is evident. Cooperation results in a larger equilibrium resource capital stock when there is a common discount rate. When there is a leading country, it exploits its power with greater short run harvest and a lower steady state population.

Fischer and Mirman (1992, 1996), also comparing Nash equilibria and global optima, extended the analysis to interacting species. Vislie (1987) extended Munro's (1979) cooperative game to examine a self-enforcing sharing agreement without strictly binding contracts. Clark (1980) considered a limited access fishery as an N-person, nonzero-sum differential game.

Stackelberg games are non-cooperative sequential move games, where the Stackelberg leader takes into account its ability to manipulate the other agent's decision. The Stackelberg follower follows the Nash non-cooperative strategy. The Stackelberg game is applied when one country has a relatively large fishing industry, and therefore the power to act as a leader, or when stocks migrate, and one country can harvest before another country (Kronbak 2005). Naito and Polasky (1997) apply this approach.

Threats can contribute to the stability of cooperative fishing games (Kronbak 2005). An efficient cooperative strategy can be supported as an equilibrium, by threat of credible punishment, provided the discount rate is low enough. If anyone deviates from a cooperative strategy, the game reverts to the Cournot-Nash one-shot non-cooperative solution. For a low enough discount rate, the short-term gain from defection is offset by the long-term foregone gains from cooperation. Kaitala (1985) examined credible threats for each player in the game, and Kaitala and Phjola (1988) introduced trigger strategies where deviation triggers a switch to play another predefined strategy and examined non-binding cooperative management. Laukkanen (2003) introduced stock uncertainty into Hannesson's (1995a) model with cooperative harvesting as a self-enforcing equilibrium supported by the threat of harvesting non-cooperatively over an infinite time horizon if defections are detected. Hannesson (1995b) considered how cooperative solutions to games of sharing fish resources can be supported by threat strategies. Schultz (1997) and Barrett (2003) discuss trade sanctions.

Munro (1979) began the consideration of a full cooperative solution and Clark (1980) and Levhari and Mirman (1980) considered a non-cooperative solution. Kronbak (2005) observes that the in-between literature addresses the formation of coalitions, where a group of players come together and form a coalition inside which they cooperate and play non-cooperatively against the players outside the coalition. This approach is applied mainly for determining the bargaining power of the players exploiting the resource (Duarte et al. 2000; Arnason et al. 2000; Lindroos and Kaitala 2000). The most recent literature, summarized by Kronbak (2005), discusses how to share the benefits among agents agreeing on joint action and is referred to as characteristic function games (c-games). These games assume that players have already agreed to cooperate and that it is a transferable utility game, and addresses how the agents distribute the benefits from cooperation. The focus is not on side payments but on the distribution of benefits. Kaitala and Lindroos (1998) introduced the theoretical framework, ignoring externalities, and concentrated on benefit sharing rules with exogenous coalition formation. Pintassilgo

(2003) continued this framework and evaluated the stabilities of the grand coalition in the presence of externalities and included endogenous coalition formation.

McKelvey (1997) introduced game theory into a sequential interception fishery where the underlying stock uncertainty is included in the model with a stochastic payoff function. Hannesson (1997) considered fishing as a supergame to analyze the importance of the number of agents exploiting a fish stock for obtaining the cooperative solution. This approach has standard information and is repeated an infinite number of times, has a closed loop, and therefore represents a situation in which a group of agents face exactly the same situation infinitely often and have complete information about each other's past behavior (Kronbak 2005).

Side payments can facilitate cooperative solutions. Munro (1979) first raised the issue. Fishery managers in Olaussen (2007) interact not by harvesting the same stock of fish, but through side payments to a third harvesting agent with stochastic survival of recruits.

Externalities and game theory, although neglected in the early literature, have since received attention. Kaitala and Lindroos (1998) established a cooperative game in characteristic function framework and determined one-point cooperative solution concepts, but their model did not incorporate externalities. Sumaila (1999) summarized research in this area at that time. Pintassilgo (2003) shows that the grand coalition is only stand-alone stable if no player is interested in leaving the cooperative agreement to adopt free-rider behavior, but does not evaluate benefits sharing inside the grand coalition. Eckmans and Finus (2004) recognize the problems with grand coalitions when externalities are present, and propose a sharing scheme for the distribution of the gains from cooperating when a solution belonging to this scheme generates the set of stable coalitions. Weikard (2005) suggests a sharing rule distributing the coalition payoff proportional to the outside-option payoff. Kronbak and Lindroos (2007) discuss the difficulty of coalition payoffs division among members in a characteristic function approach, developing a new sharing rule accounting for the stability of cooperation when externalities are present and players are heterogeneous. Moreover, Kronbak and Lindroos (2007) observe that the stability of cooperation and coalition games is affected by the way that benefits within cooperation are shared among players and that when externalities are present, that additional research is required in this area, that in fisheries coalition games the link between cooperative and non-cooperative games has received insufficient attention due to externalities not receiving attention in cooperative games. Recent research on coalitions includes Lindroos (2004, 2008), Lindroos et al. (2007), and Pintassilgo and Lindroos (2008).

Allocation of shares among the parties, especially developing nations (such as coastal states in international tuna fisheries), is critical for sustaining cooperative multilateral conservation and management and requires additional evaluation. Considerable attention has been given to sharing rules by which the economic surplus from cooperation among nations, as opposed to non-cooperation, is allocated among participating parties, as discussed above. Many models have considered two-agent games, and some authors have considered explicitly the importance of the number of agents for obtaining a cooperative solution.

Fruitful areas of research lie in empirical extensions to highly migratory species, to the area of property rights and incentives, the structure of self-enforcing treaties and agreements, repeated games, enforcement and compliance, multiple agents and the impact on the depth versus depth of cooperation, cost heterogeneity among states, the role of technical change and climate in such self-enforcing agreements (the latter is considered in on-going work by Robert McKelvey and Kathleen Miller), consideration of the rich institutional detail inherent in Regional Fishery Management Organizations, and more sophisticated and realistic population dynamics in bioeconomic applications as discussed above. Benefit sharing and the partition function game approach are important areas of current research. Important applications have been applied to whales (Amundsen et al. 1995). The importance of incentives and rights-based management are two of the most important areas of future research, as discussed in various forms by (among others) Munro (1979), Kaitala and Munro (1997), Barrett (2003), Bjørndal and Munro (2003), and various papers in Allen et al. (2006). Bjørndal and Munro have written extensively in a series of papers about the Law of the Sea, the United Nations Straddling Stock Agreement, and economics.

Another fruitful area of research is the conservation of endangered and threatened species in a transboundary context. One strand of existing literature focuses on the great whales through bioeconomic modeling (Conrad 1989; Amunbdsen et al. 1995; Allen and Keay 2001). Nonetheless, research in this area is in its infancy when taken in the broader perspective of conservation and a more disaggregated approach, and can include the role of rights, incentives, self-enforcing international agreements, mitigation and conservation investments, at-sea conservation measures, optimal regulatory instruments, and other such issues (Dutton and Squires 2008, in preparation; Gjertsen 2007a; Segerson 2007; Heberer and Stohs 2007). Segerson (2007) considers alternative policies to find the optimal balance between the goals of protected species bycatch mitigation and maintaining fishing opportunities, while Heberer and Stohs (2007) considered the choice of cleanest gear (least incidental take rate), and hence treats the regulatory policy as exogenous rather than as a choice variable. Some of the key issues include the cost-effectiveness of alternative conservation measures and interventions in different stages of the life cycle (such as nesting sites, artisanal fisheries, and high-seas fisheries for sea turtles; considerable work is underway in this area by Heidi Gjertsen [Gjertsen 2007b] and Mark Plummer); the relative importance and role of alternative conservation instruments, including technology standards such as gear changes, production standards such as quotas, forms of property rights, conservation in second-best situations such as the simultaneous presence of both transnational and resource externalities. The application of the economics of international environmental agreements to the International Whaling Convention, the Latin American sea turtle treaty, the Indian Ocean-Southeast Asian Memorandum of Understanding, the Agreement on the International Dolphin Conservation Program, and others is another potential area of research. Considerable research has been conducted on the politics and law of the International Whaling Convention (Gillespie 2005), but little formal economics has been conducted with the exception of Clark (1973), Conrad (1989), and Schneider and Pearce (2004).

Yet another fruitful area of research is learning from other international environmental agreements. Conventions and treaties dealing with transboundary pollution, water, and the atmosphere are pervasive (Barrett 2003), but research on ocean issues has only begun to scratch this surface.

The theory of international ocean agreements is still in its infancy. Considerable work in this area has been accomplished (much is reviewed in Barrett 2003; Bjørndal and Munro 2003; Munro et al. 2004; Munro 2006). Contract theory has yet to be fully mined for its insights in this area. Enforcement is a key, because of the principle of state sovereignty, so that these agreements must be self-enforcing (the contracting parties enforce their contract) rather than external enforcement (a third party takes actions as a function of verifiable information, as directed by the contract) (Barrett 2003). How to structure incentives so that parties will agree, changing the game into an induced game with the proper incentives, is a major question. Enforcement and compliance are critical factors, in which trade measures are important and about which little research has been conducted. Standard economic modeling shows that effective monitoring and verifiability are critical to identifying and sanctioning violators. However, analysis has yet to uncover the precise conditions under which cooperation can be supported when information flows in a decentralized manner. Even less understood is how institutions can deal with renegotiation, collusion, and the rescinding of agreements, all of which can interfere with schemes to punish violators. The maintenance of limited entry, dealing with free-riding by non-cooperating members, and critically, rights-based management stronger than limited entry are key areas of concern for future research. New members and IUU fishing are major issues.

32.13 Spatial Management and Marine Reserves

The recognition of the importance of the spatial dimension to management and its role in firm behavior has come relatively late to economics in general and fisheries economics is not an exception. Brown (2000, p. 876) observed:

[T]here is an essential spatial component to living resources. Biota of the same species spatially differentiate themselves and sometimes are then linked together by more or less well defined corridors, as when larvae collect from many separate sources in common pools, then disperse to separate colonies. The peripatetic nature of many renewable resources often makes it prohibitively expensive to bend them suitably into the status of private property.

The economic issues are more realistic representations of fisher behavior, metapopulations, marine protected areas, and spatial rights as a key policy question. Brown (2000, p. 903) observes: "The institutional fabric, including a suitable property rights structure, designed to achieve efficient exploitation of a metapopulation, necessarily differs from the one designed to manage many competitive firms producing for the same market with no technical interdependence." Brown (2000, p. 909) further observes: "Indisputable economies of scale and non-convexities inherent in the spatial dimension and behavior of important species invite analysis." Wilen (2004) recognizes the importance of when spatially disaggregated polices are likely to pay off, how different rights mechanisms might be designed, and exactly how altering incentives with direct or indirect instruments affects a spatially exploited bioeconomic system.

The early fisheries work in spatial behavior and management was concerned with the choice of fishing location (Hilborn and Ledbetter 1979) and has subsequently received additional research (Eales and Wilen 1986; Curtis and Hicks 2000; Holland and Sutinen 1999, 2000; Mistiaen and Strand 2000a; Smith 2002; Hutton et al. 2004; Curtis and McConnell 2004). These are all microeconomic production models, so the concerns raised by Wilen (2007) about production models may be germane here. One of the conclusions emerging from this research is that some fleets are comprised of sub-fleets defined by home port, and that some of these respond relatively quickly to profit changes and others which are more sluggish in response; and that highliners seem to be more mobile and opportunistic. Moreover, fishermen behave as economic theory suggests, adjusting high fixed costs and relatively inflexible inputs, such as vessel capital sluggishly, while adjusting other flexible inputs such as vessel days and fishing location much more quickly (Wilen 2004). Research questions in this area include (Wilen 2004, p. 14):

> For short-term participation choices, what are the relevant opportunity costs? What kinds of alternative within-season employment opportunities do skippers, crew, and owner/operators have? How do these affect decisions about whether to fish or not? How different are opportunity costs in fishing, and are these differences responsible for the kinds of heterogeneous behavior that we typically witness?

The spatial dimension has also received attention within the mathematical programming-bioeconomic framework (Holland 2003; Kjærsgaard and Frost 2007), and would benefit from research in the public good framework, thereby capturing non-market benefits such as biodiversity conservation. Important early general papers on metapopulation modeling include Tuck and Possingham (1994) and Brown and Rouhgarden (1997). When modeling spatial impacts, as with all production models, whether disaggregated and static or aggregated and dynamic, and including choice of location, problems exist with simplified production processes that do not capture the richness and complexity of fisher behavior and production and as with all bioeconomic models, simplified population dynamics that do not capture the complexity of populations.

The issue of marine reserves has emerged as one of the policy questions of keen interest, largely arising out of conservation biology. Here the focus extends beyond spatial management and single species to include biodiversity (Hanna 1999). Much of the focus has been on reserves as a form of insurance for the biomass, biodiversity, and ecosystem (Brown 2000; Murray 2007). Arguments for marine reserves often extend beyond the benefits for biodiversity and the ecosystem to increases in resource stocks outside of the reserves that lift catches outside of the reserves. Arguments tend to overlook the bunching up of the fishing capacity outside of the reserves, the length of time and management measures required to deal with the overcapacity that frequently develops, the effects of discounting, the costs of fishing and enforcement and compliance of reserves, density dependence of populations, and biological, environmental, and economic uncertainty. Empirical discussions have yet to properly account for before and after the treatment and with and without the treatment using control areas that control for habitat, environment, species abundance and density, and markets (especially distance from markets). From the perspective of research design (much coming out of the medical and public health literature), there is a glaring absence of research design.

Grafton et al. (2005a) review the recent bioeconomic literature in this area. They observe that bioeconomic models of marine reserves need to consider a number of key processes: the transfer rate and flows between reserves and harvested areas, the effect of reserves on fisher behavior and the influence of environmental stochasticity and shocks on both the reserve and fished populations. Some models assume that total fishing effort is constant and that fishing mortality in the reserve area transfers into the open area. Sanchirico and Wilen (2001, p. 206) observe that most models "consider the problem of carving out a fraction of space in an otherwise homogeneous system in which mixing is perfect, uniform, and generally instantaneous."

Important bioeconomic papers on marine reserves include Holland and Brazee (1996), Hannesson (1998, 2002), Holland (2000, 2002), Sanchirico and Wilen (1999, 2001), Sanchirico (2004, 2005), Smith and Wilen (2003), Sumaila (1998b), Sumaila et al. (2000), Grafton and Kompas (2005), Grafton et al. (2005b), and Murray (2007). Along similar lines, Holland and Schnier (2006) considered ITQs for habitat. Holland and Brazee (1996) found that whether or not increases in spawning stock biomass generate a net increase in the present value of economic benefits depends importantly on the discount rate and the pre-reserve exploitation level, as well as bioeconomic parameters. Hannesson (1998), using a model that is not spatially explicit, showed that a marine reserve in open-access equilibrium is unlikely to improve catches and will create overcapacity. Sanchirico and Wilen (2001) and Smith and Wilen (2003) demonstrate that fishers' spatial behavior is important, and that their models are open access and focus on improving net yields (but not economic welfare since there is open access) when spillover of fish and larvae is sufficient to compensate fishers for lost catches from reserve areas. Sanchirico and Wilen (2001) demonstrate economic results are highly dependent upon the type of interaction between different patches, and which patch is closed, because of complex spatial and inter-temporal effort redistribution. They find that, under open access, most reserve scenarios produce a biological benefit but that there are very few combinations of biological and economic parameters that give rise to both a harvest increase and a biological benefit. In particular, they find that harvest increases are likely only when the designated reserve patch has been severely overexploited in the pre-reserve setting. Sanchirico and Wilen (2001) assume that fishermen respond to profit opportunities by entering and exiting the fishery and by moving over space in response to spatial arbitrage opportunities and find that relative dispersal rates in a patchy system are important in choosing which patchy areas to close. Smith and Wilen (2003) observe that the typical assumptions made by biologists for analytical tractability, such as exogenous and constant effort, consistently bias the predicted impacts in a manner that makes reserves look more favorable than what they might actually be. Smith and Wilen (2003) found that reserves can produce harvest gains in an age-structured model but only when the biomass is severely overexploited and that even when steady state harvests are increased with a spatial closure, the discounted returns are often negative, reflecting slow biological recovery relative to the discount rate. Concerns over production modeling certainly apply here (Wilen 2007).

Uncertainty is important with marine reserves, since a key function is insurance. Murray (2007) observes that none of these models consider uncertainty with the exception of Lauck (1996), Lauck et al. (1998), and Grafton et al. (2005b). Sumaila (1998a) applied a Beverton-Holt bioeconomic simulation model to evaluate a shock in the fishable area outside of a reserve and demonstrated that a reserve can protect discounted rent. Conrad (1999) found that marine reserves can reduce biomass variation when there is a general shock to the system, but that reserves also reduce harvest and rent compared to private property without a reserve. Murray (2007) directly addresses this absence of uncertainty through incorporating parameter uncertainty into a bioeconomic model of marine reserves. Murray (2007) developed a surplus production model with uniform carrying capacity and intrinsic growth rates for the biomass across an open area and reserve area and risk neutrality for fishers. Murray evaluated a Schaefer yield-effort model to show that under parameter uncertainty marine reserves increase the harvest outside of the reserve area and decrease the probability of a resource stock crash in long-term steady-state equilibrium. Additional research would help clarify the uncertainty associated with marine reserves.

Smith and Wilen (2003) raise potential bioeconomic research issues, including whether oceanographic dispersal is the key driver of spatial closure impacts, or whether harvester dispersal may be equally important. Further research is warranted on Smith and Wilen's (2003) observation that whether a particular patch is a source or sink depends on its relative level of exploitation as well as its physical placement in an oceanographic system. These models also need to be extended to more realistically include the endogeneity of regulations, formulated under regulated open access rather than pure open access as argued by Homans and Wilen (1997) in a different context, and state of technology (e.g., VMS or gear restrictions which can be endogenous) as forcefully argued by Homans and Wilen (1997), reflecting further than most fisheries regulation occurs not in pure open access but regulated open access (Homans and Wilen 1997). An open question remains, nonetheless, whether there is sufficient data and analytical techniques to address these questions with any degree of resolution and certainty. The concerns raised about production models by Wilen (2007) are double here because of the oceanographic issues that compound fisheries production modeling.

Some of the critical research for the future requires (Grafton et al. 2005a, p. 173) "[m]odels that explicitly include the spatial behavior of fishers are of particular importance to managers as they emphasize the importance of economic considerations when establishing reserves, and the need to explicitly model the endogeneity of fishing effort in a decision-making framework." Bioeconomic models require allowing for negative shocks and management error or environmental stochasticity (Grafton et al. 2005a). Empirical modeling that incorporates more realistic population modeling will be required to have a substantive impact on policy at the level of the regulator and to be accepted by the practicing population biologists without whose blessings little will be accepted in the policy process. Moreover, as noted by Grafton et al. (2005a), despite many empirical studies on reserves only a few of these investigations include before and after data and spatial variation, well-designed empirical studies are necessary to separate the 'reserve effect' from the 'habitat effect', and to determine the efficacy, or otherwise, of marine reserves. As observed above, this recommendation can be taken further to encourage scientific evaluation that includes before and after treatment (reserves) and, critically, a control group without reserves that gives a without and without treatment. The experimental design should also control for the habitat and environment effect. Equally important are the uncertainties in terms of the 'connectivity' between reserves and no-take areas at the larval, juvenile, and adult stages, and also critical habitat size, renders the problem of determining the size and location of reserves a very difficult task. Moreover, fishing spillovers from reserves very much depend on their design and must consider advection, as well as diffusion processes, and an appreciation of both dispersal distance and the number of population sources. Bioeconomic modeling has yet to fully incorporate density-dependent theory in which yield-per-recruit effects that are lowest at carrying capacity, i.e., unfished populations, that compensation at lower population levels, i.e., fished populations, produces a sustainably harvestable surplus, and lower body growth rates at higher population densities. The effects of stochastic recruitment, where recruitment might be a function of the environment, have not yet been considered in a bioeconomic model. A full economic assessment of the net benefits from reserves would also have to account for the total economic value, including non-market values from biodiversity and ecosystem benefits. The full relationship between marine reserves to restore and maintain biodiversity and ecosystems, raise catches (through the spillover of adults and the export of eggs, larvae, and juveniles as the density of fish populations within reserves increases), and act as insurance and "standard" fisheries management, as discussed for example by Hilborn et al. (2006) and Roberts et al. (2005), also remains unsettled and a potential research topic. Armstrong (2006), in a review paper, states that "much work still remains with regards to the analysis of different management options than solely open and limited access outside marine reserves."

Bycatch and reserves remain an under-researched area (Armstrong 2006). Bonceur et al. (2002), in one of the few papers in this area, apply a two-species, two-area model of marine reserve implementation. Reithe (2006) and Armstrong observed that the ecological interaction also affects the possibility of obtaining a win-win situation when implementing a reserve, and also determines the optimal patch to close. What about two potential reserve areas and closing only one and one area benefits one species and the other area benefits another?

Specific potential research questions in the context of spatial management relate to (Wilen 2004, p. 16)

> management system design with mixed public and private values and mixed consumptive and non-consumptive services. What kind of spatial system might be used to manage both kinds of services? Would it be best to manage a mixed system with a mix of closed areas

and spatially regulated restricted access policies, or would it be best to move directly to clarifying and allocating a system of restricted property rights? If partial rights systems are developed, what are the implications of different degrees of specificity, transferability, and excludability?

As noted before, the underappreciated element in these schemes is the collective action problem arising with their institutional and managerial organization. Compounding this difficulty is the ability of management institutions to actually implement such fine-tuning at the margin (see Schrank 2007 for the difficulties of such fine-tuning and use of structural models in the context of TACs), and the ability to actually achieve compliance and enforcement with fine spatial scales (horizontally and vertically within the water column) and imperfect monitoring.

32.14   Risk and Uncertainty

Risk and uncertainty are pervasive in fishing industries due to multiple sources, including markets, regulatory institutions, technological change, the resource and ecosystem, environment, climate, oceanography, and other factors. The environment, climate, oceanography, resource stocks, and ecosystem are inherently stochastic, i.e. partly random. Insufficient knowledge about these future states of nature is complicated because even less is known about the social and economic impact of alternative policies, compounded by the long time horizons in climate and ecosystems (Pindyck 2007). Little is known about the current and future costs of a policy; how will fishers respond to alternative policies? Resource and environmental cost and benefit functions tend to be highly nonlinear (or even nonconvex), so that for example, environmental and ecological costs may be very low for low harvest rates but then become extremely high for higher harvest rates. Ecological resiliency may be robust until a threshold or tipping point occurs at which point the impact of a harvest policy becomes extremely severe. The precise shapes of these functions are also unknown. What discount rate should be used to calculate present values? In the broadest sense, policy-makers and scientists never really know what the benefits and costs will be from alternative policies, what discount rate is appropriate, and what the resource stocks and ecosystem will look like in the future (or even the present). Irreversibilities may also arise, such as extinction, permanent changes in species mixes from harvest policies (think of New England groundfish). Very long time horizons may also be involved, so that the effects of discounting might not fully capture the long-term effects on the ecosystem and climate; what is the full impact of the interplay between climate and harvest policies for northern cod? A long time horizon compounds the uncertainty over policy costs, benefits, and discount rates. Uncertainty also arises over the type, extent, timing, and impact of technological change that could ameliorate the economic impacts and/or reduce the policy costs. Weitzman (1974a) considered the implications of uncertainty for optimal choice of policy instruments; this has been studied extensively in environmental economics, and has received some attention in fisheries economics.

Much of the early work in bioeconomics assumed that the growth function for the resource stock is known and deterministic, when in fact the growth function is stochastic and even highly stochastic (Pindyck 2007). The stock dynamics might be better described by a stochastic differential equation. As is well known to population biologists, the actual resource stock cannot be observed, but can only be estimated subject to error (including relying on fishery dependent data). The optimal resource management problem then becomes a problem in either stochastic dynamic programming or stochastic dynamic simulation. How do stochastic fluctuations in resource growth affect the optimal regulated extraction rate (Pindyck 2007)? What is the optimal policy if there is an increase in the volatility of stochastic fluctuations in the resource stock as the stock becomes smaller and the potential for irreversible decline to extinction arises?

Although this is one of the most important topics of research, much remains to be done in fisheries economics on the choice of optimal policy instrument under uncertainty and even on understanding the very nature of the uncertainty confronting industry and policy makers. Important works include those by Reed (1979), Andersen (1982), Andersen and Sutinen (1984), Clark and Kirkwood (1986), Bockstael and Opaluch (1993), Dupont (1993), Grafton (1994), Li (1998), Mistiaen and Strand (2000b), McDonald et al. (2002), Weitzman (2002), Eggert and Tveteras (2004), Eggert and Martinsson (2004), Grafton and Kompas (2005), Saphores (2003), Herrara (2005), Sethi et al. (2005), Singh et al. (2006), Kugarajah et al. (2006), McConnell and Price (2006), Murillas and Chamorro (2006), Nøstbakken (2006), Segerson (2007), Eggert and Lokina (2007). Research topics include further examination of fishers' risk preferences, fishery management under uncertainty, including Bayesian decision analysis, irreversibilities, extinction, and accounting for rare events (e.g., sea turtles poised on the brink of extinction) in policy, closer examination of the tails of distributions, and how these may change over time rather than just central tendencies, risk, and uncertainty in ex-vessel price and ITQ markets, uncertainty of regulations, optimal policy instruments under uncertainty, causes, and likelihood of severe or catastrophic outcomes, nonlinear benefit and cost functions, and credible bioeconomic models explicitly dealing with uncertainty, perhaps through Bayesian approaches as practiced by population biologists. Many of the issues raised by Pindyck (2007) can be adapted to potential research in fisheries economics.

32.15   Bycatch and Multispecies Issues

Bycatch is simply fish or other species caught as part of a joint production process, but fish that are an undesirable output. Incidental takes can also occur for sea turtles, dolphins, other cetaceans such as whales, sea birds, and other species. In economics, some progress has been made in bioeconomic framework (Boyce 1996; Herrera 2004; Enriquez-Andrade and Vaca-Rodriguez 2004). Herrera (2004) and Jensen and Vestergaard (2002c) consider the problem of strategic interactions between catch-discarding fishermen and a regulator, a moral hazard problem. Directional distance functions are a promising approach to evaluate the microeconomic joint production process with bycatch (Färe et al. 2006), since they allow for both weak and strong disposability of target catches when bycatch is reduced, along with extensions to include the environment and its effects and the potential for underutilized capital (Kjærsgaard et al. 2007). Another promising approach, by Abbot and Wilen (2007) builds on game theory and develops a firm-level positive analysis in a second-best framework to analyze the intra-seasonal game played between fishermen and regulators even when complete and perfect information is available on all sides. Segerson (2007) developed an alternative approach within a static environment applying models originally developed in the environmental economics literature for undesirable outputs such as pollution while explicitly incorporating uncertainty. More modeling work remains to be done in both areas. Policy work remains in the area of property and use rights and economic incentives.

32.16   Buybacks

Buybacks of vessels, gear, or rights are widely used throughout the world although buybacks are largely, but not universally, deplored by economists. Considerable recent work has been done on evaluating buybacks as a policy instrument and their design by Holland et al. (1999), Groves and Squires (2007), Hannesson (2007e), and Curtis and Squires (2007), but little formal economics research has been conducted with the exception of Sun (1998), Campbell (1989), Weninger and McConnell (2000), Walden et al. (2003), Kitts et al. (2001), Guyader et al. (2004), and Clark et al. (2005, 2007). Fruitful and obvious topics of research include the design of auctions and accounting for information asymmetries including moral hazard and adverse selection, and evaluating the impact of incentives (Curtis and Squires 2007). Other than the important work of Weninger and McConnell (2000) and Clark et al. (2005, 2007), little formal modeling of investment in this context has occurred. Investment modeling would be best served with a disaggregated output vector more closely corresponding to the actual decisions made by fishers. Another important area is the role and design of buybacks for conservation purposes, especially in developing countries, such as the vaquita or for sea turtles, in which willingness to pay or accept for existence values from developed countries is expressed. Since buybacks can be viewed as a form of contract, buybacks should provide a fruitful area of research for contract theory.

32.17   Aquaculture

Increasingly, cultured fish and other seafood comprise much of the fish that is consumed, and cultured fish and other seafood can be expected to comprise ever-increasing amounts. Only a few salient issues will be touched upon on this under-researched but increasingly important topic. Rising final demand for cultured seafood raises the derived demand for a critical input in the production process, fish meal, and other fish feed often derived from small pelagic species such as anchovies and sardines. These fish in turn are often forage fish for other species in the food web, so that growing demand for small pelagics not only raises the issue of overfishing but also impacts upon the ecosystem; finding the right balance in an era of ecosystems management presents an important research topic. Just what are the economic and strictly ecological values of alternative pelagic species in the ecosystem? For example, the value of menhaden in Chesapeake Bay is under consideration, with menhaden serving as a reduction species for fish meal, for a source of Omega-3 for direct human consumption, as primary feeders and forage fish in the ecosystem, and other ecological and human purposes. The interaction and effects of cultured species, such as some species of salmon, with wild species presents another research topic, as for example, interbreeding. Wastes and diseases from cultured fish can also impact other species of fish and the surrounding ecosystem. Conflicting uses of the land and offshore areas for culture and other purposes is another concern.

32.18   Experimental Economics

Experimental economics entails laboratory experiments, i.e., formal scientific experiments with carefully constructed hypotheses and formal controls to evaluate the effects of various treatments. In contrast to most economics and population dynamics, this approach closely follows laboratory approaches routinely used by many scientific disciplines. Experimental economics can be used to evaluate many areas of economics, but is separately distinguished due to its uniqueness. This approach has been widely used in public and behavioral economics for many years, but is only now starting to be applied in fisheries economics to areas of auctions and their design, property rights, and incentives, largely by Chris Anderson of the University of Rhode Island (Anderson and Sutinen 2005, 2006; Anderson and Holland 2006). This promising approach can provide new insights into auctions, property rights, and behavioral economics applied to fisheries.

32.19   Behavioral Economics

This important area of research in finance and theoretical economics has yet to be applied to fisheries economics and will only be briefly touched upon. In essence, behavioral economics widens the behavioral hypotheses of economic agents – consumers and producers – from strict self-interested rationality of the individual economic agent. Behavioral Economics is the combination of psychology and economics that investigates what happens in markets in which some of the agents display human limitations and complications. Humans deviate from the standard economic model in several ways, including (Mullainathan and Thaler 2000):

(1) bounded rationality, which reflects the limited cognitive abilities that constrain human problem solving; (2) bounded willpower, which captures the fact that people sometimes make choices that are not in their long-term interest; and (3) bounded self-interest, which incorporates the comforting fact that humans are often willing to sacrifice their own interests to help others.

32.20 Invasive Species

Invasive species represent an increasingly important policy issue facing nations and hence is growing in research importance. The spread of exotic species into ecosystems and threats to biodiversity and sustainable resource use represent both an ecological and economic problem. Management of invasive species entails the expenditure of limited resources, which must be carried out in the most effective manner. Little will be mentioned here of this topic, other than to note its growing importance and several key extant research papers. A key question, as with pollution, is the optimal level of invasive species eradication, prevention, or ongoing management, all of which depend on the current level of invasion and the nature of the recipient ecosystem, which is essentially a cost–benefit question. Other fundamental questions include the potential importance of linking the dynamics of the invasive species with the behaviors of individuals and institutions. How do individual behaviors and decisions lead to invasion? How are externalities incorporated into models, since invasive species automatically entail social costs not adequately considered in the decisions facing individuals?

32.21 Concluding Remarks

The key ideas introduced by fisheries economics that have resonated with policy-makers and other disciplines have been the importance of property rights and incentives for the conservation and management of fisheries, ecosystems, and biodiversity. These ideas have permeated the fields of population dynamics, ecology, and conservation biology, and are routinely considered by policy makers. The importance of incentives is just becoming clear in conservation, especially of transnational global public and common resources.

The main thrust of fisheries economics is the development of normative economically optimal harvest strategies, but the results have largely failed to resonate with policy-makers, industry, and other disciplines. This is in part because of the discordance in how population dynamics and uncertainty are approached and modeled compared to the far more sophisticated population biology used in management, and in part because questions centered on prediction and distribution rather than normative are relevant for actual management. In this regard, contemporary and comprehensive population dynamics and stochastic dynamic simulations rather than normative economic optimization are critical for policy relevance. As observed by Deacon et al. (1998, p. 392), "[s]tylized optimization-based models probably have already yielded their most important insights," and (Deacon et al. 1998, p. 392), "[i]f we are interested in being useful beyond offering simple generalities, however, then it is time to begin incorporating more of the realism that already exists in the biological literature into our models." Nonetheless, there is room for the on-going incorporation of environmental services and biodviersity into a more comprehensive normative bioeconomic model. In any case, the main focus of formal fisheries economics research has by choice largely confined itself to an internal discussion of a focused topic. The seminal developments in the field were largely introduced outside of the economics profession (Warming 1911; Gordon 1954), or by economists outside of academia (Christy 1973, 1982; Sinclair 1961), with the singular exception of Scott (1955a, b).

Important ideas for future research include: (1) to better understand the workings of new forms of use and property rights as they evolve for not only fisheries management but also conservation of biodiversity and the ecosystem (especially territorial rights such as TURFs, fishing cooperatives, and the global ocean commons) and climate responses; (2) technical change and more generally, productivity growth (fishing power); (3) valuation and modeling of ecosystems and ecosystem approaches to management using the concept of natural capital and its flow of services; (4) conservation of protected species and habitat and biodiversity in general, as forms of pure and impure public goods; (5) analysis of Pareto-improving policies in a second-best world, i.e., less emphasis on a normative economic optimum under ideal conditions and greater emphasis on policies that improve economic welfare broadly defined under conditions that are less than ideal; (6) bioeconomic models closely aligned to current population dynamics and that rely less on analytical solutions and that allow for technical change (which current population dynamics allows); (7) broadening bioeconomic models to incorporate ecosystem management (as well as more realistic population dynamics and consideration of uncertainty); (8) international treaties, transboundary resource stocks, transnational fisheries, and international trade; (9) better understanding fishing industries as industries and the regulatory process as regulation of an industry under asymmetric and incomplete information; (10) better understanding the risk and uncertainty that pervades economic models and conservation and management of fisheries, biodiversity, and ecosystems; (11) aquaculture; and (12) developing country fisheries, including various scales of coastal and artisanal fisheries.

Let me finish with a few personal observations. As fisheries, especially in developed countries, evolve from purely commercial interests to concerns over the public good components – as fisheries in these countries are increasingly viewed as an environmental, ecological, and biodiversity issue rather than simply an extractive industry – the locus of research in these countries will follow suit. Environmental and public economics, ecology, and conservation biology will increasingly serve as the defining intellectual framework in what is known as ecosystems management and conservation. Traditional fisheries economics, with its focus on normative optimal harvesting strategies – pure extraction – will remain important to academic fisheries economics, but continue to languish inside management agencies where the concern is largely over predictive and distributive rather than normative concerns. The most interesting and pressing research questions will shift to those now voiced by society about how to best mix commercial fisheries and private goods with environmental, ecological, and public good concerns over biodiversity, healthy ecosystems and their services, and sustainable resource use.

Fisheries in developed countries increasingly focus on fresh, high-value fish caught by lines and traps and aimed at restaurants and "high-end" markets. Bottom trawls and ecologically harmful fishing practices are phasing out with some notable exceptions such as the Alaskan pollock fishery. Mass consumption is increasingly filled by international trade and developing country fisheries and aquaculture. In short, as emphasis shifts to environmentally sustainable fishing, healthy ecosystem and their services, and biodiversity conservation, former environmental issues are exported out of sight and out of mind. The issues do not go away; they simply reside elsewhere, which raises the important research topic of sustainable fisheries and aquaculture in developing countries along with the link of international trade as well as the growing domestic consumption within these developing countries.

Concern continues to mount over the global public good of the atmosphere, and a similar concern can be expected to grow over the global ocean commons, raising the issue of transboundary fisheries and biodiversity conservation with an environmental and public good twist, such as whales, sea turtles, coral reefs, and the like. Game theory, contract theory, and public and environmental economics will become increasingly important as intellectual frameworks. Technical change is one of the most important contributors to the growth in fishing capacity, biodiversity loss, and environmental degradation, and increasing research attention in this area is critical, including appropriate (largely second-best) policy responses. The presence of technical change obviates the notion of a steady-state equilibrium, and making normative renewable resource models relevant will require grappling with technical change and the continuing disequilibrium. Technical change has been studied extensively for other industries and this body of work suggests many important and relevant research questions. In addition, there is no substitution possible between the resource stock, man-made inputs and technical change; a lower resource stock through the impact of ongoing technical progress simply leaves less catch at some point, and theoretical and empirical research opportunities into this process are many. The effects of technical change and understanding what drives technology and its diffusion are important. Risk and uncertainty will remain as complex and critical issues of concern, especially as economists confront the limits of what are admittedly over-aggregated and highly abstract models concerned with shifting oceanography and ecosystems in the complex dynamic process of the oceans. Consideration of the tails of distributions – rare events or the "Black Swan" – is critical for conservation issues. Climate change will become an increasingly critical issue for oceans and fisheries. Fisheries and development economists have only begun to scratch the surface of the resource-based development of millions of artisanal and commercial fisheries in developing countries, and the intellectual foundation for this worthy project lies in the nexus of resource, environmental, and development economics. Much can also be learned from agriculture and terrestrial conservation economics.

Increasing interest is shown over spatial processes and territorial rights as a means to internalize or address the spatial externalities and processes and metapopulations present in the oceans. Considerable normative modeling efforts are given to this issue, with ecology and oceanography as important intellectual components. The degree of resolution and certainty of the results and the ability to manage off imperfect aggregate and largely normative models remain open questions, and care should be given not to oversell what can realistically be achieved. Much of value will undoubtedly be learned of a qualitative and general nature that can shed light on optimal policy. Concomitant with this normative modeling will be greater thought given to spatial elements in conservation and management and the nature of zoning and territorial property rights. Spatial economics, landscape analysis, and GIS provide good sources of research ideas and topics.

In sum, the central and continuing theme of economics that the proper formation and functioning of institutions – including property rights, evaluation of trade-offs and opportunity costs, and economic incentives will continue as the most important and enduring message from economics. Fisheries economics has insufficiently taken advantage of other fields of applied and theoretical economics for methods and research topics to understand fisheries as industries within either a positive or normative framework. Most importantly, the focus of economics' recurring themes will increasingly shift from the classic overfishing problem to the three great challenges facing humanity in addressing global oceanic environmental problems: sustainability of ecosystems and their services, loss of biodiversity, and the oceanic interaction with the atmosphere through climate. That is, rather than simply addressing the classic overfishing and resource stock externality issue identified by Warming (1911), Gordon (1954), and Scott (1955a, b), fisheries economics will rise to the challenge of addressing multiple externalities and a broader concept of sustainable resource use, often in a second-best rather than first-best context: the timeless resource stock externality, the transnational externality for transboundary and straddling stocks and global public goods, and those externalities arising from the public good issues of biodiversity conservation and sustainability of ecosystems and their services. The benefits of these three great challenges are largely non-market and non-use with important future values, so that their research challenge requires fisheries economics to stretch their standard dynamic models and reliance on markets and reach beyond the well-established confines of fisheries economics.

References

Abbott J, Wilen J (2007) Regulation of Fisheries Bycatch with Common-Pool Output Quotas. Working Paper, Department of Agricultural Economics, University of California, Davis

Ahmed M, Boonchuwongse P, Dechboon W, Squires D (2007) Overfishing in the Gulf of Thailand: Policy Challenges and Bioeconomic Analysis. Environment and Development Economics 12(1):145–172

Allen R, Joseph J, Squires D (2009) Conservation and Management of Transboundary Tuna Fisheries. Blackwell (in press)

Allen RC, Keay I (2001) The First Great Whale extinction: The End of the Bowhead Whale in the Eastern Arctic. Explorations in Economic History 38(4):448–477

Amundsen ES, Bjørndal T, Conrad J (1995) Open Access Harvesting of the Northeast Atlantic Minke Whale. Environmental and Resource Economics 6(2):167–185

Anderson CM, Sutinen JG (2005) A Laboratory Assessment of Tradable Fishing Allowances. Marine Resource Economics 20(1): 1–24

Anderson, CM, Sutinen JG (2006) The Effect of Initial Lease Periods on Price Discovery in Laboratory Tradable Fishing Allowance Markets. Journal of Economic Behavior and Organization 61:164–180

Anderson CM, Holland D (2006) Auctions for Initial Sale of Annual Catch Entitlement. Land Economics 82(3):333–52

Andersen LG (2007) Does Capacity Analysis Help us Meet Fishery Policy and Management Objectives? Comments. Marine Resource Economics 21(2):89–93

Andersen P (1982) Commercial Fisheries Under Price Uncertainty. Journal of Environmental Economics and Management 9(1):11–28

Andersen P, Sutinen JG (1984) Stochastic Bioeconomics: A Review of Basic Methods and Results. Marine Resource Economics 1:117–36

Armstrong C (2006) A Note on the Ecological-Economic Modeling of Marine Reserves in Fisheries. Working Paper in Economics and Management 05/06, Norwegian College of Fisheries Science, University of Tromsø, Norway

Arnason R (1990) Minimum Information Management in Fisheries. Canadian Journal of Economics 23(3):630–653

Arnason R, Magnusson G, Agnarsson S (2000) The Norwegian Spring-Spawning Herring Fishery: A Stylized Game Model. Marine Resource Economics 15(4):293–319

Asche F, Bjørndal T, Eggert H, Frost H, Gordon D, Gudmundsson E, Hoff A, Jensen C, Pascoe S, Sissener E, Tveterås R (2005) Modeling Fishermen Behavior Under New Management Regimes: Final Report. SNF Report No. 28/05, Institute for Research in Economics and Business Administration, Bergen, Norway

Asche F (1997) Dynamic Adjustment in Demand Equations. Marine Resource Economics 12(3): 221–237.

Asche F, Bremnes H, Wessells CR (1999) Product Aggregation, Market Integration and Relationships between Prices: An Application to World Salmon Markets. American Journal of Agricultural Economics 81(3):568–581

Asche F, Gordon D, Jensen C (2007) Individual Vessel Quotas and Increased Fishing Pressure on Unregulated Species. Land Economics 83(1):41–49

Baland J-M, Platteau J-P (1996) Halting Degradation of Natural Resources: Is There a Role for Rural Communities? Oxford University Press, Oxford.

Baltagi B, Griffen J (1988) A General Index of Technical Change. Journal of Political Economy 96(1): 20–41

Barbier EB, Strand I (1998) Valuing Mangrove-Fishery Linkages: A Case Study of Campeche, Mexico. Environmental and Resource Economics 12:151–56

Barrett S (2003) Environment and Statecraft: The Strategy of Environmental Treaty Making. Oxford University Press, Oxford

Barten AP, Bettendorf LJ (1989) Price Formation of Fish: An Application of an Inverse Demand System. European Economic Review 33:1509–1525

Batstone C, Sharp B (2003) Minimum Information Management Systems and ITQ Fisheries Management. Journal of Environmental Economics and Management 45(2/1):492–504

Beamish RJ, Bouillon DR (1993) Pacific Salmon Production Trends in Relation to Climate. Canadian Journal of Fisheries and Aquatic Sciences 50(5):1002–1016

Beinhocker E (2006) The Origin of Wealth: Evolution, Complexity and the Radical Remaking of Economics. Random House, London

Berck P (1979) Open Access and Extinction. Econometrica 47(4):877–882

Berck P, Perloff J (1984) An Open-Access Fishery with Rational Expectations. Econometrica 52(2):489–506

Bergland H, Pedersen PA (1997) Catch Regulation and Accident Risk: The Moral Hazard of Fisheries Management. Marine Resource Economics 12(4):281–292

Bess R, Rallapudi R (2007) Spatial conflicts in New Zealand fisheries: The rights of fishers and protection of the marine environment. Marine Policy 31(6):719–729

Beverton RJH, Holt SJ (1957) On the Dynamics of Exploited Fish Populations. Chapman & Hall, London (reprinted in 1993)

Bisack KD, Sutinen JG (2006) Harbor Porpoise Bycatch: ITQs or Time/Area Closures in the New England Gillnet Fishery. Land Economics 82(1):85–102

Bjørndal T (1987) Production Economics and Optimal Stock Size in a North Atlantic Fishery. Scandinavian Journal of Economics 89(2):145–164

Bjørndal T, Conrad J (1987) The Dynamics of an Open Access Fishery. Canadian Journal of Economics 20(1):74–85

Bjørndal T, Brasão A (2006) The East Atlantic Bluefin Tuna Fisheries: Stock Collapse or Recovery? Marine Resource Economics 21(2):193–210

Bjørndal T, Munro GR (2003) The Management of High Seas Fisheries Resources and the Implementation of the UN Fish Stocks Agreement of 1995. In Tietenberg T, Folmer H (eds) The International Yearbook of Environmental and Resource Economics 2003/2004: A Survey of Current Issues, Cheltenham, Edward Elgar, pp 1–35

Bockstael N Opaluch J (1993) Discrete Modeling of Supply Response under Uncertainty: The Case of the Fishery. Journal of Environmental Economics and Management 10(2):125–137

Bockstael N, McConnell T (2006) Environmental and Resource Valuation with Revealed Preferences: A Theoretical Guide to Empirical Models. Springer

Bonceur J, Alban F, Guyader O, Thebaud O (2002) Fish, Fishers, Seals and Tourists: Economic Consequences of Creating a Marine Reserve in a Multi-Species Multi-Activity Context. Natural Resource Modeling 15(4):387–411

Boyce JR (1992) Individual Transferable Quotas and Production Externalities in a Fishery. Natural Resource Modeling 6:385–408

Boyce JR (1995) Optimal Capital Accumulation in a Fishery: A Nonlinear Irreversible Investment. Journal of Environmental Economics and Management 28:324–339

Boyce JR (1996) An Economic Analysis of the Fisheries Bycatch Problem. Journal of Environmental Economics and Management 31:314–336

Boyce JR (2004) Instrument Choice in a Fishery. Journal of Environmental Economics and Management 47(1):183–206

Brock WA, Xepapadeas A (2003) Valuing Biodiversity from an Economic Perspective: A Unified Economic, Ecological, and Genetic Approach. American Economic Review 93:1597–1614

Bromley D (1991) Environment and Economy: Property Rights and Public Policy. Blackwell, Oxford

Brown GM (2000) Renewable Natural Resource Management and Use without Markets. Journal of Economic Literature 38:875–914

Brown G, Roughgarden J (1997) A Metapopulation Model with Private Property and a Common Pool. Ecological Economics 22:65–71

Campbell H (1989) Fishery Buy-Back Programs and Economic Welfare. Australian Journal of Agricultural Economics 33:20–31

Castilho C, Pichika D, Srinivasu N (2005) Bio-Economics of a Renewable Resource in a Seasonally Varying Environment. EEE Working Paper Series 16

Cheung S (1970) The Structure of a Contract and the Theory of a Non-exclusive Resource. Journal of Law and Economics 13(1):49–70

Christensen S, Vestergaard N (1993) A Bioeconomic Analysis of the Greenland Shrimp Fishery in the Davis Strait. Marine Resource Economics 8:345–365

Christy, Jr FT (1973) Fisherman Quotas: A Tentative Suggestion for Domestic Management. Occasional Paper 19, Law of the Sea Institute, Honolulu, HI.

Christy, Jr FT (1982) Territorial Use Rights in Marine Fisheries: Definitions and Conditions. Technical Paper 227. Food and Agriculture Organization of the United Nations, Rome.

Clark CW (1973) Profit Maximization and the Extinction of Animal Species. Journal of Political Economy 81:950–961

Clark CW (1980) Restricted Access to Common-Property Resource: A Game-Theoretic Analysis in P. Liu, editor, Dynamic Optimization and Mathematical Economics. Plenum Press, New York

Clark CW (1990) Mathematical Bioeconomics: The Optimal Management of Renewable Resources, 2nd edition. Wiley, New York

Clark CW, Munro GR (1975) The Economics of Fishing and Modern Capital Theory: A Simplified Approach. Journal of Environmental Economics and Management 2(2):92–106

Clark CW, Clarke FH, Munro GR (1979) The Optimal Exploitation of Renewable Resource Stocks: Problems of Irreversible Investment. Econometrica 47(1):25–47

Clark CW, Kirkwood GP (1986) On Uncertain Renewable Resource Stocks: Optimal Harvest Policies and the Value of Stock Surveys. Journal of Environmental Economics and Management 13:234–244

Clark CW, Munro GR, Sumaila R (2005) Subsidies, Buybacks, and Sustainable Fisheries. Journal of Environmental Economics and Management 50(1):47–58

Clark CW, Munro GR, Sumaila R (2007) Buyback Subsidies, the Time Consistency Problem, and the ITQ Alternative. Land Economics 83(1):50–58

Coase R (1960) The Problem of Social Cost. Journal of Law and Economics 3:1–44

Conrad JM (1989) Bioeconomics and the Bowhead Whale. Journal of Political Economy 97(4): 974–987

Conrad JM (1999) The Bioeconomics of Marine Sanctuaries. Journal of Bioeconomics 1:205–217

Conrad JM, Adu-Asamoah R (1986) Single and Multispecies Systems: The Case of Tuna in the Eastern Tropical Atlantic. Journal of Environmental Economics and Management 3:50–68

Costello C, Adams R, Polasky S (1998) The Value of El Niño Forecasts in the Management of Salmon: A Stochastic Dynamic Assessment. American Journal of Agricultural Economics 80(4): 765–777

Crocker TD (1966) The Structuring of Atmospheric Pollution Control Systems. In Wolozin H (ed) The Economics of Air Pollution. W.W. Norton, New York

Crutchfield JA, Zellner A (1962) Economic aspects of the Pacific halibut fishery. Fishery Industrial Research, Vol. 1, No. 1, U.S. GPO, Washington, DC

Curtis R, Hicks R (2000) The Cost of Sea Turtle Protection: The Case of Hawaii's Pelagic Longliners. American Journal of Agricultural Economics 82(5):1191–1197

Curtis R, McConnell K (2004) Incorporating Information and Expectations in Fishermen's Spatial Decisions. Marine Resource Economics 19(1):131–43

Dales JH (1968) Pollution, Property, and Prices. University of Toronto Press, Toronto

Dalton MG (2001) El Niño, Expectations, and Fishing Effort in Monterey Bay, California. Journal of Environmental Economics and Management 42(3):336–359

David PA (1969) Contribution to the Theory of Diffusion. Stanford Center for Research in Economic Growth, Memorandum #71. Stanford University, Stanford, CA

Deacon RT (1989) An Empirical Model of Fishery Dynamics. Journal of Environmental Economics and Management 27(12):167–183

Deacon RT, Kolstad C, Kneese A, Brookshire D, Scrogin D, Fisher A, Ward A, Smith K, Wilen J (1998) Research Trends and Opportunities in Environmental and Resource Economics. Environmental and Resource Economics 11(3/4):383–397

Duarte CC, Brasão A, Pintassilgo P (2000) Management of the Northern Atlantic Bluefin Tuna: An Application of C-Games. Marine Resource Economics 15(1):21–36

Dupont D (1990) Rent Dissipation in Restricted Access Fisheries. Journal of Environmental Economics and Management 19:26–44

Dupont D (1991) Testing for Input Substitution. American Journal of Agricultural Economics 73(1):155–164

Dupont D (1993) Price Uncertainty, Expectations Formation, and Fishers' Location Choices. Marine Resource Economics 8(3):219–247

Dupont D, Fox KJ, Gordon D, Grafton RQ (2005) Profit and Price Effects of Multi-Species Individual Transferable Quotas. Journal of Agricultural Economics 56(1):31–57

Dutta P (1995) Collusion, Discounting and Dynamic Games. Journal of Economic Theory 66:289–306

Dutton P, Squires D (2008) A Holistic Strategy for Pacific Sea Turtle Recovery. Ocean Development and International Law 39(2):200–222

Dutton P, Squires D (eds) (in preparation) The Conservation of Pacific Sea Turtles.

Eales J, Wilen J (1986) An Examination of Fishing Location Choice in the Pink Shrimp Fishery. Marine Resource Economics 2:331–351

Eckmans J, Finus M (2004) An Almost Ideal Sharing Scheme for Coalition Games with Externalities. Working Paper 155. Trieste, Italy: Fondazionne Eni Enrico Mattei.

Edwards S (2008) Ocean Zong, First Possession and Coasean Contracts. Marine Policy 32(1):46–54

Eggert H (1998) Bioeconomic Analysis and Management: The Case of Fisheries. Environmental and Resource Economics 11(3–4):399–411

Eggert H, Martinsson P (2004) Are Commercial Fishers Risk Lovers? Land Economics 80(4): 550–60

Eggert H, Tveteras R (2004) Stochastic Production and Heterogeneous Risk Preferences: Commercial Fishers' Gear Choices. American Journal of Agricultural Economics 86(1):199–212

Eggert H, Lokina RB (2007) Small-Scale Fishermen and Risk Preferences. Marine Resource Economics 22(1):49–68

Eggert H, Tveteras R (2008) Potential Rent and Overcapacity in the Swedish Baltic Sea Trawl Fishery for Cod (*Gadus morhua*). ICES Journal of Marine Science (in press)

Enriquez-Andrade RR, Vaca-Rodriguez JG (2004) Evaluating Ecological Tradeoffs in Fisheries Management: A Case Study for the Yellowfin Tuna Fishery in the Eastern Pacific Ocean. Ecological Economics 48:303–315

FAO (1998) Report of the Technical Working Group on the Management of Fishing Capacity. FAO Fisheries Report 586, Rome, Italy.

Färe R, Kirkley J, Walden J (2006) Adjusting Technical Efficiency to Reflect Discarding: The Case of the U.S. Georges Bank Multi-Species Otter Trawl Fishery. Fisheries Research 78(2–3):257–265

Felthoven R, Paul CM (2004) Directions for Productivity Measurement in Fisheries, Marine Policy 28(2):97–108

Finnoff D, Tschirhart J (2003) Harvesting in an Eight-Species Ecosystem. Journal of Environmental Economics and Management 45(3):589–611

Fischer RD, Mirman LJ (1992) Strategic Dynamic Interaction: Fish Wars. Journal of Economics Dynamics and Control 16:267–287

Fischer RD, Mirman LJ (1996) The Compleat Fish Wars. Biological and Dynamic Interactions. Journal of Environmental Economics and Management 30:34–42

Flaaten O (1988) The Economics of Multispecies Harvesting: Theory and Application to the Barents Sea Fisheries. Springer-Verlag, Berlin

Flaaten O (1991) Bioeconomics of Sustainable Harvest of Competing Species. Journal of Environmental Economics and Management 20:163–80

Flores-Lagunes A, Horrace WC, Schnier KE (2007) Identifying Technically Efficient Fishing Vessels: A Non-Empty, Minimal Subset Approach. Journal of Applied Econometrics 22:729–745

Fousekis P, Revell B (2004) Retail Fish Demand in Great Britain and its Fisheries Management Implications. Marine Resource Economics 19(4):495–510

Fox K, Grafton RQ, Squires D, Kirkley J (2003) Property Rights in a Fishery: Regulatory Change and Firm Performance. Journal of Environmental Economics and Management 46(1): 156–177

Fulginiti L, Perrin R (1993) The Theory and Measurement of Producer Response under Quotas. Review of Economics and Statistics 75:97–106

Gilbert B, Groves T, Squires D, Yeo BH (2007) Technical Change in an Artisanal Fishery. Draft memo, Dept. of Economics, University of California San Diego.

Gillespie A (2005) Whaling Diplomacy: Defining Issues in International Environmental Law. Edward Elgar, Northampton, MA.

Gjertsen H (2007a) Direct Incentive Approaches to Nesting Beach Protection. In Dutton P, Squires D, Ahmed M (eds) Conservation of Pacific Sea Turtles. La Jolla, CA: National Marine Fisheries Service.

Gjertsen H (2007b) Can We Improve Our Conservation Bang for the Buck? Cost-Effectiveness of Alternative Leatherback Turtle Conservation Strategies. In Dutton P, Squires D, Ahmed M (eds) Conservation of Pacific Sea Turtles.

Gonzalez E (1996) Territorial Use Rights in Chilean Fisheries. Marine Resource Economics 11:211–218

Gordon HS (1954) The economic theory of a common property resource: the fishery. Journal of Political Economy 62:124–142

Gould JR (1972) Extinction of a Fishery by Commercial Exploitation: A Note. Journal of Political Economy 80:1031–1038

Grafton RQ (1994) A Note on Uncertainty and Rent Capture in an ITQ Fishery. Journal of Environmental Economics and Management 27(3):286–294

Grafton RQ, Kompas T (2005) Uncertainty and the Active Adaptive Management of Marine Reserves. Marine Policy 29:471–479

Grafton RQ, Squires D, Fox K (2000a) Common Resources, Private Rights and Economic Efficiency. Journal of Law and Economics 43(2):679–713

Grafton RQ, Sandal L, Steinshamn SI (2000b) How to Improve the Management of Renewable Resources: The Case of Canada's Northern Cod Fishery. American Journal of Agricultural Economics 82(3):570–580

Grafton RQ, Kompas T, Schneider V (2005a) The Bioeconomics of Marine Reserves: A Selected Review with Policy Implications. Journal of Bioeconomics 7:161–178

Grafton RQ, Kompas T, Lindenmeyer D (2005b) Marine Reserves with Ecological Uncertainty. Bulletin of Mathematical Biology 67:957–971

Grafton RQ, Arnason R, Bjørndal T, Campbell D, Campbell H, Clark CW, Connor R, Dupont D, Hannesson R, Hilborn R, Kirkley J, Kompas T, Lane D, Munro D, Pascoe S, Squires D, Steinshamn S, Turris B, Weninger Q (2006) Incentive-Based Approaches to Sustainable Fisheries. Canadian Journal of Fisheries and Aquatic Sciences 63(3):699–710

Griliches Z (1957) Hybrid Corn: An Exploration in the Economics of Technological Change. Econometrica 25:501–522

Groves T, Ledyard J (1977) Optimal Allocation of Public Goods: A Solution to the "Free Rider" Problem. Econometrica 45(4):783–810

Groves T, Squires D (2007) Lessons from Fisheries Buybacks. In Curtis R, Squires D (eds) Chap 2, Fisheries Buybacks. Blackwell, Oxford, UK

Gudmundsson E, Wessells CR (2000) Ecolabeling Seafood for Sustainable Production: Implications for Fisheries Management. Marine Resource Economics 15(2):97–113

Guyader O, Daures F, Fifas S (2004) A Bioeconomic Analysis of the Impact of Decommissioning Programs: Application to a Limited-entry French Scallop Fishery, Marine Resource Economics 19(2):225–242

Hanna S (1999) From Single-Species to Biodiversity – Making the Transition in Fisheries Management. Biodiversity and Conservation 8(1):45–54

Hannesson R (1983a) The Bioeconomic Production Function in Fisheries: A Theoretical and Empirical Analysis. Canadian Journal of Fisheries and Aquatic Sciences 40(7):968–982

Hannesson R (1983b) Optimal Harvesting of Ecologically Interdependent Fish Species. Journal of Environmental Economics and Management 10:329–345

Hannesson R (1993) Bioeconomic Analysis of Fisheries. Fishing News Books, Oxford

Hannesson R (1995a) Sequential Fishing: Cooperative and Non-cooperative Equilibria. Natural Resource Modeling 9:51–59

Hannesson R (1995b) Fishing on the High Seas: Cooperation or Competition? Marine Policy 19:371–377

Hannesson R (1996) Long Term Industrial Equilibrium in an ITQ Managed Fishery. Environmental and Resource Economics 8:63–74

Hannesson R (1997) Fishing as a Supergame. Journal of Environmental Economics and Management 32:309–322

Hannesson R (1998) Marine Reserves: What Would They Accomplish? Marine Resource Economics 13(3):159–170

Hannesson R (2002) The Economics of Marine Reserves. Natural Resource Modeling 15:273 – 290

Hannesson R (2006) Sharing the Northeast Arctic Cod: Possible Effects of Climate Change. Natural Resource Modeling 19:633–654

Hannesson R (2007a) Geographical Distribution of Fish Catches and Temperature Variations in the Northeast Atlantic Since 1945. Marine Policy 31:32–39

Hannesson R (2007b) Global Warming and Fish Migrations. Natural Resource Modeling 20(2): 301–320

Hannesson R (2007c) Growth Accounting in a Fishery. Journal of Environmental Economics and Management 53(3):374–376

Hannesson R (2007d) Cheating About Cod. Marine Policy 31(6):698–705

Hannesson R (2007e) Do Buyback Programs Make Sense? In Curtis R, Squires D, (eds) Fisheries Buybacks, Chap 3, Blackwell Oxford, UK

Hannesson R (2008) ITQs for Bycatches: Lessons for the Tuna-Dolphin Issue. In Allen, Joseph, Squires, (eds) Conservation and Management of Transboundary Tuna Fisheries. Blackwell Oxford, UK (in press)

Hannesson R, Barange M, Herrick S, (eds) (2006) Climate Change and the Economics of the World's Fisheries: Examples of Small Pelagic Stocks, Edward Elgar, Cheltenham, UK

Hannesson R, Salvanes K, Squires D (2007a) The Lofoten Fishery over a Hundred Years. Working Paper, Department of Economics, Norwegian School of Economics and Business Administration, Bergen

Hannesson R, Herrick S, Field J (2007b) Ecological and Economic Considerations in the Conservation and Management of the California Sardine. Working Paper, Southwest Fisheries Science Center, La Jolla

Haraden J, Herrick S, Squires D, Tisdell C (2004) Economic Benefits of Dolphins in the United States Eastern Tropical Pacific Purse Seine Tuna Industry. Environmental and Resource Economics 28:451–468

Hatcher A (2005) On the Microeconomics of Quota Management in Fisheries. Marine Resource Economics 20(1):77–99.

Heal G (2007) Review of Environmental and Resource Economics. Review of Environmental Economics and Policy 1(1):7–25

Heaps T (2003) The Effects on Welfare of The Imposition of Individual Transferable Quotas on a Heterogeneous Fishing Fleet. Journal of Environmental Economics and Management 46(3):557–576

Heberer C, Stohs S (2007) Which Swordfish Gear is Cleanest? In Dutton P, Squires D, Ahmed M, (eds) Conservation of Pacific Sea Turtles. La Jolla, CA: National Marine Fisheries Service.

Herrera G (2004) Stochastic Bycatch, Informational Asymmetry, and Discarding. Journal of Environmental Economics and Management 49(3):463–483

Herrera G (2006) Benefits of Spatial Regulation in a Multispecies System. Marine Resource Economics 21(1):63–79

Herrick S, Norton JG, Mason JE, Bessey C (2007) Management Application of an Empirical Model of Sardine–Climate Regime Shifts. Marine Policy 31(1):71–80

Hilborn R, Ledbetter M (1979) Analysis of the British Columbia Salmon Purse Seine Fleet: Dynamics of Movement. Journal of the Fisheries Research Board of Canada 36:384–91

Hilborn R, Orensanz JM, Parma A (2005) Institutions, Incentives, and the Future of Fisheries. Philosophical Transactions of the Royal Society B 360 (No. 1453):47–57

Hilborn R, Micheli F, De Leo GA (2006) Integrating Marine Protected Areas with Catch Regulation. Canadian Journal of Fisheries and Aquatic Sciences 63:642–649

Holland DS (2000) A Bioeconomic Model of Marine Sanctuaries on Georges Bank. Canadian Journal of Fisheries and Aquatic Sciences 57(6):1307–1319

Holland D (2003) Integrating Spatial Management Measures into Traditional Fishery Management Systems: The Case f the Georges Bank Multispecies Groundfish Fishery. ICES Journal of Marine Science 60:915–929

Holland D, Gudmundsson E, Gates J (1999) Do Fishing Vessel Buyback Programs Work: A Survey of the Evidence. Marine Policy 23(1):47–69

Holland D, Brazee R (1996) Marine Reserves for Fisheries Management. Marine Resource Economics 11(3):157–171

Holland D, Sutinen J (1999) An Empirical Model of Fleet Dynamics in New England Trawl Fisheries. Canadian Journal of Fisheries and Aquatic Sciences 56(2):253–264

Holland DS, Sutinen, JG (2000) Location Choice in the New England Trawl Fisheries: Old Habits Die Hard. Land Economics 76:133–149

Holland D, Sanchirico JN, Curtis RE, Hicks RL (eds) (2004) An Introduction to Spatial Modeling in Fisheries Economics. Marine Resource Economics 19(1): 1–6

Holland DS, Schnier KE (2006) Individual Habitat Quotas for Fisheries. Journal of Environmental Economics and Management 51(1):72–92

Holloway G, Tomberlin D (2006) Bayesian Ranking and Selection of Fishing Boat Efficiencies. Marine Resource Economics 21(4):415–432

Holt M, Bishop R (2002) A Semiflexible Normalized Quadratic Inverse Demand System: Price Formation of Fish. Empirical Economics 27:23–48

Homans F, Wilen J (1997) A Model of Regulated Open Access Resource Use. Journal of Environmental Economics and Management 32:1–21

Huppert DD, Ellis GM, Noble B (1996) Do Permit Prices Reflect the Discounted Value of Fishing? Evidence from Alaska's Commercial Salmon Fisheries. Canadian Journal of Fisheries and Aquatic Sciences 53:761–67

Hutton T, Mardle S, Pascoe S, Clark R (2004) Modeling Fishing Location Choice within Mixed Fisheries: English North Sea Beam Trawlers in 2000 and 2001. ICES Journal of Marine Science 61(8):1443–1452

Janmaat J (2005) Sharing Clams: Tragedy of an Incomplete Commons. Journal of Environmental Economics and Management 49(1):26–51

Jensen C (1990) Behavioral Modeling of Fishermen in the EU. Ph.D. dissertation, University of Southern Denmark, Denmark

Jensen C (2002) Applications of the Dual Theory in Fisheries: A Survey. Marine Resource Economics 17(4):309–334

Jensen R (2007) The Digital Provide: Information (Technology), Market Performance and Welfare in the South Indian Fisheries Sector. Quarterly Journal of Economics 122(3):879–924

Jensen F (2008) Uncertainty and Asymmetric Information: An Overview. Marine Policy 32(1):89–103

Jensen F, Vestergaard N (2000) Moral Hazard Problems in Fisheries: The Case of Illegal Landings. Resource and Energy Economics 24(4):281–299

Jensen F, Vestergaard N (2001) Management of Fisheries in the EU: A Principal-Agent Analysis. Marine Resource Economics 16:277–291

Jensen F, Vestergaard N (2002a) Regulation of Renewable Resources in Federal Systems: The Case of Fishery in the EU. Marine Resource Economics 16:277–291

Jensen F, Vestergaard N (2002b) A Principal-Agent Analysis of Fisheries. Journal of Institutional and Theoretical Economics158 (2):276–285

Jensen F, Vestergaard N (2002c) Moral Hazard Problems in Fisheries Regulation: The Case of Illegal Landings and Discard. Resource and Energy Economics 24(4):281–299

Jensen F, Vestergaard N (2007) Asymmetric Information and Uncertainty: The Usefulness of Logbooks as a Regulation Measure. Ecological Economics 63(4):815–827

Jeon Y, Squires, D Reid C (2008) Is there a global market for tuna? Policy implications for tropical tuna fisheries. Ocean Development and International Law 39(1):32–50

Jin L, Thurnberg E, Kite-Powell H, Blake K (2002) Total Factor Productivity Change in the New England Groundfish Fishery: 1964–1993. Journal of Environmental Economics and Management 44:540–556

Johansen L (1987) Production Functions and the Concept of Capacity. Recherches recents sur la fonction de production. Collection Economie Mathématique et Econometrie, 2. Reprinted in Førsund FR (ed) Collected Works of Leif Johansen Vol 1. Elsevier, Amsterdam, pp 639–650

Johnson RN, Libecap GN (1982) Contracting Problems and Regulation: the Case of the Fishery. American Economic Review 72:1005–22

Kaitala V (1985) Game Theoretic Models of Dynamic Bargaining and Contracting in Fisheries Management. Helsinki Institute of Mathematics, Helsinki University of Technology, Helsinki.

Kaitala V, Pohjola M (1988) Optimal Recovery of a Shared Stock: A Differential Game with Efficient Memory Equilibria. Natural Resource Modeling 3:91–119

Kaitala V, Munro G (1997) The Conservation and Management of High Seas Fishery Resources under the New Law of the Sea. Natural Resource Modeling 10:87–108

Kaitala V, Lindroos M (1998) Sharing the Benefits of Cooperation in High Seas Fishing Game: A Characteristics Function Game Approach. Natural Resource Modeling 11:275–299

Karpoff JM (1984a) Insights from the Markets for Limited Entry Permits in Alaska. Canadian Journal of Fisheries and Aquatic Sciences 41:1160–66

Karpoff JM (1984b) Low-Interest Loans and the Markets for Limited Entry Permits in the Alaska Salmon Fisheries. Land Economics 60:69–80

Karpoff JM (1985) Non-Pecuniary Benefits in Commercial Fishing: Empirical Findings from the Alaska Salmon Fisheries. Economic Inquiry 23:159–174

Karpoff JM (1987) Suboptimal Controls in Common Resource Management: The Case of the Fishery. Journal of Political Economy 95:179–194

Keller W (2004) International Technology Diffusion. Journal of Economic Literature 42(3):752–782

Kerstens K, Vestergaard N, Squires D (2006) A Short-Run Johansen Industry Model for Common-Pool Resources: Planning a Fishery's Industrial Capacity to Curb Overfishing. European Review of Agricultural Economics 33(3):361–389

Kirkley J, Strand I (1988) The Technology and Management of Multispecies Fisheries. Applied Economics 20:1279–1292

Kirkley J, Squires D (1999) Measuring Capacity and Capacity Utilization in Fisheries. In Greboval (ed) Managing Fishing Capacity: Selected Papers on Underlying Concepts and Issues. FAO, Rome

Kirkley J, Squires D, Strand I (1995) Assessing Technical Efficiency in Commercial Fisheries: The Mid-Atlantic Scallop Fishery. American Journal of Agricultural Economics 77(3):686–697

Kirkley J, Squires D, Strand I (1998) Characterizing Managerial Skill and Technical Efficiency in a Fishery. Journal of Productivity Analysis 9:145–160

Kirkley J, Squires D, Alam F, Ishak O (2003) Capacity and Offshore Fisheries Development: The Malaysian Purse Seine Fishery. American Journal of Agricultural Economics 85(3):647–662

Kirkley J, Morrison Paul C, Cunningham S, Catanzano J (2004) Embodied and Disembodied Technical Change in Fisheries: An Analysis of the Sète Trawl Fishery, 1985–1999. Environmental and Resource Economics 29(2):191–217

Kitts A, Thunberg E, Robertson J (2001) Willingness to Participate and Bids in a Fishing Vessel Buyout Program: A Case Study of New England Groundfish. Marine Resource Economics 15(3):221–232

Kjærsgaard J, (2007) Question for the Appropriate Industry Overcapacity. Fisheries Economics and Management Division at the Institute of Food and Resource Economics, Denmark

Kjærsgaard J, Vestergaard N, Kerstens K (2007) Ecological Benchmarking to Explore Alternative Fishing Schemes: The Danish Demersal Fishery in the North Sea. Fisheries Economics and Management Division at the Institute of Food and Resource Economics, Denmark

Kjærsgaard J, Frost H (2008) Effort Allocation and Marine Protected Areas: Is the North Sea Plaice Box A Management Compromise? ICES Journal of Marine Science 65(7): 1203–1215

Klein L (1960) Some Theoretical Issues in the Measurement of Capacity. Econometrica 28(2):272–286

Kolstad C (1999) Environmental Economics. Oxford University Press, Oxford

Kronbak LG, (2005) Essays on Strategic Interaction and Behavior among Agents in Fisheries. Unpublished Ph.D. Dissertation, University of Southern Denmark, Denmark

Kronbak LG, Lindroos M (2007) Sharing Rules and Stability in Coalition Games with Externalities. Marine Resource Economics 22(2):137–154

Kugarajah K, Sandal L, Berge G (2006) Implementing a Stochastic Bioeconomic Model for the North-East Arctic Cod Fishery. Journal of Bioeconomics 8(1):35–53

Kuperan K, Ishak O, Jeon Y, Kirkley J, Squires D, Susilowati I (2002) Fishing Capacity and Fishing Skill in Developing Country Fisheries: The Kedah, Malaysia Trawl Fishery. Marine Resource Economics 16(4):293–313

Lane D (1988) Investment Decision Making by Fishermen. Canadian Journal of Fisheries and Aquatic Sciences 45:782–96

Lauck T (1996) Uncertainty in Fisheries Management. In Gordon DV, Munro GR (eds) Fisheries and Uncertainty: A Precautionary Approach to Resource Management. University of Calgary Press, Calgary, Alberta, pp. 45–58

Lauck T, Clark CW, Mangel M, Munro GR (1998) Implementing the Precautionary Principle in Fisheries Management through Marine Reserves. Ecological Applications, 8(1):72–78 Supplement: Ecosystem Management for Sustainable Marine Fisheries

Laukkanen M (2003) Cooperative and Non-cooperative Harvesting in a Stochastic Sequential Fishery. Journal of Environmental Economics and Management 45:454–473

Levhari D, Mirman LJ (1980) The Great Fish War: An Example Using a Dynamic Cournot-Nash Solution. Bell Journal of Economics 11(1):322–324

Li E (1998) Option Value of Harvesting: Theory and Evidence. Marine Resource Economics 13: 135–142

Li CZ, Löfgren KG (1998) A Dynamic Model of Biodiversity Preservation: Theory and Applications. Environment and Development Economics 3(2):157–172

Li C-Z, Lofgren KG, Weitzman ML (2001) Harvesting versus Biodiversity: An Occam's Razor Version. Environmental and Resource Economics 18:355–366

Libecap G (1989) Contracting for Property Rights.Cambridge University Press, New York

Liebcap G (2006a) Assigning Property Rights in the Common Pool: Implications of the Prevalence of First-Possession Rules. Paper prepared for Sharing the Fish 06 Conference, February 26–March 2, 2006, Fremantle, Perth, Western Australia

Libecap G (2006b) Rights-Based Management of Tuna Fisheries: Lessons from the Assignment of Property Rights on the Western US Frontier. Paper prepared for the Workshop on Regional Economic Cooperation in the Pacific Fishery for Tropical Tunas, October 10–12, La Jolla, California

Lindroos M, Kronbak LG, Kaitala V (2007) Coalitions in Fisheries Games. In Björndal T, Gordon D, Arnason R, Sumaila U (eds) Advances in Fisheries Economics – Festschrift in Honour of Professor Gordon R. Munro, Blackwell, Oxford, UK, pp 184–195

Lindroos M (2004) Restricted Coalitions in the Management of Regional Fisheries Organisations. Natural Resource Modeling 17:45–70

Lindroos M (2007) Coalitions in International Fisheries Management. Natural Resource Modeling 21(3):366–384

Lindroos M, Kaitala V (2000) Nash Equilibria in a Coalition Game of the Norwegian Spring-Spawning Herring Fishery. Marine Resource Economics 15(4):321–340

Lipton D, Strand, I Jr (1989) The Effect of Common Property on the Optimal Structure of the Fishing Industry. Journal of Environmental Economics and Management 16:45–51

Lipton D, Strand, Jr IE (1992) Effect of Stock Size and Regulations on Fishing Industry Cost and Structure: The Surf Clam Industry. American Journal of Agricultural Economics 74:197–208

Mardle S, Pascoe S (1999) A Review of Applications of Multiple-Criteria Decision-Making Techniques to Fisheries. Marine Resource Economics 14:41–63

Mardle S, Pascoe S (2002) Modeling the Effects of Trade-Offs between Long and Short-Term Objectives in Fisheries Management. Journal of Environmental Management 65:49–62

McAusland C (2005) Learning by Doing in the Presence of an Open Access Renewable Resource: Is Growth Sustainable? Natural Resource Modeling 18(1):41–68

McConnell KE, Price M (2006) The Lay System in Commercial Fisheries: Origin and Implications. Journal of Environmental Economics and Management 51(3):295–307

McDonald AD, Sandal LK, Steinshamn SI (2002) Implications of a Nested Stochastic/ Deterministic Bio-economics Model for Pelagic Fishery. Ecological Modeling 149:193–201

McKelvey R (1997) Game-Theoretic Insights into the International Management of Fisheries. Natural Resource Modeling 10(2):129–171

De Meza D, Gould JR (1992) The Social Effiency of Private Decisions to Enforce Property Rights. Journal of Political Economy 100(3):561–580

Mistiaen J, Strand, Jr IE (2000a) Spatial Models in Fisheries Economics: Location Choice of Commercial Fishermen with Heterogeneous Risk Preferences. American Journal of Agricultural Economics 82(5):1184–1190

Mistiaen J, Strand, Jr I (2000b) Location Choice of Commercial Fishermen with Heterogeneous Risk Preferences. American Journal of Agricultural Economics 82:1184–190

Moloney D, Pearse P (1979) Quantitative Rights as an Instrument for Regulating Commercial Fisheries. Journal of Fisheries Research Board of Canada 36:859–866

Mullainathan S, Thaler R (2000) Behavioral Economics. Working Paper 00–27, Department of Economics, Massachusetts Institute of Technology, Massachusetts

Munro GR (1979) The Optimal Management of Transboundary Renewable Resources. Canadian Journal of Economics 12(3):355–376

Munro GR (2006) International Allocation Issues and the High Seas: An Economist's Perspective. Paper presented to Sharing the Fish Conference 2006, Freemantle, Australia, 49p

Munro GR, Van Houtte A, Willmann R (2004) The Conservation and Management of Shared Fish Stocks: Legal and Economic Aspects. FAO Fisheries Tech Paper 465, Rome

Murillas A, Chamorro J (2006) Valuation and Management of Fishing Resources Under Price Uncertainty. Environmental and Resource Economics 33(1):39–71

Murray JH (2005) The implications of technological progress for fishery management. Social Science Association of America Annual Meeting, New Orleans LA, March 2005

Murray JH (2006) Natural Resource Collapse: Technological Change and Biased Estimation. Job Market Paper, UC San Diego

Murray JH (2007) Constrained Marine Resource Management. Unpublished Ph.D. Dissertation, University of California, San Diego, CA

Naito T, Polasky S (1997) Analysis of Highly Migratory Fish Stocks Fishery: A Game-Theoretic Approach. Marine Resource Economics 12(1):179–201

National Research Council (2005) Valuing Ecosystem Services: Toward Better Environmental Decision-Making. National Academies Press, Washington, DC,

National Science Board (2006) Science and Engineering Indicators 2006. National Science Foundation, Arlington, VA

Neary JP (1985) International Factor Mobility, Minimum Wage Rates, and Factor-Price Equalization: A Synthesis. Quarterly Journal of Economics C:551–570

Neary JP, Roberts K (1980) The Theory of Household Behavior under Rationing European Economic Review 13:25–42

Newell R, Sanchirico J, Kerr S (2005) Fishing Quota Markets. Journal of Environmental Economics and Management 49(3):437–462

Newell R, Papps K, Sanchirico J (2007) Asset Pricing in Created Markets. American Journal of Agricultural Economics 89(2):259–272

Nøstbakken L (2006) Regime switching in a fishery with stochastic stock and price Journal of Environmental Economics and Management 51(2):231–241

Ostrom E (1990) Governing the Commons: The Evolution of Institutions for Collective Action. Cambridge University Press, New York

Park H, Thurman WN, Easley, Jr JE (2004) Modeling Inverse Demands for Fish: Empirical Welfare Measurement in Gulf and South Atlantic Fisheries. Marine Resource Economics 19(3):333–351

Pearse P (1980) Property Rights and the Regulation of Commercial Fisheries. Journal of Business Administration 11:185–209

Pearse P (1981) Fishing Rights, Regulations, and Revenues. Marine Policy 5(2):135–146

Perrings C, Folke C, Mäler KG (1992) The Ecology and Economics of Biodiversity Loss: The Research Agenda. AMBIO 21(3)

Pindyck R (2007) Uncertainty in Environmental Economics. Review of Environmental Economics and Policy 1(1):45–65

Pintassilgo P (2003) A Coalition Approach to the Management of High Seas Fisheries in the Presence of Externalities. Natural Resource Modeling 16(2):175–197

Pintassilgo P, Lindroos M (2008) Coalition Formation in High Seas Fisheries: A Partition Function Approach. International Game Theory Review (in press)

Pinto da Silva P Kitts A (2006) Collaborative fisheries management in the Northeast US: Emerging initiatives and future directions. Marine Policy 30(6):832–841

Pizer W, Popp D (2007) Endogenizing Technical Change: Matching Empirical Evidence to Modeling Needs. Discussion Paper RFF DP 07–11. Resources for the Future, Washington, DC

Placenti VGR, Spagnolo M (1992) A Bio-Economic Model for the Optimization of a Multi-Species, Multi-Gear Fishery: The Italian Case. Marine Resource Economics 7:275–95

Plourde C (1970) A Simple Model of Replenishable Resource Exploitation. American Economic Review 60:518–520

Pontryagin LS, Boltyanskii VS, Gamkrelidze RV, Mishchenko EF (1962) The Mathematical Theory of Optimal Processes. New York, Wiley Interscience

Punt A, Hilborn R (1997) Fisheries Stock Assessment and Decision Analysis: The Bayesian Approach. Reviews in Fish Biology and Fisheries 7(1):35–63

Quirk JP, Smith VL (1970) Dynamic Economic Models of Fishing. In Scott A, (ed) Economics of Fisheries Management: A Symposium. Institute of Animal Research and Ecology, University of British Columbia, Canada

Reed WJ (1979) Optimal Escapement Levels in Stochastic and Deterministic Harvesting Models. Journal of Environmental Economics and Management 6:350–363

Reithe S (2006) Marine Reserves as a Measure to Control Bycatch Problems – The Importance of Multispecies Interactions. Natural Resource Modeling 19(2):221–242

Roberts CM, Hawkins JP, Gell FR (2005) The Role of Marine Reserves in Achieving Sustainable Fisheries. Philosophical Transactions of the Royal Society B 360 (1453):123–132

Rosenman RE (1986) The Optimal Tax for Maximum Economic Yield: Fishery Regulation under Rational Expectations. Journal of Environmental Economics and Management 13:348–362

Ruttan V (2000) Technology, Growth and Development: An Induced Innovation Perspective. Oxford University Press, New York and Oxford

Salvanes K, Steen F (1994) Testing for Relative Performance between Seasons in a Fishery. Land Economics 70(4):431–447

Salvanes K, Squires D (1995) Transferable Quotas, Enforcement Costs and Typical Firms: An Empirical Application to the Norwegian trawler fleet. Environmental and Resource Economics 6:1–21

Salvanes K, DeVoretz D (1997) Household Demand for Fish and Meat Products: Separability and Demographic Effects. Marine Resource Economics 12(1):37–55

Sanchirico JN (2004) Designing a Cost-Effective Marine Reserve Network: A Bioeconomic Metapopulation Analysis. Marine Resource Economics 19(1):41–65

Sanchirico J (2005) Additivity Properties in Metapopulation Models: Implications for the Assessment of Marine Reserves. Journal of Environmental Economics and Management 49(1):1–25

Sanchirico J Wilen J (1999) Bioeconomics of Spatial Exploitation in a Patchy Environment. Journal of Environmental Economics and Management 37(2):129–150

Sanchirico, James and James Wilen. 2001. A Bioeconomic Model of Marine Reserve Creation. Journal of Environmental Economics and Management 42(3): 257–276.

Sanchirico J, Wilen J (2005) Optimal spatial management of renewable resources: matching policy scope to ecosystem scale, Journal of Environmental Economics and Management 50(1):23–46

Saphores J-D (2003) Harvesting a Renewable Resource Under Uncertainty. Journal of Economics and Control 28(3):509–529

Schaefer MB (1954) Some Aspects of the Dynamics of Populations Important to the Management of Commercial Marine Fisheries. Bulletin of the Inter-American Tropical Tuna Commission 1:25–56

Schneider V, Pearce D (2004) What Saved the Whales? An Economic Analysis of 20th Century Whaling. Biodiversity and Conservation 13:543–562

Schlenker W, Carson R, Granger C, Jackson J (2007) Fisheries Management under Cyclical Population Dynamics. Department of Economics, Columbia University.

Schnier KE, Anderson CM (2006) Decision Making in Patchy Resource Environments: Spatial Misperception of Bioeconomic Models. Journal of Economic Behavior and Organization 61:234–254

Schrank W (2007) Is There Any Hope for Fisheries Management? Marine Policy 31(3):299–307

Schultz C-E (1997) Trade Sanctions and Effects on Long-Run Stocks of Marine Mammals. Marine Resource Economics 12(3):159–178

Scott AD (1955a) The Fishery: The Objectives of Sole Ownership. Journal of Political Economy 63:116–124

Scott AD (1955b) Natural Resources: The Economics of Conservation. University of Toronto Press

Scott AD (1988) Development of Property in a Fishery. Marine Resource Economics 5:289–311

Scott T, Kirkley J, Rinaldo R, Squires D (2007) Assessing Capacity in the US Northwest Atlantic Pelagic Longline Fishery for Highly Migratory Species with Undesirable Outputs. FAO Fisheries Proceedings, Rome

Segerson K (2007) Reducing Stochastic Sea Turtle Bycatch: An Efficiency Analysis of Alternative Policies. Department of Economics, University of Connecticut, Connecticut

Segerson K, Miceli (1998) Voluntary Environmental Agreements: Good or Bad News for Environmental Protection? Journal of Environmental Economics and Management 36:109–130

Segerson K, Squires D (1990) On the Measurement of Economic Capacity Utilization for Multiproduct Industries. Journal of Econometrics 44:347–361

Segerson K, SquiresD (1993) Multiproduct Capacity Utilization under Regulatory Constraints. Review of Economics and Statistics 75(1):76–85

Segerson K, Squires D (1995) A Note on the Measurement of Capacity Utilization for Revenue-Maximizing Firms. Bulletin of Economic Research 47(1):77–84

Sethi G, Costello C, Fisher A, Hanemann M, Karp L (2005) Fishery Management under Multiple Uncertainty. Journal of Environmental Economics and Management 50(2):300–318

Sinclair S (1961) License Limitation: A Method of Economic Fisheries Management. Canada Department of Fisheries, Ottawa

Singh R, Weninger Q, Doyle M (2006) Fisheries management with stock growth uncertainty and costly capital adjustment. Journal of Environmental Economics and Management 52(2):582–599

Smith M (2002) Two econometric approaches for predicting the spatial behavior of renewable resource harvesters. Land Economics 78:522–538

Smith M, Wilen J (2003) Economic Impacts of Marine Reserves: The Importance of Spatial Behavior. Journal of Environmental Economics and Management 46(2):183–206

Smith M (2006a) Bioeconometrics: Empirical Modeling of Bioeconomic Systems. Nicholas School of the Environment, Duke University

Smith M (2006b) State Dependence and Heterogeneity in Fishing Location Choice. Journal of Environmental Economics and Management 50(2):319–340

Smith VL (1968) Economics of Production from Natural Resources. American Economic Review 58:409–431

Smith VL (1969) On Models of Commercial Fishing. Journal of Political Economy 77(2):181–198

Sparre P, Vestergaard N (1990) Bioeconomic Modelling of the Shrimp Fishery of Tanzania. Proceedings of the Man. of the Shallow Water Shrimp Fishery of Tanzania, July 18–27. FAO 58:81–94

Squires D (1986) Ex-Vessel Price Linkages in the New England Fishing Industry. Fishery Bulletin 84(2):437–442

Squires D (1987a) Long-Run Profit Functions for Multiproduct Firms. American Journal of Agricultural Economics 69(3):558–569

Squires D (1987b) Fishing effort: Its testing, specification, and internal structure in fisheries economics and management. Journal of Environmental Economics and Management 14(3):268–282

Squires D (1987c) Public Regulation and the Structure of Production in Multiproduct Industries: An Application to the New England Otter Trawl Industry. RAND Journal of Economics 18(2):232–247

Squires D (1988) Production Technology, Costs, and Multiproduct Industry Structure: An Application of the Long-Run Profit Function to New England Fisheries. Canadian Journal of Economics 21(2):359–378

Squires D (1992) Productivity Measurement in Common Property Resource Industries: An Application to Pacific Fisheries. RAND Journal of Economics 23(2):221–236

Squires D (1994) Firm Behavior under Input Rationing. Journal of Econometrics 61(2):235–257

Squires D (2007) Producer Quotas and Virtual Quantities. Draft memo, Dept of Economics, University of California San Diego

Squires D, Grafton RQ (2000) Regulating the Commons: The Effects of Input Controls on Productivity Growth in the Northern Australian Prawn Industry. Paper delivered to the June 11–13 2000 Summer Workshop of the Association of Environmental and ResourceEconomics at Scripps Institution of Oceanography, La Jolla, CA

Squires D, Kirkley J (1991) Production Quota in Multiproduct Pacific Fisheries. Journal of Environmental Economics and Management 21(2):109–126

Squires D, Kirkley J (1995) Resource Rents from Single and Multispecies Individual Transferable Quota Programs. ICES Journal of Marine Science 52:153–164

Squires D, Kirkley J (1996) Individual Tradable Quotas in a Multiproduct Common Property Industry. Canadian Journal of Economics 24(2):318–342

Squires D, Kirkley J (1999) Skipper Skill and Panel Data in Fishing Industries. Canadian Journal of Fisheries and Aquatic Sciences 56:2011–2018

Squires D, Reid C (2001) Fishing Capacity and Productivity Growth in Tuna Purse Seine Fleets of the Western and Central Pacific Ocean. Papers presented to the 2001 13th Annual US Multilateral Consultation and 2nd Negotiation on US Multilateral Treaty Extension, Apia, Samoa

Squires D, Vestergaard N (2004, 2007) The paradox of Private Efficiency and Social Inefficiency with Common-Pool Natural Resources: Economic Efficiency and Bio-economics. University of California, San Diego

Squires D, Kirkley J, Tisdell C (1995) Individual Transferable Quotas as a Fishery Management Tool. Reviews in Fisheries Science 3(2):141–169

Squires D, Ishak O, Jeon Y, Kirkley J, Kuperan K, Susilowati I (2002) Excess Capacity and Sustainable Development in Java Sea Fisheries. Environment and Development Economics 8:105–127

Squires D, Reid C, Jeon Y (2008) Productivity Growth in Natural Resource Industries and the Environment: An Application to the Korean Tuna Purse-Seine Fleet in the Pacific Ocean. International Economic Journal 22(1):81–93

Steinshamn SI (1992) Economic Evaluation of Alternative Harvesting Strategies for Fish Stocks. Bergen, Centre for Fisheries Economics

Sumaila UR (1998a) Protected Marine Reserves as Fisheries Management Tools: A Bioeconomic Analysis. Fisheries Research 37:287–296

Sumaila UR (1998b) A Review of Game-Theoretic Models of Fishing. Marine Policy 23(1): 1–10

Sumaila UR, Guenette S, Alder J, Chuenpagdee R (2000) Addressing Ecosystem Effects of Fishing Using Marine Protected Areas. ICES Journal of Marine Science 57:752–760

Sumaila UR, Munro GR, Sutinen J (2007) Recent Developments in Fisheries Economics: An Introduction. Land Economics 83(1):1–5

Sun C-H (1998) Optimal Number of Fishing Vessels for Taiwan's Offshore Fisheries: A Comparison of Different Fleet Size Reduction Policies. Marine Resource Economics 13(4):275–288

Sun C-H, Chiang F-S, Tsa E, Chen M-H (2006) The Effects of El Niño n the Mackerel Purse-Seine Fishery Harvests in Taiwan: An Analysis Integrating the Barometric Readings and Sea Surface Temperature. Ecological Economics 56:268–279

Teisl M, Roe B, Hicks R (2002) Can Eco-Labels Tune a Market? Evidence from Dolphin-Safe Labeling. Journal of Environmental Economics and Management 43(3):339–522

Tilman D, Polasky S, Lehman C (2005) Diversity, Productivity and Temporal Stability in the Economies of Humans and Nature. Journal of Environmental Economics and Management 49:405–26

Thirtle CG, Ruttan VW (1987) The role of demand and supply in the generation and diffusion of technical change. Great Britain: Harwood Academic Publishers

Townsend RE (2005) Producer Organizations and Agreements in Fisheries: Integrating Regulation and Coasian Bargaining. In Leal DR, (ed) Evolving Property Rights in Marine Fisheries, Chap 7. Rowman & Littlefield, Lanham, MD

Tuck G, Possingham HP (1994) Optimal Harvesting Strategies for a Metapopulation. Bull Math Biology 56(1):107–27

Turvey R (1964) Optimization and Suboptimization in Fishery Regulation. American Economic Review 54:64–76

Vestergaard N (1999) Measures of Welfare Effects in Multiproduct Industries: The Case of Multispecies Individual Quota Fisheries. Canadian Journal of Economics Vol. 32(2):729–43

Vestergaard N, Jensen F, Jørgensen HP (2005) Sunk Cost and Entry-Exit Decisions under ITQs: Why Industry Restructuring is Delayed. Land Economics 81(3):363–378

Vislie J (1987) On the Optimal Management of Transboundary Renewable Resources: A Comment on Munro's Paper. Canadian Journal of Economics 20:870–875

Walden J, Kirkley J, Kitts A (2003) A Limited Economic Assessment of the Northeast Groundfish Fishery Buyout Program. Land Economics 79(3):426–439

Walters C, Parma AM (1996) Fixed Exploitation Rate Strategies for Coping with Effects of Climate Change. Canadian Journal of Fisheries and Aquatic Sciences 53:148–158

Warming J (1911) Om grundrente af fiskegrunde. Nationaloekonomisk Tidsskrift 112:1–8. In Andersen P, On rent of fishing grounds: a 1983 translation of Jens Warming's 1911 article. History of Political Economy 15 3 (1911):391–396

Weikard H-P (2005) Cartel Stability under an Optimal Sharing Rule. Workshop Paper 77.2005. Trieste, Italy, Fondazione Eni Enrico Mattei

Weitzman ML (1974a) Prices vs. Quantities. Review of Economic Studies 41:477–491

Weitzman ML (1974b) Free Access vs. Private Ownership as Alternative Systems for Managing Common Property. Journal of Economic Theory 8:225–34

Weitzman ML (2002) Landing Fees vs. Harvest Quotas with Uncertain Fish Stocks. Journal of Environmental Economics and Management 43(2):325–338

Weninger Q (1998) Assessing Efficiency Gains From Individual Transferable Quotas: An Application to the Mid- Atlantic Surf Clam and Ocean Quahog Fishery. American Journal of Agricultural Economics 80:750–64

Weninger Q, Just RE (1996) An Analysis of Transition From Limited Entry to Transferable Quota: Non-Marshallian Principles for Fisheries Management. Natural Resource Modeling 10:53–83

Weninger Q, McConnell KE (2000) Buyback programs in commercial fisheries: efficiency versus transfers. Canadian Journal of Economics 33(2):394–412

Wessells CR, Johnston RJ, Donath H (1999) Assessing Consumer Preferences for Ecolabeled Seafood: The Influence of Species, Certifier and Household Attributes. American Journal of Agricultural Economics 81:1084–1089

Wilen J (1976) Common Property Resources and the Dynamics of Overexploitation: The Case of the North Pacific Fur Seal. Department of Economics Programme in Natural Resource Economics, Working Paper 3, University of British Columbia, Vancouver

Wilen J (1979) Regulatory Implications of Alternative Models of Fishermen Behavior, Journal of the Fisheries Research Board of Canada 36(7):855–858

Wilen J (1985) Bioeconomics of Renewable Resource Use. Handbook of Natural Resource and Energy Economics 1:61–124

Wilen J (2000) Renewable Resources Economists and Policy: What Differences Have We Made? Journal of Environmental Economics and Management 39:306–327

Wilen J (2004) Spatial Management of Fisheries. Marine Resource Economics 19:7–19

Wilen J (2007) Thoughts on Capacity Analysis: Is Capacity Analysis Giving Policy Makers Information They Need? Marine Resource Economics 22(1):79–82

Yamamoto T (1995) Development of a Community-based Fishery Management System in Japan. Marine Resource Economics 10(1):21–34

Chapter 33
Restoring Rivers in the Twenty-First Century: Science Challenges in a Management Context

Timothy J. Beechie, George R. Pess, Michael M. Pollock,
Mary H. Ruckelshaus, and Phil Roni

Abstract Legal mandates force consideration of at least some level of river restoration in many developed nations (e.g., Clean Water and Endangered Species Act in the United States, or the Water Framework Directive in the European Union), but a lack of specifics in legislation compels decision-makers to ask three persistent management questions: (1) How much river restoration do we need? (2) How do we best achieve cost-effective river restoration? (3) How do we know we have restored enough? Moreover, the broader management context is permeated with tremendous inertia to continue development of rivers for societal and economic gain, continual application of small and fragmented restoration actions, and skepticism that river restoration can succeed in the face of climate change and steady population growth. It is in this context that we identify key science challenges for river restoration in the twenty-first century. We suggest that a fundamental shift toward restoring watershed and river processes (process-based restoration) is needed if scientists are to begin developing the tools needed to provide relevant policy answers. The basic conceptual framework of process-based restoration requires that we understand how habitat is formed and changes, how habitat changes alter biota, and how human actions alter both river habitats and the landscape processes that create river habitats. Restoration actions must then directly address human actions that caused habitat degradation, thereby addressing the root causes of biological impacts. Understanding this framework will allow scientists to better address key science challenges for advancing river restoration, including development of ecosystem models to predict what kinds of and how much restoration is needed, an expanded suite of process-based restoration techniques for large river ecosystems, and comprehensive but cost-effective suites of metrics for monitoring river health.

Keywords Ecosystem models · process-based restoration · river restoration · river health watershed processes

T.J. Beechie(✉)
Northwest Fisheries Science Center, NOAA Fisheries, Seattle, Washington
2725 Montlake Blvd. E., Seattle, Washington, 98112
e-mail: tim.beechie@noaa.gov

R.J. Beamish and B.J. Rothschild (eds.), *The Future of Fisheries Science in North America*, 697
Fish & Fisheries Series, © Springer Science + Business Media B.V. 2009

33.1   Introduction

In the last century the world's rivers have been dramatically restructured and fragmented by efforts to harness water supplies, control flooding, and produce hydroelectric power, and their condition has been severely degraded by land uses that increase erosion rates, introduce pollutants, and degrade habitat quality. Rivers have been disconnected from their floodplains and deltas (Sedell and Frogatt 1984; Ward et al. 1999; Aparicio et al. 2000; Beechie et al. 2001; Hohensinner et al. 2003), as well as from the seas into which they flow (Nilsson et al. 2005; Syvitski et al. 2005). Storage of water and sediment (Graf 1999; Dynesius and Nilsson 1994; Vörösmarty et al. 2003; Nilsson et al. 2005; Syvitski et al. 2005), and interruption of both downstream and upstream nutrient fluxes (Cederholm et al. 1999; Achord et al. 2003; Syvitski et al. 2005) have dramatically altered riverine and nearshore ecosystems. Increased intensity of land uses has raised inputs of nutrients and pollutants to the point of degrading not only riverine ecosystems, but also their receiving estuaries and marine ecosystems (Rabalais et al. 1996; Schultz 2001; Tilman et al. 2001). Steadily rising human demands for water will increasingly stress riverine ecosystems worldwide (Postel et al. 1996; Vörösmarty et al. 2000), especially in regions where climate change is expected to reduce availability of water during summer when irrigation and ecological demands are high (Mote et al. 2003). Such massive changes to the world's watersheds and rivers have left a persistent impact on riverine ecosystems (Nehlsen et al. 1991; Morita and Yamamoto 2002; Palmer and Allan 2006; Poff et al. 2007). Moreover, significant investments in river restoration over the last 20 years have failed to halt declines in habitat quality and ecosystem function, and river health continues to deteriorate (Bernhardt et al. 2005; Palmer and Allan 2006).

Recognition of these issues has led to recent calls for national and international river restoration efforts (European Commission 2000; Palmer and Allen 2006), and a push for development of ecologically sound standards for river restoration (Palmer et al. 2005). A number of recent articles have focused on suites of challenges we face in river restoration, including knowledge limitations (Wohl et al. 2005), developing consistent standards for measurement of success (Palmer et al. 1997, 2005), social and institutional barriers to river conservation and restoration (Arlinghaus et al. 2002; Nilsson et al. 2007), the need for cooperative efforts among scientists and stakeholders (Poff et al. 2003), and homogenization of riverine ecosystems by introduction and extirpation of species (Olden and Poff 2004). These and other articles describe a broad array of issues facing scientists, managers, and society as a whole, yet there remains a need to frame scientific challenges in the context of common management questions and the looming challenges posed by global climate change and population growth.

In this paper we focus on scientific challenges relevant to three persistent policy questions: (1) How do we determine how much restoration is necessary? (2) How do we implement river restoration that is both ecologically effective and sustainable? (3) How do know that we have restored enough? Without answering these questions, decision makers cannot convince themselves or others that river restoration is both necessary and cost-effective, or that reversal or modification of certain river management practices has both ecological and societal value. We identify specific science challenges embedded within each of these three questions, and illustrate how answering each of these questions is important to initiating and sustaining large-scale ecological restoration of rivers.

33.2 The Management Context

Policy makers generally understand that the condition of our rivers is vastly different than it once was, yet most major physical modifications to rivers (e.g., levees and dams) have been made precisely to benefit societies, and most habitat degradation (e.g., habitat simplification, pollution) is a byproduct of agricultural, urban, and industrial development for economic gain. Each type of development has required substantial societal and economic investment, and each continues to fuel national economies. Hence, there is considerable pressure to continue present land and water management practices, and vague, non-quantitative environmental goals are insufficient impetus to initiate large-scale ecological restoration and drive the level of action needed to restore rivers. On the other hand, legal mandates often require that degraded rivers be restored to some degree. For example, the US Clean Water Act (1972), US Endangered Species Act (1973), and the European Union Water Framework Directive (European Commission 2000) drive the need to achieve some level of river health, measured either by the status of water quality or the condition of biota. However, such legislation cannot specify goals and standards for individual rivers or sections of rivers, nor can it state how one should achieve those goals or measure progress towards them. Therefore, these acts force a set of policy and science questions aimed at the identification of goals, practices, and criteria for achieving river restoration: How much restoration do we need? How much will it cost? And how do we know if our efforts are succeeding? These questions sit squarely at the science-policy interface, and science can make important contributions to answering these questions if there is a clear separation of science and policy roles (Ruckelshaus et al. 2002a; Steel et al. 2005).

A second element of the management context is the considerable momentum behind piecemeal restoration actions that do not address the ultimate causes of habitat and species declines. That is, stream and river restoration has long focused on small and isolated actions that are based on narrowly defined restoration goals or techniques (e.g., Hunter 1991; Slaney and Zaldokas 1997) and moving beyond such actions is both technically and politically difficult (Roni et al. 2005a; Nilsson et al. 2007). This problem has been recognized in scientific critiques of traditional restoration approaches in the past decade, noting tendencies toward "one-size-fits-all" habitat standards (Bisson et al. 1997), not managing for spatial or temporal variation in habitats (Reeves et al. 1995; Bisson et al. 1997), and addressing symptoms of a disrupted ecosystem rather than the causes (Frissell and Nawa 1992; Beechie and Bolton 1999). Nevertheless, the rarity of restoration techniques that restore processes that create and maintain habitats and biological communities (process-based restoration) remains a significant obstacle to effective ecological restoration of river systems (Wohl 2005; Kondolf et al. 2006), and efforts to advance the notion that river restoration should accept the spatio-temporal dynamics of rivers is only gradually gaining acceptance.

Finally, a third aspect of the management context is the recognition that population growth will continue, especially in urban areas, and that climate change will alter river flow and temperatures in the coming decades (e.g., Mote et al. 2003; Arnell 1999; Lins and Slack 1999). Hence, there is some skepticism that river restoration can, in fact, be successful over the long term. In combination, these three components of the management context (economic pressure for development, the momentum of existing piecemeal efforts, and skepticism that restoration can succeed) comprise a substantial socio-political inertia trending towards continued degradation of rivers. It is in this context that scientists must help answer three persistent questions that arise from legal pressures for river restoration. Moreover, the science challenge is not simply to contribute to developing river restoration goals, techniques, and evaluation criteria, but to make these contributions useful where restoration to pristine conditions is rarely an option, development of landscapes and rivers to benefit people will continue, and climate change will gradually alter riverine environments.

33.3   The Science Challenges

In light of this management context, we identify key scientific challenges embedded within each of the three management questions (Table 33.1). We focus on river restoration that is intended to improve the status of fishes or biological communities, as opposed to efforts targeting water quality or aesthetic values (Wohl et al. 2005). We draw on our experiences in North America, especially those addressing recovery of Pacific Salmon (*Oncorhynchus spp.*), but also consider restoration issues and approaches worldwide.

33.3.1   Question 1: How Much Restoration Do We Need? – Tools for Identifying and Prioritizing Restoration Actions

Goal setting for river restoration is predominantly driven by stakeholder and policy input, yet determining what we are trying to achieve relies heavily on scientific analyses of ecological needs (e.g., key species, habitat quality) – particularly when the goals of restoration are to improve the status of individual aquatic species or ecological communities (Stanford and Poole 1996; Ruckelshaus et al. 2002a). We find that invariably managers ask the question "How much restoration do we need?", which contains within it the requirements of setting goals for specific biota (McElhaney et al. 2000; Tear et al. 2005) and estimating how much habitat restoration is needed to achieve biological goals (Beechie et al. 2003) (Table 33.1). This question of how much restoration is needed is at the heart of goal setting because managers and politicians must balance ecological needs for river restoration against competing needs for the river's water and land. That is, a persistent – but rarely met – policy need is to have a clear sense of how much and what kinds of river restoration are needed to achieve a specified level of river health (Ruckelshaus et al. 2002a). Only with such estimates in hand can policy makers make informed decisions to balance the needs of biota against the needs of society.

Scientists have recently taken steps to answer the first part of "How much is enough?", which is the setting of goals for riverine biota. Scientific tools to determine how many animals or populations are needed include population and metapopulation models to set goals for specific species, and multispecies metrics and focal species approaches to managing for many species simultaneously (Karr 1991; Schmutz et al. 2000; Ruckelshaus et al. 2002a; Tear et al. 2005). For endangered Pacific salmon, targets address number of fish needed within populations, productivity levels, spatial structure of populations, and life history and genetic diversity (McElhany et al. 2000), as well as how many populations are needed to sustain

Table 33.1 Scientific challenges in river restoration, listed with reference to key management questions

Management question	Scientific challenges
1. How much restoration do we need? Tools for identifying and prioritizing restoration actions	(a) Frame trade-offs managers face in balancing ecological health of rivers with human needs
	(b) Develop tools for predicting the effects of land and water use on watershed processes, habitats, and biota, as well as for predicting restoration outcomes
2. How do we get there? The need for new restoration techniques	(a) Expand repertoire of process-based restoration techniques that are compatible with human demands for land and water
	(b) Develop tools for restoration planning at scales commensurate with the scales of river processes and species life histories
	(c) Develop restoration strategies that are robust to climate change
3. How do we know when we get there? Comprehensive monitoring strategies restoration	(a) Understand and incorporate decades-long time lags between restoration implementation and stream response into monitoring programs
	(b) Develop more cost-effective and comprehensive techniques and programs for measuring river health and restoration effectiveness

a Distinct Population Segment of the species (Waples 1991; Ruckelshaus et al. 2002b). However, the challenge of setting goals for river fishes of less commercial importance is more difficult because far fewer data are available to assess the status and trends of abundance, productivity, spatial structure, and diversity. Nevertheless, many efforts have succeeded in describing key biological needs even with scant data (e.g., for Gila trout *Onchorynchus gilae,* U.S. Fish and Wildlife Service 2003), although some recovery goals have been narrative in nature and difficult to quantify (e.g., snail darter *Percina tanasi* and humpback chub *Gila cypha,* US Fish and Wildlife Service 1979, 1990). Such goals typically focus on known habitat preferences and suspected critical life stages, but are unable to quantitatively assess which habitats constrain species recovery. Multispecies goals are perhaps even more difficult to address, but methods for doing so have long been available (Fausch et al. 1990; Karr 1991), and they continue to be employed in setting targets for river health (Schmutz et al. 2000). Setting such goals is based primarily on a reference condition approach, which bases targets on community composition in relatively undamaged rivers (Karr 1991; Schmutz et al. 2000).

The second part of "how much is enough?" – estimating how much river restoration is needed to achieve the biological goals – has received far less attention and remains a daunting scientific challenge, particularly since we know that riverine habitats are highly dynamic and a product of multiple, interacting watershed processes (e.g., hydrologic regime, sediment dynamics, nutrient fluxes, organic inputs). While there is a relative abundance of models describing habitat–species relationships (Lichatowich et al. 1995; Ruckelshaus et al. 2002b; Scheuerell et al. 2006), there are virtually no models that adequately represent how land uses affect habitat change by altering the myriad processes that drive riverine ecosystems (Bartz et al. 2006). For many species, population declines are primarily driven by habitat loss (Frissell 1993; Beechie et al. 1994; Belsky et al. 1999; Wang et al. 2001; Morley and Karr 2002) or by dams that alter and fragment habitat (US Fish and Wildlife Service 1979, 1990; Filipe et al. 2004; Sheer and Steel 2006). However, the lack of suitable models has meant that it is difficult to specify how much restoration is needed with any level of accuracy or precision. Moreover, some species are also affected by hatchery practices, harvest of fish, and introduction of nonnative species (Ross 1990; Ruckelshaus et al. 2002a), adding to the complexity of science challenges in estimating how much restoration is required to achieve specific ecological goals.

Identifying what types of restoration and how much restoration is necessary for any river system first requires analysis of watershed processes and disruptions to those processes by land and water uses (Kondolf et al. 2006). Such analyses must (1) identify how habitats have changed and altered biota, (2) identify the causes of those habitat changes, and (3) identify restoration actions needed to address those causes. These assessments are based on a conceptual model of watershed-river function that illustrates how landscape and watershed processes (e.g., delivery of wood, light, water and sediment, and nutrients) drive habitat condition, which in turn drives biological responses (Fig. 33.1). The important point of this diagram is not the details of how a watershed works, but the depiction of intermediate linkages between land uses or restoration actions, changes in riverine

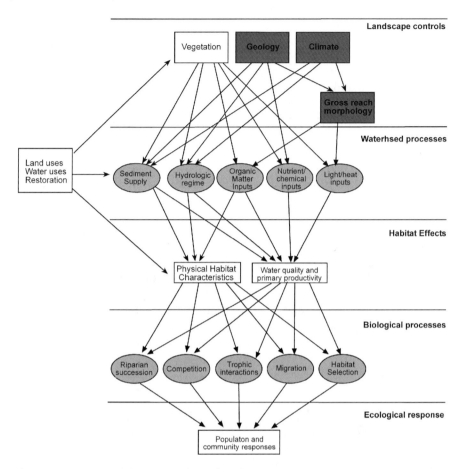

Fig. 33.1 Schematic diagram of linkages among land use and landscape controls on watershed processes, riverine habitat characteristics, and biological responses in river ecosystems. Dark gray boxes represent landscape and climate controls that limit the range of environmental conditions that can be expressed within any river reach. Open boxes are conditions or states of the ecosystem, and light gray ovals are ecosystem processes. Human uses and restoration actions can indirectly affect habitats via landscape attributes (vegetation), or by directly altering habitat-forming processes or habitat characteristics (Adapted from Beechie et al. 2003)

habitat (physical habitat, water quality, and primary productivity), and biological responses. Analysis must begin with assessment of historical or natural functions and conditions, with the explicit purpose of understanding the natural potential of river segments and habitats (Sedell and Frogatt 1984; Wohl 2005). It is not important at the analysis stage that some river segments cannot be restored to natural function. Rather, the analysis is fundamental to understanding where habitat loss and degradation have occurred (Collins and Montgomery 2001; Hohensinner et al. 2005; Wohl 2005), and the degree to which those losses explain the current state of biota in riverine ecosystems (Beechie et al. 2001).

    The greatest scientific challenge in this problem is translating land uses and
restoration actions into estimates of ecological benefit, which requires the intermediate step of predicting how land use alters habitat (Fig. 33.1). It has long been
a focus of scientists to explain species–habitat relationships in river ecosystems
and to evaluate traditional habitat restoration techniques such as wood and boulder
placement, barrier removal, and constructed floodplain habitats (e.g., Angermeier
and Schlosser 1989; Solazzi et al. 2000; Roni and Quinn 2001; Morley et al. 2005).
However, there have been few evaluations of process-based restoration actions such
as sediment reduction and riparian restoration (Roni et al. 2002). Recent attempts to
do so for rivers in northwestern United States have encountered significant obstacles in lack of data and quantitative models (Baker et al. 2004; Steel et al. 2004;
Bartz et al. 2006). Challenges include not only estimating the ultimate effectiveness
of efforts to restore landscape and watershed processes, but also understanding
recovery trajectories and lag times associated with a variety of restoration actions
(Cairns 1990; Beechie et al. 2000, 2005). Moreover, many processes interact
with each other in complex ways, and predicting restoration outcomes remains
exceedingly difficult. Scientists are therefore challenged to devise research
programs that will advance our understanding of recovery rates and pathways,
as well as models for predicting them. Moreover, such future predictions must
account not only for the future effects of habitat restoration, but also of population
growth, land use change, and climate change (Hulse et al. 2004; Bartz et al. 2006;
Scheuerell et al. 2006; Battin et al. 2007).

33.3.2   Question 2: How Do We Get There? – The Need
             for New Restoration Techniques

Perhaps more daunting than setting restoration goals is developing restoration
strategies and techniques that are likely to achieve those goals. Most broad policy
goals, such as the recovery of wide ranging ESA-listed species (e.g., Pacific salmon)
or restoration of a specific ecosystem, (e.g., Chesapeake Bay) require restoration over
a large geographical area (USEPA 1985; Feist et al. 2003). Furthermore, the causes
that led to ecosystem degradation or decline of a particular species usually occurred
over long periods of time and resulted from multiple activities (NRC 1992). Many
current restoration strategies do not succeed because they lack an understanding of
the changes in natural processes that led to system or species degradation in the first
place (e.g., Kondolf et al. 2001), they rarely address problems at the scale of environmental damage and biological needs (Palmer and Allan 2006), and they fail to
account for future population growth and climate change (Hulse et al. 2004; Battin
et al. 2007). The primary scientific challenges in developing effective river restoration strategies and techniques are thus (1) devising new restoration techniques that
allow at least partial river dynamics yet still preserve specific ecosystem goods
and services that society extracts from rivers, (2) developing tools for restoration
planning at scales commensurate with the scales of river processes and biota, and
(3) developing restoration strategies that are robust to climate change.

Many traditional restoration techniques are designed to create "stable" channels and habitat (e.g., bank stabilization techniques, creation of riffle or pool habitats, channel remeandering) (Slaney and Zaldokas 1997), yet such actions are often antithetical to the natural functioning of streams and rivers. That is, riverine habitats are created by the dynamics of sediment supply and routing, discharge variation and floods, lateral channel migration and floodplain construction, and riparian successional processes (Naiman et al. 1992). Attempts to construct stable habitats typically do not account for these dynamics (e.g., pool construction), or else attempt to control them (e.g., bank stabilization), and as a result many projects fail to accomplish their objectives of restoring habitats and species. Process-based restoration is a contrasting approach focused on understanding how natural processes in a watershed have been changed by human activities, and how they can be restored to redirect a watershed onto a recovery trajectory (see comparative examples in Table 33.2). Although this approach or various permutations of it have been advocated by us and others (Goodwin et al. 1997; Beechie and Bolton 1999; Palmer et al. 2005; Wohl et al. 2005; Kondolf et al. 2006), it has not been frequently implemented, perhaps because it is not sufficiently clear what this term means, and whether traditional restoration techniques can be used to restore processes. However, some process-based techniques such as riparian planting are combined with more traditional techniques such as wood placement to provide both short- and long-term restoration benefits.

The first key science challenge in advancing cost-effective restoration of rivers is that we lack a broad array of techniques intended to restore habitat-forming processes. Many common restoration techniques do not consider the natural potential

Table 33.2 Examples of contrasting approaches to correcting perceived habitat problems in rivers. Traditional restoration approaches often focus on creating specific desired conditions in rivers, whereas process-based approaches directly address causes of identified habitat problem. Some traditional techniques may be combined with process-based measures to produce near-term improvements and long-term restoration of processes

	Restoration approach	
Problem and cause	Traditional	Process-based
Symptom: Lack of wood and pools Cause: Wood removal and logging of riparian forest	Creation of pools by adding wood structures, sometimes substituting boulders for wood	Restore riparian wood recruitment processes, sometimes in combination with wood placement
Symptom: Increased lateral migration rate Cause: High sediment load	Bank protection measures to halt lateral movement	Address sediment supply at the source (e.g., logging road removal or reconstruction)
Symptom: Loss of floodplain habitats Cause: Bank armoring and restricted channel movement	Construct ponds or groundwater channels	Remove armoring, allow channel to migrate across its floodplain and create varied floodplain habitats
Symptom: High summer stream temperature Cause: Loss of hyporheic exchange after incision	Riparian planting to create shade	Restore sediment retention mechanisms to initiate channel aggradation

of restoration sites, and restoration actions attempt to create habitats that are geomorphologically and ecologically unsustainable (Wohl 2005; Kondolf et al. 2006). Moreover, most habitat and land use changes have been in lowland floodplains and deltas (Beechie et al. 1994; Hohensinner et al. 2003; Burnett et al. 2007), yet restoration actions most often focus on headwaters and small tributaries. Recently however, scientists and engineers have begun to develop new restoration approaches for large rivers, including levee setbacks to increase channel–floodplain interactions (Rohde et al. 2004), engineered log jams to initiate recovery of floodplain forests (Abbe et al. 2003), and environmental flow regimes that address the suite of flows required to maintain processes that support riverine ecosystems (Tharme 2003; Richter et al. 2006). These new techniques differ from many traditional restoration techniques in that they address specific causes of habitat degradation, they are designed for restoration of larger rivers that typically have not been considered for restoration, and they generally acknowledge human constraints in the design of restoration actions. Although progress has been made in developing new techniques, a key science challenge is to expand the repertoire of such techniques available to practitioners.

The second science challenge is developing tools to help devise restoration strategies that are commensurate with scale of environmental problems. Developing such strategies requires a more thorough understanding of how watershed processes interact to form and sustain habitats, as well as tractable models to predict outcomes of restoration strategies. Scientific challenges include research into recovery rates and pathways for various physical and ecological processes (e.g., Madej and Ozaki 1996; Beechie et al. 2000; Beechie 2001), and translation of this research into a model of watershed function that can predict how various combinations of restoration activities will interact to influence recovery of river ecosystems (Dietrich and Ligon 2005). Inherent in the recognition that successful long-term river restoration involves the manipulation of watershed processes is that some rivers will not be fully restored. For example, if restoration of natural sediment transport processes requires removal of a series of hydroelectric dams, it may be less likely to occur. At the same time, this approach helps clarify that restoration activities which do not address fundamental ecosystem processes will have limited biological effectiveness. This does not mean that such actions are not worth pursuing, just that they need to be pursued with a clear understanding of their limitations. Making informed decisions in such cases requires predictions of recovery trajectories and outcomes (particularly when considering only partial restoration), and scientists are again challenged to produce better tools for predicting the effects of large-scale restoration actions. Perhaps most importantly, such predictions must consider the watershed context in which restoration actions must occur, and explicitly model both physical and biological processes that will alter riverine ecosystems (e.g., Gore and Milner 1990; Poff and Ward 1990; Huxel and Hastings 1999).

The third scientific challenge to achieving effective ecological restoration is identifying restoration actions that are robust to climate change, and recognizing where climate change may fundamentally alter riverine environments (e.g., Battin et al. 2007). A process-based approach to river restoration is advantageous in that it is inherently robust to climate change. That is, restoring watershed processes and functions that allow for natural dynamics of river channels and biota will create river ecosystems that naturally adapt to climate change. However, this is not to say that all desired (or even legally mandated) outcomes will be achieved, particularly if goals are narrowly framed for specific species or community structures (Cowx and van Zyll de Jong 2004). In other words, climate change is expected to alter hydrologic and thermal regimes of rivers in this century, and biota are likely to respond to such changes by shifts in species distributions or life history strategies (e.g., Eaton and Scheller 1996; Beechie et al. 2006). Still, restoring habitat diversity through process-based restoration gives organisms greater opportunities to express diverse life history strategies, and therefore affords them greater opportunities to adjust and flourish in a climate-altered future. The key scientific challenge is to better predict how climate change may alter riverine ecosystems, and to develop river restoration strategies that accommodate such changes (Mote et al. 2003; Battin et al. 2007).

33.3.3 Question 3: How Do We Know When We Get There? – Comprehensive Monitoring Strategies

Since the early twentieth-century biologists have focused on indicator organisms to assess aspects of river health (Richardson 1929; Cairns and Pratt 1993), yet many water-quality monitoring efforts in the North America continued to focus on physiochemical measures related to toxicity (Cairns and Pratt 1993). Nevertheless, within the last two decades many monitoring efforts in the United States, Europe, and Australia have become more focused on the need to monitor biological or ecological integrity in river systems (ANZEEC 1992; European Commission 2000; Karr 2006). Hence, the number of biological and multispecies metrics that can be used to measure and monitor aquatic ecosystem health has grown rapidly (Karr 1981; Karr 1991; Schmutz et al. 2000; Welcomme et al. 2006). While these biological metrics are a critical element of monitoring and evaluating river health and quality (Karr 1999; Palmer et al. 2005), they are by themselves insufficient to assess whether restoration actions are successful in achieving multiple ecological and societal goals.

Because effective river restoration must focus on restoring the timing, magnitude, and frequency of natural processes that create and sustain riverine ecosystems, monitoring protocols must also reflect that focus and expand beyond narrow biological monitoring. That is, defining river health only in terms of aquatic biota ignores other important river functions, including supplying water for human uses, flood attenuation by healthy floodplain ecosystems, building and maintenance of beaches and coastal marshes, and connectivity among river reaches to maintain healthy populations of migratory fishes. Thus, a comprehensive definition of river health must consider physical and chemical attributes of the river (Gilliom et al. 1995; Fitzpatrick et al. 1998), as well as rates of watershed processes that drive habitat conditions such as sediment supply and transport, or wood recruitment to streams (Beechie et al. 2005; Pollock et al. 2005), measures of floodplain health (Pess et al. 2005; Florsheim et al. 2006), status of migratory fishes (McElhaney et al. 2000), and the degree to which rivers can support coastal processes and eco-systems (Palmer and Allan 2006). Such broad-ranging metrics obviously present a logistical and financial challenge, but they also challenge scientists to contribute to the development of cost-effective monitoring programs that comprehensively represent the health of riverine ecosystems.

For the purposes of this paper, we illustrate the scientific challenge of broadening how we measure river health by focusing on key watershed processes that should be incorporated into a monitoring program. These watershed processes are outside the traditional venue of aquatic system monitoring, even though they create, maintain, and alter the biological, physical, and chemical metrics used as operational targets for river health (Naiman et al. 1992, Beechie et al. 2008). Thus, monitoring water-shed processes such as sediment supply and transport, stream shading by riparian forests, hydrologic regime, the connectivity between rivers and their floodplains and tributaries, and the delivery of pollutants are all critical elements of river and watershed health (Table 33.3). Moreover, several types of monitoring are important in evaluating the long-term success of river restoration, including implementation monitoring, effectiveness monitoring, and status and trend monitoring (Roni et al. 2005b). Implementation monitoring addresses programmatic questions such as the number of restoration actions implemented, the proportion of actions that fol-lowed set plans, or the proportion of actions implemented over the specified time

Table 33.3 Examples of linking monitoring metrics for watershed processes, habitat characteristics, and biota to process-based restoration objectives

Watershed process category	Restoration objective	Watershed process metrics	Stream channel metrics	Biotic indicators
Sediment	Reduce fine sediment delivery to stream channels	Sediment input rates	Percent fine sediment (<2 mm) in streambeds	Benthic inverte-brate species composition
Hydrology	Decrease magnitude and frequency of peak flows	Peak flow frequency magnitude and duration	Frequency of streambed scour events	Salmonid egg to fry survival
Riparian functions	Restore riparian shade and wood recruitment to channels	Shade, wood recruitment rate	Temperature, in-channel wood loading	Growth and survival of juvenile fishes
Floodplain functions	Restore river–floodplain dynamics and diversity of floodplain habitats	Channel migration rate	Amount and quality of floodplain channel habitats	Proportion of floodplain channel-dependent species

period. Effectiveness monitoring evaluates how well restoration projects of specific classes or types meet their intended objectives (e.g., barrier removal, placement of in-stream structures, riparian planting), and is conducted on a statistically valid sub-set of actions to draw generalized conclusions about project effectiveness. Finally, status and trend monitoring is a common approach to tracking the condition of streams and rivers through time (Karr 2006), usually through repeated measurements at consistent sampling sites. The need for such comprehensive monitoring strategies poses unique challenges to scientists, including (1) developing monitoring approaches that explicitly account for spatial and temporal separation between watershed treatments and in-river habitat and biological responses, and (2) developing cost-effective suites of monitoring metrics for comprehensively measuring river health.

The first challenge stems from our knowledge that sampling designs vary with action type, and that spatial scales of sampling effort vary as a function of habitat-forming processes addressed by restoration actions. The scale of monitoring depends on whether a specific action type influences river processes or conditions at the scale of habitat units (on the order of 10^1–10^2 m in length), river reaches (on the order of 10^2–10^4 m in length), or river sub-basins or basins (>10^3 km^2 in area). Many biological indicators and stream channel metrics are measured at the scale of habitat units (Roni et al. 2005c), and monitoring at this scale is appropriate for restoration actions that attempt to modify the characteristics of specific habitat units, such as the construction of wood or boulder structures in streams to create pools. Effects of riparian and floodplain restoration actions are most strongly expressed at the scale of the treated reach, and monitoring of both out-of-stream (e.g., riparian processes, floodplain connectivity, lateral channel migration) and in-stream (e.g., habitat types, fish community structure) parameters should focus at this scale (Pess et al. 2003; Pollock et al. 2005). Finally, hydrologic and sediment processes occur at the watershed scale, and effectiveness monitoring of those actions should be undertaken strategically at that scale (Beechie et al. 2005). For reach and watershed level actions, the critical science challenges are devising cost-effective monitoring programs that explicitly recognize years- to decades-long time lags between certain treatments and responses, and that are designed to indicate improvements in river health long before in-stream biota express recovery. For example, riparian forests on trajectories for recovery indicate improving river health even before stream habitats and fish communities respond (e.g., Pollock et al. 2005). Similarly, the quantification of sediment input rates into stream channels with sediment budgets completed before and after upslope restoration (e.g., road decommissioning, erosion control techniques implemented, no continuing land use in high landslide hazard areas) can identify if sediment supply at the source is being reduced years to decades before changes occur in the stream channel (Madej and Ozaki 1996; Beechie et al. 2005). However, this knowledge is not explicitly incorporated into monitoring programs, and the scientist's challenges are to communicate the importance of understanding long time lags in watershed processes and to incorporate knowledge of these time lags into informative programs for monitoring river and watershed health.

The second major scientific challenge for monitoring river restoration is identifying a simplified suite of monitoring parameters that broadly represent river health. Characterizing the condition and trend of myriad attributes of riverine ecosystems is daunting, yet monitoring a narrow suite of end-point metrics is insufficient for ascertaining whether restoration actions are achieving their objectives. Therefore, scientists are challenged to help identify a small but comprehensive set of metrics to monitor river health, and to devise metrics that are diagnostic in nature and capable of detecting which aspects of river health have been improved by different restoration actions. This suite of metrics should represent physical, chemical, and biological endpoints of restoration, but should also capture landscape and watershed processes that form and sustain riverine ecosystems. Moreover, such metrics should consider monitoring parameters relevant to societal goals in order to increase the relevance of river restoration to the general public. For example, channel-floodplain restoration actions may include both ecological and societal criteria for success, including species richness of aquatic and floodplain-dependent species and the amount of land that is needed to maintain fluvial processes (Larsen et al. 2006). These additional metrics allow for balancing ecological benefits against the cost of acquiring land along rivers that naturally migrate across their floodplains.

33.4 Conclusions

Three persistent management questions frame the suite of science challenges for river restoration in the twenty-first century: (1) How much restoration do we need? (2) How do we best achieve cost-effective river restoration? (3) How do we know we have restored enough? These questions are driven by legal mandates that force varying levels river restoration (e.g., Clean Water and Endangered Species Act in the United States, or the Water Framework Directive in the European Union). However, the broader management context includes continued development of rivers for societal and economic gain, continued application of traditional restoration techniques in piecemeal fashion, and skepticism that river restoration can succeed in the face of climate change and steady population growth. It is in this context that we identify key science challenges for river restoration.

At the heart of the science challenges is developing a better understanding of how watersheds and their river ecosystems function. Each of the three questions requires pragmatic answers, and those answers cannot be achieved by continuing to rely solely on modeling, creating, and monitoring specific habitats or biota. Rather, a fundamental shift toward process-based river restoration is needed if scientists are to begin developing the tools needed to correctly identify causes of degradation and develop restoration strategies and techniques to address those causes. In its simplest form, the conceptual framework of process-based restoration requires that we understand how land uses cause habitat change, and subsequently how habitat change alters biota. This simple model clarifies the cause–effect structure through which land uses affect biota, and thereby helps scientists envision approaches to addressing the three science challenges.

The three science challenges for river restoration stem from the three management questions. The first science challenge is to develop better tools for identifying specific causes of ecosystem degradation that need to be addressed through restoration, as well as development of new tools for predicting the effects of land uses or restoration actions on watershed processes, habitats, and biota. The second science challenge is to develop new restoration strategies and techniques that address root causes of river ecosystem degradation, and to communicate that recovery of river ecosystems will be a long-term effort. And finally, the third science challenge is to develop more cost-effective and comprehensive programs for measuring river health and restoration effectiveness. Meeting each of these challenges requires not only that we advance the scientific knowledge and tools necessary for river restoration, but also that we recognize the management context for these tools and produce tools that are relevant in that context.

References

Abbe, T.B., Pess, G.R., Montgomery, D.R. and Fetherston, K.L. (2003) Integrating log jam technology into river rehabilitation. In: *Restoration of Puget Sound Rivers* (eds S. Bolton, D.R. Montgomery and D. Booth), University of Washington Press, Seattle, WA, pp. 443–482.

Achord, S., Levin, P.S. and Zabel, R.W. (2003) Density dependent mortality of in Pacific salmon: the ghost of impacts past? *Ecology Letters* **6**, 335–342.

Angermeier, P.L. and Schlosser, I.J. (1989) Species area relationships for stream fishes. *Ecology.* **70(5)**, 1450–1462.

ANZEEC (1992) Australian Water Quality Guidelines for Fresh and Marine Waters: National Water Quality Management Strategy. Australian and New Zealand Environment and Conservation Council.

Aparicio, E., Vargas, M.J., Olmo, J.M. and de Sostoa, A. (2000) Decline of native freshwater fishes in a Mediterranean watershed on the Iberian Peninsula: a quantitative assessment. *Environmental Biology of Fishes* **59**, 11–19.

Arlinghaus, R., Mehner, T. and Cowx, I.G. (2002) Reconciling traditional inland fisheries management and sustainability in industrialized countries, with emphasis on Europe. *Fish and Fisheries* **3**, 261–316.

Arnell, N.W. (1999) The effect of climate change on hydrological regimes in Europe: continental perspective. *Global Environmental Change* **9**, 5–23.

Baker, J.P., Hulse, D.W., Gregory, S.V., White, D., Van Sickle, J., Berger, P.A., Dole, D. and Schumaker, N.H. (2004) Alternative futures for the Willamette River basin, Oregon. *Ecological Applications* **14**, 313–324.

Bartz, K.L., Lagueux, K. Scheuerell, M.D., Beechie, T.J., Haas, A. and Ruckelshaus, M.H. (2006) Translating restoration scenarios into habitat conditions: an initial step in evaluating recovery strategies for Chinook salmon (*Oncorhynchus tshawytscha*). *Canadian Journal of Fisheries and Aquatic Sciences* **63**, 1578–1595.

Battin, J., Wiley, M.W., Ruckelshaus, M.H., Palmer, R.N., Korb, E., Bartz, K.K. and Imaki, H. (2007) Projected impacts of climate change on salmon habitat restoration. *Proceedings of the National Academy of Sciences* **104(16)**, 6720–6725.

Beechie, T.J. (2001) Empirical predictors of annual bed load travel distance, and implications for salmonid habitat restoration and protection. *Earth Surface Processes and Landforms* **26**, 1025–1034.

Beechie, T.J. and Bolton, S. (1999) An approach to restoring salmonid habitat-forming processes in Pacific Northwest watersheds. *Fisheries* **24**:6–15.

Beechie, T.J., Beamer, E. and Wasserman, L. (1994) Estimating coho salmon rearing habitat and smolt production losses in a large river basin, and implications for habitat restoration. *North American Journal of Fisheries Management* **14**, 797–811.

Beechie, T.J., Collins, B.D. and Pess, G.R. (2001) Holocene and recent geomorphic processes, land use and salmonid habitat in two north Puget Sound river basins. In: *Geomorphic Processes and Riverine Habitat* (eds J.B. Dorava, D.R. Montgomery, F. Fitzpatrick and B. Palcsak), Water Science and Application Volume 4, American Geophysical Union, Washington, DC, pp. 37–54l.

Beechie, T.J, Pess, G., Kennard, P., Bilby, R.E. and Bolton, S. (2000) Modeling recovery rates and pathways for woody debris recruitment in northwestern Washington streams. *North American Journal of Fisheries Management* **20**, 436–452.

Beechie, T.J., Pess, G., Beamer, E., Lucchetti, G. and Bilby, R.E. (2003) Role of watershed assessments in salmon recovery planning. In: *Restoration of Puget Sound Rivers* (eds D.R. Montgomery, S. Bolton, D. Booth and L. Wall), University of Washington Press, Seattle, WA, pp. 194–225.

Beechie, T.J., Veldhuisen, C.N., Beamer, E.M. Schuett-Hames, D.E., Conrad, R.H. and DeVries, P. (2005) Monitoring Treatments to Reduce Sediment and Hydrologic Effects from Roads. In: *Monitoring Stream and Watershed Restoration* (ed P. Roni), American Fisheries Society, Bethesda, MD, pp. 35–65.

Beechie, T.J., Ruckelshaus, M., Buhle, E., Fullerton, A. and Holsinger, L. (2006) Hydrologic regime and the conservation of salmon life history diversity. *Biological Conservation* **130**, 560–572.

Beechie, T., Pess, G., Roni, P. and Giannico, G. (2008) Setting river restoration priorities: a review of approaches and a general protocol for identifying and prioritizing actions. *North American Journal of Fisheries Management* **28**, 891–905.

Belsky, A.J., Matzke, A. and Uselman, S. (1999) Survey of livestock influences on stream and riparian ecosystems in the western United States. *Journal of Soil and Water Conservation* **54**, 419–431.

Bernhardt, E.S., Palmer, M.A., Allan, J.D., et al. (2005) Synthesizing U.S. river restoration efforts. *Science* **308**, 636–637.

Bisson, P.A., Reeves, G.H., Bilby, R.E. and Naiman, R.J. (1997) Watershed management and Pacific salmon: desired future conditions. In: *Pacific Salmon and Their Ecosystems: Status and Future Options* (eds D.J. Stouder, P.A. Bisson and R J. Naiman), Chapman & Hall, New York, pp. 447–474.

Burnett, K.M., Reeves, G.H., Miller, D.J., Clarke, S., Vance-Borland, K. and Christiansen, K. (2007) Distribution of salmon-habitat potential relative to landscape characteristics and implications for conservation. *Ecological Applications* **17**, 66–80.

Cairns, J. (1990) Lack of a theoretical basis for predicting rates and pathways of recovery. *Environmental Management* **14(5)**, 517–526.

Cairns, J. and Pratt, J.R. (1993) A history of biological monitoring using benthic macroinvertebrates. In: *Freshwater Biomonitoring and Benthic Macroinvertebrates* (eds D.M. Rosenberg and V.H. Resh), Chapman & Hall. New York, pp. 10–40.

Cederholm, C.J., Kunze, M.D., Murota, T., and Sibatani, A. (1999) Pacific salmon carcasses: essential contributions of nutrients and energy for terrestrial and aquatic ecosystems. *Fisheries* **24**, 6–15.

Collins, B.D. and Montgomery, D.R. (2001) Importance of archival and process studies to characterizing pre-settlement riverine geomorphic processes and habitat in the Puget Lowland. In: *Geomorphic Processes and Riverine Habitat* (eds J.B. Dorava, D.R. Montgomery, F. Fitzpatrick and B. Palcsak), Water Science and Application Volume 4, American Geophysical Union, Washington, DC, pp. 227–243.

Cowx, I.G. and van Zyll de Jong, M. (2004). Rehabilitation of freshwater fisheries: tales of the unexpected? *Fisheries Management and Ecology* **11**, 243–249.

Dietrich, W.E. and Ligon, F.K. (2005) "Desktop watersheds" in river restoration: static to dynamics digital terrain-based modeling. *Eos, Transactions of the American Geophysical Union* **86(18)**, Joint Assembly Supplement, Abstract NB21G-04.

Dynesius, M. and Nilsson, C. (1994) Fragmentation and flow regulation of rivers systems in the northern third of the world. *Science* **266**, 753–762.

Eaton, J.G. and Scheller, R.M. (1996) Effects of climate warming on fish thermal habitat in streams of the United States. *Limnology and Oceanography* **41**, 1109–1115.

European Commission. (2000) Directive 2000/60/EC, Establishing a Framework for Community Action in the Field of Water Policy. European Commission PE-CONS 3639/1/100 Rev. 1, Luxembourg.

Fausch, K.D., Lyons, J., Karr, J.R. and Angermeier, P.L. (1990) Fish communities as indicators of environmental degradation. *American Fisheries Society Symposium* **8**, 123–144.

Feist, B.E., Steel, A.E., Pess, G.R. and Bilby, R.E. (2003) The influence of scale on salmon habitat restoration priorities. *Animal Conservation* **6**, 271–282.

Filipe, A.F., Marques, T.A., Seabra, S., et al. (2004) Selection of priority areas for fish conservation in Guadiana River basin, Iberian Peninsula. *Conservation Biology* **18**, 189–200.

Fitzpatrick, F.A., Waite, I.R., D'Arconte, P.J., Meador, M.R., Maupin, M.A., Gurtz, M.E. (1998) *Revised methods for characterizing stream habitat in the National Water-Quality Assessment Program.* (USGS Water-Resources Investigations Report 98–4052) U.S. Geological Survey, Raleigh, NC.

Florsheim, J.L., Mount, J.F. and Constantine, C.R. (2006) A geomorphic and adaptive management monitoring framework to assess the effects of lowland floodplain river restoration on channel-floodplain sediment continuity. *River Research and Applications* **22**, 353–375.

Frissell, C.A. (1993) Topology of extinction and endangerment of native fishes in the Pacific Northwest and California (U.S.A.). *Conservation Biology* **7**, 342–354.

Frissell, C.A., and Nawa, R.K. (1992) Incidence and causes of physical failure of artificial habitat structures in streams of western Oregon and Washington. *North American Journal of Fisheries Management* **12**, 182–197.

Gilliom, R.J., Alley, W.M. and Gurtz, M.E. (1995) *Design of the National Water-Quality Assessment Program: Occurrence and distribution of water-quality conditions.* (U.S. Geological Survey Circular 1112) U.S. Geological Survey, Washington, DC.

Goodwin, C. N., Hawkins, C.P. and Kershner, J.L. (1997) Riparian restoration in the western United States: Overview and perspective. *Restoration Ecology* **5**, 4–14.

Gore, J.A. and Milner A.M. (1990) Island biogeographical theory: can it be used to predict lotic recovery rates. *Environmental Management* **14**, 737–753.

Graf, W.L. (1999) Dam nation: a geographic census of American dams and their large-scale hydrologic impacts. *Water Resources Research* **35**, 1305–1311.

Hohensinner S., Habersack, H., Jungwirth, M. and Zauner, G. (2003) Reconstruction of the characteristics of a natural alluvial river-floodplain system and hydromorphological changes following human modifications: the Danube River (1812–1991). *River Research and Applications* **20**, 25–41.

Hohensinner, S., Jungwirth, M., Muhar, S. and Habersack, H. (2005) Historical analysis: a foundation for developing and evaluating river-type specific restoration programs. *International Journal of River Basin Management* **3**, 1–10.

Hulse, D.W., Branscomb, A. and Payne, S.G. (2004) Envisioning alternatives: using citizen guidance to map future land and water use. *Ecological Applications* **15**, 325–341.

Hunter, C. J. (1991) *Better trout habitat: a guide to stream restoration and management.* Island Press, Washington, DC, 320 p.

Huxel, G.R. and Hastings, A. (1999) Habitat loss, fragmentation, and restoration. *Restoration Ecology* **7**, 309–315.

Karr, J.R. (1981) Assessment of biotic integrity using fish communities. *Fisheries* **6(6)**, 21–27.

Karr, J.R. (1991) Biological integrity: a long-neglected aspect of water resource management. *Ecological Applications* **1**, 66–84.

Karr, J.R. (1999) Defining and measuring river health. *Freshwater Biology* **41**, 221–234.

Karr, J. (2006) Seven Foundations of Biological Monitoring and Assessment. *Biologia Ambientale* **20**, 7–18.

Kondolf, G.M., Smeltzer, M.W. and Railsback, S.F. (2001) Design and Performance of a Channel Reconstruction Project in a Coastal California Gravel-Bed Stream. *Environmental Management* **28**, 761–776.

Kondolf, G.M., Boulton, A.J. O'Daniel, S., et al. (2006) Process-based ecological river restoration: visualizing three-dimensional connectivity and dynamic vectors to recover lost linkages. *Ecology and Society* **11(2)**, Art. 5, [online] URL: http://www.ecologyandsociety.org/vol11/iss2/art5/

Larsen, E.W., Girvetz, E.H. and Fremier, A.K. (2006) Assessing the affects of alternative setback channel constraint scenarios employing a river meander migration model. *Environmental Management* **37**, 880–897.

Lichatowich, J., Mobrand, L., Lestelle, L. and Vogel, T. (1995) An approach to the diagnosis and treatment of depleted Pacific salmon populations in Pacific Northwest watersheds. *Fisheries* **20**, 10–18.

Lins, H.F. and Slack, J.R. (1999) Streamflow trends in the United States. *Geophysical Research Letters* **26(2)**, 227–230.

Madej, M.A. and Ozaki, V. (1996) Channel response and partial recovery after large sediment inputs to Redwood Creek, California. *Earth Surface Processes and Landforms* **21**, 911–927.

McElhany, P., Ruckelshaus, M., Ford, M., Wainwright, T. and Bjorkstedt, E. (2000) *Viable salmonid populations and the recovery of evolutionarily significant units.* U.S. Dept. Commerce, NOAA Tech, Memo. NMFS-NWFSC-42, Seattle, Washington.

Morita, K. and Yamamoto, S. (2002) Effects of habitat fragmentation by damming on the persistence of stream-dwelling charr populations. *Conservation Biology* **16**, 1318–1323.

Morley, S.A., Garcia, P.S., Bennett, T. and Roni, P. (2005) Juvenile salmonid use of constructed and natural side channels in Pacific Northwest Rivers. *Canadian Journal of Aquatic and Fishery Sciences* **62**, 2811–2821.

Morely, S.A. and Karr, J.R. (2002). Assessing and restoring the health of urban streams in the Puget Sound basin. *Conservation Biology* **16**, 1498–1509.

Mote, P.W., Parson, E.A., Hamlet, A.F., et al. (2003) Preparing for climatic change: the water, salmon, and forests of the Pacific Northwest. *Climatic Change* **61**, 45–88.

Naiman, R.J., Beechie, T.J., Benda, L.E., et al. (1992) Fundamental elements of ecologically healthy watersheds in the Pacific Northwest coastal ecoregion. In: *Watershed Management: Balancing Sustainability and Environmental Change* (ed. R.J. Naiman), Springer-Verlag, New York, pp. 127–188.

Nehlsen, W., Williams, J.E. and Lichatowich, J.A. (1991) Pacific salmon at the crossroads: Stocks at risk from California, Oregon, Idaho, and Washington. *Fisheries* **16**, 4–21.

Nilsson, C., Reidy, C.A., Dynesius, M. and Revenga, C. (2005) Fragmentation and flow regulation of the world's large river systems. *Science* **308**, 405–408.

Nilsson, C., Jansson, R., Malmqvist, B. and Naiman, R.J. (2007) Restoring riverine landscapes: the challenge of identifying priorities, reference states, and techniques. *Ecology and Society* **12(1)**, 16 [online] URL: http://www.ecologyandsociety.org/vol12/iss1/art16/

NRC (1992) *Restoration of aquatic ecosystems.* National Academy Press, Washington, DC.

Olden, J.D. and Poff, N.L. (2004) Ecological processes driving biotic homogenization: testing a mechanistic model using fish faunas. *Ecology* **85(7)**, 1867–1875.

Palmer, M. A. and Allan, J. D. (2006) Restoring rivers. *Issues in Science and Technology* **winter**, 40–48.

Palmer, M.A., Ambrose, R.F. and Poff, N.L. (1997) Ecological theory and community restoration ecology. *Restoration Ecology* **5**, 291–300.

Palmer, M.A., Bernhardt, E.S., Allan, J.D., et al. (2005) Standards for ecologically successful river restoration. *Journal of Applied Ecology* **42**, 208–217.

Pess, G.R., Morley, S.A. and Roni, P. (2005) Evaluating fish response to culvert replacement and other methods for reconnecting isolated aquatic habitats. In: *Monitoring stream and watershed restoration* (ed P. Roni), American Fisheries Society, Bethesda, MD, pp. 267–276.

Pess, G.R., Beechie, T.J., Williams, J.E., Whitall, D.R., Lange, J.I. and Klochak, J.R. (2003) Watershed assessment techniques and the success of aquatic restoration activities. In: *Strategies* for restoring river ecosystems: Sources of variability and uncertainty in natural and managed systems (eds R.C. Wissmar and P.A. Bisson), American Fisheries Society, Bethesda, MD, pp. 185–201.

Poff, N.L. and Ward, J.V. (1990) Physical habitat template of lotic systems" recovery in the context of historical pattern of spatiotemporal heterogeneity. *Environmental Management* **14**, 629–645.

Poff, N.L., Allan, J.D., Palmer, M.A., et al. (2003) River flows and water wars: emerging science for environmental decision making. *Frontiers in Ecology and the Environment* **1**, 298–306.

Poff, N.L., Olden, J.D., Merritt, M.D. and Pepin, D.M. (2007) Homogenization of regional river dynamics by dams and global biodiversity implications. *Proceedings of the National Academy of Sciences* **104(14)**, 5732–5737.

Pollock, M.M., Beechie, T.J., Chan, S. and Bigley, R. (2005) Monitoring and evaluating riparian restoration efforts. In: *Monitoring Stream and Watershed Restoration*. (ed P. Roni) American Fisheries Society, Bethesda, MD, pp. 67–96.

Postel, S.L., Daily, G.C. and Ehrlich, P.R. (1996) Human appropriation of renewable freshwater. *Science* **271**, 785–788.

Rabalais, N.N., Turner, R.E., Justice, D. and Dortch, Q. (1996) Nutrient changes in the Mississippi river and systems responses on adjacent continental shelf. *Estuaries* **19**, 386–407.

Reeves, G.H., Benda, L.E., Burnett, K.M., Bisson, P.A. and Sedell, J.R. (1995) A disturbance-based ecosystem approach to maintaining and restoring freshwater habitats of evolutionarily significant units of anadromous salmonids in the Pacific Northwest. In: *Evolution and the aquatic ecosystem: defining unique units in population conservation* (ed J.L. Nielsen), *American Fisheries Society Symposium* **17**, 334–349.

Richardson, R.E. (1929) The bottom fauna of the Middle Illinois River 1913–1925: its distribution, abundance, valuation, and index value in the study of stream pollution. *Bulletin of the Illinois Natural History Survey* **17**, 387–475.

Richter, B.D., Warner, A.T., Meyer, J.L. and Lutz, K. (2006) A collaborative and adaptive process for developing environmental flow recommendations. *River Research and Applications* **22**, 297–318.

Rohde, S., Kienast, F. and Burgi, M. (2004) Assessing the restoration success of river widenings: a landscape approach. *Environmental Management* **34**, 574–589.

Roni, P., Beechie, T.J., Bilby, R.E., Leonetti, F.E., Pollock, M.M. and Pess, G.R. (2002) A review of stream restoration techniques and a hierarchical strategy for prioritizing restoration in Pacific Northwest watersheds. *North American Journal of Fisheries Management* **22**, 11–20.

Roni, P., Fayram, A.H. and Miller, M.A. (2005c) Monitoring and evaluating instream habitat enhancement. In: *Monitoring Stream and Watershed Restoration* (ed P. Roni), American Fisheries Society, Bethesda, MD, pp. 209–236.

Roni, P., Hanson, K., Beechie, T., Pess, G., Pollock, M. and Bartley, D. (2005a). Habitat rehabilitation for inland fisheries: global review of effectiveness and guidance for rehabilitation of freshwater ecosystems. (*FAO Fisheries Technical Paper* No. 484) U.N. Food and Agriculture Organization, Rome.

Roni, P., Liermann, M.C., Jordan, C. and Steel, E.A. (2005b). Design considerations for monitoring restoration effectiveness. In: *Monitoring stream and watershed*. (ed P. Roni), *restoration*, American Fisheries Society, Bethesda, MD, pp. 13–34.

Roni, P. and Quinn, T.P. (2001) Effects of artificial wood placement on movements of trout and juvenile coho salmon in natural and artificial channels. *Transactions of American Fisheries Society* **130**, 675–685.

Ross, S.T. (1990) Mechanisms structuring stream fish assemblages: are there lessons from introduced species. *Environmental Biology of Fishes* **30**, 359–368.

Ruckelshaus, M.H., Levin, P., Johnson, J. and Kareiva, P.M. (2002a) The Pacific salmon wars: What science brings to the challenge of recovering species. *Annual Review of Ecology and Systematics* **33**, 665–706.

Ruckelshaus, M., McElhany, P. and Ford, M.J. (2002b). Recovering species of conservation concern: Are populations expendable? In: *The importance of species – perspectives on expendability and triage* (eds P.M. Kareiva and S.A. Levin), Princeton University Press, Princeton, NJ, pp. 305–329.

Scheuerell, M.D., Hilborn, R., Ruckelshaus, M.H., Bartz, K.K., Lagueux, K.M., Haas, A., D. and Rawson K. (2006) The Shiraz model: a tool for incorporating anthropogenic effects and fish–habitat relationships in conservation planning. *Canadian Journal of Fisheries and Aquatic Sciences* **63**, 1596–1607.

Schmutz, S., Kaufmann, M., Vogel, B., Jungwirth, M. and Muhar, S. (2000) A multi-level concept for fish-based river-type-specific assessment of ecological integrity. *Hydrobiologia* **422/423**, 279–289.

Schultz, R. (2001) Comparison of spray drift- and runoff-related input of azinphos-methyl and endosulfan from fruit orchards into the Lourens River, South Africa. *Chemosphere* **45**, 543–541.

Sedell, J.R. and Froggatt, J.L. (1984) Importance of streamside forests to large rivers: the isolation of the Willamette River, Oregon, U.S.A., from its floodplain by snagging and streamside forest removal. *Verhandlung Internationale Vereinigung Limnologie* **22**, 1828–1834.

Sheer, M.B. and E.A. Steel. (2006) Lost watersheds: barriers, aquatic habitat connectivity, and species persistence in the Willamette and Lower Columbia basins. *Transactions of the American Fisheries Society* **135**:1654–1669.

Slaney, P.A. and Zaldokas, D. (1997) *Fish Habitat Rehabilitation Procedures.* (Watershed Restoration Technical Circular No. 9) Ministry of Environment, Lands and Parks) Vancouver, BC.

Solazzi, M. F., Nickelson, T.E., Johnson, S.L, Rodgers, S.D. (2000). "Effects of increasing winter rearing habitat on abundance of salmonids in two coastal Oregon streams." *Canadian Journal of Fisheries and Aquatic Sciences* **57**: 906–914.

Stanford, J.A., and Poole, G.C. (1996) A protocol for ecosystem management. *Ecological Applications* **6**, 741–744.

Steel, E.A., Feist, B.E., Jensen, D., Pess, G.R., Sheer, M., Brauner, J. and Bilby, R.E. (2004) Landscape models to understand steelhead (*Oncorhynchus mykiss*) distribution and help prioritize barrier removals in the Willamette basin, OR, U.S.A. *Canadian Journal of Fisheries and Aquatic Sciences* **61**: 999–1011.

Steel, E.A., Beechie, T.J., Ruckelshaus, M., Fullerton, A.H., McElhany, P. and Roni, P. (2005) Mind the gap: Uncertainty and model communication between managers and scientists (eds H. Michael, C. Steward and E. Knudsen). American Fisheries Society Symposium, Anchorage, AK.

Syvitski J.P.M., Vörösmarty, C.J., Kettner, A.J. and Green, P. (2005) Impact of humans on the flux of terrestrial sediment yield to the global coastal ocean. *Science* **308**, 376–380.

Tear, T.H., Kareiva, P., Angermeier, P.L., et al. (2005) How much is enough? The recurring problem of setting measurable objectives in conservation. *BioScience* **55**, 835–849

Tharme, R.E. (2003) A global perspective on environmental flow assessment: emerging trends in the development and application of environmental flow methodologies for rivers. *River Research and Applications* **19**, 397–441.

Tilman, D., Fargione, J., Wolff, B., et al. (2001) Forecasting agriculturally driven global environmental change. *Science* **292**, 281–284.

U.S. Clean Water Act. (1972) Title 33 U.S. Code, Section 1251.

U.S. Endangered Species Act. (1973) Title 16 U.S. Code, Section 1531.

U.S. Fish and Wildlife Service. (1979) *Snail Darter Recovery Plan.* U.S. Fish and Wildlife Service, Asheville, NC.

U.S. Fish and Wildlife Service. (1990) *Humpback Chub Recovery Plan.* U.S. Fish and Wildlife Service, Denver, CO.

U.S. Fish and Wildlife Service. (2003) *Gila Trout Recovery Plan (Third revision).* U.S. Fish and Wildlife Service, Albuquerque, New Mexico.

U.S. EPA. (1985) *Chesapeake Bay restoration and protection plan.* United States Environmental Protection Agency, Washington, DC.

Vörösmarty, C.J., Green, P., Salisbury, J. and Lammers, R.B. (2000) global water resources: vulnerability from climate change and population growth. *Science* **289**, 284–288.

Vörösmarty, C.J., Meybeck, M., Fekete, B., Sharma, K., Green, P. and Syvitski J.P.M. (2003) Anthropogenic sediment retention: major global impact from registered river impoundments. *Global and Planetary Change* **39**, 169–190.

Wang, L., Lyons, J., Kanehl, P. and Bannerman, R. (2001) Impacts of urbanization on stream habitat and fish across multiple spatial scales. *Environmental Management* **28**, 255–266.

Waples, R.S. (1991) Definition of "species" under the Endangered Species Act: application to Pacific Salmon. National Marine Fisheries Service, *NOAA Technical Memorandum* NMFS F/ NWC-194, Seattle, Washington.

Ward, J.V., Tockner, K. and Schiemer, F. (1999). Biodiversity of floodplain river ecosystems: ecotones and connectivity. *Regulated Rivers: Research and Management* **15**, 125–139.

Welcomme, R.L., Winemiller, K.O. and Cowx, I.G. (2006) Fish environmental guilds as a tool for assessment of ecological conditions of rivers. *River Research and Applications* **22**, 377–396.

Wohl, E. (2005) Compromised rivers: understanding historical impacts on rivers in the context of restoration. *Ecology and Society* **10(2)**, 2 [online] URL: http://www.ecologyandsociety.org/ vol10/iss2/art2/

Wohl, E., Angermeier, P.L., Bledsoe, B., et al. (2005) River restoration. *Water Resources Research* **41**, W10301.

Index